STRUCTURAL ENGINEERING, MECHANICS AND COMPUTATION

Proceedings of the International Conference on Structural Engineering, Mechanics and Computation

Volume 2

Elsevier Science Internet Homepage
http://www.elsevier.nl (Europe)
http://www.elsevier.com (America)
http://www.elsevier.co.jp (Asia)

Consult the Elsevier homepage for full catalogue information on all books, journals and electronic products and services.

Elsevier Titles of Related Interest

CHAN & TENG
ICASS '99, Advances in Steel Structures.
(2 Volume Set)
ISBN: 008-043015-5

DUBINA
SDSS '99 - Stability and Ductility of Steel Structures.
ISBN: 008-043016-3

FRANGOPOL, COROTIS & RACKWITZ
Reliability and Optimization of Structural Systems.
ISBN: 008-042826-6

FUKUMOTO
Structural Stability Design.
ISBN: 008-042263-2

MAKELAINEN
ICSAS '99, Int. Conf. on Light-Weight Steel and Aluminium Structures.
ISBN: 008-043014-7

RIE & PORTELLA
Low Cycle Fatigue and Elasto-Plastic Behaviour of Materials.
ISBN: 008-043326-X

KO & XU
Advances in Structural Dynamics (2 Volume Set)
ISBN: 008-043792-3

USAMI & ITOH
Stability and Ductility of Steel Structures.
ISBN: 008-043320-0

VOYIADJIS *ET AL*
Damage Mechanics in Engineering Materials.
ISBN: 008-043322-7

WANG, REDDY & LEE
Shear Deformable Beams and Plates.
ISBN: 008-043784-2

Related Journals
Free specimen copy gladly sent on request. Elsevier Science Ltd., The Boulevard, Langford Lane, Kidlington, Oxford, OX5 1GB, UK

Advances in Engineering Software
CAD
Composite Structures
Computer Methods in Applied Mechanics
 & Engineering
Computers and Structures
Construction and Building Materials
Engineering Failure Analysis
Engineering Fracture Mechanics
Engineering Structures
Finite Elements in Analysis and Design

International Journal of Mechanical Sciences
International Journal of Plasticity
International Journal of Solids and Structures
International Journal of Fatigue
Journal of Applied Mathematics and Mechanics
Journal of Constructional Steel Research
Journal of Mechanics and Physics of Solids
Mechanics of Materials
Mechanics Research Communications
Structural Safety
Thin-Walled Structures

To Contact the Publisher
Elsevier Science welcomes enquiries concerning publishing proposals: books, journal special issues, conference proceedings, etc. All formats and media can be considered. Should you have a publishing proposal you wish to discuss, please contact, without obligation, the publisher responsible for Elsevier's civil and structural engineering publishing programme:

Mr Ian Salusbury
Senior Publishing Editor
Elsevier Science Ltd,
The Boulevard, Langford Lane,
Kidlington, Oxford,
OX5 1GB, UK

Phone: + 44 1865 843425
Fax: +44 1865 843920
E.mail: i.salusbury@elsevier.co.uk

General enquiries, including placing orders, should be directed to Elsevier's Regional Sales Offices – please access the Elsevier homepage for full contact details (homepage details at the top of this page).

STRUCTURAL ENGINEERING, MECHANICS AND COMPUTATION

Proceedings of the International Conference on Structural Engineering, Mechanics and Computation

2- 4 April 2001, Cape Town, South Africa

Edited by

A. Zingoni

Department of Civil Engineering, University of Cape Town, Rondebosch 7701, Cape Town, South Africa

Volume 2

2001

ELSEVIER

AMSTERDAM - LONDON - NEW YORK - OXFORD - PARIS - SHANNON – TOKYO

ELSEVIER SCIENCE Ltd
The Boulevard, Langford Lane
Kidlington, Oxford OX5 1GB, UK

© 2001 Elsevier Science Ltd. All rights reserved.

This work is protected under copyright by Elsevier Science, and the following terms and conditions apply to its use:

Photocopying
Single photocopies of single chapters may be made for personal use as allowed by national copyright laws. Permission of the Publisher and payment of a fee is required for all other photocopying, including multiple or systematic copying, copying for advertising or promotional purposes, resale, and all forms of document delivery. Special rates are available for educational institutions that wish to make photocopies for non-profit educational classroom use.

Permissions may be sought directly from Elsevier Science Global Rights Department, PO Box 800, Oxford OX5 1DX, UK; phone: (+44) 1865 843830, fax: (+44) 1865 853333, e-mail: permissions@elsevier.co.uk. You may also contact Global Rights directly through Elsevier's home page (http://www.elsevier.nl), by selecting 'Obtaining Permissions'.

In the USA, users may clear permissions and make payments through the Copyright Clearance Center, Inc., 222 Rosewood Drive, Danvers, MA 01923, USA; phone: (+1) (978) 7508400, fax: (+1) (978) 7504744, and in the UK through the Copyright Licensing Agency Rapid Clearance Service (CLARCS), 90 Tottenham Court Road, London W1P 0LP, UK; phone: (+44) 207 631 5555; fax: (+44) 207 631 5500. Other countries may have a local reprographic rights agency for payments.

Derivative Works
Tables of contents may be reproduced for internal circulation, but permission of Elsevier Science is required for external resale or distribution of such material.
Permission of the Publisher is required for all other derivative works, including compilations and translations.

Electronic Storage or Usage
Permission of the Publisher is required to store or use electronically any material contained in this work, including any chapter or part of a chapter.

Except as outlined above, no part of this work may be reproduced, stored in a retrieval system or transmitted in any form or by any means, electronic, mechanical, photocopying, recording or otherwise, without prior written permission of the Publisher.
Address permissions requests to: Elsevier Science Global Rights Department, at the mail, fax and e-mail addresses noted above.

Notice
No responsibility is assumed by the Publisher for any injury and/or damage to persons or property as a matter of products liability, negligence or otherwise, or from any use or operation of any methods, products, instructions or ideas contained in the material herein. Because of rapid advances in the medical sciences, in particular, independent verification of diagnoses and drug dosages should be made.

First edition 2001

Library of Congress Cataloging in Publication Data
A catalog record from the Library of Congress has been applied for.

British Library Cataloguing in Publication Data
A catalogue record from the British Library has been applied for.

ISBN: 0-08-043948-9

The paper used in this publication meets the requirements of ANSI/NISO Z39.48-1992 (Permanence of Paper).
Printed in The Netherlands.

FOREWORD

**by
Professor Patrick Dowling CBE DL FREng FRS
Vice-Chancellor and Chief Executive
University of Surrey, UK
Chairman, Steel Construction Institute, UK**

These Proceedings will be regarded in time as seminal in that they record the enormous progress which has been made in Structural Engineering, Mechanics and Computation in the later half of the 20th century, and point the way to the future agenda for research in those areas for the 21st century.

The two volumes record the collected work of some of the most able researchers working on the world stage at this moment in time, and who are laying the foundations to some exciting new developments in the future. In that respect, the Proceedings should prove essential reading for those newly entering the field.

It is also most appropriate that the SEMC 2001 International Conference be held in South Africa where there is a New Dawn unfolding, and the spirit of cooperation is infusing all sectors of society, not least places of learning such as universities and colleges which are at the very heart of the New Knowledge Economy and Life Long Learning Agenda.

I salute all of those involved in the organisation, preparation and presentation of the Conference, as well as the delegates for their contribution to the shaping of these Proceedings, and assure all potential readers of the high quality and usefulness of its contents.

PREFACE

The International Conference on Structural Engineering, Mechanics and Computation was held in Cape Town (South Africa) from 2 to 4 April 2001. Organised by the University of Cape Town, the conference (SEMC 2001) aimed at bringing together from around the world academics, researchers and practitioners in the broad area of structural engineering and allied fields, to review the achievements of the past 50 years in the advancement of structural engineering, structural mechanics and structural computation, share the latest developments in these areas, and address the challenges that the future poses.

The Proceedings contain, in two volumes, a total of 180 papers written by Authors from around 40 countries worldwide. The contributions include 6 Keynote Papers and 12 Special Invited Papers. In line with the aims of the SEMC 2001 International Conference, and as may be seen from the List of Contents, the papers cover a wide range of topics under a variety of themes. There is a healthy balance between papers of a theoretical nature, concerned with various aspects of structural mechanics and computational issues, and those of a more practical nature, addressing issues of design, safety and construction. As the contributions in these Proceedings show, new and more efficient methods of structural analysis and numerical computation are being explored all the time, while exciting structural materials such as glass have recently come onto the scene. Research interest in the repair and rehabilitation of existing infrastructure continues to grow, particularly in Europe and North America, while the challenges to protect human life and property against the effects of fire, earthquakes and other hazards are being addressed through the development of more appropriate design methods for buildings, bridges and other engineering structures.

I would like to thank all Authors for preparing their work towards this compilation which, on account of the wealth of information it contains in just two volumes, will undoubtedly serve as a useful reference to practitioners, researchers, students and academics in the areas of structural engineering, structural mechanics, computational mechanics, and allied disciplines. Special thanks are due to Members of the International Scientific/Technical Advisory Board of SEMC 2001, who gave their time in advising on the selection of material contained in these Proceedings. The financial support of the Sponsoring Organisations is gratefully acknowledged. I am indebted to my colleagues in the Organising Committee, for the hard work they put into the preparations for the conference. Last but not least, I would like to thank my wife Lydia for the immense contribution she made towards making the SEMC 2001 International Conference a success.

A. Zingoni

INTERNATIONAL CONFERENCE ON STRUCTURAL ENGINEERING, MECHANICS AND COMPUTATION

Local Organising Committee

A. Zingoni, University of Cape Town (Chairman)
M. Latimer, Joint Structural Division of SAICE & IStructE
B.D. Reddy, University of Cape Town
J. Retief, University of Stellenbosch
F. Scheele, University of Cape Town
A.R. Kemp, University of the Witwatersrand
A. Masarira, University of Cape Town
M.G. Alexander, University of Cape Town
N.J. Marais, University of Cape Town
D. Douglas, University of Cape Town
G.N. Nurick, University of Cape Town

International Scientific/Technical Advisory Board

- P.J. Dowling, University of Surrey, *UK*
- Y.K. Cheung, University of Hong Kong, *China*
- P.L. Gould, Washington University, *USA*
- M. Bradford, University of New South Wales, *Australia*
- M. Papadrakakis, National Technical University of Athens, *Greece*
- L.A. Clark, University of Birmingham, *UK*
- B.D. Reddy, University of Cape Town, *South Africa*
- Y. Fujino, University of Tokyo, *Japan*
- D.A. Nethercot, Imperial College of Science, Technology & Medicine, *UK*
- O. Buyukozturk, Massachusetts Institute of Technology, *USA*
- A.R. Kemp, University of the Witwatersrand, *South Africa*
- V. Tvergaard, Technical University of Denmark, *Denmark*
- M.G. Alexander, University of Cape Town, *South Africa*
- H. Nooshin, University of Surrey, *UK*
- S.N. Sinha, Indian Institute of Technology at Delhi, *India*
- D.R.J. Owen, University of Wales at Swansea, *UK*
- G.N. Nurick, University of Cape Town, *South Africa*
- R. Zandonini, University of Trento, *Italy*
- S. Shrivastava, McGill University, *Canada*
- V. Marshall, University of Pretoria, *South Africa*
- J.M. Ko, Hong Kong Polytechnic University, *China*
- J.B. Obrebski, Warsaw University of Technology, *Poland*
- H.G. Schaeffer, University of Louisville, *USA*
- S. Heyns, University of Pretoria, *South Africa*
- D. Muir Wood, University of Bristol, *UK*
- R. de Borst, Delft University of Technology, *Netherlands*
- Y. Ballim, University of the Witwatersrand, *South Africa*
- A. Ghobarah, McMaster University, *Canada*
- A. Nowak, University of Michigan, *USA*
- N.M. Hawkins, University of Illinois at Urbana-Champaign, *USA*
- U. Schneider, Technical University of Vienna, *Austria*
- B.W.J. van Rensburg, University of Pretoria, *South Africa*
- P. Moss, University of Canterbury, *New Zealand*
- J. Retief, University of Stellenbosch, *South Africa*
- P. Dunaiski, University of Stellenbosch, *South Africa*
- O. Vilnay, Technion Israel Institute of Technology, *Israel*
- J.J.R. Cheng, University of Alberta, *Canada*
- S.K. Bhattacharyya, University of Durban-Westville, *South Africa*
- H. Adeli, Ohio State University, *USA*

Sponsoring Organisations

- *Joint Structural Division of the South African Institution of Civil Engineering (SAICE) and the UK Institution of Structural Engineers (IStructE)*
- *The Southern African Institute of Steel Construction (SAISC)*
- *The Cement and Concrete Institute (CCI) of South Africa*
- *The South African Association for Theoretical and Applied Mechanics (SAAM)*

CONTENTS

VOLUME 1

Foreword — v

Preface — vii

Local Organising Committee, International Scientific/Technical Advisory Board, Sponsoring Organisations — ix

KEYNOTE PAPERS

Y.K. CHEUNG, Y.S. CHENG and F.T.K. AU
Vibration analysis of bridges under moving vehicles and trains — 3

D.A. NETHERCOT
The importance of combining experimental and numerical study in advancing structural engineering understanding — 15

B.D. REDDY and O. EHL
Enhanced strain finite elements for Mindlin-Reissner plates — 27

P.L. GOULD
Recent advances in local-global FE analysis of shells of revolution — 39

A.R. KEMP
A new mixed flexibility approach for simplifying elastic and inelastic structural analysis — 51

P.J. PAHL and M. RUESS
Eigenstates of large profiled matrices — 63

INVITED PAPERS

F.M. MAZZOLANI and A. MANDARA
Advanced metal systems in structural rehabilitation of monumental constructions — 75

R. HARTE and W.B. KRATZIG
Lifetime-oriented analysis and design of large-scale cooling towers — 87

J.T. KATSIKADELIS
The BEM for vibration analysis of non-homogeneous bodies — 99

J.M. KO, Z.G. SUN and Y.Q. NI
A three-stage scheme for damage detection of Kap Shui Mun cable-stayed bridge   111

S.A. SHEIKH
Rehabilitation of concrete structures with fibre reinforced polymers   123

D.R.J. OWEN, Y.T. FENG, P.A. KLERCK and J. YU
Computational strategies for discrete systems and multi-fracturing materials   135

S.A. TIMASHEV
Optimal control of structure integrity and maintenance   147

J.B. OBREBSKI
On the mechanics and strength analysis of composite structures   161

H. PASTERNAK, S. SCHILLING and S. KOMANN
The steel construction of the new Cargolifter airship hangar   173

T. VROUWENVELDER
The fundamentals of structural building codes   183

R.T. SEVERN
Earthquake engineering research infrastructures   195

STEEL STRUCTURES: GENERAL CONSIDERATIONS

L.H. TEH and G.J. HANCOCK
Beam elements in structural analysis and design of steel frames   213

J.M. DAVIES
Second-order elastic-plastic analysis of plane frames   221

J. STUDNICKA
Steel structures in Czech Republic   231

A.N. GERGESS and R. SEN
Inelastic response of simply supported I-girders subjected to weak axis bending   243

M.M. TAHIR and D. ANDERSON
Performance of flush end-plate joints connected to column web   251

R.J. CRAWFORD and M.P. BYFIELD
A numerical model for predicting the bending strength of Larssen steel sheet piles   259

A. BURKHARDT
Practical use of probabilistic analysis for steel structures   267

J.A. KARCZEWSKI, M. GIZEJOWSKI, S. WIERZBICKI and E. POSTEK
Double butt, bolted connections: Influence of prestressing   275

A. MASARIRA
Joint type and the behaviour of frame beams   283

G.J. KRIGE and M.M. KHAN
Effect of rock movements on the integrity and performance of mine shaft steelwork   291

J.A. MWAKALI
Plasticity enhancement in axially compressed members   301

CONCRETE STRUCTURES: GENERAL CONSIDERATIONS

M.A. MANSUR, K.H. TAN and W. WENG
Analysis of reinforced concrete beams with circular openings using strut-and-tie model   311

R.V. JARQUIO
True parabolic stress method of analysis in reinforced concrete beams   319

S.H. CHOWDHURY
Crack width predictions of reinforced and partially prestressed concrete beams: A unified formula   327

H.Y. LEUNG and C.J. BURGOYNE
Analysis of FRP-reinforced concrete beam with aramid spirals as compression confinement   335

R.V. JARQUIO
Ultimate strength of CFT circular and square columns   343

H.D. BEUSHAUSEN, R.D. KRATZ and M.G. ALEXANDER
The contribution of screed to the structural behaviour of precast prestressed concrete elements   351

L.F. BOSWELL
Serviceability criteria for the vibration of post tensioned concrete flat slab floors   359

I. ISKHAKOV
Quasi-isotropic ideally elastic-plastic model for calculation of RC elements without empirical
coefficients   367

STEEL-CONCRETE COMPOSITE CONSTRUCTION

B. McKINLEY and L.F. BOSWELL
Large deformation performance of double skin composite construction using Bi-Steel   377

H.J.C. GALJAARD and J.C. WALRAVEN
Behaviour of different types of shear connectors for steel-concrete structures   385

M.P. BYFIELD
An analysis of inter shear-stud slip in composite beams   393

D. LAM and E. El-LOBODY
Finite-element modelling of headed stud shear connectors in steel-concrete composite beam   401

MASONRY, GLASS AND TIMBER STRUCTURES

G. de FELICE
Overall elastic properties of brickwork via homogenization   411

E.A. BASOENONDO, R.S. GILES, D.P. THAMBIRATNAM and H. PURNOMO
Response of unreinforced brick masonry wall structures to lateral loads   419

H.C. UZOEGBO
Lateral loading tests on dry-stack interlocking block walls   427

M.M. ALSHEBANI
Cyclic residual strains of brick masonry   437

F.A. VEER, G.J. HOBBELMAN and J.A. van der PLOEG
The design of innovative nylon joints to connect glass beams   447

G.J. HOBBELMAN, G.P.A.G. van ZIJL, F.A. VEER and C.N. TING
A new structural material by architectural demand   455

N. BOCCHIO, J.W.G. van de KUILEN and P. RONCA
The impact strength of timber for guard rails   463

S.J. FICCADENTI, G.C. PARDOEN and R.P. KAZANJY
Experimental and analytical studies of diaphragm to shear wall connections   473

PLATE, SHELL AND CONTAINMENT STRUCTURES

N. HASEBE and X.F. WANG
Green's functions for the thin plate bending problem under various boundary conditions   483

H. SHIN and D. REDEKOP
Nonlinear analysis of a storage tank by the DQM   491

P.D. AUSTIN, D. BUTLER, A.M. NASIR and D.P. THAMBIRATNAM
Dynamics of axisymmetric shell structures   499

E.S. MELERSKI
Analysis for temperature change effects in circular tanks   507

A. ZINGONI
On the possibility of parabolic ogival shells for egg-shaped sludge digesters   515

P.E. TRINCHERO
Field testing of column-supported silos and an introduction to the SAISC silo guideline   525

F. SHALOUF
Influence of anti-dynamic tube on reduction of dynamic flow pressure and elimination of
pulsation and vibration in grain silo   533

LAMINATED COMPOSITE PLATES AND SHELLS

H. MATSUNAGA
Vibration of cross-ply laminated composite plates   541

A. BENJEDDOU and S. LETOMBE
Free vibrations of piezoelectric sandwich plates: A two-dimensional closed solution   549

S.C. SHRIVASTAVA
Plastic buckling of spherical sandwich shells under external pressure   557

P.K. PARHI, S.K. BHATTACHARYYA and P.K. SINHA
Hygrothermal effects on the bending behaviour of multiple delaminated composite plates   565

A. SECU, R. BOAZU and D.P. STEFANESCU
New methods to establish the elastic characteristics of the fabric reinforced laminae   573

BRIDGES, TOWERS AND MASTS

M. SAMAAN, K.M. SENNAH and J.B. KENNEDY
Comparative structural behaviour of multi-cell and multiple-spine box girder bridges   583

C. GENTILE
Full-scale testing and system identification of a steel-trussed bridge   591

K.M. SENNAH, M.H. MARZOUCK and J.B. KENNEDY
Horizontal bracing systems for curved steel I-girder bridges   599

K.H. RESAN and I. OTHMAN
Torsional moments in Y-beam bridge deck under Malaysian abnormal load   607

X. LIANG, G.J. JUN and J.J. JING
Experimental modal analysis of the HuMen suspension bridge   613

M. IORDANESCU
Dedicated software for the structural analysis of guyed antenna towers   621

B. BEIROW and P. OSTERRIEDER
Dynamic investigations of TV towers   629

FINITE ELEMENT FORMULATIONS

M. BARIK and M. MUKHOPADHYAY
A new stiffened plate element for the analysis of arbitrary plates — 639

S. GEYER and A.A. GROENWOLD
A new 24 d.o.f. assumed stress finite element for orthotropic shells — 647

D. SONG, H. WANG, P.K. BANERJEE and D.P. HENRY, Jr
Finite element analysis of material and geometry nonlinearities with remeshing — 655

A. ZINGONI
Subspace formulation for symmetric finite elements — 663

G. TABAN-WANI and S.S. TICKODRI-TOGBOA
Finite element formulations in the design of underground structures — 675

FINITE ELEMENT AND NUMERICAL MODELLING

X.J. YU and D. REDEKOP
FEM computation of dynamic properties of a structure using fuzzy set theory — 687

M.A. GIZEJOWSKI, J.A. KARCZEWSKI, E. POSTEK and S. WIERZBICKI
Development of semi-rigid frame model assisted by testing — 695

S.H. LO
Analysis of building structures using solid finite elements — 703

M.K. APALAK, R. GUNES and E.S. KARAKAS
Geometrically non-linear thermal stress analysis of an adhesively bonded tee joint with double support — 711

A.T. McBRIDE and F. SCHEELE
Validation of discontinuous deformation analysis using a physical model — 719

CRACKING AND FRACTURE MECHANICS

G.P.A.G. van ZIJL
The time scale in quasi-static fracture of cementitious materials — 729

Z. KNESL, L. NAHLIK and Z. KERSNER
Calculation of the critical stress in two-phase materials — 737

G.P.A.G. van ZIJL
A discrete crack modelling strategy for masonry structures — 745

SOIL-STRUCTURE AND FLUID-STRUCTURE INTERACTION

B.B. BUDKOWSKA and A. ELMARAKBI
The assessment of shear effect of soil in analysis of laterally loaded models of the piles   755

B.F. COUSINS and E.S. MELERSKI
Numerical analysis of laterally loaded piles under conditions of elasticity   763

G.A. MOHAMMED and S. BAYOUMI
Flexible pipe sewer failure: Numerical analysis   771

G.A. MOHAMMED
Experimental and numerical analyses of multi-storey cracked frames with loss of support   779

S.K. BHATTACHARYYA and D. MAITY
Evaluation of stresses of a flexible structure exposed to fluid considering fluid-structure interaction   787

S.K. BHATTACHARYYA and O.O. ONYEJEKWE
A Green element computational technique applied to a fluid-structure interaction problem   793

Author Index   S 1

Keyword Index   S 5

VOLUME 2

Foreword   v

Preface   vii

Local Organising Committee, International Scientific/Technical Advisory Board, Sponsoring Organisations   ix

STABILITY OF THIN-WALLED MEMBERS

P. OSTERRIEDER and J. ZHU
Interaction buckling design concepts for thin-walled members   803

S. SENSOY
Degenerate Hopf bifurcation phenomena of a cantilever beam on elastic foundation   811

N. BJELAJAC
Simplified computational procedure for postbuckling equilibrium branches in ideal and imperfect plates   821

Q. WANG and Y. LUO
Dynamic stability of thin-walled members — 829

M. DJAFOUR, A. MEGNOUNIF and D. KERDAL
The compound spline finite strip method for the elastic stability of U and C built-up columns — 835

VIBRATION AND DYNAMIC ANALYSIS

L.F. YANG, Q.S. LI, J.Z. ZHANG and A.Y.T. LEUNG
Stochastic transient variational principle in vibration analysis — 845

J. FARJOODI and A. SOROUSHIAN
Efficient automatic selection of tolerances in nonlinear dynamic analysis — 853

T.U. AHMED, L.S. RAMACHANDRA and S.K. BHATTACHARYYA
An elasto-plastic free-free beam subjected to pulse load at tip — 861

J. FARJOODI and A. SOROUSHIAN
Robust convergence for the dynamic analysis of MDOF elastoplastic systems — 867

N. MUNIRUDRAPPA and V.A. KUMAR
Free vibration analysis of slantlegged skew bridge — 875

VIBRATION CONTROL AND SEISMIC ANALYSIS

A. HENRY, A. RICHARDSON and M. ABDULLAH
Placement and elimination of vibration controllers in buildings — 887

Y. RIBAKOV and J. GLUCK
Viscous damping system for optimal structural seismic design — 897

N.F. du PLOOY and P.S. HEYNS
Reducing vibratory screen structural loading using a vibration absorber — 905

C.F. de ANDRADE, J.C. de ANDRADE FILHO and J.C. de ANDRADE
Acceptability vibration criterion for floors with walking occupants — 913

N.A. ALEXANDER, N. GOORVADOO, F.A. NOOR and A.A. CHANERLEY
A comparative study of a storey vs. element hysteretic nonlinear model for seismic analysis of buildings — 919

S.T. VASSILEVA
Predicting earthquake ground motion descriptions through artificial neural networks for testing the constructions — 927

SEISMIC DESIGN OF STEEL STRUCTURES

G. De MATTEIS, R. LANDOLFO and F.M. MAZZOLANI
Contributing effect of cladding panels in the seismic design of MR steel frames — 937

M. MOESTOPO, I. IMRAN, R. RENANSIVA and A. SUDARSONO
Ductility formulations of steel structural members — 947

G. DELLA CORTE, G. De MATTEIS, R. LANDOLFO and F.M. MAZZOLANI
Seismic analysis of MR steel frames accounting for connection topology — 955

B. FAGGIANO, G. De MATTEIS, R. LANDOLFO and F.M. MAZZOLANI
A survey of ductile design of MR steel frames — 965

SEISMIC DESIGN OF CONCRETE STRUCTURES

S.S.E. LAM, B. WU, Z.Y. WANG, Y.L. WONG and K.T. CHAU
Behavior of rectangular columns with low lateral confinement ratio — 977

H. YIN, P. IRAWAN, T.C. PAN and C.H. LIM
Behavior of full-scale lightly reinforced concrete interior beam-column joints under reversed cyclic loading — 985

C.H. HAMILTON, G.C. PARDOEN, R.P. KAZANJY and Y.D. HOSE
Experimental and analytical assessment of simple bridge structures subjected to near-fault ground motions — 993

S.M. ELACHACHI, M. BENSAFI and D. NEDJAR
Seismic response of reinforced concrete frames using nonlinear macro-element behaviour — 1001

E. ATIMTAY and M.E. TUNA
Designing the concrete dual system — 1009

ANALYSIS AND DESIGN FOR BLAST AND IMPACT

K.H. LOW, K.L. LIM, K.H. HOON, A. YANG and J.K.T. LIM
Parametric study on the drop-impact behaviour of mini hi-fi audio products — 1019

S.C.K. YUEN and G.N. NURICK
Deformation and tearing of uniformly blast-loaded quadrangular stiffened plates — 1029

P. BIGNELL, D. THAMBIRATNAM and F. BULLEN
Non linear response and energy absorption of vehicle frontal protection structures — 1037

F. du TOIT, K. COMNINOS and P.J. KRUGER
Non-linear design of blast/containment reinforced gunite walls for coal mines in SA — 1043

R.D. KRATZ
A philosophy for blast resistant design								1051

FIRE SAFETY AND FIRE RESISTANCE

P.J. MOSS and G.C. CLIFTON
The effect of fire on multi-storey steel frames							1063

W. SHA and N.C. LAU
Fire safety design and recent developments in fire engineering					1071

W. SHA, N.C. LAU and T.L. NGU
Fire resistance of steel floors constructed with experimental fire resistant steels			1079

K.S. AL-JABRI, I.W. BURGESS and R.J. PLANK
The influence of connection characteristics on the behaviour of beams in fire			1087

W. SHA and T.L. NGU
Heat transfer in steel structures and their fire resistance					1095

W. SHA and N.C. LAU
Temperature development during fire in slim floor beams protected with intumescent coating	1103

STRUCTURAL SAFETY AND RELIABILITY

S. CADDEMI, P. COLAJANNI and G. MUSCOLINO
On the non stationary spectral moments and their role in structural safety and reliability		1113

K. RAMACHANDRAN
Developments in structural reliability bounds							1121

J.O. AFOLAYAN and A. OCHOLI
Isosafety parameters for fink-type steel roof trusses						1129

J.O. AFOLAYAN
Cost-modelling for the economic appraisal of joint details in steel trusses			1137

STRUCTURAL OPTIMISATION

M. KAHRAMAN and F. ERBATUR
A GA approach for simultaneous structural optimization					1147

J. MAHACHI and M. DUNDU
Genetic algorithm operators for optimisation of pin jointed structures				1155

K.H. LOW and H.P. SIN
Use of a stopper for the stress reduction in beam-block switching systems of audio products	1163

DAMAGE PREDICTION AND DAMAGE ASSESSMENT

M. ZHAO, Y. ZHAO and F. ANSARI
Fiber optic assessment of damage in FRP strengthened structures — 1175

J.L. HUMAR and M.S. AMIN
Structural health monitoring — 1185

C.L. MULLEN, P. TULADHAR, B. LeBLANC and S. SHRESTHA
3D seismic damage simulations for an existing bridge substructure using nonlinear FEM calibrated with modal NDT — 1195

H.A. RAZAK and F.C. CHOI
Damage assessment of corroded reinforced concrete beams using modal testing — 1203

M.M. KHAN and G.J. KRIGE
Evaluation of the structural integrity of an aging mine shaft — 1217

R. MASIH
Problems in measuring strains in the remaining parts of partially demolished bridge — 1225

REPAIR, REHABILITATION AND STRENGTHENING

A. GHOBARAH
Seismic rehabilitation of beam-column joints — 1235

C. ARYA, J.L. CLARKE, E.A. KAY and P.D. O'REGAN
TR 55: Design guidance for strengthening concrete structures using fibre composite materials - A review — 1243

K.P. GROSSKURTH and W. PERBIX
Force transmitting filling of wet and water filled cracks in concrete structures by means of crack injection with newly developed epoxy resins — 1251

M.K. RAHMAN
Numerical simulation of moisture diffusion in a concrete patch repair — 1259

A.I. UNAY
Analytical modelling of historical masonry structures for the evaluation of strength capacity of their vulnerable elements — 1269

LOADINGS AND CODE DEVELOPMENTS

F. WERNER and P. OSTERRIEDER
Actual problems of steel-design – Future of the codes — 1279

T.R. Ter HAAR, J.V. RETIEF and A.R. KEMP
Calibration of load factors for the South African Loading Code — 1289

J.V. RETIEF, P.E. DUNAISKI and P.J. de VILLIERS
An evaluation of imposed loads for application to codified structural design — 1297

A.M. GOLIGER, R.V. MILFORD and J. MAHACHI
South African wind loadings: Where to go — 1305

J.L. HUMAR and M.A. MAHGOUB
Seismic design provisions based on uniform hazard spectrum — 1313

P.E. DUNAISKI, H. BARNARD, G. KRIGE and R. MACKENZIE
Review of provision of loads to structures supporting overhead travelling cranes — 1321

A.M. GOLIGER and J.V. RETIEF
Background to wind damage model for disaster management in South Africa — 1329

CONCRETE AND CONCRETE MATERIALS

A.S. NGAB
Structural engineering and concrete technology in developing countries: An overview — 1339

T. YEN, K.S. PANN and Y.L. HUANG
Strength development of high-strength high-performance concrete at early ages — 1349

H.Y. LEUNG and C.J. BURGOYNE
Compressive behaviour of concrete confined by aramid fibre spirals — 1357

C.P. LAI, Y. LIN and T. YEN
Behavior and estimation of ultrasonic pulse velocity in concrete — 1365

C.W. TANG, K.H. CHEN and T. YEN
Study on the rheological behavior of medium strength high performance concrete — 1373

M.F.M. ZAIN, T.K. SONG, H.B. MAHMUD, Md. SAFIUDDIN and Y. MATSUFUJI
Influence of admixtures and quarry dust on the physical properties of freshly mixed high performance concrete — 1381

A.S. NGAB and S.P. BINDRA
Towards sustainable concrete technology in Africa — 1391

S. MOHD, C.K. WAH and P.Y. LIM
Development of artificial lightweight aggregates — 1399

F. FALADE
A comparative study of normal concrete with concretes containing granite and laterite fine aggregates — 1407

A.R.M. RIDZUAN, A.B.M. DIAH, R. HAMIR and K.B. KAMARULZAMAN
The influence of recycled aggregate on the early compressive strength and drying shrinkage of concrete — 1415

H.J. CHEN and H.C. CHAN
Numerical prediction on the elastic modulus of aggregate — 1423

K. RAMACHANDRAN and A. KARIMI
Estimation of corrosion time with observed data — 1431

CONSTRUCTION TECHNOLOGY AND METHODS

A. KASA, F.H. ALI and N. NASIR
Construction and instrumentation of a concrete modular block wall — 1441

B.B. BUDKOWSKA and J. YU
Analysis of multilayer system with geosynthetic insertion - Sensitivity analysis — 1449

L. FLISS
The scale pits of Saldanha Steel: An innovative solution to a complex problem — 1457

T.A.I. AKEJU and F. FALADE
Utilization of bamboo as reinforcement in concrete for low-cost housing — 1463

J. KANYEMBA
Enhancing housing delivery using a simple precast construction method — 1471

I. WEISER
Process Chains: A base for effective project management — 1481

J.O. AFOLAYAN
Analysis of placement errors of bars in reinforced concrete construction — 1489

STRUCTURAL ENGINEERING EDUCATION

B.W.J. van RENSBURG
Teaching structural analysis: A curriculum for an undergraduate civil engineering degree and learning issues — 1497

M.Y. RAFIQ and D.J. EASTERBROOK
Interactive use of computers to promote a deeper learning of the structural behaviour — 1505

LATE PAPERS

STEEL STRUCTURES: GENERAL CONSIDERATIONS

K.F. CHUNG and M.F. WONG
Experimental investigation of cold-formed steel beam-column sub-frames: Enhanced performance — 1515

CONCRETE STRUCTURES: GENERAL CONSIDERATONS

C.T. MORLEY and S.R. DENTON
Modified plasticity theory for reinforced concrete slab structures of limited ductility — 1523

MASONRY, GLASS AND TIMBER STRUCTURES

F.S. CROFTS and J.W. LANE
Accidental damage on unreinforced masonry structures — 1531

LAMINATED COMPOSITE PLATES AND SHELLS

S. MOHAMMADI and A. ASADOLLAHI
A contact based dynamic delamination buckling analysis of composites — 1539

BRIDGES, TOWERS AND MASTS

S.H. CHENG and D.T. LAU
Modelling of cable vibration effects of cable-stayed bridges — 1551

S.H. CHENG and D.T. LAU
Parametric study of cable vibration effects on the dynamic response of cable-stayed bridges — 1559

FINITE ELEMENT AND NUMERICAL MODELLING

T.C.H. LIU, K.F. CHUNG and A.C.H. KO
Finite element modelling on Vierendeel mechanism in steel beams with large circular web opening — 1567

CRACKING AND FRACTURE MECHANICS

R.A. FODHAIL
Loading parameters at cracks and notches — 1575

STABILITY OF THIN-WALLED MEMBERS

E.P. DJELEBOV
Investigations on local stability of compressed wall of hollow reinforced concrete bridge
pier with rectangular cross section — 1583

REPAIR, REHABILITATION AND STRENGTHENING

S.A. EL-REFAIE, A.F. ASHOUR and S.W. GARRITY
Strengthening of reinforced concrete continuous beams with CFRP composites — 1591

CONCRETE AND CONCRETE MATERIALS

G.C. FANOURAKIS and Y. BALLIM
Assessment of a range of design models for predicting creep in concrete — 1599

S.J. FICCADENTI
Effects of cement type and water to cement ratio on concrete expansion caused by sulfate attack — 1607

VIBRATION CONTROL AND SEISMIC ANALYSIS

G.C. PARDOEN, R. VILLAVERDE, R. TAVARES and S. CARNALLA
Improved modelling of electrical substation equipment for seismic loads — 1615

Author Index — S 1

Keyword Index — S 5

STABILITY OF THIN-WALLED MEMBERS

INTERACTION BUCKLING DESIGN CONCEPTS FOR THIN-WALLED MEMBERS

P. Osterrieder and J. Zhu

Department of Civil Engineering, BTU Cottbus, FRG
PO Box 101344, D-03013 Cottbus

ABSTRACT

In the study principles of load carrying behaviour in case of interaction between local plate buckling and global beam buckling and solution procedures within a numerical approach are discussed. Three different engineering design concepts which include simultaneous treatment of combined loading are developed and compared against each other and also versus numerical results. Concept 1 is easily applicable and leads to both safe and economic results. Concept 2 especially yields a smooth transition between all limit cases. To improve application in engineering practise however there is further research need.

KEYWORDS

Interaction Buckling, Buckling Capacity, Design Concepts, Local and Global Buckling, Stability Design

INTRODUCTION

Thin-walled beams and columns are susceptible to interaction buckling with local plate buckling and overall member buckling. For sake of simplification of the stability design procedure the overall problem in general is decomposed into two separate parts which exclude the other instability phenomena. This approach is justified only when bifurcation loads for both problems differ essentially.
The design concepts presented in this study are based on the following criterions.
- The estimated design capacity must be both economic and safe.
- The method has to reflect the problem in a way that the requirements of structural mechanics are satisfied.
- The expense for analysis must be reasonable with respect to practical engineering application.

LOAD CARRYING BEHAVIOUR IN CASE OF INTERACTION BUCKLING

Local plate buckling essentially reduces cross section stiffness and subsequently member buckling stiffness. If local buckling occurs compression stresses will shift to stiffer and less loaded section parts. The governing quantity for member buckling is the member slenderness λ_S and for plate buckling the plate slenderness λ_P. Post buckling behaviour for either instability phenomena however differs essentially and should be taken into account in modern design concepts within an ultimate load approach.

Interaction buckling will occur when bifurcation loads for member and plate buckling are close together. In this case member slenderness λ_S and plate slenderness λ_P differ only to a minor extent. Table 1 classifies failure modes depending on λ_S and λ_P.

TABLE 1
FAILURE MODES DEPENDING ON SLENDERNESS

		Member slenderness λ_S		
		small	medium	large
Plate slenderness λ_P	small	Yielding	Member Buckling	*Member Buckling*
	medium	*Plate Buckling*	Interaction	*nearly pure member*
	large	*Plate Buckling*	*nearly pure plate buckling*	Interaction

If plate slenderness λ_P is very small local plate buckling will not occur. The load carrying capacity of such a member is described in figure 1 by the solid line. For short members the capacity will be N_P, which leads to yield stresses in the whole cross section. If members length and subsequently member slenderness λ_P increases the load carrying capacity decreases. The reduction compared to the plastic capacity is specified via a reduction factor. Both the German Steel Stability code DIN 18800 Part 2 (1990) and the Eurocode EC3 (1992) employ for this purpose the European Buckling Curves in Eq.(1).

$$\kappa = \begin{cases} 1 & \bar{\lambda}_S \leq 0.2 \\ \dfrac{1}{k + \sqrt{k^2 - \bar{\lambda}_S^2}} & \bar{\lambda}_S > 0.2 \\ k = 0.5 \cdot [1 + \alpha \cdot (\bar{\lambda}_S - 0.2) + \bar{\lambda}_S^2] \end{cases} \quad (1)$$

The parameter α depends both on the type of cross section and the buckling direction.

If plate slenderness λ_P increases even short members will due to local plate buckling not reach the fully plastic internal force N_P. This fraction is indicated in figure 1 by Q (Q≤1). If the length of a member with a cross section susceptible to local buckling increases interaction between local and global buckling will be observed (see dashed line in figure 1). The load carrying capacity will decrease further till global member buckling governs the problem.

A proper design concept should account for a smooth transition between the different regions. Unfortunately neither the German code nor the American code are able to do so.

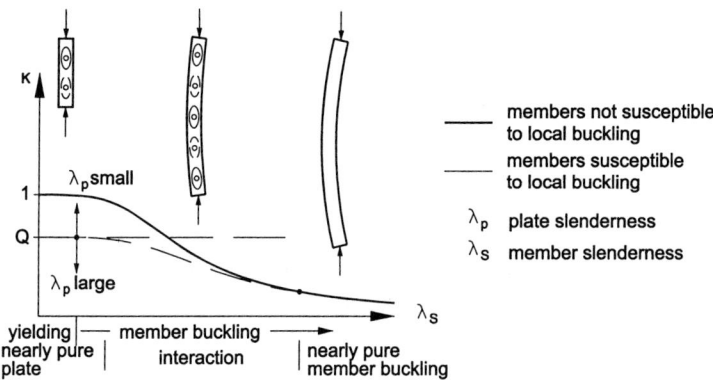

Figure 1 Interaction local and member buckling

DESIGN CONCEPTS

In German Steel Stability codes two different concepts are offered for interaction buckling. In DIN 18800 Part 3 (1990) the buckling capacity is calculated by stress reduction. The buckling capacity of the whole section is limited by the buckling capacity of the weakest section part. This is conservatively demanded for bridge structures. On the opposite DIN 18800 Part 2 (1990) is limiting the capacity of the entire cross section. The buckling capacity is evaluated using method of effective width. For light weight steel structures this approach is widely recognized.

First the method of stress reduction to allow for local buckling is discussed. For plate buckling Winter derived based on Karman´s plate theory the following Eq.

$$\kappa_p = \begin{cases} 1 & \overline{\lambda}_p \leq 0.673 \\ \dfrac{1}{\overline{\lambda}_p} - \dfrac{0.22}{\overline{\lambda}_p^2} & \overline{\lambda}_p > 0.673 \end{cases} \qquad (2)$$

This modification of v.Karman´s theory is based on experimental investigations and is used in stability codes all over the world.

Design Concept 1

The simplest way to deal with interaction between member and plate buckling is to consider firstly both phenomena independent. The final reduction factors is then evaluated by the product of the member and plate buckling reduction factors.

$$\begin{cases} \kappa^* = \kappa_s(f_y) \cdot \kappa_p(f_y) = \kappa_s \cdot \kappa_p \\ \kappa_s(f_y) \quad \text{reduction factor member buckling} \\ \kappa_p(f_y) \quad \text{reduction factor plate buckling} \end{cases} \qquad (3)$$

This approach is used in DIN 18800 part 3 (Element 503). In fig. 2 the load carrying capacity according to this method is plotted over the related member slenderness. It can be observed that there is no smooth transition between nearly pure member buckling and nearly pure plate buckling. This way the real load carrying behaviour is approximated very conservatively. This is especially true if normal stress σ_x are resulting primarily from bending moments rather than from axial compression forces and if at the same time the reduction factor κ_s is rather small. The reason for this is due to the fact that at present there are no results available for cases with pure bending and pure compression respectively.

To improve this design concept for problems with combined loading a modification of the reduction factors for member buckling is suggested as follows

$$\kappa_s = \frac{\text{load factor allowing for member buckling}}{\text{load factor neglecting for member buckling}} \quad (4)$$

The load factor μ_s allowing for member buckling in case of combined loading by N and M is evaluated according to DIN 18800 part 2 so, that the internal forces $\mu_s \cdot N$ and $\mu_s \cdot M$ satisfy Eqn.24 in DIN 18800 part 2. Thus the reduction factor for member buckling is larger.

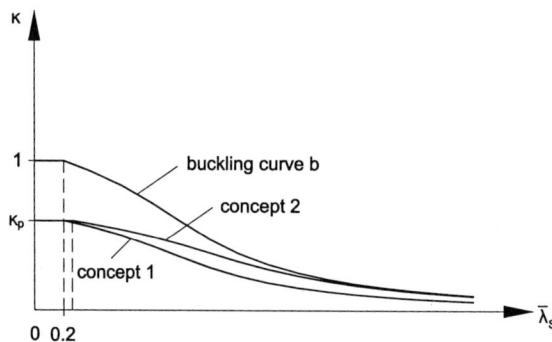

Figure 2   Comparison of design concepts 1 and 2 with european buckling curve b

Design Concept 2

Design concept 1 may be improved essentially by replacing in the equation for the reduction factor accounting for member buckling the yield stress f_y by the limit buckling stress $\kappa_p \cdot f_y$. This leads to a smaller slenderness and subsequently to less reduction for member buckling.

$$\begin{cases} \kappa^* = \kappa_s(\kappa_p(f_y) \cdot f_y) \cdot \kappa_p(f_y) \\ \kappa_s(\kappa_p(f_y) \cdot f_y) \quad \text{reduction factor member buckling} \\ \qquad\qquad\qquad \text{with limit buckling stress } \kappa_p(f_y) \cdot f_y \\ \kappa_p(f_y) \quad \text{reduction factor plate buckling} \\ \qquad\qquad\qquad \text{with yield stress } f_y \end{cases} \quad (5)$$

For concentric compression this method agrees with the approach in EC 3 and also with the out dated american standard AISI. Fig. 2 shows the load carrying capacity according to this concept depending on the related member slenderness. Especially in the medium slender area the reduction is less than in case of concept 1. For very slender members there is no difference between concepts 2 and buckling curve b. Thus the requested smooth transition is guaranteed.

In case of combined loading the reduction factor for member buckling should be - like for concept 1 - calculated applying Eqn. 4. Concept 2 can be summarized as follows. Firstly plate buckling is analyzed. Due to local buckling the member can only carry the limit buckling stress $\kappa_p \cdot f_y$ instead of the yield stress f_y. This reduces the section capacities from N_P and M_P to $\kappa_p \cdot N_{el} = \kappa_p \cdot N_P^*$ and $\kappa_p \cdot M_{el} = \kappa_p \cdot f_y \cdot W_{el}$ respectively. These capacities are to be used for member buckling design.

Great benefit for concept 1 and 2 comes from the fact that no interaction relations are required for evaluation of cross sections susceptible to local buckling.

Design Concept 3

While for concept 2 the reduction factor for member buckling is raised for concept 3 the capacity for plate buckling will be increased according to Eqn. 6.

$$\left\{ \begin{array}{ll} \kappa^* = \kappa_s(f_y) \cdot \kappa_p(\kappa_s(f_y) \cdot f_y) & \\ \kappa_s(f_y) & \text{reduction factor member buckling} \\ & \text{with yield stress } f_y \\ \kappa_p(\kappa_s(f_y) \cdot f_y) & \text{reduction factor plate buckling} \\ & \text{with member buckling stress } \kappa_s(f_y) \cdot f_y \end{array} \right. \quad (6)$$

Again the design is checked firstly with respect to pure member buckling resulting in an ultimate member buckling stress. The effect of local plate buckling is taken into account by limiting the capacity for plate buckling instead by the yield stress f_y by the ultimate member buckling stress. Thus the plate buckling slenderness is reduced and subsequently the reduction for buckling is smaller.

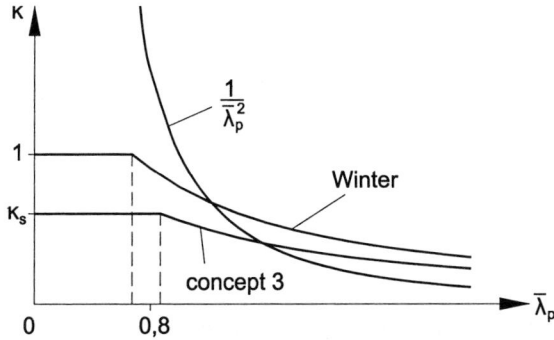

Figure 3 Comparison of concept 3 with *Winter*

For concentric compression this concepts agrees with the design approach in the current American Standard AISI (1980). Using in case of combined loading Eqn. 4 for the evaluation of the reduction factor due to member buckling no further interaction is required within this procedure. From fig. 3 it is to observe, that with Winter´s formula for plate buckling the associated reduction factor does not converge to the one for plate buckling if plate slenderness is large. This means that there is no transition to the case of nearly pure plate buckling.

BACKGROUND OF NUMERICAL INVESTIGATIONS

To account for interaction buckling in thin-walled beams and columns the numerical model must allow for geometrically and materially nonlinear behaviour and geometric imperfections. Geometric imperfections must be applied so that both instability modes - member buckling and plate buckling - are initiated. Thus the member itself and the cross section must be subjected to geometric imperfections.

While the shape of the member imperfection is quite obvious, the leading plate buckling imperfections are unknown. One way to proceed is to run in the first place a linear eigenvalue analysis and subsequently add the eigenmode associated to the lowest eigenvalue to the member buckling mode. If doing so, still the magnitude of the imperfections need to be determined.

In the present study a more practically related method is used to consider geometric imperfections within a nonlinear finite-strip approach. First the column with a geometric member imperfection is loaded with a small compression force which creates an compression stress due to axial compression and bending of 1 N/mm². The next step consists of a linear eigenvalue analysis to determine to lowest eigenvalue and the associated mode. The later is then scaled and added to the total Lagrangian displacements.

For very short columns these eigenmodes reflect pure plate buckling. For very long columns the lowest eigenvalue is due to member buckling. In this case local plate buckling does not essentially affect the member buckling behaviour. Therefore plate buckling imperfections can be omitted for this problems and the additional member imperfection will lead to conservative ultimate load capacities. Within the analysis the eigenmode is considered a member buckling mode if the critical member stress is less than the critical buckling stress of the cross section plates.

The amplitude of a member buckling imperfection is chosen according to EC 3 as

$$w_s = \alpha \, (\overline{\lambda}_s - 0.2) \, \frac{W_P}{A}$$

The amplitude of the plate buckling imperfection is taken from DIN 18800 part 3 as

$$w_p = \frac{b}{250}$$

The geometric imperfections taken into account in the analysis include also structural imperfections from residual stresses and variation of strength.

Another problem in numerical analysis comes from local stresses under the applied load. Using shell elements in the program NIPL for discretization stiffeners are arranged along the centrelines of the end sections in the numerical model to prevent stress concentration at either ends. This is also done in the experimental investigations.

NUMERICAL EXAMPLES

Examples are taken from the commentary of Lindner, Scheer & Schmidt (1993) on DIN 18800. The column consists of a thin-walled box section under axial compression (see fig. 4) with steel of grade St52 and a yield stress $f_y = 360$ N/mm².

Firstly the problem of concentric compression ($e_z = 0$) is investigated. In this case the slenderness for flanges and webs is identical (b = 390 mm, t =10 mm, $\sigma_{cr} = 499$ N/mm²). The reduction factor according to Winter is 0.872. Neglecting projecting ends both the method of reduced stresses and the method of effective widths lead to the buckling capacity of $F_{up} = 0.872 \cdot 36 \cdot 163 = 5120$ kN. Applying program NIPL for a column of length L = 249 cm (in this case $\overline{\lambda}_s$=0.2 and the effect of member buckling according DIN 18800 Part 2 and EC3 is neglectible) the ultimate load is $F_u = 0.846 \cdot 36 \cdot 163 = 4967$ kN. Though Winter´s results are known to be somewhat unconservative in this slenderness area the agreement is quite good.

In table 2 ultimate loads obtained from previously discussed concepts and from program NIPL analysis are compared to member buckling loads for column lengths up to L = 2000 cm ($\overline{\lambda}_s$=1.606). Results are scaled with respect to the section capacity of $N_P = 36 \cdot 163 = 5868$ kN. Load carrying capacities calculated according to concept 1 are conservative with respect to the numerical results. Agreement between Concepts 2 and NIPL is satisfying which in return justifies the approach of dealing with geometric imperfections in NIPL. Concept 3 leads compared to concept 1 and 2 to the highest capacities.

Next columns loaded eccentrically with $e_z = 65$ mm are investigated. Due to eccentric compression the higher compressed flange will buckle first. Using method of reduced stress the reduction factor will be 0.872 to be applied to the elastic limit load $F_{el} = 3926$ kN. Method of effective width leads to considerably higher loads since it allows for stress redistribution in the post-buckling range. Program NIPL calculates for L = 249 cm an ultimate capacity of $F_u = 0.974 \cdot 3926 = 3824$ kN.

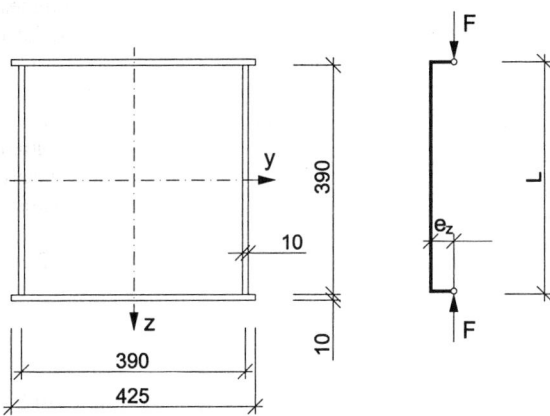

Figure 4   Thin-walled column with axial compression

TABLE 2
ULTIMATE LOADS FOR CONCENTRICALLY LOADED COLUMN OF VARIABLE LENGTH RELATED TO F=5868 KN

L (cm)	Member Buckling	Concept 1	Concept 2	Concept 3	NIPL
249	1.000	0.872	0.872	0.872	0.846
400	0.956	0.834	0.841	0.846	0.844
800	0.815	0.711	0.730	0.758	0.752
1200	0.620	0.541	0.577	0.620	0.597
1600	0.434	0.379	0.417	0.434	0.414
2000	0.306	0.267	0.298	0.306	0.296

In table 3 again ultimate loads based on concept 3 are except from one value higher than those from concepts 1 and 2. Agreement with NIPL is best for concept 2. It is to note that if ultimate buckling capacities are calculated with a more precisely (e.g. method of effective widths) than the ultimate interaction capacity with respect to concept 2 can be somewhat unconservative compared to the results of NIPL. As before ultimate loads from concept 1 are conservative.
In fig. 5 the ultimate load according to DIN 18800 part 2 for a column of length L = 800 cm is calculated as the product of the reduction factors for member buckling of 0.815 and for plate buckling of 0.872 to 0.711. However since the column is eccentrically compressed Eqn. 4 may be applied leading to a minor reduction of 0.850. Multiplying this value with the reduction factor 0.872 for plate buckling gives 0.741 which is 4.2% more capacity. For L = 2000 cm thus the reduction is 0.306 as compared to 0.267 which means an increase of 28.1%. The improvement in terms of capacity is higher the longer the column and the larger the bending moment.

TABLE 3
ULTIMATE LOADS FOR ECCENTRICALLY LOADED COLUMN OF VARIABLE LENGTH RELATED TO F=3926 KN

L (cm)	Member Buckling	DIN18800 part 3	Concept 1	Concept 2	Concept 3	NIPL
249	1.000	0.872	0.872	0.872	0.872	0.974
400	0.992	0.834	0.866	0.872	0.868	0.926
800	0.850	0.711	0.741	0.760	0.780	0.783
1200	0.681	0.541	0.594	0.624	0.666	0.622
1600	0.520	0.379	0.453	0.488	0.520	0.488
2000	0.392	0.267	0.342	0.375	0.392	0.370

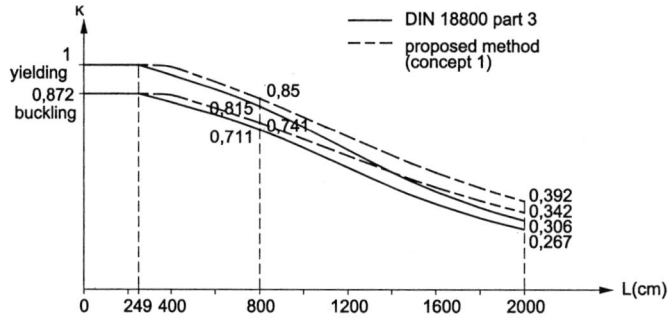

Figure 5 Comparison of DIN 18800 part 2 and concept 1

References

DIN 18800 Part 2: Stahlbauten: Stabilitätsfälle; Knicken von Stäben und Stabwerken, November 1990
DIN 18800 Part 3: Stahlbauten: Stabilitätsfälle; Plattenbeulen, November 1990
EC 3 (ENV 1993): Design of steel structures; Part 1.1: General rules and rules for buildings, 1992
AISI: Specification for the design of cold-formed steel structural members, 1980
Lindner, J., Scheer, J. und Schmidt, H.: Stahlbauten, Erläuterungen zu DIN 18800 Teil 1 bis Teil 4, Beuth Verlag GmbH Berlin, Köln, Ernst & Sohn Berlin, 1993
Winter, G.: Commentary on the 1968 edition of the specification for the design of cold-formed steel structural members, AISI, march 1970
Beedle, L.S. (Editor-in-Chief): Stability of metal structures, a world view, second edition, U. S. International Development Cooperation Agency, 1991
Fischer, M., Priebe, J. und Zhu, J.: Zum gegenwärtigen Stand des Einsatzes der Methode der wirksamen Breiten, Der Stahlbau, 10/1995, S. 326-335
Fischer, M., Zhu, J. und Priebe, J.: Die Methode der wirksamen Breiten, Forschungs-bericht aus dem Fachgebiet Stahlbau, Nr. 21, Universität Dortmund, Dezember 1996
Zhu, J. und Fischer, M.: Die Methode der wirksamen Schnittgrößen, Forschungsbericht aus dem Fachgebiet Stahlbau, Nr. 20, Universität Dortmund, Dezember 1996

Degenerate Hopf Bifurcation Phenomena of a Cantilever Beam on Elastic Foundation

S. Sensoy

Eastern Mediterranean University, Department of Civil Engineering, Gazimagusa, North Cyprus, Mersion 10 Turkey

ABSTRACT

Symmetries and/or other conditions may lead to so-called degenerate Hopf bifurcations in engineering systems. In this paper, dynamic bifurcation phenomena of a cantilever beam on elastic foundation are studied. Attention is focused on two degenerate cases, namely symmetric and flat Hof bifurcations. In the former case, the main Hopf condition is violated and a pair of complex conjugate eigenvalues touches the imaginary axis without crossing it. In the later case, another significant coefficient associated with the non-linear terms vanishes and leads to so-called flat Hopf bifurcation. In both cases, harmonically excited non-autonomous system is analyzed conveniently, with reference to the corresponding autonomous system. Bifurcation and stability analysis are performed via the intrinsic harmonic balancing (IHB) technique. The technique yields the bifurcation equation as an integral part of the perturbation procedure. It is demonstrated that unlike the symmetric Hopf bifurcation associated with the corresponding autonomous system, the harmonically excited non-autonomous system exhibits an exchange of stability at a shifted critical point where bifurcations from a family of periodic motions to a family of quasi-periodic motions on a torous take place. A symbolic computer language, namely MAPLE, facilitates the analysis as well as verification of the ordered approximations to the solutions via the IHB technique.

KEYWORDS

Degenerate, bifurcation, oscillations, flutter, perturbation, stability, instability, periodic, quasi-periodic.

INTRODUCTION

Engineering systems affected by several independent parameters may loose stability and bifurcate into another state as these parameters are varied slowly. Therefore, it is important for engineers to predict and design such dangerous developments.

Instability analysis of such systems can best be handled by understanding basic concepts and characteristics of these systems and using appropriate systematic methods. It should be noted that for post instability analysis non-linear formulation is essential. On the other hand, conservative and non-conservative systems may be identified as two board classes of systems for instability analysis. Conservative systems can exhibit divergence instabilities whereas non-conservative systems may exhibit both divergence and dynamic (flutter) instabilities. Under the influence of non-conservative forces, flutter instabilities may be observed and systems bifurcates from an initially stable state to a dynamic mode, limit cycles (e.g. Hopf bifurcation). Tacoma Narrows Bridge might be an illustration of flutter instability in structural engineering as mentioned in El Nashie (1990) and Scanlan (1998).

Degenerate Hopf bifurcations occurs when basic hypothesis in the classical Hopf bifurcation theorem are violated. Thus, symmetric Hopf bifurcation case might be observed when a pair of complex conjugate eigenvalues, as parameter of the system varies slowly, touches the imaginary axis without crossing it unlike in the case of classical Hopf bifurcation and so called flat Hopf bifurcation occurs when the curvature coefficient in the solution of Hopf bifurcation is violated. It should be emphasized that most significant contributions in solving for the local dynamics in the vicinity of critical points of degenerate Hopf bifurcation phenomena of autonomous systems, works of Andronov at. all (1973), Takens (1973), Flockerzi (1979), Kielhöfer (1979), Vanderbauwhede (1980) and Golubitsky & Langford (1981) can be shown. On the other hand, Atadan & Huseyin (1984, 1986) and Huseyin (1986) employed the intrinsic harmonic balancing technique (a perturbation method) effectively for the analysis of non-linear oscillations and bifurcation problems of systems having degeneracy.

Recently, research in this field led to the consideration of non-autonomous systems, i.e. harmonically excited systems [see for example, Wang and Huseyin (1993) and Yu et. al. (1996)]. In this paper, harmonically excited cantilever on elastic foundation, subjected to follower force, is studied and instability phenomena in the vicinity of a critical point are discussed. Attention here is focused on the degenerate cases and effect of harmonic excitation on the corresponding autonomous system is emphasized. In all the analysis, intrinsic harmonic balancing (IHB) technique is used effectively. The methodology lends itself to symbolic computer language, namely MAPLE readily, which in turn, enhances the applicability of the technique.

FORMULATION

Consider a cantilever beam on a non-linear elastic foundation over its length, l, which is subject to a concentrated follower type force and a *small* disturbing moment at its tip as shown in Figure 1.

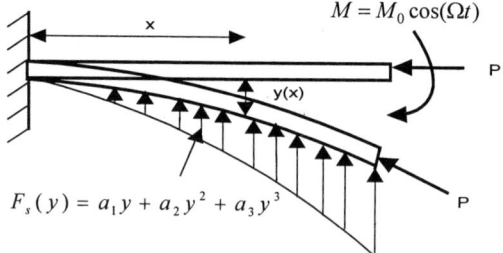

Figure 1: Cantilever beam on elastic foundation

Let S denote the bending stiffness of the beam and assume spring modulus is not a constant value and it changes with $y(x)$, such that the force exerted by the foundation to the beam is given by

$$F_s(y) = a_1 + a_2 y + a_3 y^2 \qquad (1)$$

where a_1, a_2 and a_3 are certain coefficients. Kinetic energy of the system may be written as

$$T = \frac{1}{2}m\int_0^l \left(\frac{\partial y(x;t)}{\partial t}\right)^2 dx, \qquad (2)$$

where m is the mass/length along the beam length. Strain energy of this continuum problem can be expressed as

$$U = \frac{1}{2}S\int_0^l \left[\left(\frac{\partial^2 y(x;t)}{\partial x^2}\right)^2 - 3\left(\frac{\partial^2 y(x;t)}{\partial x^2}\right)^2 \left(\frac{\partial y(x;t)}{\partial x}\right)^2\right]dx$$
$$+ \int_0^l \left[\frac{1}{2}k_1 y(x;t)^2 + \frac{1}{2}k_2 y(x;t)^3 + \frac{1}{4}k_3 y(x;t)^4\right]dx \qquad (3)$$

where $k_1 = a_1$, $k_2 = \frac{2}{3}a_2$ and $k_3 = a_3$. Damping of the system is defined by

$$D = \int_0^l \frac{1}{2}d\left(\frac{\partial y(x;t)}{\partial t}\right)^2 + \frac{1}{6}d_1\left(\frac{\partial y(x;t)}{\partial t}\right)^3 dx \qquad (4)$$

where d and d_1 are damping coefficients. Work done by the external loads during bending is expressed by

$$W = \frac{1}{2}\int_0^l P_1\left(\frac{\partial^2 y(x;t)}{\partial x^2}\right)^2 - \frac{1}{4}P_1\left(\frac{\partial^2 y(x;t)}{\partial x^2}\right)^4 dx - P_2 y(l;t) - M\frac{\partial y(x;t)}{\partial x}\bigg|_{x=l}, \qquad (5)$$

where

$$P_1 = P\cos\left(\frac{\partial y(x;t)}{\partial x}\bigg|_{x=l}\right) \text{ and } P_2 = P\sin\left(\frac{\partial y(x;t)}{\partial x}\bigg|_{x=l}\right).$$

The continuum represented above, may be discretized by assuming a displacement function as

$$y(x;t) = q_1(t)\left(x^2 - \frac{1}{2}\frac{x^4}{l^2} + \frac{1}{5}\frac{x^5}{l^3}\right) + q_2(t)\left(x^3 - \frac{x^4}{l} + \frac{3}{10}\frac{x^5}{l^2}\right) \qquad (6)$$

where q_1 and q_2 are the generalized coordinates and assumed displacement function (6) satisfy all the *natural* and *forced* boundary conditions.

In order to carry out the analysis easily, that's assign numerical values as $l = 4$, $m = 0.5$ and $d = 0.8$ where specific units are not given and units assumed to be consistent. Using the Lagrange equation

$$\frac{d}{dt}\left(\frac{\partial T}{\partial \dot{q}_i}\right) - \frac{\partial T}{\partial q_i} + \frac{\partial U}{\partial q_i} + \frac{\partial D}{\partial \dot{q}_i} = \frac{\partial W}{\partial q_i}, \qquad (7)$$

one can derive a set of second order differential equations where $\dot{q}_i = \frac{dq_i}{dt}$ and upon writing the second order differential equations in the form of first order differential equations, one obtains;

$$\frac{dz_1}{dt} = z_2,$$

$$\frac{dz_2}{dt} = (-2k_1 - 3.307P + 0.703S)z_1 - 1.2z_2 + (7.5S - 8.125P)z_3 + 47.823k_2z_1^2$$

$$+ 15.94d_1z_2^2 + 185.787k_2z_3^2 + 61.929d_1z_4^2 + 194.621k_2z_1z_3$$
$$+ 64.874d_1z_2z_4 + (520.364k_3 + 148.726P - 67.611S)z_1^3$$
$$+ (1214.599P + 2827.183k_3 - 710.769S)z_3^3$$
$$+ (1813.666P + 4868.963k_3 - 1010.874S)z_1z_3^2$$
$$+ (2772.624k_3 + 901.237P - 467.832S)z_1^2z_3 + 2.0508M_0\cos(\Omega t),$$

$$\frac{dz_3}{dt} = z_4 \quad (8)$$

$$\frac{dz_4}{dt} = (2.006P - 0.505S)z_1 + (4.888P - 2k_1 - 4.830S)z_3 - 1.2z_4 - 45.744k_2z_1^2$$

$$- 15.248d_1z_2^2 - 158.432k_2z_3^2 - 52.811d_1z_4^2 - 173.357k_2z_1z_3$$
$$- 57.786d_1z_2z_4 + (420.407k_3 - 98.063P + 43.983S)z_1^3$$
$$+ (-794.795P - 2196.270k_3 + 459.648S)z_3^3$$
$$+ (-1189.403P - 3821.174k_3 + 653.959S)z_1z_3^2$$
$$+ (-2203.782k_3 - 592.491P + 303.049S)z_1^2z_3 - 1.334M_0\cos(\Omega t),$$

upto third order terms, where $z_1 = q_1$, $z_2 = \dot{q}_1$, $z_3 = q_2$, $z_4 = \dot{q}_2$.

Fundamental equilibrium surface of the corresponding autonomous system ($M_0 = 0$) of Eqn. 8. is $z_i = f_i(\eta) = 0$ and the Jacobian matrix of the corresponding autonomous system evaluated at the fundamental equilibrium surface takes the form

$$J = \begin{bmatrix} 0 & 1 & 0 & 0 \\ (0.703S - 3.307P - 2k_1) & -1.2 & 7.5S - 8.125P & 0 \\ 0 & 0 & 0 & 1 \\ 2.006P - 0.505S & 0 & 4.888P - 2k_1 - 4.830S & -1.2 \end{bmatrix}, \quad (9)$$

where P, S and k_1 can be assumed to be the parameters of the system. The characteristic polynomial of the Jacobian can be written as

$$C(\lambda) = \lambda^4 + A_1\lambda^3 + A_2\lambda^2 + A_3\lambda + A_4, \quad (10)$$

where

$$A_1 = 2.4,$$
$$A_2 = 4.127S + 4k_1 - 1.581P + 1.44,$$
$$A_3 = 4.8k_1 - 1.897P + 4.9521S,$$
$$A_4 = 8.2534k_1S + 0.38924S^2 + 0.130P^2 + 0.2669PS + 4k_1^2 - 3.162Pk_1.$$

Using the Hurwitz criterion and recalling that $S > 0$, $k_1 > 0$ and $P > 0$ are the physical requirements, one obtains critical boundaries where divergence and/or flutter instabilities occurs as

$$0.172P^2 + 144.630 - 12.6517P + k_1 = 0 \quad \text{(flutter)}, \tag{11}$$

and

$$0.395P - k_1 - 10.317 \pm \sqrt{0.1237P^2 - 8.823P + 96.706} = 0 \quad \text{(divergence)} \tag{12}$$

after assigning a numerical value to one of the parameters, say $S = 10$.

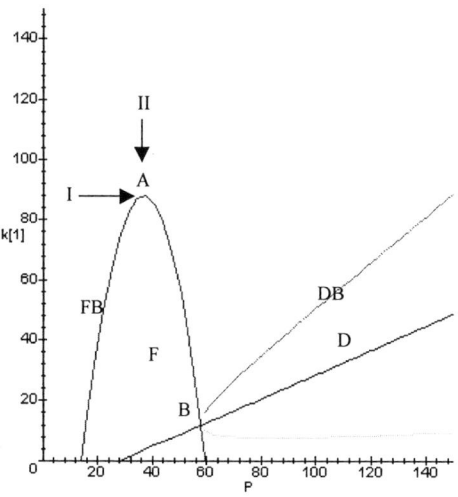

Figure 2: Flutter and divergence boundaries for beam on elastic foundation when $S = 10$

A plot of P versus k_1 for both Eqn. 11. and Eqn.12. is shown in Figure 2. In Figure 2, FB denotes the flutter boundary, DB denotes the divergence boundary and F and D represent regions where dynamic and static instabilities occurs, respectively. Point A (36.785, 88.023) on Figure 2 is on the flutter boundary and represents the extreme point of Eqn. 12. On the other hand, at point B (57.743, 12.506) flutter boundary and divergence boundary intersects and the Jacobian evaluated at that point has double zero eigenvalues where one may observe interactive static and dynamic bifurcations. Here, attention is focused on the critical point A. It is obvious that if the parameters vary in the vicinity of the critical point A, then the corresponding autonomous system exhibits Hopf bifurcation. In order to consider all the possible behavior of the system in the vicinity of the critical point A, now introduce the transformation

$$P = 36.785 + \mu_1, \text{ and } k_1 = 88.023 + \mu_2 \tag{13}$$

into Eqn. 8. $\mu_1 = \mu_2 = 0$, now identifies the critical point A. After applying certain transformations, Eqn. 8. takes the form

$$\frac{dw_i}{dt} = W_i(w_j, \mu_\alpha) + G_i(w_i, \Omega t) \quad i, j = 1,2,..,4 \text{ and } \alpha = 1,2, \tag{14}$$

which facilitates the bifurcation and stability analysis via the IHB technique.

BIFURCATION AND STABILITY ANALYSIS

IHB technique yields the ordered approximate solution and bifurcation equation as an integral part of the perturbation procedure (see e.g. Sensoy and Huseyin, 1998). Now, let us assume the solution of Eqn. 14. as

$$w_i(\tau_1,\tau_2;\varepsilon) = \sum_{\substack{m=0 \\ m_1+m_2=m}}^{M} p_{i,m_1,m_2}(\varepsilon)\cos(m_1\tau_1+m_2\tau_2)+r_{i,m_1,m_2}(\varepsilon)\sin(m_1\tau_1+m_2\tau_2) \qquad (15)$$

where $\tau_1 = \Omega t$ and $\tau_2 = \omega(\varepsilon)t$ and m_1 and m_2 may be chosen positive or negative; M is an arbitrary positive integer. It should be noted that Eqn. 15. embraces equilibrium solution, periodic solution as well as quasi-periodic motions which enables one to identify bifurcations from one solution to the other. Thus, substituting Eqn. 15. into the series of perturbation equations and balancing all the harmonics one can construct the solution in the form

$$w_i(\tau_1,\tau_2;p_{101}) = p_{101}w_i'(\tau_1,\tau_2;0)+\frac{1}{2}p_{101}^{2}w_i''(\tau_1,\tau_2;0)+O(p_{101}^{3}) \qquad (16)$$

Bifurcation equation, which identifies bifurcations from one solution to the other, is obtained as an integral part of the perturbation procedure as discussed in the following section. On the other hand, non-resonance relationship is satisfied by assuming $\Omega = 3$ since $\omega(0) = \omega_c = 12.95$

Symmetric Bifurcation

Following path I (i.e. $\mu_2 = 0$ and μ_1 varies slowly in the vicinity of A), one may observe the symmetric Hopf bifurcation phenomena of the corresponding autonomous system. Thus, applying the procedure described above

$$\frac{dp_{101}}{dt} = \frac{1}{2}p_{101}\left[-0.00123\mu^2 - \frac{1}{2}\Delta_F - \frac{1}{6}\Delta_P p_{101}^2\right] \qquad (17)$$

and

$$\frac{d\theta}{dt} \equiv \omega(p_{101}) = 12.95 - 0.0305\mu_1 \qquad (18)$$

where

$$\Delta_P = 0.27k_3 + 1.934 - 0.492 d_1 k_2 - 11.310 d_1^2 - 0.0088 k_2^2,$$

$$\Delta_F = (8.711\times10^{-7}k_3 + 5.274\times10^{-4} + 6.4\times10^{-6}d_1 k_2 + 2.133\times10^{-6}k_2^2 - 3.230\times10^{-4}d_1^2)M_0^2,$$

and ε is replaced by p_{101}, amplitude of the first harmonics, after first perturbation.

Equating right-hand side of Eqn. 17. to zero one obtains two steady-state solutions as

$$p_{101} = 0 \qquad (19)$$

and

$$p_{101}^{2} = \frac{6}{\Delta_P}\left[-0.00123\mu^2 - \frac{1}{2}\Delta_F\right], \qquad (20)$$

A careful inspection on the solution with the aid of amplitudes (solution is not given here explicitly), reveals that Eqn. 19. indicates a periodic solution with frequency Ω where Eqn. 20. represents quasi-periodic motions on an invariant torous with frequencies Ω and $\omega(p_{101})$.

which indicates symmetric bifurcation. Depending on the signs of Δ_P and Δ_F, several bifurcation phenomena may be observed. The bifurcating family of quasi-periodic solutions (if exist) is expressed as

$$\mu_1 = \sqrt{-\left[\frac{1}{2}\left(\frac{\Delta_F}{0.00123}\right) + \frac{1}{6}\left(\frac{\Delta_P}{0.00123}\right)p_{101}^2\right]} \tag{21}$$

It should be recalled that solution given by Eqn. 21. may or may not exist, depending on whether the terms under square root is positive or negative, respectively. Depending on the values of k_2, k_3 and d_1 the coefficients Δ_P and Δ_F can take positive or negative values. Therefore, the following phenomena may be observed.

a) For $\Delta_P < 0$ and $\Delta_F < 0$

As μ_1 varies in the vicinity of the critical point $\mu_{1c} = \pm\sqrt{\frac{1}{2}\left|\frac{\Delta_F}{0.00123}\right|}$, a family of periodic solution with the frequency of the harmonic excitation looses (gains) its stability at $\mu_{1c} = +\sqrt{\frac{1}{2}\left|\frac{\Delta_F}{0.00123}\right|}$ ($\mu_{1c} = -\sqrt{\frac{1}{2}\left|\frac{\Delta_F}{0.00123}\right|}$) while the bifurcating family of quasi-periodic solutions from the critical points is unstable as shown in Figure 3a. It should be noted that the corresponding autonomous ($M_0 = 0$) system in this case exhibits symmetric bifurcation where a pair unstable intersecting bifurcation paths, which represents a family of limit cycles, bifurcates from the critical point $\mu_c = 0$.

b) For $\Delta_P < 0$ and $\Delta_F > 0$

In this case, the family of periodic solutions with the frequency of the harmonic excitation retains its stability and one cannot observe any bifurcation into a family of quasi-periodic motions. The existing bifurcation path defines an unstable quasi-periodic motion as shown in Figure 3b.

c) For $\Delta_P > 0$ and $\Delta_F < 0$

The family of periodic solution looses (gains) its stability at the critical points $\mu_{1c} = \pm\sqrt{\frac{1}{2}\left|\frac{\Delta_F}{0.00123}\right|}$ and a stable bifurcating family of quasi-periodic solutions is described by a loop emanating from the lower and upper critical points, as shown in Figure 3c.

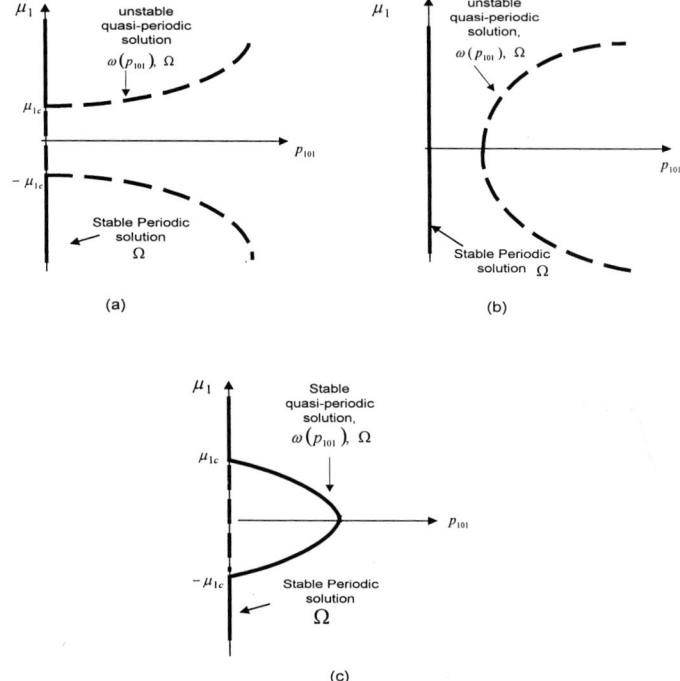

Figure 3: Symmetric bifurcation phenomena of the beam on elastic foundation.

Flat Hopf Bifurcation

Following path I (i.e. $\mu_1 = 0$ and μ_2 varies slowly in the vicinity of A), the corresponding autonomous system exhibits Hopf bifurcation and in addition if $\Delta_p = 0$, i.e. slope information is vanished the so-called flat Hopf bifurcation is observed. In this case, higher order perturbations should be carried out and information on local dynamics in the vicinity of the critical point can be obtained after four successive perturbations. Thus, the first order approximation for the bifurcation path is obtained as

$$\mu_2 = (0.00864 - 0.000839k_2 - 0.0015k_2^2)M_0^2 \\ - \frac{1}{24}(0.00435k_2^4 + 1.006k_2^3 + 91.449k_2^2 + 3621.01k_2 + 44324.114)p_{101}^4 \quad (22)$$

after introducing $d_1 = \frac{3}{2}$ into the system.

Bifurcation path associated with the flat Hopf bifurcation represents a flat curve bifurcating from the critical point $\mu_{2c} = (0.00864 - 0.000839k_2 - 0.0015k_2^2)M_0^2$, which represents a family of quasi-periodic motions. It can be shown that for $\mu_2 < \mu_{2c}$ ($\mu_2 > \mu_{2c}$) the family of periodic solution is unstable (stable). On the other hand, assuming $k_2 > 0$, then the bifurcation solution is stable as shown in Figure 4.

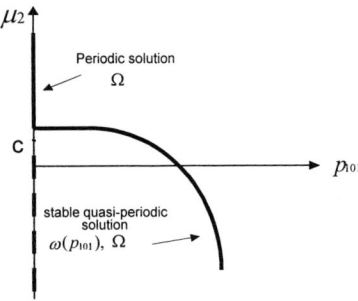

Figure 4: perturbed flat Hopf bifurcation phenomena for the beam on elastic foundation

CONCLUSIONS

It is demonstrated that a cantilever model of a beam on elastic foundation can exhibit degenerate Hopf bifurcations under the influence of a concentrated follower type force. It is observed that, choosing an appropriate path in the parameter space through a specific critical point, various interesting bifurcation phenomena associated with the flutter instability may be observed. Effects of harmonic excitation on the bifurcation and stability behaviour of the system in the vicinity of the critical point are explored with reference to the corresponding autonomous system. It is observed that bifurcations from a family of periodic motions to a family of quasi-periodic motions may be occurred in the vicinity of a pair of critical points in symmetric Hopf bifurcation case. On the other hand, when the curvature coefficient associated with Hopf bifurcation is violated, flat Hopf bifurcation phenomena is observed. In this case, critical point, where bifurcation from a family of periodic motions to the family of quasi-periodic motions occurs, is shifted due to harmonic excitation.

Symbolic computer language, MAPLE, is used extensively to obtain ordered approximation for the solution and bifurcation equations. It is expected that the exposition presented in this paper will enhance the applications to a variety of specific problems especially in wind induced oscillation problems.

REFERENCES

Andronov, A. A., Lentovich, E. A., Gordon, I. I. And Maier, A. G. (1973). *Theory of Bifurcations of Dynamical Systems in the Plane*, Israel Program for Sci. Transl., Halstead Press, Jhon Wiley, New York

Atadan, A. S. and Huseyin, K. (1984). Symmetric and Flat Bifurcations: An Oscillatory Phenomenon. *Acta Mechanica* **53**, 213-232

Atadan, A. S. and Huseyin, K. (1986). A Double Hopf Bifurcation Phenomenon. *Meccanica* **21**, 123-129.

El Nashie, M. S. (1990). *Stress, Stability and Chaos in Structural Engineering: An Energy Approach*, Mc Graw-Hill, Berkshire.

Flockerzi, D. (1979). Existence of Small Periodic Solutions of Ordinary Differential Equations in R^2. *Arch. Math.* **33**, 263-278

Golubitsky, M. and Langford, W. F. (1981). Classification and Unfoldings of Degenerate Hopf Bifurcations. *J. of Diff. Eqn.* **41**, 375-415.

Huseyin, K. (1986). *Multiple Parameter Stability, Theory and Its Applications*, Oxford University Press, Oxfored, UK.

Kielhöfer, H. (1979). Generalized Hopf Bifurcation in Hilbert Space. *Math. Meth. Appl. Sci.* **1**, 498-513.

Scanlan, R. H. (1998). Bridge Flutter Derivatives at Vortex Lock-in. *J. of Struct. Eng.* **124:4**, 450-458.

Sensoy, S. and Huseyin, K. (1998). On the Application of IHB Technique to the Analysis of Non-linear Oscillations and Bifurcations. *J. of Sound and Vibration* **215:1**, 35-46.

Takens, F. (1973). Unfoldings of Certain Singularities of Vector Fields: Generalized Hopf Bifurcations. *J. of Diff. Eqns.* **14**, 476-493.

Vanderbauwhede, A. (1980). An Abstract Setting for the Study of Hopf Bifurcation. *Nonlinear Anal.* **4**, 547-566.

Wang, S. S. and Huseyin, K. (1993). Bifurcation and Stability Properties of Non-linear Systems with Symbolic Software. *Mathl. Compt. Modeling* **18:8**, 21-38.

Yu, P., Taneri, U. and Huseyin, K. (1996). Forced Oscillations Bifurcations and Stability of Molecular Systems (Part I): non-resonance. *Int. J. Systems Science* **27**, 1339-1350.

SIMPLIFIED COMPUTATIONAL PROCEDURE FOR POSTBUCKLING EQUILIBRIUM BRANCHES IN IDEAL AND IMPERFECT PLATES

Nina Bjelajac

Department for Technical Mechanics
Faculty of Civil Engineering
University of Zagreb,
Fra A. Kačića Miošića 26, Croatia

ABSTRACT

A simplified computational method to determine post-buckling equilibrium branches for thin elastic plates subjected to in-plane loading, considering ideal and imperfect plate geometry as well is presented. The discretisation of the problem is carried out by means of the finite-difference method. Formulation of the problem and approximations involved are discussed, together with a range of algorithmic issues related to the application of the method. Plates with variable boundary and variable load conditions are considered and several illustrative examples are included. The obtained secondary paths correspond to analytical solutions. Finite difference method is very suitable for programming and sufficiently accurate, as it tends to an exact solution when the node density is increased.

KEYWORDS

Stability, critical loading, imperfection sensitive, post-critical behaviour, finite difference, equilibrium paths

1. INTRODUCTION

It is well known that the load bearing capacity of thin elastic plates, (which exhibit buckling within elastic range), is not exhausted in the post-critical load regime. This can be explained by the character of the equilibrium at the branching point and by the shapes of the post-buckling equilibrium paths, which have a stable character. Taking advantage of this phenomenon is mainly confined to metal structures, especially so in the aircraft industry and naval architecture.

In recent years a number of articles reported various methods of defining critical loads for various types of plates and various boundary conditions [11,15,17]. Although the shapes of post-buckling equilibrium branches are well understood, methods for their determination are still being developed, specially when an imperfection is taken into account. Most solutions are based on the perturbation techniques or the finite element method [1,4,6,10,19].

Here, a novel numerical procedure for the determination of post-critical equilibrium paths for thin elastic ideal and imperfect plates using the finite difference method is presented. The versatility of this method is clearly limited in comparison to finite elements for known reasons, but for regular-shaped plates of uniform thickness the procedure is very relevant. Therefore, the intention is to use the ordinary finite difference method and to develop a simplified procedure for the above mentioned problem, which could at the same time serve as a comparison scale with other numerical methods.

2. PROBLEM FORMULATION AND ASSOCIATED ITERATIVE PROCEDURES

For plates with ideal geometry, mathematical interpretation of the problem involves equilibrium equations which define the plane stress (eqn. 2.1) and plate bending problem(eqn.2.2).For small displacement assumption those equations are independent of each other and can be solved separately.

$$\frac{\partial n_x}{\partial x} + \frac{\partial n_{yx}}{\partial y} = 0 \quad , \quad \frac{\partial n_{xy}}{\partial x} + \frac{\partial n_y}{\partial y} = 0 \tag{2.1}$$

$$D\nabla^4 w = (n_x \frac{\partial^2 w}{\partial x^2} + n_y \frac{\partial^2 w}{\partial y^2} + 2n_{xy} \frac{\partial^2 w}{\partial x \partial y}) \tag{2.2}$$

At the critical load level, with small displacement theory still valid, eqn. (2.2) can be expressed as:

$$D\nabla^4 w = \lambda (n_{x0} \frac{\partial^2 w}{\partial x^2} + n_{y0} \frac{\partial^2 w}{\partial y^2} + 2n_{xy0} \frac{\partial^2 w}{\partial x \partial y}) \tag{2.3}$$

In this form it defines a critical buckling load as an eigen value problem and can be solved for critical load factor λ

$D = \frac{Eh^3}{12(1-v^2)}$ .... bending stiffness of the plate

h .... plate thickness
E .... Young's modulus
n_x, n_y, n_{xy} .... membrane forces/unit length (constant over plate thickness)
n_{x0}, n_{y0}, n_{xy0} .... membrane forces due to prescribed edge loads
w .... out of plane displacements
λ .... critical load factor
v .... Poisson's ratio

For the post-buckling regime, the theory of small strains and moderately large displacements is assumed. Only the geometric non-linearity is taken into account. Equations (2.1) and (2.2) are in this case dependent of each other and can be solved by iteration.

Taking this into account the expressions for strains are as follows:

$$\varepsilon_x \cong \frac{\partial u}{\partial x} + \frac{1}{2}\left(\frac{\partial w}{\partial x}\right)^2, \quad \varepsilon_y \cong \frac{\partial v}{\partial y} + \frac{1}{2}\left(\frac{\partial w}{\partial y}\right)^2, \quad \gamma_{xy} \cong \frac{\partial u}{\partial y} + \frac{\partial v}{\partial x} + \frac{\partial w}{\partial x}\frac{\partial w}{\partial y} \tag{2.4}$$

and the relationship between stresses and displacements becomes:

$$\sigma_x = \frac{E}{1-v^2}\left[\frac{\partial u}{\partial x} + \frac{1}{2}\left(\frac{\partial w}{\partial x}\right)^2 + v\left(\frac{\partial v}{\partial y} + \frac{1}{2}\left(\frac{\partial w}{\partial y}\right)^2\right)\right]$$

$$\sigma_y = \frac{E}{1-v^2}\left[\frac{\partial v}{\partial y} + \frac{1}{2}\left(\frac{\partial w}{\partial y}\right)^2 + v\left(\frac{\partial u}{\partial x} + \frac{1}{2}\left(\frac{\partial w}{\partial x}\right)^2\right)\right] \quad (2.5)$$

$$\sigma_{xy} = \frac{E}{2(1+v)}\left(\frac{\partial u}{\partial y} + \frac{\partial v}{\partial x} + \frac{\partial w}{\partial x}\frac{\partial w}{\partial y}\right)$$

For the plates with imperfection, equilibrium eqn.(2.2) becomes of the form:

$$D\nabla^4 w = n_x \frac{\partial^2(w+w_{oi})}{\partial x} + n_y \frac{\partial^2(w+w_{oi})}{\partial y} + n_{xy}\frac{\partial^2(w+w_{oi})}{\partial y \partial x} \quad (2.6)$$

w_{oi} .... displacement due to imperfection
w .... displacement due to the applied loading

Imperfection influences the relationship between strains and displacements as follows:

$$\varepsilon_x = \frac{\partial u}{\partial x} + \frac{1}{2}\left(\frac{\partial w}{\partial x}\right)^2 + \frac{\partial w}{\partial x}\frac{\partial w_{oi}}{\partial x}$$

$$\varepsilon_y = \frac{\partial v}{\partial y} + \frac{1}{2}\left(\frac{\partial v}{\partial y}\right)^2 + \frac{\partial w}{\partial y}\frac{\partial w_{oi}}{\partial y} \quad (2.7)$$

$$\gamma = \frac{\partial u}{\partial y} + \frac{\partial v}{\partial x} + \left(\frac{\partial w}{\partial x} + \frac{\partial w_{oi}}{\partial x}\right)\left(\frac{\partial w}{\partial y} + \frac{\partial w_{oi}}{\partial y}\right) - \frac{\partial w_{oi}}{\partial x}\frac{\partial w_{oi}}{\partial y}$$

and further the relationship between stresses and displacements:

$$\sigma_x = \frac{E}{1-v^2}\left[\frac{\partial u}{\partial x} + \frac{1}{2}\left(\frac{\partial w}{\partial x}\right)^2 + \frac{\partial w_{oi}}{\partial x}\frac{\partial w}{\partial x} + v\left(\frac{\partial v}{\partial y} + \frac{1}{2}\left(\frac{\partial w}{\partial y}\right)^2 + \frac{\partial w_{oi}}{\partial y}\frac{\partial w}{\partial y}\right)\right]$$

$$\sigma_y = \frac{E}{1-v^2}\left[\frac{\partial v}{\partial y} + \frac{1}{2}\left(\frac{\partial w}{\partial y}\right)^2 + \frac{\partial w_{oi}}{\partial y}\frac{\partial w}{\partial y} + v\left(\frac{\partial u}{\partial x} + \frac{1}{2}\left(\frac{\partial w}{\partial x}\right)^2 + \frac{\partial w_{oi}}{\partial x}\frac{\partial w}{\partial x}\right)\right] \quad (2.8)$$

$$\tau_{xy} = \frac{E}{2(1+v)}\left[\frac{\partial u}{\partial y} + \frac{\partial v}{\partial x} + \left(\frac{\partial w}{\partial x} + \frac{\partial w_{oi}}{\partial x}\right)\left(\frac{\partial w}{\partial y} + \frac{\partial w_{oi}}{\partial y}\right) - \frac{\partial w_{oi}}{\partial x}\frac{\partial w_{oi}}{\partial y}\right]$$

Eqn.(2.6) is coupled with equations (2.1) and is to be solved by iteration.

The computational procedure for evaluation of post-buckling equilibrium branches, presented in this paper, is based on the fact that the equilibrium state of the buckled plate is defined with equations which basically define the plane stress (membrane) problem and plate bending problem. So, the idea is that for each post-buckling load level the problem is divided into two parts - the *membrane problem* and the *plate bending* problem.

By applying the finite central difference operator to equilibrium eqn. (2.1), a system of algebraic equations is obtained which, for plane stress, has the following form:

$$\overline{\mathbf{K}}\mathbf{q} = \mathbf{p} \qquad (2.9)$$

$\overline{\mathbf{K}}$ .... stiffness matrix for the plane stress problem
$\mathbf{q}$ .... in plane displacement vector $(u_{i,j}, v_{i,j})$
$\mathbf{p}$ .... edge load vector (in plane)

and by applying the finite difference operator to equilibrium eqn. (2.2), a system of algebraic equations for plate bending is obtained as follows:

$$\mathbf{K}\mathbf{w} = \mathbf{A}\mathbf{w} \qquad (2.10)$$

$\mathbf{K}$ .... stiffness matrix for bending problem
$\mathbf{w}$ .... out of plane displacement vector

For determination of critical load factor a system of algebraic equations of the form $\mathbf{K}\mathbf{w} = \lambda\,\mathbf{A}\,\mathbf{w}$ is solved.

The stiffness matrix $\mathbf{K}$ for the plate bending problem and the stiffness matrix $\overline{\mathbf{K}}$ for the membrane problem, remain constant throughout the procedure. For plates with ideal geometry, first the critical load factor is determined. In post-buckling regime the resulting out of plane displacement vector $\mathbf{w}$ for a given load level, and the corresponding membrane forces, are evaluated by iteration, starting with an assumed initial displacement vector $\mathbf{w}^{(0)}$, following an iterative procedure described below:

1. Definition of the desired load level by multiplying the critical load vector with an arbitrary factor.
2. As the out of plane displacement vector $\mathbf{w}$ is not known, an initial displacement vector $\mathbf{w}^{(0)}$, proportional to the first buckling form is assumed.
3. In addition, the plane stress problem $\overline{\mathbf{K}}\mathbf{q} = \mathbf{p}$ is solved for a given load level, as if the plate is not deformed out of its plane.
4. The obtained solution does not correspond to the real state, because the given load gives rise to the out of plane displacements, which can no longer be considered small. For this reason, new expressions for internal forces must be derived, however in this form they do not satisfy the equilibrium equations for plane stress. The difference between the internal forces for the plane stress problem and for the deformed shape case (taking out of plane displacements into account) is defined with the following expressions:

$$\Delta n_x = \frac{Eh}{1-v^2}\left[\frac{1}{2}\left(\frac{\partial w}{\partial x}\right)^2 + \frac{1}{2}v\left(\frac{\partial w}{\partial y}\right)^2\right]$$

$$\Delta n_y = \frac{Eh}{1-v^2}\left[\frac{1}{2}\left(\frac{\partial w}{\partial y}\right)^2 + \frac{1}{2}v\left(\frac{\partial w}{\partial x}\right)^2\right] \qquad (2.11)$$

$$\Delta n_{xy} = \frac{Eh}{2(1+v)}\frac{\partial w}{\partial x}\frac{\partial w}{\partial y}$$

These forces can be understood as additional out-of-balance load vector $\Delta \mathbf{p}^{(0)}$

5. The next step is to solve the plane stress problem again, for the defined load and an additional part $\Delta \mathbf{p}^{(0)}$

$$\overline{\mathbf{K}}\,\mathbf{q}^{(1)} = \mathbf{p} + \Delta\,\mathbf{p}^{(0)} \qquad (2.12)$$

6. With the new values of internal forces, the load vector $\mathbf{A}\mathbf{w}^{(0)}$ for the system of finite difference equations which approximate the plate bending problem $\mathbf{K}\,\mathbf{w}^{(1)} = \lambda\,\mathbf{A}\,\mathbf{w}^{(0)}$ is defined. The solution $\mathbf{w}^{(1)}$ represents the new displacement vector, which is an improved approximation of the true displacement vector for the prescribed load as compared to the first one, $\mathbf{w}^{(0)}$.

Stages 1 to 6 represent one cycle of the iterative procedure and the procedure is repeated until the desired convergence criterion is achieved. In this way, an equilibrium position of the plate for the prescribed load in the post critical regime is derived, or in other words, one point on the secondary equilibrium path is found for the chosen location. The procedure is repeated for the new load level, until the convergence is no longer possible. Numerical instability that occurs at a certain load level may represent a new branching point, however an analysis of causes for such instability is beyond the scope of this investigation.

The iterative procedure for the determination of equilibrium paths for thin elastic plates with initial imperfection is basically the same as the procedure for plates with an ideal geometry, if the influence of imperfection is taken into account in appropriate expressions for strains and stresses ((2.7),(8.8)) . The difference is that there is no branching point. Imperfection is defined with a displacement vector $\mathbf{w}_{oi}$ in the form of the first buckling mode, multiplied by an arbitrary chosen factor (denoted with "rimp" on the charts).

3. EXAMPLES

All examples concern a model plate problem: 50 x 50 cm, h = 1 cm , mesh 13 x 13 , $\Delta x = \Delta y = 4.16$ cm, $E = 6.9 \times 10^7$ kN/m^2, $v = 0.25$
Out of plane and in plane boundary conditions are distinguished. Out of plane: simply supported, clamped, free. In plane boundaries: free, fixed in x direction, fixed in y direction, completely fixed.

Results for all model problems with the corresponding edge loads are illustrated on Figs 1-3, indicating the post-critical equilibrium paths for an ideal model and a model with imperfection, as well as the buckling modes for corresponding load case.

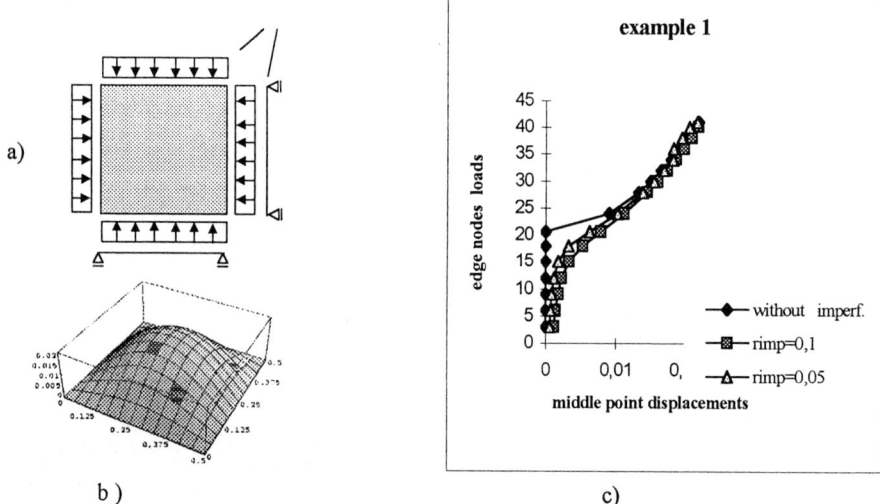

Figure 1 : Model plate 1 : a)boundary and load conditions 1; b)Equilibrium paths for perfect end imperfect model ; c) buckling mode

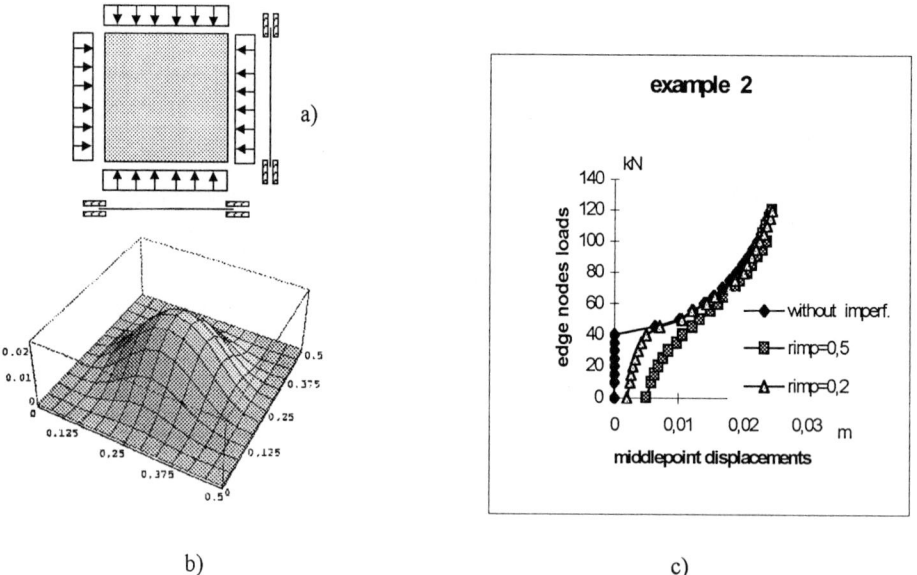

Figure 2 : Model plate 2 ; a)boundary and load conditions 1; b)Equilibrium paths for perfect end imperfect plate model ; c) buckling mode

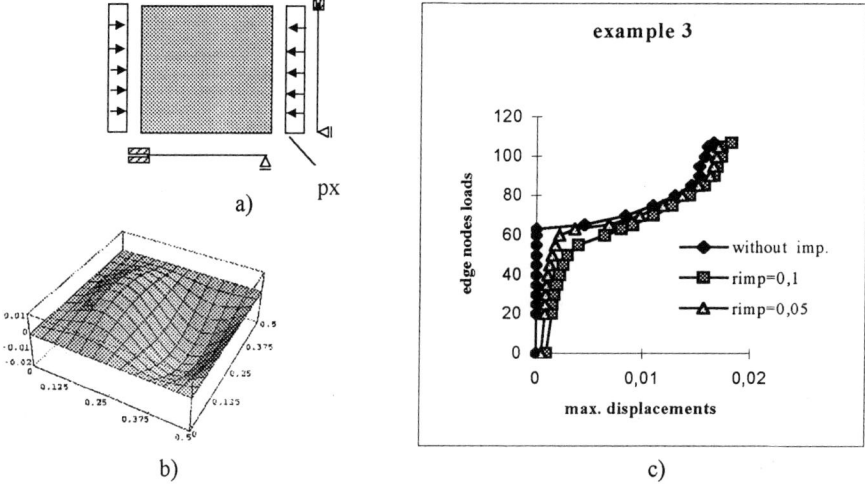

Figure 3: Model plate 3; a)Boundary and load conditions1; b) Equilibrium paths for perfect and imperfect plate model ; c) buckling mode

4. CONCLUSIONS

From the presented methodology and numerical results the following conclusions can be drawn:

- the obtained secondary equilibrium paths correspond to analytical solutions

- the effect of imperfection influences the critical load-level

- the load limit in the post-critical area, for which the plate is still in a stable equilibrium, depends on the boundary conditions

- the finite difference method is very suitable for programming and sufficiently accurate, as it tends to an exact solution when the node density is increased

- results from the formulation presented could serve as a comparison with other numerical methods

REFERENCES

[1] Casciaro R.,G.Garcea,G.Attanasio,F.Giordano,(1998) Perturbation approach to elastic postbuckling analysis, Computers and Structures,Vol.66.,N0.5, 585-595.

[2] Croll J.,G.,A.,Walker A.,C.,:The Finite Difference and Localised Ritz Methods,Int.J. for Numerical Methods in Engineering,Vol.3, 155-160,1971.

[3] Croll,J.,G.,A.,Walker, A.,C.,(1972) Elements of Structural Stability, McMillan Press Ltd, London Basinstoke,

[4] Dombourian,E.,M., Smith,C.,V., Carlson,R.,L., A Perturbation Solution to a Plate Postbuckling Problem,Int.J. Non-Linear Mechanics,Vol.11 .49-58. Pergamon Press , 1976

[5] Esben Byskov : Smooth postbuckling stresses by a modified finite element method,Int.,J.for Numerical Methods in Engineering, Vol.28, 2877-2888, 1989 .

[6] Lanzo D.,A., Garcea G.,Casciaro R.,(1999), Asymptotic Post-Buckling Analysis of rectangular Plates by HC Finite elements, Int.Journal for Numerical Methods in Engineering, Vol.38, 2325-2345 .

[7] Matkovsky Bernard J. and L.J.Putnick, (1974),: Multiple buckled states of rectangular plates,Int.J. Non-Linear Mechanics,Vol.9, 89-103.

[8] Nishino F., Hartono W.,(1989)Influental Mode of Imperfection on Carrying Capacity of Structures, Journal of Engineering Mechanics , Vol. 115, No. 10,2150-2165.

[9] Noor A. K., Peters J. M.,(1989), Buckling and Postbuckling analysis of laminated anisotropic structures, Int. Journal for numerical methods in engineering,Vol. 27 , 383-401.

[10] Palassopoulos G. V.,(1993) New Approach to buckling of imperfection- sensitive structures , Journal of Engineering Mechanics,Vol. 119, No. 4, 850-869.

[11] Riks E., (1982) : Bifurcation and Stability, A Numerical Aproach,National Aerospace Laboratory Laboratory, The Netherlands , 313-344.

[12] Ruff Daniel C.,U. Schulz : Der einfluss von Imperfektionen auf das Tragverhalten von Platten Stahlbau 68 ,Heft 10, 829-834 ,1999.

[13] K.R. Rushton ,(1971,Large deflection of plates with initial curvature.,Int.J. Sci.,vol 12, 1037-1051.

[14] Swartz E.S: O'Neill J.R.(1995), Linear Elastic Buckling of Plates Subjected to Combined Loads, Elsevier Science Limited, 1-15.

[15] Szilard, R. : Theory and Analysis of Plates, (1974)Prentice-Hall,INC.,Englewood Cliffs,New Jersey,

[16] Thompson, J.,M.,T., Hunt, G.,W., (1973):A General Theory of Elastic Stability, John Wiley & Sons Ltd.,London

[17] Thompson, J.,M.,T., Hunt, G.,W., (1984)Elastic Instability Phenomena, John Wiley & Sons,Ltd., London,

[18] Timoshenko,S.P. Gere J.,M., (1961):Theory of Elastic Stability,McGraw - Hill Company, Inc., New York,

[19] Ye,J.(1994) Large deflection of imperfect plates by iterative BE-FE method, J.of Eng. Mechanics, Vol.120,No.3, 431-442 .

[20] Wagner W. , Wriggers P. A simple method for the calculation of post-critical branches, Engineering Computations, Vol.5 , June 1988.

[21] Williams, D.,G., Aalami, B., (1979)Thin Plate Design for In-Plane Loading, Granada Publishing Limited, Crosby Locwood Staples,

DYNAMIC STABILITY OF THIN-WALLED MEMBERS

Quanfeng Wang and Yi Luo

Department of Civil Engineering, Huaqiao Univ., 362011 Quanzhou, Fujian, China

ABSTRACT

The dynamic stability of thin-walled structures subjected to periodically alternating axial forces is discussed in this paper. A system of second-order differential equations with period coefficients of the Mathieu type describes the dynamic stability of thin-walled members without damping. To solve the problem, the finite element method is applied, which can greatly reduce the process of simplification. Using MATLAB package, a computer program is developed to calculate the region of dynamic instability without damping corresponding to bending vibration, torsion and warping coupling vibration. The results prove to be more efficient and credible if compared with other calculational methods.

KEYWORDS

dynamic stability, variation stiffness, thin-walled member, excited vibration

INTRODUCTION

If a thin-walled member is subjected to a periodic longitudinal load, and if the amplitude of the load is less than that of the static buckling value, then in general, the member experiences only longitudinal vibrations. However, it can be shown that for certain relationships between the disturbing frequency θ and the natural frequency of transverse vibration ω, a thin-walled member becomes dynamically unstable and transverse vibrations occur; the amplitude of these vibrations rapidly increases to large values. A thin-walled member subjected to periodically alternating axial force is essentially variation stiffness. Linear stiffness matrix of thin-walled member remains constant, while nonlinear geometry stiffness matrix changes with periodically alternating axial force (Zhu 1998). So the problem being discussed is on dynamic stability of variation stiffness thin-walled member. Dynamic stability is one of the three criteria to dynamic design of structures (Zhang and Zhu 1996). In this paper the stability problem is solved by Finite Element Method. The results prove to be more efficient and credible if compared with those of Yang and Tong (1998 and 1992).

BASIC ASSUMPTIONS

We will make the usual assumptions in the field of strength of materials, i.e. that Hooke's law holds and plane sections remain plane. As in the case of the applied theory of vibrations, we will not include the influence of longitudinal inertia forces and the inertia forces associated with the rotation of the cross sections of the rod with respect to its own principal axes.

MATHEMATICAL MODEL

If the damping of the system is so small that it can be disregarded, then, variation stiffness thin-walled member can be represented by an assembly of finite elements connected together at the nodes. The matrix equation for axially loaded discretized system is

$$Mx'' + Kx = 0 \qquad (1)$$

where M is global mass matrix and K is global stiffness matrix. For thin-walled member subjected to a periodic longitudinal force $P = P_o + P_t Cos\theta t$, where θ is the disturbing frequency, the static and time dependent components of the load P_o and P_t can be represented as a fraction of the fundamental static buckling load P^*. Hence, putting $P = \alpha P^* + \beta P^* Cos\theta t$, with α and β as percentages of the static buckling load P^*. A periodic longitudinal force P is used to modify nonlinear geometry stiffness matrix of thin-walled member in Eq.(1), thus oscillation equation of variation stiffness thin-walled member is obtained:

$$Mx'' + \{K_e - (\alpha P^* + \beta P^* Cos\theta t)K_G\}x = 0 \qquad (2)$$

Eq.(2) is essentially a second-order differential equation with periodic coefficients, where K_e is linear stiffness matrix which reflects strain energy and K_G is nonlinear geometry stiffness matrix which reflects the influence of P_o and P_t. I representing the unit matrix, Eq.(2) may be written as:

$$x'' + M^{-1}\{K_e - (\alpha P^* + \beta P^* Cos\theta t)K_G\}x = 0$$

$$x'' + M^{-1}\{K_e - \alpha P^* K_G - \beta P^* Cos\theta t K_G\}x = 0$$

$$x'' + M^{-1}\{K_e - \alpha P^* K_G\}\{I - (K_e - \alpha P^* K_G)^{-1} * \beta P^* K_G * Cos\theta t\}x = 0$$

$$x'' + M^{-1}\{K_e - \alpha P^* K_G\}\{I - 2*(2K_e - 2\alpha P^* K_G)^{-1} * \beta P^* K_G * Cos\theta t\}x = 0$$

where

$$M^{-1}\{K_e - \alpha P^* K_G\} = \Omega^2 \qquad (3)$$

$$(2K_e - 2\alpha P^* K_G)^{-1} * \beta P^* K_G = \mu \qquad (4)$$

Eq.(1) becomes a second-order differential equation with periodic coefficients of the Mathieu type.

$$x'' + \Omega^2(I - 2*\mu*Cos\theta t)x = 0 \qquad (5)$$

Mathieu equation is called to be periodic in the sense that it satisfies Eq.(5) for every positive $T = \frac{2\pi}{\theta}$, $2T = \frac{4\pi}{\theta}$.

We seek the periodic solution with a period 2T in the form

$$x_i(t) = \sum_{k=1,3,5}^{\infty}\left(a_k \sin\frac{k\theta t}{2} + b_k \cos\frac{k\theta t}{2}\right) \qquad (6)$$

Substituting the series Eq.(6) into Eq.(5), and equating the coefficients of identical $\sin\frac{k\theta t}{2}$ and $\cos\frac{k\theta t}{2}$ leads to the following system of linear homogeneous algebraic equations in terms of a_k and b_k:

$$\begin{cases} \left(I+\mu-\dfrac{\theta^2}{4\Omega^2}\right)a_1-\mu a_3=0 \\ \left(I-\dfrac{k^2\theta^2}{4\Omega^2}\right)a_k-\mu(a_{k-2}+a_{k+2})=0 \\ \left(I-\mu-\dfrac{\theta^2}{4\Omega^2}\right)b_1-\mu b_3=0 \\ \left(I-\dfrac{k^2\theta^2}{4\Omega^2}\right)b_k-\mu(b_{k-2}+b_{k+2})=0 \end{cases} \quad (k=3,5,7,----)$$

The necessary condition for the existence of the periodic solution of Eq.(5) is that the obtained determinants of the homogeneous systems be equal to zero. Combining the two conditions under the ± sign, we obtain

$$\begin{vmatrix} I\pm\mu-\dfrac{\theta^2}{4\Omega^2} & -\mu & 0 & \cdots \\ -\mu & I-\dfrac{9\theta^2}{4\Omega^2} & -\mu & \cdots \\ 0 & -\mu & I-\dfrac{25\theta^2}{4\Omega^2} & \cdots \\ \cdots & \cdots & \cdots & \end{vmatrix}=0 \qquad (7)$$

This equation relating the frequencies of external loading with the natural frequency of the rod and the magnitude of the external force makes it possible to find regions of instability that are bounded by the periodic solutions with a period 2T. To determine the regions of instability bounded by the periodic solutions with a period T, we proceed in an analogous manner. Substituting the series

$$x_i(t)=b_{i0}+\sum_{k=2,4,6}^{\infty}\left(a_{ik}\sin\dfrac{k\theta t}{2}+b_{ik}\cos\dfrac{k\theta t}{2}\right) \qquad (8)$$

in Eq.(5), we obtain the following systems of algebraic equations:

$$\begin{cases} \left(I-\dfrac{\theta^2}{\Omega^2}\right)a_2-\mu a_4=0 \\ \left(I-\dfrac{k^2\theta^2}{4\Omega^2}\right)a_k-\mu(a_{k-2}+a_{k+2})=0 \\ b_0 I-\mu b_2=0 \\ \left(I-\dfrac{\theta^2}{\Omega^2}\right)b^2-\mu(2b_0+b_4)=0 \\ \left(I-\dfrac{k^2\theta^2}{4\Omega^2}\right)b_k-\mu(b_{k-2}+b_{k+2})=0 \end{cases} \quad (k=4,6,8,----) \qquad (9)$$

Equating the determinants of the homogeneous system to zero, we arrive at the following equations for the boundary frequencies respectively:

$$\left| \begin{array}{ccccc} I - \dfrac{\theta^2}{\Omega^2} & -\mu & 0 & \cdots \\ -\mu & I - \dfrac{\theta^2}{4\Omega^2} & -\mu & \cdots \\ 0 & -\mu & I - \dfrac{9\theta^2}{4\Omega^2} & \cdots \\ \cdots & \cdots & \cdots & \cdots \end{array} \right| = 0 \qquad (10)$$

and

$$\left| \begin{array}{ccccc} I & -\mu & 0 & 0 & \cdots \\ -2\mu & I - \dfrac{\theta^2}{\Omega^2} & -\mu & 0 & \cdots \\ 0 & -\mu & I - \dfrac{\theta^2}{4\Omega^2} & -\mu & \cdots \\ 0 & 0 & -\mu & I - \dfrac{9\theta^2}{4\Omega^2} & \cdots \\ \cdots & \cdots & \cdots & \cdots & \cdots \end{array} \right| = 0 \qquad (11)$$

DYNAMIC STABILITY OF THIN-WALLED MEMBER WITH OPEN CROSS SECTION y

Numerical Examples

Fig.1 shows an open I-shaped cross-section of thin-walled member with both ends simply supported. The following properties were taken for numerical computations: Length $L = 2m$, cross-sectional dimension: $t1 = 11.05mm$, $b1 = 203mm$, $t2 = 7.24mm$, $b2 = 203mm$, $h = 191.95mm$.
Young's modulus: $E = 2.0e11 N/m^2$; Shear modulus: $G = 0.8e11 N/mm^2$; Poisson's ratio: $\gamma = 0.25$.

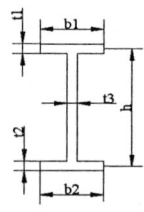

Fig. 1: I-shaped section

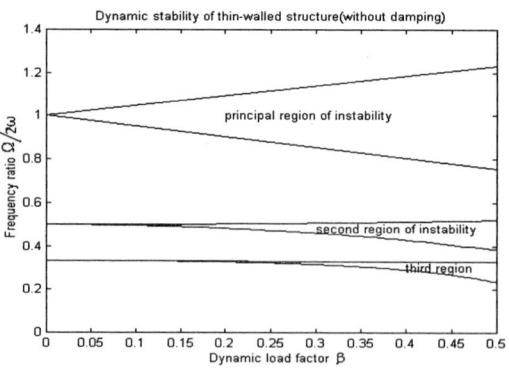

Figure 2: Dynamic stability of thin-walled structure

We are much obliged to Kiipornchai and Chan (1988) for their nonlinear thin-walled open section finite element. The MATLAB program developed here adopts their element but eliminates axial deformation, that is, we take into account of bending coupled with twist and warping deformation of thin-walled member. By using MATLAB programs to work out graph of dynamic instability regions whose Y-coordinate is the proportion of disturbing frequency to fundamental natural frequency, a graph of the distribution of the regions of instability is presented in Fig.2 :

SUMMARY AND CONCLUSION

(a). Instability of thin-walled member is a parametric excitation problem (Liu and Chen 1998). Fig.2, which is a distribution graph of the instability regions for variation stiffness thin-walled member, reveals some interesting features. First, regions of stability are larger than regions of instability. Second, judging from the magnitude of the relative width parameter, the first region of instability is large and reduces rapidly for the second and third regions, which indicates that the instability is highly dominant in the first region. Therefore the first region is always called the principal region and is generally most important while the second and third instability regions are of much less practical importance. (b). With the foregoing research, it is to be noted that the upper boundary and lower boundary of instability regions for variation stiffness thin-walled member change slightly with α. From the calculation we may see clearly that the static compressive load is most pronounced in lowering the frequencies and enlarging the width parameters of the instability regions. As α increases, i.e., the static axial load approaches the critical buckling load for the thin-walled member, the instability regions appear at lower disturbing frequencies and the size of the instability regions increases. This is due to the fact that nonlinear geometry stiffness matrix decreases as the static axial load increases, though linear stiffness matrix which reflects strain energy remains invariable. (c). We can also see from computing the example that for a certain value of static load factor α beyond a particular value, as dynamic load factor β increases beyond the limit point, the imaginary eigenvalues of the boundary of the instability region imply that the periodic solution of the Mathieu equation does not exist in that region. The stability behavior in that region is definitely unstable and this may be due to the large lateral displacement of the thin-walled member due to increasing values of static and dynamic load factors. Indeed, the thin-walled member loses its stability because nonlinear geometry stiffness matrix decreases to a lowest point with static and dynamic load on it. (d). At $\beta = 0.0$, $\mu = 0.0$, the dynamic load takes no action on the thin-walled member, which should be considered as a critical state. (e). Both parametrically excited vibrations and forced vibrations could lead to instability phenomena, and they are similar in appearance. But they are not the same dynamic response. Forced vibrations occurs when disturbing frequency is close to the natural frequency of thin-walled member, while parametrically excited vibrations take place at many cases such as: disturbing frequency is 2.0, 1.0 or 0.6 times the natural frequency. Thus, it can be concluded that to avoid dynamic instability of thin-walled member is more difficult than to prevent it from sympathetic vibrating. (f). The length, size of thin-walled members as well as boundary conditions have no influence upon graph of dynamic instability regions. The programs are also valid for variable cross section thin-walled members, even if it is not revised and enlarged. What's more, fundamental static buckling load fundamental natural frequency and frequency under axial force are obtained once done and for all. If different model of static buckling and vibrating are to be considered, changing finite element is not a tough issue.

REFERENCES

Kiipornchai S. and Chan S.L. (1988). Finite Element Analysis of Thin-Walled Structures: Stability and Non-linear Finite Element Analysis of Thin-Walled Structures, Elsevier Applied Science, London, UK

Liu Y.Z., Chen W.L. and Chen L.Q. (1998). Mechanics of Vibration, Advanced Education Press, Beijing, China (in Chinese)

Tong L.W. and Zhou G.L. (1992). The Dynamic Stability of Thin-walled Members With Closed Cross-Section Under Periodic Loading. Shanghai Journal of Mechanics **13:2,** 41~48 (in Chinese)

Yang P. And Shun L. (1998). The Dynamic Stability of Thin-walled Members Under Periodic Loading. Journal of Wuhan Transport Science and Technology University **22:4,** 403~407 (in Chinese)

Zhang A.Z. and Zhu D..P.. (1996). Practically Vibration Engineering, Aviation Industry Press, Beijing, China (in Chinese)

Zhu B.F. (1998). Theory and Application of Finite Element Method, China Water and Electricity Publishing House, Beijing, China (in Chinese)

THE COMPOUND SPLINE FINITE STRIP METHOD FOR THE ELASTIC STABILITY OF U AND C BUILT-UP COLUMNS

M. DJAFOUR[1], A. MEGNOUNIF[1] & D. KERDAL[2]

1. Civil Engineering department, University of Tlemcen, B.P 230, Tlemcen Algeria
2. Civil Engineering department, UST Oran, B.P 1505 El M'Naouer, Oran Algeria

ABSTRACT

Using the basis of the spline finite strip method, we propose a "compound" spline method to study the elastic stability of axially compressed battened columns composed of U and C shaped channels. These channels are modelled with spline finite strips while the connecting elements are represented by a twelve degrees of freedom beam finite element which its stiffness matrix is modified in order to ensure complete compatibility with the strips.

First of all, the effect of the batten plates number on slender U and C columns has been examined. Then, fixing the number of batten plates, a stability analysis was conducted. The aim is to study the variation of the critical stress and nature of the associated buckling mode with the column length.

Finally, elastic stability curves for U and C built-up columns were proposed for practical use.

KEYWORDS

Batten plate, Built-up column, Elastic stability, Spline finite strip.

INTRODUCTION

Structures such as cold formed thin walled sections usually rely upon the in-plane stiffness of the individual thin plates from which they are made up. However, these plate elements present a relatively high out-of-plane flexibility which makes them susceptible to the various types of buckling modes, local, flexural, torsional or mixed ones. [e.g, Hancock (1985), Al Bermani & al (1990)]

Sometimes, for economical point of view, these cold formed elements were assembled by connecting elements to form a built-up column as shown in figure 1. The few works done to predict the buckling load of axially compressed built-up columns proposed some theoretical formulae which are very limited in use [e.g, Engesser (1891), Müller-Breslau (1910), Timoshenko (1961)]. For these structures powerful numerical methods are available. [Niazi (1993)]

While the finite element method is in theory able to solve these problems, in practice, the writing of the program and the data preparation is often formidable and expensive. It is found that the spline finite strip method is very powerful, easier to set-up and gives accurate results. [e.g, Fan & al (1983), Hancock (1985)]

In this method, the elements are longitudinal strips of plate which are joined to one another along nodal lines running the full length of the structure. The use of B3 spline fits perfectly the displacement functions in the longitudinal direction, which have then a C^2 continuity. This method can model complex boundary conditions and local effects such as intermediate supports and is also easily adapted to the buckling analysis of built-up columns. Using the basis of this method, the authors propose a "Compound" spline finite strip method to study the elastic stability of these structures. The channels are modelled with spline finite strips while the connecting elements are represented by a twelve degrees of freedom beam finite element which its stiffness matrix is modified in order to ensure complete compatibility with the strips.

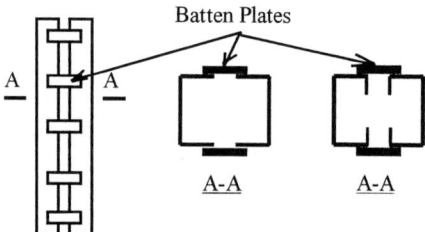

Figure 1: A built-up column

The aim of this work is to contribute to the elastic stability analysis of axially compressed U and C shaped built-up columns. First, the behaviour of long battened columns which fail by overall buckling has been examined. The number of batten plates has been varied and the resulting critical loads compared to the Engesser's values. Then, fixing the number of batten plates, a stability analysis was conducted for both U and C built-up columns in the aim of studying the variation of the critical stress and the associated of buckling load with the column length.

THE "COMPOUND" SPLINE FINITE STRIP METHOD

In the spline finite strip method, the thin walled structures are subdivided into longitudinal strips connected along nodal lines. n+1 sections, called nodal sections, divide them into n intervals. At each longitudinal end, an additional nodal section is required to handle all kinds of boundary conditions. They are located outside the strip as shown in figure 2. Each nodal section, say k, has four degrees of freedom, the flexural displacements $(v_k; \theta z_k)$ and the membrane displacements $(u_k; w_k)$. A B3-spline interpolation is used in the longitudinal direction (Z) and polynomial displacement functions are used in the transverse direction (x). For each nodal line (q), four displacement functions having C^2 continuity can be evaluated. In the global axis (X,Y,Z), the notation used is, $U_q(Z)$, $V_q(Z)$, $W_q(Z)$ and $\theta z_q(Z)$.

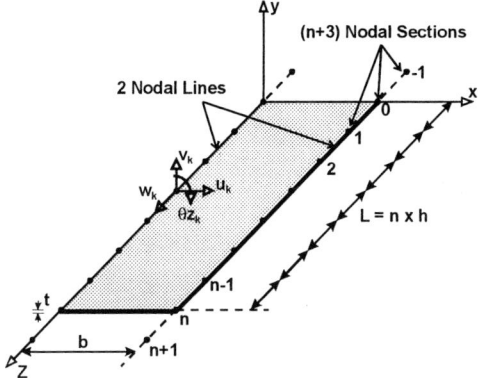

Figure 2: A typical B3-spline strip

A beam element modelling a batten plate has two nodes i and j, and six degrees of freedom per node as shown in figure 3. Its ends i and j are connected to two nodal lines, q and r, at $Z = Z_i$ and $Z = Z_j$, respectively (see figure 4). For the node i, the displacement compatibility in the global axis requires to write:

$$U_i = U_q(Z_i) \, , \, V_i = V_q(Z_i) \, , \, W_i = W_q(Z_i) \text{ and } \theta z_i = \theta z_q(Z_i) \qquad (1)$$

The remaining two degrees of freedom can be related to the nodal line displacements by:

$$\theta x_i = -\left.\frac{dV_q(Z)}{dZ}\right|_{Z=Z_i} \text{ and } \theta y_i = \left.\frac{dU_q(Z)}{dZ}\right|_{Z=Z_i} \qquad (2)$$

Equations (1) and (2), when applied to nodes i and j, define a matrix relating the beam's degrees of freedom to those of the nodal sections located at lines q and r. This is used to transform the beam stiffness matrix, and the resulting one can be easily assembled to the system stiffness matrix.

A computer program based on this technique has been developed. Its characteristics are described elsewhere, [Djafour & al (1996)]. The program calculates lowest buckling stresses and draws the corresponding failure modes.

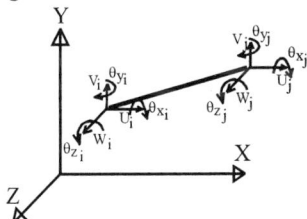

Figure 3: A beam element

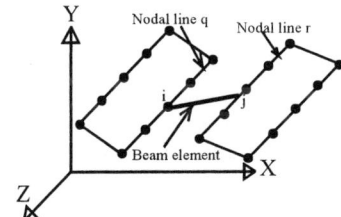

Figure 4: Two strips connected by a beam

ANALYSED BUILT-UP COLUMNS

The developed computer program has been used to study the buckling of an axially compressed and simply supported built-up column composed of two U and C shaped channels. The cross section and some characteristics are shown in figure 5.

In all examples, the U and C channels were studied with their flanges divided into two strips, web into four strips and stiffeners into one strip (see figure 6). The number of nodal sections was always greater or equal to 35.

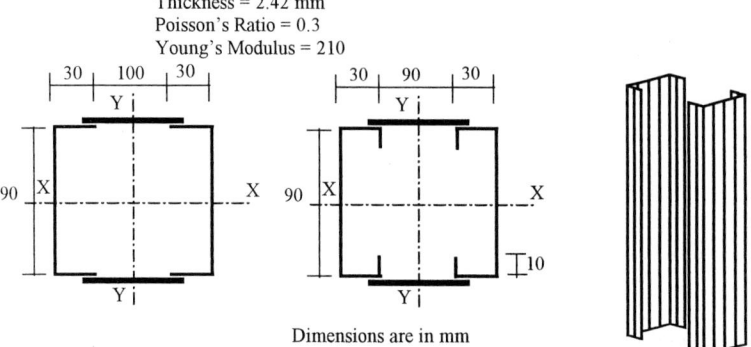

Figure 5: Built-up columns cross section   Figure 6: Finite strip model

EFFECT OF THE BATTEN PLATES NUMBER

The buckling of slender columns connected by a variable number of batten plates is first analysed and the results compared to the predictions of Engesser's formula.

In Engesser's theory, the batten plates and the channels are represented by beams. The distance "a" between two plates is constant. Hinges are supposed at mid span of the plates and, in the columns, at mid points between two plates. The critical load is:

$$P_{cr} = \frac{\pi^2.E.I}{L^2} \frac{1}{1+\frac{\pi^2.E.I}{L^2}\left(\frac{a.b}{12.E.I_b}+\frac{a^2}{24.E.I_c}\right)} \quad (3)$$

Where b is the length of the plates, E is the material Young's modulus, I, I_c and I_b are the flexural moments of inertia of the built-up column, the channels and the plates, respectively, and L is the length of the built-up column set equal to 6.0 m..

Varying the number of batten plates from 2 to 11, the elaborated program was used to calculate the lowest critical loads corresponding to different buckling modes. In order to make a direct comparison with the prediction of equation 3, only the loads corresponding to flexural buckling in the X-X direction and their respective positions have been given in table 1 for both U and C shapes. The limit cases corresponding to the isolated channels and the closed box column were also calculated with "a" infinite and "a=0", respectively.

From table 1, it is seen that the flexural buckling load given by the program depends on the number of batten plates. The difference between the results of the finite strip method and the predictions of equation 3 varies from 1.24% to 11.70% for the U shape and 0.43% to 11.90% for the C shape. The case of a=6000mm, where the difference is important, is due to the assumptions of equation 3. Assuming a=L means that there is only one batten plate located at mid span whereas in the finite strip model there is one plate at each longitudinal end. Due to symmetry, and as shown by the program results for a=6m and a=3m, for both the U and C shapes, the central batten plate does not affect the buckling load. This explains also why Engesser's prediction for a=6m is closer to the buckling load of isolated channels.

Furthermore, table 1 shows that from six batten plates for the U shape and from five batten plates for the C shape, the flexural buckling in the X-X direction is no longer critical. Instability occurs then by flexural buckling in the Y-Y direction.

TABLE 1
COMPARISON OF BUCKLING LOADS OF THE U AND C BUILT-UP COLUMNS

Nbr of batten plates	a (mm)	U Shape				C Shape			
		XX-Flexural mode Spline Finite Strip Method		Engesser	Diff (%)	XX-Flexural mode Spline Finite Strip Method		Engesser	Diff (%)
		Pcr (KN)	Mode's position	Pcr (KN)		Pcr (KN)	Mode's position	Pcr (KN)	
0	∞	3.55	1st	0	----	6.38	1st	0	----
2	6000	13.59	1st	3.76	----	23.31	1st	7.37	----
3	3000	13.62	1st	14.27	-4.56	23.31	1st	26.46	-11.90
4	2000	29.38	1st	29.75	-1.24	46.74	1st	51.32	-8.90
5	1500	48.95	1st	48.10	1.77	72.41	2nd	76.83	-5.75
6	1200	71.48	2nd	67.46	5.96	97.36	2nd	100.07	-2.71
7	1000	94.03	2nd	86.47	8.75	118.52	2nd	119.95	-1.20
9	750	134.55	2nd	120.46	11.70	148.99	2nd	149.96	-0.65
11	600	163.01	2nd	147.59	10.45	170.73	2nd	170.00	0.43
∞	0	254.86	2nd	255.69	-0.32	215.22	2nd	229.97	-6.41

CRITICAL STRESSES VERSUS COLUMN LENGTH

To study the effect of the length column on the critical stress and the associated buckling mode, several values of length column were considered, for both U and C shapes. Only three batten plates were used in this case, two are located at L/30 from each longitudinal end and the third one is positioned at mid span. The program calculates, then the ten lowest critical loads and draw the failure modes.

From these results the variation of the buckling stress with the column length is given in figures 7 and 8 for the U and C built-up columns respectively. We note that curves 1 to 10 correspond to a local buckle with a number of half waves varying from 1 to 10. Figure 9 shows some the local buckling, for both U and C shaped columns.

In figures 7 and 8 we note also that curves 11 stops the local buckling and the flexural buckling begins. There is a continuous transformation from local to flexural buckling. Some of these configurations are given in figure 10.

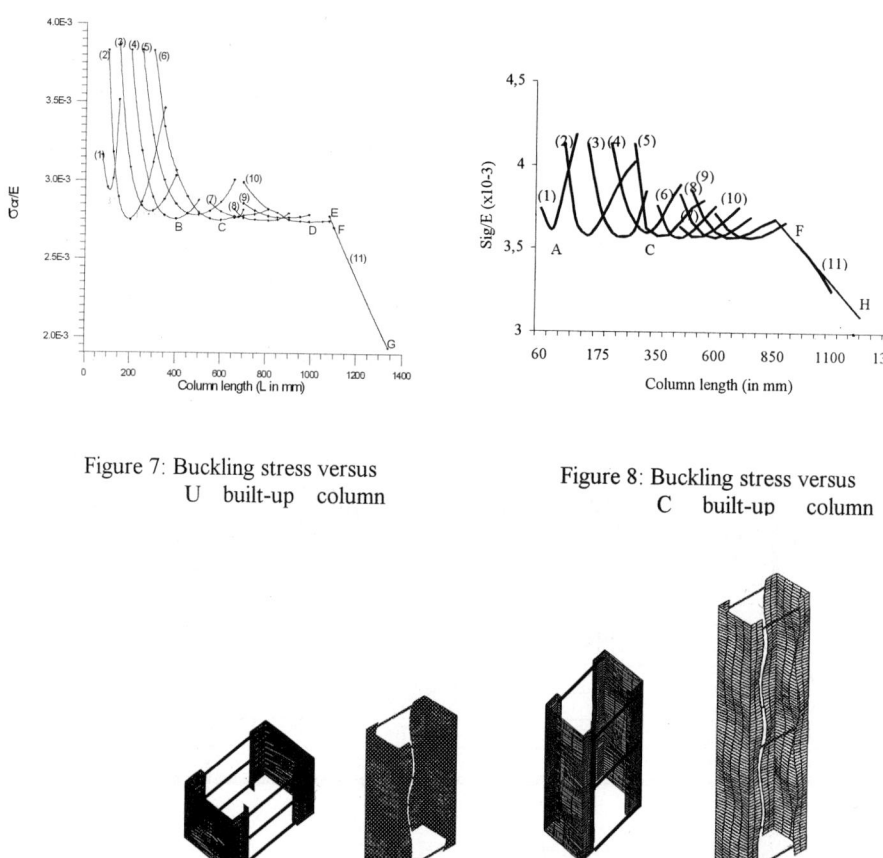

Figure 7: Buckling stress versus U built-up column

Figure 8: Buckling stress versus C built-up column

Figure 9: Some of local-buckling modes of the U and C battened columns.

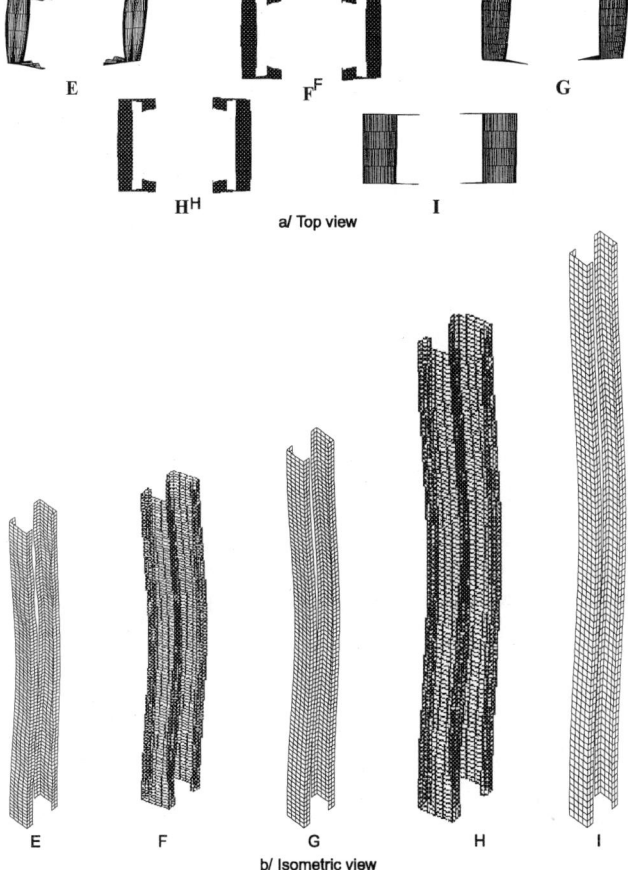

Figure 10: Some of the flexural buckling modes of the U and C built-up columns

Every local buckling curves in figure 7 and 8 displays a minimum. In table 2, the corresponding critical stresses are compared with that obtained by the finite strip method for the U and C shaped channels [Kerdal & al (1995); Kerdal & al (1995)].

For the U built-up column, it is seen that odd number of half waves gives a more significant difference with the finite strip prediction than the even ones. This is result of the restraining effect of the mid-span batten plate, which is more solicited in buckling with odd number of half-waves. But for the C built-up columns the two results are closer for all values of m.

Thus, for this example we can conclude that the batten plates have an influence on the critical loads and the failure modes of the short built-up columns.

TABLE 2
MINIMUM LOCAL BUCKLING STRESSES

Number of half-waves		m=1	m=2	m=3	m=4	m=5	m=6	m=7	m=8	m=9	m=10
σcr/E (×10⁻³)	Built-up U column	2.957	2.747	2.871	2.752	2.852	2.756	-	2.747	-	2.741
	Single channel	2.724	2.724	2.724	2.724	2.724	2.724	2.724	2.724	2.724	2.724
Difference (%)		8.55	0.84	5.40	1.03	4.70	1.17	-	0.84	-	0.62
σcr/E (×10⁻³)	Built-up C column	3.611	3.575	3.568	3.571	3.577	3.567	3.572	3.567	3.567	3.567
	Single channel	3.565	3.565	3.565	3.565	3.565	3.565	3.565	3.565	3.565	3.565
Difference (%)		1.30	0.30	0.10	0.60	0.34	0.05	0.20	0.05	0.05	0.05

CONCLUSION

Elastic stability analysis of U and C built-up column is held using a technique based on the spline finite strip method. The channels can have any arbitrary cross section and any boundary conditions. The method is able to predict the buckling load incorporating all the possible failure modes (local, flexural, torsional and mixed modes).

The proposed method has been applied to study the buckling of axially compressed battened columns composed of U and C shaped channels. First, the effect of the batten plates number on slender columns has been examined and the results compared with the predictions of a theoretical formula. The difference between the two predictions varies from 1.24% to 11.70% for the U shape and 0.43% to 11.90% for the C shape.

Then, a stability analysis of a column connected by only three batten plates has been performed. The results show that the batten plates have an influence on the critical loads and the failure modes of the short built-up columns.

REFERENCES

Al-Bermani, F. G. A., and Kitipornchai, S. (1990). Non-linear analysis of thin-walled structures using least element/number. *J. Struct. Eng., ASCE*, **116:1**, 215-234.

Djafour, M. and Megnounif, A. (1996). Etude de la stabilité des colonnes à étrésillons par la méthode des spline bandes finies. *Rapport de recherche*, Université de Tlemcen.

Engesser, F. (1891). Die knickfestigkeitgerader stabe zentralblatt der bauverwaltung. Berlin.

Fan, S. C., and Cheung, Y. K. (1983). Analysis of shallow shells by spline finite strip method. *Engrg. Structures*, **5:4**, 255-263.

Hancock, G. J., and Lau, C. W. (1985). Buckling of thin flat walled structures by a spline finite strip method. *Research report N°R487*, University of Sidney.

Kerdal, D., Djafour, M. and Megnounif, A. (1995). Etude du voilement de profils à parois minces en U soumis à la compression. *Revue Algérie Equipement*, **22**, 7-11.

Kerdal, D., Djafour, M. and Megnounif, A. (1995). Stabilité élastique des profils en C soumis à la compression. *Revue Algérie Equipement*, **21**, 13-17.

Müller-Breslau, H. (1910). Uber exzentrisch gedruckte gigliederte stabe. *Stizungsberichte d.k. preuss, Akad. d. wissenschaften*, Berlin.

Niazi, A. (1993). Contribution à l'étude de la stabilité des structures composées de profils à parois minces et section ouverte de type C. *Thèse de doctorat*, Université de Liège.

Timoshenko, S. P., and Gere, J. M. (1961). *Theory of elastic stability*. Mc Graw-Hill, New York

VIBRATION AND DYNAMIC ANALYSIS

STOCHASTIC TRANSIENT VARIATIONAL PRINCIPLE IN VIBRATION ANALYSIS

L.F.Yang[1*]   Q.S.Li[1]   J.Z.Zhang[2]   A.Y.T.Leung[1]

[1]Department of Building and Construction, City University of Hong Kong,
Tat Chee Avenue, Kowloon, Hong Kong, China
[2]Department of Civil Engineering, Wuhan University of Technology, Hubei, China

ABSTRACT

This paper demonstrates the dependence of virtual displacements on real displacements, hence on the material properties and loading conditions of the structures. For stochastic structures involved with physical and/or geometrical stochastic parameters, the virtual displacements are regarded as stochastic quantities. Therefore, a general purposes version of stochastic transient variational principle (STVP) is developed for vibration analysis of linear continuum. Probabilistic distributions of random parameters are consistently incorporated. Stochasticity of all random quantities, as well as their variations, is taken into consideration. The second-order perturbation techniques are employed to expand all the random quantities involved in the energy functional. Accordingly, a set of deterministic recursive equations is obtained as the alternative expressions of the stochastic transient variational principle (STVP). On the basis of the STVP, the stochastic finite element method (SFEM) is developed for vibration analysis, so that the roundabout procedures for formulations of the SFEM are avoided.

KEYWORDS

Stochastic, transient variational principle, perturbation technique, stochastic finite element method, virtual displacements, and structural dynamics

INTRODUCTION

In the field of numerical analysis of uncertain structures, the stochastic finite element methods (SFEM) were widely investigated. However, most of the formulations of them are based on the framework of the direct stiffness method, which imposed the probabilistic distributions into the characteristic equations of the FEM directly. In order to deal with the different sorts of randomness in a natural and consistent manner, some stochastic variational principles were proposed, and overcome some drawbacks of the direct stiffness method. Liu et al. proposed the single and the three field stochastic variational principles by means of minimum potential energy principle (Liu, etc. 1986) and Hu-

---

[*]Corresponding Author. On leave from Guangxi University, Dept. of Civil Engineering

Washizu principle (Liu, etc. 1988), respectively. Hien and Kleiber (1990) developed the stochastic Hamilton principle for linear dynamic problems. Elishakoff etc. (1996) developed a variational principles for stochastic beams. The last one is not based on the perturbation techniques, however is limited to some specific cases. Liu et al. (1986,1988) and Hien and Kleiber (1990) took the virtual displacements as deterministic quantities, and developed the stochsatic variational principles (SVP) of general purposes version. This paper demonstrates the dependence of virtual displacements on real displacements, hence on the material properties and loading conditions of the structures. Therefore, a general purposes version of STVP is developed for vibration analysis. Probabilistic distributions of random parameters are consistently incorporated into the transient energy functional, and the roundabout procedures for formulation of the SFEM are avoided..

THERECICAL DRAWBACKS OF THE SFEM

For SFEM formulated within the framework of direct stiffness matrix, the randomness are incorporated into the characteristic equations of the FEM directly. Considering the formulation of deterministic FEM based on the deterministic variational principle, it is evident that the direct stiffness matrix approach was developed through a roundabout procedure of two steps: step 1, the functional of transient potential energy was established neglecting stochasticity invovled, and read:

$$\Pi = \int_V [A - \{u\}^T (\{F\} + \{F_I\} + \{F_D\})]dV - \int_{S_\sigma} \{u\}^T \{T\}dS \qquad (1)$$

in which, $\{u\}$, $\{\varepsilon\}$ are displacement components and strain components, respectively; $\{F\}$ and $\{T\}$ denote body force and tractions, respectively; V denotes the domain of interest; the boundary of the domain S is simply divided into two parts: traction boundary S_σ and the prescribed boundary S_u, and satisfy: $S_\sigma \cup S_u = S$ and $S_\sigma \cup S_u = 0$. A, $\{F_I\}$ and $\{F_D\}$ denotes the strain energy, inertia force and damping force vector, respectively. For linear structural systems, they are given by:

$$A = \frac{1}{2}\{\sigma\}^T\{\varepsilon\} = \frac{1}{2}\{\varepsilon\}^T[D]\{\varepsilon\} \qquad (2)$$

$$\{F_D\} = -c\{\dot{u}\} \quad \{F_I\} = -\rho\{\ddot{u}\} \qquad (3a,b)$$

in which, ρ and c denote the density and damping coefficients, respectively.

In FEM analysis, the domain of interest is discretized by the finite element idealization. For a typical element, its displacement model shows:

$$\{u\} = [N]\{a\} \qquad (4)$$

where, $[N]$ is the shape function matrix, $\{a\}$ denotes the nodal displacement vector, $\{u\}$ and $\{a\}$ are both functions of time t. Based on the transient variational energy principle, the dynamic equations of motion for the deterministic FEM is obtained, and given in matrix form as:

$$[K]\{a\} + [M]\{\ddot{a}\} + [C]\{\dot{a}\} = \{P\} \qquad (5)$$

where, $[K]$, $[M]$ and $[C]$ denote the stiffness matrix, mass matrix and damping matrix, respectively; $\{P\}$ denotes the nodal force vector; the dot and double dot represent the first and second derivatives with respect to time t, respectively.

Step 2, stochastic properties of parameters and variables involved are incorporated into Eqn.4, so that matrices $[K]$, $[M]$, $[C]$ and $\{P\}$ may be of stochastic quantities. Therefore, Eqn.4 is a stochastic equation. Based on the second order perturbation techniques, the stochastic matrix $[K]$, $[M]$, $[C]$, $\{P\}$ and $\{a\}$ can be expanded about the mean value of the random field $\{X\}$, denoted by $\{\overline{X}\}$, via Taylor series, and retained only up to second order terms as:

$$(\bullet) = (\overline{\bullet}) + \sum_{m=1}^{r} \alpha_m (\bullet)_{,m} + \frac{1}{2} \sum_{m,n=1}^{r} \alpha_m \alpha_n (\bullet)_{,mn} \qquad (6)$$

where, $(\bullet)$ denotes any of the stochastic quantities $[K]$, $[M]$, $[C]$, $\{P\}$ or $\{a\}$; $(\overline{\bullet})$ denotes $[\overline{K}]$, $[\overline{M}]$, $[\overline{C}]$, $\{\overline{P}\}$, and $\{\overline{a}\}$, representing values of $[K]$, $[M]$, $[C]$, $\{P\}$, and $\{a\}$ evaluated at $\{\overline{X}\}$, respectively; similarly, $(\bullet)_{,m}$ denotes $[K]_{,m}$, $[M]_{,m}$, $[C]_{,m}$, $\{P\}_{,m}$, and $\{a\}_{,m}$, representing the first order partial derivatives of $[K]$, $[M]$, $[C]$, $\{P\}$, and $\{a\}$ with respect to X_m, evaluated at $\{\overline{X}\}$, respectively; $(\bullet)_{,mn}$ denotes $[K]_{,mn}$, $[M]_{,mn}$, $[C]_{,mn}$, $\{P\}_{,mn}$, and $\{a\}_{,mn}$, representing the second order partial derivatives of $[K]$, $[M]$, $[C]$, $\{P\}$, and $\{a\}$ with respect to X_m and X_n, evaluated at $\{\overline{X}\}$, respectively; α_i, $i=1,2,\cdots,r$, are small parameters, and can be expressed by $\alpha_i = X_i - \overline{X}_i$.

Substituting the expanded expressions of $[K]$, $[M]$, $[C]$, $\{P\}$ and $\{a\}$, given by Eqn.6, into Eqn.4, and equating coefficients of like powers of α_i, the dynamic equations of motion for SFEM are obtained:

$$[\overline{K}]\{\overline{a}\} + [\overline{M}]\{\ddot{\overline{a}}\} + [\overline{C}]\{\dot{\overline{a}}\} = \{\overline{P}\} \qquad (7a)$$

$$[\overline{K}]\{a\}_{,n} + [\overline{M}]\{\ddot{a}\}_{,n} + [\overline{C}]\{\dot{a}\}_{,n} = \{P\}_{,n} - [K]_{,n}\{\overline{a}\} - [M]_{,n}\{\ddot{\overline{a}}\} - [C]_{,n}\{\dot{\overline{a}}\} \qquad (7b)$$

$$[\overline{K}]\{a''\} + [\overline{M}]\{\ddot{a}''\} + [\overline{C}]\{\dot{a}''\} = \{P''\} \qquad (7c)$$

in which,

$$\{a''\} = \frac{1}{2}\sum_{i,j=1}^{r} \text{Cov}(X_i, X_j)\{a\}_{,ij} \qquad (8a)$$

$$\{P''\} = \frac{1}{2}\sum_{i,j=1}^{r} \text{Cov}(X_i, X_j)\big(\{P\}_{,mn} - [K]_{,m}\{a\}_{,n} - [\overline{M}]_{,m}\{\ddot{a}\}_{,n} - [\overline{C}]_{,m}\{\dot{a}\}_{,n}$$
$$- [K]_{,n}\{a\}_{,m} - [\overline{M}]_{,n}\{\ddot{a}\}_{,m} - [\overline{C}]_{,n}\{\dot{a}\}_{,m} - [K]_{,mn}\{\overline{a}\} - [M]_{,mn}\{\ddot{\overline{a}}\} - [C]_{,mn}\{\dot{\overline{a}}\}\big) \qquad (8b)$$

Once $\{\overline{a}\}$, $\{\dot{\overline{a}}\}$ and $\{\ddot{\overline{a}}\}$ are solved by Eqns.7a, $\{a\}_{,n}$, $\{\dot{a}\}_{,n}$ $\{\ddot{a}\}_{,n}$ and, $\{a''\}$ $\{\dot{a}''\}$, $\{\ddot{a}''\}$ can be determined recursively. Then the first and second order moments of the stochastic nodal displacement vector $\{a\}$ are given by:

$$E[\{a\}] = \{\overline{a}\} + \{a''\} \; ; \quad \text{Var}[\{a\}] = \sum_{i,j=1}^{r} \text{Cov}(X_i, X_j)\{a\}_{,i}\{a\}_{,j} \qquad (9)$$

In this two-step procedure, the stochastic properties of parameters and variables involved in the transient potential energy functional, as given by Eqn.1, are neglected, such that the deterministic dynamic equations of motion for the deterministic FEM are established by the deterministic transient variational principle. Then, the stochastic properties are incorporated directly into these dynamics equations of motion to yield the characteristic equations of the SFEM. Therefore, it is not too much to say that the direct stiffness method takes a roundabout course of "neglecting at first and then incorporating". Might the neglect of stochasticity cause any errors in formulation of dynamic equation? The direct stiffness matrix method could not answer it theoretically. On the other hand, the random properties, as pointed out by Liu, etc. (1986), were not incorporated consistently into the SFEM within the framework of the direct stiffness matrix method, where only the equilibrium is satisfied in a weak sense while constitutive laws and prescribed boundary conditions are satisfied in a strong sense.

In order to deal with a number of classes of randomness in a natural and concise manner, and to lay a theoretical basis for stochastic numerical methods, Liu et al. (1986, 1988), Hien and Kleiber (1990) proposed the stochastic variational principles (SVP) and the stochastic Hamilton principle based on the virtual displacement principle. In these versions of SVP, the deterministic functional of the transient potential energy Π, as expressed by Eqn.1, was established neglecting all the uncertainties involved. Hence, the first variation of Π read:

$$\delta \Pi = \int_V [\delta A - (\{F\} + \{F\}_I + \{F\}_D) \delta\{u\}] dV - \int_{S_\sigma} \{T\} \delta\{u\} dS \tag{10}$$

Then, the stochasticity of material property and the loads are incorporated into Eqn.10, so that Π, A, $\{F\}$, $\{F_I\}$, $\{F_D\}$, $\{T\}$, $\{u\}$, and the stain vector $\{\varepsilon\}$ are all treated as stochastic quantities, and can be expanded about $\{\overline{X}\}$ via Taylor series, as shown by Eqn.6.

By Eqn.2, the first variation of A shows:

$$\delta A = \delta\{\varepsilon\}^T [D]\{\varepsilon\} \tag{11}$$

It is noteworthy that, in this version of SVP, though the displacement vector $\{u\}$ is assumed to be stochastic, the first variation of $\{u\}$, denoted by $\delta\{u\}$, is assumed to be a deterministic quantity, and independent of any uncertainties, since $\delta\{u\}$ is arbitrary and independent of material properties or the external load.

According to the relationship between $\{\varepsilon\}$ and $\{u\}$, one has:

$$\delta\{\varepsilon\} = [L]\delta\{u\}; \quad \{\overline{\varepsilon}\} = [L]\{\overline{u}\}; \quad \{\varepsilon\}_{,m} = [L]\{u\}_{,m}; \quad \{\varepsilon\}_{,mn} = [L]\{u\}_{,mn} \tag{12a,b,c,d}$$

in which, $[L]$ denotes a differential operator matrix, it is independent of the material properties and loading conditions. Therefore, $\delta\{\varepsilon\}$ may also be assumed independent of any uncertainties, as $\delta\{u\}$ does, though the strain vector $\{\varepsilon\}$ is stochastic. Substituting the expanded expressions of δA, $[D]$ and $\{\varepsilon\}$ into Eqn.11, and equating the like power of $\alpha_{,m}$, one has:

$$\delta\overline{A} = \delta\{\varepsilon\}^T [\overline{D}]\{\overline{\varepsilon}\} \tag{13a}$$

$$\delta A_{,m} = \delta\{\varepsilon\}^T ([\overline{D}]\{\varepsilon\}_{,m} + [D]_{,m}\{\overline{\varepsilon}\}) \tag{13b}$$

$$\delta A_{,mn} = \delta\{\varepsilon\}^T ([\overline{D}]\{\varepsilon\}_{,mn} + [D]_{,m}\{\varepsilon\}_{,n} + [D]_{,n}\{\varepsilon\}_{,m} + [D]_{,mn}\{\overline{\varepsilon}\}) \tag{13c}$$

Similarly, according to Eqn.4, one has:

$$\{\overline{u}\} = [N]\{\overline{a}\} \quad \{u\}_{,m} = [N]\{a\}_{,m} \quad \{u\}_{,mn} = [N]\{a\}_{,mn} \tag{14a,b,c}$$

Substituting Eqns.12 into Eqns.13 leads to:

$$\delta \overline{A} = \delta \{u\}^T [L]^T [\overline{D}][L]\{\overline{u}\} \tag{15a}$$

$$\delta A_{,m} = \delta \{u\}^T [L]^T \left([\overline{D}][L]\{u\}_{,m} + [D]_{,m}[L]\{\overline{u}\} \right) \tag{15b}$$

$$\delta A_{,mn} = \delta \{u\}^T [L]^T \left([\overline{D}][L]\{u\}_{,mn} + [D]_{,m}[L]\{u\}_{,n} + [D]_{,n}[L]\{u\}_{,m} + [D]_{,mn}[L]\{\overline{u}\} \right) \tag{15c}$$

Analogously, substitution of the expanded expressions of δA, $\delta \Pi$, $\{F\}$, $\{F_I\}$, $\{F_D\}$ and $\{T\}$ into Eqn.10, equating the same power of coefficients $\alpha_{,m}$, the zeroth, first and second order potential energy functional are shown as:

$$\delta \overline{\Pi} = \int_V [\delta \overline{A} - \delta \{u\}^T (\{\overline{F}\} + \{\overline{F}_I\} + \{\overline{F}_D\})] dV - \int_{S\sigma} \{\delta \{u\}^T \{\overline{T}\} dS \tag{16a}$$

$$\delta \Pi_{,m} = \int_V [\delta A_{,m} - \delta \{u\}^T (\{F\}_{,m} + \{F_I\}_{,m} + \{F_D\}_{,m})] dV - \int_{S\sigma} \delta \{u\}^T \{T\}_{,m} dS \tag{16b}$$

$$\delta \Pi_{,mn} = \int_V [\delta A_{,mn} - \delta \{u\}^T (\{F\}_{,mn} + \{F_I\}_{,mn} + \{F_D\}_{,mn})] dV - \int_{S\sigma} \delta \{u\}^T \{T\}_{,mn} dS \tag{16c}$$

in which

$$\{\overline{F}_I\} = -\overline{\rho}\{\ddot{\overline{u}}\}; \quad \{F_I\}_{,m} = -\overline{\rho}\{\ddot{u}\}_{,m} - \rho_{,m}\{\ddot{\overline{u}}\}; \quad \{F_I\}_{,mn} = -\overline{\rho}\{\ddot{u}\}_{,mn} - \rho_{,m}\{\ddot{u}\}_{,n} - \rho_{,n}\{\ddot{u}\}_{,m} - \rho_{,mn}\{\ddot{\overline{u}}\} \tag{17}$$

$$\{\overline{F}_D\} = -\overline{c}\{\dot{\overline{u}}\}; \quad \{F_D\}_{,m} = -\overline{c}\{\dot{u}\}_{,m} - c_{,m}\{\dot{\overline{u}}\}; \quad \{F_D\}_{,mn} = -\overline{c}\{\dot{u}\}_{,mn} - c_{,m}\{\dot{u}\}_{,n} - c_{,n}\{\dot{u}\}_{,m} - c_{,mn}\{\dot{\overline{u}}\} \tag{18}$$

Substituting Eqns.15 into Eqns.16, the stationary conditions, shown as $\delta \overline{\Pi} = 0$, $\delta \Pi_{,m} = 0$ and $\delta \Pi_{,mn} = 0$, lead to:

$$\int_V \delta \{u\}^T [[L]^T [\overline{D}][L]\{\overline{u}\} - (\{\overline{F}\} + \{\overline{F}_I\} + \{\overline{F}_D\})] dV - \int_{S\sigma} \{\delta \{u\}^T \{\overline{T}\} dS = 0 \tag{19a}$$

$$\int_V \delta \{u\}^T \left([L]^T [\overline{D}][L]\{u\}_{,m} + [L]^T [D]_{,m}[L]\{\overline{u}\} - \{F\}_{,m} - \{F_I\}_{,m} - \{F_D\}_{,m} \right) dV$$
$$- \int_{S\sigma} \delta \{u\}^T \{T\}_{,m} dS = 0 \tag{19b}$$

$$\int_V \delta \{u\}^T \left([L]^T [\overline{D}][L]\{u\}_{,mn} + [L]^T [D]_{,m}[L]\{u\}_{,n} + [L]^T [D]_{,n}[L]\{u\}_{,m} \right.$$
$$\left. + [L]^T [D]_{,mn}[L]\{\overline{u}\} - \{F\}_{,mn} - \{F_I\}_{,mn} - \{F_D\}_{,mn} \right) dV - \int_{S\sigma} \delta \{u\}^T \{T\}_{,mn} dS = 0 \tag{19c}$$

Substitution of Eqns.14 into Eqns.19 yields the same recursive characteristic equations of motion for SFEM, as shown in Eqns.7 and Eqns.8, where:

$$[\overline{K}] = \int_V [B]^T [\overline{D}][B] dV \quad [\overline{M}] = \int_V \overline{\rho}[N]^T [N] dV \quad [\overline{C}] = \int_V \overline{c}[N]^T [N] dV$$

$$[K]_{,m} = \int_V [B]^T [D]_{,m}[B] dV \quad [M]_{,m} = \int_V \rho_{,m}[N]^T [N] dV \quad [C]_{,m} = \int_V c_{,m}[N]^T [N] dV$$

$$[K]_{,mn} = \int_V [B]^T [D]_{,mn}[B] dV \quad [M]_{,mn} = \int_V \rho_{,mn}[N]^T [N] dV \quad [C]_{,mn} = \int_V c_{,mn}[N]^T [N] dV$$

$$\{\overline{P}\} = \int_V [N]^T \{\overline{F}\} dV + \int_{S\sigma} [N]^T \{\overline{T}\} dS \; ; \quad \{P\}_{,n} = \int_V [N]^T \{F\}_{,n} dV + \int_{S\sigma} [N]^T \{T\}_{,n} dS$$

$$\{P\}_{,mn} = \int_V [N]^T \{F\}_{,mn} dV + \int_{S\sigma} [N]^T \{T\}_{,mn} dS$$

$$\tag{20}$$

in which, $[B] = [L][N]$.

Though the principle of virtual displacement gives the first order variation of the total potential functional, Sometimes it is the functional of total potential energy, rather than its first variation, that is required for computation and analysis of some types of structures. In this case, the formulation of the SVP in this section, though overcomes some drawbacks of the direct stiffness matrix method, still follows the roundabout course of "neglecting stochasticity in Eqn.1, and then incorporating it into Eqn.10", which is similar to the conventional SFEM

The more important is that this version of SVP is based on the assumption that $\delta\{u\}$, $\delta\{a\}$ and $\delta\{\varepsilon\}$ are independent of stochasticity, which might be questioned.

THE STOCHASTICITY OF THE VIRTUAL DISPLACEMENTS

In the SVP stated in the previous section, $\delta\{u\}$, $\delta\{a\}$ and $\delta\{\varepsilon\}$ are all assumed to be independent of any stochasticity, so that they cannot be expanded via Taylor series, as given by Eqn.4. This issue may be questioned for two reasons: first, the virtual displacement vector should satisfy the boundary conditions. In actuality, there exist imperfection and uncertainty, including stochasticity embedded in structural boundary conditions; they affect significantly the static and dynamic behavior of structures. Some articles have addressed this issue. Ohira (1961) pointed out that different boundary conditions might cause large variations in the critical buckling loads for isotropic thin-walled shells. Arbocz (2000) investigated the effect of imperfect boundary conditions on the collapse behavior of anisotropic shells. Gutierrez and De Borst (2000) imposed the boundary constraint imperfection by Lagrange multipliers to evaluate the statistical properties of localization phenomena. Now that the boundary conditions can affect the displacements, hence affect the virtual displacements in a significant manner, the stochastic properties of the boundary conditions will inevitably result in the stochasticity of $\delta\{u\}$. The second reason, $\delta\{u\}$ is the arbitrary increment vector based on the real displacement vector $\{u\}$, as shown in Figure 1, and equal to the admissible displacement vector $\{u\}^p$ minus the real displacement vector $\{u\}$. That's to say, $\delta\{u\}$ is dependent on the real displacement vector $\{u\}$, hence the stochastic property of $\{u\}$ inevitably cause the randomness of $\delta\{u\}$. On the other hand, if $\delta\{u\}$ has nothing to do with $\{u\}$, how could one derive the relationship between the virtual displacement vector $\delta\{u\}$ and the virtual nodal displacement vector $\delta\{a\}$, given by:

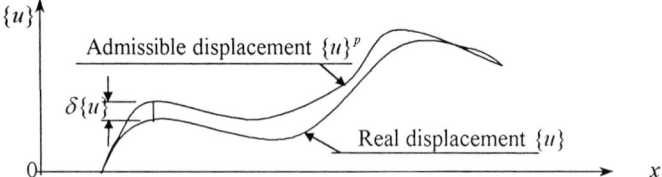

Figure 1: Real displacement, virtual displacement, and admissible displacement

$$\delta\{u\} = [N]\delta\{a\} \qquad (21)$$

via the displacement model defined by Eqn.4. Therefore, a conclusion could be arrived that the virtual displacement vector $\delta\{u\}$, involved with the real displacement vector $\{u\}$, is dependent on material properties and loading conditions as $\{u\}$ does, though the geometrically admissible displacement vector $\{u\}^p$ may be regarded as being independent of material properties and loading conditions. This relationship among $\delta\{u\}$, $\{u\}$ and $\{u\}^p$ may also be demonstrated by the following expression:

$$\delta\{u\} = \{u\}^p - \{u\} \tag{22}$$

where, if any two vectors among $\delta\{u\}$, $\{u\}$ and $\{u\}^p$ are independent of material property and the loading conditions of the structure of interest, and hence independent of stochasticity, the third one should also exhibit the same property.

THE STOCHASTIC TRANSIENT VARIATIONAL PRINCIPLE

In order to avoid the roundabout course of "neglecting at first, and then incorporating", the stochasticity should be incorporated in the functional of transient potential energy, as defined by Eqn.1, at the beginning so that Π, A, $\{F\}$, $\{F_I\}$, $\{F_D\}$, $\{T\}$, $\{u\}$ ρ and c are all dealt with as stochastic quantities (Yang, 1998), and can be expanded via Taylor series. Substituting their expanded expressions into Eqn.1, and equating the like powers of coefficients α_m, the zeroth, first and second order perturbation expressions of the transient functional are given by:

$$\overline{\Pi} = \int_V [\overline{A} - \{\overline{u}\}^T (\{\overline{F}\} + \{\overline{F_I}\} + \{\overline{F_D}\})]dV - \int_{S_\sigma} \{\overline{u}\}^T \{\overline{T}\}dS \tag{23a}$$

$$\Pi_{,m} = \int_V [A_{,m} - \{\overline{u}\}^T (\{F\}_{,m} + \{F_I\}_{,m} + \{F_D\}_{,m}) - \{u\}_{,m}^T (\{\overline{F}\} + \{\overline{F_I}\} + \{\overline{F_D}\})]dV$$
$$- \int_{S_\sigma} (\{\overline{u}\}^T \{T\}_{,m} + \{u\}_{,m}^T \{\overline{T}\})dS \tag{23b}$$

$$\Pi_{,mn} = \int_V [A_{,mn} - \{\overline{u}\}^T (\{F\}_{,mn} + \{F_I\}_{,mn} + \{F_D\}_{,mn}) - \{u\}_{,m}^T (\{F\}_{,n} + \{F_I\}_{,n} + \{F_D\}_{,n})$$
$$- \{u\}_{,n}^T (\{F\}_{,m} + \{F_I\}_{,m} + \{F_D\}_{,m}) - \{u\}_{,mn}^T (\{\overline{F}\} + \{\overline{F_I}\} + \{\overline{F_D}\})]dV$$
$$- \int_{S_\sigma} (\{\overline{u}\}^T \{T\}_{,mn} + \{u\}_{,m}^T \{T\}_{,n} + \{u\}_{,n}^T \{T\}_{,m} + \{u\}_{,mn}^T \{\overline{T}\})dS \tag{23c}$$

It should be pointed out that the requirements of stationary condition of $\Pi_{,mn}$, given by $\delta\Pi_{,mn} = 0$, include the requirements of stationary conditions of $\overline{\Pi}$ and $\Pi_{,m}$, given by $\delta\overline{\Pi} = 0$, and $\delta\Pi_{,m} = 0$, respectively (Yang, etc. 1999). Accordingly, when one addresses the stochastic analysis of engineering structures via the stochastic transient variational principle (STVP), only the highest order expanded functional, i.e. $\Pi_{,mn}$ herein, is necessary and sufficient to derive all the governing equations. Therefore, $\overline{A}$ and $A_{,m}$ are not necessary for formulation. $A_{,mn}$ shows:

$$A_{,mn} = \{\overline{\varepsilon}\}^T [\overline{D}]\{\varepsilon\}_{,mn} + \{\overline{\varepsilon}\}^T [D]_{,m}\{\varepsilon\}_{,n} + \{\overline{\varepsilon}\}^T [D]_{,n}\{\varepsilon\}_{,m} + \{\varepsilon\}_{,m}^T [\overline{D}]\{\varepsilon\}_{,n} + \frac{1}{2}\{\overline{\varepsilon}\}^T [D]_{,mn}\{\overline{\varepsilon}\} \tag{24}$$

Substituting $\overline{\varepsilon}$, $\varepsilon_{,m}$, $\varepsilon_{,mn}$, given by Eqns.12, into Eqn.24, then replacing $A_{,mn}$ in Eqn.23c by that in Eqn.24, the first-order variation of $\Pi_{,mn}$ with respect to $\{\overline{u}\}$, $\{u\}_{,m}$ and $\{u\}_{,mn}$ leads to:

$$\delta\Pi_{,mn} = \int_V [\delta\{\overline{u}\}^T ([L]^T[\overline{D}][L]\{u\}_{,mn} + [L]^T[D]_{,m}[L]\{u\}_{,n} + [L]^T[D]_{,n}[L]\{u\}_{,m} + [L]^T[D]_{,mn}[L]\{\overline{u}\}$$
$$- \{F\}_{,mn} + \rho_{,mn}\{\ddot{\overline{u}}\} + \rho_{,m}\{\ddot{u}\}_{,n} + \rho_{,n}\{\ddot{u}\}_{,m} + \overline{\rho}\{\ddot{u}\}_{,mn} + c_{,mn}\{\dot{\overline{u}}\} + c_{,m}\{\dot{u}\}_{,n} + c_{,n}\{\dot{u}\}_{,m} + \overline{c}\{\dot{u}\}_{,mn})$$

$$+\delta\{u\}_{,m}^T([L]^T[\overline{D}][L]\{u\}_{,n}+[L]^T[D]_{,n}[L]\{\overline{u}\}-\{F\}_{,n}+\rho_{,n}\{\ddot{\overline{u}}\}+\overline{\rho}\{\ddot{u}\}_{,n}+c_{,n}\{\dot{\overline{u}}\}+\overline{c}\{\dot{u}\}_{,mn})$$
$$+\delta\{u\}_{,mn}^T([L]^T[\overline{D}][L]\{\overline{u}\}-\{\overline{F}\}+\overline{\rho}\{\ddot{\overline{u}}\}+\overline{c}\{\dot{\overline{u}}\}))dV$$
$$-\int_{S_\sigma}(\delta\{u\}^T\{T\}_{,mn}+\delta\{u\}_{,m}^T\{T\}_{,n}+\delta\{u\}_{,n}^T\{T\}_{,m}+\delta\{\overline{u}\}^T\{T\}_{,mn})dS \tag{25}$$

As Eqn.4 is adopted as the displacement model, one has:

$$\{\dot{\overline{u}}\}=[N]\{\dot{\overline{a}}\} \qquad \{\dot{u}\}_{,m}=[N]\{\dot{a}\}_{,m} \qquad \{\dot{u}\}_{,mn}=[N]\{\dot{a}\}_{,mn} \tag{26a,b,c}$$
$$\{\ddot{\overline{u}}\}=[N]\{\ddot{\overline{a}}\} \qquad \{\ddot{u}\}_{,m}=[N]\{\ddot{a}\}_{,m} \qquad \{\ddot{u}\}_{,mn}=[N]\{\ddot{a}\}_{,mn} \tag{27a,b,c}$$

Substituting Eqns.14, Eqns.26 and Eqns.27 into Eqn.25, the stationary condition, $\delta\prod_{,mn}=0$, yields:

$$\frac{\partial\prod_{,mn}}{\partial\{\overline{a}\}}=\frac{\partial\prod_{,mn}}{\partial\{a\}_{,m}}=\frac{\partial\prod_{,mn}}{\partial\{a\}_{,mn}}=0 \tag{28}$$

Then the recursive equations for dynamic SFEM are obtained, and are given by Eqns.7-8 and Eqns.20.

CONCLUSION

The stochastic transient variational principle is developed based on the second order perturbation techniques. The stochastic properties of all the involved quantities are incorporated into the functional of total potential energy so that the roundabout course of "neglecting at first, then incorporating" is overcome.
The virtual displacement vector is based on the real displacement vector, they both are dependent on the material properties and loading conditions of the system, and hence are not independent of randomness. Though the geometrically admissible displacement vector may be independent of the material properties and loading conditions.

REFERENCE

Arbocz J., etc. (2000). The effect of imperfect boundary conditions on the collapse behavior of anisotropic shells. *International Journal of Solids and Structures*, **37**, 6891-6915.
Elishakoff I., Ren Y.J. (1996). and Shinozuka. Variational principles developed for and applied to analysis of stochastic beams. *Journal of Engineering Mechanics*, **122:6**, 559-565.
Hien T.D. and Kleiber M. (1990). Finite element analysis based on stochastic Hamilton variational principle. *Computers and Structures*, **37**, 893-902.
Liu W.K., Besterfield G.H. and Belyschko T. (1988). Variational approach to probabilistic finite elements. *Journal of engineering mechanics*, **114**, 2115-2133.
Liu W.K., et al. (1986). A variational approach for probabilistic mechanics. In *Finite Element Method for Plate and Shell Structures*, Vol.2, Edited by Hughes T.J.R. and Hinton E., Pineridge Press, Swansea, U.K., 285-311.
Ohira H. (1961). Local buckling theory of axially compressed cylindrical shells. *Proc. 11th Japan National Congress of Applied Mechanics*, 37-40
Yang L.F. (1988). Fuzzy stochastic finite element methods and their applications. *Dissertation of PhD*, Wuhan University of Technology.
Yang L.F., Gao D.X., and Li G.Q. (1999). Stochastic functional of potential energy developed for stochastic structures under stochastic loads. *Chinese Journal of Applied Mechanics*, **16:4**, 35-39.

EFFICIENT AUTOMATIC SELECTION OF TOLERANCES IN NONLINEAR DYNAMIC ANALYSIS

J. Farjoodi and A. Soroushian

Department of Civil Engineering, Faculty of Engineering,
University of Tehran, Tehran 11365, Iran

ABSTRACT

Direct time integration is the most popular method for solving nonlinear equations of motion. In nonlinear problems, tolerances are additional parameters that should be determined in advance. These parameters are almost always essential for terminating the iterations, when non-linearity occurs. However the sensitivity of responses, to these tolerances may be as high as instability, forcing the selection of small tolerances. On the other hand, unstudied selection of very small tolerances may considerably increase the computational cost. To overcome this drawback, the authors have succeeded to introduce a new technique in this paper. According to the suggested technique, the non-linearity tolerances are being automatically computed so as to remove or at least highly decrease the effects of non-linearity on stability. To attain this goal, the total relative error of the non-linearity detected time steps is forced to be less than the same error in the last previous linear time steps. By this technique, without any additional parameter as input, the effect of residual errors, while localizing non-linearity, is being controlled so as to have a total error, mainly dependent on the time step size. The efficiency of the suggested technique is revealed within some highly nonlinear examples.

KEYWORDS

Time Integration, Nonlinear, Dynamic, Tolerance, Stability, Truncation error, Nau's Method

1. INTRODUCTION

Time integration is one of the most popular and simplest methodologies suggested for dynamic analysis of structural systems. However only methods based on assumptions in their formulation can practically analyze the most general problems. To be able to control the effect of these assumptions that appear as the responses' error, a time integration method should be stable and consistent, Wood (1991) i.e., convergent, Richtmyer & Morton (1967). Emphasizing on the stability of a time integration method, the following familiar equation,

$$\left\{\begin{array}{c} u/\Delta t^2 \\ \dot{u}/\Delta t \\ \ddot{u} \end{array}\right\}_n = [A] \cdot \left\{\begin{array}{c} u/\Delta t^2 \\ \dot{u}/\Delta t \\ \ddot{u} \end{array}\right\}_{n-1} + \{L\} \cdot P_n \qquad (1)$$

Belytschko & Hughes (1983), Wood (1991) is very effective. This equation reveals the step-by-step computation of displacements, velocities and accelerations in an arbitrary DOF of an arbitrary vibration mode of linear systems, (or an arbitrary instantaneous vibration mode, of nonlinear systems). In other words, u, $\dot{u}$ and $\ddot{u}$ are the displacement, velocity and acceleration in one of the DOFs in a special vibration mode. P_n is the RHS of the corresponding equation of the motion at t_n, [A] and {L} are the 3×3 amplification matrix and a 3 member vector, representing the characteristics of the time integration method and instantaneous vibration mode, t is the time step size and finally n denotes the time step being studied. Mathematically, for a method to be stable, the eigenvalues of [A] should be at most equal to one, Bathe (1996). However in nonlinear problems the discussion is a little different. In these problems, employing tolerances for detection/localization non-linearity, is almost obligatory, Bernal (1991) and it is yet practically impossible to use zero tolerances in general problems. Thus in presence of non-linearity, going from the time station t_{n-1} to t_n, an error will appear at t_n even if the response is accurate at t_{n-1}. In other words, when non-linearity is detected in time step n, because of the non-zero tolerance, a non-zero residual error will remain in the LHS of Eqn. 1, showing a non-zero error in [A] and {L} at the RHS of Eqn. 1. Besides, almost in all of the non-linearity detection/localization techniques, Bathe (1996), Belytschko & Hughes (1983) the non-zero error at the RHS of Eqn. 1 and thus the error of [A] can not be determined before time integration. Therefore, in presence of non-linearity, the eigenvalues of [A] and so the stability can not be studied in advance.

Researches on instability of time integration methods in nonlinear regimes returns at least to mid-seventies, Belytschko & Schoeberle (1975). Later the phenomena was reported within simple and even some practical actual problems, Xie and Steven (1994), Cardona & Geradin (1989), Rashidi & Saadeghvaziri (1997), and as explained above, incorrect refinement of time steps in localization of non-linearity is recognized as its main reason, Hughes (1987), Kardestuncer (1987). As time integration is by itself time consuming and in presence of non-linearity this drawback highlights, and more, an unstable response is completely useless, it is practically very important to have stable responses. Therefore the amount of researches in this regard is still considerably high. The main idea in most of these studies is controlling the energy or/and momentum during time integration, Kuhl & Ram (1996), Laursen & Chawla (1997).

In this study, paying attention to the mathematical aspects of stability, a technique is suggested that can highly decrease the instability problem in nonlinear regimes, without requiring any additional parameter, as an advantage. In the next section, the suggested technique is explained in detail. Then applying the technique to some highly nonlinear problems, its efficiency is studied and finally the conclusions are mentioned briefly.

2. THE SUGGESTED TECHNIQUE

Replacing the exact response in Eqn. 1, and neglecting the round off error, the following equation,

$$\left\{\begin{array}{c} u(t_n)/\Delta t^2 \\ \dot{u}(t_n)/\Delta t \\ \ddot{u}(t_n) \end{array}\right\} = [A] \cdot \left\{\begin{array}{c} u(t_{n-1})/\Delta t^2 \\ \dot{u}(t_{n-1})/\Delta t \\ \ddot{u}(t_{n-1}) \end{array}\right\} + \{L\} \cdot P_n + \{\tau_n^L\} \qquad (2)$$

is obtained. In this equation, $\{\tau_n^L\}$ is the local truncation error for step n i.e., the error of the response at t_n when the response is accurate at t_{n-1}, and is defined for linear regimes. Also, $\{\tau_n^L\}$ depends only on the time integration method and time step size. Subtracting Eqn. 1 from Eqn. 2 and assuming the zero error at the initial instant, $\{\tau_n^L\}$ can be accepted as a representative of the responses' error (the non-zero $\{\tau_n^L\}$ causes the non-zero difference of $u(t_n)$ and u_n, and vice versa.). Consequently, in order to prevent instability in linear regimes, $\|\{\tau_n^L\}\|$ should be small enough, satisfying,

$$\|\{\tau_n^L\}\| \leq C \tag{3}$$

where C can be interpreted as an upper limit to prevent instability. Now considering nonlinear regimes, as the structural properties are assumed constant within each time step, Eqn. 2 would be rewritten as,

$$\begin{Bmatrix} u(t_n)/\Delta t^2 \\ \dot{u}(t_n)/\Delta t \\ \ddot{u}(t_n) \end{Bmatrix} - \{\delta_n^{NL}\} = [A] \cdot \begin{Bmatrix} u(t_{n-1})/\Delta t^2 \\ \dot{u}(t_{n-1})/\Delta t \\ \ddot{u}(t_{n-1}) \end{Bmatrix} + \{L\} \cdot P_n + \{\tau_n^L\} \tag{4}$$

in order to consider the residual error while detecting/localizing non-linearity. In Eqn. 4, $\{\tau_n^L\}$ is again the local truncation error, assuming linear behavior, and $\{\delta_n^{NL}\}$ is the deviation of $\{\tau_n^L\}$ from its actual value. Thus in presence of non-linearity the error of response corresponding to time step n is,

$$\{\tau_n^L\} + \{\delta_n^{NL}\} \tag{5}$$

Considering the two adjacent $n-1$ and n^{th} time steps, where non-linearity is detected in the second step, if we force $\{\delta_n^{NL}\}$ to satisfy Eqn. 6,

$$\|\{\tau_n^L\} + \{\delta_n^{NL}\}\| \leq \|\{\tau_{n-1}^L\}\| \tag{6}$$

and assume continuous variation of error, similar to RHS, the LHS of Eqn. 6 will satisfy Eqn. 3, and thus will be small enough not to cause instability. Now, if non-linearity is detected in the $n+1^{th}$ step, Eqns. 7,

$$\|\{\tau_{n+1}^L\} + \{\delta_{n+1}^{NL}\}\| \leq \|\{\tau_n^L\} + \{\delta_n^{NL}\}\| \tag{7}$$

6 & 3 show that instability sources are removed from time step $n+1$. Hence, besides error's continuity, the only assumption essential in employing Eqns. 6 & 7 is the existence of at least one time step with linear behavior before the time steps showing non-linearity. This assumption is very slight and usually true in actual problems. It is worth noting that maintaining stability for displacement, velocity and acceleration will be stable too. So instead of Eqns. 6 & 7, Eqn. 8 would rather be used,

$$\left| \tau_{u_n}^L + \delta_{u_n}^{NL} \right| \leq \left| \tau_{u_{n-1}}^L \right| \tag{8.a}$$

$$\left| \tau_{u_{n+1}}^L + \delta_{u_{n+1}}^{NL} \right| \leq \left| \tau_{u_n}^L + \delta_{u_n}^{NL} \right| \tag{8.b}$$

In Eqn. 8 all of the errors, are displacement errors relating to an arbitrary vibration mode. Also as the local truncation error is in fact the error caused in computing the increment of response, in order to have a better comparison of errors when the adjacent time steps are not equal, Eqn. 8 is replaced with

$$\frac{\left|\tau^L_{u_n} + \delta^{NL}_{u_n}\right|}{\left|u_n - u_{n-1}\right|} \leq \frac{\left|\tau^L_{u_{n-1}} + \delta^{NL}_{u_{n-1}}\right|}{\left|u_{n-1} - u_{n-2}\right|} \tag{9}$$

where $\{\delta^{NL}_{u_n}\}$ is zero for linear time steps. Eqn. 9 is the criteria for terminating iterative calculations when non-linearity is detected in time step n. As Eqn. 9 is resulted from Eqn. 1, for MDOF systems, it should govern the errors relation in all of the instantaneous vibration modes. But almost all kinds of non-linearity are defined in the actual physical space. So paying attention to the fact that the frequency space is only another special definition for the actual space, Eqn. 9 is interpreted consistent with the detected non-linearity as,

$$\frac{\left|\tau^L_{d_n} + \delta^{NL}_{d_n}\right|}{\left|d_n - d_{n-1}\right|} \leq \frac{\left|\tau^L_{d_{n-1}} + \delta^{NL}_{d_{n-1}}\right|}{\left|d_{n-1} - d_{n-2}\right|} \tag{10}$$

The definition of the new variable d_n in Eqn. 10 is problem dependent. For example in truss elements made of elastoplastic material, d_n is,

$$d_n = u^e_n - u^s_n \tag{11}$$

where u^e_n and u^s_n are the displacements at the two ends of the member in the direction of member, at t_n. In Eqn. 10, $\tau^L_{d_n}$, the local truncation error of d in the end of time step n according to linear properties, can be computed by different approximate equations. In this paper the following equation,

$$\tau^L_{d_n} = \frac{1}{12} \cdot \Delta t^2 \cdot \left(\ddot{d}_{n+1} - \ddot{d}_n\right) \tag{12}$$

Zienkiewicz & Xie (1991) is employed for the original Newmark method, Wiberg & Li (1993) and because of shortcomings of the above equation, the following equation,

$$\tau^L_{d_n} = \frac{1}{6} \cdot \Delta t^2 \cdot \left(\ddot{d}_{n+1} - \ddot{d}_n\right) \tag{13}$$

Zienkiewicz & Xie (1991), when employing the linear acceleration method. To compute $\delta^{NL}_{d_n}$, regarding to Eqn. 4, $\delta^{NL}_{d_n}$ is not only the deviation of $\tau^L_{d_n}$, but comparing Eqns. 2 & 4, is also the deviation of the calculated displacement from a constant specific value. For example the yield limits in elastoplastic material, or a specified distance in contact/impact problems. To complete the discussion for truss elements, using Nau's technique in detection/localization of non-linearity, Nau (1983), Mahin & Lin (1983), $\delta^{NL}_{d_n}$ is automatically computed during the localization procedure.

So in brief, according to the suggested technique when detecting non-linearity in a time step we can control the steps' error by Eqn. 10, if acceptable continue the time integration and otherwise divide the time step to sub-time-steps, repeat the analysis until the non-linearity is detected again and once more use Eqn. 10 and continue the procedure until Eqn. 10 is satisfied. This is the same as Nau's technique,

Nau (1983), Mahin & Lin (1983), except that Eqn. 10 is replaced instead of an ordinary tolerance controlling equation. To lessen the possibility of any numerical shortcoming in the dominators of Eqn. 10, only the error and response of an original time step and not divided time steps will be employed, in the RHS of this equation.

3. NUMERICAL EXAMPLES

In the first example, an elastoplastic SDOF system with 5 percent damping, .16 slugs mass, 13.1595 lb/in initial stiffness, and .802028 inches yield limit is subjected to the 10 seconds North-South component of the Elcentro(1940) strong motion, Chopra (1996). Figure 1.a shows the exact response.

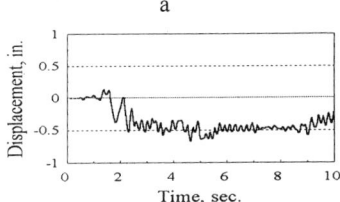 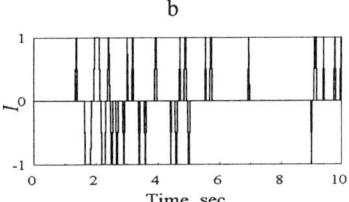

Figure 1:The exact response of the elastoplastic system defined in example one

In Figure 1.b , I indicates an index which is 0 when the system is in the elastic range and ±1 when behaves plastically. Figure 1.b clearly shows that the behavior of the system is highly nonlinear.

Applying the average acceleration time integration method, considering .0025 sec. time steps, Nau's technique, Mahin & Lin (1983), and a constant non-linearity tolerance equal to .005, causes the unstable response shown in Figure 2.a. Repeating the analysis, employing the suggested technique and

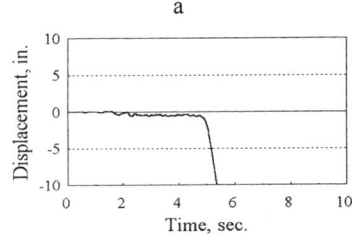 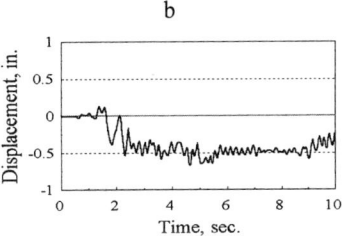

Figure 2:The response of example one, employing average acceleration, .0025 sec. time steps, a) non-linearity tolerance equal to .005 b) the suggested technique

the result is shown in Figure 2.b. Comparing Figures 1.a, 2.a & 2.b, apparently reveals the significant effect of the suggested technique and the adequacy of the attained response.

As a second example, an eight story plane shear building with 5 percent modal damping and the other characteristics mentioned in Table 1 is excited by the first 10 seconds of the North-South component of the Elcentro(1940) strong motion, Chopra (1996). Using the average acceleration method,

TABLE 1
PROPERTIES OF THE MEMBERS OF AN EIGHT STORY PLANE SHEAR BUILDING

Level	Mass	Stiffness	Yield Deformation
1	.518 k-sec^2/in	860 k/in	.36 in
2	.517 k-sec^2/in	840 k/in	.37 in
3	.516 k-sec^2/in	820 k/in	.38 in
4	.515 k-sec^2/in	700 k/in	.40 in
5	.514 k-sec^2/in	680 k/in	.41 in
6	.513 k-sec^2/in	660 k/in	.42 in
7	.512 k-sec^2/in	640 k/in	.43 in
8(top)	.511 k-sec^2/in	620 k/in	.44 in

step size equal to .02 sec. and a non-linearity tolerance equal to .001 for Nau's technique, causes the response shown in Figure 3.a for the top displacement. In this figure, instability is apparent.

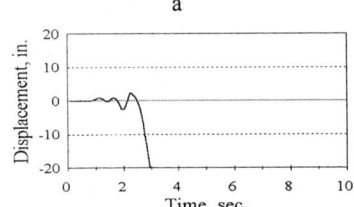
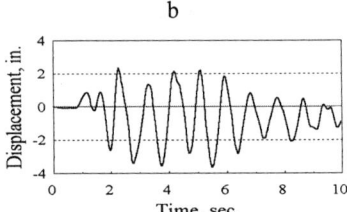

Figure 3: Top displacement in the 2nd example, employing average acceleration, .02 sec. time steps, a) non-linearity tolerances equal to .001 b) the suggested technique

In order to have a better response, the time integration is repeated with half time steps several times, Clough & Penzien (1993). By this methodology a stable response is attained only after four repetitions, and a total number of time steps equal to 16260. Now employing the suggested technique while analyzing this system with a time step size equal to .02 sec., the response shown in Figure 3.b is obtained. Opposite to Figure 3.a, Figure 3.b shows a stable response, after 2275 time steps, which is considerably less than 16260 when we employ non-linearity tolerances. As a closure for this example, Figures 4.a & 4.b reveal the convergence of the top floor's maximum displacement when we repeatedly analyze the system, halving the time steps each time, with a constant tolerance equal to .001 and applying the suggested technique, respectively.

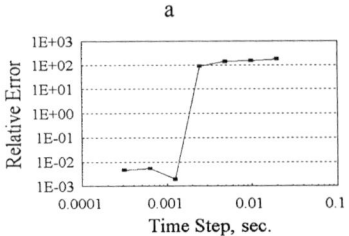
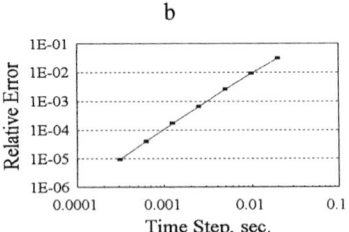

Figure 4: Convergence of the maximum top disp. in the 2nd example, employing average acceleration, a) non-linearity tolerances equal to .001 b) the suggested technique

4. CONCLUSION

Suggesting a technique for automatic selection of non-linearity tolerances, the problem of appropriate tolerances in nonlinear dynamic analysis seems to be removed. Also, applying the suggested technique, the error of nonlinear responses will mainly depend on time step size and time integration method, causing a considerable reliability in practical analyses. Also according to this study, the computational effort is considerably reduced by the suggested technique.

REFERENCES

Bathe K.J. (1996). *Finite Element Procedures*, Prentice-Hall, USA
Belytschko T. and Hughes T.J.R. (1983). *Computational Methods for Transient Analysis*, North-Holland, Elsevier, USA
Belytschko T. and Schoeberle D.F. (1975). On the Unconditional Stability of an Implicit Algorithm for Nonlinear Structural Dynamics. *Journal of Applied Mechanics, ASME* 42:4, 865-869.
Bernal D. (1991). Locating Events in Step-By-Step Integration of Equations of motion. *Journal of Structural Engineering, ASCE* 117:2, 530-545.
Cardona A. and Geradin M. (1989). Time Integration of the Equations of Motion in Mechanism Analysis. *International Journal Computers and Structures* 33:3, 801-820.
Chopra A.K. (1995). *Dynamics of Structures: Theory and Application to Earthquake Engineering*, Prentice-Hall, USA
Clough R.W. and Penzien J. (1993). *Dynamics of Structures*, McGraw-Hill, Singapore
Hughes T.J.R. (1987). *The Finite Element Method: Linear Static and Dynamic Finite Element Analysis*, Prentice-Hall, USA
Kardestuncer H. (1987). *Finite Element Handbook*, McGraw-Hill, USA
Kuhl D. and Ram E. (1996). Construction of Energy and Momentum for implicit single step time integration schemes. *Proc. Dynamics-Eurodyn'96*, Augusti, Borri & Spinelli, Balkema, 349-356
Laursen T.A. and Chawla V. (1997). Design of Energy Conserving Algorithms for Frictionless Dynamic Contact Problems. *International Journal for Numerical Methods in Engineering* 40:5, 863-886.
Mahin S.A. and Lin J. (1983). *Construction of Inelastic Response Spectra for Single Degree-of-Freedom Systems*, Report No. UCB/EERC-83/17, Earthquake Engineering Research Center, University of California, Berkeley, USA
Mohraz B. and Elghadamsi F.E. and Chang C.J. (1991). An Incremental Mode Superposition for Non-Linear Dynamic Analysis. *Earthquake Engineering and Structural Dynamics* 20:5, 471-481.
Nau J.M. (1983). Computation of Inelastic Spectra. *Journal of Engineering Mechanics, ASCE* 109:1, 279-288.
Rashidi S. and Saadeghvaziri M.A. (1997). Seismic Modeling of Multispan Simply-Supported Bridges Using Adina. *International Journal Computers and Structures* 64:5/6, 1025-1039.
Richtmyer R.D. and Morton K.W. (1967). *Difference Methods For Initial-Value Problems*, Interscience Publishers, USA
Wood W.L. (1990). *Practical Time-Stepping Schemes*, Oxford, USA
Xie Y.M. and Steven G.P. (1994). Instability, Chaos, and Growth and Decay of Energy of Time-Stepping Schemes for Nonlinear Dynamic Equations. *Communications in Numerical Methods in Engineering* 10:5, 393-401.
Wiberg N.-E. and Li X.D. (1993). A post-processing Technique and an A Posteriori Error Estimate for the Newmark Method in Dynamic Analysis. *Earthquake Engineering and Structural Dynamics* 22:6, 871-887.
Zienkiewicz O.C. and Xie Y.M. (1991). A Simple Error Estimator and Adaptive Time Stepping Procedure for Dynamic Analysis. *Earthquake Engineering and Structural Dynamics* 20:9, 871-887.

AN ELASTO-PLASTIC FREE-FREE BEAM SUBJECTED TO PULSE LOAD AT TIP

T.U.Ahmed[1], L.S. Ramachandra[1] and S.K.Bhattacharyya[2]

[1]Civil Engineering Department, Indian Institute of Technology, Kharagpur, INDIA.
[2]Civil Engineering Discipline, University of Durban-Westville, S. Africa.

ABSTRACT

The dynamic response of an isotropic strain-hardening free-free beam subjected to transverse rectangular pulse load is presented. Large displacement Lagrangian finite element formulation has been adopted together with Newmark's time integration algorithm. The onset of yielding is identified by von Mises yield criterion. The response of beam in terms of displacement, curvature profile and energy dissipation in the form of plastic work done have been studied. Numerical results reveal that the interaction between the faster reflected waves and slower progressive waves determine the final deformed configuration of the free-free beam. The propagation of travelling plastic front (hinge) in the early transient response is also discussed. The energy used for the rigid body displacement and the elasto-plastic deformation has been evaluated.

KEYWORDS

Impact, elasto-plastic, free-free beam, pulse load, finite element technique

INTRODUCTION

The response of a free-free beam subjected to intense dynamic loading is of continuing interest in the aerospace engineering. The aircraft structures are often subjected to landing impacts, high wind pressure, and variety of other loading. The governing equations of motion based on an elasto-plastic material law are highly non-linear and complex in nature and it is difficult to obtain an analytical solution even for a simple beam of finite length. However, the governing equations based on a rigid, perfectly-plastic material law can often allow the non-linear equations of motion to reduce to a set of first-order differential equations that are relatively easy to solve analytically/numerically for given initial and boundary conditions.

Lee and Symonds [1952] studied the dynamic behaviour of a free-free beam subjected to different types of pulse load adopting rigid, perfectly-plastic idealisation, which greatly simplified their analysis. The authors could predict the location of the plastic hinges and also discussed the different phases of beam behaviour in respect to hinge formation with the application of loads. Later Symonds [1953] and Symonds and Leth [1954] worked on the same idealisation. Jones and Wierzbicki [1987] studied the dynamic plastic collapse of a free-free beam of uniform/stepped cross-section subjected to rectangular pulse/triangular distributed impulsive load. Authors have presented a closed form solution of the problem using rigid, perfectly-plastic material idealisation for two different cases. They have demonstrated the failure mechanism of free-free beams due to the formation of stationary plastic hinge at centre of the beam. According to their predictions, in one particular case of ideal impulse, only 25% of the external energy are converted into plastic deformations. However the energy used for the plastic deformation is overestimated in their work as elastic energy is not considered. It is found that rigid-plastic theory provides a better approximation for short beams than compared to long beams.

On the other hand, in relation to the wave propagation problem, several authors [Symonds et.al.(1984), Reid et.al. [1987] have demonstrated the behaviour of cantilever beam subjected to dynamic load in the form of step/impulsive load to explain the experimental findings of Hall [1971]. Finite element procedure or code has been adopted in their analysis. Most recently, Yu et al. [1996] have demonstrated the effect of reflected elastic wave on the plastic wave of a cantilever beam subjected impulsive/step load using finite difference technique with elastic-perfectly plastic material properties. Authors have confirmed the observation of Reid and Gui [1987]. Authors have discussed more clearly the oscillatory nature of the plastic hinge due to interaction of reflected elastic wave.

In case of free-free beams, Yu et al. [1996] have presented the elasto-plastic response of free-free beams subjected to a single point impact at the centre. The governing partial differential equations of motion have been solved using finite difference technique. They have demonstrated the effect of faster reflected elastic waves on the plastic wave front, which helps to understand how elastic-perfectly plastic model is different from that of rigid, perfectly-plastic model. Also the authors have discussed the energy dissipation mechanism of free-free beams.

However Yang, et al. [1998] have demonstrated the behaviour of free-free beams subjected to various magnitude of concentrated step-loading at any cross-section along its span adopting rigid, perfectly-plastic material idealisation. Authors have discussed the rate of dissipation of energy in the form of plastic work done for different types of hinge mechanisms. They have shown that the plastic dissipation energy is always less than 33.33% of input energy.

In the present paper, the response of a free-free beam subjected to transverse rectangular pulse load at tip is studied. Total Lagrangian formulation has been adopted to obtain the behaviour of the beam considering large displacement. The Newmark's time integration algorithm is used to solve the non-linear equations of motion of the beam. Elasto-plastic strain-hardening material properties are used in the analysis and the onset of yielding is identified by von Mises yield criterion. Linear scaling down procedure is adopted to satisfy the natural boundary conditions of the beam in drawing the stress profile. Authors have studied the response of beam in terms of displacement, curvature profile and energy dissipation in the form of plastic work done. The propagation and formation of plastic fronts are also studied.

FINITE ELEMENT FORMULATION

Equation of motion for the target (beam without damping) at time $t+\Delta t$ is expressed as

$$[M]\{\ddot{d}\}_{t+\Delta t} + \{p\}_{t+\Delta t} = \{F\}_{t+\Delta t} \tag{1}$$

where $[M]$, $\{\ddot{d}\}_{t+\Delta t}$, $\{p\}_{t+\Delta t}$ and $\{F\}_{t+\Delta t}$ are the mass matrix, acceleration, internal force and external force vector of beam at time $t+\Delta t$ respectively.

An eight-noded quadrilateral plane stress element is adopted here to discretise the beam. In the present formulation large displacement small strain is considered in the analysis. The internal resisting force vector at time $t_n + \Delta t$ is given by

$$\{p\}_{t+\Delta t} = \int_\Omega [B]^T_{t+\Delta t} \sigma_{t+\Delta t} \, d\Omega \tag{2}$$

$$[B]^T_{t+\Delta t} = [B_L]_{t+\Delta t} + [B_{NL}]_{t+\Delta t} \tag{3}$$

where $[B_L]_{t+\Delta t}$ and $[B_{NL}]_{t+\Delta t}$ are linear and nonlinear strain–displacement matrices at time step $t+\Delta t$.

In the present work, von Mises yield criterion is used to identify the onset of plastic deformation. The initial conditions of the problem are

$$\{d\}_{t_0=0} = \{0\}, \qquad \{\dot{d}\}_{t_0=0} = \{0\}$$

Solution procedure

The governing equations of motion of the beam are solved using Newmark's constant-average acceleration technique. At each time step increment, the residual forces are obtained due to the application of the external force. Also since the governing equations are nonlinear, iterations are carried out at each time step to satisfy the equilibrium equation. The residual forces (ψ) are obtained from the following equation at time $t+\Delta t$ in i-th iteration

$$\{\psi\}^{(i)} = \{F\}_{t+\Delta t} - [M]\{\ddot{d}\}^{(i)}_{t+\Delta t} - \{p\}^{(i)}_{t+\Delta t} \tag{4}$$

where subscript 'i' stands for the iteration number.
The effective stiffness matrix is obtained by the following equation

$$[\bar{K}]_{t+\Delta t} = \frac{1}{\Delta t^2 \beta}[M] + [K_T]^{(i)}_{t+\Delta t} \tag{5}$$

where $[K_T]^{(i)}_{t+\Delta t}$ is tangent stiffness matrix at time step $t + \Delta t$ at i-th iteration.

The additional displacement due to residual forces are obtained from the following relation

$$[\bar{K}]_{t+\Delta t}\Delta d^{(i)} = \psi^{(i)} \tag{6}$$

The displacement, velocity and acceleration are corrected after obtaining the incremental displacement from equation (6). All the stress quantities are checked at each Gaussian point for yielding. As a result, an element may behave in partly elastic and partly elasto-plastic manner.

RESULTS AND DISCUSSION

A computer code is developed based on the above formulation and it is validated with the existing results available in the literature. The transverse deflections of a free-free beam calculated by the present method are compared with the numerical results of Yu, et al. [1997]. The transient behavior of the free-free steel beam with the following geometrical and material properties is discussed in the following sections.

Half length of beam $L = 0.13970$ m;   Yield stress $Y = 0.2075$ GPa
Width of beam $b = 0.01270$ m; Modulus of elasticity $E = 199.00$ Gpa
Depth of beam $h = 0.00318$ m; Hardening parameter $H = 0.5700$ GPa
Mass density $\rho = 7850$ kg/m^3 ; Pulse load $F = 1000$ N
Duration of pulse $T = 1.0$ ms.

The instantaneous deflection profiles of the beam are shown in Figs.1 & 2. At the initial stage, the rate of deflection of the tip is higher due to the presence of pulse. The tip continues to move at a lesser rate in the positive sense after the withdrawal of pulse. The point having zero slope also gradually proceeds towards the far end of the beam. This signifies that the positive wave front moves towards the free end.

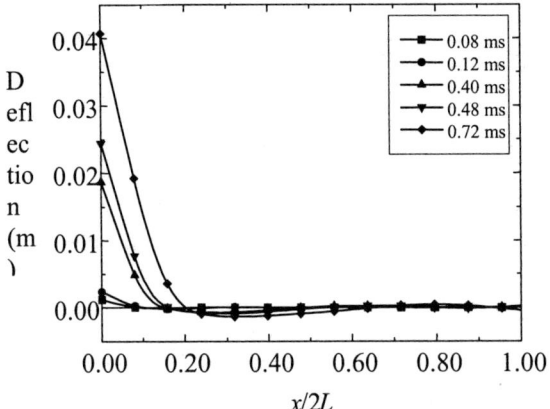

Fig. 1 The deflection profiles of the free-free beam

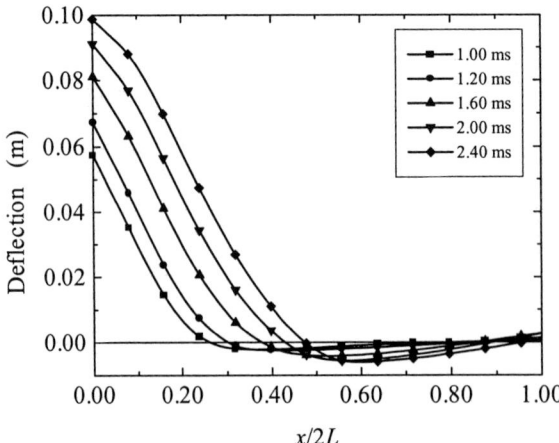

Fig. 2 The deflection profiles of the free-free beam

The variation of stress (σ_x), evaluated at gauss points, along the length of the beam at various instants of time are shown in Figs.3 & 4. At the initial stage, the oscillation of the stress profiles is uniform in nature. The stress distribution suffers a disturbance for the first time created by the interaction of reflected flexure wave. The stress profile suffers more and more disturbance with time as the interaction increases.

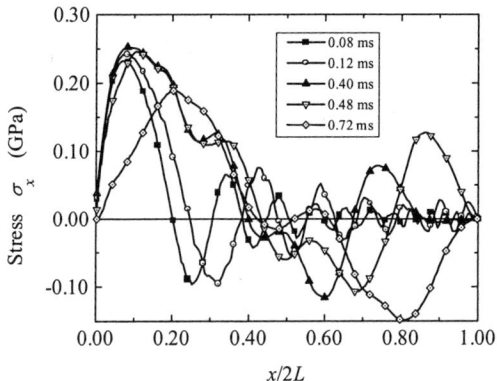

Fig. 3: Stress distribution along the length of beam

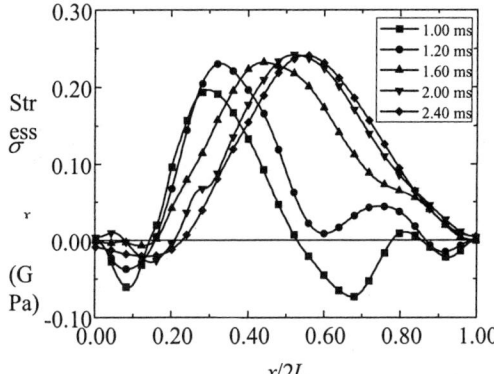

Fig. 4: Stress distribution along the length of beam

For isotropic hardening material, the yield stress at that particular time instance is much higher than the initial yield stress. Depending on the current yield stress, hinge will form at a distance 0.02402 m away from load point. In the present analysis, the first such hinge is developed at a distance of 0.0229m away from the load point.

The energy used for the rigid body motion of the beam (kinetic energy of the beam), the work done by the applied pulse, the energy absorbed by the beam and the elastic energy with respect to time are evaluated.

CONCLUSIONS

Elasto-plastic response of a free-free beam subjected to rectangular pulse load has been investigated. At the initial stage, the interaction between the faster reflected waves and slower progressive waves is predominant factor that governs the response of beam. The speed of the plastic front is much higher than that for rigid, perfectly plastic idealization. But variation in the locations of instantaneous plastic hinge is less than the variation in speed. The maximum rate of energy absorbed and gain in kinetic energy by the beam is equal to 30% and 72% respectively. Also it is observed that the total energy absorbed by the beam and the total kinetic energy remained in the beam are 45.076% and 54.924% of the total input energy due impulse.

REFERENCES

Hall,R.G., Al-Hassani,S.T.S. and Johnson,W.(1971), The impulsive loading of cantilever. *Int. J. Mech. Sci.*, 13, pp. 415-430.

Jones, N. and Wierzbieki,T.(1987), Dynamic plastic failure of a free-free beam. *Int. J. Impact Engng.*, 6, pp. 225-240.

Lee,E.H.,and Symonds,P.S.(1952), Large plastic deformations of beams under transverse impact. *J. Appl. Mech.*, 19, pp. 308.

Reid,S.R. and Gui,X.G. (1987), On the elasto-plastic deformation of cantilever beams subjected to tip impact. *Int. J. Impact Engng.*, 6, pp. 109-127.

Stronge, W. J. and Yu, T. X.(1993), *Dynamic Models for Structural Plasticity,* Springer-Verlag,.

Symonds,P.S.(1953), Dynamic load characteristics in plastic bending of beam. *J. Appl. Mech.*, 20, pp. 475.
Symonds,P.S. and Leth,C.F.A (1954), Impact of finite beam of ductile metal. *J. Appl. Mech..*, 2, pp. 92.

Symonds,P.S.and Fleming,, W.T. (1984), Jr. Parkes revised: on rigid-plastic and elasto-plastic dynamic structural analysis, *Int. J. Impact Engng.*, 2, pp. 1-36.

Yu,T.X., Yang,J.L. and Reid,S.R (1997). Interaction between reflected flexural waves and a plastic 'hinge' in the dynamic response of pulse loaded beams. *Int. J. Impact Engng.*, Vol. 19, No 5-6, pp. 457-475.

Yu, T.X., Yang, J. L., Reid, S.R. and Austin, C.A.(1996), Dynamic behaviour of elastic-plastic free-free beams subjected to impulsive loading. *Int. J. of Solids Structures.*, Vol.33, No 18, pp. 2659-2680.

Yang, J. L.,Yu,T.X. and Reid,S.R.(1998), Dynamic behaviour of a rigid, perfectly plastic free-free beam subjected to step-loading at any cross-section along its span. *Int. J. Impact Engng.*, Vol.21, No 3, pp. 165-175.

Zienkiewiez, D.C. & Taylor, R.L. (1991), The finite element method, *4th edition,* Vol – 2 , *McGraw-Hill Int. Ed.*

ROBUST CONVERGENCE FOR THE DYNAMIC ANALYSIS OF MDOF ELASTOPLASTIC SYSTEMS

J. Farjoodi and A. Soroushian

Department of Civil Engineering, Faculty of Engineering,
University of Tehran, Tehran 11365, Iran

ABSTRACT

Direct time integration is the most popular method for solving nonlinear equations of motion. Because of the inherent error in most of these methods, in order to use the responses caused by analyses in design, it is customary to repeat them with smaller time steps, Clough & Penzien (1993). However except for some complicated, new and yet not practical time integration methods, this methodology does not always work in nonlinear regimes. To overcome this drawback, the authors have recently suggested a methodology and succeeded to cause robust convergence for dynamic analysis of SDOF elastoplastic systems, Farjoodi & Soroushian (2000). The main idea of the suggested methodology is to choose the non-linearity detection tolerance such as to cause the truncation inherent error to have the most contribution in the total error. In this paper after a brief study of the non-convergence problem in nonlinear dynamic analysis, a generalized methodology is suggested for robust convergence of MDOF elastoplastic systems. The efficiency of the suggested methodology is then studied using different numerical examples analyzed by different time integration methods. The numerical results reveal the considerable effect of this methodology on practical dynamic analysis.

KEYWORDS

Nonlinear, Elastoplastic, Dynamic, Time Integration, Truncation Error, Convergence, Tolerance, Nau's method, Newmark's method

1. INTRODUCTION

In spite of methods such as modal analysis and some new methods based on combination of modal analysis and time integration, Mohraz et al. (1991), direct time integration is still the most popular methodology in practical dynamic analysis. Besides simplicity, the set of these methods has the capability of analyzing problems with every kind of non-linearity, non-classical damping, close natural frequencies and sophisticated excitations. Hopelessly, only time integration methods that consider assumptions and approximations in their formulations are versatile in nonlinear MDOF problems.

The errors caused by these approximations depend on the time step size. Thus in practical dynamic analysis it is being recommended to repeat the analyses with smaller time steps and compare the results, Clough & Penzien (1993). However, this approach may not work for nonlinear problems. This phenomenon, which seems considerably important in practice, was already recognized at mid seventies, Belytschko & Schoeberle (1975). Since then, it is discussed, Belytschko & Schoeberle (1975), Belytschko & Hughes (1983), Kardestuncer (1987) and shown by simple examples, Cardona & Geradin (1989), Low (1991), Xie (1996) and even implicitly in actual problems, Rashidi & Saadeghvaziri (1996). Incorrect refinement of time steps in the detection of non-linearity is known as the main reason of this event, Hughes (1987), Kardestuncer (1987). To briefly explain why the response of step-by-step analysis of nonlinear problems do not always converge, we can pay attention to the following familiar equation,

$$\left\{ \frac{u}{\Delta t^2} \quad \frac{\dot{u}}{\Delta t} \quad \ddot{u} \right\}_n^T = [A] \cdot \left\{ \frac{u}{\Delta t^2} \quad \frac{\dot{u}}{\Delta t} \quad \ddot{u} \right\}_{n-1}^T + \{L\} \cdot P_n \qquad (1)$$

Belytschko & Hughes (1983), Wood (1991) for each of the instantaneous vibration modes, attained using instantaneous properties. In Eqn. 1, P_n is the RHS of the equation of motion at t_n, u, $\dot{u}$ and $\ddot{u}$ indicate the response, $[A]$ and $\{L\}$ are the 3×3 amplification matrix and a 3×1 vector representing the characteristics of the time integration method, Δt is the time step size and finally n denotes the time step being studied. For step-by-step method to be stable, the error; i.e. deviation from the exact response, should not be increased from t_{n-1} to t_n for a given n, Bathe (1996). Applying the deviation operator to Eqn. 1, opposite to linear problems, in nonlinear problems $[A]$ has also a deviation. This is so, because, except for the exact Event-to-Event method(which is not practical and efficient for nonlinear MDOF systems, Bernal (1991)), the tolerances used in localizing stiffness changing instants can not be zero and thus regarding to the RHS of Eqn. 1, $[A]$ and $\{L\}$ should have deviations from their exact values. The non-zero deviation of $[A]$ is an error besides the inherent error caused by the assumptions made in the formulation of time integration method and depends not only on the time step and tolerance but also on the non-linearity detection/localization method. Therefore when non-linearity happens, the stability of a special problem will depend both on the time step size and also on other parameters such as the tolerances and the non-linearity detection/localization methods. Also, replacing exact values of the response in Eqn. 1, for the sake of equality a term such as $\{\tau(t_n)\}$, named local truncation error, need to be added to the RHS. For consistency it is essential to,

$$\Delta t \to 0 \qquad \|\{\tau(t_n)\}\| \le C \cdot \Delta t^{q+1} \qquad q > 0 \qquad (2)$$

Belytschko & Hughes (1983), where C and q are respectively the constant error and convergence rate of the integration method. As in nonlinear regimes the deviations of $[A]$ and $\{L\}$ depend on the nonlinear behavior, by definition, $\{\tau(t_n)\}$ also depends on non-linearity localization tolerances and methods and thus Eqn. 2 and consistency, may be untrue. Therefore in nonlinear problems, neither stability nor the consistency of the direct time integration methods can be guaranteed and so according to the Lax's Equivalence theorem, Richtmyer & Morton (1967) convergence of the response can neither be guaranteed. Considering SDOF elastoplastic systems and Nau's technique in detecting/localizing yielding/unloading, the authors have recently suggested the following equations,

$$^{i+1}\overline{\eta} = r \cdot \text{Min}_t(^i\eta) \qquad (3.a)$$

$$r = \left(^{i+1}\Delta t / ^i \Delta t\right)^q \qquad (3.b)$$

Farjoodi & Soroushian (2000). According to these two equations, if we start an analysis with the time step size $^{i}\Delta t$ and tolerance $^{i}\eta$, after terminating the analysis, we can attain a more accurate response, starting another analysis with the time step size $^{i+1}\Delta t$ and non-linearity localization tolerance $^{i+1}\eta$. In fact Eqn. 3 causes the time steps to be refined such that the additional non-linearity based error decreases faster than the truncation inherent error and thus the effect of the additional error lessens gradually and becomes negligible during iterative analyses. In this paper, generalizing Eqn. 3, a methodology is suggested to maintain convergence for time integration of MDOF elastoplastic systems. The efficiency of the suggested methodology is also studied by examples analyzed with different time integration methods and some variation of other parameters.

2. THE SUGGESTED METHODOLOGY

First of all it is essential to mention that paying attention to the shortcomings of the Newton Raphson technique in Dynamic analysis of elastoplastic systems, Belytshcko & Hughes (1983), in this study, the fractional time stepping technique, Nau (1983) has been used to detect and localize the yielding and unloading instants. According to this technique, when the stiffness in the start and end of a time step is recognized different, the time step should be divided by a natural number, the calculations of the last step should be redone with the refined step size until non-linearity is detected again, and this procedure should be repeated continuously until the specified tolerance is satisfied and then, the response at the instant when the non-linearity has been detected first, should be calculated. This operation should be followed for all non-linearity detection instants. Besides the above mentioned natural number, the maximum number of repetitions of the above procedure in localizing a single stiffness change instant is a parameter of the fractional time stepping technique. This parameter is being determined in advance in order to prevent any computational difficulties. Similar to the usual practice, Mahin and Lin (1983), these two parameters are selected to be respectively 10 and 5. Thus, as mentioned before in nonlinear regimes, besides step size, the response depends on the errors remained while localizing non-linearity

$$u = u\left(\Delta t, \eta_{j=1,2,\ldots}\right) \qquad (4)$$

The error caused by Δt is inherent for time integration methods, and for small Δt is controlled by q and C, denoted in Eqn. 2. The effect of q and C is such that in absence of non-linearity, if we have two time integration analyses for a unique problem with different time steps, Δt and $\Delta t'$, the errors of these analyses in a special time instant, E and E', with a given norm will be such that

$$\frac{E}{E'} = \left(\frac{\Delta t}{\Delta t'}\right)^q \qquad (5)$$

Now if we disregard the inherent error, and consider a single yielding or unloading instant, paying attention to the fact that in fractional time stepping technique the tolerance equation can be stated linearly in terms of the response, Mahin and Lin (1983), Eqn. 6

$$\frac{E}{E''} = \frac{\eta}{\eta''} \qquad (6)$$

will govern the relation between the errors in two assumed time integration analyses without inherent error, E' and E'',(for example when Δt is very small and the round off error negligible) and with different remained errors in non-linearity detection/localization instants, η' and η''.

Besides, paying attention to Eqn. 4, both of these errors can be considered together, in Eqn. 7,

$$\delta u = \frac{\partial u}{\partial (\Delta t)} \cdot \delta(\Delta t) + \sum_j \frac{\partial u}{\partial \eta_j} \cdot \delta \eta_j \qquad (7)$$

Now, as the number of yielding/unloading instants in a special analysis is limited, if we force

$$\frac{\eta}{\eta'} \leq \left(\frac{\Delta t}{\Delta t'}\right)^q \qquad (8)$$

for all of the non-linearity detection instants in all of the spatial non-linearity controlling locations, it seems, Eqn. 5 will govern the variation of error either in linear or in nonlinear problems. However, as in the Nau's technique only the upper limit of the non-linearity based error can be controlled by the tolerance $\bar{\eta}$, practically, Eqn. 8 would rather be replaced with

$$\frac{\bar{\eta}}{\eta'} = \left(\frac{\Delta t}{\Delta t'}\right)^q \qquad (9)$$

Compared to Eqn. 8, Eqn. 9 requires an additional computational cost. Also as in the analysis of an actual MDOF system in a long duration, a considerably high number of loading/unloading may happen, Eqn. 9 will be computationally expensive. Instead if, for each spatial section under non-linearity control, the non-linearity localization tolerance is determined by

$$\bar{\eta} = \left(\mathop{Min}_t(\eta')\right) \cdot \left(\frac{\Delta t}{\Delta t'}\right)^q \qquad (10)$$

the additional saving of information, will be reduced to a single vector with members calculated easily from Eqn. 10. The dimension of this vector will be equal to the number of spatial non-linearity controlling sections. In other words within each time integration analysis the value of $Min(\eta')$ for each non-linearity controlling location can be renewed cheaply after each time step ($\Delta t'$ or the corresponding refined sub-steps) and gradually cause the result of Eqn. 10 in the end of the time duration. These tolerances will be then used for the time integration with step size equal to Δt. Of course this reduction of computational cost may be balanced by the additional cost caused by attaining smaller tolerance in Eqn. 10, but it is selected here because it may be effective in a highly nonlinear problem. To reduce the computational cost more, it is decided, to start each of the second, third, forth, ... analyses with a constant time step smaller than their previous analysis' time step and leave the reduction of the sub-time steps, that are created by Nau's technique within the previous analysis, to be controlled by Eqn. 10. This seems in contradiction with the presented justification for Eqn. 10, however as shown in the next section, selecting practical values for the Nau's technique, the inherent error of sub-time steps is small, causing little effect on the desired results.

In the closure of this section it is worth repeating that according to the suggested methodology, to become sure of the correctness of the responses attained from the time integration of an MDOF elastoplastic dynamic system, besides the usual methodology, Clough & Penzien (1993), it should be remembered to change the non-linearity localization tolerances according to Eqn. 10 after each analysis and separately for different non-linearity controlling sections.

3. NUMERICAL EXAMPLES

In the first example a three story undamped plane shear building with characteristics mentioned in Table 1 is subjected to the 30 second North-South component of the Elcentro (1940) strong motion,

TABLE 1
PROPERTIES OF THE MEMBERS OF A THREE STORY PLANE SHEAR BUILDING

Level	Mass	Stiffness	Yield Deformation
1	2.0 k-sec^2/in	180 k/in	.64 in
2	1.5 k-sec^2/in	120 k/in	.65 in
3(top)	1.0 k-sec^2/in	60 k/in	.67 in

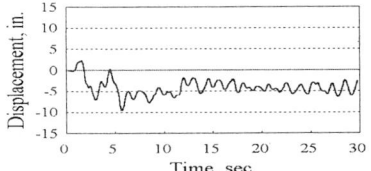 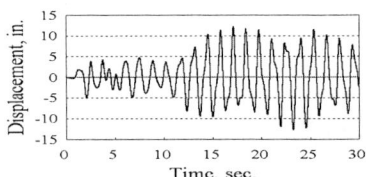

Figure 1: Exact top disp. for the 2D three story shear building; a) elastoplastic, b) linear elastic, system

Chopra (1996). The exact response for the top floor displacement is shown in Figure 1.a. Also using very high values for the yield deformation, the exact top floor response is attained for the linear elastic system, corresponding to the elastoplastic system. This response is shown in Figure 1.b. Now starting with the Newmark's average acceleration method, time step size equal to .02 seconds and a localization tolerance equal to .001 and then, after complete time integration, repeat it several times, each time, i, with half steps, the maximum top displacement error of these analyses varies such as shown in Figures 2.a & 2.b. In these figures and also the rest of figures, the errors have become

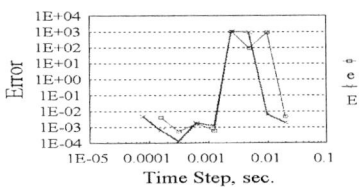 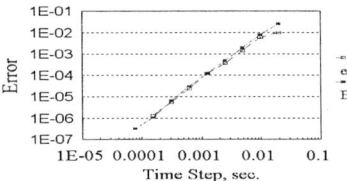

Figure 2: The error in the 2D three story shear building in a) elastoplastic, b) linear elastic, system

dimensionless after being divided by exact values. Besides the heavy line that shows the error compared to the exact response, ^iE, the light line shows the error compared to the same response attained from the next analysis. We call this error the apparent error and for the i th time integration analysis use to show it with ^ie. Though ^ie is not the true error, it is the only error that can be computed in practice. Comparing Figures 2.a & 2.b, similar to the SDOF case, Farjoodi and Soroushian (2000):
1. Opposite to the linear elastic case, in the elastoplastic case, using smaller time steps, a less error can not be guaranteed.
2. Opposite to the linear elastic case, in the elastoplastic case, the apparent error's variation is not a good representative for the true error's variation.

3. As shown in Figure 2.b, opposite to the elastoplastic case, in the linear elastic case, the equation,

$$^{i+1}E \leq {}^i e \tag{11}$$

is true. This equation which can be helpful in practical analyses is also in agreement with Eqn. 5. Now using the suggested methodology, if in the original analysis, we employ the same tolerance, .001 for all members and then for the second analysis with half time steps, change members' tolerances individually according to Eqn. 10, and similarly go on for the third, forth, ... analyses, Figure 3 will be attained. The high resemblance of Figures 3 & 2.b shows the significant effect of Eqn. 10.

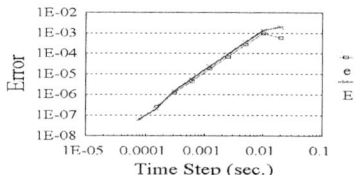 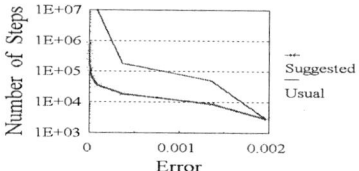

Figure 3: Error variation in the EP case of example one, using the suggested methodology

Figure 4: The total number of steps, essential to gain a given accuracy, in example one

In fact as apparent in Figure 3, all of the above mentioned difficulties are highly diminished. To completely study this example, Figure 4 shows the more efficiency of the suggested methodology compared to the usual methodology, Clough & Penzien (1993), where Eqn. 10 is not used.

As a second example, an eight story plane shear building with 5 percent modal damping and the other characteristics such as mentioned in Table 2 is excited by the first 10 seconds of the North-South

TABLE 2
PROPERTIES OF THE MEMBERS OF AN EIGHT STORY PLANE SHEAR BUILDING

Level	Mass	Stiffness	Yield Deformation
1	.518 k-sec^2/in	860 k/in	.36 in
2	.517 k-sec^2/in	840 k/in	.37 in
3	.516 k-sec^2/in	820 k/in	.38 in
4	.515 k-sec^2/in	700 k/in	.40 in
5	.514 k-sec^2/in	680 k/in	.41 in
6	.513 k-sec^2/in	660 k/in	.42 in
7	.512 k-sec^2/in	640 k/in	.43 in
8(top)	.511 k-sec^2/in	620 k/in	.44 in

component of the Elcentro(1940) strong motion, Chopra (1996). Starting with an average acceleration analysis, .02 second time step and .001 tolerance and halving the time step for each new analysis, once with and once without applying Eqn. 10, the errors of the maximum top floor displacement for the two cases will be such as shown in Figure 5. Figure 5 clearly shows the considerable effect of the suggested methodology. Repeating the convergence analyses with the tolerance .0001 and then redoing all computations with the linear acceleration method, the results shown in Figures 6 & 7 & 8 are attained. The efficiency of these examples is similar to what mentioned in Figure 4.

In the above examples, some data from other studies, Mohraz et al. (1991), have been employed.

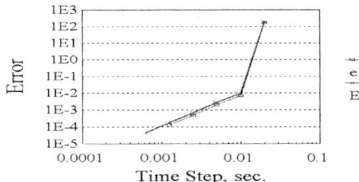 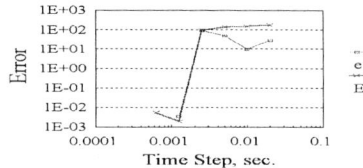

Figure 5: Convergence for the response of an elastoplastic system, employing average acceleration, $\bar{\eta} = .001$ in the first analysis, using a) the suggested methodology b) the usual methodology

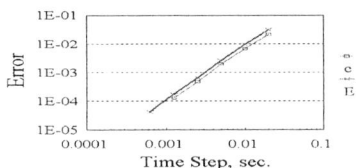 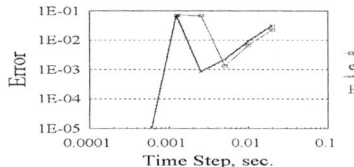

Figure 6: Convergence for the response of an elastoplastic system, employing average acceleration, $\bar{\eta} = .0001$ in the first analysis, using a) the suggested methodology b) the usual methodology

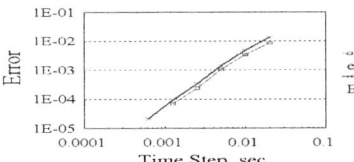 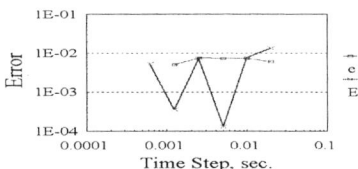

Figure 7: Convergence for the response of an elastoplastic system, employing linear acceleration, $\bar{\eta} = .001$ in the first analysis, using a) the suggested methodology b) the usual methodology

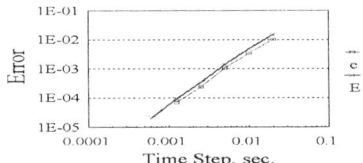 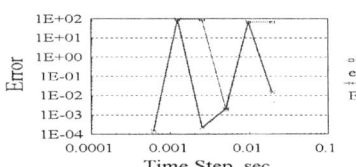

Figure 8: Convergence for the response of an elastoplastic system, employing linear acceleration, $\bar{\eta} = .0001$ in the first analysis, using a) the suggested methodology b) the usual methodology

4. CONCLUSION

In this paper, considering elastoplastic dynamic problems, the logical reduction of tolerances accompanied with the methodology mentioned in the literature, Clough & Penzien (1993), is introduced as a robust methodology to attain responses with less error. This methodology can also be employed in an iterative convergence analysis to gain responses with more accuracy. Regardless of the time integration method and the tolerances used in the first analysis, the methodology is efficient, however it can be improved.

REFERENCES

Bathe K.J. (1996). *Finite Element Procedures*, Prentice-Hall, USA

Belytschko T. and Hughes T.J.R. (1983). *Computational Methods for Transient Analysis*, North-Holland, Elsevier, USA

Belytschko T. and Schoeberle D.F. (1975). On the Unconditional Stability of an Implicit Algorithm for Nonlinear Structural Dynamics. *Journal of Applied Mechanics, ASME*, 42:4, 865-869.

Bernal D. (1991). Locating Events in Step-By-Step Integration of Equations of motion. *Journal of Structural Engineering, ASCE* 117:2, 530-545.

Cardona A. and Geradin M. (1989). Time Integration of the Equations of Motion in Mechanism Analysis. *International Journal Computers and Structures* 33:3, 801-820.

Chopra A.K. (1995). *Dynamics of Structures: Theory and Application to Earthquake Engineering*, Prentice-Hall, USA

Clough R.W. and Penzien J. (1993). *Dynamics of Structures*, McGraw-Hill, Singapore

Farjoodi J. and Soroushian A. (2000). More Accuracy in Step-By-Step Analysis of Nonlinear Dynamic Systems. *Proc. 5th International Conference on Civil Engineering*, Ferdowsi University of Mashhad, 222-229, (In Persian)

Hughes T.J.R. (1987). *The Finite Element Method: Linear Static and Dynamic Finite Element Analysis*, Prentice-Hall, USA

Kardestuncer H. (1987). *Finite Element Handbook*, McGraw-Hill, USA

Low K.H. (1991). Convergence of the Numerical methods for Problems of Structural Dynamics. *Journal of Sound and Vibration* 150:2, 342-349.

Mahin S.A. and Lin J. (1983). *Construction of Inelastic Response Spectra for Single Degree-of-Freedom Systems*, Report No. UCB/EERC-83/17, Earthquake Engineering Research Center, University of California, Berkeley, USA

Mohraz B. and Elghadamsi F.E. and Chang C.J. (1991). An Incremental Mode Superposition for Non-Linear Dynamic Analysis. *Earthquake Engineering and Structural Dynamics* 20:5, 471-481.

Nau J.M. (1983). Computation of Inelastic Spectra. *Journal of Engineering Mechanics, ASCE* 109:1, 279-288.

Rashidi S. and Saadeghvaziri M.A. (1997). Seismic Modeling of Multispan Simply-Supported Bridges Using Adina. *International Journal Computers and Structures* 64:5/6, 1025-1039.

Richtmyer R.D. and Morton K.W. (1967). *Difference Methods For Initial-Value Problems*, Interscience Publishers, USA

Wood W.L. (1990). *Practical Time-Stepping Schemes*, Oxford, USA

Xie Y.M. (1996). An Assessment of Time Integration Schemes for Nonlinear Dynamic Equations. *Journal of Sound and Vibration* 192:1, 321-331.

FREE VIBRATION ANALYSIS OF SLANTLEGGED SKEW BRIDGE

N.MUNIRUDRAPPA[1] & V.ANIL KUMAR[2]

[1]Professor Dept. of Civil Engg. (U.V.C.E) Bangalore University, Jnanabharathi Campus Bangalore-560056, India.
[2]P.G. Student (Structures) Dept. of Civil Engg. (U.V.C.E) Bangalore University, Jnanabharathi Campus Bangalore-560056, India

ABSTRACT

Knowledge of natural modes and frequency are basic to understand the dynamic response of a structure under any kind of excitation. In the present study, free vibration characteristics of slant legged skew bridge is determined by considering it as a skew type. The bridge is modeled as a space bar element with six degrees of freedom at each node, i.e. displacements in x, y, and z directions and rotations in x, y, and z directions. The element stiffness matrix, for the space bar is given the stiffness matrix K_1 and geometric stiffness matrix K_g (i.e. $K = K_1 + K_g$). The frequency of the slant legged bridge is studied by Jacobi method which represents standard eign value problem, the solution of which gives 'n' values of 'λ' where 'n' is the degree of freedom and 'λ' represents the natural frequency of the system. Corresponding to each value of 'λ' a vector $[a_n]$ can be evaluated representing the natural mode or principal mode of vibration. Further various parameters like skew angle, span of the bridge and geometric stiffness has been considered and their effect on the bridge was studied. Software has been developed in C++ to perform the analysis, which basically involves in obtaining the static deflections, shear force, bending moment, natural frequencies and mode shapes.

KEYWORDS

Free Vibration, Slant Legged, Skew Bridge, Parametric Study, C++, Static Deflection, Natural Frequencies, Modal Shapes.

INTRODUCTION

The slant legged rigid frame is one type of the most widely used highway bridges treating the super structure and the sub structure as one unit. This type of construction eliminates the need for concrete piers and positions the supports away from the lower roadway, thus giving safe structure. The rigorous analysis of the slant legged bridge is very complicated, time consuming and the structure will be having infinite degrees of freedom. Hence, from the review of literature [5,7,8,9] it is seen that in many cases the bridge is idealized as an open grid or as plane frame with masses lumped at nodes. The error involved in the calculation of the response due to this transformation is generally small. In general for all practical considerations, only first three modes of vibrations are important. Hence the finite degree of freedom system is derived out of plane frame by lumped mass idealization. The aim of the present study is to investigate the static deflections, shear force, bending moment, natural frequencies and mode shapes.

The slant bridge is modeled as a space bar system with six degrees of freedom at each node i.e. displacements in x, y and z directions and rotations in x, y and z directions respectively. The element stiffness matrix for the space bar is given by the sum of the stiffness matrix K_1 and geometric stiffness matrix K_g (i.e. $K=K_1+K_g$). This element stiffness matrix, which is developed for its local axis, is transferred to the global axis by multiplying with the transformation matrix. This transformation matrix will account for the slant leg and also skew angle. By combining the entire element stiffness matrix the global stiffness matrix is obtained, which in turn will be used to determine the static deflection at each node. Further the free vibration of the slant legged skew bridge is studied by Jacobi method which represents standard eign value problem. Further to make the study more meaningful, parametric study has been done under the following heads (i) skew angle (ii) span of the bridge and (iii) geometric stiffness.

MATHEMATICAL IDEALIZATION OF BRIDGE MODEL

The systems generally encountered in practice have continuously distributed mass and elasticity [1,2,3,4,6]. To specify the position of every particle in the elastic body, infinite number of co-ordinates are necessary and such bodies therefore posses an infinite degree of freedom. However in order to arrive at a definite solution it is desirable to limit the number of degree of freedom to a finite value. Also sufficient accuracy can be obtained, if we consider only the first few modes of vibration in the analysis. This is made possible, by replacing the continuous mass distribution of the real structure by an equivalent set of masses concentrated at the joints. The error involved in the calculation of the response of the structure due to this transformation is generally small. The accuracy of the results can be improved by having finer mesh systems for grid at which the masses are lumped, but at the expense of costly computer time. By neglecting the action of the deck slab, it is reasonable to assume that masses are lumped at the joints of main and cross beams.

Figure 1: Mathematical Model Of The Bridge

Fig 1, illustrates the analytic bridge chosen for the analysis wherein the bridge is of rigid frame with three spans. The two interior supports are inclined columns casted integrally with the girders. The ends of the bridge are simply supported, and the bridge has a road width of 12m (as per IRC standards for N.H) and 20cm thick deck slab.

The bridge is divided into 16 elements with 17 nodes and the node-to-node length is varied by 4m, 6m and 8m. The cross section of the bridge is varied as 0.4m x 0.7m, 0.4m x 0.8m respectively. The inclination of slant legs are taken as 45^0 for all the span lengths of the bridge and it is tested up to a skew angle of 60^0 to see the effect of the inclination on the natural frequencies at the related modal values.

FORMATION OF STIFFNESS MATRIX

Formulation Of Member Stiffness Matrix

In the present study the bridge is modeled as a space beam element considering three dimensional prismatic beam element with the member axis as shown in fig 2. the X-axis coincides with the centroidal axis of the member and is positive from i to j. the y and z axis are chosen such that X-Y and X-Z are principal planes of bending. The member axis x, y and z are parallel to the global axis and the degrees of freedom of global and member axis system are the same, as shown in Fig 2. For an arbitrarily inclined three-dimensional beam, the member stiffness matrix can be transformed into the global system.

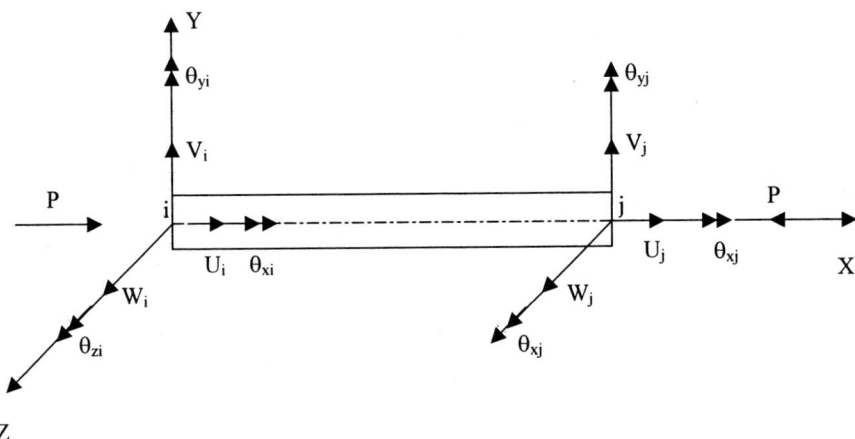

Figure 2: Space Frame Member

The element stiffness matrix of a prismatic beam can be written in the form

$$K_m = K_1 + K_g \quad (1)$$

In which K_1 is the standard linear matrix for space frame and K_g represents the effect of axial effect on the bending stiffness of the element.

TRANSFORMATION MATRIX

The stiffness matrix obtained from equation 1 is with reference to its member axis. In order to transform the member stiffness matrix from member axis to structure axis, the rotation transformation matrix R_T is given as

$$R_T = \begin{pmatrix} R & 0 & 0 & 0 \\ 0 & R & 0 & 0 \\ 0 & 0 & R & 0 \\ 0 & 0 & 0 & 0 \end{pmatrix} \quad (2)$$

Where

$$R = \begin{pmatrix} Cx & Cy & Cz \\ (-CxCy\cos\alpha - Cz\sin\alpha)/Cxz & Cxz\cos\alpha & (-CyCz\cos\alpha + Cx\sin\alpha)/Cxz \\ (CxCy\sin\alpha - Cz\cos\alpha)/Cxz & -Cxz\sin\alpha & (CyCz\sin\alpha + Cx\cos\alpha)/Cxz \end{pmatrix} \quad (3)$$

Where

$$Cx = (Xj-Xi)/L \qquad Cy = (Yj-Yi)/L \qquad Cz = (Zj-Zi)/L$$

$$L = \sqrt{(Xj-Xi)^2 + (Yj-Yi)^2 + (Zj-Zi)^2}$$

α = Angle measured from principal axis of the cross section to that plane.

The stiffness matrix of the element with reference to global system of axes is given by

$$[K] = [R_T]^T [K_m] [R_T] \quad (4)$$

Where
 $[K_m]$ is the stiffness matrix with reference to local member axis obtained from equation 1

Total Structural Stiffness Matrix

The stiffness matrix obtained in equation 4 is for one member, the total stiffness matrix is formulated by assembling all the elemental stiffness matrices by considering the node connectivity. The total structural stiffness matrix is given by

$$K = \sum_{i=1}^{m} K_{Mis} \quad (5)$$

Where the symbol K_{Mis} represents the ith elemental stiffness matrix with end actions and displacements (for both ends) taken in the direction of structural axis.

STATIC ANALYSIS

The set of simultaneous equations that arise in the static analysis are given by:

$$[K][U] = [F] \quad (6)$$

Where
 [K] is the total structural stiffness matrix (equation 6) of order n x n where n is the degree of freedom of the structure.
 [U] is the displacement vector and
 [F] is the load vector

The above equation is solved for displacement vector by using Cholesky method of decomposition. The obtained displacement vector will give the displacements at each node in six directions (i.e. u, v, w, θ_n, θ_y, θ_z). By reverse substituting this displacement vector the shear force and bending moment in each member are obtained.

FREE VIBRATION ANALYSIS

Derivation of Characteristic Equation

The basic dynamic equilibrium equation derived by using Newton's second Law of Motion can be shown equal to

$$[M][\ddot{y}] + [C][\dot{Y}] + [K][Y] = [F(t)] \tag{7}$$

Where
  [M] is the diagonal mass matrix of size n x n where n is the number of degrees of freedom of the structure.
  [K] is the total structural stiffness matrix of size n x n given in the equation---where n is the degree of freedom.
  [C] is the damping coefficient
  [F (t)] is the column matrix of the applied time dependent forces.

To determine the frequencies and mode shapes of any vibrating system subjected to action of time dependent forces, first of all it is necessary to understand its behavior under no load conditions. When a system is displaced from its equilibrium position by giving a small initial displacement and there after allowed to vibrate under its own weight without applying any external force, the system is said to be in free vibration. So if [F (t)] = 0 and since damping effect is neglected, equation 8 will become

$$[M][\ddot{y}] + [K][Y] = 0 \tag{8}$$

Further, it was shown by French mathematician J.Forier (1768-1830)[6] that any periodic motion can be represented by a series of sines and cosines that are harmonically related.

Hence [Y] can be written as $[a_n] \sin \omega_n t$

$$[\ddot{y}] = -[a_n]\omega_n^2 \sin \omega_n t \tag{9}$$

Substituting equation 10 in 9 we get

$$[[M] - \lambda[K]][a_n] = 0 \tag{10}$$

The above equation represents generalized eign value problem (i.e. $A_x = \lambda B_x$). This equation will have a non-trivial solution only if the determinant corresponding to

$$|[M] - \lambda[K]| = 0$$

The above determinant is called as "the frequency determinant", the solution of which gives n values of λ where n is the degree of freedom and λ represents the natural mode or principal mode of vibration.

NUMERICAL COMPUTATIONS

To determine the static deflections and natural frequencies computer programs were developed by considering only the self-weight of the structure. The programs were developed taking into account the elements in slant legged rigid frame bridge as beam elements in space with uniform cross section and the skew angle is induced by declaring the necessary z-coordinates. The programs were developed in C++ and run on Pentium based PC.

EFFECT OF SKEW ANGLE ON STATIC DEFORMATION

From figure 3 (i.e. when the node distance was 4m and cross section 0.4m x 0.7m) it was observed that, as skew angle increases, the vertical deflection at node 8 are increasing marginally. This indicates

1. Skew induces more flexibility to the structure
2. The induced flexibility is marginal

EFFECT OF NODE-TO-NODE LENGTH ON STATIC DEFORMATIONS

It was observed from the results that significant increase in deflection and hence the structure becomes more flexible. As the spans are increasing the effect of skew angle is producing considerable increase in torsional forces (fig 5&6)

EFFECT OF SKEW ANGLE ON SHEAR FORCE AND BENDING MOMENT

As the skew angle increases for a node-to-node distance of 4m (fig 4) it was noticed that the shear force is remaining constant where as the moment in x-direction is increasing considerably. this indicates that as the skew angle increases, torsional moments are increasing predominantly and the structure has to be designed for torsion.

EFFECT OF GEOMETRIC STIFFNESS

It was observed that for a skew angle of 30^0 and node-to-node length of 4m, the static deformations in both the cases (with and without stiffness matrix) remained the same. This confirmed that the structure was rigid. It was also observed that for a skew angle of 30^0 and node-to-node length of 6m, the static deformation decreased. This is because as the span increases, the structure will become more flexible and hence the effect of geometric stiffness matrix is more

EFFECT OF SKEW ANGLE ON NATURAL FREQUENCIES

From the results it was observed that the fundamental mode always is predominantly a normal mode, it is further noticed that the fundamental frequency is not changing significantly when the skew angle is increased. When the skew is 0^0 the contribution of torsional modes is virtually zero. As the skew is increased the contribution of torsional modes are increasing slowly. Fig 7a & 7bshows the mode shapes for a skew of 30^0.

CONCLUSIONS

1) The static deflections remain almost same irrespective of the skew angle in Y-direction. Whereas, in the other directions there is a gradual increase in deflections.

2) The shear force has remained constant irrespective of the skew angle. Whereas, the torsional moment has increased with the increase in the skew angle

3) The addition of geometric stiffness matrix has induced a minor increase in the deflections and showing that the structure is rigid.

4) The fundamental frequency is not changing significantly when the skew angle is increased.

5) The modal values are predominantly in Y and Z directions when the angle is less, but as the skew increases, the modal values are in θ_x and θ_y directions showing that the torsion is playing an important role and the structure has to be designed for torsion.

REFERENCES

1) A.S.Veletsos and Hung (1970). "*Analysis Of Dynamic Response Of Highway Bridges*", J.Engrg Mechanics Division, ASCE, 96 (10), pp 593-619

2) Bleich.H.H., (1950). "*Frequency Analysis Of Beam And Girder Floors*", Transactions, ASCE, vol.115, paper No.2416, pp. 1023-1061.

3) Biggs.J.M and Suer.H.S and Louw.J.M (1956). "*Vibration of single span highway bridge*", transactions, ASCE, Vol.124, paper No.2979, pp.291-318

4) Cox A.L and Denke.P.H, (1956). "*Stress distribution instability and free vibrations of beams grid works on elastic foundation*", J.Aeronautical Sciences, Vol.23 (2), pp.173

5) Fernando Venanico Filno. (1966). "*Dynamic influence lines of beams and frames*", J.Struct. Engrg., ASCE, 92(4), pp.371-386

6) Mario Paz. (1991). "*A text book of Structural Dynamics*". CBS Publishers and Distributors, India.

7) Munirudrappa.N.(1969). "*Dynamic response of orthogonal bridge grid under moving forces*". M.Tech. Thesis, Indian Institute Of Bombay, India

8) Thein Wah. (1963). "*Natural Frequencies of Uniform Grillages*", J. Applied Mech, Tranactions, ASME, No.4, pp.571-578

9) Ton LO Wang, Danghou Huang and Mohsen Shahaway, (1994). "*Dynamic Behivour of slant legged rigid frame highway bridge*", J. Strut. Engrg. ASCE, 120(3), pp.885-902.

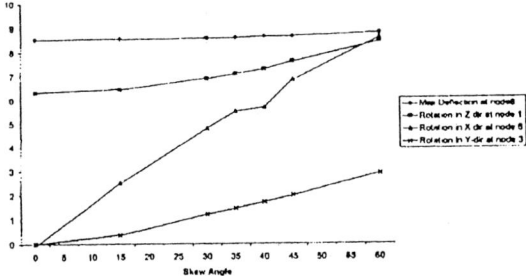

Fig 3   Effect of Skew Angle on Max Deflection and Rotation

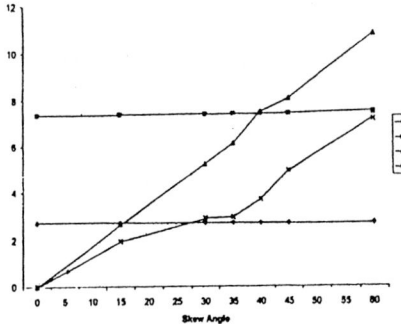

Fig 4   Effect of Skew Angle on Forces and Moments

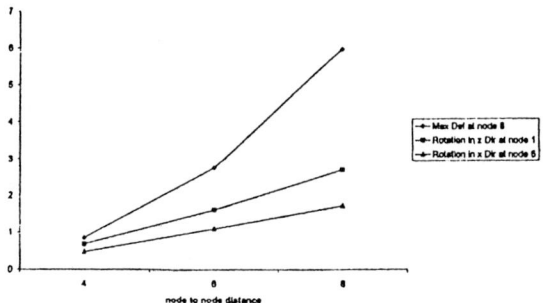

Fig 5   Effect of Node to Node Distance on Maximum Deflection and Rotations

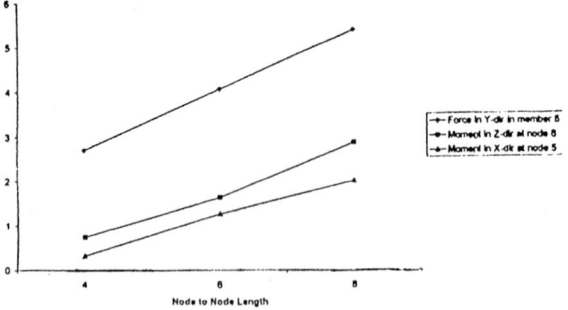

Fig 6   Effect of Node to Node Distance on Forces and Moments

Fig 7 Model Shapes

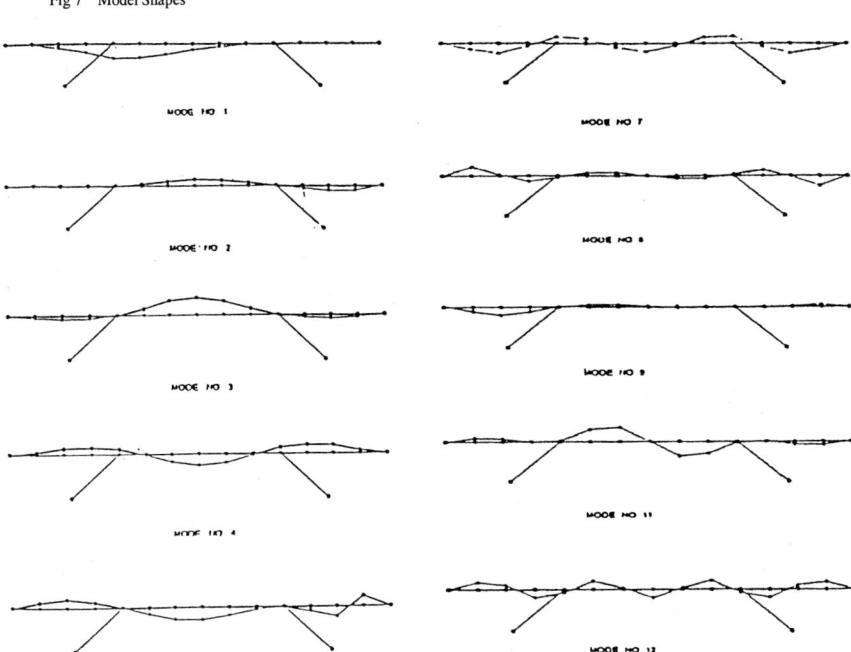

VIBRATION CONTROL AND SEISMIC ANALYSIS

PLACEMENT AND ELIMINATION OF VIBRATION CONTROLLERS IN BUILDINGS

A. Henry, A. Richardson, M. Abdullah

Department of Civil Engineering, Florida Agricultural and Mechanical
University / Florida State University, Tallahassee, Fl 32310, U.S.A.

ABSTRACT

Dynamic loads on civil structures due to earthquakes can cause excessive vibrations, which can lead to serious structural instability resulting in building damage and/or collapse. Building vibrations can be reduced using controllers (sensor/actuators), however these control devices are quite expensive. It is therefore necessary for structural engineers to determine not only the optimal placement and design of these controllers, but also determine the number of controllers required to achieve a predetermined performance. Presently there are several methods of placing controllers, many of which only consider the placement of a predetermined number of sensor/actuators. In this paper, a new method is introduced to simultaneously determine the number of necessary controllers and the optimal placement and design of these controllers needed to retrofit existing structures. To do so, it will be assumed that active tendon controllers are placed between every floor of the structure. Controllers will then be eliminated in part on their gains and their effect on the building's response. Controllers with smaller gains will be eliminated first, as their effect on the building's response will be insignificant and therefore deemed unnecessary. Controllers with larger gains will be eliminated according to the building's response with and without that controller in question. The proposed method will be compared to similar work where a finite number of sensor/actuators are placed sequentially and simultaneously.

KEYWORDS

Earthquake design, optimal placement, elimination, gains, feed back control, tall buildings

INTRODUCTION

Dynamic loads on civil structures due to earthquakes can cause excessive vibrations which often create serious structural instability, leading to damage or collapse. Civil structures are designed based on static loads and thus have a significant amount of stiffness. However because of the building materials used, civil engineering structures provide very little damping, thus they dissipate vibration energy poorly. Controllers are therefore used to help dissipate vibration energy.

Presently there are several types of controllers some of the more commonly used include; base isolation, active tuned mass dampers, and active tendon mechanisms. In base isolation, the foundation is isolated from the superstructure using nonlinear absorbers, allowing the entire structure to function as a single degree of freedom system (Rofooei & Tadjbakhsh 1993). In this control method, the relative displacement between the base of the structure and the top floor can be greatly reduced. Active tuned mass dampers are by far the most popular control device presently being used. They consist of a mass, a spring, a viscous damper and an actuator. The underlying passive mass damper is tuned to the natural frequency of the structure, and is designed to dissipate vibration energy imparted on the structure through its own motion. The force exerted by the actuator in this system limits excessive movement of the mass damper, and improves the efficiency of the passive system. Active tuned mass dampers have been found to be effective in reducing building vibration caused by wind and earthquake forces (Abdullah 1991; Abe & Fujino 1994). Active tendon mechanisms consist of an actuator connected to cables, which are rigidly connected between floors. The control force is produced by the actuator and is transmitted through the cables to the floors of the structure. The magnitude of the force exerted is based on a predetermined control design.

In this research task, the optimal number, placement and design of active tendon mechanisms to retrofit existing structures will be considered. To design these controllers, direct velocity output feedback control is used since this control law is guaranteed to be stable, provided that sensor/actuator dynamics is neglected (Balas 1979). Output feedback control was chosen as it is more efficient than state feedback, as it requires less sensor outputs and is therefore less computationally expensive (Chung & Lin & Chu 1993). To determine the number of necessary controllers, it was first assumed that active tendon controllers are placed between every floor of the structure. Controllers will then be eliminated in part on their gains and their effect on the building's response. Controllers with smaller gains will be eliminated first, as their effect on the building's response will be insignificant and therefore deemed unnecessary. Controllers with larger gains will be eliminated according to the building's response with and without that controller in question. The effectiveness of the design will be demonstrated by comparing it to similar work where controllers are placed sequentially (Abdullah 1999) and placed simultaneously (Richardson 2000).

EQUATIONS OF MOTION

The dynamic equations used to describe the motion of a building are given as follows:

$$\mathbf{M}\ddot{\mathbf{x}}(t) + \mathbf{C}\dot{\mathbf{x}}(t) + \mathbf{K}\mathbf{x}(t) = \mathbf{b}\mathbf{u}(t) + \mathbf{h}w(t), \tag{1}$$

where $\mathbf{M}, \mathbf{C}$, and $\mathbf{K}$ are $n \times n$ mass, damping and stiffness matrices respectively, n denotes the number of floors, $\mathbf{u}$ is the controlled input to the system, w is the ground acceleration (earthquake excitation), $\mathbf{b}$ is a participation matrix for the controlled input, and $\mathbf{h}$ is a participation matrix which consists of a vector of floor masses. Equation (1) can be decoupled into madal space and expressed in state space form as:

$$\mathbf{I}\ddot{\mathbf{q}}(t) + \mathbf{C}_d \dot{\mathbf{q}}(t) + \Lambda \mathbf{q}(t) = \Phi^T \mathbf{b}\mathbf{u}(t) + \Phi^T \mathbf{h}w(t), \tag{2}$$

$$\dot{x}(t) = A_0 x(t) + Bu(t) + Hw(t), \quad y(t) = C_s x(t), \tag{3}$$

where

$$A_0 = \begin{bmatrix} 0 & I \\ -\Lambda & -C_d \end{bmatrix}, \quad x(t) = \begin{bmatrix} q(t) \\ \dot{q}(t) \end{bmatrix}, \quad B = \begin{bmatrix} 0 \\ \Phi^T b \end{bmatrix}, \quad H = \begin{bmatrix} 0 \\ \Phi^T h \end{bmatrix}, \quad C_s = B^T \tag{4}$$

In this paper, the sensor and actuator dynamics for the tendon controllers will be neglected. For the direct velocity feedback design, the control force is proportional to the sensor output $y(t)$.

$$u(t) = -Fy(t) \tag{5}$$

OPTIMIZATION TECHNIQUE

To determine the optimal placement and design of an active control system, a performance function is required and was chosen so as to include both the building response and the control effort.

$$J = \frac{1}{2}\int_0^\infty (x^T Qx + u^T Ru) dt, \tag{6}$$

where Q and R are weighting matrices with respect to the building's response and the required control force,

$$Q = Q\begin{bmatrix} \Lambda & 0 \\ 0 & I \end{bmatrix}, \quad R = R\begin{bmatrix} I & 0 \\ 0 & I \end{bmatrix} \tag{7}$$

The performance function can be expanded in terms of the fundamental transition matrix $x = e^{At}$, by substituting u as a function of x, where A is the matrix of the closed loop plant:

$$A = (A_0 - BFC_s), \quad \hat{J} = x^T(0)\left[\frac{1}{2}\int_0^\infty e^{A^T t}(Q + C_s^T F^T RFC_s)e^{At} dt\right]x(0). \tag{8}$$

A design procedure which uses this performance function will require discrete values for the initial state, $x(0)$ (Choe & Baruh 1992). Herein, this dependence on $x(0)$ is eliminated by using a performance function proposed by Levine and Athans (1970). The initial state is modeled as a random vector uniformly distributed on the surface of the $2n$-dimensional unit sphere. It has been shown that the average or expected value of $\hat{J}$, scaled by $2n$, is

$$\hat{J} = \frac{1}{2}\int_0^\infty tr\left[e^{A^T t}(Q + C_s^T F^T RFC_s)e^{At}\right]dt. \tag{9}$$

This performance function which is dependent on the actuator and sensor placements embedded in the matrices C_s and B, and the feedback gain matrix F, is used hereafter. Levine and Athans (1970) also showed that the gradient of the performance function could be determined. However two matrix integrals are needed:

$$P = \int_0^\infty e^{A^T t}(Q + C_s^T F^T R F C_s) e^{At} dt, \quad L = \int_0^\infty e^{At} e^{A^T t} dt \qquad (10)$$

The matrix integral P is related to the standard performance function by

$$J = \frac{1}{2} \text{tr}[P]. \qquad (11)$$

To reduce the computational effort and time, the matrix integrals in (10) can be easily solved by finding the solution to the following two Lyapunov equations:

$$PA + A^T P + Q + C_s^T F^T R F C_s = 0, \quad LA + A^T L + I = 0. \qquad (12)$$

The use of single loop collocated sensors and actuators with direct velocity feedback gains, eliminates the effects of residual modes causing instabilities in the control system (Balas 1982). The gradient of the gain matrix can be expressed as such:

$$\frac{\partial \hat{J}}{\partial F} = -RFC_s LC_s^T + C_s PLC_s^T. \qquad (13)$$

ELIMINATION TECHNIQUE

Initially it is assumed that sensor/actuators are placed on every floor. To determine the optimal gains for these controllers, a random search algorithm is used to evaluate random test points from the specified range space, of these, the point (vector) that yields the lowest objective function value is used as the starting point for the Davidon-Fletcher-Powell (DFP) algorithm. The DFP method is used to update the Hessian matrix, which is always maintained to be positive definite so that the search direction is always in the decreasing direction. Once the search direction vector is computed, a cubic polynomial line search procedure is used to locate a one-dimensional minimum. When this point is found, another search direction vector is computed and the procedure continues until a local minimum is found.

Once the optimum gains are determined, the building's response is simulated, and its RMS response as well as the control effort is noted for each of the given inputs. During the design process, the building response is simulated for each of the chosen inputs, while decreasing the number of sensor/actuators. Thus in the first simulation for a chosen input, it is assumed that all controllers are present. In each successive run thereafter, the controller with the lowest gain is eliminated. The process of eliminating the controller with the lowest gains after each simulation will continue until one controller is left.

Using the data obtained, a quantitative comparison is made between the required control effort and the RMS building response to determine the number of necessary sensor/actuators.

BUILDING PARAMETERS

In this research task, a 40-story structure with identical floors is used. The floor mass, m, is 1290 tons, the elastic stiffness of each floor, k, is 10^6 KN/m and the external damping constant, β, is 1000 KN/m/s and the internal damping constant, c, is 0. For this structure, $\mathbf{Q}$ in (6) is chosen to be a diagonal matrix with $\mathbf{Q}_{ii} = 13000$ for $i = 1,........8$, and $\mathbf{Q}_{ii} = 0$ for $i = 9,........16$. The weighting matrix $\mathbf{R}$ is chosen to be the identity.

RESULTS

Using the building parameters, the elimination method is implemented for El Centro, Kern County, and North Ridge earthquakes. The results obtained are compared to similar work, where sensor/actuators are placed sequentially (Abdullah 1999) and simultaneously (Richardson 2000). Figure 1 shows the RMS top floor building response vs the number of sensor/actuators. Figure 2 shows the sum of the RMS control effort required by each controller vs the number of sensor/actuators. From figure 1, it is seen that the top floor building response is smallest when there are 40 controllers, and as this number decreases, the RMS response of the top floor increases. Figure 2 shows that the control effort is practically the same when there are 40 or only 20 controllers, suggesting that half of the controllers are exerting very little force to improve the building response. From figure 1, it is also evident that for all three inputs, there is not a marked increase in the RMS top floor response when there are 40 or 20 sensor/actuators. This again suggests that half of the controllers are not helping to improve the building response and thus should be eliminated. With 20 sensor/actuators eliminated, the elimination process becomes more rigorous. Figure 1 shows that the change in the RMS response from twenty sensor/actuators to eight is less than 3cm, while figure 2 shows that this change reduces the RMS control effort by at least 5000 KN for all three earthquakes. If less than eight controllers are used, the RMS response will be much larger than the case with 40 controllers. Thus eight sensor/actuators are chosen as the minimum number of controllers necessary to effectively reduce the top floor RMS response of the structure.

The resulting sensor/actuators placement and their corresponding gains are compared to similar work where the sensor/actuators are placed sequentially (Abdullah 1999) and simultaneously (Richardson 2000) in table 1. It should be noted that the controller gains are comparable to those obtained by the other two placement methods. Tables 2 show the RMS response of the top floor of the structure respectively. From table 2, it is seen that for all three earthquakes, the proposed elimination technique reduced the top floor building response more than the sequential method, while having comparable percent reductions as the simultaneously placed method. Table 3 compares the sum of the RMS control effort required by each of the three-placement methods. This table shows that the control effort required by the elimination method is less than that required by the sequential method, while having percent reductions within 10% of that of the simultaneously placed method. Figure 3 shows the top floor building response of the uncontrolled structure as compared to the structure with the proposed control design. Figure 4 shows the top floor response for the three-placement methods.

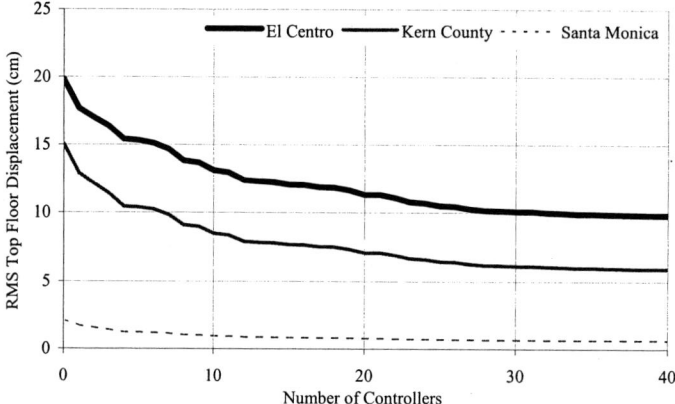

Figure 1: RMS Top Floor Displacements

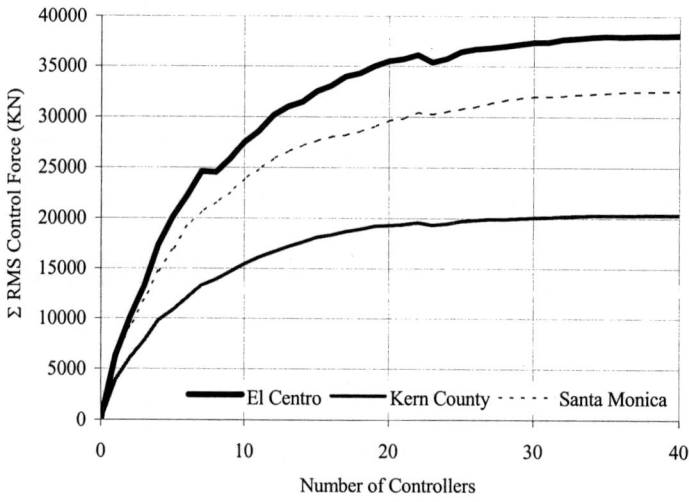

Figure 2: Sum of RMS Control Forces

Table 1: Summary of the Controller Location and Gains for the Forty-Story Structure

Tendon location (Simultaneously placed)	1	2	3	4	8	9	12	14	26	36
Gains/m_i (1/s)	206	496.3	491.9	111.7	385.4	181.1	182.2	289.4	64	32
Tendon location (Sequentially placed)	3	4	7	10	11	15	19	24	30	36
Gains/m_i (1/s)	2199	0.01	1108	1864	70.7	1797	1442	1743	1395	874.6
Tendon location (Elimination)	1	2	4	18	20	21	31	37		
Gains/m_i (1/s)	365.0	466	417.6	426.4	460.7	376.7	388.4	410.5		

Table 2: Summary of the RMS Top Floor Displacement of the Forty-Story Structure

	Elcentro		Kern County		Northridge at Santamonica	
	RMS Displacement (cm)	% Reduction in RMS Disp	RMS Displacement (cm)	% Reduction in RMS Disp	RMS Displacement (cm)	% Reduction in RMS Disp
Uncontrolled	19.81		15.02		21.57	
Simultaneously placed	13.31	32.81	8.58	42.88	9.49	56.00
Sequentially placed	13.55	31.60	9.55	36.42	11.46	46.87
Elimination	13.80	30.34	9.08	39.55	10.18	52.8

Table 3: Summary of RMS Control Effort

	El Centro	Kern County	Northridge at Santa Monica
Σ RMS control force (10^3 KN) (Simultaneously placed)	20.45	11.79	18.06
Σ RMS control force (10^3 KN) (Sequentially placed)	45.70	29.00	46.51
Σ RMS control force (10^3 KN) (Elimination)	24.53	13.89	21.50
%Reduction of RMS control force (Simultaneously placed)	55.25	59.34	61.17
%Reduction of RMS control force (Elimination)	46.32	52.10	53.77

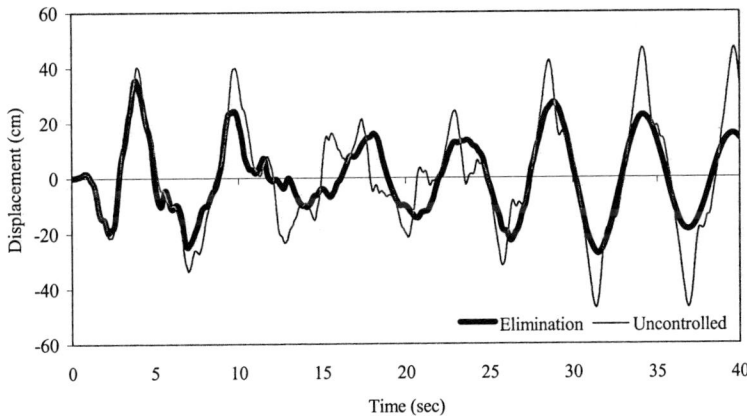

Figure 3: Forty-Story Top Floor Displacement of Uncontrolled vs Elimination Method

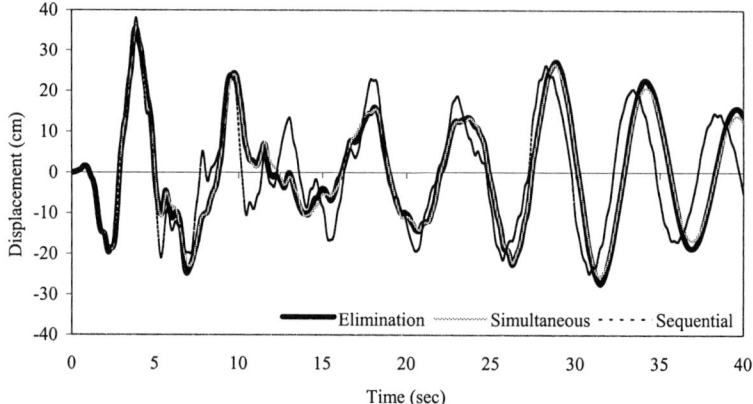

Figure 4: Forty-Story Top Floor Displacement of all Three-Control Methods

CONCLUSION

The proposed control sensor/actuator elimination method has proved to be an effective means of determining the optimal number, placement and design of sensor/actuators. Using the proposed method a similar building response was achieved while using fewer sensor/actuators and less control effort than the sequential method. The newly proposed method required a similar control effort to that of the simultaneous method. Such savings in-terms of the number of sensor/actuators needed and the total control effort required essentially mean more saving to the building owner.

ACKNOWLEDGEMENTS

The authors are grateful for funding provided by the National Science Foundation (NSF), Grant 9703020, and the National Aeronautic and Space Administration (NASA), Grant NAG4-164. The cognizant NSF program officials are William Anderson and Allison Flatau, and the cognizant NASA program official is Dr. Kajal Gupta. All opinions expressed or implied herein are solely those of the writers.

REFERENCES

Rofooei, F.R., Tadjbakhsh, I.G. 1993, 'Optimal control of structures with acceleration, velocity and displacement feedback', *Journal of Engineering Mechanics, ASCE*, vol. 119, no. 10, pp. 1993-2010.

Abdullah, M. 1991, Active control of tall buildings, Masters Thesis, Northwestern University.

Abe, M., Fujino, Y. 1994, 'Dynamic characterization of multiple tuned mass dampers', *Earthquake Engineering and Structural Dynamics*, vol. 23, no. 8, pp. 813-835.

Balas, M. J. 1979, 'Direct velocity feedback control of large space structures', *Journal of Guidance Control Dynamics*, vol. 2, no. 3, pp. 252-253.

Chung, L. L., Lin, C. C., Chu, S. Y. 1992, 'Optimal direct output feedback of structural control', *Journal of Engineering Mechanics. ASCE,* vol. 119, no. 11, pp. 2157-2173.

Abdullah, M. 1999, 'Optimal Placement of DVFC controllers on buildings subjected to earthquake loading', *Earthquake Engineering and Structural Dynamics*, vol. 28, pp. 127-141.

Richardson, A. 2000, Placement of sensor/actuators on civil structures using genetic algorithm, Masters Thesis, Florida Agricultural and Mechanical University.

Choe, K., Baruh, H. 1992, 'Actuator placement in structural control', *Journal of Guidance Control,* vol. 15, no. 1, pp. 40-48.

Levine, W. S., Athans, M. 1970, 'On the determination of the optimal constant output feedback gains for the linear multivariable systems', *IEEE Transactions on Automatic Control*, vol. AC-15, no. 1, pp. 44-48.

Balas, M. J. 1982, 'Trends in large space structure control theory: fondest hopes, wildest dreams', *IEEE Transactions on Automatic Control*, vol. AC-27, no.3, pp. 522-536.

Viscous Damping System for Optimal Structural Seismic Design

Y. Ribakov[1], J. Gluck [2]

[1] Civil Engineering Division, Department of Engineering,
College of Judea and Samaria, Ariel, 44837, Israel
[2] Structural Engineering Division, Department of Civil Engineering,
Technion - Israel Institute of Technology,
Technion City, Haifa 32000, Israel.

ABSTRACT

A method for design a passive control system for multistory structures is presented. Viscous dampers installed at each story of the building are used to improve the response of the structure during earthquakes. Optimal control theory is used to obtain the properties of the devices. Optimization leads to different level of damping at each story. A method that enables the use of standard devices that are available off-the-shelf is proposed. Numerical analysis of a seven story shear framed structure is represented as an example. It is shown that by using the proposed method the response of structures with devices that are available off-the-shelf is close to that of structures with viscous dampers, selected according to optimal control theory. Significant improvement was obtained in the behavior of the controlled structure compared to the uncontrolled one.

KEYWORDS

Viscous damping system, optimal passive control.

INTRODUCTION

Viscous dampers are known as effective devises improving structural response to earthquakes. The damping force developed by the viscous damper depends on the physical properties of the fluid used, on the pattern of flow in the device and on its size (Constantinou et al. 1992). Gluck et al. (1996) demonstrated that structures dominated by a single mode of vibration, provided with optimally designed supplemental viscous dampers, have a response close to that of an actively controlled structure. An Optimal Control Theory (OCT) using a linear quadratic regulator is adapted to design the linear passive viscous devices in accordance to their deformation and velocity. The main difficulty of method is that the optimal solution requires different levels of damping at each story which is inconvenient and can be expansive.

This paper proposes a method by which off-the-shelf Viscous Damping System (VDS) which is connected between the floor diaphragm and chevron brace (see Fig.1) is used at all stories of the structure.

CONSTRUCTION OF VDS

A VDS (Fig. 2) consists of a main damper (1), damping correctors (2), and a cylinder (3), to which the dampers are connected. The main damper is connected at one end to cylinder, and at the other end to the activating bar. The damping correctors are connected between the cylinder and the activating bar. The angle between the damping correctors and the vertical plane is chosen so that the energy dissipated in it is equal to that dissipated in a viscous damper with optimal properties.

The device may be installed in the structure by connecting the cylinder to chevron braces, which are fixed to the lower floor, and the activating bar to the upper floor diaphragm. The force developed by the device, $F_D(x)$, is given by:

$$F_D = F_M + nF_C \sin\alpha \qquad (1)$$

where F_M is the force in the main spring, F_C is the force in a damping corrector, n is the number of damping correctors, and α is the angle between the damping correctors and the vertical plane (Fig. 2). Expressing the forces in the dampers in terms of their damping coefficients and velocities, the expression for the force in the VDS (Eq. 1) takes the form:

$$F_D(x) = C_M \dot{u} + nC_C \dot{u} \sin\alpha \qquad (2)$$

where C_M and C_C are the damping coefficients of the main damper and of the corrector, respectively, and a is the diameter of the cylinder. An important feature is that it is possible to choose the parameters of the device in such a way that for a limited range of the activating bar's displacement, the force, produced by the device, would depend only on the angle α.

The idea which is examined in this study is to obtain the required properties of the optimal viscous devices by using the Optimal Control Theory (OCT). Then the angle between the damping correctors and the vertical plane is chosen so that the VDS consisting of standard viscous dampers will produce optimal control forces.

OPTIMAL DESIGN OF PASSIVE ENERGY DISSIPATORS

The response of a structure provided with supplemental dissipating devices is described by the following dynamic equation of equilibrium (Soong 1990):

$$M\ddot{u}(t) + C\dot{u}(t) + Ku(t) = Lf_e(t) + Df_c(t) \qquad (3)$$

where M, C, and K are the mass, damping, and stiffness matrices, respectively; $u(t)$, $\dot{u}(t)$, and $\ddot{u}(t)$ are the displacement, velocity and acceleration vectors, respectively; f_e is the vector of forces in the supplemental devices, f_e is the external excitation vector. D and L are the control and excitation forces-location matrices, respectively.

The force produced by a linear fluid viscous device is proportional to the velocity of the piston in a viscous environment (Constantinou et al., 1992):

$$F_{vd} = C_{vd}\dot{u}(t) \quad (4)$$

where F_{vd} is the force in the device; C_{vd} is the viscous characteristic of the device; $\dot{u}(t)$ is the velocity of the piston in the fluid viscous medium.

Gluck et al. (1996) proposed a method to obtain the optimal properties of viscous dampers connected to chevron braces at each floor, i, which takes the form:

$$\Delta c_i = \frac{\int_0^{t_f} \sum_j g_{ij,d} \dot{d}_j(t) dt}{\int_0^{t_f} \dot{d}_i(t) dt} \quad (5)$$

where t_f is the duration time of the seismic event, and Δc_i is the damping coefficient of the damper installed at the ith floor. In the case when one mode, m, is relevant to the structural response, Eq. (3) may be further simplified as follows:

$$\Delta c_i = \frac{\sum_j g_{ij,d} \Phi_{jm}}{\Phi_{im}} \quad (6)$$

where Φ_{im} is the element of the eigenvector corresponding to mode m and degree of freedom i. Noting that the damping coefficients Δc_i are independent of the earthquake history, but only on the characteristics of the structure, Eq. (6) may be used to set the viscous characteristics of the dampers connected to chevron braces.

NUMERICAL EXAMPLE

To investigate the effectiveness of the proposed design technique simulations of a seven story shear framed structure with stiff beams (Fig. 3) were carried out. The responses to four different seismic excitations (specified below) were computed. All simulations were using routines written in MATLAB (MATLAB, 1993). The structure is characterized by the following matrices:

$$M = 8.75 \times 10^4 \, I_{7 \times 7} \, [\text{kg - mass}]$$

$$K = \begin{bmatrix} 29.28 & -14.64 & & & & & 0 \\ -14.64 & 31.59 & -16.95 & & & & \\ & -16.95 & 30.96 & -14.01 & & & \\ & & -14.01 & 28.02 & -14.01 & & \\ & & & -14.01 & 25.13 & -11.12 & \\ & & & & -11.12 & 22.24 & -11.12 \\ 0 & & & & & -11.12 & 11.12 \end{bmatrix} \times 10^7 \ [\text{N/m}]$$

where M is the mass matrix of the structure, $I_{7 \times 7}$ is a unit diagonal matrix, and K is the structural stiffness matrix.

The optimization was carried out using the OCT, which yields the following optimal values of the viscous coefficients:

$$\Delta C_k^T = \begin{bmatrix} 1.7219 & 1.7219 & 1.9936 & 1.6478 & 1.6478 & 1.3079 & 1.3079 \end{bmatrix} \times 10^6 \ \frac{N \times s}{m}$$

The matrix represents the optimal solution of a structure provided by supplemental viscous dampers. The solution requires different damping levels at each story, which may be significantly simplified by application of the proposed VDS technique, with $C_M = 1.07 \times 10^6 \ \frac{N \times s}{m}$ and $C_C = 0.36 \times 10^6 \ \frac{N \times s}{m}$. Four damping correctors were used in each VDS. According to the procedure proposed above the vector of angles between the damping correctors and the vertical plane is:

$$\alpha^T = \begin{bmatrix} 30.0° & 30.0° & 43.6° & 26.7° & 26.7° & 12.3° & 12.3° \end{bmatrix}$$

The following four seismic excitations were used in the analysis: El-Centro S00E, 1940, Taft N21E, 1952, Kocaeli EW, 1999, and Eilat EL1226NS, 1995.

The roof displacement time histories are shown in Fig. 4. Peak roof displacements and base shear forces for the uncontrolled, optimally designed structure with viscous dampers and for the structure with the proposed VDS, respectively, are presented in Tables 1, 2 and 3, respectively. Tables 2 and 3 show that VDS yields very similar results to those assumed to be the optimal solution.

Using the proposed technique the peak displacements and base shear forces in the controlled structure were reduced up to 50% (see Tables 1, 2, 3).

CONCLUSIONS

A procedure was developed for optimal design of passive controlled viscous damped structures with Viscous Damping Systems (VDS). First the optimal viscous properties of the dampers were selected using the OCT design, and then the angles of the damping correctors to the vertical plane were obtained for application of devices that are off-the-shelf available.

A numerical simulation of a seven-story structure showed that its performance with VDS including standard dampers is close to the optimal design. Reductions of up to 50% of the peak displacements and of the base shear forces were obtained and the accelerations were reduced up to 60%. These results show a promising method of using off-the-shelf viscous dampers in VDS. Thus, the damping at each floors is controlled and provides an improved behavior of the structure during an earthquake.

REFERENCES

Constantinou, M.C., and Symans, M.D. (1992). "Experimental and Analytical Investigation of Seismic Response of Structures with Supplemental Fluid Viscous Dampers". Tech. Rep. NCEER-92-0027, National Center of Earthquake Engrg. Res. State Univ. of New York (SUNY) at Buffalo N.Y.

Gluck, N., Reinhorn, A .M., Gluck, J., and Levy, R.(1996). "Design of Supplemental Dampers for Control of Structures". Journal of Structural Engineering, **122:12**, 1394-1399.

Soong, T.T. (1990). "*Active Structural Control: Theory and Practice*". John Wiley & Sons, Inc., New York, N.Y.

MATLAB - High Performance Numeric Computation and Visualization Software. User's Guide (1993). The Math Works Inc.

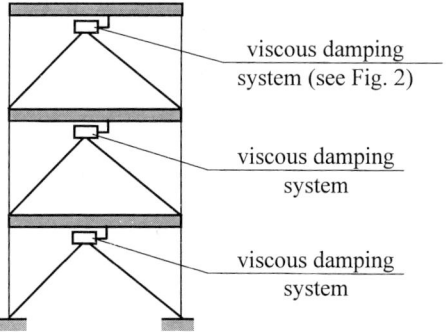

Figure 1: Three-story viscous damped structure

TABLE 1
PEAK RESPONSE OF THE UNCONTROLLED STRUCTURE

	El - Centro	Taft	Kocaeli	Eilat
Roof displacements, cm	4.33	5.18	8.32	4.18
Base shearforce, kN	1220.7	1383.0	1992.7	1139.6

TABLE 2
PEAK RESPONSE OF THE STRUCTURE CONTROLLED BY DAMPERS
SELECTED ACCORDING TO OPTIMAL CONTROL THEORY

	El - Centro	Taft	Kocaeli	Eilat
Roof displacements, cm	2.76	2.39	6.52	2.68
Base shear force, kN	900.9	663.7	1994.6	810.1

TABLE 3
PEAK RESPONSE OF THE STRUCTURE CONTROLLED BY VDS

	El - Centro	Taft	Kocaeli	Eilat
Roof displacements, cm	2.85	2.62	6.52	2.67
Base shear force, kN	853.876	802.9	1994.5	739.3

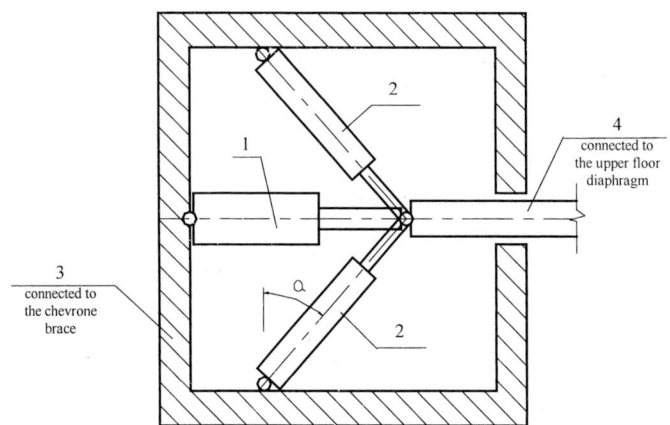

Figure 2: VDS construction scheme

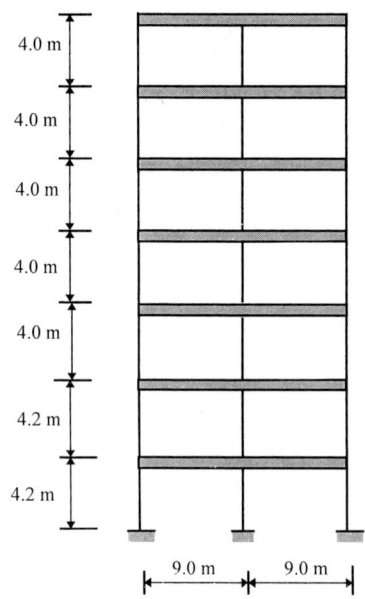

Figure 3: Uncontrolled seven-story structure

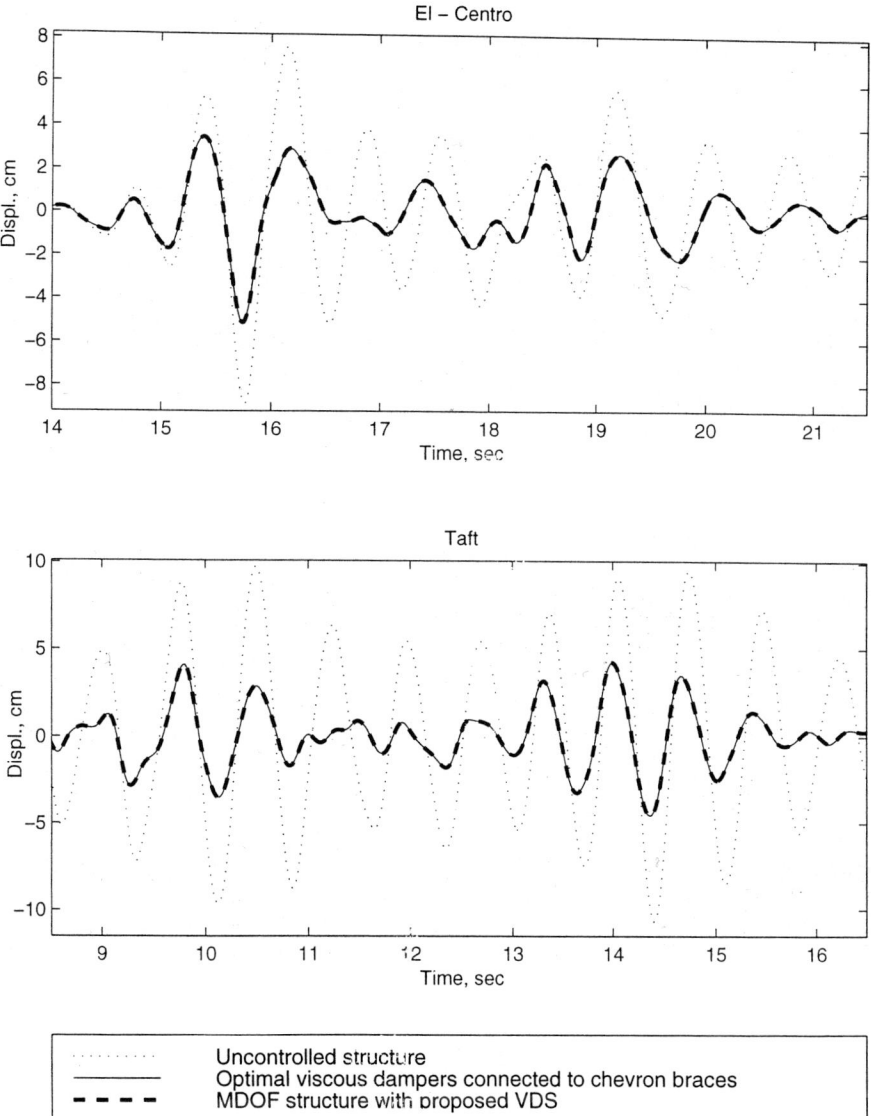

Figure 4. Roof displacements of the structure under different earthquakes

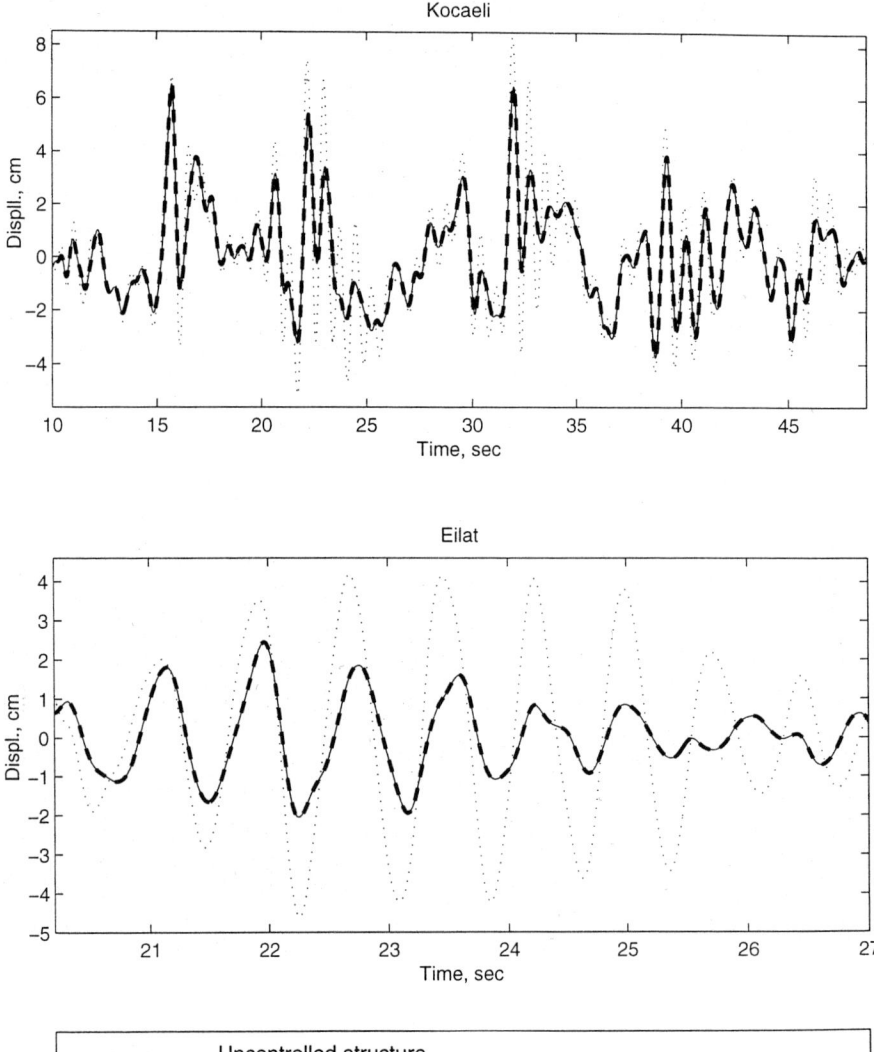

Figure 4 (cont.).

REDUCING VIBRATORY SCREEN STRUCTURAL LOADING USING A VIBRATION ABSORBER

N.F. du Plooy and P.S. Heyns

Dynamic Systems Group
Department of Mechanical and Aeronautical Engineering
University of Pretoria, Pretoria, 0002

ABSTRACT

In the nature of their operation, vibratory screens are subject to high levels of acceleration and may impart significant dynamic loads on plant buildings, especially in cases where large numbers of these screens are employed in process applications. To avoid premature failure and reduce plant building construction costs in critical cases, a variety of techniques are applied in practice. Most of these require substantially increased system mass, which is undesirable. With this work the use of a liquid inertia tuned vibration absorber to reduce the structural input loading without significantly affecting the screen acceleration, is investigated. A mathematical model is developed and applied to the design of an experimental vibration absorber. This absorber was built and tested on a Schenck Hydropuls machine and the mathematical model was validated. The model was subsequently employed in a design study on a typical sand washing plant screen. Different absorber configurations are considered and it is demonstrated that substantial benefits may be realised by using a vibration absorber for reducing structural input loading in such applications.

KEYWORDS

Liquid inertia absorber, Sand washing plant, Structural input loading, Tuned absorber, Vibration absorber, Vibratory screen.

INTRODUCTION

Many modern industrial machines rely on vibration for their operation. Oscillatory motion is essential for screens and feeders. Their proper operation demands high levels of vibration, usually to the detriment of support structures. The combined forces imparted by the vibration of several vibratory screens simultaneously, may for example cause fatigue failure of the supporting structures and plant buildings, or may alternatively require much more expensive civil engineering infrastructure than is really required (Greenway, 1983).

Various techniques exist that can be employed to reduce this vibration. The most common approach is the use of flexible isolators. This method is limited by the permissible static deflection of the isolating springs. Other methods include the use of sub-frames and counterweights. These methods add significant mass and cost to the structure. Sub-frames will also increase the transient response of the system and counterweights will require more force to achieve the desired acceleration. This work investigates the use of vibration absorbers to reduce the vibration transmitted to the support structure.

Vibration absorbers are devices used to attenuate the vibration of a primary system by adding a secondary spring and mass system. The absorber reduces the response at a tuned isolation frequency. At the isolation frequency the inertia of the secondary system is utilised to cancel the forces normally transmitted to the support structure. Such devices can therefore be used on machines that are primarily operated at a dominant frequency of excitation, which is the case for vibratory screens.

The Liquid Inertia Vibration Eliminator (Halves, 1980, 1981) provides an absorber configuration that is believed to be fundamentally suitable for vibratory screen applications. Based on this configuration an absorber was developed and tested.

VIBRATION ABSORBER

The absorber comprises an inner housing with a tuning passage connected to an outer housing through an elastomeric spring (figure 1). These two housings are respectively connected to points A and B between which the force transmission should be minimised. Under the influence of the relative vibration between these two housings, an incompressible, high-density, low viscosity and high surface tension fluid is accelerated through the tuning port from chamber to chamber. This generates amplified inertial forces on the inner and outer housings. At a tuned frequency, the inertia forces of the fluid match the spring forces and the net effect is that only the damping force is transmitted.

Since the fluid is simultaneously used as the hydraulic fluid as well as the absorber mass, the overall size of the unit is inversely proportional to the square of the fluid density. Mercury and selenium bromide has been suggested as working fluids. In this work treated water is however used as the working fluid, to avoid the risks of dangerous substances in plants.

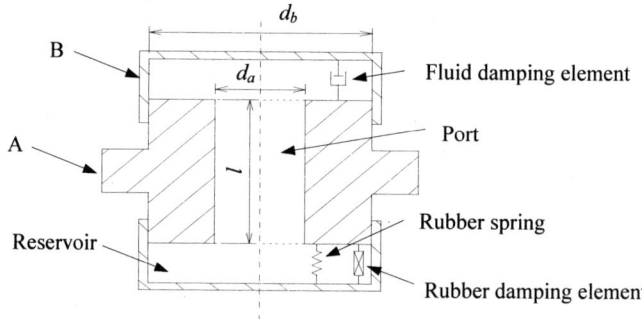

Figure 1 Schematic representation of vibration absorber

Considering the balance of forces on the system, an equation of motion can be formulated from which it follows that the force transmissibility is (Du Plooy, 1999):

$$\frac{F_o}{F_i} = \frac{k(1+i\eta)+i\omega c + \omega^2 m_B\left(1-\frac{A_b}{A_a}\right)\frac{A_b}{A_a}}{k(1+i\eta)+i\omega c - \omega^2\left[m+m_B\left(1-\frac{A_b}{A_a}\right)^2\right]} \qquad (1)$$

Once a working fluid has been selected, the response of the system is governed by the stiffness and three geometric variables:
- Stiffness k
- The reservoir diameter d_b. (The area of the reservoir is denoted A_b)
- The port diameter d_a. (The area of the port is denoted A_a)
- The tuning port length l.

Obviously, the objective of the absorber is to minimise the force transmitted at a specified tuning frequency. This can be accomplished through an optimisation process, with design variables $x_1=k$, $x_2=d_b$,

$x_3 = d_a$ and $x_4 = l$. Using these variables and substituting the viscous damping in terms of a damping ratio and critical damping, the objective function becomes (Du Plooy, 1999):

$$f(\bar{x}) = \frac{\left| x_1(1+i\eta) + i\omega_a 2\zeta \sqrt{x_1 \left[m + \frac{\pi}{4}\rho x_4 x_2^2 \left(1 - \frac{x_3^2}{x_2^2}\right)^2 \right]} + \frac{\pi}{4}\omega_a^2 \rho x_4 x_3^2 \left(1 - \frac{x_3^2}{x_2^2}\right) \right|}{\left| x_1(1+i\eta) + i\omega_a 2\zeta \sqrt{x_1 \left[m + \frac{\pi}{4}\rho x_4 x_2^2 \left(1 - \frac{x_3^2}{x_2^2}\right)^2 \right]} - \omega_a^2 \left[m + \frac{\pi}{4}\rho x_4 x_2^2 \left(1 - \frac{x_3^2}{x_2^2}\right)^2 \right] \right|}$$

subject to (2)

$$g_1(\bar{x}) = -x_1 + k^{min} \leq 0$$
$$g_2(\bar{x}) = x_2 - d_b^{max} \leq 0$$
$$g_3(\bar{x}) = -x_3 + d_a^{min} \leq 0$$
$$g_4(\bar{x}) = \sqrt{\frac{x_1}{m + \frac{\pi}{4}\rho x_4 x_2^2 \left(1 - \frac{x_3^2}{x_2^2}\right)^2}} - \omega_a \leq 0$$

The first three constraints impose minimum limits on the relevant design variables, whereas the last constraint ensures that the design is in a feasible region where the natural frequency is smaller than the design isolation frequency. Without this constraint the solution can get stuck at a local minimum, which is larger than the global minimum.

This procedure was applied to the design of an absorber tuned to 50 Hz with water as the absorber fluid and a polyurethane spring. The absorber was fabricated and tested on a 100 kN Schenck Hydropuls servo-hydraulic testing machine. Transmissibilities of as low as 0.16 were achieved. The transmissibility model also fitted the measured data very well, provided that careful attention was given to the characterisation of the elastomeric material.

SCREEN ABSORBER DESIGN STUDY

Once confidence is established in the mathematical model, the model may be used for design studies to investigate the feasibility of using vibration absorbers for vibratory screen applications. The design methodology described in the previous paragraph is used, and in this case design parameters typical of a sand washing plant screen are considered.

First a single degree-of-freedom screen model is analysed. An absorber is then added. Thereafter a two degree-of-freedom screen model is designed according to SDRC guidelines (Riddle et al., 1984). An absorber is again added between the sub-frame and the ground to illustrate the improvement possible. To simplify comparison between the different configurations, hysteretic damping is assumed for both the normal screen suspension and with the added absorber. The screen is fitted on steel coil springs with a loss factor of 0.01. It is further assumed that the combined viscous and material damping of the absorber can be represented by a loss factor of 0.1.

Single degree-of-freedom simple screen model

Two rotating eccentric mass excitation motors transfer 27 000 kg of force to the simple screen schematically depicted in figure 2. The motors rotate at 750 rpm. The spring rate is given as 30.4 kg/mm and there are 4 springs on each corner. The screen has a 2-second run-up time, which is too short for any problematic transient response amplification. Run-down can take as long as 2 minutes and damping can therefore become important to reduce the response at resonance. Tables 1 and 2 summarise the physical properties of the screen and the operating characteristics calculated according to the suggested analysis procedure.

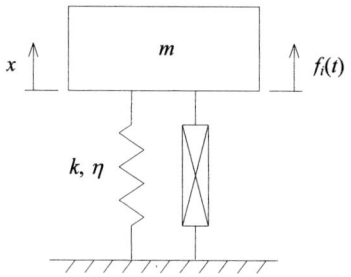

Figure 2 A single degree-of-freedom screen model

TABLE 1
PHYSICAL PROPERTIES OF THE SCREEN

Mass [kg]	m	13×10^3
Force amplitude [N]	F_i	$27000 \times 9.81 = 264.87 \times 10^3$
Stiffness [N/m]	k	$4 \times 4 \times 30.4 \times 10^3 \times 9.81 = 4.77 \times 10^6$
Spring loss factor	η	0.01
Operating speed [Hz]	f	12.5

TABLE 2
OPERATING CHARACTERISTICS OF THE SCREEN

Natural frequency [Hz]	f_n	3.05
Frequency ratio	ω/ω_n	4.1
Stroke amplitude [mm]	X	3.5
Acceleration [m/s^2]	$\omega^2 X$	2.21g
Static deflection [mm]	x_S	26.7
Dynamic force transmitted [N]	F_o	16758.55
Transmissibility [%]	$|T_r|$	6.327

Figures 3(a) and 3(b) show the receptance and force transmissibility of this design. The low damping of the spring can cause high amplitudes at the natural frequency during run-down.

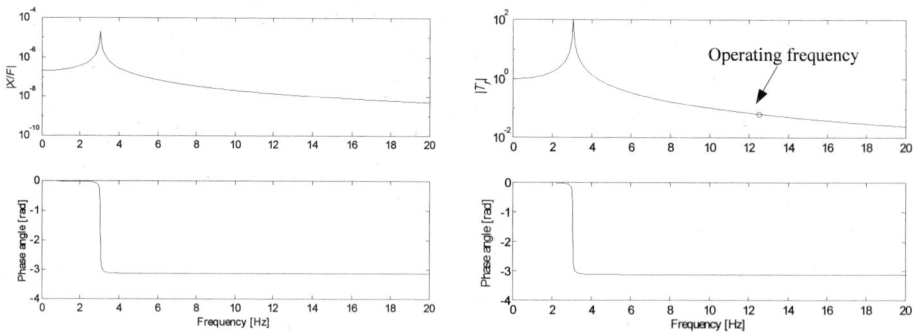

Figures 3 (a) Receptance (b) Force transmissibility

The design of the screen is in accordance with current practice and therefore the operating frequency is 4 times the resonance frequency. Although the transmissibility is only 6%, a total of 16.8 kN is still transmitted to the support structure.

Single-degree-of-freedom screen with absorber model

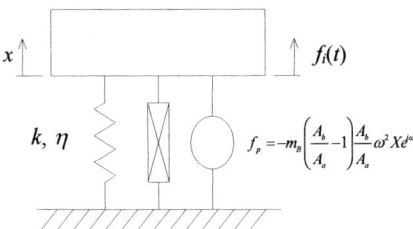

Figure 4 Screen fitted with absorber

The force f_p in figure 4 is the force generated by the absorber. The best isolation would be achieved when the stiffness of the spring is as low as possible. For the screen application low stiffness would result in high static deflection. For a steel spring the static deflection will become problematic at about 125 mm. The absorber cannot accommodate high static deflection easily. The first problem is the available space, but more importantly the polyurethane's ultimate tensile strength must not be exceeded. Both these problems can be overcome if the static deflection is reasonably low. The experimental absorber discussed in the previous paragraph was built to deflect only 10 mm, but with a softer material and a different geometry it will be possible to attain larger deflections. It is believed that a 20 mm deflection can reasonably be accommodated. The results are therefore listed for 20 mm deflection (best practically possible) as well as for 26.7 mm (ideal) in order to compare it to the previous result for the screen. The absorber was again designed according to the procedures outlined in the previous paragraph. The mass, force and forcing frequency is the same as before.

TABLE 3
PHYSICAL PROPERTIES OF THE SYSTEM

		$x_s = 20$	$x_s = 26.7$
Mass [kg]	m	13×10^3	13×10^3
Force amplitude [N]	F_i	264.87×10^3	264.87×10^3
Stiffness [N/m]	k	6.38×10^6	4.78×10^6
Spring loss factor	η	0.1	0.1
Operating speed [Hz]	f	12.5	12.5
Area ratio	A_r	36	36
Reservoir diameter [mm]	d_b	180	180
Absorber liquid density [kg/m^3]	ρ	1000	1000

TABLE 4
DESIGN RESULTS AND OPERATING CHARACTERISTICS OF THE SYSTEM

		$x_s = 20$	$x_s = 26.7$
Port length [mm]	l	290	217
Port diameter [mm]	d_a	30	30
Natural frequency [Hz]	f_n	3.52	3.05
Frequency ratio	ω/ω_n	3.55	4.1
Stroke amplitude [mm]	X	3.64	3.64
Acceleration [m/s^2]	$\omega^2 X$	2.29g	2.29g
Dynamic force transmitted [N]	F_o	2322	1738
Transmissibility [%]	$|T_r|$	0.797	0.597

At a static deflection of 20 mm the force transmitted to the support structure is 7.2 times less than a screen fitted with conventional isolators. Figures 5(a) and (b) show the receptance and transmissibility for the design.

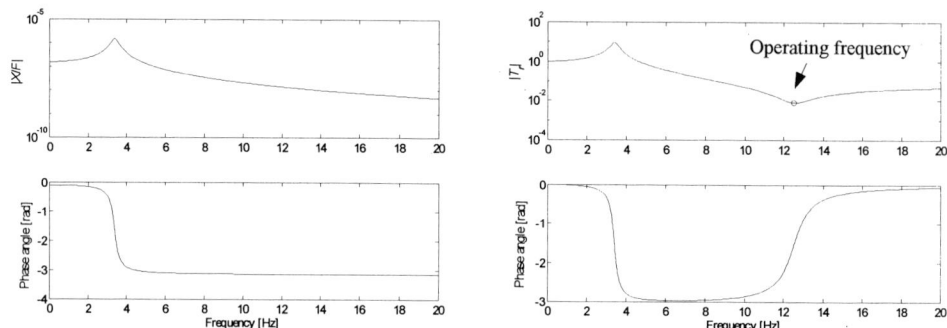

Figures 5(a) Receptance (b) Force Transmissibility of screen with an absorber fitted ($x_s = 20$ mm)

Two degree-of-freedom screen with sub-frame model

The screen mass, stiffness, excitation force and frequency is the same as before. The sub-frame mass is 1.3 times the screen mass (Riddle *et al.*, 1984). The sub-frame stiffness was calculated using the permissible static deflection of a rubber spring, which was given as 15% of its free length (37.5 mm for a 250 mm spring) by Riddle *et al.* (1984).

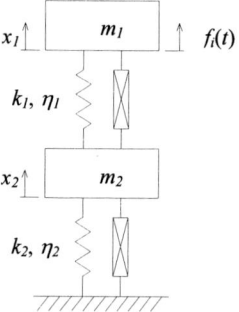

Figure 6 Screen with sub-frame

TABLE 5
PHYSICAL PROPERTIES OF SCREEN WITH SUB-FRAME

		$X_2 = 20$ mm	$x_2 = 37.5$ mm
Screen mass [kg]	m_1	13×10^3	13×10^3
Sub-frame mass [kg]	m_2	16.9×10^3	16.9×10^3
Mass ratio	m_2/m_1	1.3	1.3
Screen stiffness [N/m]	k_1	4.77×10^6	4.77×10^6
Sub-frame stiffness [N/m]	k_2	14.67×10^6	7.82×10^6
Loss factor of the screen spring	η_1	0.01	0.01
Loss factor of the absorber spring	η_2	0.1	0.1
Static deflection of the screen [mm]	x_1	46.7	64.22
Stiffness ratio	k_2/k_1	3.07	1.64

TABLE 6
OPERATIONAL CHARACTERISTICS OF SCREEN WITH SUB-FRAME

		$x_2 = 20$ mm	$x_2 = 37.5$ mm
Natural frequencies [Hz]	f_1	2.53	2.15
	f_2	5.66	4.85
Frequency ratios	ω/ω_1	2.21	5.8
	ω/ω_2	4.94	2.57
Static deflection of the screen [mm]	x_1	46.7	64.2
Dynamic force transmitted [N]	F_o	2922	1442
Stroke amplitude [mm]	X_1	3.52	3.52
	X_2	0.198	0.183
Acceleration [m/s^2]	$\omega^2 X_1$	2.216g	2.215g
Transmissibility [%]	$|T_r|$	1.1	0.544

It will still be possible to achieve low transmissibility without using the upper bound of 130% of the screen mass for the sub-frame mass. It can be shown that the gain in isolation will be less at high mass ratios. A 50% mass ratio will already provide good isolation while not adding too much weight to the assembly.

Screen with sub-frame and absorber

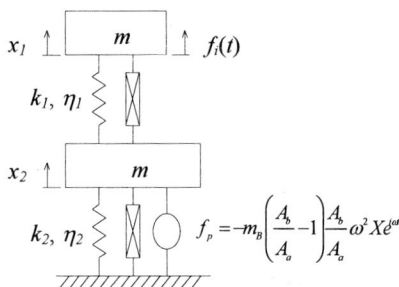

Figure 7 Screen with sub-frame and absorber

TABLE 7

		$x_2 = 20$ mm	$x_2 = 37.5$ mm
Port length	l	337	179
Port diameter	d_a	30	30
Natural frequencies [Hz]	f_1	2.33	1.99
	f_2	10.1	9.54
Frequency ratios	ω/ω_1	5.36	6.29
	ω/ω_2	1.23	1.31
Dynamic force transmitted [N]	F_o	1808	955
Stroke amplitude [mm]	X_1	3.67	3.67
	X_2	2.47	2.44
Acceleration [m/s^2]	$\omega^2 X_1$	2.3g	2.3g
Transmissibility [%]	$|T_r|$	0.679	0.36

PHYSICAL PROPERTIES OF THE SYSTEM

TABLE 8
DESIGN RESULTS AND OPERATING CHARACTERISTICS OF THE SYSTEM

		$x_2 = 20$ mm	$x_2 = 37.5$ mm
Screen mass [kg]	m_1	13×10^3	13×10^3
Secondary mass [kg]	m_2	1950	1950
Mass ratio	m_2/m_1	0.15	0.15
Screen stiffness [N/m]	k_1	4.77×10^6	4.77×10^6
Absorber stiffness [N/m]	k_2	7.33×10^6	3.91×10^6
Stiffness ratio	k_2/k_1	1.54	0.82
Loss factor of the screen spring	η_1	0.01	0.01
Loss factor of the absorber spring	η_2	0.1	0.1
Static deflection of the screen [mm]	x_1	46.7	64.22
Area ratio	A_r	36	36
Reservoir diameter [mm]	d_b	180	180
Absorber liquid density [kg/m^3]	ρ	1000	1000

In both these cases the response is lower than for a conventional sub-frame while only adding 15% of the screen mass to the assembly. The additional weight of m_2 will require higher absorber stiffness and therefore a longer port. A sub-frame with the same mass (15%) will have a transmissibility of 8%, which is even higher than a screen with conventional isolators.

CONCLUSIONS

From the analysis presented it can be concluded that:
- A single degree-of-freedom screen fitted with an absorber will offer the most benefit since good isolation can be achieved and there is no addition of mass to the system.
- A sub-frame fitted with an absorber will reduce the mass ratio for this type of design significantly. For applications where more isolation is needed than achievable with the single degree-of-freedom system this method is recommended.
- The damping of the absorber should be as low as possible.
- It is important that the absorber spring should be as soft as possible. It seems as if it will be difficult to design such a spring, but the anticipated problems can probably be overcome.
- The acceleration of the screen, which is the most important parameter affecting its operation, is not altered significantly using any of these methods.

Overall, the use of absorbers for screen applications has shown enough advantages over existing isolation methods for screens, to warrant further development.

REFERENCES

Du Plooy, N.F. The development of a vibration absorber for vibrating screens, M.Eng. dissertation, *University of Pretoria*, 1999.

Greenway, M. 1983. Vibration in screening plants: designing out maintenance, *Proceedings of the 1983 Maintenance Management Convention*, October 17-19.

Halwes, D.R., Simmons, W.A., 1980. *Vibration suppression system*, U.S. patent no. 4,236,607.

Halwes, D.R., 1981. Total main rotor isolation system analysis, Bell Helicopter Textron, *NASA Contractor Report No. 165667*. Langley Research Center, Hampton, Virginia, June.

Riddle, T., Bates, G. & Thomas, G. 1984. Summary report on the design on nine isolation system subframes systems, *Structural Dynamics Research Corporation*, Ohio: SDRC.

ACCEPTABILITY VIBRATION CRITERION FOR FLOORS WITH WALKING OCCUPANTS

C. F. de Andrade[1], J. C. de Andrade Filho[2], and J. C. de Andrade[1]

[1] Departamento de Engenharia Estrutural e Construção Civil
Universidade Federal do Ceará
Rua Pascoal de Castro de Alves, 718
60155-420 Papicu, Fortaleza, Ceará, Brazil
[2] Curso de Engenharia Civil
Universidade de Fortaleza
Fortaleza, Ceará, Brazil

ABSTRACT

Vibrations imposed by the action of normal occupant activities may affect the structural serviceability of many modern building structures, in terms of causing physical or psychological discomfort and general distress to occupants. During the past years, a number of research programs have been directed toward the establishment of serviceability criteria for the evaluation of the dynamic behavior of floor systems under the action of human activities. Several studies have been performed attempting to establish and relate levels of structural vibration to human comfort and acceptable or tolerable levels, trying to identify imperceptible, perceptible, tolerable, annoying, intolerable or uncomfortable vibrations. However, it is generally recognized that most of these procedures using either large general or small specialized available dynamic analysis computer programs are often time-consuming, not easily understood and promptly applied by inexperienced engineers, and, in many instances, inappropriate in routine design. This paper discusses a simple and straightforward procedure to evaluate the dynamic behavior of floor systems under normal occupant walking activities, based on a design criterion that the fundamental frequency of the floor system be greater than a specified constant limiting value. The proposed serviceability criterion provides evidence that rational simplified design rules for the evaluation of the performance of a given floor system to an acceptable and confident degree can be developed without the need for elaborate computations. In addition, it can be inferred that a design criterion limiting the fundamental frequency of a floor system is tantamount to limiting its static deflection to a specified constant deflection limit.

KEYWORDS

Acceptability criterion, floor vibrations, walking activity, objectionable vibrations, dynamic behavior, human tolerance.

INTRODUCTION

Extensive experimental and analytical studies have been carried out, in past years by numerous investigators from around the world, in attempts to establish acceptable vibration design criteria for the evaluation of the structural performance of floor systems under normal human walking activities. However, it is generally recognized that most of these procedures using either large general or small specialized available dynamic analysis computer programs are often time-consuming, not easily understood and promptly applied by inexperienced engineers, and, in many instances, inappropriate in routine design, particularly during the preliminary design phase.

Man-induced vibrations imposed by the action of normal occupant activities may affect the structural serviceability of many modern building structures, which are becoming lighter and more flexible, in terms of causing physical or psychological discomfort and general distress to occupants (see, *e.g.*, Allen & Rainer 1976; Ebrahimpour *et al.* 1996; Galambos & Ellingwood 1986; Murray 1991).

In general, people expect that buildings or other habitable installations, which they have to use, will remain apparently stationary under certain ordinary conditions. In many instances, however, they are psychologically prepared to tolerate some level of motion and unexpected or excessive structural movements may be taken as an indication that the installation is unsafe and the occupants may become annoyed, discomfortable or alarmed.

During the last few years, several studies have been performed attempting to establish and relate levels of structural vibration to human comfort and acceptable or tolerable levels, trying to identify imperceptible, perceptible, tolerable, annoying, intolerable or uncomfortable vibrations. In addition, relations have been established between displacement, velocity, acceleration and the fundamental natural frequency of the structural system, for different damping levels.

Correcting troublesome floor systems is, in many instances, expensive and extremely difficult. Alternative approaches to control or reduce objectionable floor vibrations have been proposed using either passive or active control techniques to improve the dynamic characteristics of the floor system (see, *e.g.*, Hanagan & Murray 1996). Nevertheless, the subject has received limited attention and only a restricted amount of literature can be found pertaining to structural serviceability criteria for floor systems under normal occupant activities. There appears, for this reason, to exist a definite need to further develop and establish structural serviceability criteria and design rules to control or reduce excessive and objectionable floor vibrations and deflections of modern flexible structures imposed either by occupant walking activities or by in-situ human activities such as spectators at a sporting event.

Therefore, from the point of view of the practicing structural designer, it is recommended that structural serviceability checks continue to rely on the establishment of clear, basic and simplified rules to enable the designer to arrive, without excessive speculation on psychological, physiological, pathological or subjective factors, at an acceptable, rapid and confident evaluation of the performance of simple floor systems under normal human walking activities, without the need for elaborate computations. Furthermore, one should not have to rely on public opinion to determine the adequacy of a given floor system once it has been designed and constructed.

Many researchers agree that the acceleration is an appropriate indicator of vibration perception and tolerance. For this reason, acceptable or tolerable levels of acceleration have been established and related to the occupancy of the building. Recent studies (see Dolan *et al.* 1999) validate that if the stiffness of the floor system is sufficient to keep the fundamental frequency of the floor system above a certain specified value, acceptable vibration will be obtained. Man-induced objectionable floor vibrations are more likely to occur in offices, walkways and public assembly areas having relatively long spans. The main source of floor vibration of such occupancies is normal walking activity.

STATEMENT OF THE PROBLEM

In many instances, it may be appropriate to use a dynamic analysis method based on the heel-drop impact test for avoiding annoying floor vibrations due to normal walking activities in residential and office buildings. The method may be particularly appropriate to evaluate the dynamic behavior of concrete slab steel joist systems and it has also been used to evaluate normal concrete slab and beam construction systems.

The heel-drop impact test consists of a person designated as the impactor weighing about 780 N raising his heels approximately 65 mm on his toes and suddenly dropping them to the floor, inducing a dynamic excitation to the floor system. The procedure can be used both as a design tool and for testing the adequacy of existing floors for acceptability against walking vibrations.

The heel-drop impact test may be quite meaningful for evaluating the dynamic response of floor systems exposed to activities that cause impact forces such as jumping in place. The method appears to be a particularly useful tool for screening certain potentially troublesome floor systems.

It needs to be emphasized, however, that the heel-drop impact test produces an isolated, single, in situ, pulse-type dynamic excitation, suddenly applied to the floor system, whereas the act of normal walking constitute a dynamic and repetitive process, which occurs with a definite coordinated sequence of events that take place progressively during a complete gait cycle. The gait cycle, in terms of foot placement, begins with the foot's initial heel-strike with the floor, followed by opposite toe-off, opposite heel-strike, toe-off, and terminates when the same foot strikes the floor again. These significant events of the walking cycle determine the footstep force-time history.

The heel-drop impact test produces an isolated pulse-type dynamic excitation, which is useful and convenient in evaluating the transient behavior of floor systems. Normal walking, however, causes a dynamic motion, which may approach, in many instances, a steady-state condition. Under a steady-state condition, the annoyance threshold level is lower than under transient vibration. Therefore, the structural vibration produced by a heel-drop impact test may be more easily tolerated than the dynamic motion caused by normal pedestrian walking.

In a finite-element idealization, the impactor may be modeled as a mass-spring-dashpot single-degree-of-freedom system, in which the mass of the impactor is subjected to an initial velocity corresponding to the velocity of the heel when it impacts the floor system.

As the body of the impactor is elevated during the heel-drop impact test, energy is supplied by muscle action. Just before the heel-drop impact occurs, the muscles of the limb may be further contracted concentrically to provide extra motor power to the body, or they may be contracted eccentrically to perform as shock absorbers. At the moment the heel strikes the floor, the downward velocity of the body drops to zero during a very short time interval (approximately 0.05 s), and this high degree of deceleration creates a considerable inertia force that determines the impact action and induces a pulse excitation to the floor.

The impact of a person with a mass of 80 kg, performing a heel-drop test from a height of 65 mm, causes an impulse of approximately 65 N.s (see, *e.g.*, Ellingwood & Tallin 1984). As the response of the floor is highly dependent on the excitation, it appears to be quite accurate to use the experimentally measured impulsive excitation as the input to the finite-element simulation model, and, thus, discrepancies in the results may be due solely to possible inaccuracies in the simulating analytical model rather than to simplification in the excitation force. Therefore, by using the well-known classical impulse-momentum relationship, it seems that a realistic value for the velocity of the heel, when it impacts the floor, may be adopted when modeling the impactor as a mass-spring-dashpot single-degree-of-freedom system in a finite-element idealization.

The fundamental frequency of simple floor systems may be estimated assuming that the floor behaves as simply supported flexural members. Therefore, the fundamental frequency, f, of the floor system can be estimated using

$$f = \frac{\pi}{2l}\sqrt{\frac{gEI}{Wl}}, \qquad (1)$$

where g = acceleration of gravity; E = modulus of elasticity; I = moment of inertia of the floor system; l = span of the floor; $W = W_D + pW_L$ = total load on the floor, in which W_D = total dead load; W_L = total live load; and p = the percentage of the nominal live load that actually participates in the floor dynamic response.

By combining Eqn. 1 and the familiar expression

$$\delta = \frac{5Wl^3}{384EI}, \qquad (2)$$

which gives the static midspan deflection, δ, of a simply supported one-way flexural member under a uniformly distributed load, the following requirement may be established

$$\delta < \frac{5\pi^2 g}{1536 f_{lim}^2}, \qquad (3)$$

based on a design criterion that the fundamental frequency of the floor system be greater than a specified constant limiting value f_{lim}. The heel-drop impact test is used to establish the acceptable level of vibration in terms of a limiting deflection value, thus eliminating all of the floor systems with unacceptable and marginal vibrational performance.

NUMERICAL EXAMPLE

To illustrate the proposed design criterion, consider the limiting value of 14 Hz (see Dolan et al. 1999), for occupied floors. For this simple illustrative numerical example, the requirement given by Eqn. 3 results in $\delta <$ 1.6 mm (0.063 in.).

CONCLUSION

The proposed serviceability criterion provides evidence that rational simplified design rules for the evaluation of the performance of a given floor system to an acceptable and confident degree can be developed without the need for elaborate computations. In addition, it can be inferred from the above considerations that a design criterion limiting the fundamental frequency of a floor system is tantamount to limiting its static deflection to a specified constant deflection limit.

References

Allen D. E. and Rainer J. H. (1976). Vibration Criteria for Long-Span Floors. *Canadian Journal of Civil Engineers* **3:2**, 165-173.

Dolan J. D., Murray T. M., Johnson J. R., Runte D., and Shue B. C. (1999). Preventing Annoying Wood Floor Vibrations. *Journal of Structural Engineering,* ASCE **125:1**, 19-24.

Ebrahimpour A., Hamam A., Sack R.L., and Patten W.N. (1996). Measuring and Modeling Dynamic Loads Imposed by Moving Crowds. *Journal of Structural Engineering,* ASCE **122:12**, 1468-1474.

Ellingwood B. and Tallin A. (1984). Structural Serviceability: Floor Vibrations. *Journal of Structural Engineering,* ASCE **110:2**, 401-418.

Galambos T. V. and Ellingwood B. (1986). Serviceability Limit States: Deflection. *Journal of Structural Engineering,* ASCE **112:1**, 67-84.

Hanagan L. M. and Murray T. M. (1996) Active Control Approach for Reducing Floor Vibrations. *Journal of Structural Engineering,* ASCE **123:11**, 1497-1505.

Murray T. M. (1991) Building Floor Vibrations. *Engineering Journal,* AISC **Third Quarter**, 102-109.

A COMPARATIVE STUDY OF A STOREY vs. ELEMENT HYSTERETIC NONLINEAR MODEL FOR SEISMIC ANALYSIS OF BUILDINGS

N. A. Alexander[1], N. Goorvadoo[1], F. A. Noor[1] and A. A., Chanerley[2]

[1]Department of Civil Engineering, University of East London,
Dagenham, Essex, RM8 2AS, UK
[2]Department of Electrical Engineering, University of East London,
Dagenham, Essex, RM8 2AS, UK

ABSTRACT

This paper introduces a non-linear inelastic hysteretic storey model to study the behaviour of buildings under bi-directional seismic actions. This model describes the behaviour of the storey as a whole rather than modelling the assemblage of non-linear discrete elements. The model is formulated from an extension of the linear elastic equations for a single storey using multi-dimensional Taylor series expansion. A quadratic hypersurface is used to model the stiffness action/displacement relationship, the characteristics of which are related to the reduction factor and a natural frequency parameter. In addition a hyperplane is defined for the undeforming process. Only one nonlinear parameter is required to describe the storey system, and it is shown to have some physical significance. As a comparison, the system is also modelled by an assemblage of discrete elements. These elements exhibit a bilinear, bi-axial, hysteretic behaviour. A nonlinear timehistory analysis is carried out using both the storey and the element model for a selection of corrected earthquake records and the adequacy of the storey model is investigated. The total acceleration response and ductility demand spectra are compared and discussed.

KEYWORDS

Non-linear, hysteretic, torsion, seismic, bi-axial, ductility, response spectra, earthquake

INTRODUCTION

Traditionally dynamic studies of structures make use of element-based non-linearity model. This requires setting up the positions of the resisting elements and assembling their stiffnesses to model the behaviour of the structure as a whole. The number of parameters is greatly increased as the model is refined, e.g. if stiffness degradation, pinching and other effects are included and if bi-lateral seismic loading is considered. However if the behaviour of the structure is modelled using a storey formulation, the number of parameters are greatly reduced as the location of resisting elements and their individual yield parameter need not be specified. Thus storey model lends itself more easily to general parametric study. This paper describes the storey hysteretic non-linearity model that is formulated from only five linear system parameters and one nonlinear parameter. A nonlinear time-history analysis is carried out for a system acted upon by bi-directional ground motions, using both the storey model and an element model. The results are discussed and compared.

ELEMENT HYSTERETIC NONLINEAR MODEL

This model is used in various forms by researchers, Jiang *et al* (1996), Chandler *et al.* (1994), Tso *et al* (1992), Corderoy *et al.* (1983), etc. In all these model types the location, dimensions and nonlinear characteristics (including hysteretic and biaxial behaviour) for every vertical, resisting element, need to be defined. This does, obviously, provide a rather large number of system parameters and thus it does challenge the generality of the results derived from these element model types. In the present study, it is conjectured that the number of resisting elements will effect the transition from elasticity to inelasticity. Too few resisting elements may make the structure heavily dependent on an individual element's behaviour. Thus the system could have abrupt changes in stiffness as a result of yielding in any individual element. Hence a structure with 16 elements as shown in figure 1(a) is chosen. The reference axes are centred at centre of mass (CM) of the floor. The element model has a square plan of 24×24 m, with elements 1-13 having cross-sections of 0.32×0.32m and elements 14-16 having cross-sections 0.45×0.45 m. Thus the system is given an predefined eccentricity in stiffness.

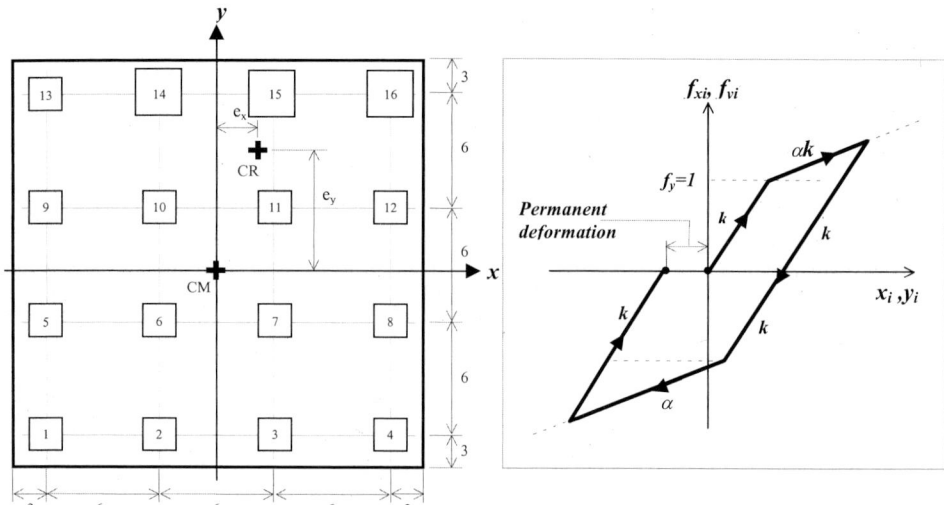

(a) Plan view of structure with 16 elements   (b) Bi-Linear Hysteretic element model
Figure 1: Element model

In this study, the force-displacement relationship is described using a bi-linear model where the stiffness decreases after yielding. The secondary post-yielding stiffness slope is fixed with $\alpha = 0$ for the present study. This model, as shown in figure 1(b), allows plastic deformation and hence hysteresis in both x and y direction. Yielding of an individual element is controlled by interaction formula when the yield parameter

$$f_y = \left(\frac{f_{xi}}{f_{yx}}\right)^2 + \left(\frac{f_{yi}}{f_{yy}}\right)^2 \geq 1 \; , \qquad (2)$$

where f_{xi} & f_{yi} are the *ith* element stiffness actions. The global system equation is in terms of three degrees of freedom $\underline{u} = [x \; y \; \varphi]^T$ defined at the CM, where (x, y) are sway displacement co-ordinates and θ is the structure rotation. The angular displacement is defined $\varphi = r_m \theta$ where r_m is the radius of gyration of the floor mass about the (CM), thus,

$$\underline{\ddot{u}} + [\mu]\underline{\dot{u}} + \underline{F} = -\underline{\ddot{u}}_g, \tag{3}$$

where the global nonlinear stiffness actions vector $\underline{F}$ is assembled in the standard way from the contributions of all the resisting element stiffness actions which would include the hysteretic behaviour defined in figure 1(b) and biaxial interaction equation (2). The orthogonal damping matrix $[\mu]$ is identical to that used in the storey hysteretic nonlinear model. The system of equations (3) is solved by a classical Runge-Kutta method with appropriate time step to maintain numerical accuracy. A spectrum of structures is obtained by varying the heights of all resisting elements. This does not vary the eccentricity ratios $\varepsilon_{mx}, \varepsilon_{my}$ or frequency ratios λ_y, λ_{mm} but will vary ω_x.

STOREY HYSTERETIC NONLINEAR MODEL

The parametric linear model is well known and adequately characterised by, Chopra et al. (1987), and has recently been extended to introduce nonlinear inelastic behaviour, Alexander et al. (1999,2000). In it a three-degrees of freedom idealisation of a building is conjectured. The equation of motion, for the system, is set up in the standard way. This is also defined using the CM as the reference axis. The parameterised equation at the CM would use frequency parameters $(\omega_x^2, \omega_y^2, \omega_{mm}^2)$. The eccentricity ratios are $(\varepsilon_{mx}, \varepsilon_{my})$. The degrees of freedom are $\underline{u} = [x \; y \; \varphi]^T$ where (x, y) are sway displacement co-ordinates and θ is the structure rotation. The angular displacement is defined $\varphi = r_m \theta$ where r_m is the radius of gyration of the floor mass about the (CM).

$$\underline{\ddot{u}} + [\mu]\underline{\dot{u}} + \underline{F} = -\underline{\ddot{u}}_g, \quad \underline{F} = \omega_x^2 \underline{f}, \quad \underline{f} = \begin{bmatrix} f_1 \\ f_2 \\ f_3 \end{bmatrix} = [\Omega]\underline{u}, \quad [\Omega] = \begin{bmatrix} 1 & 0 & -\varepsilon_{my} \\ 0 & \lambda_y^2 & \lambda_y^2 \varepsilon_{mx} \\ -\varepsilon_{my} & \lambda_y^2 \varepsilon_{mx} & \lambda_{mm}^2 \end{bmatrix} \tag{4}$$

where $(\lambda_y = \omega_y/\omega_x, \lambda_{mm} = \omega_{mm}/\omega_x)$ are the frequency ratio parameters. The ground acceleration vector is $\underline{\ddot{u}}_g = [\ddot{x}_g \; \ddot{y}_g \; 0]^T$. The non-dimensional system frequency matrix is $[\Omega]$ and classical orthogonal damping matrix is $[\mu]$. The general linear elastic stiffness action / deflection relationship is $\underline{F}$. The normalised stiffness action $\underline{f}$ is replaced by a nonlinear formulation. A multidimensional Taylor series expansion is used to produce a general quadratic hypersurface which models nonlinear elastic behaviour and this is described in detail in Alexander et al. (2000). For the process of deforming this general hypersurface is used to describe the nonlinear, biaxial, stiffness action / deflection behaviour of the entire storey. In general the three quadratic hypersurfaces for deforming in the (x, y, φ) directions are given by equation (5). The co-ordinates $(x_{ip}, y_{ip}, \varphi_{ip})$ are conceptually the unrecoverable permanent deformations and the current origin for the quadratic hypersurface. k_{in} is the element of the system frequency matrix $[\Omega]$ defined at row i column n. The nonlinearity parameter β, in equation (5), is the deflection(s) $\underline{u}$ at which the maximum normalised action(s) $\underline{f}$ is applied.

$$f_i = k_{i1}x_i^* + k_{i2}y_i^* + k_{i3}\varphi_i^* - \frac{1}{2\beta}\{k_{i1}x_i^*|x_i^*| + k_{i2}y_i^*|y_i^*| + k_{i3}\varphi_i^*|\varphi_i^*|\}$$
$$+ \frac{1-\sqrt{2}}{2\beta}\{k_{i1}[x_i^*|y_i^*| + y_i^*|x_i^*|] + k_{i2}[\varphi_i^*|y_i^*| + y_i^*|\varphi_i^*|] + k_{i3}[x_i^*|\varphi_i^*| + \varphi_i^*|x_i^*|]\}, \quad (5)$$
$$x_i^* = x - x_{ip}, \quad y_i^* = y - y_{ip}, \quad \varphi_i^* = \varphi - \varphi_{ip}$$
$$i \in \{1,2,3\}$$

$$\beta = \Delta_1 \left(\frac{1}{\omega_x}\right)^{4/3} \qquad \Delta_1 = \frac{C}{R}. \qquad (6)$$

The value of β is given by equation (6) that has previously been derived from a (UBC 97) type design spectrum and ultimate limit state considerations. The parameter R a classical stiffness action reduction factor due to inelastic behaviour, C is a design spectrum parameter. The constant Δ_1 would normally be in the range of approximately 0.1 to 2. Thus in equation (6) β would be dependant on the reduction factor provided in the structure as well as the fundamental circular frequency parameter.

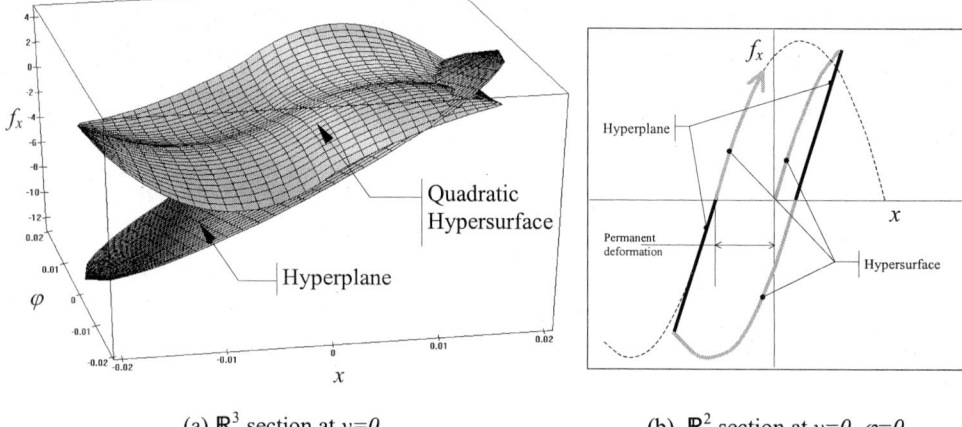

(a) $\mathbb{R}^3$ section at $y=0$ \qquad\qquad (b) $\mathbb{R}^2$ section at $y=0$, $\varphi=0$
Figure 2: Intersection of undeforming hyperplane & deforming quadratic hypersurface (f_x)

The use of an equation of the type of (6) seems consistent with a general seismic design philosophy. This equation does represent a conceptual difference between this storey hysteretic nonlinear model and the element hysteretic nonlinear model. Basically it would seem structurally reasonable for the yield force in resisting elements to be a function of structural period which is itself a function of the resisting element dimensions. When comparing small and tall buildings, normally taller buildings have larger fundamental periods and larger columns, shear walls, cores and beams. Hence there is an argument for element moment capacities, of some structural elements, to increase with increasing period. This does bring into question a nonlinear analyses based period independent, constant, values for element yield forces. Hysteretic behaviour is modelled by the introduction of a hyperplane parallel to the linear $\underline{F} = \omega_x^2[\Omega]\underline{u}$ hyperplane, given by,

$$f_i = k_{i1}x_i + k_{i2}y_i + k_{i3}\varphi_i + f_{ip},$$
$$f_{ip} = f_{i1} - k_{i1}x_{i1} - k_{i2}y_{i1} - k_{i3}\varphi_{i1} \qquad (7)$$
$$i \in \{1,2,3\}$$

where $(x_{i1}, y_{i1}, \varphi_{i1}, f_{i1})$ is the point at which the system changes from deforming to undeforming. The algorithm that is needed to process the transition from the deforming, equations (5,6), and undeforming, equation (7) has been described in detail previously and is an integral part of the algorithm. Figure 2(a) shows graphically a section through the hypersurface and hyperplanes in $\mathbb{R}^3$ space. The intersection of the hyperplane and the quadratic hypersurface is shown in the Figure 2(a). Figure 2(b) shown a section in $\mathbb{R}^2$ where a typical stiffness action / deflection path is pictorially described.

SELECTION OF EARTHQUAKE RECORDS

The availability of earthquake records has greatly increased in the last decade as a result of more instrumentation and dissemination. As with most theoretical/numerical studies in earthquake engineering, the validity and accuracy of the findings depends immensely on the quality of the records used. Hence the choice of the records should be based on the nature of their use. In this study records have been extracted from the NOAA database, which contains digitised and corrected records. This is important so as to obtain records that are as near as possible to a true representation of the event. In order to reduce the influence of the local soil conditions, Veletsos et al. (1989), on the response of the structure, only records from rock sites (UBC 97 Soil Class 1) are used. The amount of energy released during an earthquake is normally dissipated within an area close to the event, referred to as near-field. Several definitions of near-field exist, Martinez et al.(1998); in this study the Campbell definition described in Kramer (1996), is used. This is where the near field records are classified as those within a region described by the epicentral distance and the focal depth of the event. The selection of records in this study are hence restricted to rock sites and the near-field. A scaling procedure is required to compare the chosen records, each having a different peak value. Choosing which accelerogram the x component and which the y component acceleration is problematic. Thus each accelerogram pair is used twice. Initially the first accelerogram is the x component and the second accelerogram is the y component. Then the accelerogram pair is swapped. The first accelerogram becoming the y component and second is the x component. The normalisation of the records is also problematic as the amplitude ratio of x component to y component needs to be maintained. Hence the x component is normalised to $1 m/s^2$ and the y component is normalised to maintain the original $x:y$ amplitude ratio. Given that for a particular accelerogram pair both orthogonal ground components are granted the opportunity to be the x component any bias is minimised. The description of the each record is summarised in table 1.

TABLE 1
DATABASE OF CORRECTED ACCELEROGRAMS.

Event Name	Date	Country	Mag. (M_L)	Slant Dist. (km)
Michoacan Aftershock, Aeropuerto	21.09.85	Mexico	7.6	36
Whittier Narrows Earthquake, Mt. Wilson	01.10.87	USA	5.9	20
Nahanni Aftershock, Battlement Creek NWT	25.12.85	Canada	5.7	19
Westmorland Eathquake, Superstition Mt. Ca	26.04.81	USA	5.6	23
San Fernando Earthquake, Lake Hughes	09.02.71	USA	6.5	29

NUMERICAL STUDIES

An exact comparison between the two models (a) element nonlinear and (b) storey nonlinear is difficult primarily because of the different formulations of each method. The elastic system parameters are matched exactly, the eccentricity ratios $\varepsilon_{mx} = 0.11, \varepsilon_{my} = 0.32$ and the frequency ratios $\lambda_y = 1.0, \lambda_{mm} = 1.014$. However the difficulty comes with the nonlinear parameters. For the element model, each resisting element is symmetric in stiffness and strength, and the overall model has no predefined strength eccentricity. Thus the main nonlinear parameter is the element yield force $f_{yx} = f_{yy}$. For the storey model the Δ_1 is the single nonlinear parameter. For a comparison these two parameters f_{yx} and Δ_1 have to be set at levels that will produce similar levels of peak ductility demand at the CM. For the present study (a) element nonlinear $f_{yx} = f_{yy} = 35\text{kN}$, (b) storey nonlinear $\Delta_1 = 0.35$, are compared for a spectrum of structures.

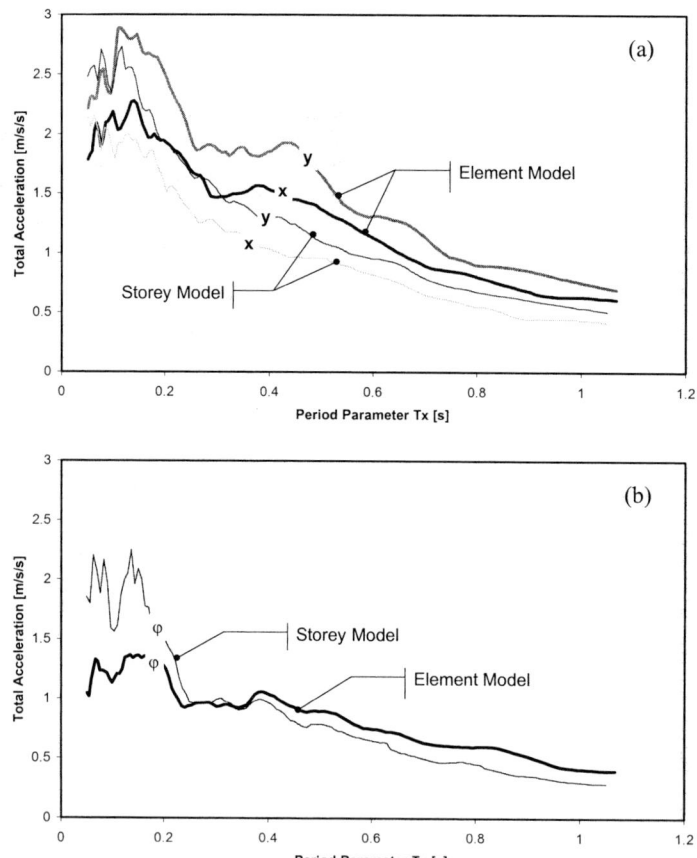

Figure 3: 95% Total acceleration spectra (a) x & y accelerations (b) φ accelerations

The frequency ratio parameter ω_x is varied while the eccentricity ratios, frequency ratios and nonlinear parameters are kept constant. The mean and sample standard deviation spectra are calculated from the results. Then 95% total acceleration spectra (at CM) are produced for both methods shown in figure 3. The element model produces larger total accelerations in the x and y directions, figure 3(a), though the peak total accelerations for the two models are generally comparable. In the case of the building torsion, the storey model produces larger torsional accelerations for low period. In figure 4 the ductility demand spectra are displayed. It is clear from figure 4(a) that the element model produces a narrower band, in period, of large ductility behaviour, while the storey model shows large ductility demand over a broad range of structural periods. It is difficult, at this stage, to say which method produces results that are compatible with a real structure. The broad band of large ductility demands, for the storey model, corresponds to the lower total accelerations that this model exhibits over a similar range in period. Figure 4(b) is the 95% ductility demand spectra for element model's worst resisting element. Note that these spectra are very similar in shape to the spectra at the CM but are of the over of 2 to 3 times larger. Thus it could be argued that the values of the ductility demand at the CM derived for the storey model would correspond to much larger, of the order of 2 to 3 times larger, ductility demand at a corner/edge element.

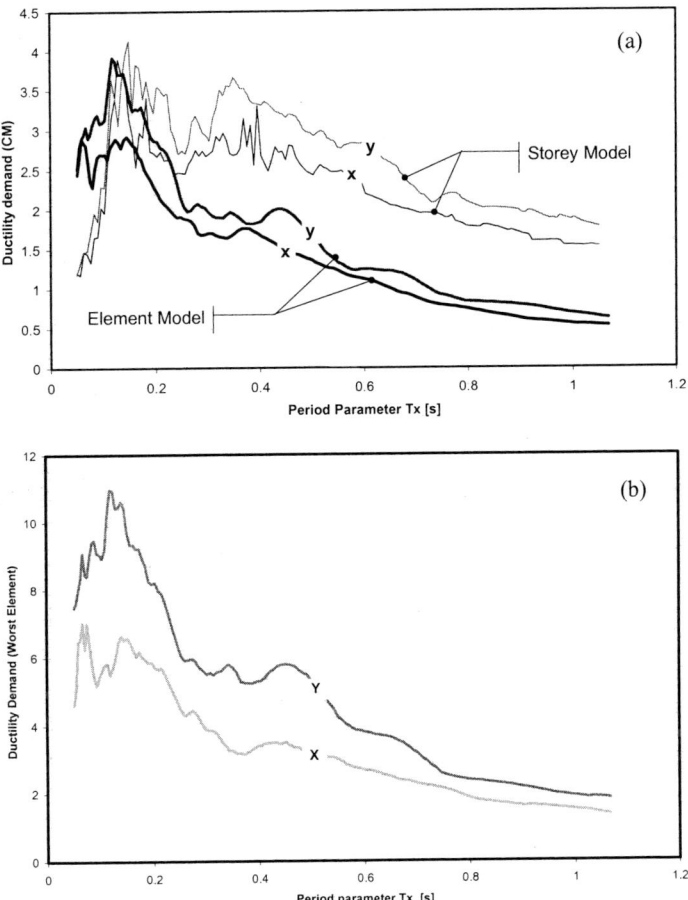

Figure 4: 95% Ductility demand spectra (a) defined at CM (b) for the worst element

CONCLUSIONS

The storey nonlinear model and the more traditional element nonlinear model have been compared by nonlinear timehistory analysis. However, as each method has a different nonlinear formulation, there are differences under inelastic conditions. The element model has the advantage (disadvantage?) of being able to model each idealised resisting element with different force / deflection relationships. This has enabled various researchers to introduce the concept of strength eccentricity. However this does lead to a large number of system parameters and so it is difficult to assess the generality of studies based on this formulation. The storey nonlinear model as presented has the advantage (disadvantage?) of modelling the entire buildings nonlinear behaviour without the need to recourse to individual idealised resisting elements. It does this by the introduction of a single nonlinear parameter. Biaxial, hysteretic and stiffness softening behaviour are all modelled. In the present study the comparison shows that the classical nonlinear hysteretic energy dissipation is present in both models: ie. reduced total accelerations with increased ductility demand. However for the element model the range in structural period over which this energy dissipation is displayed is much smaller than for the storey model. The reasons for this difference are not clear at present, but one possible explanation is that the element model has no falling branch of the force / deflection relationship ie. no degradation in capacity of the structure. The storey model characteristics are quadratic, parabolic, hence have both increase and decrease in capacity of the structure after yield. The storey model only produces information at the CM of the structure while the element model gives information about the behaviour of critical edge/corner elements. However the present study shows that the values of the ductility demand at the CM derived for the storey model would correspond to much larger, of the order of 2 to 3 times larger, ductility demand at a corner/edge element. Thus information about critical edge/corner elements can still be assessed using the storey model. The storey model is a useful tool in the parametric analysis of building subject to seismic action.

REFERENCES

1. Alexander, N.A., Goorvadoo, N., Noor, F.A. & Chanerley, A.A. (2000). A new parametric storey non-linearity including hysteretic behaviour in the dynamic analysis of buildings under bi-directional seismic action. *The 5th International Conference on Computational Civil & Structural Engineering, Belgium* **2, 169-176**
2. Alexander, N.A., Javed, K., Noor, F.A. & Goorvadoo, N. (1999). Parametric study of torsional response of phased accelerograms. *Proceedings of the 8th Canadian Conference on Earthquake Engineering* **,751-756**
3. Corderoy, H.J.B., Thambiratnam, D.P (1983). Microcomputer analysis of torsionally coupled multistorey buildings for earthquakes. *Computers and Structures* **46, 593-602**
4. Chandler, A.M., Corerenza, J.C. & Hutchinson, G.L. (1994) Period-dependant effects in siesmic torsional response of code systems. *Journal of Structural Engineering* **120 (12), 3418-3434**
5. Hejal, R. & Chopra, A.K. (1987). Earthquake response of torsionally coupled buildings, Earthq. Eng. Research Center, Uni. of California at Berkeley **UCB/EERC-87/20**
6. Jiang, W., Hutchinson, G.L. & Wilson, J., L. (1996). Inelastic torsional coupling of building model. *Engineering structures* **18(4), 288-300**
7. Kramer S. L. (1996). Geotechnical Earthquake Engineering, Prentice Hall.
8. Martinez A. and Bommer J. (1998). What is near-field?, *The 6th SECED Conference on Seismic Design Practice into the Next Century*, Ed. Booth.E. **245-252**
9. Tso, W.K., & Zhu, T.J. (1992) Design of torsionally unbalancedstructural systems based on code provisions I: Ductility demand. *Earthquake engineering and structural dynamics* **21(7), 609-627**
10. Veletsos A. S. And Prasad A. M. (1989). Seismic Interaction of Structures and Soils: A Stochastic Approach. *Journal of structural engineering* **115(4),935-956**

PREDICTING EARTHQUAKE GROUND MOTION DESCRIPTIONS THROUGH ARTIFICIAL NEURAL NETWORKS FOR TESTING THE CONSTRUCTIONS

S. T. Vassileva[1]

[1] Department of Computer Aided Engineering, University of Architecture Civil Engineering and Geodesy, Hypodrouma 20A ent. A, fl. 9, 1612 Sofia, Bulgaria

ABSTRACT

The paper presents results from a number of investigations into the problem of the predicting ground motions using an artificial neural network for testing the constructions. For that purpose, a number of previous earthquake accelerograms referring to a 100 different earthquakes occurring within a reasonably small geographic area in California have been acquired and processed in order to extract some of the features that could describe an earthquake more synthetically than the full records. The proposed method for generating spectrum compatible accelerograms uses the learning capabilities of neural networks to develop the knowledge of the inverse mapping from the response spectra to earthquake accelerogram. In this work is proposed a two stage approach. In the first stage a replicator neural network is used as a data compression tool to compress the vector of the discrete Fourier spectrum of the earthquake accelerogram to a vector of much smaller dimension. In the second stage a multi layer feed forward neural network learns to relate the response spectrum to the compressed Fourier spectrum. As a result of combining these two neural networks, the artificial earthquake accelerograms are generated directly from the given response spectra. With this approach can be received as well different descriptors of the earthquake, such as the magnitude or the peak accelerations. The number of attributes, which allows receiving the artificial earthquake accelerogram, can be changed in such a manner to receive artificial generation of the accelerogram with expected characteristics. The received results were used for testing different parts of constructions.

KEYWORDS

Predicting ground motion, Construction, Neural network, Knowledge based systems, Fourier spectrum of the earthquake accelerogram, Artificial earthquake accelerogram.

INTRODUCTION

Building design is a complex process in which various components should be cared for in a constructive and consistent way throughout the duration from the very beginning up to the realization phase. The components include the persons in charge for realization of the project, the materials used

during the construction, and the information about the process of construction, which demands appropriate information processing tools. Building design involves multi-dimensional aspects to be considered with conflicting criteria and as a result of this many types of expertise are required. One of requirements is connected with eventual occurrence of an earthquake and the effects upon existing structures. These effects are a function of a number of attributes characterizing the ground motion acting the foundations of the structure. The characteristics of such ground motion depend on:
- the seismic source and the magnitude of the epicenter;
- the propagation path and distance from epicenter;
- the local site conditions and the geotechnical site characteristics.

The statistical characterization of "predictable earthquakes" can be used by the engineering seismology as a guide to the writing of design codes. If there exists a typical earthquake motion depending of certain parameters, characterizing selected area then a structure which will be located in a given seismic region should endure on eventual ground motion in the area. According to EUROCODE 8, the seismic action can be defined in terms of response spectra whose shape and maximum ordinates are characterised by means of the seismic zone and subsoil conditions. EUROCODE 8 allows as well the use of adequate response spectrum compatible accelerograms.

Connected with this, it would be interesting to have possibility of mapping the knowledge about the occurrence of expected with certain probability earthquakes with a given magnitude, duration and epicenter, to a ground motion described in the form of an accelerogram. If this accelerogram refers to a location not coinciding with the epicenter, it would be pertinent to map the expected earthquake to the accelerogram of the ground motion reaching the location where the structure is situated. Such mapping would enable the establishment of the relationships described in Figure 1.

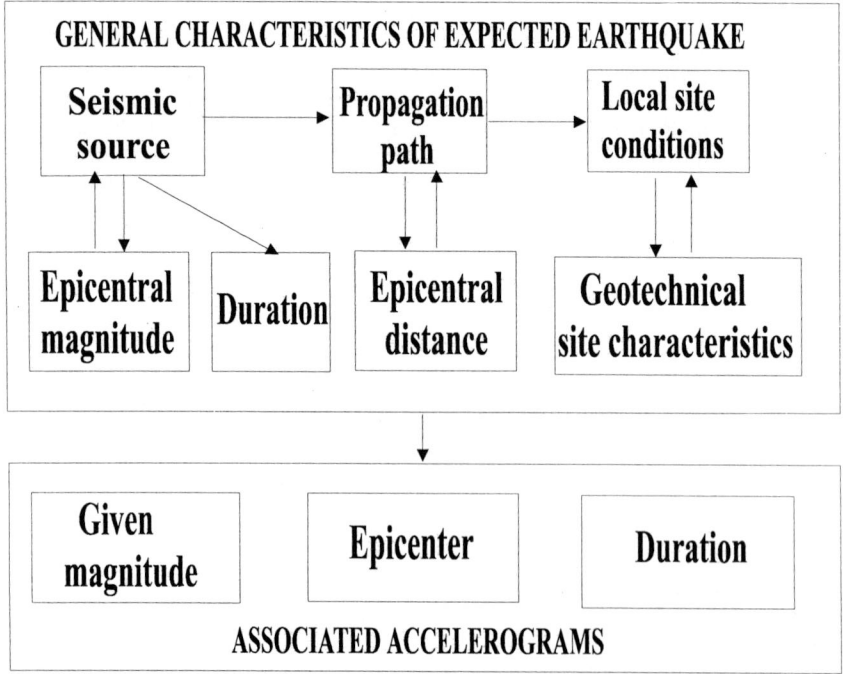

Figure 1: The relationships between general earthquake characteristics and associated accelerograms

The most relevant data describing expected earthquakes can be given by a few parameters. These parameters can be magnitude and location. A corresponding accelerogram, measured at a given distance of the epicenter of the earthquake, is generally described, on a diagram, shown on Figure 2.

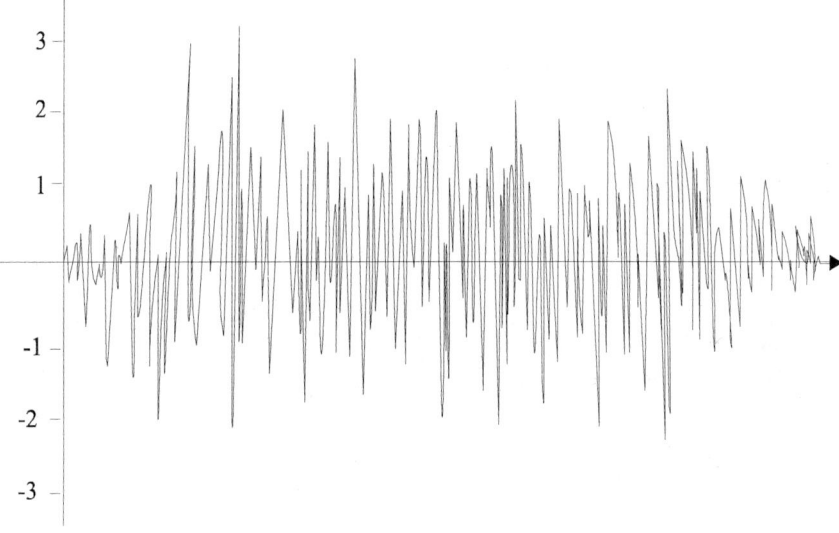

Figure 2: Accelerogram from Northridge earthquake

In structural design and analysis often is used earthquake response spectrum. Developing an artificial earthquake accelerogram, or selecting an existing recorded accelerogram, compatible with a given response spectrum is desirable for earthquake analysis of the constructions. The increasing recorded accelerograms permit the development of systematic processing and utilization of the massive volume of the accelerogram records in analysis and design of constructions and suggesting methods for generating spectrum compatible accelerogram.

As determining the spectra from accelerograms is a direct problem, then, determining the accelerograms from their spectra is an inverse problem. In the case of Fourier spectra, the mapping is reversible and the inverse problem can be uniquely solved, by performing an Inverse Fourier Transform. However, the mapping of response spectrum is not reversible since significant amount of information in the accelerogram gets lost when computing the response spectrum. Nevertheless, it is possible to develop accelerograms whose response spectrum are closed to a given response spectrum.

On the base of actual recorded earthquake accelerograms neural network can be trained to learn to associate the response spectrum with the corresponding accelerograms. This can be made in two stages:
1. A replicator neural network is used to compress the Fourier spectra of the accelerograms.
2. A multi layer feed forward neural network is used to relate the discretized response spectrum to the compressed Fourier spectra.

The proposed method gives possibilities to generate earthquake accelerograms directly from response spectrum. This method for generating spectrum compatible accelerograms uses the learning capabilities of neural networks to develop the knowledge of the inverse mapping from the response spectra to earthquake accelerogram.

MAPPING EARTHQUAKES INTO ACCELEROGRAMS

For mapping earthquakes into accelerograms there should be a model, capable to describe this mapping. Such a model can be based on statistical approaches or on the use of machine learning technologies. The most perspective approach is use of a supervised training technique which is based on descriptions of existing data about past earthquakes and their accelerograms.

On the base of actual recorded in past earthquake accelerograms neural network can be trained to learn to associate the response spectrum with the corresponding accelerograms. For defining the process of training the network, which will be capable to provide the mapping between expected earthquakes and corresponding accelerograms, it's necessary to define the attributes, characterising earthquakes and accelerograms. As it was shown, the earthquakes can be characterised by magnitude, distance, duration etc. All existing accelerograms for certain area should be processed for synthesis of their main characteristics. This can be done using Fourier analysis, and in the case of Fourier spectra, the mapping is reversible. This determining of the accelerograms from their spectra can be solved, by performing an Inverse Fourier Transform. The mechanism for mapping between characteristics of earthquakes, accelerograms and associated descriptors is shown on Figure 3.

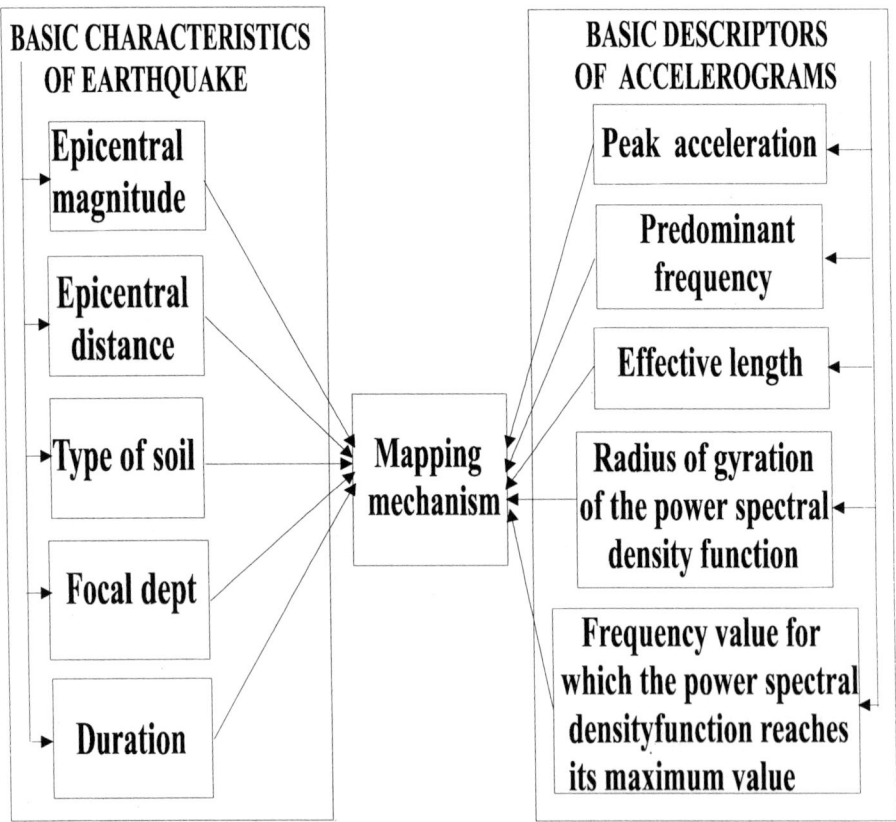

Figure 3: The mechanism for mapping between characteristics of earthquakes, accelerograms and associated descriptors

NEURAL NETWORK DEFINITION AND PRE-PROCESSING OF DATA

Neural networks are massively parallel computational models, which has the learning capabilities to develop neural networks based methods for certain mathematically intractable problems. They are assemblages of interconnected artificial neurons, which receives signals, performs some simple operations, and sends signals along its output connections. The knowledge learned by neural network is stored in its connection weights. In multi layer feed forward neural network, used in this study, the artificial neurons are arranged in layers. The neurons in each layer are fully connected to all the nodes in the next layer. During the training, while the connection weights have not fully matured, the output of the neural network will differ from the desired output. In this work is used a combined form of the quick pomp algorithm Fahlman (1988) and local adaptive learning rate algorithm Cichocki & Unbehauen (1993) to speed up the rate of learning in neuron networks.

The notification, that the vector F is the output of the neural network can be written like Eqn.1, where NN architecture denotes neural network architecture and $Q_{(output)}$ denotes the output of the multi layer feed forward neural network.

$$F = Q_{(output)}(\{ \text{input parameters} \} : \{ \text{NN architecture} \}) \qquad (1)$$

The first argument of Eqn.1 describes the input of the neural network and the second describes the neural network architecture and its training history.

The proposed method of generating artificial earthquake accelerograms is based on developing a neural network which tales the discretized ordinates of the response spectrum as input, and produces the real and imaginary parts of the Fourier spectra of the generated earthquake accelerograms as the output. Such a neural network will be trained with the response spectrum and Fourier spectra, which are discretized with a large number of discrete values since a reasonable accuracy should be maintained, of a number of actual recorded accelerograms. As the resulting neural network will be very huge and very difficult to train, this problem will be overcome with two-stage method. First the neural network learns to relate the discretized response spectrum to a compressed form of the Fourier spectra, and then, it learns to relate the compressed Fourier spectra to the actual Fourier spectra.

The architecture of proposed replicator neural network for data compression consist of four layers with identical input and output layers and a much smaller number of processing units in the middle hidden layer. These neural networks are trained to replicate in their output layer the vector given at their input layer. Going from the input layer to the output layer the signals pass through small number of processing unit in the middle hidden layer, thus, the activations of the middle hidden layer form an internal compressed representation of the input data. The replicator neural networks were first invented by Kohonen (1976). Ackley (1985) studied the replicator neural networks in the context of the encoder problem. The multi layer feed forward replicator neural network were developed by Cottrell (1987). Hecht Nielson (1996) clarified certain fundamental aspects of the operation of replicator neural networks. The replicator neural networks perform a mapping from the n dimensional input vector space to a unit cube in the k dimensional vector space of middle hidden layer, where k is much smaller than n. In his study Hecht Nielson (1996) shows that the middle hidden layer of the replicator neural network produce optimal source codes.

For effective work of the neural network very important is pre-processing activity in order to establish the required parameters, as effective strength, type of soil, predominant frequency and radius of gyration of the power spectra and the maximum spectral acceleration. The pre-processing begun with the conversion of the existing coordinates to longitudinal and transversal components. A brief description of the parameters computed or extracted for training the replicator neural network can be given as follows:

- **Magnitude** – normalised measure of the amplitude of the earthquake waves measured at a given distance of the epicenter. The most commonly used is the Richter scale, which provides a 1 to 10 scale. The values of the magnitude were directly extracted from the data source.

- **Epicentral distance** – distance between the epicenter and the station where the motion was recorded, and constitutes one of the attributes that was used to sharacterise each eatrthquake. The value of this attribute is given in km and was calculated as Eqn.2.

$$D_{EP} = \sqrt{\left[(Lat_{EP} - Lat_{ST}) \times 111\right]^2 + \left\{\cos\left[(Lat_{EP} + Lat_{ST})/2\right] \times (Long_{EP} - Long_{ST}) \times 111\right\}^2} \qquad (2)$$

- **Focal dept** – the actual dept of the hypocenter. Hypocenter or focus is the region of the globe where the earthquake occurs. Epicenter is its projection on the surface of the earth. The focal dept is usually given in km and this parameter was directly extracted from the data source.

- **Peak acceleration** – may be obtained from the record of the ground motion after conversion to longitudinal and transversal components. As turned out, the peak acceleration was refer to the longitudinal component and this component was used for training the network.

- **Type of soil** – where a structure is located is extremely relevant for the definition of the ground motion. For consideration of this parameter it's necessary to have a classification of the soil where each base station is located.

- **Effective duration** – the period corresponding to occurrence of the higher components of the motion. For providing a better description of the duration of the earthquake it is necessary to remove the initial and final parts of the accelerogram, during which the values of the acceleration show lesser significance. The computation of effective duration was made, by defining an accumulated function of the acceleration. This function was used for establishing the beginning and the end of the effective period.

- **Radius of gyration** – of the power spectral density function is a relevant factor for the generation of artificial accelerograms. Defining the radius of gyration id necessary to provide a measure of the concentration or dispersion of the frequency contents around the central value or around the main frequency.

- **Predominant frequency** – is given in order to describe an earthquake in the frequency domain. Description of the earthquake in the time domain is given by an accelerogram. Predominant frequency describes the contents of an earthquake in each of its constituent frequencies. The function of spectral density is named power spectrum and is shown on Figure 4.

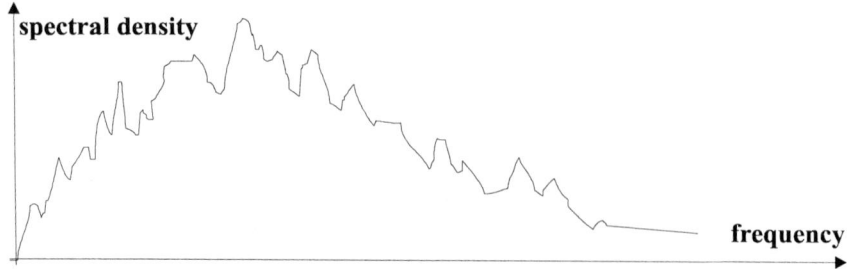

Figure 4: Spectral density function

The attributes used for training for predominant frequency was obtained from the spectral density function. Predominant frequency represents the central value of this function, which had to be computed for each of the records used for training.

♦ **Maximum spectral frequency** – represents the maximum of the spectral density function. Sometimes the central frequency of a strong motion earthquake does not coincide with the central value and then is better to train the network with maximum spectral frequency for receiving better results. This parameter allows taking into account some eventual skewing of the spectra around the central value.

REPLICATOR NEURAL NETWORK FOR DATA COMPRESSION

The proposed method of generating artificial earthquake accelerograms is based on developing a neural network which tales the discretized ordinates of the response spectrum as input, and produces the real and imaginary parts of the Fourier spectra of the generated earthquake accelerograms as the output. Such a neural network was trained with using the mentioned above parameters extracted from real accelerograms for a small geographic area in California. The replicator neural network was used as a data compression tool to compress the vector of the discrete Fourier spectrum of the earthquake accelerogram to a vector of much smaller dimension. In the second stage a multi layer feed forward neural network learns to relate the response spectrum to the compressed Fourier spectrum. As a result of combining these two neural networks, the artificial earthquake accelerograms are generated directly from the given response spectra.

For training the replicator neural network were used Fourier spectrum taken from accelerograms of different earthquakes for the certain small geographic area. The replicator neural network learns to relate the discretized response spectrum to a compressed form of the Fourier spectra. Going from the input layer to the output layer, the replicator neural networks channels the information through the few neurons at the middle hidden layer. Therefore, the activation of these neurons encode a compressed version of the Fourier spectra. The data compression in replicator neural networks trends to retain the essential and important features of the Fourier spectra, while discarding the less important features. In fact there are two separate replicator neural networks, one replicator neural network processes the real part while the second one processes the imaginary part of the Fourier spectra of the same accelerograms.

MULTI LAYER FEED FORWARD NEURAL NETWORK TO RELATE THE RESPONSE SPECTRUM TO THE COMPRESSED FOURIER SPECTRUM

For training the multi layer feed forward neural network were used parameters as follows: for input layer were used the parameters on the left side of Figure 3, and for output layer – the parameters on right side. These basic descriptors were used for artificial generating of an accelerogram to a given earthquake. For this generation procedure first on the base of the descriptors of last three parameters from the left were defined a power spectral density function. For this purpose were used a combination of Gaussian functions. The mean value of these functions corresponds to descriptor of the predominant frequency, whose standard devitation was related to descriptor of the radius of gyration of the power spectral density, and whose maximum value corresponds to the frequency value for which the power spectral density function reaches its maximum.

The overall agreement of the generation process can be assessed at following levels:

♦ Comparing the obtained descriptors with similar values gathered from the used accelerograms.

♦ Making a similar comparison in terms of the power spectral density functions, assuring that there is a good description of the frequency content of the ground motion.

- Comparison in terms of the accelerograms, allowing assuring a good description of the time domain evolution of the earthquake, that is especially important in case the generated accelerograms. This comparison was used for the purpose of non-linear structural analysis.

- Assessment by means of elastic response spectra for comparison in terms of their effects on structures between the trained and the real systems. These spectra, which are currently used for structural design purposes, were measures of the maximum response of a generic single degree of freedom system to the ground motion described by means of an accelerogram.

CONCLUSIONS

The purpose of this work was to determine whether the trained with actual recorded earthquake accelerograms neural network is capable of generating reasonable looking accelerograms from design spectra. The proposed neural network method for generating artificial earthquake accelerograms gives possibility after training with the different initial connection weight to receive several different accelerograms with almost the same response spectrum as input. It was demonstrated that proposed strategy would lead to interesting results. The method was tested as well by using design spectra as input and generating accelerograms compatible with those design spectra.

The generated accelerogram has similar characteristics as those in the training set and its response spectrum is very close to the input design spectra. This is the useful property of the neural network based method, because it will enable generation of accelerograms compatible with any specified design spectra. The generated accelerograms can be used for analysis of linear and nonlinear structures.

With this method was possible to receive a good results even for a small number of earthquake records. The method gives possibility for training the network to predict the parameters required for the generation of artificial accelerograms, almost equivalent to those that had been actually recorded. Comparison of the input and output accelerograms and their response spectrum indicates that the trained neural networks has been learnt the training cases very well.

As a result of combining two neural networks, the artificial earthquake accelerograms are generated directly from the given response spectra. With this approach can be received as well different descriptors of the earthquake, such as the magnitude or the peak accelerations. The number of attributes, which allows receiving the artificial earthquake accelerogram, can be changed in such a manner to receive artificial generation of the accelerogram with expected characteristics. The received results were used for testing different parts of constructions.

REFERENCES

Ackley D.H., Hinton G.E. and Sejnowski T.J. (1985). A Learning algorithm for boltzmann machines, *Journal of Cognitive Science,* Vol. 9., pp. 122 - 147.

Chen L., Xu L. and Chi H. (1999), Improved learning algorithms for mixture of experts in multiclass classification. *Neural Networks 12,* pp. 1229 – 1252.

Cichocki A. and Unbehauen R. (1993). *Neural networks for optimization and signal processing,* John Wiley &Sons, Chichester.

Fahlman S. E. (1988). Faster learning variations on back propagation: an empirical study *Proceeding of 1988 Connectionist Models Summer School,* Morgan Kaufmann Los Altos CA, pp. 38 – 51.

Hertz J., Krogh A. and Palmer R.G. (1991). *Introduction to the theory of neural computations,* Addison Wesley, Redwood City, California, USA.

Hecht Nielsen R. (1996). Data manifolds, natural coordinates, replicator neural networks, and optimal source coding, ICONIP 96, pp. 126 - 142.

Kohonen T. (1976). A principle of neural associative memory, *Neuroscience,* vol.2, pp. 1065 – 1088.

SEISMIC DESIGN OF STEEL STRUCTURES

CONTRIBUTING EFFECT OF CLADDING PANELS IN THE SEISMIC DESIGN OF MR STEEL FRAMES

G. De Matteis [1], R. Landolfo [2] and F. M. Mazzolani [1]

[1] University of Naples Federico II, P.le Tecchio, 80, I 80125 Napoli, ITALY
[2] University of Chieti G. D'Annunzio, V.le Pindaro 42, 65127 Pescara, Italy

ABSTRACT

Current paper is dealing with the influence of light-weight cladding panels on the seismic design of multi-storey steel moment resisting (MR) frames. To this purpose, the possibility to profit of the contributing effect of shear walls on the serviceability limit state of the structure is analysed. In particular, a new design procedure is stated and discussed. It is essentially based on the separation between strength and drift requirements. Therefore, the stiffening diaphragm action provided by cladding panels is taken into account in order to increase the lateral stiffness of the bare frame, once the latter has been designed according to resistance and ductility requirements only. In the study, which is referred to three different structural typologies, namely two-, five- and seven-storey frames, different strategies to chose cladding panels characteristics are considered. In general, obtained results show that the proposed procedure leads to more rational and economic solutions, allowing for an important reduction of structural member sizes. Besides, it is noticed that obtained requirements for cladding panels in terms of both strength and stiffness are never prohibitive, they being in the same range of ones provided by commercial products.

KEYWORDS

Cladding panels, Overstrength, Sandwich panels, Seismic design, Serviceability limit state, Shear walls, Steel frames, Stressed skin design.

INTRODUCTION

Since many years it has been recognised that moment resisting steel frames provide high strength, ductility and energy dissipation capacities. Therefore, their use for seismic purpose has been widely promoted worldwide (Mazzolani and Piluso, 1996). On the other hand, such structural typology is often characterised by a very low lateral stiffness. The latter aspect strongly influences the usefulness of the structure, because the respect of serviceability limit state requirements gives rise to member sizes noticeably larger then the ones strictly related to strength and ductility requirements. The problem is particularly important in case of seismic design, due to the emphasis given by building codes to "stiffness checks", which in same cases appear to be particularly compelling and conservative. This means that the high potential of structural performance of both the material itself and the structural typology under consideration cannot be fully profited. On the contrary, it is wasted by this inherent inconsistency between strength and stiffness capacity of the structure. Therefore, it is a common opinion that something in the design procedure of MR steel frames should be revised, aiming at improved solutions, where the "strength" and "stiffness" criteria have a similar impact on the determination of the member sizes.

To this purpose, the effect of cladding panels, as contributing for the lateral stiffness of multi-storey steel MR frames, seems to represent a useful chance. In fact, previous studies have shown that even steel lightweight cladding panels, commonly adopted in practice as enclosure of the building, if suitably connected to the bearing members, may provide considerable in-plane strength and stiffness, due to their interaction with main bearing members. Hence, aiming at more efficient and cost-effective solutions, recent design philosophies consider the enclosure surfaces acting as diaphragms in assistance or substitution of other lateral load resisting systems (Miller and Serag, 1978; De Matteis, 1998). Adequate allowance for enclosure panel effects should be therefore considered both in predicting actual seismic actions, due to their stiffening effect, and in the sharing of reaction forces between members and infill panels. This means that the actual structural properties of infill panels, namely stiffness, strength and dissipative behaviour, have to be preventively determined and then controlled, in order to assess correctly the contribution they can supply to the main structure under earthquake loading.

A suitable design method has been therefore pursued by the authors. It is based on a twofold analysis. Firstly, cladding panels are not accounted for in evaluating the response of the structure at the ultimate limit state, and the bare steel frames are designed according to "strength" and "ductility" criteria only. Then, in order to fulfil "stiffness" criteria also, the cladding action is accounted for, it being considered as a complementary action to that provided by the bare frame structure. Obviously, some questions arise. The first one is concerned with the actual possibility to cover such stiffness requirements by means of ordinary systems, compatible with commonly adopted cladding typologies. The second one is concerned with the potential contribution of cladding panels to the global performance of the whole structure under high-intensity earthquake loading, assessing the possibility to account for cladding action at the ultimate limit state as well.

The current paper essentially focuses on the first of the above aspects, emphasising, on one hand, the beneficial effect that could be obtained by profiting of the contribution effect provided by cladding panels, and showing, on the other hand, if ordinary existing lightweight cladding panels could be actually compatible with design purposes. Such results are related to different typical frame configurations, for which the applications of the above design procedure is presented and discussed.

THE DESIGN PROCEDURE

In order to increase the lateral stiffness of moment resisting steel frames, the contributing effect of cladding panels, acting as a collaborating system in absorbing horizontal actions, may be favourably taken into account. External lateral forces are therefore resisted by a composite action. In fact, besides to the intrinsic stiffness provided by the bare frame itself, the whole system may benefit of the complementary stiffness provided by cladding panels acting as shear diaphragms. In particular, according to previous studies, in order to simplify the structural modelling of the latter contribution, such an effect may be schematised through equivalent pinned diagonal members (De Matteis and Landolfo, 2000).

In the first phase of the design procedure, in order to minimise structural weight and taking in mind the complementary stiffness provided by cladding bracing action, which will by taken into account successively, frames may be designed according to ultimate limit states only, on the basis of design seismic and vertical loading. Then, in case that the obtained bare structure is not able to fulfil serviceability limit state checks by itself, in the second phase of the design procedure, cladding panels are properly designed in order to provide the necessary stiffening action which permits covering the gap with the considered design interstorey drift limits. The key of the procedure is how evaluating both stiffness and strength of panels in order to fulfil the code provisions concerning with interstorey drift limits. Obviously, elastic stiffness of cladding panels could be computed at every storey in such a way the deformability checks of the frame are satisfied. By assuming an elastic structural response, the global external horizontal action can be simply shared between the bare frame and the shear walls, as a function of the relative elastic stiffness ratio.

On the basis of the above simple consideration as well as of several numerical results related to different multi-storey frame configurations (De Matteis and Landolfo, 1999a), for each single storey (i), the stiffness increment of the whole system due to the single cladding panel may be expressed by the following ratio:

$$\psi_i = \frac{K_{p,i}}{V_i / \Delta_i} \quad (1)$$

where $K_{p,i}$ is the cladding panel stiffness at storey i, V_i is the design storey shear load and Δ_i is the interstorey drift computed on the bare structure under design loads.

Therefore, it is possible to synthesise the increment of the structural performance at each storey i of the infilled structure ($P_{fl,i}$) as respect to the bare structure ($P_{br,i}$) by using the following improving ratio:

$$R_i = \frac{P_{fl,i}}{P_{br,i}} = 1 + c_i \psi_i \quad (2)$$

where c_i is the number of bays at storey i where cladding panels are located to act as a bracing system. Consequently, the required stiffness for cladding panel $K_{p,i,req}$ at each storey i may be simply determined by imposing the following criterion:

$$R_i = \frac{\Delta_i}{\Delta_{i,lim}} = \phi_i \quad (3)$$

where $\Delta_{i,lim}$ is the design interstorey drift limit at storey i, and ϕ_i is the corresponding ratio with respect to the actual displacement value at that storey Δ_i.

Once initial stiffness of cladding panels has been fixed at all the storeys, then elastic strength of each cladding panel may be determined in order to avoid any damage of the panel itself under moderate earthquakes. To this purpose, the internal force distribution among structural members and cladding panel equivalent diagonal members can be easily determined by means of an elastic analysis under the pre-defined external action. The latter may be defined as the one corresponding to the elastic response spectrum given by the code for high return period divided by the service factor v. Finally, in the fourth phase of the procedure, once required elastic stiffness and strength of cladding panels have been computed, the problem to choose a convenient and appropriate cladding system should be overcame. To this purpose, the method proposed by Bryan and Davies (1982) could be applied. It permits to determine the global response of cladding panels starting from the characteristics of basic components, namely panel sheeting, connections and external members. As an alternative, sophisticated finite element models could be considered, allowing all panel typologies and connecting systems to be accurately schematised and taken into account (De Matteis and Landolfo, 1999d).

THE STUDY CASES

Frame Schemes and basic Assumptions

In order to verify the possibility to profitably apply the above design procedure to actual cases, some frame configurations have been selected. The study has been referred to a typical residential building with a rectangular lay-out constituted by 8 different frames in the transversal direction and 3 frames in the longitudinal direction, which is the same analysed in (De Matteis et al., 1998). Here, the analysis is carried out with reference to an internal 2-bay transversal steel rigid MR frame. Aiming at investigating on the effect of cladding panels in case of different conditions of the bare structure in terms of global lateral rigidity, in the following three different solutions are considered, namely a seven-storey, five-story and two-storey building. In all the cases, the basic geometric dimensions of the frame are the same. In

Particular, a storey height equal to 4.00 m and 3.5 m have been assumed for the first storey and for all the other storeys, respectively. The characteristic yield strength of the members is 275 N/mm², corresponding to steel grade Fe 430. IPE profiles and HEB profiles have been considered for beam and column members, respectively. Live loads have been assumed equal to 3 kN/m² for both typical floors and roof. Dead loads, comprising steel structure weight, have been evaluated in 5.1 kN/m² and 4.3 kN/m² for typical floor and roof, respectively. Sub-soil class B, corresponding to a deposit of medium density sand, and a peak ground acceleration $a_{g,d} = 0.2\ g$ have been hypothesised, g being the gravity acceleration. Further details on the deign criteria are given in De Matteis et al. (1998).

Design Procedure for Frames

Steel frame member sizes have been determined by means of a simplified response spectrum analysis, where the effect of seismic action on the structure has been evaluated according to EC1 and EC8 provisions. Second order (P-Δ) effects have been accounted for by carrying out a non-linear 2nd order analysis. Structural regularity in plan and in elevation has been hypothesised. In order to quantify the contributing effect provided by cladding panels in different design conditions, several cases have been considered. First of all and generally speaking, three different design methodologies have been followed. The first two series of frames (strength-type) have been designed according to ultimate limit state only and both not complying (frames type 1) and complying (frames type 2) with second order effect limitation $\theta_i \leq 0.3$, where parameter θ_i is considered according to EC8. In addition, a third frame series (deformability-type), which accounts for serviceability limit state also, has been considered (frames type 3). Obviously, frames type 1 provide the lightest solution, where, possibly, the contribution provided by cladding panels is very important and profitable, while frames type 3 represent the reference case, i.e. frames fully complying with code provisions.

Besides, three design q-factors, namely $q_d = 4, 6, 8$, have been accounted for, while interstorey drifts have been referred to three different limits Δ_{lim}, namely 0.004 h_i, 0.006 h_i and 0.008 h_i, h_i being the storey height. The first two interstorey limits correspond to the ones suggested by EC8, while the latter represents a less severe condition, which has been recently proposed in the draft of the new Italian seismic code for steel structures. Finally, in order to define seismic events according to serviceability loading conditions, the values of reduction coefficient v, which accounts for the lower return period respect to high-intensity seismic events, have been fixed equal to 2 and 3. The first limit is suggested by EC8 for ordinary buildings, while the latter has been introduced for the same building category in the draft of the new Italian seismic code for steel structures.

Design Procedure for Bracing Panels

As already stated, once the bare structure has been dimensioned, the elastic stiffness of cladding panels may be evaluated at each storey level via eq. (1), (2) and (3). On the other hand, as far as their distribution throughout the frame height is concerned, it should be considered that several procedures could be followed, giving rise to a different increment of the global performance of the structure. In the current study, three procedures have been analysed. All of them are based upon interstorey drift ratios ϕ_i evaluated on the bare frame, by means of a preventive elastic analysis.

In the first procedure, which in the following is labelled as "$\Psi = \Psi_{max}$", panels are assumed to be the same at every storey, allowing both the procedure and its practical application to be simplified. Besides, in this way the same strength and stiffness increment towards the height of the frame as respect to the bare structure is conferred. Therefore, once structural displacements have been determined at all the storey levels, the maximum interstorey drift ratio ϕ_{max} is considered. Then, the corresponding panel stiffness $K_{p,max,req}$, which is the maximum among all the storeys, is computed at that storey. Finally, for all the other storeys, the same panel configuration is considered.

The second procedure, which in the following is labelled as "$\Psi = \Psi_{const}$", aims at adopting the same Ψ value at every storey, allowing for the same percentage increment of the lateral stiffness at every storey as respect to the bare frame. Therefore, once interstorey drift ratios ϕ_i have been determined, then cladding panels are designed at each storey in such a way to obtain the same value Ψ_i. Obviously, this value should be the one related to the storey presenting the highest value of drift ratio ϕ_{max}.

The third procedure, which in the following is labelled as "$\Psi = \Psi_i$", aims at the optimisation of the complementary stiffening system, whose characteristics are selected as the ones strictly necessary at each storey. Therefore, once interstorey drift ratios ϕ_i have been determined, cladding panels may designed at every storey in such a way to obtain the interstorey drift ratio ϕ_{lim} provided by the considered code. In this way, it is assured a linear deformed shape of the structure under the design external force distribution.

THE APPLICATION OF THE PROPOSED DESIGN PROCEDURE

General

The different combinations of all the above defined parameters would lead to several different frame configurations to be considered. In order to reduce the number of study cases, these parameters have been not ubiquitously varied, considering only some possible combinations. All the analysed case are synthesised in Table 1, where the corresponding obtained frame weight values are also devised. It is possible to recognise easily that the selected frame configurations allow all the effects of all the parameters to be suitably investigated.

TABLE 1
FRAME WEIGHT [t] FOR ANALYSED CASES

Frame geometry	q_d	Strength-type		Deformability-type					
				$\nu = 2$			$\nu = 3$		
				Δ_{LIM}/h			Δ_{LIM}/h		
		$\vartheta_i > 0.3$	$\vartheta_i \leq 0.3$	0.004	0.006	0.008	0.004	0.006	0.008
7-storey	4	8.73			15.35				
	6	7.98		17.97	15.35	13.36	15.35	12.39	10.59
	8	7.56			15.35				
5-storey	4	5.82	5.82		9.28				
	6	4.83	5.4		9.28				
	8	4.63	6.16		9.28				
2-storey	4	2.12			3.43				
	6	1.79			3.43				
	8	1.79			3.43				

Obtained Frame Configurations

According to the above defined design procedures, member sizes for every selected frame configuration have been determined. They are synthesised in Table 2, where, for all the analysed cases, results are reported separately for 7-storey, 5-storey and 2-storey frames, providing the required profile for beams, external columns and internal columns at each storey level. It is important to notice that in all the cases, member sizes have been obtained aiming at optimising the total weight of the structure, within the respect of the design conditions. Nevertheless, some other criteria have been preserved. In particular, according to EC8 capacity design rules, the strong column-weak beam criterion has been observed; besides, changing of column size at each storey has been avoided; finally, in case of frame designed according to stiffness, the fulfilment of the interstorey drift limitation has been assured by increasing progressively both column and beam sections, aiming at the minimum structural weight.

TABLE 2
MEMBER SIZES FOR ANALYSED FRAME CONFIGURATIONS

	Strength-type $\vartheta_i > 0.3$						Deformability-type					
							$\nu=2$ ($q_d=6$)			$\nu=3$ ($q_d=6$)		
	$q_d=4$		$q_d=6$		$q_d=8$		$\Delta_{LIM}/h=0.004$	$\Delta_{LIM}/h=0.006$	$\Delta_{LIM}/h=0.008$	$\Delta_{LIM}/h=0.004$	$\Delta_{LIM}/h=0.006$	$\Delta_{LIM}/h=0.008$

Storey No.	Ext. Col HEB	Beam IPE	Int. Col HEB	Ext. Col HEB	Beam IPE	Int. Col HEB	Ext. Col HEB	Beam IPE	Int. Col HEB	Ext. Col HEB	Beam IPE	Int. Col HEB	Ext. Col HEB	Beam IPE	Int. Col HEB	Ext. Col HEB	Beam IPE	Int. Col HEB	Ext. Col HEB	Beam IPE	Int. Col HEB	Ext. Col HEB	Beam IPE	Int. Col HEB	Ext. Col HEB	Beam IPE	Int. Col HEB
7	180	270	200	180	270	200	180	270	200	340	400	360	320	300	360	280	300	320	320	300	360	260	270	300	220	270	260
6	180	270	200	180	270	200	180	270	200	340	400	360	320	360	360	280	300	320	320	360	360	260	270	300	220	270	260
5	220	300	260	200	300	240	200	300	220	400	400	450	360	400	400	320	360	360	360	400	400	300	330	340	260	300	300
4	220	330	260	200	300	240	200	300	220	400	400	450	360	450	400	320	360	360	360	450	400	300	330	340	260	300	300
3	220	330	260	200	300	240	200	300	220	450	450	500	400	450	450	320	400	400	400	450	450	340	400	360	300	360	240
2	260	330	300	240	300	280	220	300	240	450	450	500	400	450	450	360	400	400	400	450	450	340	400	360	300	360	340
1	260	330	300	240	300	280	220	300	240	450	450	500	400	450	450	360	400	400	400	450	450	340	400	360	300	360	340

	Strength-type $\vartheta_i > 0.3$						Strength-type $\vartheta_i \leq 0.3$						Deformability-type
	$q_d=4$		$q_d=6$		$q_d=8$		$q_d=4$		$q_d=6$		$q_d=8$		$\nu=2$ ($q_d=4,6,8$) $\Delta_{LIM}/h=0.006$

Storey No.	Ext. Col HEB	Beam IPE	Int. Col HEB	Ext. Col HEB	Beam IPE	Int. Col HEB	Ext. Col HEB	Beam IPE	Int. Col HEB	Ext. Col HEB	Beam IPE	Int. Col HEB	Ext. Col HEB	Beam IPE	Int. Col HEB	Ext. Col HEB	Beam IPE	Int. Col HEB	Ext. Col HEB	Beam IPE	Int. Col HEB
5	180	270	200	160	270	180	160	270	180	180	270	200	160	270	180	180	270	200	240	300	240
4	180	300	200	160	270	180	160	270	180	180	300	200	160	270	180	180	300	200	280	360	280
3	220	300	220	180	270	200	160	270	200	220	300	220	200	300	220	220	300	260	320	400	320
2	220	330	240	180	300	220	180	270	220	220	330	240	200	330	240	260	330	260	320	450	320
1	240	330	240	200	300	220	180	300	220	240	330	240	240	330	240	260	330	260	320	450	320

	Strength-type $\vartheta_i > 0.3$						Deformability-type
	$q_d=4$		$q_d=6$		$q_d=8$		$\nu=2$ ($q_d=4,6,8$) $\Delta_{LIM}/h=0.006$

Storey No.	Ext. Col HEB	Beam IPE	Int. Col HEB	Ext. Col HEB	Beam IPE	Int. Col HEB	Ext. Col HEB	Beam IPE	Int. Col HEB	Ext. Col HEB	Beam IPE	Int. Col HEB
2	200	270	200	180	240	180	180	240	180	300	300	300
1	200	270	200	180	240	180	180	240	180	300	300	300

The obtained results are also summarised in Figure 1, where for each analysed case the structural weight normalised as respect to the number of storeys is depicted. The comparison of both member sizes and structural weight per storey allows many useful considerations to be drawn. Firstly, it is to be observed that frames designed according to stiffness are much heavier than the ones designed according to strength. In particular, in Figure 2a the percentage steel saving corresponding to the lightest solutions, i.e. frames designed without complying with second order effect limitation (frames type 1), is reported. It is shown that the structural weight saving of these frame typology respect to the ones designed according to stiffness and complying with the less severe conditions stated by EC8 ($\Delta_{LIM}/h = 0.006$ and $\nu = 2$) ranges between 38% and 50%, it still increasing as far as both design q-factor and number of storey increases. Similarly, in Figure 2b histogram of steel saving is depicted with respect to the other influencing parameters, showing that even in case of less severe conditions, namely $\Delta_{LIM}/h = 0.008$ and $\nu = 3$, frames designed according to strength are not able to comply with deformability checks, giving rise to a steel saving of about 25%.

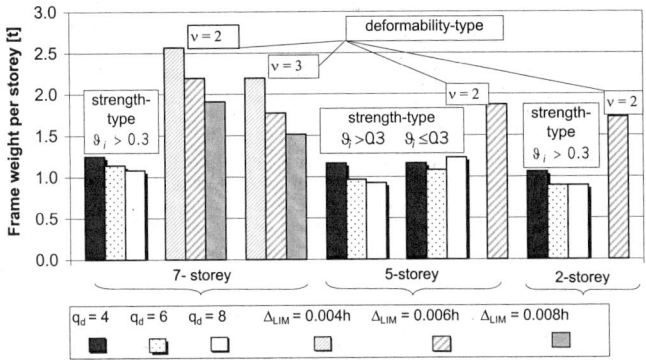

Figure 1: Histogram of normalised frame weight respect to the number of storeys [t]

Finally, it should be noticed that the limitation provided by EC8 related to second order effects is particularly influencing in case of frames designed with high values of q_d. This is shown in Figure 1 for 5-storey frames, where it is evident that, in case of $\theta_i \leq 0.3$, the frame corresponding to $q_d = 8$ is substantially heavier than the ones corresponding to $q_d = 6$ and $q_d = 4$. Obviously this is a nonsense that is due to the dependence of the coefficient θ_i from the q-factor value, which leads to a more compelling limitation as far as the behaviour q-factor value increases.

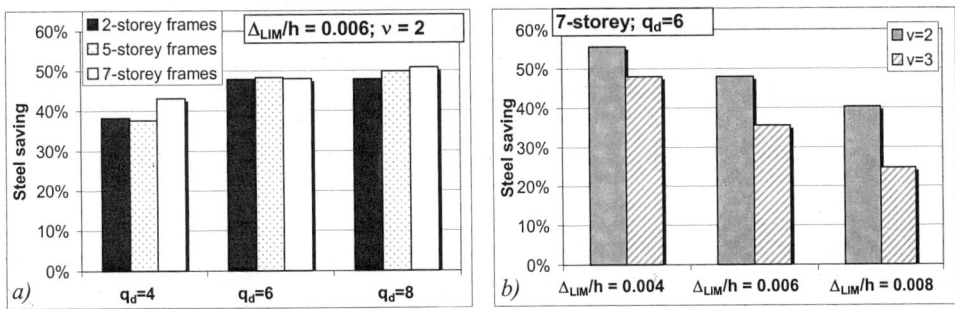

Figure 2: Histograms of steel saving for frames designed according to strength only ($\theta_i > 0.3$)

Evaluation of Cladding Panel Properties

As it is clear by comparing relative structural weights, frames designed according to strength are very far to respect common interstorey drift limits imposed by seismic code under earthquakes related to relatively small return period. In the current study, for such frames the complementary lateral stiffness necessary to comply with these displacement limits are supposed to be supplied by a purposely conceived cladding system. Therefore, for all the analysed cases, the required initial stiffness and elastic strength for panel equivalent diagonal members have been evaluated according to the design procedure previously defined. Comparison in terms of panel characteristics due to the applied procedure to define the panel distribution throughout the frame is shown in Figure 3a, where one case related to the 7-storey frame configuration is analysed. In general, it is evident as the procedure "$\Psi = \Psi_{max}$" provides the highest requirements for panels, while the other two procedures furnishes panel stiffness values decreasing from bottom storeys to top storeys. In particular, the procedure "$\Psi = \Psi_i$" seems to be the most effective, providing minimum stiffness requirements. As far as strength requirements are concerned, the three adopted procedures provide results, linearly increasing from top storeys towards bottom ones. Also in this case, the procedure "$\Psi = \Psi_i$" minimises the demand to shear walls. In Figure 3b, with reference to the procedure "$\Psi = \Psi_i$", bracing panel requirements are depicted for different values of the design q-factor. It is clearly shown that requirements in terms of strength decrease with q_d, while the required stiffness of cladding panels increases with q_d. In Figures 3c and 3d the effects of both interstorey drift limit (Δ_{LIM}/h) and factor v are devised. As it could be expected, in case of less severe limits ($\Delta_{LIM}/h = 0.008$ and v = 3), panel requirements are very small, they being also equal to zero for some storeys. Finally, in Figures 3e and 3f the effect of frame configuration is shown. Also in this case, as it could be expected, requirements for cladding panels in terms of both strength and stiffness increase with the number of storeys in the frame. In particular, such a variation is quite regular for the strength, while it is somewhat irregular for the stiffness, emphasising the dependence of results on the configuration of the bare structure, which is actually influenced by several design parameters.

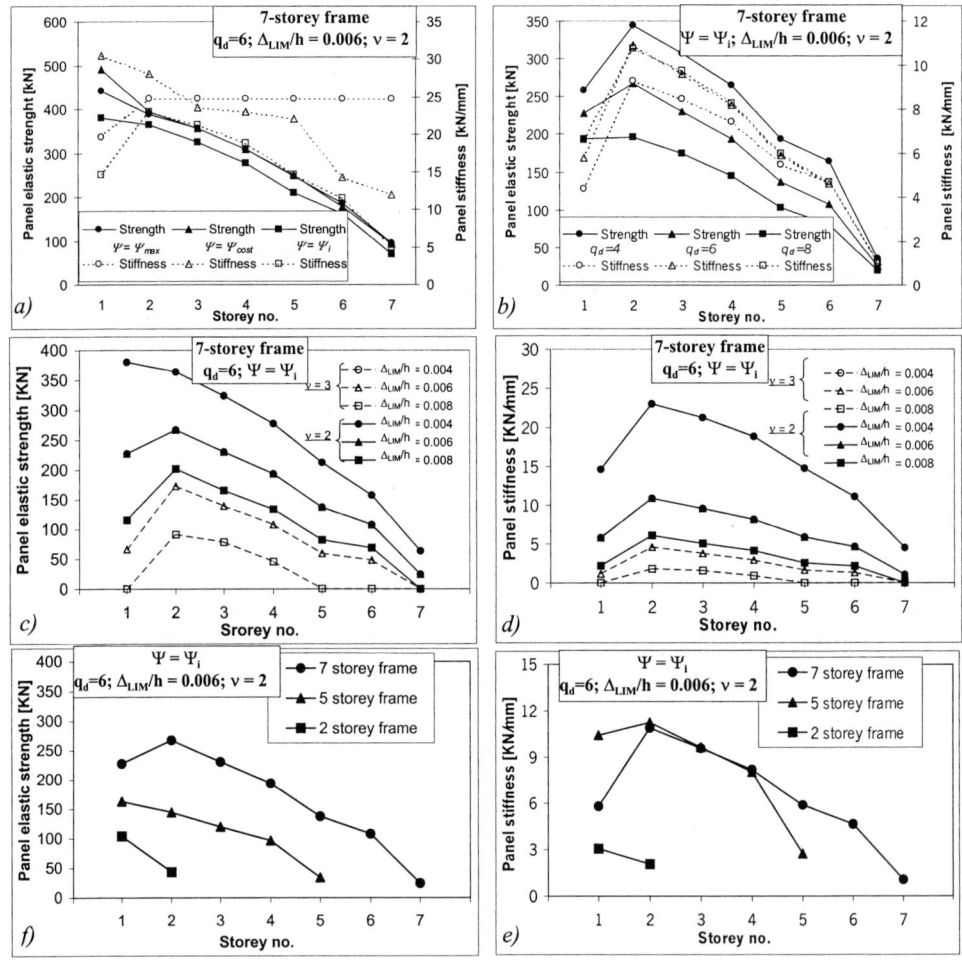

Figure 3: Strength and stiffness requirements for cladding panels

THE DESIGN OF BRACING PANELS

It is interesting to verify if the above panel requirements in terms of both strength and stiffness, which have been obtained for typical European steel MR frame configurations, might be covered by realistic and available cladding panels typologies. For the analysed cases and for the procedure "$\Psi = \Psi_i$", the obtained demand in terms of both stiffness and strength is up to about 400 kN and to 25 kN/mm for the elastic strength and the initial stiffness, respectively. Anyway such requirements are often less severe, especially in case of two- and five-storey frames. Therefore, at a first glance, it seems that obtained demands are not prohibitive, they being in the range of usually adopted light-weight shear walls, constituted of thin metal sheeting connected to each other and to the bearing structure by means of simple mechanical fasteners (self-drilling screws or steel bolts). For instance, previous experimental results carried out by authors, which concern with sandwich panels with polyurethane foam core, have provided elastic strength and initial stiffness equal to 78 kN and 4.6 kN/mm, respectively (De Matteis and Landolfo, 1999b), they being therefore already able to cover many of the situations under consideration. On the other hand, it is to be considered that tested specimen was only slightly connected to the bearing structure. In fact, it was connected to the external frame on two sides only, it did not have any internal girt and the spacing of both panel-to-panel and panel-to-frame connections was quite large. Therefore, it is expected that higher value of both strength and stiffness could be compatible with available mechanical characteristics of already existing panels and may be accomplished by adopting simple technological details.

In the following, in order to verify the above possibility, an application of the stressed skin design method (ECCS, 1995) suggested by Bryan and Davies (1982) is carried out. According to such a method, the strength and stiffness of the whole shear panel are determined by means of a sort of component method. In fact, by setting all the individual components, their own strength and flexibility are firstly computed. The strength of the panel is then obtained as the one corresponding to the lowest value among all the possible failure modes, while its flexibility is determined by summing the contributions of all the components.

In particular reference has been made to the seven storey building, designed according to $q_d = 6$, $\Delta_{LIM}/h = 0.004$, $v = 2$, and by considering the procedure "$\Psi = \Psi_i$" for determining the panel distribution throughout the frame. In such a case, according to Figure 3c and 3d, elastic strength and stiffness demands to the panel located at the second storey are equal to 365 kN and 23 kN/mm, respectively. These values practically correspond to the maximum requirements among all the analysed cases, when applying the procedure "$\Psi = \Psi_i$", which has been demonstrated to be the most suitable one. Panel geometry is shown in Figure 4. It is related to a four side fastened panel with two internal girt members. In such a practical application, the basic components of the panel system have been supposed to be the same of the tested specimens. Therefore, the stiffness and the strength of the seam connections and panel-to-external frame connections have been assumed according to the available experimental test results (De Matteis and Landolfo, 1999c), which are referred to self-drilling screws of 6.3 mm in diameter and steel bolts of 8 mm in diameter. In particular, the elastic strength of connections has been assumed equal to 0.8 times the ultimate strength value determined by tests. Furthermore, the properties of shear connectors have been assumed the same as the ones related to panel-to-member connections.

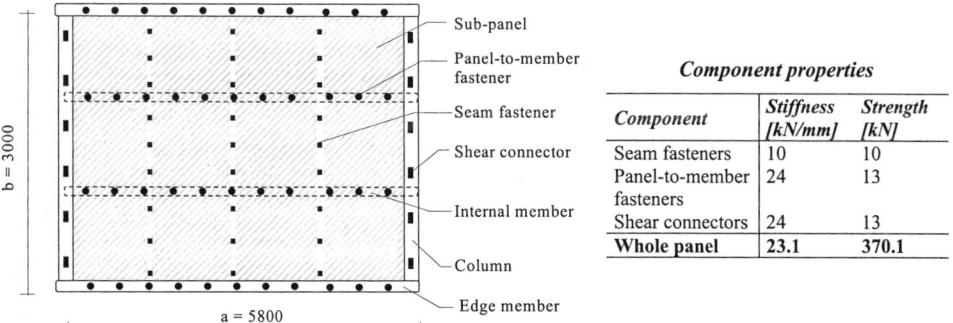

Figure 4: Cladding panel geometry and assumed property values for components

In order to determine the panel strength, as possible collapse modes, both failure along the lines of seam fasteners and failure of panel-to-member fasteners have been considered. Panel global buckling has not been regarded, since the stabilising action of the internal core of the sandwich system should allow for a high shear buckling strength. As far as the shear stiffness is concerned, the total flexibility of the panel has been computed by deeming the following component contributions: shear strain in the sheet, axial strain in the edge members, movement in the panel seams, movement in the panel-to-member connections and movement in the shear connectors. By assuming for all the connecting systems a fastener pitch equal to 105 mm and by considering the properties of each single fastener as shown in the table of Figure 4, the stiffness and the elastic strength of the panel system has been evaluated. Results are reported in the same figure. It should be observed that they are practically coincident with the panel requirements obtained by the application of the proposed procedure, demonstrating the suitability of the chosen panel typology.

CONCLUDING REMARKS AND FURTHER DEVELOPMENTS

On the basis of the presented results, it can be concluded that the proposed design philosophy, which is based upon stiffening action provided by cladding panels, appears to be very effective and rational, leading to a remarkable lightening of steel moment resisting frames. In fact, with reference to several steel moment resisting frame configuration and influencing parameters, it has been shown that serviceability limit conditions have an important impact on member sizes. Therefore, if frames are designed according to ultimate limit states only, allowing for a noticeable steel saving, then cladding panels could be taken into account providing the complementary stiffness, acting as shear walls also rather than as enclosure elements only. According to this procedure, cladding panel requirements, in terms of both elastic strength and rigidity, have been evaluated for several cases. By applying the well known stressed skin design method, it has been also demonstrated that such requirements are non prohibitive if compared with the ones provided by lightweight steel sheeting. Further developments are now in progress. In fact, it is expected that cladding panels could also contribute to the global response of the structure at ultimate limit state. To this purpose, starting from the actual disipative capacity of cladding panels, as it has been experimentally stated (De Matteis and Landolfo, 1999*b*), non linear time history dynamic analyses will be carried out, aiming at quantifying the global structural performance of analysed building as well as the improvement of the frame global performance endowed with shear walls.

References

Davies, J.M. and Bryan, E.R. (1982). *Manual of Stressed Skin Diaphragm Design*, Granada Publishing, London.

De Matteis, G. (1998). *The Effect of Cladding Panels in Steel Structures under Seismic Actions*, PhD Thesis, University of Naples Federico II.

De Matteis, G., Landolfo, R., Mazzolani, F.M., (1998). Dynamic Response of Infilled Multistorey Steel Frames. Proc. of *XIth European Conference on Earthquake Engineering,* Paris la Defense, France, 6-11 September.

De Matteis, G. and Landolfo, R. (1999*a*). *Cladding Panel Action in Drift Control of Seismic Resistant Steel Frames*, in Proc. of 2nd European Conference on Steel Structures, (Eurosteel 99), Praha, Czech Republic, CD-ROM, Paper no. 163.

De Matteis, G. and Landolfo, R. (1999*b*). Structural Behaviour of Sandwich Panel Shear Walls: an Experimental Analysis. *Materials and Structures,* **32**, 331-341.

De Matteis, G. and Landolfo, R. (1999*c*). Mechanical fasteners for cladding sandwich panels: interpretative models for shear behaviour. *Thin-Walled Structures,* **35**, 61-79.

De Matteis, G. and Landolfo, R. (1999*d*). Modelling of lightweight shear diaphragms for dynamic analysis. *Journal of Constructional Steel Research,* **53: 1**, 33-61.

De Matteis, G. and Landolfo, R. (2000). Diaphragm Action of Sandwich Panels in Pin-Jointed Steel Structures: a Seismic Study. *Journal of Earthquake Engineering,* Imp. College Press, **4:3**, 251-275.

ECCS (1995). *European Recommendation for the Application on Metal Sheeting acting as a Diaphragm*, European Convention for Constructional Steelwork, Brussels.

Mayes, R.L., (1995). Interstory Drift Design and Damage Control Issue. *The Structural Design of Tall Buildings*, **4**, 15-25.

Mazzolani, F.M. and Piluso, V. (1996). *Theory and Design of Seismic Resistant Steel Frames*, E & FN SPON.

Miller, C.J. and Serag, A.E. (1978). Dynamic Response of Infilled Multi-Storey Steel Frames. In Proc. of *Fourth Int. Spec. Conf. on Cold-formed Steel Structures*, Univ. of Missouri, Rolla, 557-586.

DUCTILITY FORMULATIONS
OF STEEL STRUCTURAL MEMBERS

Muslinang Moestopo, Iswandi Imran, Revi Renansiva, Atin Sudarsono

Structural Mechanics Laboratory
Inter University Center for Research in Engineering Sciences
Institute of Technology Bandung, 10 Ganesha, Bandung 40132, INDONESIA

ABSTRACT

Formulations to quantify the ductility of steel structural members are proposed in this paper. Parametric study has been carried out by numerical and experimental work to investigate some ductility parameters on I-shaped steel members, i.e. steel grade, flange width-thickness, and member slenderness. Inelastic behaviour and local buckling as well as member buckling are considered in a steel member subject to uniform bending moment. Assumed stress-strain relationship with nominal values of yield stress and tensile stress is shown to provide lower ductility ratio. The influences of flange width-thickness and member slenderness as major factors affecting the member ductility, are formulated. Variation due to these ductility parameters on the flexural moment capacity and the curvature ductility are also presented. Based on this study, simplified formulas involving these parameters are presented to determine curvature ductility, rotational ductility and displacement ductility of I-shaped steel members. The formulas could be further extended to quantify the ductility provided by various dimensions of steel member as compared to the ductility that implicitly assumed in the seismic resistant design specification.

KEYWORDS

Ductility formulation, I-shaped steel member, curvature ductility, rotational ductility, displacement ductility, flange width-thickness ratio, slenderness ratio, seismic resistant design.

INTRODUCTION

The recent development in seismic resistant structure design has set the structural ductility as the very important parameter to determine the level of seismic load to be carried by a certain type of structure. The Seismic Provisions of the 1997 Uniform Building Code [6], for example, defines the seismic force reduction factor to account for the structural over-strength and the ductility of the structures. This factor is used in a single value for a particular structure (e.g. special moment resisting frames, eccentrically braced-frames, etc.).

Previous studies [4, 7, 8] had investigated a number of parameter affecting the ductility characteristic of steel structures, among others are the fundamental period of the structure and the slenderness ratios of the members. While the relationship between the ductility ratios and the fundamental period had been well accepted, the formulation to quantify the ductility by the member slenderness ratio has not been clearly determined. It was only assumed that the use of compact sections could prevent the occurrence of inelastic buckling of I-shaped steel member and could ensure the ductility performance of the structure.

A ductility formulation, which depends on the slenderness ratios of the steel members being used in the real construction, needs to be proposed to ensure the validity of the seismic design factor used in design as compared to the one assumed in the specification. This paper presents numerical and experimental results on the ductility with respect to the slenderness ratios as well as the ductility formulation of steel member subjected to pure bending moment.

MEMBER DUCTILITY

The term ductility has been widely understood as the capacity of structural member to undergo significant inelastic deformation without considerable loss of strength. Among many definition of ductility of a structural member, three definitions of ductility, are presented in this paper, i.e. curvature ductility ratio (μ_ϕ), rotational ductility ratio (μ_θ), and displacement ductility ratio (μ_δ). These definitions refer to the ratio between the value of the displacement being considered at the ultimate member strength and the value of such displacement at the member yielding.

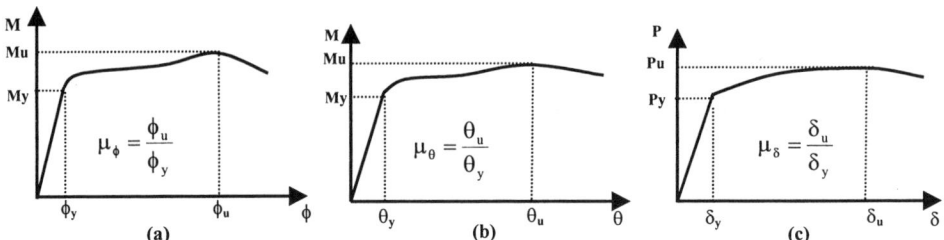

Figure 1 (a). Curvature Ductility, (b) Rotational Ductility, (c) Displacement Ductility

Previous studies [4, 8] had shown that the ductility of steel structural member are significantly reduced by the instability occurrence in the section or member. The critical loads due to local buckling of the I-shaped steel section and lateral-torsional buckling of the member are governed by the flange width-thickness ratio, web width-thickness ratio, and beam slenderness ratio. To ensure the ductility performance of the structure via the flexural-yielding mode of the specified member, the AISC Seismic Provisions [1] limits those ratios stricter than the ones specified in AISC-LRFD Specification [2]. For I-shaped steel member, those limits are the flange width-thickness ratio, $b/2t_f \leq 52\sqrt{f_y}$, the web width-thickness ratio, $h/t_w \leq 520\ [1-1.54P_u/P_y]/\sqrt{f_y}$ (for member with less dominant compression), and the member slenderness ratio, $\lambda = L/r_y \leq 2500/f_y$ (for member of a special moment resisting frame); where : b = overall flange width (in.), t_f = flange thickness (in.), h = web height (in.), t_w = web thickness (in.), L = laterally unsupported length of beam, r_y = radius of gyration with respect to weak-axis, f_y = nominal yield stress (ksi), P_u = factored compression force (lbs), and P_y = nominal yield force (lbs).

PARAMETRIC STUDY

The influences of those parameters on ductility of steel member, were investigated on an I-shaped steel beam subject to a uniform bending moment. Parametric study was conducted by both numerical work using a non-linear finite element program (NASTRAN) and experimental work on a number of member specimens which are considered for seismic resistant structures.

a. Numerical Work

Numerical work was conducted on different steel sections, different member lengths and two different steel constitutive models. The first model used the commonly assumed stress-strain relationship with the value of nominal yield stress, tensile stress, modulus of elasticity, and modulus in hardened region are 240 MPa, 370 MPa, 200 GPa and 160 GPa respectively. The second model used stress-strain relationship obtained from the laboratory tensile-testing which showed higher strength yet lower elongation. Both geometric and material non-linearity were considered in the analysis. The yielding was modeled by the von Mises yield criteria for associative flow-rule with isotropic hardening. The ductility ratios were calculated based on the displacement attained at the beam ultimate strength.

The member was represented by shell-elements in each cross-section along the member-length with aspect ratio of approximately equals 1.0 as shown in Figure 2. The shell element is a quadrilateral isoparametric element with four nodes and three translational degrees of freedom per-nodes with two rotational degrees of freedom. This enables the program to detect instability due to local buckling as well as member buckling.

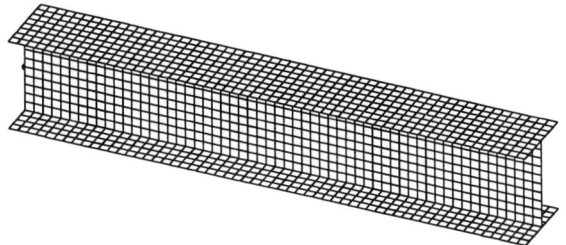

Figure 2 : Modeling beam member by shell element

b. Experimental work

Two-point bending test was conducted on a number of I-shaped steel beam subject to incremental loading applied by a Loading Frame as shown in Figure 3. Various steel member sections tested in the laboratory are from mild-steel with the average yield-stress, f_y = 385 Mpa and the average tensile stress, f_u = 480 MPa. Three sections were used in the entire tests: IWF 100x100x6x8, 150x75x5x7 and 200x100x4.5x7 (dimension in milimeter). In addition, the laterally-unsupported length (L) of the beam as well as the distance between the point loads were varied to represent different member slenderness ratios. The beam curvature, ϕ, was calculated from the data of strain-gauges mounted on the top and bottom flanges at the beam mid-span. The beam rotation, θ, was measured using the inclinometer, and the displacement, δ, was measured by LVDT at the beam midspan.

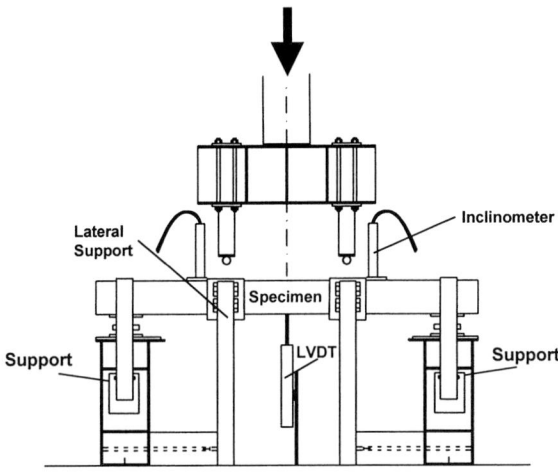

Figure 3 : Testing set-up of beam specimen

RESULTS AND DISCUSSIONS

Three curvature ductility relationships for various R_f (=$b/2t_f$) were obtained from the numerical analysis using two different constitutive equations and from the testing results (Figure 4). The first equation prescribed by the nominal values gives lower ductility ratios than the others. It confirms the use of assumed constitutive equations with nominal values for determining the level of ductility of steel member. Figure 4 is also shown that all relationships well describe the curvature ductility, μ_ϕ, to be proportional to $(R_f)^{-2}$ or $(1/R_f^2)$.

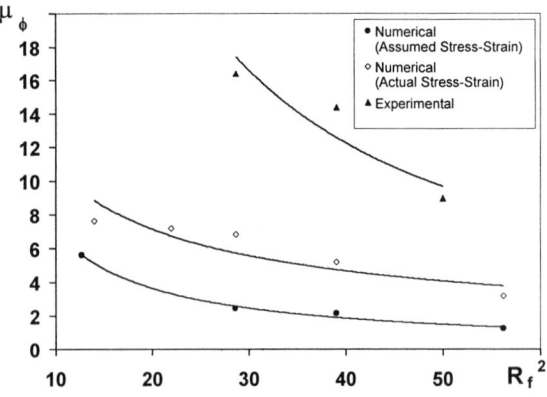

Figure 4 : Numerical and experimental result of curvature ductility ratio

The variations of curvature ductility ratio with respect to the flange width-thickness ratio ($R_f = b/2t_f$) for different member slenderness ratio ($\lambda = L/r_y$) are shown in Figure 5 and 6. The variation of ductility ratio with respect to R_f is found to be insensitive for $\lambda \geq 40$, and becomes more sensitive for more compact sections ($\lambda < 40$). For more slender member ($\lambda \geq 40$) the ratio is considered constant and close to 1.0. Figure 6 is also shown that all relationships well describe the curvature ductility, μ_ϕ, to be proportional to λ^{-1}. The curvature ductility ratio more significantly increases as λ or R_f approaches zero.

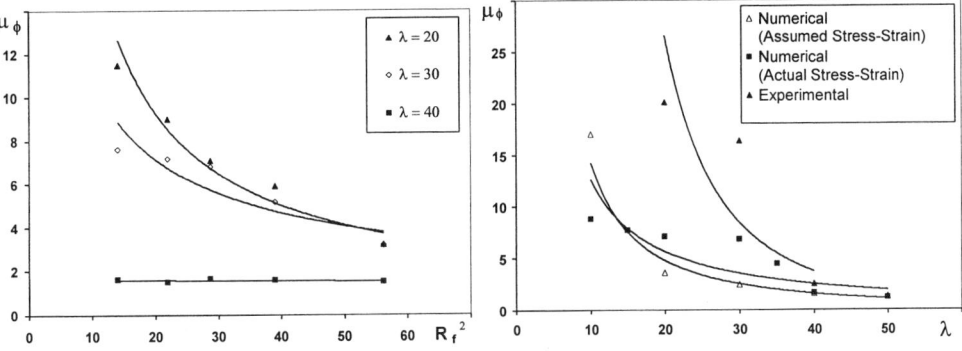

Figure 5 : Curvature ductility ratio, μ_ϕ vs R_f^2

Figure 6 : Curvature ductility, μ_ϕ vs λ

The curvature ductility of I-shaped steel member is more sensitive to the flange width-thickness ratio than to the web width-thickness ratio, R_w, as shown in Figure 7. Although the moment capacity increases for section with thicker web and constant flange thickness, the member could posses lower ductility since the flexural yielding of such section is resisted mostly by the yielding of the flange and not the web.

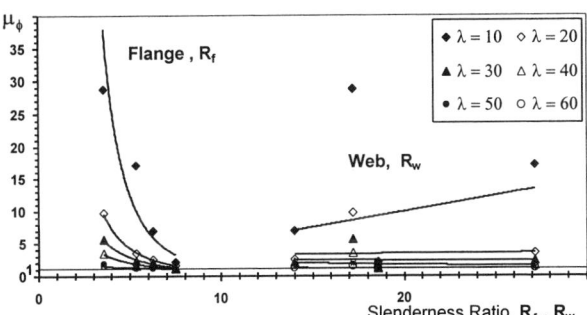

Figure 7 : Numerical results of curvature ductility, μ_ϕ vs R_f and R_w

As far as the strength of the member is also concerned, the changes of curvature ductility ratio and ultimate bending moment due to variation on R_f and λ were presented in Figure 8. The ordinate describes M^*/M_y, where M^* is the ultimate bending strength determined by yielding or buckling of the member, and M_y is the yield moment. The curvature ductility ratio is shown to be more sensitive than the over-strength of bending moment. This result confirms the importance of the ductility ratio that should be quantified more accurately for design purposes such that the single-valued design parameter specified in the Code could be more appropriately used.

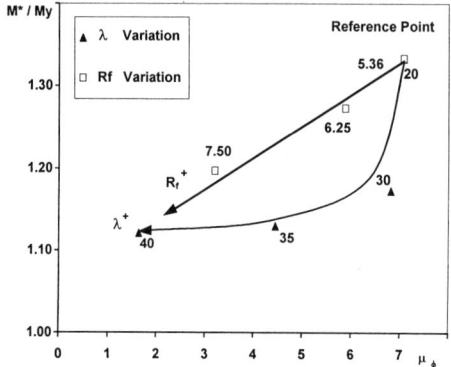

Figure 8 : Changes of curvature ductility ratio, μ_ϕ and ultimate bending moment, M^*/M_y due to variation in R_f and λ

DUCTILITY FORMULATIONS

Figure 5 and 6 shows the influences of parameter R_f^2 and λ on the curvature ductility, which are well described as a function of $(1/R_f^2)$ and $(1/\lambda)$. Since the member slenderness ratio is found to be a more dominant factor of ductility ratio, a normalized value is used by eliminating the effects of steel grade, $\overline{\lambda} = \lambda/(\pi\sqrt{E/f_y})$. Using two parameters R_f and $\overline{\lambda}$, a simplified curvature ductility ratio is proposed as follows:

$$\mu_\phi = \frac{30}{\overline{\lambda} R_f^2} \qquad (1)$$

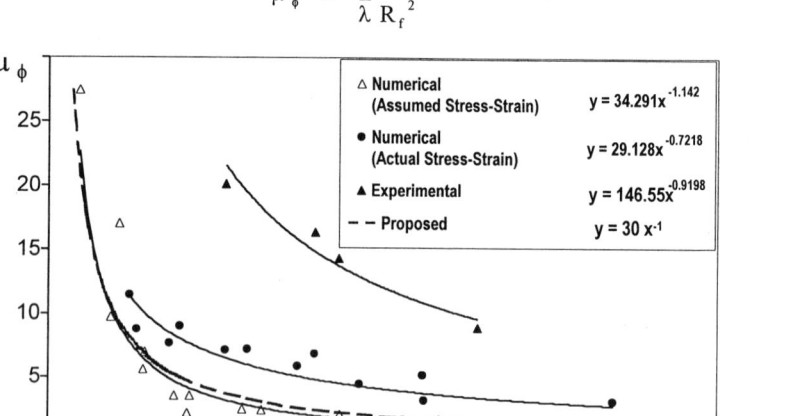

Figure 9 : Results for curvature ductility

The similar results were also obtained for the rotational ductility ratio and the displacement ductility ratio. Both ratios could be well described to be proportional to $(R_f)^{-2}$ and λ^{-1}. The rotational ductility ratio, μ_θ, and the displacement ductility ratio, μ_δ, are proposed by simplified formula as follows:

$$\mu_\theta = \frac{22}{\bar{\lambda} R_f^2} \quad (2)$$

$$\mu_\delta = \frac{22}{\bar{\lambda} R_f^2} \quad (3)$$

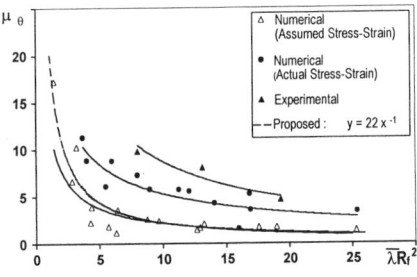 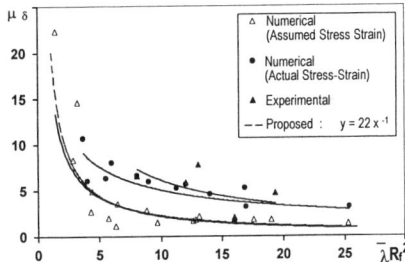

Figure 10 : Results for rotation ductility   Figure 11 : Results for displacement ductility

The simplified formulation could be used to quantify the ductility provided by various dimensions of I-shaped steel member as compared to the ductility assumed in the seismic resistant design specification. The formulation could also be further extended to include more ductility parameters and to cover very compact sections where the ductility ratio would become more sensitive to the change of R_f^2 and λ.

CONCLUSIONS

Numerical and experimental results to study some ductility parameters of I-shaped steel member have been presented:
1. The assumed stress-strain relationship using nominal values of yield-stress and tensile-stress of steel gives more conservative ductility ratios compared with the ones obtained by using the values from laboratory tensile-test results.
2. For fairly compact I-shaped steel section specified for seismic resistant member, the curvature ductility ratio decreases inversely with respect to the member slenderness ratio, λ, and to the square of the flange width-thickness ratio, R_f^2. As λ or R_f approaches zero, the curvature ductility increases more significantly. This is also true for rotational and displacement ductility ratios.
3. Simplified formulas to quantify curvature ductility as well as rotational ductility and displacement ductility of I-shaped steel member are proposed as a function of the member slenderness ratio and flange width-thickness ratio.

ACKNOWLEDGEMENT

The authors wish to thank URGE Project – Ministry of National Education the Republic of Indonesia for providing financial support through the University Research for Graduate Education (URGE) Project Batch IV 1999-2001, and the graduate students involved in this research project.

REFERENCES

1. American Institute of Steel Construction (1997). *Seismic Provisions for Structural Steel Buildings*, AISC, Chicago, USA.
2. American Institute of Steel Construction (1994). *Load and Resistance Factor Design*, AISC, Chicago, USA.
3. Bruneau, M. ,et.all (1998). *Ductile Design of Steel Structures*, McGraw Hill, New York,USA
4. Earls, C.J. (1999) On the Inelastic of High Strength Steel I-shaped beams, *Journal of Constructional Steel Research* **49**, 1-24.
5. Gioncu,V.(2000). Framed structures.Ductility and seismic response. General Report. *Journal of Constructional Steel Research* **55**, 125-154.
6. International Council of Building Officials (1997). *Uniform Building Code*, Whittier,CA
7. Miranda,E. and Bertero,V.V (1994). Evaluation of Strength Reduction Factors for Eathquake-resistant Design, *Earthquake Spectra* **10:2**, 357-379.
8. Moestopo,M. (2000) *On Ductility Characteristic of Steel Structural Members*, URGE Project Seminar, Jakarta, Indonesia.
9. Nakashima,M (1994). Variation of Ductility Capacity of Steel Beam-Columns. *Journal of Structural Engineering* **120:7,** 1941-1960.

SEISMIC ANALYSIS OF MR STEEL FRAMES BASED ON REFINED HYSTERETIC MODELS OF CONNECTIONS

G. Della Corte[1], G. De Matteis[1], R. Landolfo[2], F. M. Mazzolani[1]

[1]Department of Structural Analysis and Design, University of Naples "Federico II"
P.le V. Tecchio, Napoli 80125, Italy
[2]University of Chieti "Gabriele D'Annunzio", V.le Pindaro 42, Pescara 65127, Italy

ABSTRACT

Evaluation of the seismic performance of MR steel frames is usually carried out assuming the elastic-perfectly-plastic type of hysteresis model for plastic zones (in particular for connections). This type of model is associated with conventional ultimate plastic rotations giving indication of the maximum deformation beyond which strength degradation is likely to occur. This methodology leads to conservative results and the elastic-perfectly-plastic type of hysteresis model is definitely adequate to obtain reliable information on deformation demands when the above maximum values are not exceeded. However, the safety of the frame at 'collapse' remains unknown. The analyses presented in this paper show that safety margins on the global collapse of the structure can be predicted only by considering more realistic hysteresis behaviours, i.e. basing the numerical results on models able to take into account strength degradation. Moreover, it is shown that the design of conventional steel building systems (like the fully welded highly redundant frame schemes) according to EC8 leads to strongly over-resistant structures, due to the limitation of story drift angles for non-structural damage control under frequent earthquakes. It is also shown that this result is unavoidable due to the indications provided by the European seismic code for admissible story drift angles associated to non-structural elements and for the assumed seismic base shear force demand under frequent earthquakes. Finally, aiming at obtaining more rational structures, i.e. structures satisfying both the serviceability and ultimate limit states with the same level of confidence, some possible solutions are proposed and discussed.

KEYWORDS

Collapse, Connections, Damage, Modelling, Moment Resisting Frames, Seismic Behaviour, Steel Structures, Structural Codes.

INTRODUCTION

There is complete agreement, within the international scientific community, about the importance of the explicit modelling of connections in order to have a realistic and reliable prediction of the global frame behaviour, both under static and dynamic external actions. Many papers have shown also that the panel zone of the column web at the intersection of columns and beams may significantly affect stiffness, strength and ductility of the whole frame. Thus, MR frames have to be conceived as the assemblage of three main structural components: beams, columns and joints (Bruneau et al., 1998). There is a large agreement about the most significant parameter for measuring damage of the frame, too. It is the story drift angle (Δ/h), i.e. the ratio between the displacement Δ of one floor relative to the adjacent lower one and the story height h. The story drift angle could be related, through a simplified approach, to local plastic deformations (Gupta et al., 2000). On the contrary, there is no complete agreement about the correct methodology for evaluating frames seismic performance. We have to recognise that this is a major and more general problem, which involves evaluation and analysis of every structural system. There are both seismological and structural aspects that can significantly affect the frame seismic performance.

From the seismological point of view, the main difficulty arises from a still partial understanding of the parameters that should be used for measuring the ground motion damage potential. Currently, the most dependable parameter we can use is the spectral elastic pseudo-acceleration ($S_{a,e}$, usually computed with reference to a 5% damping ratio and for the first vibration mode of the structure) normalised by means of the gravity acceleration (g). Thus we obtain the ratio between the maximum base shear that would act on the structure if it behaves linearly elastic and the seismic weight. This is the parameter usually adopted by seismic codes for establishing seismic hazard at a given site.

From the structural point of view, one major uncertainty is related to the type of hysteresis model used in the numerical analysis. In fact, looking at the experimental results, it is apparent that the usually assumed elastic-perfectly-plastic hysteresis model is far from faithfully representing the cyclic response of connections. Thus, it seems proper to wonder which is the effect of the type of hysteresis model adopted in the numerical analysis of the whole frame on the predicted response at different performance levels.

A synthetic picture of the frame seismic performance can be obtained by relating the ground motion damage potential ($S_{a,e}/g$) and the maximum damage (maximum Δ/h ratio computed during the time history throughout the building height) produced at that level of earthquake intensity. In the following, this type of relationship will be referred to as "performance curve". This curve is usually obtained by scaling a given ground acceleration time-history at increasing values of the peak ground acceleration and computing the relevant maximum story drift angle. By fixing limit values to the maximum story drift angle, in relation to the selected limit state (serviceability, damage control, structural collapse), it is possible to evaluate the earthquake intensity ($S_{a,e}/g$) inducing that damage level and compare this intensity with that fixed by the code, for that limit state and in relation with the site seismic hazard. It could be useful to remark that if the two parameters, which have been chosen as representative of seismic damage potential on one hand and of frame actual damage on the other hand, are the correct ones, then an increase of the former should induce an increase of the latter. In other words, we should expect a monotonically increasing relationship between the ground motion damage potential parameter and the structural damage parameter. As it will be shown, this not always happens. The following numerical analysis will show that, sometimes, an increase in the normalised spectral acceleration is associated with a constant or even decreasing maximum story drift angle. We must conclude that, in those cases, the parameters chosen are not completely representative either of the earthquake damage potential or of the structure damage sensitivity.

The "performance curve", as above defined, is a very useful tool, by which it is possible to highlight the influence on the predicted seismic response both of the hysteresis model adopted and of the seismic input chosen. This is shown in the following, where some numerical analyses are presented and discussed.

EXPERIMENTAL BEHAVIOUR AND MATHEMATICAL MODELLING OF BEAM-TO-COLUMN CONNECTIONS

A number of beam-to-column connection types are currently available in practice and their mechanical response is strongly different from each other, both under monotonic and cyclic loading conditions. Connections exhibit hardening not only during the monotonic loading phase but also during cyclic loading at a given deformation amplitude. Furthermore, under cyclic loading conditions, geometrical damages (changes in the geometry of the connection due to plastic deformations) induce a reduction of stiffness and strength. Finally, in all types of connections fatigue-related damages induce a reduction of the available ductility, so that at the end of a given deformation history the residual plastic deformation capacity of the connection could be very small.

The first major step towards the development of reliable methodologies for evaluation of MR steel frames seismic performance is the development of mathematical models able to capture correctly the connection hysteretic behaviour. The model developed by the authors is semi-empirical. It is based on some parameters to be calibrated by the comparison with experimental results. The main features of the model are discussed in details in Della Corte et al. (2000), where it is also shown, on the basis of existing experimental results, that the proposed model is able to represent quite well both hardening, softening and pinching phenomena. It is essential to remark that it has been developed looking at the need to understand the effects of the model itself on the predicted response of the whole frame. Therefore, it has not predictive aims, but its main quality is flexibility and simplicity for the implementation in numerical codes.

Discussion is necessary about some assumptions made within the model, in particular with reference to cyclic local buckling effects on the moment-rotation relationship of plastic hinges. Cyclic buckling phenomena occurring in the plastic hinges induce strength degradation, accompanied by negative stiffness branches in the moment-rotation relationship. The accurate simulation of the hysteretic behaviour in this case is quite difficult, mainly because the point of beginning of the softening branch moves during the deformation history. In the numerical analyses presented in this paper, a simplified approach has been undertaken. It has been assumed that, during the generic deformation history, the rotational stiffness becomes negative (i.e. a softening branch in the moment-rotation relationship takes place) when the plastic excursion in one direction becomes equal to the value of the rotation producing local instability in case of monotonic loading. In the following, this limit value of rotation will be referred to as "softening rotation". After its first activation, the softening branch will remain fixed defining a boundary for the moment-rotation curve. Internal flexural strength will be decreasing along the softening branch, being equal to the value reached on this branch. In order to simulate the cyclic strength degradation effect, another type of degradation rule has been superimposed to the one related to the reaching of excessive plastic excursions in one direction. This cyclic degradation law is based on the use of a "damage factor" β very similar to the one introduced by Park and Ang (1985). Denoting by M_0 the initial (undamaged) conventional plastic strength of the connection, by φ_{ult} the conventional ultimate rotation for monotonic loading conditions and by E_h the dissipated plastic energy, the reduced value of plastic strength ($M_{0,red}$) is given by:

$$M_{0,red} = M_0(1-\beta) = M_0(1- E_h/M_0\varphi_{ult}) \qquad (1)$$

Then, strength degradation is obtained by the combination of the two damage rules, the one related to the excessive maximum excursion and the one related to the dissipated energy. Such a model is deemed to be adequate in case of few numbers of plastic excursions of large amplitude. Some improvements are currently under development, in order to obtain a more general and robust representation of the cyclic nature of local buckling during seismic action, when rotational stiffness may become negative due to the repetition of small plastic rotations.

A STUDY CASE

The Frame under Study

Some numerical analyses have been performed with reference to a regular plane frame extracted from a regular plan civil building. The frame has been designed according to EC3 and EC8. In particular, sub-soil class B (medium density sand), peak ground acceleration equal to 0.20g and a design q-factor equal to 6 were assumed in the design procedure (De Matteis et al., 1998). The story drift angle capacity at the serviceability limit state has been supposed to be equal to 0.006 rad. The stiffness requirement has conditioned the design, such that the final solution is significantly over-resistant. It may be useful to remark that beam-to-column joints were assumed to be full-strength and rigid in the design procedure, even if their mechanical properties have subsequently changed in the analysis phase, considering also semi-rigid and partial-strength connections.

Figure 1 shows the geometry of the analysed frame and the different modelling assumptions made in the evaluation phase. As can be seen from Figure 1, column web panels have been assumed to be always rigid and resistant enough to remain elastic. Beam-to-column connections have been modelled by means of lumped elasto-plastic springs. Two types of connections have been considered:

1 Rigid and full-strength connections characterised by moment-rotation relationships with strength degradation.
2 Semi-rigid and partial-strength connections characterised by moment-rotation relationships with or without pinching of hysteresis loops.

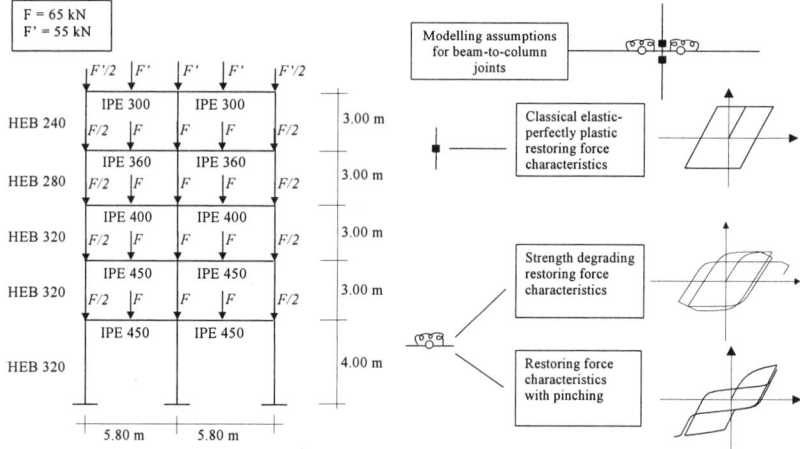

Figure 1: The frame under study and the modelling assumptions

The Seismic Input

Three ground acceleration time-histories have been selected for the numerical analyses to be carried out. The first two of them (Rinaldi record from Northridge earthquake, 1994, and JMA record from Kobe earthquake, 1995) have been chosen as representative of near-fault recordings. The last one (Llollelo record from Chile earthquake, 1985) has been chosen as characteristic for long duration strong earthquakes. Accordingly, the first two acceleration records are characterised by a relatively small value of the Trifunac duration (7÷8 s) with respect to Llollelo record (36 s) (see Table 1, Elgamal et al. 1998). This criterion of subdivision derives from the intention to emphasise the effect of earthquake duration on the seismic performance when strength degradation is taken into account. It is apparent, in fact, that simulating cyclic strength degradation might be more important for long duration earthquakes than for pulse-type ones.

TABLE 1
FUNDAMENTAL CHARACTERISTICS OF THE CHOSEN EARTHQUAKES

Earthquake	Magnitude	Record name	a_{max} (g)	Arias Intensity (cm/s)	Predominant Period (s)	Trifunac duration (s)
Northridge (17 January 1994)	6.7	Rinaldi 228	0.84	14.86	1.36	7.04
Kobe (17 January 1995)	6.9	JMA NS	0.83	71.48	0.69	8.36
Chile (3 March 1985)	8.0	Llolleo 10	0.71	46.16	0.53	35.91

ANALYSIS OF NUMERICAL RESULTS

General

In the analysis of the structural response, the acceleration records have been scaled with peak ground accelerations starting from $0.2PGA_r$ and with increasing step equal to $0.2PGA_r$, where PGA_r is the recorded peak ground acceleration. The generic 'performance curve' (elastic spectral acceleration at 5% damping and at the first period of elastic vibration vs. maximum story drift angle) has been stopped when dynamic instability occurred. This condition is deemed to be representative of the frame 'collapse', i.e. the achievement of a limit situation where the structure capacity of bearing vertical loads is lost (Hamburger et al., 2000), or unacceptably reduced. However, as will be shown, when non-degrading hysteresis models are adopted, the "performance curve" is still increasing for very high values of the story drift angle and/or spectral acceleration. In those cases the analyses have been stopped when a very large story drift angle and/or spectral acceleration occurred.

The 5% damped elastic spectral pseudo-acceleration given by EC8 for the ultimate limit state of the frame under investigation is indicated in the next sections by $S_{a,e}$(ULS-EC8) (the return period of this spectral acceleration is equal to 475 years).

In the following some moment-rotation relationships computed during the integration of time-histories will be also shown. They refer to a pre-selected beam-to-column connection, corresponding to the one located between the first column on the left of the frame and the beam at the first floor of the frame. These curves give to the reader a clearer picture of the connection hysteretic behaviours assumed in the numerical analyses.

Continuous Frame with Strength Degrading Beam-to-Column Connections

The first examined case is that of rigid and full-strength connections. The effects of strength degradation related to local softening phenomena and to cyclic fatigue-related damage phenomena are here separately investigated.

Figure 2 relates the influence on the computed seismic performance of the strength degradation concerned with local softening phenomena, i.e. with the achievement of large plastic deformation excursions leading to a negative rotational stiffness. Two values of the 'softening rotation' are taken into account: 0.03 and 0.01 rad. Figure 2d illustrates the moment-rotation history computed for the chosen reference connection, in case of JMA record with a peak ground acceleration equal to 1.6 times the historical value. This figure refers to the case of a softening rotation equal to 0.01 rad.

The first thing evident from Figure 2 is that the "performance curve" is insensitive to the type of model for relatively small values of the drift demand (until 0.02÷0.03 rad is achieved). 0.02÷0.03 rad is exactly the story drift angle usually adopted as ultimate deformation capacity in conventional numerical analyses based on elastic-perfectly-plastic types of models. From Figure 2 it can be also observed that at larger drift amplitudes the correct evaluation of the frame performance should be necessarily based on realistic hysteretic models, that is on models able to consider local softening. It is interesting to notice that at the seismic damage potential level equal to $S_{a,e}$(ULS-EC8) the drift demand is around 0.01 rad, and at $3S_{a,e}$(ULS-EC8) the demand is around 0.02÷0.03. This result is the necessary consequence of the design methodology, as will be deepened in the following.

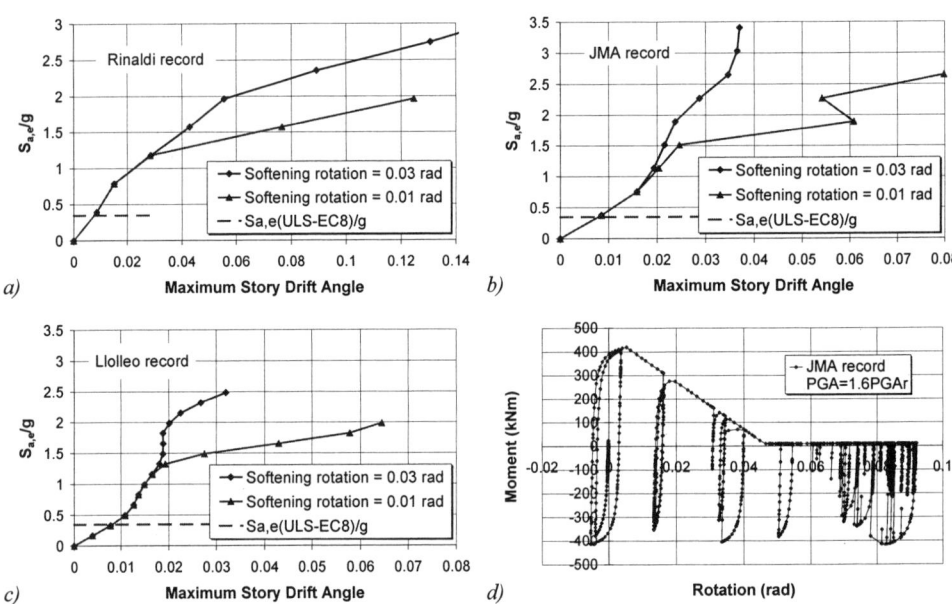

Figure 2: Effect of local softening phenomena on the seismic performance of the whole frame

Figure 3 relates the influence of the cyclic strength degradation, showing the modifications of the frame "performance curve" with increasing values of the damage factor β. In particular, a damage factor equal to 0.025 indicates the practical absence of strength degradation, while β = 0.30 indicates a very strong strength degradation. Figure 3d provides the hysteretic behaviour computed for the reference connection in case of Llollelo record scaled up to 1.8 times the recorded peak ground acceleration and with reference to a damage factor equal to 0.15.

It is again apparent that strength degradation may strongly affect the frame seismic performance at the ultimate limit state (collapse). In fact, the performance curve is quite insensitive to the level of strength degradation when low levels of drift demand are involved, while the type of model becomes very important for assessing the safety at larger drift demands. Once again, it can be noticed that safety at collapse is quite large in Europe (compare the performance curve with $S_{a,e}$(ULS-EC8)), when low levels of strength degradation are considered. It is also interesting to notice the different impact that cyclic strength degradation can have in relation with the earthquake duration. In fact, in case of Llollelo record the drift limit at which modelling of strength degradation is important is around 0.02 rad, while this drift limit is around 0.05÷0.06 rad for shorter pulse-type earthquakes, like the Kobe and Northridge ones.

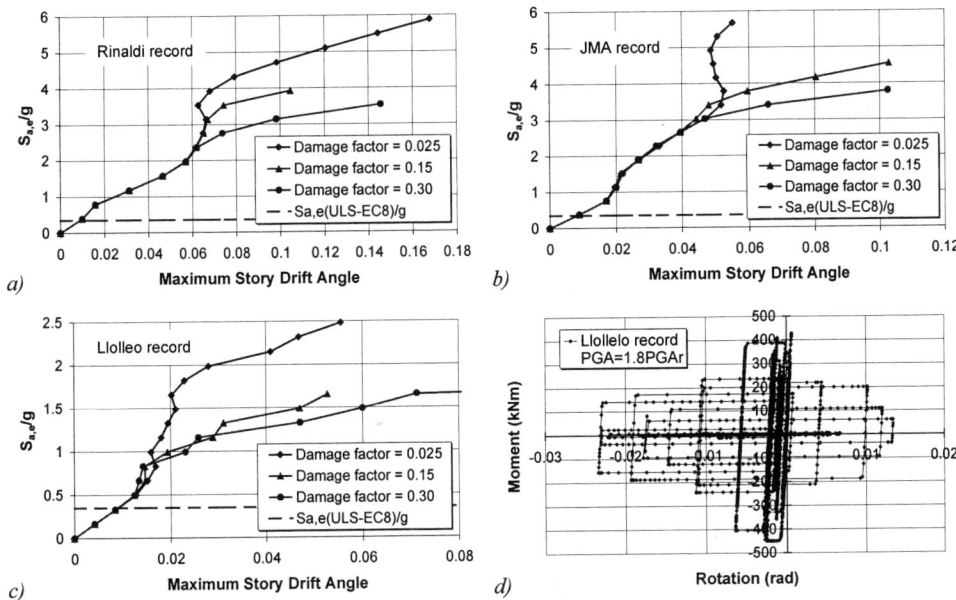

Figure 3: Effect of cyclic fatigue-related strength degradation phenomena on the seismic performance of the whole frame

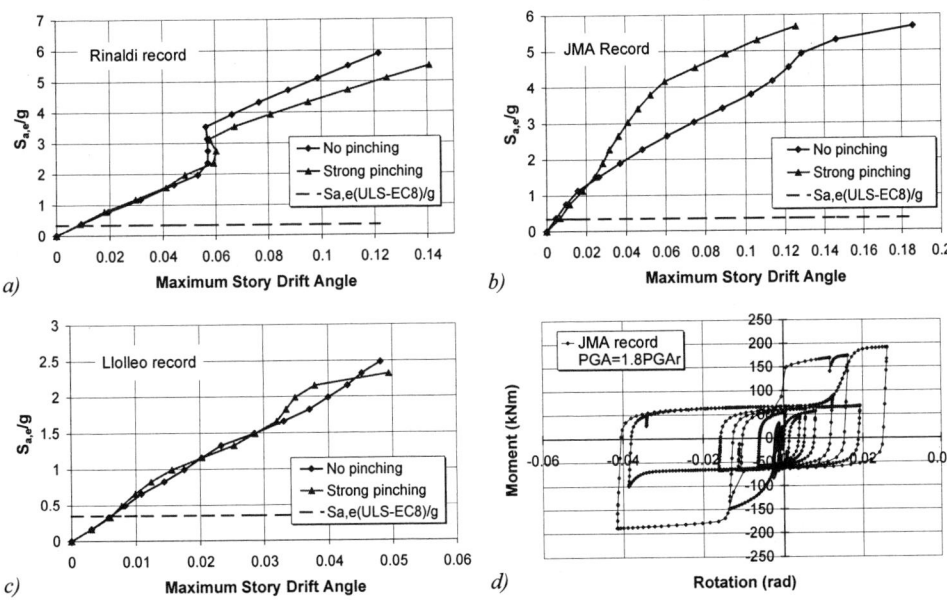

Figure 4: Effect of pinching of hysteresis loops on the seismic performance of the whole frame

Semi-Continuous Frame with Beam-to-Column Slipping Connections

The seismic performance of the above frame, without changing the dimensions of structural members, is also investigated in case of semi-rigid and partial-strength connections. In particular, the ratio between connection and beam flexural stiffness has been assumed equal to the half of the limit value related, according to EC3, to the boundary with rigid connections, while the moment capacity of the connection has been hypothesised equal to the half of the beam plastic strength. Since this type of connections usually exhibit hysteresis loops with a slip-type behaviour, the influence of the modelling of pinching on the computed frame performance is analysed. Two types of hysteresis models are considered, assuming the same skeleton curve but a different cyclic behaviour (with and without pinching of loops). Figure 4*d* illustrates the moment-rotation relationship computed for the reference connection when analysing the structure under the JMA ground acceleration time-history scaled up to 1.8 times the historical peak ground acceleration.

Unfortunately, based on these few numerical results, it is not possible to determine a general trend for the effects of pinching. In fact, in case of Rinaldi and Llollelo records, there is a very small impact of pinching on the required drift values for every level of spectral acceleration. On the contrary, in case of JMA record, the model with pinching of loops produces a strong reduction of structure drift demands. Based on these results, the conclusion could be that pinching is not a dangerous phenomenon, the story drift angle demands being practically insensitive or even decreasing with the level of pinching assumed in the hysteresis model. Moreover, the results presented are based on the adoption of non-degrading hysteresis models. Thus, accordingly to observations made in previous sections, it is not possible to delineate, with a sufficient confidence, the performance at collapse of the semi-continuous frame. However, it can be noticed that under the EC8 prescribed seismic intensity at ultimate limit state ($S_{a,e}$(ULS-EC8)) the maximum story drift angle demand is around 0.01 rad, which may be considered as a level of deformation easily sustainable by this type of connections.

CONCLUSIONS AND FURTHER DEVELOPMENTS

Seismic evaluation of MR steel frames is usually carried out using simplified hysteresis models, such as the elastic-perfectly-plastic type, which are associated to the assumption of story drift angle capacities having the meaning of upper limits beyond which strength degradation is likely to take place. In this way, the ultimate limit state of the frame is practically identified with the achievement of the level of story drift angle for which the numerical validity of the model itself is lost. Numerical results shown both in current paper and in a previous one (Landolfo et al., 2000) demonstrate that for low levels of story drift angle demand the type of model used in the numerical analysis is definitely not influent. As a consequence, the adoption of simplified hysteresis models for levels of story drift angles demand lesser than those corresponding to a first significant appearing of strength degradation seems to be adequate to obtain reliable information on the expected drift demand.

However, if simplified hysteresis models are used, the 'real' safety at collapse of the building remains unknown. The numerical results shown in this paper demonstrate that the elastic-perfectly-plastic type of model (without strength degradation) is absolutely not adequate to obtain this information. Using realistic hysteresis models allows also the shift of the attention towards a more physically meaningful ultimate limit state, that corresponding to the achievement of a limit situation where the building capacity of carrying vertical loads is lost, or unacceptably reduced.

The methodology for the evaluation of the frame seismic performance is still an open question. The use of elastic spectral pseudo-acceleration as ground motion damage potential parameter and of maximum story drift angle as damage measure for the frame gives rise to not always monotonically increasing relationships. It has been noted that this inconsistency disappears when more realistic types of hysteresis models are taken into account. Anyway, this aspect deserves further investigation.

Besides, the obtained numerical results demonstrate that, for typical steel building systems (such as the fully welded tri-dimensional highly redundant frame schemes), the conventional ultimate limit state (achievement of story drift angle demands around 0.03 rad) is far from being reached under the design earthquake. In fact, the normalised spectral elastic pseudo-acceleration corresponding to drift demands of 0.03 rad has been found to be always about 3 times the codified value $S_{a,e}(ULS-EC8)/g$. This result is well expected and could be mathematically established through a simple reasoning. To this purpose, in Figure 5, a generic push-over curve for a multi-story frame, relating the base shear seismic coefficient ($C = V/W$, V being the total base shear and W the seismic weight) versus the maximum story drift angle (Δ/h, Δ being the displacement of one floor relative to the adjacent lower one and h the story height) is considered (full line curve). It is assumed that the maximum story drift angle producing the ultimate limit state is $\Delta_c/h = 0.03$, while the admissible inter-story drift under frequent earthquakes is $\Delta_s/h = 0.005$. In Figure 5, $C_{e,u}$ indicates the base shear seismic coefficient assumed by the code for the ultimate limit state, while $C_{e,u}/q$ is the design base shear. Under the latter level of lateral force the system ductility should be equal to q (the approximation $\mu = q$ is made). Consequently, the story drift angle demand under the design base shear will be around $0.03/q$, that is equal to $0.03/6 = 0.005$ for $q = 6$. This is exactly equal to the assumed admissible story drift angle under frequent earthquakes. On the other hand, according to EC8, the base shear requirement under frequent earthquakes is $C_{e,u}/v$, where the suggested value for v is about 2. Thus, re-applying the ductility approximation, the drift requirement under frequent earthquakes, for the structure optimally designed at the ultimate limit state, will be $0.03/v = 0.015 >> 0.005$! It is now necessary for the designer to re-design the structure providing a higher stiffness, in order to satisfy the story drift angle limitation under frequent earthquakes. The simplest way of stiffening the structure is by increasing the dimensions of the structural members. But, this will induce an increase of the system strength and the final design solution will be strongly over-resistant. Looking at Figure 5, it can be seen that the ultimate story drift angle (0.03 rad) will be consequently reached, for the final design solution, under a base shear demand equal to $q/vC_{e,u} = 6/2C_{e,u} = 3C_{e,u}$.

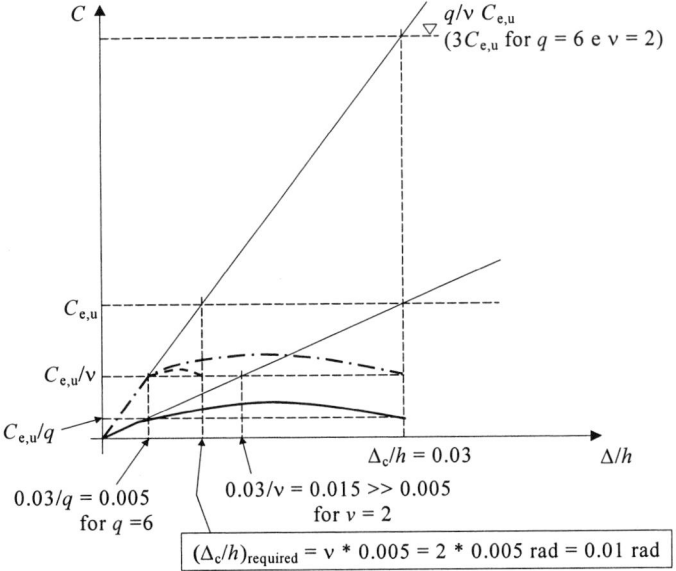

Figure 5: Some features of codified design rules of EC8

Therefore, either there is something wrong in the codified design rules or we have to pursue different design strategy implementing new construction technologies. For example, it might be supposed that the specified "admissible" levels of story drift angle for damage control of non-structural elements under frequent earthquakes are too small and/or the specified level of strength demand for frequent earthquakes is too large. Assuming that both these two parameters are correctly specified by EC8, an optimal design solution (i.e. a structure satisfying both the ultimate and the serviceability limit states with the same level of confidence) could be achieved in two ways. One solution is the stiffening of the structure without significantly increasing the system strength, which could be achieved by using the stiffening effect of cladding panels (De Matteis et al., 1998). A second solution is the adoption of less ductile structural systems. In this direction, the dashed line in Figure 5 shows that an ultimate limit state story drift angle capacity equal to 0.01 rad could be adequate for European seismic regions. In particular, the second solution could be pursued by adopting member sections characterised by low ductility classes (for instance thin-walled cold-formed members), or by using partial-strength and partial-ductile connection types (bolted connections). The latter have been already proposed by several Authors (Astaneh, 1994; Leon, 1995; Kasai et al., 2000). It is important to emphasise that in all the above cases the proposed solutions could lead to a remarkable reduction of the cost of the structure, giving more competitiveness to steel constructions. On the other hand, it should be considered that the adoption of new a-seismic building technologies, which are not based on a very long experience like the traditional solutions, requires a very careful examination. In particular, the availability of rational procedures for reliably evaluating the structural behaviour up to collapse is needed.

REFERENCES

Astaneh-Asl, A. (1994). Seismic Behaviour and Design of Semi-Rigid Steel Frames. *Proc. of STESSA 1994 – International Conference on the Behaviour of Steel Structures in Seismic Areas*, F. Mazzolani and V. Gioncu (eds.), E&FN Spon, Chapman & Hall, London.
Bruneau M., Uang C-M. and Whittaker A. (1998). *Ductile Design of Steel Structures*. McGraw Hill.
Della Corte G., De Matteis G. and Landolfo R. 2000. Influence of Connection Modelling on Seismic Response of Moment Resisting Steel Frames. In F.M. Mazzolani (ed.), *Moment resistant connections of steel building frames in seismic areas*. London, E & FN SPON.
De Matteis G., Landolfo R. and Mazzolani F.M. (1998). Dynamic Response of Infilled Multi-Story Frames. *Proc. of XI European Conference on Earthquake Engineering*, Paris 6-11 Sept. CD-ROM.
Elgamal A., Ashford S. and Kramer S. (eds.) (1998). Proc. of the 1st PEER Workshop on Characterization of Special Source Effects, *University of California, San Diego 20-21 July 27-28*.
Park Y.-J. and Ang A.H.-S. (1985). Mechanistic Seismic Damage Model for Reinforced Concrete. *Journal of Structural Engineering* **111**:4, 722-739.
Gupta A. and Krawinkler H. (2000). Relating Ground Motion Spectral Information to Structural Deformation Demands. *Proc. of STESSA 2000 – International Conference on the Behaviour of Steel Structures in Seismic Areas*, F. Mazzolani & R. Tremblay (eds.), E&FN Spon, Chapman & Hall, London, 687-694.
Hamburger R.O., Foutch D.A. and Cornell C.A. (2000). Performance Basis of Guidelines for Evaluation, Upgrade and Design of Moment-Resisting Steel Frames. *Proceedings of the 12th World Conference on Earthquake Engineering*, Oakland, New Zealand, 30 January - 4 February.
Kasai K., Xu Y. and Mayangarum A. (2000). Experiment and Analysis of Bolted Semi-Rigid Beam-Column Connections Part I: Cyclic Loading Experiment. *Proc. of STESSA 2000 – International Conference on the Behaviour of Steel Structures in Seismic Areas*, F. Mazzolani & R. Tremblay (eds.), E&FN Spon, Chapman & Hall, London, 199-206.
Landolfo R., Della Corte G. and De Matteis G. (2000). The Effect of Connection Hysteretic Behaviour on Seismic Damage to MR Steel Frames. *Proc. of STESSA 2000 – International Conference on the Behaviour of Steel Structures in Seismic Areas*, F. Mazzolani & R. Tremblay (eds.), E&FN Spon, Chapman & Hall, London, 601-610.
Leon R. (1995). Seismic Performance of Bolted and Riveted Connections. *Background Reports: Metallurgy, Fracture mechanics, Welding, Moment Connections and Frame Systems Behavior*. SAC Steel Project Report No. 95-09.

A SURVEY ON DUCTILE DESIGN OF MR STEEL FRAMES

B. Faggiano[1], G. De Matteis[1], R. Landolfo[2], F.M. Mazzolani[1]

[1] University of Naples Federico II, P.le Tecchio, 80, I 80125 Napoli, Italy
[2] University of Chieti G. D'Annunzio, V.le Pindaro 42, 65127 Pescara, Italy

ABSTRACT

The paper focuses on the current methodologies for seismic design of moment resisting steel frames, paying special attention to the specific prescriptions provided by the European codes. Therefore, some regular steel framed structures are investigated by analysing the efficiency of different design methods in conferring to the structure the aimed prerequisites, in terms of ductility, resistance, stability and economy. To this purpose, the seismic structural response of the analysed cases is evaluated by means of both static pushover and dynamic time-history analyses, allowing for the determination of different levels of structural performance, in the perspective of the Performance-based approach. Finally, on the basis of the obtained results and in relation to both Ultimate and Serviceability Limit States, a critical assessment of current design procedures is presented and discussed.

KEYWORDS

Overstrength, performance-based design, q-factor, seismic design, serviceability limit state, dynamic analyses, steel frames.

INTRODUCTION

During the last years much awareness of the seismic behaviour of steel Moment-Resisting (*MR*) frames has been achieved through the extensive research activities for either damage inspections, or experimental and numerical elaborations, ensuing worldwide from the recent seismic events (De Matteis et al., 2000). The acquired knowledge has been already or is going to be introduced into structural design provisions in all earthquake prone Countries, giving rise to a new generation of seismic codes. In Europe, the present phase of conversion from *ENV* to *EN* of Eurocode 8 moves towards this direction (Mazzolani, 2000). The effort has been addressed mainly to the revision of the existing design concepts, of the current design procedures and of the structural details, in order to eliminate or, at least, to reduce the lack of correspondence between the design requirements and the actual structural response. In particular, aiming at improving the design process, the Performance-based design approach has been set up, by making explicit the association between the performance objectives that a seismic resistant structure has to attain and the codified design decisions (*SEAOC* Vision, 1995). Furthermore, as an alternative to the traditional Force-based approach, the potentiality of the Displacement-based design has been established and the appropriate design procedure adjusted (Priestley, 2000).

In this general context, the current paper is devoted to analyse the structural performance of steel MR frames in relation to the design procedures founded on the traditional Force-based approach and mainly focusing on the provisions provided by the present European codes for both Ultimate and Serviceability Limit States. Moreover, several methods for the application of "capacity design" rules, which aims at conferring to the structure a ductile behaviour, are considered and their impact on the determination of member sizes and on the structural behaviour of the whole structure is investigated. To this purpose, the seismic response of several frame configurations is assessed by using both static pushover and dynamic time-history analyses, while the interpretation of the corresponding results is carried out by means of a number of different parameters related to resistance, stability, ductility and economy of the structure as well as at the light of the Performance-based approach. In such a way, several important conclusions are drawn, allowing a better understanding of the main influencing factors on the seismic performance of steel MR frames.

BASIC PRINCIPLES FOR THE SEISMIC DESIGN OD STEEL MR FRAMES

Existing General Design Procedures

The limit state philosophy was introduced after the 2nd World War, changing the classical deterministic approach to the safety conception and check method. It basically identifies three design objectives, which should be achieved for three different increasing intensity levels of the earthquake actions: (1) to avoid damage in structural and non-structural elements for frequent (minor) ground motions; (2) to minimize non-structural damage during occasional (moderate) seismic events; (3) to avoid collapse or serious damage in rare (major) earthquakes (Mazzolani et al., 1995; Bertero, 1996). Actually, in present codes, the two first levels are assembled in only one, they corresponding to the limitation of damage level, by checking the Serviceability Limit State (*SLS*). On the other hand, the 3rd objective is accomplished by checking the Ultimate Limit State (*ULS*). Anyway, it has to be observed that the current procedure for realizing such objectives is strongly simplified compared to the complexity of the problem.

In general, the present seismic design approach for most building typologies is based on the evaluation of the earthquake required strength (base shear), which is determined using a Smoothed Linear Elastic Design Response Spectrum (*SLEDRS*) corresponding to a strong seismic event, whose damage potential is in general associated to an assumed return period. Profiting of the energy dissipation due to plastic deformation of ductile structures, in seismic codes a reduction factor (called q-factor in Europe, response modification factor R or structural system factor R_w in U.S.A.) is introduced, leading to the definition of the Smoothed Inelastic Design Response Spectra (*SIDRS*). In such a way even though the inelastic resources of structural systems are not neglected, the simplicity of code design procedure is preserved, linear elastic methods of analysis being allowed. Furthermore an economical benefit is gained due to the fact that structural design leads to lighter constructions. On the other hand, the evaluation of the reduction factor as representative of structural ductility is a critical task. Generally, in code provision such a coefficient is assigned as a function of the structural typology, but a correct estimation should account for many other factors, such as the fundamental period of the structure, the actual structural ductility, the soil conditions at the site and the dynamic characteristics of the ground motions (Miranda et al., 1994). Moreover, a direct correlation between the reduction factor and the structural ductility is possible only for *SDOF* systems, whereas for *MDOF* systems it still represents an open problem (Santa-Ana & Miranda, 2000; Mazzolani, 2000).

In the current design procedure, the *SLS* is usually checked *a posteriori*, once the structure has been preliminary designed at the *ULS*. Generally, only the interstorey drift check is carried out under a lower return period earthquake, while any strength check is not required under the serviceability seismic actions.

The limitation of interstorey drift should avoid or restrict the damage to structural elements as well, but it does not allow the structural damage control, which is expected to occur because for usual values of the strength reduction factor, structural members are sized for a lower strength than that required by a serviceability earthquake. In a better design procedure, two different limit states should be considered under moderate earthquakes: the first one being referred to the preservation of functionality of the building, while the second one to the structural damage.

In order to overcome some of the above difficulties, the Performance-based design approach has been recently proposed in U.S.A. and is now developing worldwide. The relevant concept is the multi-level design aimed to completely achieve fixed design objectives. The coupling of a structural performance level with a specific level of ground motion defines a performance design objective. The purpose is to provide the designer with the criteria for selecting the appropriate structural system and its layout and for proportioning and detailing both structural and non-structural components, so that for specified levels of earthquake intensity the structural damage will be constrained within acceptable extents. In this way, the minimisation of the cost/benefit ratio, which takes into account the construction cost and the expected losses, can be attained for all the limit states occurring during the life of the structure.

A detailed definition of the Performance-based design has been supplied within the activities of *SEAOC* Vision 2000 Committee (1995). The acceptability of different levels of damage should be determined on the basis of the consequences of this damage to the user community and the frequency with which such damage occurs. Four performance levels have been proposed: Fully operational (no damage), the consequences to the building user community are negligible; Operational (moderate damage to non-structural elements and contents and light damage to structural elements), the damage does not compromise the safety of the building for occupancy; Life safe (damage state, moderate damage to structural and non-structural elements), the structure lateral stiffness and ability to resist additional lateral loads is reduced, but some margins against collapse remain; Near collapse (extreme state), in which the lateral and vertical load resistance of the building are substantially compromised, aftershocks could result in partial or total collapse of the structure. It is clear that structures fail at successive more severe stages with increasingly less probability that the seismic actions will reach the corresponding intensity levels.

It is apparent that current seismic design approach does not permit the structural performance at each of the above levels to be predicted accurately. Therefore, in order to allow for a multi-level design approach, the common design practice should be based on a simplified procedure, relating the considered seismic performance objectives to different design criteria. Towards this direction, in recent times the interest in Displacement-based design is remarkably increasing (Priestley, 2000), this approach being more direct, feasible and reliable to attain some fixed performance objectives. The Displacement-based design is already included in the New Zealand seismic codification for the assessment of primary lateral force resisting elements, as an alternative method to the Force-based design (*NZSS* Draft, 1996). Recently, in Europe a big effort has been addressed to this new design procedure (Faccioli et al., 1998), but the correct definition of the basic parameters should be still definitively stated in order to allow for an easy and convenient application of the method.

European Practice for the Seismic Design of Steel Frames

The European seismic design code (*EC*8, 1994) is based on the traditional design principles. It identifies two performance levels: (1) limited damage, for frequent moderate seismic events; (2) no collapse, for rare major earthquakes. Their acquisition requires a check against the Serviceability and the Ultimate Limit States, respectively.

With regard to the *ULS* check, seismic resistant dissipative structures are designed to withstand severe earthquakes by means of a proper combination of strength and energy dissipation capacity. This design goal is pursued by provisions aimed at the development of a minimum level of strength, accompanied by detailing rules for obtaining the required energy dissipation capacity. In the case of moment resisting steel frames, in order to optimise the exploitation of the plastic resources of the structural scheme, it is suggested a design methodology based on capacity design for members, which has been introduced for the 1st time in New Zealand seismic codes almost 20 years ago. The general principle is the strong column-weak beam (*SC-WB*) concept, aiming at forcing inelastic deformations in beams rather than in columns. In fact due to the presence of significant axial forces a strong reduction of the columns member ductility arises (Krawinkler, 1995). The aim of the design method is to avoid that structure can attain poor dissipative failure modes, such as the so-called soft story, and to favour the development of a global type mechanism, which represents the ideal collapse condition because the energy input is dissipated at all the floors by inelastic bending cyclic deformations involving all the beams of the structures and the columns at the first floor. As a consequence of design purposes, dissipative regions have to be designed to resist the design loads and due to the high ductility demand, they have to be accurately detailed. This means that the beam cross-section must be of ductile type, for avoiding the local buckling phenomena that compromise the member ductility, whereas beam-to-column connections have to be full-strength, for allowing the development of plastic hinges at the end of the adjacent beam, instead of in the connection, because the connection welds hinder large inelastic strains. On the other side, non-dissipative elements have to be designed to resist the internal forces and moments transmitted by dissipative zones. Nowadays, different member design procedures for hierarchy criterion have been proposed and it has been observed that only the most refined approaches are able to assure the collapse attainment by the activation of a kinematic mechanism of global type. In particular, new revised versions of the existing procedure adopted by the European Recommendation (*ECCS*, 1988), proposed by the authors (in the following indicated as *RAFmin* and *RAFmax*), have been demonstrated to be suitable, they representing a compromise between simplicity of application and efficiency (Faggiano et al., 2000).

As far as *SLS* check is concerned, in order to guarantee the construction functionality building structures under moderate earthquake, a minimum level of lateral stiffness is required. For this reason, EC8 prescribes that structural lateral displacements evaluated by means of an elastic analysis have to be contained within some deformation limits. The seismic action is defined by means of a response spectrum, which is defined on the basis of the elastic one by using the reduction coefficient ν, which accounts for a lower return period of moderate earthquakes respect to major earthquakes and for the civil importance of the building. Furthermore, it is worth mentioning that not any strength check is required under the serviceability load condition.

STUDY CASES AND DESIGN CRITERIA

In order to investigate the influence of design methods on the seismic response of steel moment frames, a wide numerical study has been carried out. All examined structural schemes are multistoreys steel *MRFs* with 2 and 4 bays (*B*), 2, 5 and 8 storeys (*S*), whose dynamic behaviour is exhaustively represented by the 1st vibrational mode. The geometry of the frames is given in Figure 1.

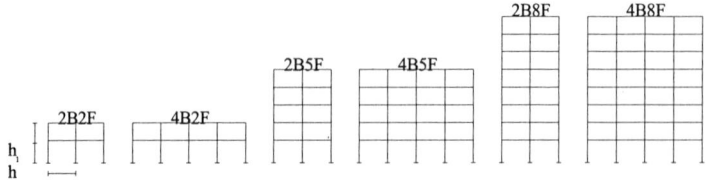

Figure 1: Examined structural typologies

Bays		Storeys		
	L		h	h_1
n°	(m)	n°	(m)	(m)
2	6	2	3.5	4.5
4		5		
		8		

The structural members have been designed according to *EC3* and *EC8* at *ULS*. Two different values of the design q-factor have been assumed, namely q_d=4 and 6. Steel grade is *Fe* 430. Dead and live loads have been assumed equal to 4.75 and 2.00 kN/m², respectively. They have been applied as concentrated vertical forces on principal beams at 1/3 and 2/3 of the bay span. The equivalent seismic action has been defined assuming a soil profile *A*, a viscous damping factor equal to 5% and *PGA*=0.35g. The 2nd order (*P-Δ*) effects have been considered by using the simplified method proposed by *EC8* (1994). It consists on amplifying the internal forces and moments by the factor ψ, which is a function of the interstorey drift sensitivity coefficient θ, which in turn represents the rate of the critical load that is accepted acting on the structure. The capacity design rule has been applied according to different criteria. In particular, besides to the criterion suggested by *EC8*, the two methods, namely *RAFmin* and *RAFmax*, previously proposed by the authors (Faggiano, 2000), have been considered. Also the case in which the hierarchy resistance criterion is not used has been considered, it being labelled with *N*. Finally, as reference cases, the same structures have been designed by means of the "mechanism curve" method, labelled as *GM*, according to Mazzolani and Piluso (1997a) method. In order to verify the impact of interstorey drift limitation on the design of the considered frames, the obtained structures have been checked for *SLS* according to *EC8* prescriptions, by assuming the reduction coefficient ν equal to 2 and an interstorey drift limit (δ_{lim}) equal to 0.006h. Besides some other less severe conditions have been considered, such as ν=3 and δ_{lim} equal to 0.008h, 0.006h, 0.004h. However, structural member sizes have not been changed for satisfying the stiffness requirements. The summary of design data is presented in Table 1. In total 60 different structures have been designed.

TABLE 1
DESIGN DATA

Structural typology	Ultimate Limit States (ULS) design			Study case No.	Serviceability Limit States (SLS) check
	Reference case	Member Hierarchy Resistance Criterion	q-factor (q_d)		Interstorey drift limit (δ_{lim})
2B2F	GM*	N**	4	30	0.004h (ν=3)
2B5F		EC8			0.006h (ν=2, 3)
2B8F		RAFmin***			0.008h (ν=3)
4B2F		RAFmax			
4B5F			6	30	
4B8F					

* GM= Global Mechanism; ** N=None; ***RAF=Revised Amplification Factor

Both static pushover analyses and dynamic time-history analyses have been performed by means of the computer program Drain-2DX (Prakash et al., 1993). In particular, time-history analyses have been carried out with reference to a selection of 3 acceleration records, whose mean response spectra approximates the Class *A* elastic spectrum provided by *EC8*. *P-Δ* effects are taken into account and the inelastic behaviour of dissipative zones is defined by means of concentrated elastic-perfectly plastic hinges.

CRITICAL ANALYSIS OF THE OBTAINED RESULTS

General

The wide study performed on Moment Resisting Frames allows a critical review of the current seismic design prescriptions. In fact on the basis of obtained results, a number of useful indications can be drawn, with regard to the effectiveness of the modern procedures in the attainment of the performance requirements in terms of strength, stiffness and ductility. For the sake of brevity, in the following the general trends of the most important topics are focused, while for a more comprehensive and detailed analysis of outcomes the reader is referred to Faggiano (2000).

The Influence of Design q-factor

As shown in Figure 2, where the structural weight (t) of considered frames is depicted, the value of the design strength reduction factor q_d does not influence the design of 2S structures. This can be due to the fact that the resistance to vertical loads governs the frame design.

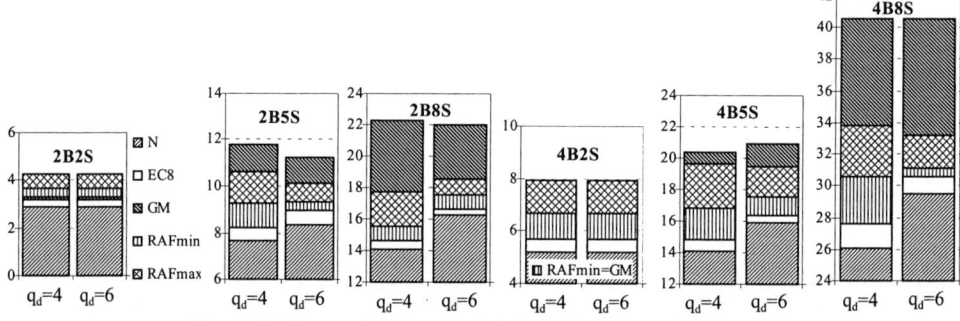

Figure 2. The results of structural design in terms of weigth (t)

As far as 5S and 8S structures are considered, it can be observed that using $q_d = 6$ is not convenient. In fact, from Figure 2 it is apparent that often the structures related to $q_d = 6$ are heavier than the ones related to $q_d = 4$ (almost 5% in average). This is a consequence of the limitation of the 2nd order effects (≤ 0.3) according to EC8 (1994), whose impact increases with higher values of design q-factor (q_d). Finally, it should be remarked that the number of bays has a slight effect on the influence of the design q-factor.

The Influence of the Design Criteria

As it is apparent from Figure 2, the adoption of different design methods for capacity design induces a structural weight increment as respect to the case where no criterion is applied (case N). Usually such a weight increment is in the following sequence: *EC8, RAFmin, RAFmax, GM*. Only for 2S frames the structures are heavier when the *RAFmax* method is used. For the other frames, the weight increment ranges from 2%, for 2B8SQ6-EC8 case to 55%, for 2B8SQ6-GM case. In particular, in 5S and 8S cases the increment is greater for structures designed by assuming $q_d = 4$ as respect to $q_d = 6$. Moreover, it is worth mentioning that as respect to the *EC8* method, *RAFmin, RAFmax, GM* methods produce a remarkable weight increment in all the analysed cases.

The Influence of the Serviceability Limit State

The *EC8* prescriptions concerning the *SLS* check appear to be excessively restrictive. In fact, for the study cases, the interstorey drift limitation always governs the structural design. In particular, when the reduction coefficient ν is assumed equal to 2 and the interstorey drift limit is equal to 0.006h, none of the study cases, which have been designed at the *ULS*, satisfies the lateral stiffness check. This occurrence has also the important consequence to make vain the design at *ULS*, partially loosing the advantage of the inelastic design with reduced seismic forces. Moreover, the application of capacity design rules, aiming at a ductile structure, cannot be fully exploited.

The assumption of $\nu = 3$ and using higher interstorey drift limits ($\delta_{lim}=0.008h$ and $0.006h$) lead to the fulfilment of the *SLS* check for each examined structural typology. On the contrary the interstorey drift limit $\delta_{lim}=0.004h$ appears to be sever even in case of $\nu = 3$, mainly for $q_d = 6$. Nevertheless the possibility to accept less restrictive interstorey drift levels could be justified only by guaranteeing that structural and non-structural elements are able to sustain larger deformations, without intolerable damage. Alternatively, frame structures should be firstly designed according to strength and stiffness requirements under the "serviceability" forces rather than to strength and ductility requirements under the "ultimate" forces. In such a case, the ULS check should be performed *a posteriori*, allowing for the necessary structural ductility as a function of the available strength in the structure.

The Overstrength Factor

The overstrength factor is here intended as the ratio between the actual yielding strength and the one related to the design seismic forces. Design provisions refer to overstrength only concerning capacity design, where the uncertainties associated to the actual material and element strength evaluation are compensated by an overstrength factor, which is applied to the strength of elements to be "protected" from plastic deformation occurrence. However the overstrength concept is more complex, because it depends on several sources. For instance, the contribution of the following factors can be simply quantified by means of a static inelastic pushover analysis, as the ratio between the actual base shear at yielding and the design one: (1) redistribution of internal forces in the case of redundant structures; (2) effects of discrete member sizes, due to the use of commercial profiles; (3) effects of stiffness requirements; (4) effects of unification of members for constructibility. Nevertheless it is quite impossible to separate the roles of each factor, due to the complexity of the structural response of a MDOF system.

From the analysis of the large number of study cases, it has been possible to point out a recurring design aspect, which seems to strongly and mostly contribute to structural overstrength. By observing the variation of the member structural dimensions due to the application of different design methods, for a given geometry, the only recognizable feature that influences overstrength appears to be the beam cross-section size, the overstrength in beams having a stronger impact than the one in columns. Furthermore, Figure 3 evidences that in the case $q_d = 6$ the overstrength value is higher than in the case $q_d = 4$. This result was expectable, because in the former 5S and 8S structural typologies have generally stronger beams than the ones designed by assuming $q_d = 4$. As above stated, this is a consequence of the prescriptions related to the limitation of 2nd order effects (≤0.3). Changing q_d from 4 to 6, the increment in q_d of 1.5 is fully reflected in the increment of overstrength. Moreover, it appears that the member hierarchy criterion usually has a slight influence on the overstrength factor and a direct correspondence between the overstrength and the structural weight cannot be established. This depends on the fact that the member hierarchy criterion does not affect the column sizes only, but also the beam sizes that can be reduced due to the different distribution of internal forces and moments. It has been also observed that for $q_d = 4$ the overstrength factor decreases as the number of storeys increases, both for 2B and 4B typologies, whereas a precise trend is not recognizable in the case $q_d = 6$. This difference can be ascribed to the stronger influence of the 2nd order effect rules for $q_d=6$.

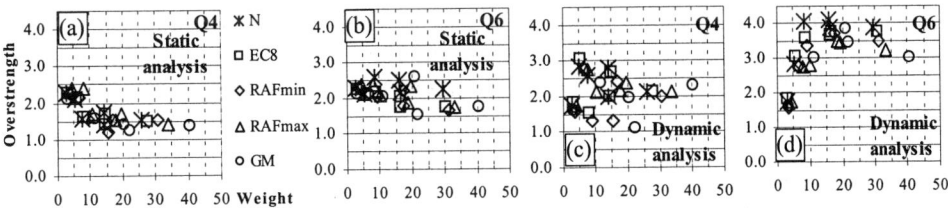

Figure 3. The overstrength factor versus structural weigth (t)

Definitively, with regard to the overstrength evaluation also in terms of comparison between «static» (Figures 3a,b) and «dynamic» (Figures 3c,d) results the following trends can be highlighted: (1) in all the cases for $q_d=6$ the overstrength values are higher than in the case $q_d=4$, the increment being of almost 70% in the dynamic case and 40% in the static case; (2) the overstrength is higher if the capacity design criterion is not applied. Nevertheless if comparing the «static» and the «dynamic» relationships between the overstrength factor and the structural weight, it appears that the trends are dissimilar, which means that the type of analysis influences the assessment of the first yielding, it occurring for larger *PGA* values in case of dynamic analysis as respect to the static analysis.

The Evaluation of the Structural Performance

In order to evaluate by a synthetic way the performance of structures in terms of available ductility and to verify the accordance with the design assumptions, a factor (q^*) could be defined as a strength reduction factor evaluated *a posteriori* to be suitably compared to the design one (q_d). Coherently with the meaning of the design q-factor, q^* should be related to the experienced structural ductility. With regard to the pushover static analysis, within the approximation of the ductility factor theory, the «static» value of the equivalent reduction factor q^* may be evaluated by the behavioural curve (α-Δ in Figure 4a) as the ductility factor μ, i.e. the ratio between top-sway displacements corresponding to the predefined ultimate limit state and to yielding, it also representing an approximation of the ratio between the *PGA* corresponding to the above two target limit states. With regard to the dynamic time-history analysis the «dynamic» value of factor q^* could be evaluated by the incremental dynamic curve (*a*-Δ in Figure 4b) as the ratio of the above defined *PGA* levels. Anyway, for a correct interpretation of the results, it should be reminded that the *a*-Δ curve obtained from dynamic incremental analyses has a different meaning with respect to the behavioural curve α-Δ obtained from the static pushover analysis: the latter represents the response history in terms of top-sway displacements under increasing monotonic horizontal loads; while the former represents the relationship between the *PGA* of a given earthquake accelerogram and the corresponding maximum top-sway displacement experienced by the examined structure during a whole acceleration history. This means that the *a*-Δ curve does not represent the evolution of the structural response under the growing actions, but the envelope of results of different histories.

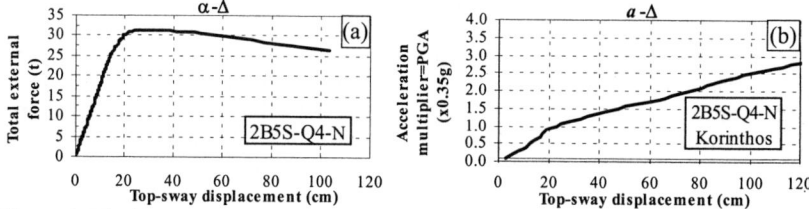

Figure 4. The incremental static α-Δ curve versus the incremental dynamic a- Δ curve

In Figure 5, the comparison between the values of factor q^* obtained from static and dynamic analyses is presented for the 2B5S-Q4-N frame typology. In both cases, the ultimate condition has been conventionally identified with the attainment of a plastic rotation equal to θ_{pl}=0.03rad in the most engaged plastic hinge. In case of dynamic analyses, in order to emphasise the important role played by the seismic input, results are presented both for all the chosen earthquake records and as mean value. However, as a general trend, it can be observed that the comparison between static and dynamic results is quite satisfactory, also in relation to the assumed design q-factor value (q_d= 4). Besides, as it could be expected, the structural performance is generally increasing when more refined design methods (*GM* and *RAFmax*) are used, while it is unsatisfactory (q^*< q_d) when no member hierarchy criterion is applied (case N).

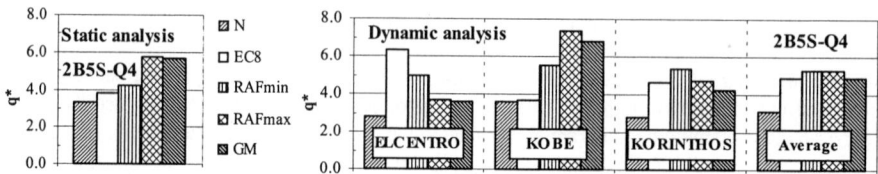

Figure 5. The «static» and «dynamic» reduction factor q^*

Nevertheless, since the evaluation of the q^* factor is affected by the assumed conventional ultimate limit state condition, for a better interpretation of the structural performance, different situations should be taken into account. In the perspective of the Performance-based design philosophy, such limit states could be intended as different seismic performance objectives. In the following, for one of the study cases, the performance levels defined according to *SEAOC* Vision have been quantitatively determined on the basis of the procedure proposed by Mazzolani and Piluso (1997b). Therefore, the corresponding limit states have been firstly fixed on the basis of the pushover curve and then transferred in terms of either target displacements or plastic rotations to the dynamic a-Δ curve in order to evaluate the *PGA* level inducing the occurrence of the considered limit conditions. In particular, with reference to the trilinear modelling of the pushover curve, the Fully Operational (*FO*) performance level is related to the most restrictive condition between the ones corresponding to the attainment of some interstorey drift (δ/h) limits, a top-sway displacement limit (Δ/h) and the occurrence of first structural damage; the Operational (*O*) performance level is related to the intersection between the linear elastic behaviour and the horizontal line at the maximum strength; the Life Safe (*LS*) performance level is related to the intersection between the horizontal line at the maximum strength and descending branch; the Near Collapse (*NC*) performance level is conventionally related to an interstorey drift value equal to 0.10 rad, which should represent the limit for compatible realistic deformation of the whole structure. Furthermore, some typical limit states corresponding to different extension and amount of damage within the structures have been considered, namely: the attainment of a plastic rotation equal to 0.03 rad in the most engaged plastic hinge; the attainment of an interstorey drift equal to $0.03h$ at the weakest floor; the attainment of a global drift equal to $0.03H$ (*H* being the building height); the mechanism activation. For the sake of example, in Figure 6, the above procedure is applied to the *2B5S-Q4-N* frame typology. From the examination of the obtained results, some significant information can be acquired: the *FO* performance level is conditioned by the limitation of non-structural damage instead of structural damage; for a *PGA* equal to the design one (0.35g) the structure overcomes the *O* performance level, it having already experienced both non-structural and structural damage ($\Delta/H=\theta_{pl} \cong 0.01$, $\delta/h \cong 0.02$); the Life Safe (*LS*) performance level is in good agreement with a plastic rotation demand $\theta_{pl} \cong 0.03$rad and it is attained at about 1.5 times the design *PGA* value; the *NC* performance level is attained for a very high value of *PGA*, it being out of the practical range of interest.

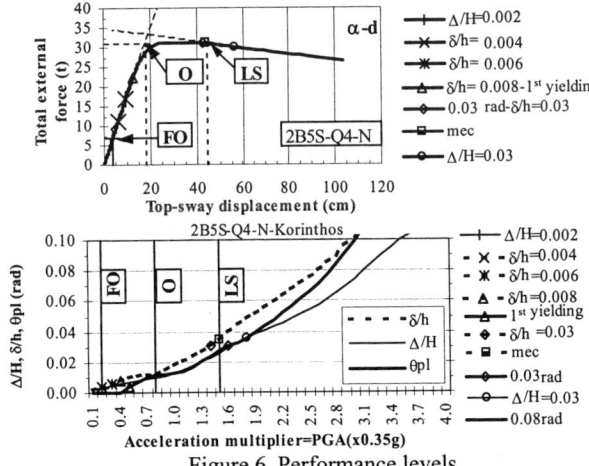

Figure 6. Performance levels

CONCLUSIVE REMARKS

The general study on the influence of design methods on the seismic response of steel MR frames presented in this paper allows several important conclusions to be drawn. Firstly, it has been emphasised the important role to be ascribed to stiffness requirements, in terms of limitation of either second order effects in column members or damage in non-structural elements, at least for the investigated frame typologies. In fact, it has been shown that both these aspects have a remarkable impact on the determination of frame member sizes, conferring to the structure a significant overstrength. For this reason all the analysed frames, even without the application of any refined design procedure aimed at a ductile structural behaviour, have experienced a satisfactory seismic performance, showing that the inherent ductility of the whole structure is not fully exploited under design earthquakes.

As far as the evaluation of the structural performance is concerned, the comparison of results coming from the application of static pushover and dynamic time-history analyses has emphasised a substantial agreement, showing that in both cases the estimated available structural ductility is aligned with the common design assumptions related to the force reduction factor, at least with reference to the conventional ultimate limit state corresponding to plastic rotation demand equal to 0.03 rad. Moreover, as it could be expected, the structural performance is generally increasing when more refined design methods are adopted and it is somehow influenced by the considered ground motion. Finally, aiming at a further generalisation, some of the obtained results have been interpreted in the perspective of the performance-based approach, where the design objectives have been firstly defined and then assessed by using a procedure based on the combination of pushover curve and dynamic time-history analysis results. The outcomes are encouraging, stating a rational correspondence between the damage states identified by typical conventional deformation limits and the predefined performance levels.

References

Bertero V.V. (1996). The need for multi-level seismic design criteria. *Proc. of the 11th WCEE*. Acapulco, Mexico.
De Matteis, G., Landolfo, R., Mazzolani, F.M. (2000). The Behaviour of Connections in Steel MR-Frames under High-Intensity Earthquake Loading. *Abnormal Loading on Structures: Experimental and Numerical Modelling*, K.S. Virdi, R.S. Mattehews, J.L. Clarke and F.K. Garas eds, E & FN SPON, London (UK), 59-73.
ECCS (1988). *European recommendations for steel structures in seismic zones*. Publication No.54.
Faccioli E., Tolis S.V., Borzi B., Elnashai A.S. and Bommer J.J. (1998). Recent developments in the definition of the design seismic action in Europe. *Proc of 11th ECEE*, Paris, France, Balkema, Rotterdam.
Faggiano B. (2000). *Earthquake resistant steel frames: a new method for ductile design*. PhD Thesis in Structural Engineering, University of Naples "Federico II".
Faggiano B., De Matteis G. and Landolfo R. (2000). Comparative study on seismic design procedures for steel MR frames according to the force-based approach. *Proc. of 3rd STESSA International Workshop*, Montreal, Canada.
Krawinkler H. (1995). *Systems behaviour of structural steel frames subjected to earthquake ground motion*. Background Reports SAC 95-09.
Mazzolani F.M. editor (2000). *Moment resistant connections of steel frames in seismic areas: design and reliability*. E & FN SPON, London.
Mazzolani F.M and Piluso V. (1997a). Plastic design of seismic resistant steel frames. *Earthquake Engineering and Structural dynamics*, **26**, 167-191.
Mazzolani F.M and Piluso V. (1997b). A simple approach for evaluating performance levels of Moment-resisting steel frames. *International Workshop on Seismic design methodologies for the new generation of codes*, Bled, Slovenia, June, Balkema, Rotterdam.
Mazzolani F.M., Georgescu D. and Astaneh-Asl A. (1995). Safety levels in seismic design. *Proc.of the 1st STESSA International Workshop*, Timisoara, Romania, June 1994, Mazzolani F. M. and Gioncu V. editors, published by E & FN SPON an Imprint of Chapman & Hall, London.
Miranda E., EERI M. and Bertero V.V (1994). Evaluation of strength reduction factors for earthquake-resistant design. *Earthquake Spectra*, **10:2**, 357-379.
Priestley M. J. N. (2000). Performance-Based seismic design. *Proc. of the 12th WCEE*. Auckland, New Zealand.
Santa-Ana P.R. and Miranda E. (2000). Strength reduction factors for multi-degree of freedom systems. *Proc. of the 12th WCEE,* Auckland, New Zealand.
SEAOC Vision 2000 (1995). *A framework for performance based design*. Volume I, Structural Engineers Association of California, Vision 2000 Committee, Sacramento, California.

SEISMIC DESIGN OF CONCRETE STRUCTURES

BEHAVIOR OF RECTANGULAR COLUMNS WITH LOW LATERAL CONFINEMENT RATIO

S.S.E. LAM[1], B. WU[2], Z.Y. WANG[2], Y.L. WONG[1] and K.T. CHAU[1]

[1]Department of Civil and Structural Engineering, The Hong Kong Polytechnic University, Hong Kong
[2]School of Civil Engineering, Harbin Institute of Technology, Harbin, China

ABSTRACT

Main objective of the study is to examine the seismic capacity of rectangular reinforced concrete columns having non-seismic detailing. Six column specimens in 1:3 scale have been tested under cycles of lateral load. The column specimens were characterised by high axial load (40%-65% of the axial load-carrying capacity), low lateral confinement ratio (0.001-0.003) and low to moderate shear span ratio (1.5-3.0). An ultimate drift ratio is introduced to quantify the seismic capacity. Poor performances were observed on column specimens with low shear span when subjected to cyclic lateral loads. Four column specimens failed under shear, and flexural failures were observed in the other two column specimens with lateral confinement ratio at 0.003 and moderate shear span. An increase in the axial load ratio within the range 0.4-0.65 resulted in a reduction in the lateral load-carrying capacity by about 10% and a reduction in the ultimate drift capacity by about 30%. Lateral confinement has significant influence on the ultimate drift, of which an increase from 0.001 to 0.003 could increase the ultimate drift capacity by up to 100%. At low to moderate shear span, contributions of shear and bond slip to the deformation must be considered. However, deformation outside the plastic region could be neglected for a column with low shear span ratio.

KEYWORDS

Reinforced concrete, rectangular column, ultimate drift ratio, lateral confinement, shear span, non-seismic detailing

INTRODUCTION

The response of reinforced concrete columns under seismic action has always been an area of both research and practical interest. In regions of low to moderate seismic risk for instance Hong Kong, columns could be designed with no seismic provisions (Lam *et al* 2000). So far, little information is available on the seismic capacity of such columns. Among others, Wehbe *et al* (1999) have examined the ductility of rectangular columns with moderate lateral confinement ratio (0.0037-0.0048) and moderate axial load (10%-25% of the axial load-carrying capacity). Mo and Hwang (1999) conducted experimental tests on rectangular columns under low axial load. Special reinforcement detail was considered with the transverse reinforcements at 50-100mm spacing and having moderate lateral confinement ratio (0.0048-0.0124). Inoue *et al* (2000) examined the seismic performance of a reinforced concrete frame. The lightly loaded columns were poorly confined with transverse reinforcements at 300mm spacing. Shear failures were observed with ultimate drift at 1/117. Contrast to these studies, typical columns in Hong Kong and in areas with no seismic provisions are characterised by low lateral confinement, high axial load and low to moderate shear span. It is the objective of this study to carry out cyclic load tests on representative column specimens having such properties, and to quantify the seismic capacity through the ultimate drift ratio.

SPECIMEN DESIGN AND TEST SETUP

Six column specimens in 1:3 scale have been designed to test under combined axial load and cyclic shear. The specimens represent prototype of reinforced concrete columns in buildings with no seismic provisions. Geometry and reinforcement details of the specimens can be divided into two types as shown in Figure 1.

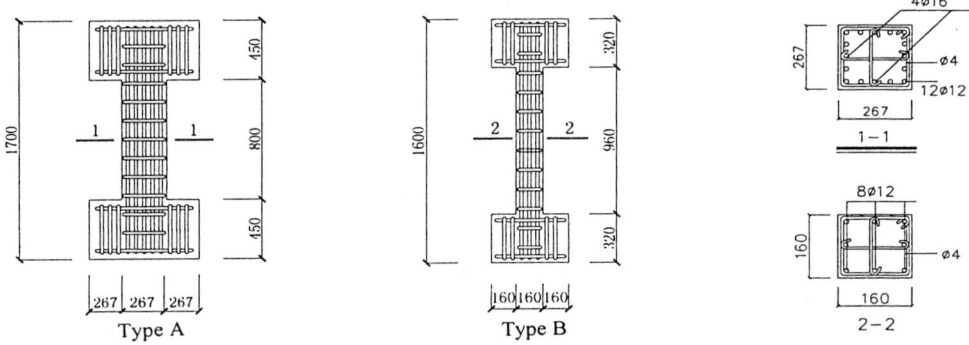

Figure 1: Geometry and reinforcement details of column specimens

Normal strength concrete with compressive strength f_{cu} around 42MPa was used. Yield strength and ultimate strength of the reinforcements were 273-370MPa and 410-520MPa respectively. Basic properties of the specimens are tabulated in Table 1. Here, m and n are the resepctive shear span ratio (defined as half clear-height divided by the depth of a column with fixed ends) and axial load ratio (or the ratio of axial load to axial load-carrying capacity). The lateral confinement ratio ρ_s is defined as $A_{sh}/S_t h_c$. Here, A_{sh} is the total area of transverse reinforcements in the transverse direction of bending, S_t is spacing of the transverse reinforcements along the axis of the member, and h_c is the cross-sectional dimension of the column core measured center-to-center of the confining reinforcements.

TABLE 1
BASIC PROPERTIES OF THE COLUMN SPECIMENS

No	Type	Dimension (mm)	Clear height (mm)	m	n	ρ_s	Main Reinforcements	Links
X-1	A	267x267	800	1.50	0.60	0.001	12ϕ12+4ϕ16 (3.03%)	ϕ4@164
X-2	A	267x267	800	1.50	0.40	0.001	12ϕ12+4ϕ16 (3.03%)	ϕ4@164
X-3	A	267x267	800	1.50	0.40	0.003	12ϕ12+4ϕ16 (3.03%)	ϕ4@55
X-4	B	160x160	960	3.00	0.65	0.003	8ϕ12 (3.53%)	ϕ4@88
X-5	B	160x160	960	3.00	0.65	0.001	8ϕ12 (3.53%)	ϕ4@264
X-6	B	160x160	960	3.00	0.45	0.003	8ϕ12 (3.53%)	ϕ4@88

Tests were conducted using an existing loading frame in the Harbin Institute of Technology. Figure 2 shows the loading apparatus. The loading frame is typical and similar to those used in other tests, for instance Nakata et al (1978) and Xiao et al (1986). Constant axial load is applied at the top stub of the test specimen by two manually controlled hydraulic jacks. A pantograph system connects the loading arm and the reaction floor to prevent the ends of the columns from rotating. The lateral loading system provides a double bending condition with the point of inflection at mid-height.

Load cells were installed to measure the axial load and lateral load. Transducers were used to measure the relative displacements between the top and bottom stubs. Transducers were installed to measure the horizontal, vertical and diagonal displacements in the vicinity of the plastic region in order to estimate the deformation due to flexural, shear and bond slip. Electric resistance strain gauges were mounted on the surfaces of the main reinforcements and transverse reinforcements to monitor the strains in the reinforcements.

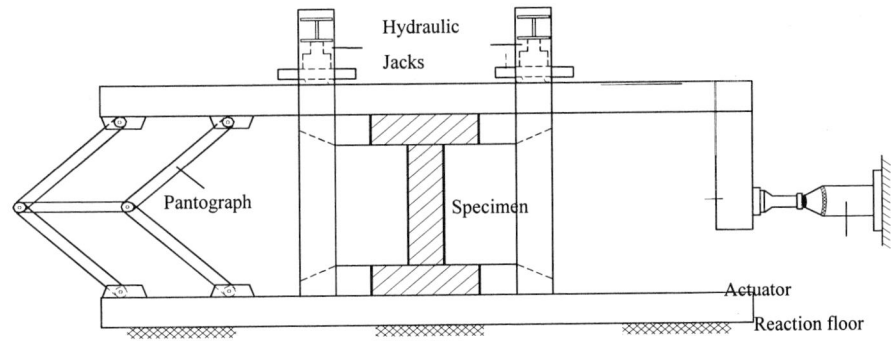

Figure 2: Setup of the loading apparatus

Axial load was introduced manually and maintained through the two hydraulic jacks. Afterwards, a series of symmetric positive and negative loading cycles (push-pull loading) was applied under load control. Cycles of lateral load were applied until obvious reduction in stiffness was observed in the hysteresis loops. The corresponding lateral displacement was U_1. Subsequent loading cycles were carried out under displacement control. Two loading cycles were applied with the maximum lateral displacement at $2U_1$, $3U_1$, $4U_1$, and so on. The loading cycles terminated when a 20 percent drop in the lateral load-carrying capacity from the maximum was observed.

EXPERIMENTAL RESULTS AND DISCUSSIONS

Figure 3 shows the ultimate conditions of the column specimens at failure. The principal mode of failure of column specimens X-1, X-2, X-3 and X-5 was shear, whereas column specimens X-4 and X-6 exhibited flexural failure. The main observations are as follows.

Column specimens X-1, X-2, X-3 and X-5

Column specimens X-1 and X-2 are characterized by low shear span ratio and low lateral confinement. For these two column specimens, failures were initiated and occurred almost instantaneously by the propagation of a large shear crack in the diagonal direction. Rupture of the transverse reinforcements and distortion of the main reinforcements were observed.

When the lateral load was around 250kN in the negative loading cycle of column specimen X-3, a shear crack was found running in the diagonal direction and a flexural crack appeared at 90mm from a column end. On progressive load reversals, the main reinforcements yielded. Subsequently, more shear cracks and flexural cracks were observed, and the transverse reinforcements at the two column ends yielded. Failure occurred when a large shear crack was developed, accompanied by rupture of the transverse reinforcements and distortion of the main reinforcements. Although the lateral confinement ratio of column specimen X-3 is three times more than X-1 and X-2, this is still very small as compared with the seismic requirements worldwide. As a result of low shear span, the principal mode of failure of column specimen X-3 is still shear.

For column specimen X-5, the first flexural crack was observed at 90mm from a column end. When the loading was reversed, more flexural cracks were found and the main reinforcements yielded. Failure occurred after the initiation of a shear crack. This shear crack propagated instantaneously and the main reinforcements buckled. The abrupt failure in shear is probably due to low lateral confinement.

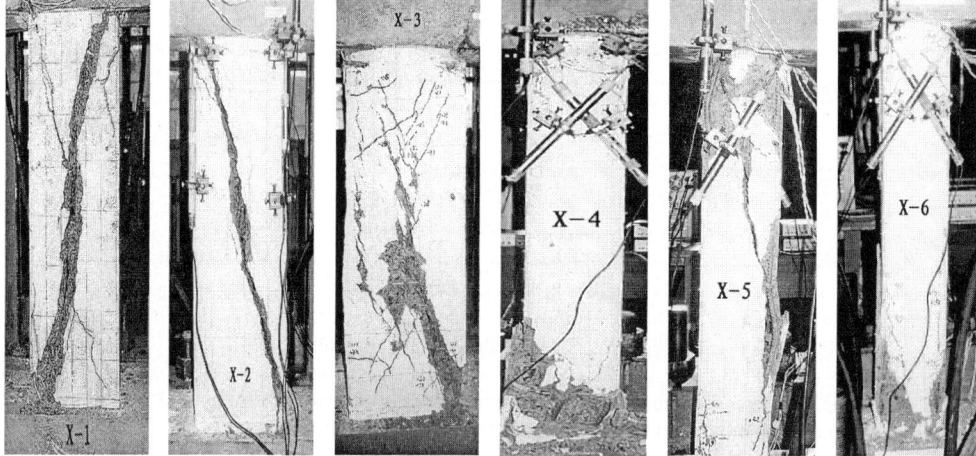

Figure 3: Modes of failure of the column specimens

Column specimens X-4 and X-6

The lateral confinement ratios of column specimens X-4 and X-6 are three times more than X-5. These two column specimens were able to sustain more cycles of load reversals after yielding of the main reinforcements. Failures are characterized by considerable spalling of concrete.

The first flexural crack appeared in column specimen X-4 at about 50kN at the positive loading cycle, and was 60mm from a column end. Another flexural crack at 120mm from a column end was observed at the following negative loading. These two cracks were not symmetrical. On progressive load reversals, the main reinforcements yielded. Subsequently, inclined cracks appeared and the transverse reinforcements in plastic region started to yield. No shear crack was found near the mid-height of the column specimen throughout the test. Failure occurred with spalling of concrete in the plastic region and buckling of the main reinforcements.

At 53.5kN in positive loading cycle, the first flexural crack appeared in column specimen X-6. A second flexural crack was found at the next negative loading, and was symmetrical to the first one. The main reinforcements yielded on subsequent load reversals. The mode of failure was similar to column specimen X-4 with some shear cracks near the mid-height. Application of cyclic load was disrupted in several occasions due to power failure. The column specimen was exposed to high axial load over considerable time, and the incomplete loading cycles had to be restarted again. This could have some adverse effects on the hysteresis response, for instance possible reduction in the ultimate drift capacity.

HYSTERESIS BEHAVIOR AND DEFORMATION

Hysteresis loops of the tests are plotted in Figure 4.

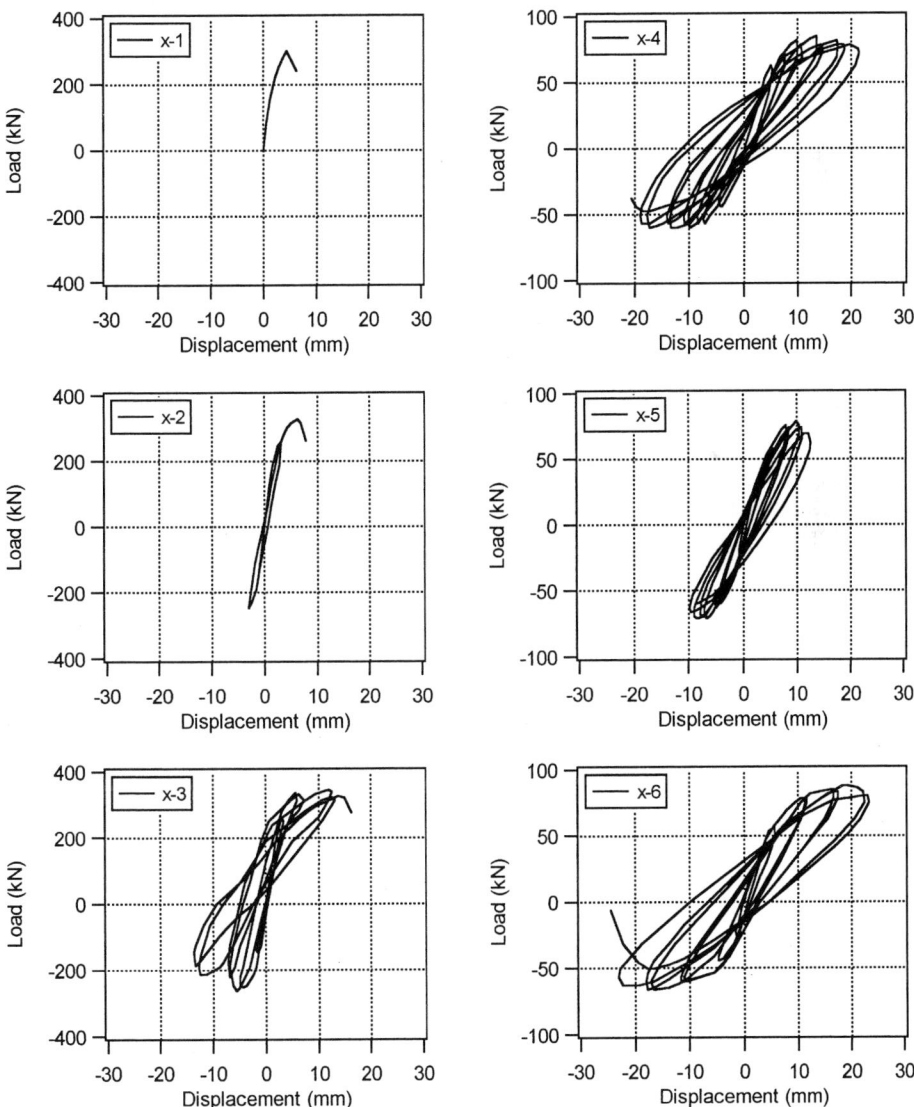

Figure 4: Hysteresis loops of lateral load against displacement

The principal values are compiled in Table 2. Maximum lateral load and corresponding displacement are obtained by averaging the data at the positive and negative loading cycles. Failure load and displacement at failure are the corresponding values at 20 percent drop in the lateral load-carrying capacity from the maximum. Ultimate drift ratio is the ratio of the relative displacement at the two ends of the column at failure against the clear height of column.

TABLE 2
PRINCIPAL DATA OF THE COLUMN SPECIMENS

Parameter	X-1	X-2	X-3	X-4	X-5	X-6
Maximum lateral load (kN)	302.4	327.6	305.6	71.8	75.6	77.18
Displacement at maximum load (mm)	4.38	6.25	7.50	13.61	8.45	17.50
Failure load (kN)	241.9	262.1	245.0	57.46	60.48	61.74
Displacement at failure (mm)	6.25	7.813	14.33	20.91	11.48	24.05
Ultimate drift ratio	1/128	1/102	1/56	1/46	1/84	1/40

As indicated in Table 2 (X-1 vs X-2; X-4 vs X-6), an increase in the axial load ratio within the range 0.4-0.65 resulted in a reduction in the lateral load-carrying capacity by about 10% and a reduction in the ultimate drift capacity by about 30%. Lateral confinement has significant influence on the ultimate drift (X-2 vs X-3; X-4 vs X-5), of which an increase from 0.001 to 0.003 increased the ultimate drift capacity by up to 100%.

The lateral load-carrying capacity of a column is influenced by many factors. In this study, particular attention is focused on the effects due to shear span and lateral confinement. Figures 5(a)-(b) compare the lateral load-displacement envelopes of the column specimens. The influence of shear span ratio is demonstrated in Figure 5(a) (X-1 vs X-5; X-3 vs X-6). Reduction in the shear span ratio has led to some reduction in the ultimate drift capacity. Figure 5(b) demonstrates the influence of lateral confinement (X-2 vs X-3; X-5 vs X-6). There were significant improvements in both the ductility and ultimate drift capacity when the lateral confinement ratio increased from 0.001 to 0.003.

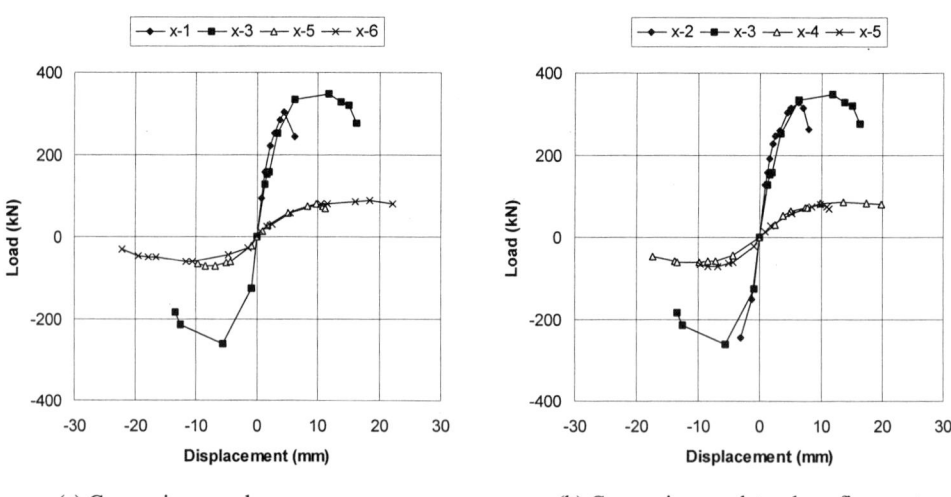

(a) Comparison on shear span   (b) Comparison on lateral confinement

Figure 5: Envelopes of lateral load against displacement

Transducer measurements taken at the top of the column specimens in the horizontal, vertical and diagonal directions have been analyzed to estimate the components of deformation, namely flexural, shear and bond slip. It is assumed that deformations at the top and the bottom are identical, and the total deformation within the plastic region is simply twice the value obtained from the estimation. Figures 6(a)-(d) show the relative deformation (expressed as fraction of the total deformation measured separately) of the various components within the plastic region against the number of positive loading cycles of column specimens X-2, X-3, X-4 and X-6 respectively. While conducting the test on column specimen X-4, movements were found in a few reference points of the transducers. As a result, summation of the deformations due to flexural, shear and bond slip was greater than the total deformation obtained separately from other transducers.

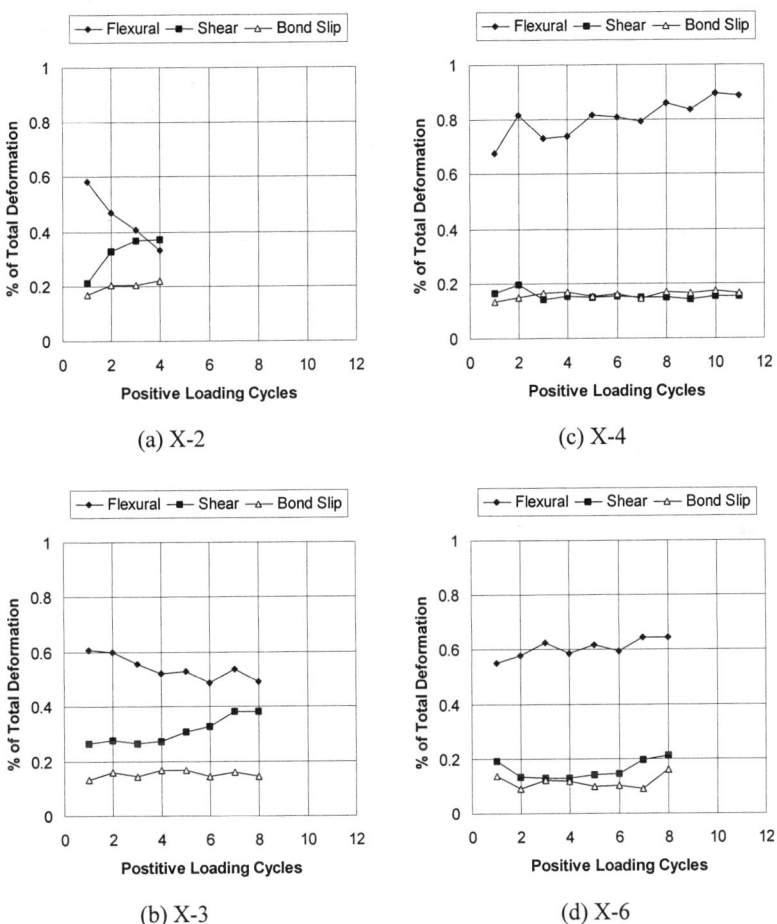

Figure 6: Variation of components of deformation against number of positive cycles

For column specimens with low shear span ratio, majority of the deformation occurred in the plastic region. Therefore, deformation outside the plastic region could be neglected for a column with low shear span ratio.

At low shear span ratio, the relative shear deformation increases with the loading cycles. Although the lateral confinement ratio of column specimen X-3 is three times more than X-2, the relative shear deformations at failure in both cases are similar and about 40%. At low shear span ratio, the relative deformation due to shear is insensitive to the range of lateral confinement ratio ($0.001 < \rho_s < 0.003$) considered in this study.

In Figures 6(a)-(b) i.e. low shear span ratio, the relative flexural deformation reduces with the loading cycles. However, in Figures 6(c)-(d) i.e. moderate shear span ratio, the relative flexural deformation increases with the loading cycles. The influence of shear to the deformation is significant at low shear span ratio. When the shear span ratio is within the range 1.5-3.0, the relative deformation due to bond slip remains the same throughout, and is significant to the deformation.

CONCLUSIONS

Six column specimens in 1:3 scale having non-seismic reinforcement details were tested under cyclic lateral load. The column specimens were characterized by high axial load, low lateral confinement and low to moderate shear span. An ultimate drift ratio is introduced to quantify the seismic capacity. Shear failures were observed in four column specimens, whereas the other two with the lateral confinement ratio at 0.003 and shear span ratio at 3.0 failed under flexure. An increase in the axial load ratio within the range 0.4-0.65 reduced the ultimate drift by about 30%. When the shear span ratio is reduced, the relative shear deformation within the plastic region increases. Lateral confinement has significant influence on the ultimate drift, of which an increase from 0.001 to 0.003 could increase the ultimate drift capacity by up to 100%. At low to moderate shear span, contributions of shear and bond slip to the deformation must be considered. However, deformation outside the plastic region could be neglected for a column with low shear span ratio.

ACKNOWLEDGEMENTS

The authors would like to thank The Hong Kong Polytechnic University for the financial assistance provided through the Area of Strategic Development Fund.

REFERENCES

Inoue N., Inai E., Wada A., Kuramoto H., Fujimoto I. and Ilba M. (2000). A Shaking Table Test of Reinforced Concrete Frames Designed Under Old Seismic Regulations in Japan. 12th World Conferene on Earthquake Engineering, Auckland, New Zealand.
Lam S.S.E., Wu B., Wong Y.L., Chau K.T., Chen A and Li C.S. (2000). *Seismic Study for Hong Kong: Ultimate drift ratio of rectangular reinforced concrete columns in Hong Kong when subjected to earthquake action*. CIDSRC Report, July 2000.
Mo Y.L. and Hwang M.L. (1999). New Configuration of Lateral Steel for RC Columns. ASCE, *Practice Periodical on Structural Design and Construction* **4(3)**, 111-118.
Nakata S., Sproul T. and Penzien J. (1978). *Mathematical Modelling of Hysteresis Loops for Reinforced Concrete Columns*. Report No. UCB/EERC-78/11.
Wehbe N., Saiidi M. and D. Sanders (1999). Seismic Performance of Rectangular Columns with Moderate Confinement. ACI, *Structural Journal* **96(2)**, 248-258.
Xiao Y. and Martirossyan A. (1998). Seismic Performance of High-Strength Concrete. ASCE, *J. of Structural Engineering* **124(3)**, 241-251.
Xiao Y., Tomli M. and Sakino K. (1986). Experimental Study on Design Method to prevent Shear Failure of Reinforced Concrete Short Circular Columns by Confining in Steel Tube. *Trans. Japan Concrete Inst.* **8**, 535-542.

Behavior of Full-Scale Lightly Reinforced Concrete Interior Beam-Column Joints under Reversed Cyclic Loading

Hui Yin[1], Paulus Irawan[1], Tso-Chien Pan[1], and Chee Hiong Lim[2]

[1] Protective Technology Research Center, School of Civil and Structural Engineering,
Nanyang Technological University, Singapore
[2] Buildings and Infrastructure Division, Defense Science and Technology Agency, Singapore

ABSTRACT

This paper presents four full-scale testing on lightly reinforced interior beam-column joints under reversed cyclic loading. The main characteristic of the specimens is that no transverse reinforcement was provided in the joint panel and the beam reinforcement had large diameter resulting in relatively poor bond condition. All the specimens failed in joint panel with gradual strength deterioration, severe stiffness and bond degradation. Anchorage of the beam reinforcement passing through the joint panel played a prominent role on the column shear strength, while column axial load had relatively stronger effect on the deformational behavior of the beam-column sub-assemblage.

KEYWORDS

Beam-column joints, lightly reinforced, reversed cyclic loading, bond deterioration, column axial load.

INTRODUCTION

Most of the reinforced concrete buildings in Singapore are designed according to the British Standard BS8110, in which the design load includes the dead load of the structure and imposed load in the vertical direction, as well as small notional horizontal load for the robustness of the structural system. However, there is no provision for seismic effect in this design code. Therefore, these reinforced concrete buildings typically with reinforcing details that are in contrast to the modern seismic design standards[1-3], are susceptible to possible ground shocks resulting from earthquakes or underground explosion. Lightly reinforced beam-column joints, where no transverse reinforcement is provided, are generally considered one of the most critical parts of a reinforced concrete frame structure under ground shocks. However, most of the studies on beam-column joints in the literature have been focusing on ductile details. There is a scarcity of information on lightly reinforced joints[4,5]. Therefore, there is a need to study the behavior of the lightly reinforced concrete beam-column joints under simulated earthquake loading, i.e. reversed cyclic loading.

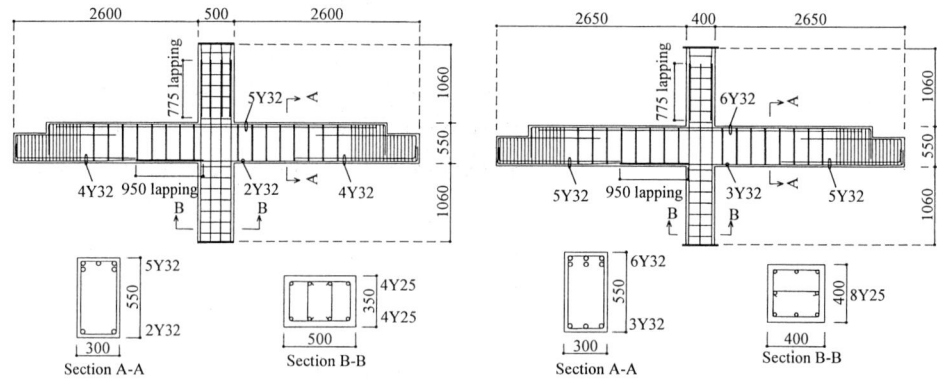

Figure 1 (a) Elevation of specimen C1   Figure 1 (b) Elevation of specimen C4

This paper presents an experimental testing on four full-scale lightly reinforced concrete interior beam-column joints under reversed cyclic loading. The specimens represented typical beam-column joint details from the existing low-rise industrial building in Singapore. The main characteristics of the specimens were that no transverse reinforcement was provided in the joint panel and the beam reinforcement had relatively large diameter resulting in very poor bond condition. The effect of bond deterioration along beam reinforcing bars and column compressive axial load equal to 15% column squash capacity was particularly studied.

EXPERIMENTAL TESTING

Specimen Descriptions

There are four full-scale specimens comprising two types of structural details denoted as C1 and C4. The dimensions and reinforcement details for each specimen are shown in Figures 1 (a) and (b). The column compressive axial load equal to 15% squash capacity was considered in specimens C1A and C4A, while no axial load was applied in specimens C1B and C4B. The nominal concrete cover of the columns was 40mm and that of the beams was 25mm. The compressive strength of concrete f_c' was 23.7 MPa, and the Young's modulus of concrete E_c was 26.2 GPa. The yielding strength f_y of main reinforcement was 500 MPa and 510 MPa for Y32 and Y25 bars, respectively. The anchorage length across the joint panel in lightly reinforced beam-column joints are normally far below the modern code requirement for ductile joints as compared in Table 1. Therefore, severe bond deterioration across the joint panel is very likely to occur under the reversed cyclic loading.

Loading Arrangement, Instrumentation and Measurement

Figure 2 presents the testing arrangement, where the reversed cyclic loading expressed in terms of story drift ratio as shown in Figure 3 was applied at the ends of the beams. The compressive column axial load was applied at the top of the column by a hydraulic jack. Figure 4 depicts the measuring arrangement. A total of twelve Linear Variable Displacement Transducers (LVDT) were used to measure, respectively, the total story drift at beam ends, the rigid body rotation at column ends due to the elastic deformation of the testing rig, and the plastic rotation at the beam-joint and column-joint interfaces mainly due to concrete cracking and bond loss between reinforcement and concrete. In the joint panel, the joint shear deformation was captured by installing inclinometers which had a

Table 1: Anchorage ratio across the joint panel

	Column depth to beam bar diameter (h_c/d_b)	Beam depth to column bar diameter (h_b/d_c)
Specimen C1	15.6	22
Specimen C4	12.5	22
ACI 318-99	20	20
NZS 3101-1995	30.8	31.4
AIJ-1997	33	33

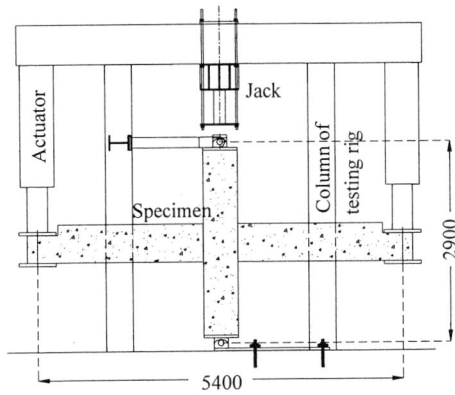

Figure 2: Loading arrangement

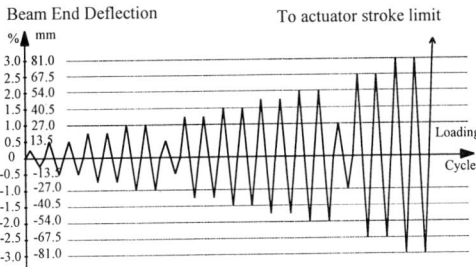

Figure 3: Loading history

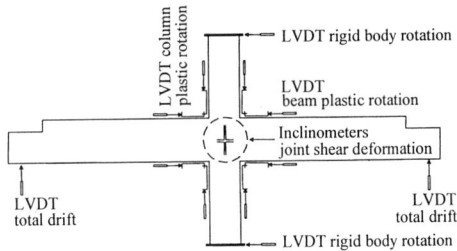

Figure 4: Measurement arrangement

measuring range of ±1.5 degrees. Electrical strain gauges were installed along the main reinforcing bars passing through the joint panel and at a minimum spacing of 150mm.

DISCUSSION OF TESTING RESULTS

Cracking Pattern and Failure Mode

In general, all specimens experienced similar cracking pattern and ultimate failure mode. All four specimens displayed joint panel failure with large block of concrete spalling and reinforcement bar exposure at the end of the testing. The framing members – beams and columns – remained relatively intact throughout the test. Figure 5 shows the typical cracking patterns at 2% story drift and the ultimate failure states for specimens C4A and C4B. When the effect of column axial load is concerned, specimen with column axial load C4A experienced less amount of cracking in the columns compared to C4B without axial load, since compressive axial load helped the closure of cracks. Meantime, when the column axial load was present, the joint diagonal cracks were mostly confined within the joint panel itself, whereas they propagated further into the column chord when the axial load was not present. At the final loading stage, it was observed that the area of concrete spalling even extended into the columns in specimens C4B where the column axial load was absent. In addition, the diagonal cracks in specimens C4A tended to be steeper than those in C4B by a range of 5 to 10 degrees, respectively. This indicates a relatively larger and more vertical concrete compressive strut had taken part in the shear resisting system when the compressive axial load was present.

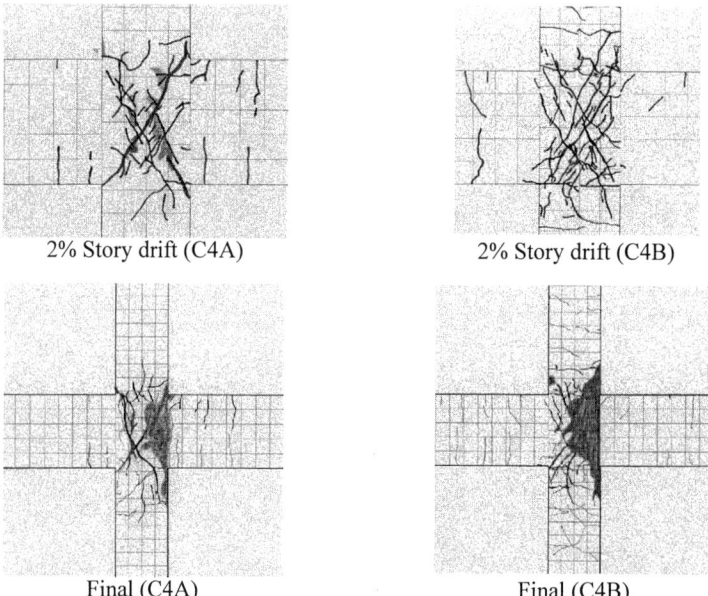

Figure 5: Typical cracking pattern and failure mode

Degradation of Column Shear V_c

The column shear force versus total story drift hysteresis loops for all specimens are shown in Figure 6. On one hand, it is interesting to observe that all the hysteresis loops display relatively gradual strength deterioration. This is different from what had been expected that the column shear strength would drop catastrophically once it reached the peak because the joint shear resisting capacity mainly contributed by the concrete in the joint core had been expected to deteriorate suddenly at the peak. On the other hand, very severe pinching is observed in the hysteresis loops especially after the drift ratio of 1.5%. This is in contrast to the behavior of a ductile beam-column joint, for which the hysteresis loops depicting the behavior of column shear versus story drift are supposed to have a spindle shape representing more desirable energy dissipation behavior. The pinching or stiffness degradation is caused by inelastic deformations due to shear and bond mechanisms after extensive diagonal cracking in the joint panel. Under reversed cyclic loading, when the applied force is reversed in direction, and before the previous cracks are closed, the compressive stress within the joint panel should have been resisted by the longitudinal reinforcement if sufficient bond strength is available across the joint width. However, if the bond strength deteriorates severely across the joint, the reinforcing bar will slip until the cracks close so that the concrete is able to resist the compression. This causes the column shear versus story drift curve to be relatively flat when the force is reversed, and it remains to be flat until the closing of concrete cracks.

The column shear strength envelopes are plotted in Figure 7 (a) for the comparison between specimens with axial load C1A and without axial load C1B. Figure 7 (b) shows the difference between specimens C1B and C4B, where C1B had a 16% larger effective joint volume and a 25% longer anchorage length for beam main bars passing through the joint panel.

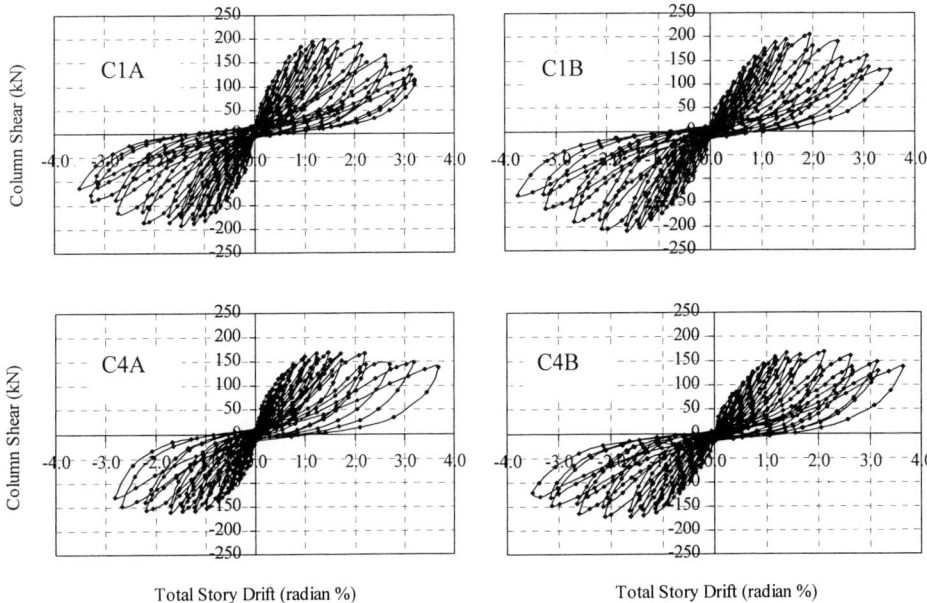

Figure 6: Column shear versus total story drift

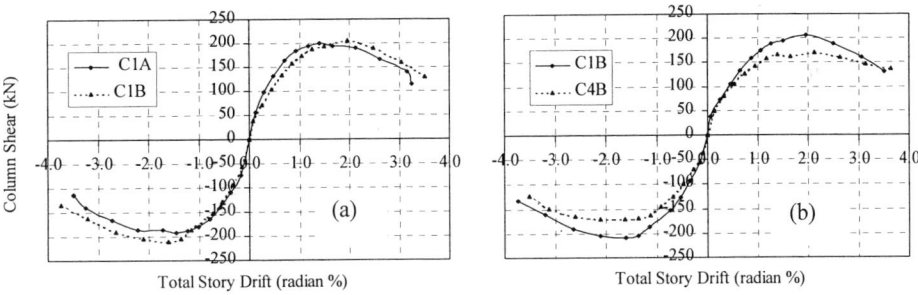

Figure 7: Comparison of column shear strength

In Figure 7 (a), it is observed that, initially C1A displays higher strength, at around 1.5% story drift, C1B surpasses C1A and remains to be stronger than C1A at the same drift level until the end of the loading history. During the early loading stage and before reaching the peak capacity when the joint core itself was still relatively strong, the column axial load stiffened the overall beam-column sub-assemblage. Hence, at the same drift level before the peak capacity, specimens with column axial load resulted in a larger shear resistance. After reaching the peak capacity, the concrete compressive strength f_c' had probably been deteriorated very much due to excessive diagonal cracking in the joint panel, the column axial load at this moment only added more load to the diagonal strut, so it ultimately undermined the column shear capacity. The magnitude of the difference on the column shear strength between specimens with column axial load (C1A) and without column axial load (C1B) was very minimal. However, there was a large disparity between the drift ratios at which the peak values occurred. The column compressive axial load in specimen C1A had accelerated the peak to occur by 45% compared to C1B. The above observation was equally true for specimens C4A and C4B. Hence it proved that column compressive axial load could lead to higher stiffness in the early loading stage. On the other hand, Figure 7 (b) shows that specimen C1B had significantly 20% higher peak column shear strength than specimen C4B. However, C1B degraded faster than C4B after the peak strength. This is believed to be due to the relatively larger bond stress built up in specimen C1B. A higher bond stress between concrete and reinforcing bar definitely leads to a higher bond demand. Although C1B had a relatively better anchorage condition than C4B, the anchorage length in both of these lightly reinforced joints were far below the code requirement. Therefore, after the peak column shear strength was reached, the available bond capacity in both C1B and C4B was believed to be very low, and C1B with the higher bond demand resulted in a more rapid strength degradation.

Stiffness Degradation

The stiffness degradation for specimens C1A and C1B is presented in Figure 8. For the convenience of comparison, stiffness at each loading stage has been converted to a ratio of the initial stiffness in the column shear versus total drift ratio plots. The initial stiffness exists when the sub-assemblage is still under the linear-elastic range. Graphically, it is the tangent value of the initial stages before the column shear-total drift relationship becomes non-linear. The stiffness at the later loading stages is obtained from the average secant stiffness which is the slope of the peak to peak values within one loading cycle.

It is observed that the stiffness degraded very rapidly especially before the drift ratio of 0.5%. At this stage, the stiffness had dropped to about 40% of the initial stiffness for specimen C1B without column axial load, and 50% for specimen C1A with column axial load. At 1.5% story drift, in every specimen, the remaining stiffness was only about 20% of the initial. Beyond 3% drift ratio, the specimens had been very severely damaged with a stiffness less than 10% of the initial. It is well known that even though the strength of a structure is high, the excessive deformation caused by the extremely low stiffness will also render the structure to be out of service.

The effect of column compressive axial load can also be clearly identified from Figure 8. Before the drift ratio of 1.5% to 2%, the compressive axial load helped the structure to retain its stiffness. With a relatively slower degradation process, the specimens with axial load showed a higher stiffness value by up to 15%. However, the effect of axial load diminished beyond the drift ratio of 2% and all specimens had about the same stiffness ultimately. The same observation was realized between specimens C4A and C4B.

Beam, Column and Joint Deformation Contributions to the Total Story Drift

Generally, the total story drift comprises beam and column elastic deformation, beam and column plastic deformation, and joint shear deformation. In the modern seismic design models, there is no specific requirement for joint shear deformation in ductile beam-column joints. Actually, all these design models implicitly assume that the joint shear deformation is negligibly small if the design is strictly carried out according to the code specification, and most of the story drift should be accumulated from the beam plastic deformation. In the current local design practice in Singapore, no seismic loading is considered, and the beam-column joint panel itself is always treated as rigid under all types of loading.

The typical progress of beam, column and joint deformation contributions as the total story drift increased is shown in Figure 9. The following observations typical to every specimen are summarized here. Firstly, the joint shear deformation contribution suddenly jumped high at 0.25-0.50% drift. This was when diagonal cracking started to form in the joint panel, and the joint shear deformation

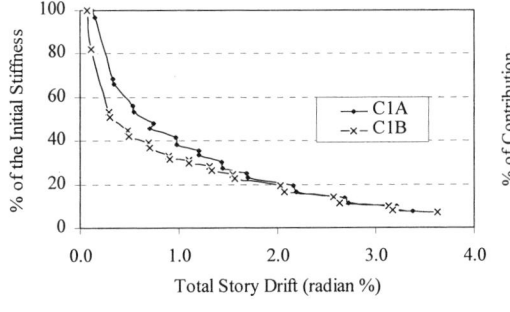
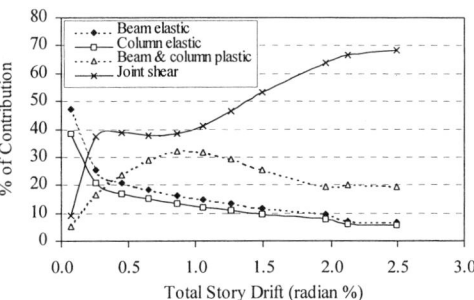

Figure 8: Stiffness degradation

Figure 9: Deformation contributions

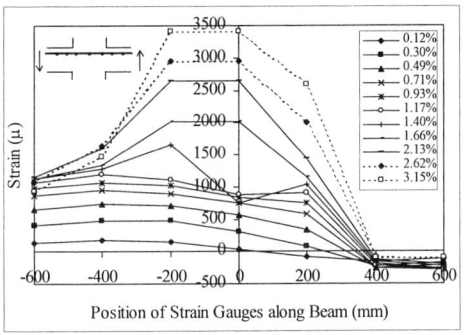
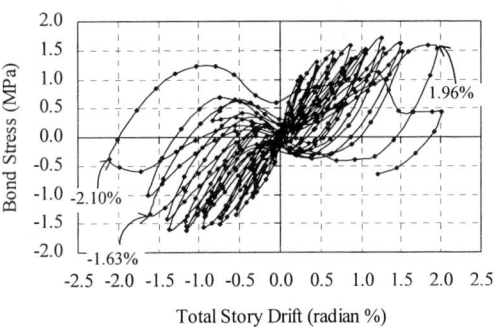

Figure 10: Beam top reinforcement stress distribution

Figure 11: C1B beam top reinforcement bond stress deterioration

contribution was about 35-40% of the total drift at this moment. After that, it increased rapidly all the way until it prevailed the other deformations and reached 60-70% of the total drift at 2.0-2.5% drift level. Secondly, the total plastic deformation of the beams and columns was no more than 20% of the total drift at the end of the collected data. Among this plastic deformation, the contribution from the columns was suspected to be the majority since the columns had much more cracking throughout the test, some of the major diagonal cracking in the joint panel even propagated into the upper columns. Lastly, the elastic deformation contribution of the beams and columns was descending all the time, especially in the initial stage before 0.5% drift, the elastic deformation contribution drastically dropped. Ultimately, each of them accounted for no more than 5~10%.

Bond Deterioration across Joint Panel

The typical stress distribution along the main reinforcing bars across the joint panel is plotted in Figure 10. It is observed that the reinforcement was in tension throughout the joint panel regardless of the loading direction except during the very initial loading stage, i.e. before about 0.2% story drift when the first diagonal crack was formed. Outside the joint panel, the length of the reinforcement in compression was also reducing as the story drift increased. All these facts clearly indicated that bond deterioration had occurred along the beam and column reinforcement.

The bond stress degradation process for the beam top reinforcement in specimen C1B is plotted in Figure 11. When the pattern of the bond stress deterioration is compared with the process of the column shear strength degradation as presented previously in Figure 7, a very close correlation could be established. At story drift -1.63%, the bond stress had degraded by 40% and it suddenly dropped very severely at drift -2.10%. Specimen C1B achieved its peak strength at drift -1.63%, and from drift -1.63% to -2.10%, the column shear strength degraded negligibly small, i.e. from -209.1kN to -205kN. As the loading direction was reversed, the remaining bond strength was far below that at the previous story drift 1.96%, and this was exactly the story drift when the peak column strength 205.4kN occurred. Hence, it is concluded that the column shear strength was directly correlated with the bond condition of the beam reinforcement. The degradation of the column shear strength was actually due to the sudden deterioration of the beam bond within the joint panel.

CONCLUSIONS

1. This series of lightly reinforced beam-column joints all failed in joint panel with severe stiffness degradation, but relatively gradual strength degradation under reversed cyclic loading.
2. The column axial load with the magnitude of 15% column squash capacity stiffened the beam-column sub-assemblage before the peak column shear strength was reached, however it had very insignificant influence on the magnitude of the peak column shear strength. In addition, the existence of column compressive axial load resulted in more concentrated diagonal cracking within the joint panel and prevented the diagonal cracking in the joint panel from propagating into the columns.
3. The behavior of the column shear strength was directly correlated with the bond condition of the beam reinforcement. The degradation of the column shear strength was due to the bond deterioration of the beam bars across the joint panel. The peak column shear occurred when the bond condition had dropped very much or when the bond suddenly collapsed. Specimens C1 with better anchorage condition reached significantly higher column shear strength than specimens C4.

ACKNOWLEDGEMENTS

The financial support by the Lands & Estates Organization of Ministry of Defense, Singapore, is gratefully acknowledged.

REFERENCES

1. ACI Committee 318, "*Building Code Requirements for Structural Concrete (ACI 318M-95) And Commentary – ACI318RM – 95*", American Concrete Institute, Farmington Hills, Michigan, 1995.
2. Architectural Institute of Japan, "*Design Guidelines for Earthquake-Resistant Reinforced Concrete Buildings Based on Inelastic Displacement Concept*" *(draft)*, 1997. (In Japanese)
3. Standards New Zealand, "*Concrete Structures Standard, NZS3101: Part 1 – The Design of Concrete Structures*", Wellington, New Zealand, 1995.
4. Blaikie, E.L., "*Behavior of Unreinforced and Lightly Reinforced Concrete Beam-Column Joints*", Proceedings of Pacific Concrete Conference, New Zealand 8-11 November 1988, pp.181-193.
5. Hakuto, S., Park, R. and Tanaka, H., "*Seismic Load Tests on Interior and Exterior Beam-Column Joints with Substandard Reinforcing Details*", ACI Structural Journal, Vol.97, No.1, January-February 2000, pp.11-25.

EXPERIMENTAL AND ANALYTICAL ASSESSMENT OF SIMPLE BRIDGE STRUCTURES SUBJECTED TO NEAR-FAULT GROUND MOTIONS

C. H. Hamilton[a], G. C. Pardoen[a], R. P. Kazanjy[a], and Y. D. Hose[b]

[a]Department of Civil and Environmental Engineering, University of California, Irvine, California 92697-2175 USA
[b]Department of Structural Engineering, University of California, San Diego, California 92093-0085 USA

ABSTRACT

Recent laboratory experiments and analytical simulations conducted at UC Irvine and UC San Diego relating to response of simple bridge structures in the near-field of seismic events are presented in the context of variability of demand characteristics between near-field and far-field time-histories. Using quasi-static and pseudo-dynamic displacement protocols to simulate near-fault loading conditions, the performance of a series of reinforced concrete columns was assessed. The results of these tests, and a related series of tests at UC San Diego, were used to calibrate a finite element model for near-fault structural response of bridge columns. Models of simple bridge structures are under construction using this information for assessment of system response to near-fault ground motion records. Preliminary conclusions include a likely reduction in plastic hinge length under pulse-type ground motions and little change in ultimate load and displacement capacities for ductile-detailed columns.

KEYWORDS

bridges, earthquake engineering, near-fault ground motion, reinforced concrete columns, analytical modeling, experimental testing

ACKNOWLEDGMENTS

This project was supported by grant 5111999 from the Pacific Earthquake Engineering Research Center, a National Science Foundation supported Engineering Research Center. The collaborative environment fostered by the PEER Center greatly contributed to the success of this project.

INTRODUCTION

It has always been clear that structures close to the epicenter of earthquakes suffer an increased risk of damage compared to those further away. In relatively recent times, it has become clear that there are significant differences between near-fault and far-field ground motions. These differences give rise to substantial variation in the response of near-fault structures, especially those in the direction of rupture propagation. The 1994 Northridge and 1995 Kobe earthquakes highlighted this issue when several structures expected to withstand ground motion from similarly-sized earthquakes collapsed, Comartin et al. (1995); Buckle (1994). As a result, recent research efforts have begun to focus on the effects of the near-source strong velocity pulse or "fling" event on structural response.

A "fling" event is characterized by a strong pulse, usually asymmetric, in the velocity domain. The pulse results from directionality of the fault rupture which is "visible" to the ground immediately adjacent to the fault in the direction of rupture. As the fault rupture propagates, newly-generated shear waves reinforce the existing wavefront, increasing its amplitude, Hall and Aagaard (1998). This results in large, directional velocity pulses in the ground motion, giving rise to sharp accelerations and large ground displacements near the fault. For more detailed information and a literature survey on the topic, the reader is referred to Bell (1998) and Hall and Aagaard (1998).

This project employed a multi-campus collaborative approach to address several issues which impact the development of testing and design protocols for near-fault ground motions showing pulse-type effects. Work performed at UC San Diego for this project led to the development of a pulse-type displacement time history of approximately 3-second duration. A reversed-cyclic, drift-based sequential loading protocol, similar to that recommended in ATC-24, was used as the far-field baseline against which response to pulse-style loadings was compared. Calibrated numerical models have been developed for column behavior which will enable more realistic system behavior modeling under simulated pulse-type loads.

EXPERIMENTAL PROCEDURES

The primary goal of this project was to develop a comparison between the effects of near-field seismic motion on bridge structures and those of far-field seismic motion. To do so, it was necessary to develop loading protocols to simulate each of these conditions. Four loading histories were used for this project. The first is a modified ATC-24-style protocol in use at USC prior to the beginning of this project, ATC (1992). This protocol was used to simulate far-field motion. The second protocol, developed by Professor Scott Ashford at UCSD for this project, represents a pulse-type time history generated by combination of several strong-motion records. Ashford (1999) A modified version of this protocol was developed and used at UCI. Finally, a monotonic loading protocol was employed to enable comparison with pushover analyses.

To maximize the applicability of these results, as well as the range of available comparative data, single-column bents with circular sections were selected as the test specimens.

Test specimen

The column test sample was designed in accordance with current California Department of Transportation (Caltrans) specifications, Caltrans (2000a,b,c). Use of these design criteria limited this test series to consideration of ductile-detailed columns. The sample design had a 406 mm diameter with 123 mm clear cover to reinforcing spirals. The concrete for the column was specified to have f'_c of 28 MPa, not to exceed 38 MPa. It contained 9.5 mm aggregate and had a specified slump of 102 ± 25 mm.

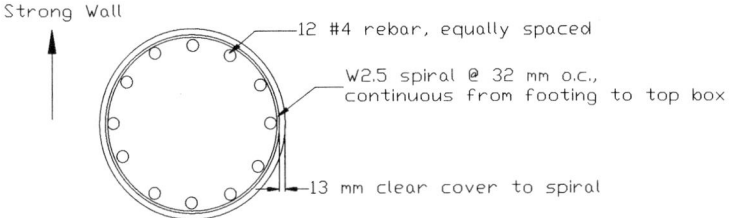

Figure 1: Section schematic of sample column

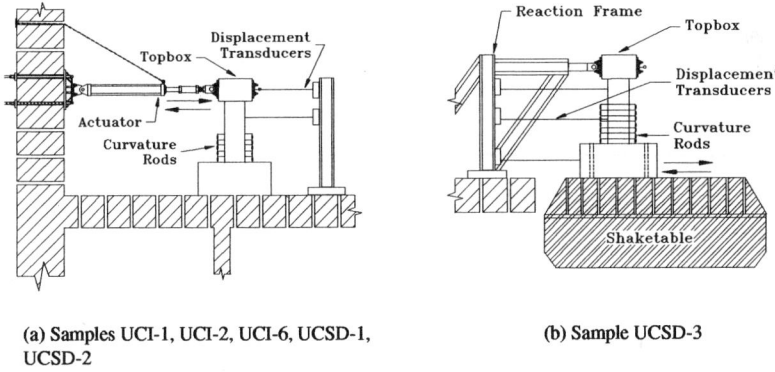

(a) Samples UCI-1, UCI-2, UCI-6, UCSD-1, UCSD-2

(b) Sample UCSD-3

Figure 2: Apparatus layout for column tests

Longitudinal reinforcing bars were CRSI #4 bars (13 mm diameter) A706 Grade 60, specified yield of 414 MPa $\leq F_y \leq$ 517 MPa. Reinforcing spirals were A82 Grade 80 W2.5 smooth wire (4.5 mm diameter, specified minimum yield $F_y =$ 552 MPa). Refer to Figure 1 for a schematic of the column section.

The tests were intended to show a column configuration with fixed (at the base)-pinned boundary conditions, similar to what is seen in seismic loading transverse to the long axis of highway bridges. Therefore, all samples were rigidly attached to the testing surface (strong floor/shaketable) by high-strength post-tensioning rod to ensure a fixed-base condition and had a clevis connection to the loading ram or reaction frame.

Test apparatus

Samples UCI-1, UCI-2, and UCI-6 were loaded using a manually-controlled hydraulic ram at low rates. Displacements were monitored using precision linear potentiometers attached to the column topbox. Samples UCSD-1 and UCSD-2 were loaded using servo-controlled hydraulic actuators slaved to LVDTs attached to the column topboxes. Sample UCSD-3 was tested using a shaketable to provide high-rate input of the UCSD pulse-sequential protocol to the column footing. These configurations are shown in Figure 2.

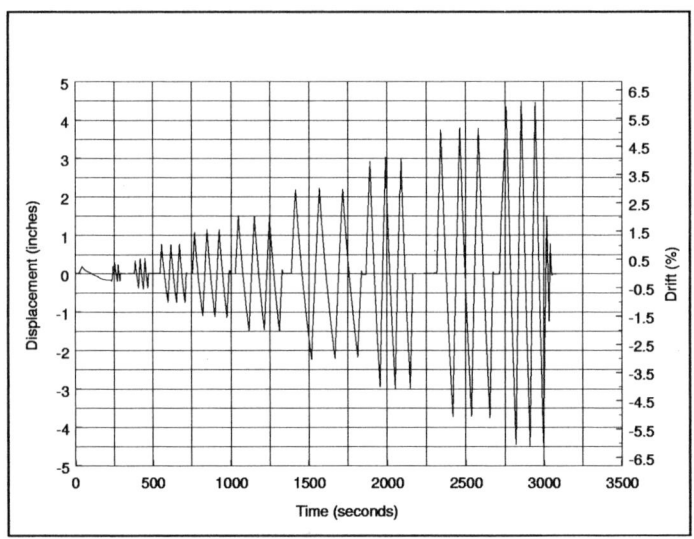

Figure 3: Modified ATC-24 loading protocol

Loading histories

All loading protocols used in this series of tests were drift-based, rather than using the traditional yield-basis. Thus, columns of identical height, regardless of section geometry or reinforcing ratios, will undergo the same set of loading cycles. This allows greater ease of comparison between elements of dissimilar section properties, an effect which will be useful in comparisons with future and past research.

Modified ATC-24-style loading protocol

Professor Yan Xiao at USC has developed a modified version of the ATC-24 reversed cyclic loading protocol. Xiao (1999) It can be distinguished from the ATC-24 protocol, ATC (1992), by several major features. First, it employs a drift-based displacement time history, where all levels of displacement are related to element height rather than to a computed yield displacement. Second, the protocol uses a different displacement level strategy from the ATC-24 protocol. The steps are by 0.25 percent increments to 1 percent drift. This is followed by increments of 0.5 percent steps up to 2 percent drift. From 2 percent to 6 percent, steps of 1 percent drift are used. Thus, the overall displacement time history imposed is as shown in Figure 3. The column tested using this protocol was designated UCI-1.

"Pulse" loading protocols

Professor Scott Ashford at UCSD developed a loading protocol based on the combination of 34 near-fault ground motions. The resulting "average pulse" was prefixed to the reversed cyclic protocol to generate the drift time history shown in Figure 4. All UCSD samples were tested using variants of this protocol. These variants are:

1. Initial pulse amplitude of +6.5 percent/-5.25 percent at dynamic rate (Sample UCSD-1)

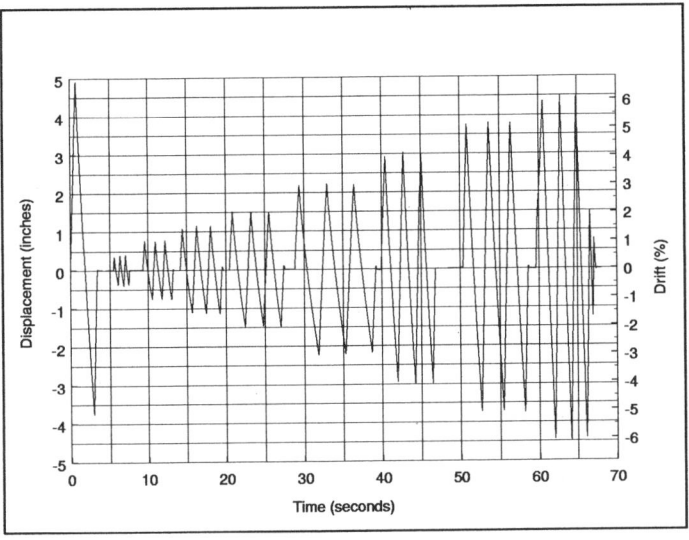

Figure 4: UCSD pulse-sequential loading protocol

2. Initial pulse amplitude of +6.5 percent/-5.25 percent at quasi-static rate (Sample UCSD-2)

3. Initial pulse amplitude of +5.5 percent/-3.8 percent at quasi-static rate (Sample UCSD-3)

The value of simply combining a "representative pulse" with the existing cyclic protocol, rather than developing en entirely different protocol as was done in Krawinkler et al. (2000), is that the effect of the pulse can be isolated from other effects such as frequency content shift in the non-pulse segment of the test protocol. Substantially more complex protocols are outside the scope of this research at this time.

UCI developed a modified version of this protocol in which the initial pulse pattern was repeated three times before applying the reversed-cyclic protocol, all at a quasi-static rate. The first repetition's amplitude will be referred to as "displacement level 1". The second repetition increased the displacement levels by 50 percent over level 1. The third repetition doubled level 1. However, due to failure of the test sample, the protocol was terminated before commencing the cyclic protocol. The resulting displacement time history is shown in Figure 5. The sample tested using this protocol was designated UCI-2.

RESULTS

Comparison of experimental results with response to typical loading protocols

Limited changes were found when comparing response to loading histories containing large-displacement cycles at the outset with response to histories lacking these large amplitude pulses. No substantial changes were observed in the load-displacement backbone curves from history to history, as can be seen in Figure 6. Minor increases in initial stiffness were observed, but nothing outside the normal range of test variability or that expected due to rate effects.

However, a substantial reduction in effective plastic hinge length below that predicted by Priestley et al. (1996) was observed in three columns subjected to large initial displacement cycles. The plastic hinge

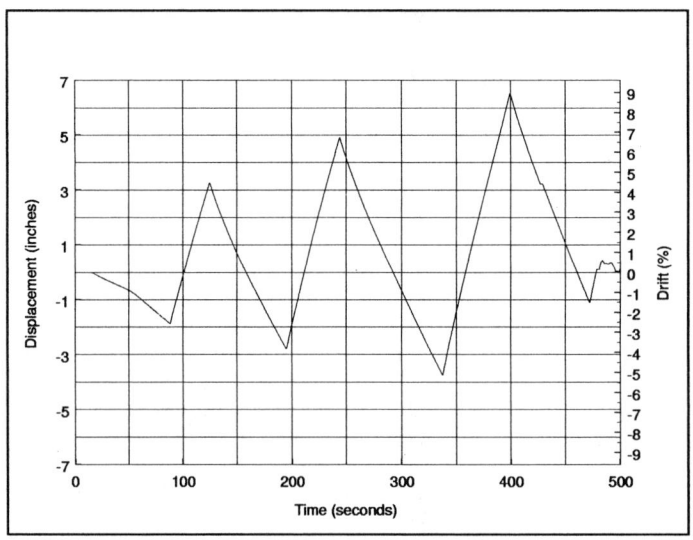

Figure 5: UCI pulse loading protocol

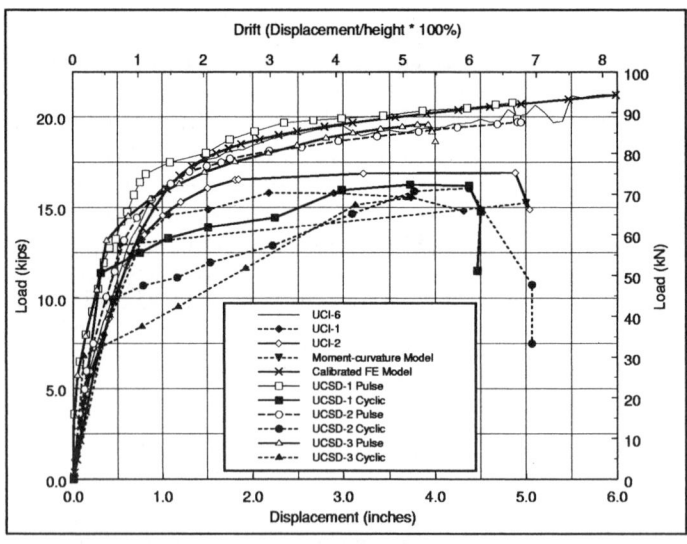

Figure 6: Comparison of load-displacement hysteresis backbones

TABLE 1
PLASTIC HINGE LENGTHS OF EXPERIMENTAL SAMPLES

Sample	UCI-1		UCI-2		UCI-6		UCSD-1		UCSD-2		UCSD-3	
Column face	Push[1]	Pull	Push	Pull	Push	Pull	Push	Pull	Push	Pull	Push	Pull
L_p (mm)	267	229	186	203	51	152	356	254	121	191	152	152

"Push" indicates direction of initial loading

length, L_p, predicted by the equation of Priestley et al. is

$$L_p = 0.08L + 0.022F_y d_{bl} \geq 0.044F_y d_{bl} = 264 \text{ mm} \qquad (1)$$

where L is the length of the sample from inflection point to the critical section, d_{bl} is the diameter of the longitudinal reinforcing bars, and all other symbols are as previously identified. The experimentally observed plastic hinge lengths are shown in Table 1. Note that UCI-1 shows a plastic hinge length similar to that predicted by Eqn. 1. This is expected as Eqn. 1 is based on reversed-cyclic loading response, albeit under a slightly different protocol. However, samples UCI-2, UCSD-2 and UCSD-3 show plastic hinge lengths substantially shorter than that predicted above. Their plastic hinge lengths are between the average plastic hinge length obtained for the monotonic test, UCI-6, and that predicted by Priestley et al. This is consistent with the researchers' expectations as the large displacement pulses, even when reversed, tend to concentrate damage around the initial weaknesses and flaws in the sample, similar to what is seen in monotonic tests. The results for sample UCSD-1 appear to contradict this result, although being only one sample, it may be a unique result. Additional testing is planned to confirm or deny this reduction of plastic hinge length under pulse loadings.

Analytical modeling of pulse response

Three phases of analytical modeling are being performed for this project. First, the monotonic pushover behavior of the sample must be modeled. Second, reversed cyclic behavior must be matched, including degradation characteristics. Third, dynamic behavior of the member must be modeled, including degradation and rate effects. The platform being used for this modeling is the OpenSees platform under development by the PEER Center. McKenna and Fenves (2000)

The first stage of the modeling has been completed and its results have been incorporated into Figure 6. The model also replicates effectively the plastic hinging behavior of the monotonic sample.

Currently, work is underway to complete modeling of reversed cyclic response of the column. This model does not yet properly show strength degradation, but its modeling of plastic hinge behavior is promising.

Experimental tests being performed at UC Berkeley are intended to serve as a source of data for calibration of the model's response to dynamic excitation.

CONCLUSIONS

Although far from exhaustive, this test program allows certain preliminary conclusions to be drawn regarding response of reinforced concrete columns to large displacement cycles. The ultimate load and displacement characteristics of ductile-detailed columns do not seem to be significantly affected by the inclusion of an initial large-displacement cycle representing a pulse or by application of large-displacement cycles only. They are subject, however, to increased curvature ductility demands based on the apparent reduction of the plastic hinge region under low-cycle large-displacement loading. In extreme cases, this could lead to failure of even ductile-detailed columns early in an earthquake containing a strong pulse component to its motion.

Research efforts are ongoing, with additional columns scheduled to be tested at UC Irvine under high-rate quasistatic and pseudodynamic loading protocols. These columns will allow for additional tests to verify the apparent reduction of plastic hinge length under pulse-type loadings, as well as enlarge the sample data base against which the finite element models may be verified. UCSD is also continuing this project with tests of reinforced concrete joint assemblies. Additional work funded by the PEER Center is underway at UC Berkeley comparing long-duration and pulse-type ground motion effects. All of this work is expected to contribute to the body of knowledge developed through the PEER Center regarding loading pattern effects on seismic response of structures.

REFERENCES

Ashford S. Private communication (1999).

ATC. Guidelines for Cyclic Seismic Testing of Components of Steel Structures. ATC-24, Applied Technology Council (1992).

Bell L. Defining the Near-Field Velocity Pulse. Peer Summer Internship Report, University of California, San Diego (1998).

Buckle I.G. The Northridge, California Earthquake of January 17, 1994: Performance of Highway Bridges. Technical Report NCEER-94-0008, National Center for Earthquake Engineering Research (1994).

Caltrans. Bridge Design Specifications. Technical report, California Department of Transportation (2000a).

Caltrans. Bridge Memo to Designers Manual. Technical report, California Department of Transportation (2000b).

Caltrans. Seismic Design Criteria. Technical report, California Department of Transportation (2000c).

Comartin C.D., Greene M. and Tubbesing S.K. The Hyogo-Ken Nanbu Earthquake: Great Hanshin Earthquake Disaster January 17, 1995. Preliminary Reconaissance Report 95-04, Earthquake Engineering Research Institute (1995).

Hall J. and Aagaard B. Fundamentals of the Near-Source Problem. *In Fifth Caltrans Seismic Research Workshop*. California Department of Transportation (1998). Session 1, Paper 4.

Krawinkler H., Parisi F., Ibarra L., Ayoub A. and Medina R. Development of a Testing Protocol for Wood Frame Structures. Draft, CUREE (2000).

McKenna F. and Fenves G.L. The OpenSees Command Language Primer. Version 1.0, Pacific Earthquake Engineering Research Center (2000).

Priestley M., Seible F. and Calvi G.M. *Seismic Design and Retrofit of Bridges*. John Wiley & Sons, New York (1996).

Xiao Y. Private communication (1999).

SEISMIC RESPONSE OF REINFORCED CONCRETE FRAMES USING NON LINEAR MACRO-ELEMENT BEHAVIOR

S.M. ELACHACHI [1], M. BENSAFI [1], D. NEDJAR [1]

[1] Department of Civil Engineering, University of Sciences and Technology of Oran
B.P. 1505 elmenouar, Oran, Algeria

ABSTRACT

This paper deals with the computation of reinforced concrete frame structures with a simplified method, using non linear macro-elements. A multi-level approach based on the notion of macro-element is presented. This approach brings the rapidity of an analysis with a reduced number of degrees of freedom (spatial and dynamic) while keeping relatively the refinement of a local analysis. Concerning the dynamic aspect, a method of analysis which uses a step-by-step integration method and whose objective is an evaluation of the non linear seismic response of structural systems, using a modal superposition technique which greatly decreases the required computational effort, is passed.
The assumption that is given out is that the non linear terms of the equations of motion can be considered, without any simplification, as additional external forces, and thus, the resulting equations of motion can be physically interpreted as the equations of motion of a conventional linear system excited in its base by a modified ground motion. Some numerical examples are presented which demonstrate the accuracy as well as the effectiveness of the method.

KEYWORDS

Reinforced concrete, simplified method, macro-elements, seismic analysis, non linear behavior

INTRODUCTION

Reinforced concrete frames are widely used for buildings placed in high seismicity zones. The inelastic incursions imposed during earthquakes require efficient numerical methods to carry out a non-linear analysis of the whole structure. This analysis is needed to correctly identify the cyclic degradation process and/or the imminent failure of the structural system as well as the coupling between acting loading components. Numerical modelling of reinforced concrete (RC) frame structures for non linear seismic analysis is commonly carried out by means of local analysis in order to compute the representative behavior of structures, where one must detail as precisely as possible the phenomena occurring at the microscopic level. For example, the constitutive law of a material is expressed on a representative elementary volume which size depends on the continuous media theory. The use of complex constitutive laws in the structural computing correctly describe the test observations in most of the cases. However this accurate local analysis leads to very expensive (if feasible) computations : great number of degrees of freedom, iterative solvings which are very consuming time.

In an other hand, global analysis which uses global element models aiming at describing the behavior of each structural element (beam, column or inter-storey portion of wall) with just a single element in the numerical model, appears as a too simplified modelling through which the structural response is ill described, and is not consistent with general bending moment distributions along the element, being given that the non linearity is restricted to the fixed plastic zones, both the plastic length development and the spread of cracking effects inside the internal linear elastic element cannot be taken into account, thus preventing an adequate simulation of the global structural stiffness.

An intermediate level consists in studying the response of an intermediate representative elementary volume whose size is between the size of the local representative elementary volume and the size of the structure. This intermediate representative elementary volume is called a 'macro-element' (Elachachi, 1992). To construct its modelisation, one needs to go throw three steps:

First step : identifying the response of the macro-element, called 'working law', as the result of the accurate local analyses. The behavior of the concrete is modelled with a damage law. A detailed local analysis is made once for each kind of concrete geometry of the whole structure at the beginning.

Second step : synthesizing the working laws under the form of explicit or numeric expressions relating averaging variables (forces and displacements for example) and including geometrical and material parameters. The responses of these macro-elements are stored in a data base. Those two steps constitute the local analysis.

Third step : assembling the macro-elements to make the global computation of the structure. The global analysis of the structure made at this third step, uses the responses data base, without coming back to the local level.

It is to note that the local analysis is uncoupled with the global analysis. From an industrial point of view, only the third step is needed, the working laws being stored in a data base.

LOCAL ANALYSIS

The choice of the macro-element and the associated macro-variables is first done. For the study of beams and columns, the simplest macro-element is a portion of beam described in figure 1. The macro-variables are the generalized forces (bending moment and normal force) and the displacements and rotation of the ending faces. It will thus be easy to introduce this macro-element in a finite element code because it uses the same variables. To study the response of the macro-element and to build the working laws, it is discretised into layers (Davenne & Elachachi 1992) in which the variables are constant (strain, stress, internal variables of the constitutive law, ...). The response of the macro-element must be obtained without knowing the structure where it will be.

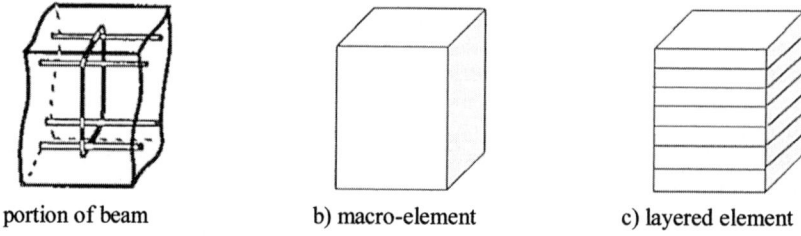

a) portion of beam   b) macro-element   c) layered element

Figure 1: beam macro-element

Modelling the behavior of the concrete

The behavior of the concrete is modelled with a continuous damage law evolved by Mazars (1984). The progressive degradation of the concrete (generation, growth and coalescence of microcracks leading to macrocraks) is described using a scalar variable D wich affects the stiffness of the material:

$$\sigma = E_o (1 - D) \varepsilon \quad (1)$$

where σ is the tensor of constraints, E_o the initial stiffness tensor, ε the strain tensor and D measures the damage, that evolves between values 0 (state initial virgin) and 1 (broken state).
The progression of the damage is guided by the evolution of an equivalent strain, calculated from the positive main strains $<\varepsilon_i>+$:

$$\tilde{\varepsilon} = \sqrt{<\varepsilon_i>_+^2} \quad (2)$$

where

$$<x>+ = (|x| + x)/2 \quad (3)$$

Under any sollicitation state, the variable of damage results from a combination of an owing damage to the D_t traction and an owing damage to the D_C compression:

$$D = \alpha D_t + (1-\alpha) D_C \quad (4)$$

The definition of a threshold function permits to start the evolution of the damage. The expression of this function is:

$$f = \tilde{\varepsilon} - k(\tilde{\varepsilon}) \quad (5)$$

where $k(\varepsilon)$ is equal to the maximal value reached by the equivalent strain during the loading and is worth initially ε_o. Laws of evolution of D_t and D_C are identified from experimental tests.

Modelling the behavior of the reinforcement

The model proposed by Dodd & Cooke (1994) is simplified to predict the monotonous behavior as well that the hysteretic behavior of steel. The cyclic effect is accounted for through the degradation of the reloading stiffness, depending on the maximum strain in each direction. A pow-function describes the response from zero stress to the maximum strain attained in the opposite direction. This model was validated on steel coupons tested by Dodd & Cooke (1994).

Obtaining of the working laws

Considering Euler Bernoulli assumption (shear influence not taken into account), the strain of a layer is obtained from the displacements:

$$\varepsilon = -\frac{\Omega}{\Delta L} + \frac{u}{\Delta L} = -\chi y + \eta \quad (6)$$

where y is the position of the layer (the reference axis is the middle of the depth), χ is the curvature, η the axial generalized strain, Ω the rotation, u the axial displacement and ΔL the length of the macro-element.
The contribution of the concrete and of the steel are separated. The whole macro-element is treated as a superposition of the concrete and of the steel. The macro-element is subjected to a proportionnal loading path, i.e. the variables Ω and u growth with a given ratio. The strain in each layer is computed

with the Eqn. 6. Then the stress in each layer is obtained with Eqn. 1. By integrating it is possible to compute the bending moment (at the reference axis) and the normal force. In order to keep efficiency in the global computation (the global matrix will be inverted only one time) and to have less storage, they are expressed as :

$$\{F\} = [K_e]\{d\} + \{F_{an}\} \tag{7}$$

$$\{F\} = \begin{Bmatrix} N \\ M \end{Bmatrix} \qquad [K_e] = \begin{bmatrix} K_N & K_C \\ K_C & K_M \end{bmatrix} \qquad \{d\} = \begin{Bmatrix} \eta \\ \chi \end{Bmatrix} \qquad \{F_{an}\} = \begin{Bmatrix} N_{an}(\eta,\chi) \\ M_{an}(\eta,\chi) \end{Bmatrix} \tag{8}$$

where K_N, K_M, K_C represent the initial stiffness, N_{an} and M_{an} the anelastic respectively normal and flexural forces which depend directly on the kinematic variables Ω and u, variables which evolve with the damage evolution of the concrete.
The functions N and M of the expression (8) are the working laws. These functions depend not only of Ω and u, but also on the path followed to arise this couple (Ω, u). It is not possible to store all the paths that could be encountered for each macro-element during the structure computation (there is an infinity of paths). In the case of monotonic loading, a data base of responses is built with a limited number of radial loading paths. Practically, the data base of one working law (i.e. a concrete geometry) is made with 20 different loading paths for a non symmetric section (T beam for example) and 10 for a symmetric section.

CYCLIC ANALYSIS

Presently, the most widely used hysteretic models are those based on a refined phenomenological law, fitted by Takeda *et al.* (1970) to the experimental load-displacement results of testing RC sub-assemblage on earthquake simulator. This law consists of tri-linear spines for monotonic loading, separately for positive and negative loading, with change of slope at cracking and yielding. It has 16 rules for unloading and reloading, covering all possible sequences. However, simpler versions, Q-hyst model by Saiidi *et al.* (1981), operating on a symmetric bilinear primary curve with simplified description of cyclic behaviour, provide in almost the same fit to non linear dynamic response results as the much more complicated original Takeda model (Saiidi 1982).

Modelisation of cyclic behavior

The complexity of the overall involved phenomena has been eliminated by simplified decoupled phenomenological models disable to consider a true interaction of normal force, shear and flexion. For more general loading histories, this approach is no longer reliable. An analogy with the theory of plasticity can then be used in order to take into account the interaction of these generalised variables.

To take into account the cyclic loading, a threshold surface has to be introduced to know if the macro-element is under loading or under unloading. In the case where the ratio Ω/u is kept constant (radial loading path), there is only two additional parameters to introduce, Ω^+ and Ω^-, wich are the maximum and minimum values reached by Ω during the loading. Over Ω^+ and under Ω^- the stored working laws are used. Between Ω^+ and Ω^-, the damage of the macro-element is supposed constant. In the reality, with the used concrete model, the damage evolves even in case of unloading (there are redistribution of stress between the layers).
Since the working laws are built with radial loading paths, they should be used only for this kind of loading. But in real structures, the path followed by an element is not always radial (Ω/u is not constant during all the non linear computation of the structure) because of the adaptation of the internal

forces when the degradation occurs in near elements or because of a non radial loading of the structure. However, this way of doing is kept to have less storage. It can be seen that this assumption is valid on the done apllications.

Application 1

To test the cyclic performances of the proposed method, a frame submitted to a horizontal load applied at the upper corner is presented on the figure 2 a). It has 4 stories and 2 bays and it is discretized into 100 elements. Framing, member sizes and required reinforcement were chosen according to the Eurocode 2 and with traditional limits on sizes and proportions. The load path follows a cyclic evolution given in figure 2 b).

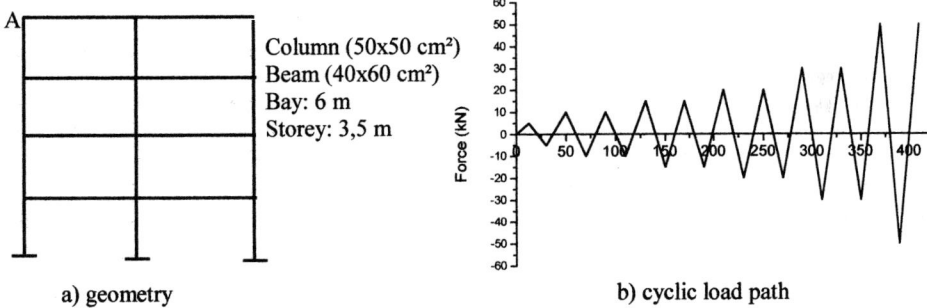

a) geometry   b) cyclic load path

Figure 2: Geometry of the frame and cyclic load path

On the figure 3 one presents the responses (horizontal displacement at point A versus load) given by the model and experiment. One can see that the difference is not great. On the loading parts the stiffness is higher in the simulations than in the test possibly because of the representation of the boundary conditions. On the unloading parts, the differences are due to the assumption on proportionnality. The damage state of the cross section is supposed to remain constant during unloading. This affects also the non elastic strains since they are dependent on damage. Nevertheless, the difference is under ten percent and is reasonable compare to the test scatter.

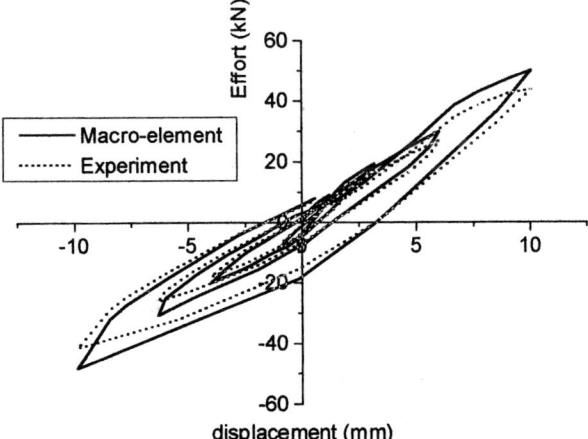

Figure 3 : experimental and simulated load-displacement curves

DYNAMIC ANALYSIS

Actually, for the knowledge of the physical non-linear behavior of structural systems, the most of the authors use dynamic analysis based in classic step-by-step direct integration methods to solve the equations of motion of structures like, Newmark method (Newmark, 1959), Houbolt method (Houbolt, 1950) or Wilson method (Wilson, Farhoomand & Bathe, 1973). Although this procedure can be a powerful analysis technique, the application to large structural systems (more than 1000 d.o.f.) is, in general, very expensive and can only be performed in main frames computers.

Another approach to the seismic analysis of structures with non-linear behavior, can be conducted through incremental techniques with modal superposition. The assumption is that the physical non linearities that occur in time in the structural system can be considered, without any simplification, as additional external forces (pseudo-forces) that make the difference between a hypothetical linear response and the real non-linear response of the structure.

This procedure allows to maintain during the whole dynamic analysis, the frequencies and the mode shapes of the structure that were computed in the beginning of the analysis for the elastic-linear system, becoming this method computationally more efficient than the classic step-by-step direct integration methods. The method proposed by Fragoso & al (1998) is an example of this approach. This approach seems to be a powerful technique of analysis to do the evaluation of the seismic non-linear response of large structural systems.

Knowing that the dynamic balance of a non linear structure of n dynamic degrees of freedom to an acceleration of soil $\{ug\}$ is expressed by:

$$[M]\{\ddot{d}(t)\} + [C]\{\dot{d}(t)\} + [K_e]\{d\} = \{\tilde{F}\} \tag{9}$$

And being given that:

$$\{\tilde{F}\} = -[M]\{J\}\{\ddot{u}_g(t)\} + \{F_{an}(d(t))\} \tag{10}$$

Where $[M]$, $[C]$ and $[K_e]$ represent the matrices of mass, damping and linear stiffness respectively of the structure, $\{J\}$ the influence vector of the soil movement and $\{F_{an}\}$ the vector of anelastic efforts defined in Eqn. 7.

One can note that the first member of the expression (Eqn. 10) represents the standard formulation of an elastic structure, while the second member constitutes the vector of a pseudo-strength acting on the structure. As it was seen, it is more interesting to transform the equation (Eqn. 9) in one system to modal coordinates, permitting to reduce of drastic manner the number of necessary operations to do to every step of time.

While putting:

$$\{d(t)\} = [\Phi]\{Y(t)\} \tag{11}$$

Where $[\Phi]$ represents the modal matrix and $\{Y(t)\}$ the generalized coordinates (modal), and while using properties of orthogonalisation, one gets:

$$[M_G]\{\ddot{Y}(t)\} + [C_G]\{\dot{Y}(t)\} + [K_G]\{Y(t)\} = -[\alpha]^T\{\ddot{u}_g(t)\} + [\Phi]^T\{F_{an}(t)\} \tag{12}$$

Where $[M_G]$, $[C_G]$ and $[K_G]$ represent the generalized matrices of mass, damping and stiffness generalized of the structure respectively, $[\alpha]$ the matrix of factors of modal participation. With properties of orthogonalisation, this system of n equations to n degrees of freedom becomes a system of n equations to only one degree of freedom and (Eqn. 12) becomes:

$$\ddot{Y}_i(t) + 2\beta\omega_i \dot{Y}_i(t) + \omega_i^2 Y(t) = -\alpha_i \ddot{u}_{gi}(t) + \frac{\{\varphi_i\}}{M_{Gi}}\{F_{an}(t)\} \tag{13}$$

Being given that a seismic action excites the first modes, particularly structures where damping is included between 5 and 10%, it is sufficient to value the first m modes. The method proposed by Clough and Penzien (1986) consisting in calculating the dynamic displacement of a structure $\{d(t)\}$ from the contribution of the m first modes while the contribution of them (n-m) remaining modes can be gotten with a good precision from a pseudo-static analysis, once inertial effects of them (n-m) modes can be neglected. One gets:

$$\{d(t)\} = \sum_{i=1}^{m}\{\varphi_i\}\left(Y_i(t) - \frac{\{\varphi_i\}^T \{\tilde{F}(t)\}}{\omega_i^2}\right) + [K_e]^{-1}\{\tilde{F}(t)\} \tag{14}$$

The first term of the second member represents the answer of the structure of the lowest m modes considered at the time of the analysis and the second and third terms represent the static correction of inertial effects relative to the (n-m) omitted modes.

Algorithm For The Modal Superposition Method

The modal superposition method for inelastic analysis of structural systems can be formulated summarily in the following algorithm (according to Fragoso & all, 1998):
1. Modal Analysis of the Structure:
    - Compute the elastic-linear stiffness matrix and the mass matrix of the structure, [K] and [M],
    - Compute the frequencies and mode shapes of the structure for the lowest m modes, solving the eigenvalue problem,
    - Compute the modal participation factors of the structure for the first m modes considered,
2 - Recurrence Matrix for the Modal Response:
    For each one of the m modes considered in the analysis, compute the respectively recurrence matrix for the i-th linear oscillator of 1 d.o.f,
3 - Inelastic Response of the Structure:
    For each time step do:
    - Compute the modal response of the generalized system for each m modes considered in time step s+1
    - Compute the displacements of the structure for the time step s+1
    - Compute de corrective force vector for time step s+1
    - Verify the structure equilibrium state
    - Using the displacements vector $\{u(t_{s+1})\}$, actualize the inelastic stiffness matrix
    - Compute the inelastic restoring force of the structure for time step s+1
    - Go to the next time step.

Application 2

The same frame presented on figure 2 which its fondamental period is equal to 0,45 s, is now subjected to a seismic acceleration given at the bottom (Elcentro S00E, 1940). Only the ten first seconds of the response are presented on the figure 4. After, the concrete of the frame is full damaged. During the first second, the responses given by the macro-element method is strictly identical to those of experiments. The concrete remain elastic. After it begins to occur some differences that grows as the damage increases in the structure. There are differences in amplitude and in frequency. This is due to the differences of damage evolution already seen in the previous application. It is to note that 2% of damping is given in this simulation. Considering the great saving of time in the macro element method, one can say it gives good enough results.

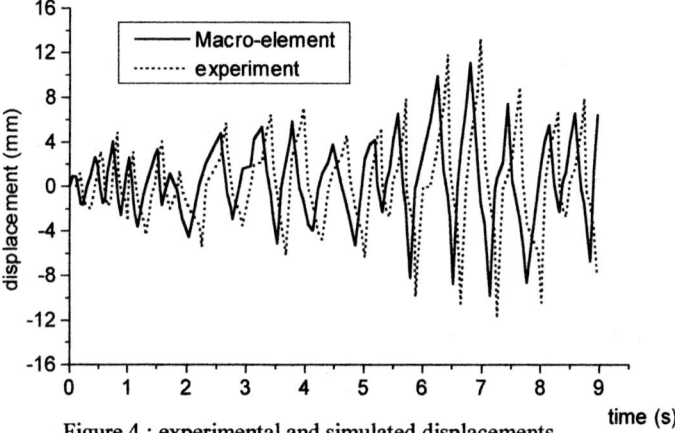

Figure 4 : experimental and simulated displacements

CONCLUSION

The general technique of conception of simplified methods is detailed in a particular application. A numerical macro-element is presented. The method has been tested under monotonic and cyclic loading on portal frames. A seismic simulation is also presented in this paper. All the presented results have a good accuracy and the computations are very efficient in time terms. The approach proves effective for cyclic as well seismic loadings and allows to describe correctly the evolution of degradations.

REFERENCES

Davenne L., Elachachi S.M. (1992). A simplified method to model the shear bending interaction of reinforced concrete cross sections, *First European Conference on Numerical Methods in Engineering, Brussels, 7-11 september 1992.*
Dodd, L. L. & Cooke, N. 1994. The Dynamic Behaviour of Reinforced-Concrete Bridge Piers Subjected to New Zealand Seismicity. *Research Report No. 92-04*, Department of Civil Engineering, University of Canterbury, Christchurch, New-Zealand.
Elachachi S.M. (1992). A simplified analysis method of civil engineering structures by macro-elements adapted to composite and damaged constructions, Doctorate thesis, ENSCachan/CNRS/University Paris VI, France (in french).
Fragoso M.R, Coelho Serra E., (1998).Non-linear analysis of structures by mode superposition method.*11th European Conference on Earthquake Engineering*,Balkema.
Houbolt, J. C. (1950). A Recurrence Matrix Solution for the Dynamic Response of Elastic Aircraft. *Journal of Aeronautical Science*, **V. 17**.
Mazars J. (1984)). application of damage mechanics to the non linear behavior and to the rupture of concrete structures, Doctorate thesis, ENS Cachan/CNRS/University Paris VI, France (in french).
Newmark, N. M. 1959. A Method of Computation of Structural Dynamics. *Journal of Mechanics Division*,ASCE,**V. 85**.
Saiidi, M. & Sozen, M.A. 1981. Simple Non-linear Seismic Analysis of RC Structures, *J. of Struct. Div.*,ASCE, **V. 107, ST5**, 937-952.
Saiidi, M. 1982. Hysteresis Models for RC, *J. of Struct. Div.*, ASCE, **V. 108, ST5**, 1077-1087.

DESIGNING THE CONCRETE DUAL SYSTEM

E. Atımtay[1] and M.E. Tuna[2]

[1] Department of Civil Engineering, Middle East Technical University
Ankara, TURKEY
[2] Department of Architecture, Gazi University
Ankara, TURKEY

ABSTRACT

Assessment of building failures in recent earthquakes have shown that shear walls are mandatory to make a structure earthquake resistant. Also, the hybrid "shear wall-moment resisting frame" should satisfy the criteria of strength, ductility and stiffness.

The structure should be designed as a dual system as defined by the Uniform Building Code. In the dual system, the shear walls are to resist the total design base shear as dictated by a proper structural analysis. Additionally, the moment resisting frame is to resist 25% of the total base shear, independently.

The amount of shear wall area as a percentage of the floor plan area is derived by an upper-bound calculation of the total design base shear and a lower-bound calculation of shear strength of shear walls containing minimum steel percentage.

The dual structure should also possess enough ductility, as expressed by the displacement ductility ratio, $\mu_\Delta = 4-5$. Considering the difficulties and ambiguousness of displacement calculations in reinforced concrete structures, a more reliable measure of ductility is employed, as the curvature ductility ratio, μ_ϕ. A plastic analysis is performed to relate the displacement ductility ratio to the more readily obtainable curvature ductility ratio.

The shear walls can be designed for a displacement ductility ratio of $\mu_\Delta = 4-5$, which in turn necessitates a curvature ductility that can be easily calculated. The cross-sections of the shear walls are to be consecutively designed to provide the curvature ductility demands which are calculated.

KEYWORDS

Earthquake resistant, dual system, shear walls, displacement ductility, curvature ductility, shear strength, minimum reinforcement.

INTRODUCTION

In the last ten (10)years, four (4) major earthquakes have hit Turkey: Erzincan (1992, R= 6.8), Dinar (1995, R= 5.9), Adana – Ceyhan (1997, R= 5.9) and Marmara (1999, R= 7.4).

Observations and assessment of building failures in these earthquakes have brought out one fact: buildings which contain shear walls have survived the above mentioned earthquakes without collapse, regardless of how poorly shear walls may have been placed in the floor plan.

In framed structures, the total earthquake force is resisted by columns alone. Degradation of stiffness of the framed structure due to reversing seismic forces, anchorage failures at beam – column joints, lack or inadequate confinement at end zones of beams and columns, unfortunate occurrence of plastic hinges at column ends, and lateral sway out of control, finally lead to structural collapse. The above mentioned defects of the structural system manifest themselves as weak columns – strong beams, formation of soft stories, short columns and excessive second order effects.

In light of the above observations, the prudent structural engineer is forced to the following design strategy: an earthquake resistant building should not be a bare framed structure, but should be a hybrid system containing shear walls as well.

Having reached this conclusion, a second question immediately follows: use shear walls, fine, but of what amount?

SATISFYING THE STRENGTH DEMAND

The structural system composed of shear walls and the moment resisting frame should be capable of resisting the effects of the earthquake force by satisfying three major design criteria as strength, ductility and stiffness.

Necessary Amount of Shear Walls to Meet the Strength Demand

What is the correct amount of shear walls to be used, necessary to make a building earthquake resistant?

In order to answer this question, the following design strategy will be adopted.

a) The total design base shear must be resisted by shear walls.
b) Because seismic action occurs in all directions, equal amounts of shear walls must be placed in both orthogonal directions of the structure.
c) The moment resisting frame elements (beams and columns) must independently be able to resist 25 % of the total design base shear.

The Uniform Building Code (UBC, 1999) defines the structure which possesses the above mentioned properties as the "dual system". The authors believe that all buildings to be built in severe earthquake zones must conform to the requirements of the dual system.

Determination of the Total Design Base Shear

The total design base shear can be determined by using the acceleration response spectrum, as defined in the UBC (1999) and shown in Figure 1.

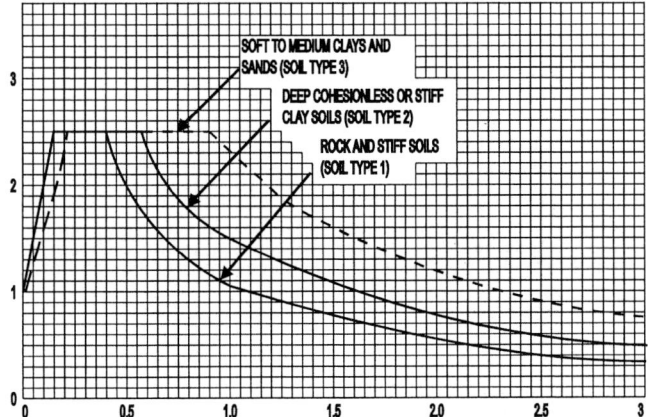

Figure 1: Acceleration Response Spectrum (UBC, 1999)

A study of the acceleration response spectrum reveals that the effective ground acceleration is magnified by a factor of 2.5, for natural periods that are most typical of building structures commonly employed in practice: 0.2-1.0 seconds. Based on this observation, the total design base shear, V, can be determined by Eqn. 1.

$$V = S(T)(A_0)(I_p)(W)/R_w \tag{1}$$

$S(T)$ = ratio of spectral acceleration to effective peak ground acceleration, its maximum value being 2.5

A_0 = effective peak acceleration, its maximum value being 0.4

I_p = seismic importance factor, which varies between 1.0-1.5, and taken as 1.0 in Eqn. 1

W = total weight of the building, as expressed by Eqn. 2, where A_p is the area of the floor plan

R_w = seismic force reduction factor

n = number of stories

$$W = \sum_{i=1}^{n} w_i A_{pi} \tag{2}$$

Assuming an average value of 10 kN/m² for w_i and considering a building n-stories high, the total design shear base of Eqn.1 becomes as expressed in Eqn.3 and 4. The force reduction factor R_w is taken as 7, which is an average typical value.

$$V = 2.5(0.4)(1.0)(10)(nA_p)/7 \tag{3}$$

$$V = 1.43(n)(A_p) \tag{4}$$

Determination of Shear Strength of Walls

A lower-bound assessment of the shear strength of the total number of walls in one orthogonal direction of the building floor plan can be done per the Uniform Building Code (1999), Eqn.5.

$$V_n = \phi \sum A_{cv}(0.166\sqrt{f'_c} + \rho_n f_y) \times 10^3 \tag{5}$$

Considering $f'_c = 20\,\text{MPa}$, $\rho_n = 0.0025$, $f_y = 420\,\text{MPa}$, $\phi = 0.7$, and equating the total design base shear to the total shear resistance provided by all shear walls in one direction, the ratio of the total area of shear walls to the area of the floor plan can be obtained, Eqn. 6 and 7.

$$1.43(n)(A_p) = 1.25 \sum A_{cv} \times 10^3 \tag{6}$$

$$\frac{\sum A_{cv}}{A_p} = 0.00114(n) \tag{7}$$

The ratio of Eqn. 7 can be expressed in tabular form as a function of number of stories, Table 1.

TABLE 1
REQUIRED AREA OF SHEAR WALLS

Number of Stories, n	$\left(\dfrac{\sum A_{cr}}{A_p}\right)^*$
1	0.0014
3	0.0042
5	0.0070
8	0.011
10	0.014
15	0.021
20	0.028

*Multiply ratio by Building Importance Factor, I_p.

It can be seen from Table 1 that about 1 % of shear wall area containing minimum steel percentage will resist the total design base shear for a building that is 8 - stories high.

SATISFYING THE DUCTILITY DEMAND

Under the total acting design base shear, it is expected that plastic hinges form at the base of shear walls. During the formation of plastic hinges it is further required that the structure behave ductile.

Ductility is a qualitative term, and it needs to be quantified. To quantify ductility, the commonly accepted measure is the displacement ductility ratio, μ_Δ, as given in Eqn. 8.

$$\mu_\Delta = \frac{\Delta_u}{\Delta_y} \tag{8}$$

Δ_y = displacement at top of the structure at initiation of yielding at base of shear wall

Δ_u = displacement at top of the structure at ultimate stage

In practice, $\mu_\Delta = 4-5$ is considered to provide enough ductility. At ultimate stage, plastic hinges form at beam ends, in addition to hinges at base of shear walls.

However, the above criterium necessitates the calculation of top sway of a reinforced concrete structure, which is rather tedious and uncertain. It is much easier to calculate the cross-sectional curvature, as commonly expressed by the axial load-moment-curvature relationship (N–M–ϕ). Therefore, it will be most convenient if an expression could be developed to relate displacement ductility ratio, μ_Δ, to curvature ductility ratio, μ_ϕ, Figure 2.

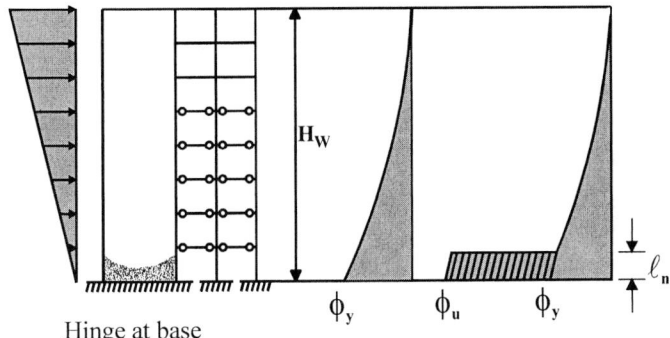

Hinge at base

Figure 2: Relating top sway of building to cross-sectional curvature of shear wall

$$\Delta_y = \frac{1}{3}\phi_y (H_w)(\frac{9}{10}H_w) \tag{9}$$

$$\Delta_u = \Delta_y + (\phi_u - \phi_y)(\ell_n)(H_w - 0.5\ell_n) \tag{10}$$

Consider a shear wall-frame structure, where a plastic hinge of height ℓ_n has formed at the base of the shear wall, Figure 2. The sway at the top of the structure can be calculated at the time of initiation of yielding and concrete crushing as Δ_y and Δ_u, respectively, Eqn. 9 and Eqn. 10. Consequently, displacement ductility can be readily calculated, as in Eqn. 11.

$$\mu_\Delta = \frac{\Delta_u}{\Delta_y} = 1 + \frac{30}{9H_w^2}\left(\frac{\phi_u}{\phi_y}-1\right)(\ell_p)(H_w - 0.5\ell_p) \tag{11}$$

From Eqn. 11, the curvature ductility ratio μ_ϕ can be readily solved for, Eqn. 12.

$$\mu_\phi = \frac{\phi_u}{\phi_y} = \frac{9}{30}\frac{H_w^2}{\ell_p}\frac{(\mu_\Delta - 1)}{H_w - 0.5\ell_p} \tag{12}$$

The displacement ductility ratio, μ_Δ, is thus expressed in terms of the more familiar curvature ductility ratio, μ_ϕ, which solely depends on cross-sectional properties of the shear wall and the axial load on it.

DESIGNING THE SHEAR WALLS

The following steps should be followed for the design of the ductile shear wall.

a) The seismic demand on each shear wall should be determined by a proper structural analysis.

b) The minimum amount of reinforcement $(\rho_n = 0.0025)$ should be uniformly distributed in the cross section. The end zone details should also be carefully applied, as dictated by the valid seismic code.

c) The (N–M–ϕ) relationship should be developed, as commonly done by an available computer program. The developed (N–M–ϕ) relationship will give the yield curvature, ϕ_y and the ultimate curvature, ϕ_u. Thus the curvature ductility ratio, μ_ϕ, can be readily calculated, $\mu_\phi = \phi_u / \phi_y$

d) The curvature ductility demand μ_ϕ, corresponding to the displacement ductility demand of $\mu_\Delta = 4-5$ can be calculated by using Eqn. 12.

The supplied μ_ϕ as obtained from the (N–M–ϕ) relationship must be greater than demanded; otherwise the cross-section must be revised.

DESIGNING THE MOMENT RESISTING FRAME

In accordance with the definition of the dual system, the moment resisting frame must be designed to resist 25 % of the total design base shear. The following steps are in order:

a) Calculate 25 % of the total design base shear, 0.25V

b) Distribute 0.25V to individual columns by a proper method of structural analysis, e.g. the Muto Method or the Smith Method.

c) The permitted redistribution among the column shear forces, per the valid seismic code can be applied.

d) The columns and beams are to be designed, considering the proper load combination of gravity and seismic loads. Magnification of column moments to account for second order effects (P - Δ effects) should be considered.

e) Of course, the confinement requirements at ends of columns and beams should be carefully applied.

CONCLUSION

1. A method has been proposed to design the earthquake resisting building composed of shear walls and moment resisting frame.

2. The "dual system" concept has been applied with a slight modification, in which shear walls are asked to resist 100 % of the total design base shear and the moment resisting frames to additionally resist 25 % of the total design base shear.

3. A plastic analysis of the shear wall has been performed to relate displacement ductility ratio and curvature ductility ratio.

4. The shear wall should be designed to meet both the strength demand and the curvature ductility demand. This can be done by developing the axial load-moment-curvature relationship of the shear wall cross-section.

REFERENCES

Atimtay, E., (1991). Seismic Design of Bridges to Relate Ductility and Sway. *Evaluation and Rehabilitation of Concrete Structures and Innovations in Design, American Concrete Institute, SP-128.*

Paulay, T. and Priestley, M.J.N.(1992). *Seismic Design of Reinforced Concrete and Masonry Buildings* John Wiley & Sons, Inc., USA.

International Conference of Building Officials *Code* (1999*) Uniform Building,* Ca., USA.

ANALYSIS AND DESIGN FOR BLAST AND IMPACT

PARAMETRIC STUDY ON THE DROP-IMPACT BEHAVIOUR OF MINI HI-FI AUDIO PRODUCTS

K. H. Low[1], K. L. Lim[2], K. H. Hoon[1], Aiqiang Yang[1] and Judy K. T. Lim[2]

[1]School of Mechanical and Production Engineering
Nanyang Technological University, Singapore 639798, Republic of Singapore
[2]Philips Singapore Pte. Ltd., 620A Lorong 1 Toa Payoh, Singapore 319762,
Republic of Singapore

ABSTRACT

Most electronic devices suffer impact-induced failure in their usage. Drop/impact performance of these products is one of the important concerns in product design. It is often time and cost consuming and thus undesirable to conduct drop tests for every design to physically detect any sign of failure and identify their drop behaviours. Finite element analysis provides a vital and powerful tool to solve the problems. The methodology of finite element modeling, simulation, and basic experimental validation should be developed to effectively study the drop-impact effects. Modeling and transient drop analysis for a mini Hi-Fi audio product are considered in this work. The impact analysis of the base-drop with buffer hitting the concrete floor is the main interest and concern of this work. The model is created with Pro-E, and the analysis is carried out with an existing software package. The analysis is focused at the deformation of the bottom plate that is carrying a transformer. The deflection measured using a 3D-probe measurement is compared to the simulated results. Experimental data have also been obtained for drop simulation correlation for the plate and buffer material properties. The G-force and the permanent deformation of the bottom plate during drop are noted. The effects of the material properties to the plate deflection under drop/impact shock have also been investigated.

KEYWORDS

Drop modeling/simulation, drop/impact behaviour, impact-induced effects, permanent deformation, buffer, G-force, plates.

INTRODUCTION

Feed-drop/impact induced damage is one of the dominant failure modes of portable products: cellular phones, pagers, radios, TVs, irons, audio products, etc. The drop/impact-induced damage includes housing fracture, cover cracking, joint breaking, connector disconnection or component failure [1-7]. Traditionally, the product reliability against this kind of damage is determined empirically by designers.

The method could be quite time-consuming without scientific insight. It was thus suggested that two basic approaches should be adopted to obtain detailed understanding of product drop performance and the drop-induced damage: testing and computer simulation [4, 7].

A computer simulation is used to analyze at every design stage in timely and efficient manner. The simulation is able to pick up complete mechanical information at any location of an analyzed object. To obtain reliable prediction in computer simulation, the correct modeling and the analyst's experience are essential. A frequent question is how accurate are the simulation model and results produced. Some basic test validations are then required and necessary to verify the model and its results. The combination of simulation and testing should therefore be the best approach to obtain convincing analysis to support product design.

Drop/impact simulation is basically an analysis of a product impacting with a surface. Products such as electronics are becoming lighter in weight and customers are demanding increased functionality. The resulting analysis information is therefore critical in their manufacturing. A major area in manufacturing is to design these products wherein certain components act as energy absorbers on impact, thereby protecting vital components. The damaged components manufactured as energy absorbers can then be repaired or replaced. Computer simulations are being used to demonstrate drop impact on products by setting critical drop/impact parameters and configurations. Among the features needed for drop/impact analysis simulation is, for example, to produce a model capable of predicting stresses and strains on the components [2, 6, 8].

This paper presents a preliminary study of the drop/impact effect on mini Hi-Fi audio products. The results with different material properties on the drop-induced damage are demonstrated and compared. Finally, some observation from the modeling and results produced are provided and discussed.

ISSUES OF COMPUTER SIMULATION

The governing equation of a dynamic system can be expressed as [9-13]

$$M\ddot{x} + C\dot{x} + K x = F, \qquad (1)$$

where x is the nodal displacement vector and the dots denote its differentiation with respect to time. Equation (1) expresses a dynamic equilibrium state due to the load vector F, including those impulse forces. Note that M, C and K are the mass, damping and stiffness matrices, respectively.

For a transient analysis, a dynamic process is divided into time steps. The time-step integration is an algorithm to predict an unknown state, at time step $t+dt$ from a known state at moment t. The initial state is defined by given initial conditions, while the entire dynamic process is calculated step by step. With finite element discretion in space and central difference in time domain, equation (1) is written as [9]

$$M\ddot{x}_t = F_t - C\dot{x}_t - K x_t, \qquad (2)$$

if an equilibrium state is considered at moment t. The algorithm based on equation (2) is the so-called *explicit* formulation. In this formulation, only the acceleration term in equation (2) is required to satisfy equilibrium state. The velocity and displacement are all calculated from acceleration at the previous step by using central difference formulation.

In contrast, if the equilibrium state is considered at $t+dt$, the governing equation can be expressed in the form [9],

$$M \ddot{x}_{t+\Delta t} + C \dot{x}_{t+\Delta t} + K x_{t+\Delta t} = F_{t+\Delta t}. \tag{3}$$

Equation (3) is called *implicit* formulation, in which the three terms of acceleration, velocity and displacement are all undetermined in an equilibrium state. By comparing equations (2) and (3), it is seen that the solution of equation (2) is much simpler than that of equation (3) in each time step. No global matrix manipulation of equation (2) is needed, and the memory space requirement is thus small. Furthermore, no extra complication in the explicit formulation even if material and geometric non-linearity is introduced. On the other hand, matrix manipulation is necessary in the implicit formulation, equation (3). Also, iteration has to be conducted in consideration of non-linearity.

It is worth mentioning that the explicit formulation is by nature well suited to solve impact, wave propagation and impact problems [11, 12]. This is due to two main reasons: (i) impact occurs and the consequential response decays within a very short time. Usually, the drop/impact process of electronic products needs only several milliseconds to be analyzed; (ii) two physical factors limit the step time in impact analysis. The first restriction is the high deformation rate in impact cases. If a larger step time is adopted, it will introduce too much strain increase in one time step, and the large strain change will cause divergence in such a large deformation analysis. The second restriction comes from the contact surface algorithm, as contact is a basic phenomenon in impact cases. The contact force is calculated as proportional to the penetration of two contact bodies. In a large step time, the contact force will be too large at the contact area, which causes local distortion and analysis failure. In view of the two stated reasons, the advantage of larger step time allowable in implicit formulation vanishes. With such a tiny step, the implicit is much time-consuming than the explicit because of its matrix integration and inversion. The explicit formulation becomes the optimal algorithm in cases of drop/impact contact.

Researchers in structural dynamics have extensively studied appropriate time step and the selection criteria for reliable and stable solutions [9, 14-16]. For PAM-CRASH codes, the standard time step criteria for different elements are based on the elastic wave travel time along the respective characteristic length [17]. The time step size is then determined by taking the minimum value over all elements. For example, the time step to ensure solution stability for shell elements is determined as follows:

$$\Delta t_{\text{shell}} = k \min\left(L/\sqrt{E/\rho}\right), \tag{4}$$

in which k is a scale factor for stability reasons, L the characteristic length defined for different types of shell elements [17], E the Young's modulus, and ρ the mass density.

Besides the above-mentioned analysis code, another issue crucial to industrial application of FEM is modeling skill. The industrial application is often focused at products with complex geometry, shape and structure. For this purpose, powerful modeling tools and sophisticated modeling skill are essential prerequisites for successful engineering application. With several existing generation tools, a user may easily create, check and edit finite element mesh in both automatic and manual manners. Exact geometry, right mass distribution and interaction among parts are the basic requirements to get correct modeling and prediction for the drop/impact simulation. Examples of new generation tools suitable to drop/impact modeling are I-DEAS, Pro-E, HYPERMESH, FEMB, PATRAN, etc.

MODELING, SIMULATION AND TEST CORRELATION

The model considered here is a mini Hi-Fi audio set. As shown in Figure 1, the set includes bottom plate, transformer, back plate, heat sink, CD changer module, front and rear cabinets. Figure 2 shows the flowchart of simulation process. All parts/components created and meshed with Pro-E are transferred to PAM-CRASH GENERIS for the final model assembly (see Figure 3). PAM-CRASH Solver executes drop simulation after the model generation is completed. The resultant time history curves and animation can then be viewed through the post-processor tool, PAM-VIEW.

The complete system model considered, as shown in Figure 3, contains 4814 shell elements, 15026 solid elements and 7705 nodes. The total mass of the audio set is about 10 kg. The mass of the buffer is 0.166 kg (with a density of 25 kg/m^3). The base drop configuration considered is the most concerned case when the rear cabinet with buffer hit the floor (see Figure 3). The drop speed when the set hits the ground is defined 3.962 m/s, which is equivalent to a 1-m free drop. The problem concerned is now the deformation of the bottom plate due to the pulling force of a 3.783-kg transformer (see Figure 1a).

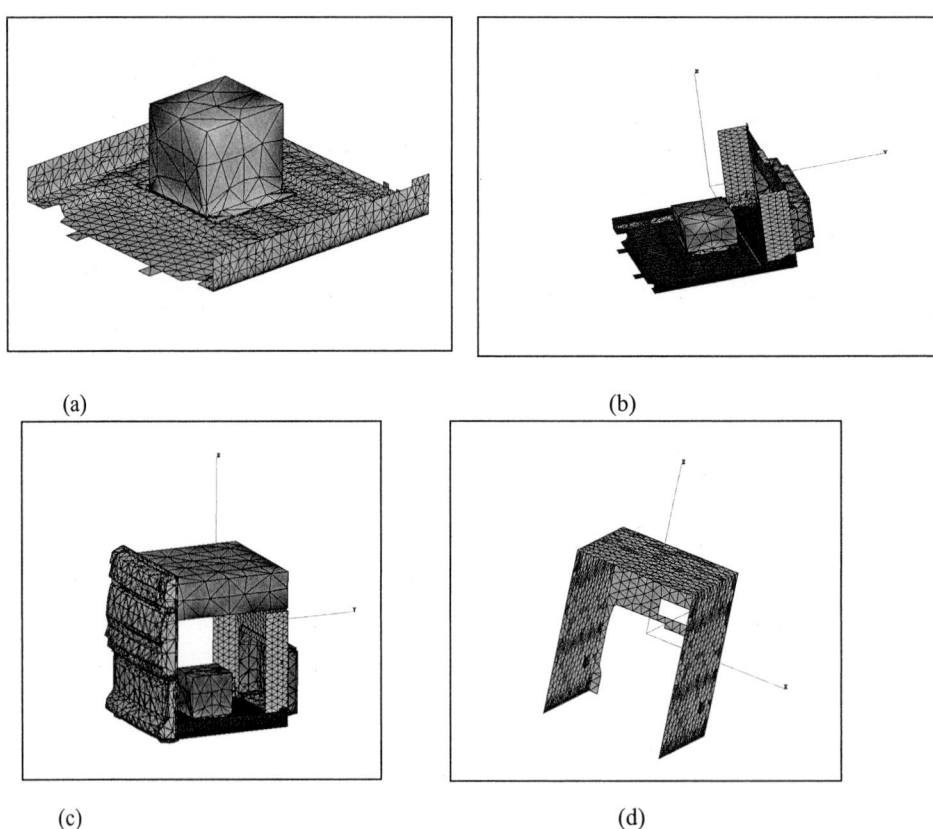

(a) (b)

(c) (d)

Figure 1 Step-by-step model assembly: (a) bottom plate with transformer, (b) addition of back plate and heat sink, (c) addition of front cabinet and CD changer module, and (d) rear cabinet

The measured properties of the bottom plate were taken in the simulation as: $E = 210$ GPa, $\sigma_y = 0.27$ GPa, $\upsilon = 0.28$, etc. The permanent deformation of the bottom plate with 0.8-mm thickness after two different drop heights was each measured using a 3D-probe measurement. The results are listed in Table 1. The points measured are in the region where the transformer was fixed onto the bottom plate. As depicted in Figure 4, the region was divided into 5 × 8 points, with a distance of 1 cm between points. It can be concluded from Table 1 that the deflection is larger for a higher drop height. Also, the deflection is increasing for the point moving away from the fixed end.

TABLE 1 MEASURED PERMANENT DEFORMATION IN mm AT DESIGNATED POINTS ON THE BOTTOM PLATE AFTER DROP

	A	B	C	D	E	F	G	H
Drop height = 80 cm								
1	0.5479	0.5771	0.6091	0.6446	0.6751	0.6930	0.7177	0.7479
2	0.6107	0.7010	0.7732	0.8364	0.8897	0.9296	0.9641	0.9787
3	0.6634	0.7669	0.8478	0.9138	0.9713	1.0200	1.0448	1.0505
4	0.5115	0.5462	0.5690	0.5970	0.6405	0.6668	0.7025	0.7202
5	0.5603	0.5974	0.6600	0.7087	0.7609	0.8033	0.8286	0.8692
Drop height = 120 cm								
1	1.1188	1.1667	1.2277	1.2806	1.3217	1.3572	1.3638	1.4317
2	1.3139	1.4433	1.5784	1.6830	1.7761	1.8419	1.8835	1.8822
3	1.2949	1.4600	1.5995	1.7218	1.831	1.9152	1.9239	1.9018
4	1.2311	1.3137	1.3855	1.4582	1.5342	1.6040	1.6394	1.6480
5	1.2403	1.3048	1.3998	1.4842	1.550	1.6221	1.6620	1.7014

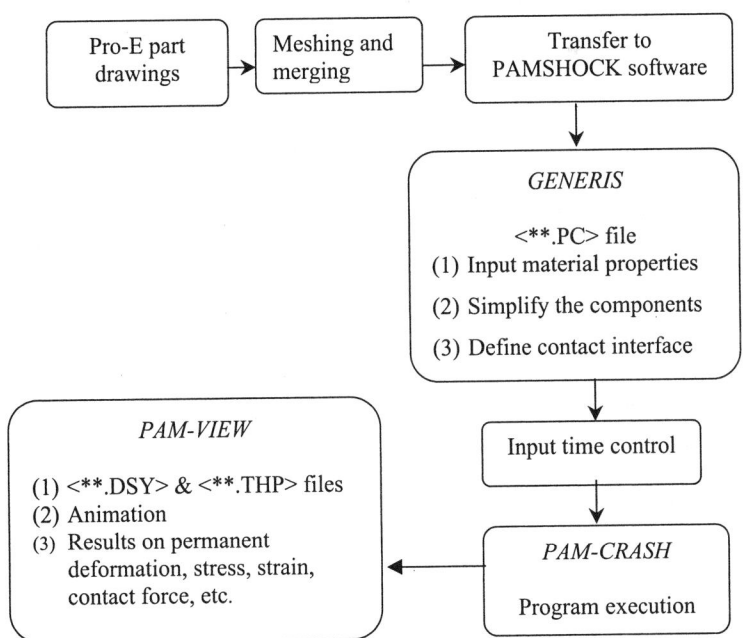

Figure 2   Flowchart of computer simulation procedures

PARAMETRIC STUDY OF BUTTOM PLATE

Another main concern in this work is to study the effects of individual material properties on the plastic deformation of the bottom plate (thickness = 0.8 mm). The reference point on the bottom plate is the point connecting the plate and the centroid of transformer (refer to Point R in Figure 4). It is hoped that the results can be useful to engineers in the design of bottom plate and the selection of its materials.

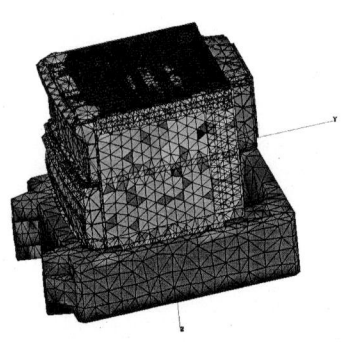

Figure 3 Inverted audio set resting on the buffer ready for base drop

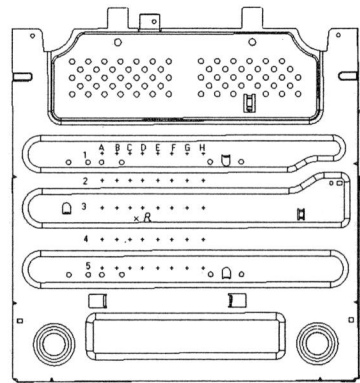

Figure 4 Designated positions of the bottom plate for deformation measurement

In comparison, the respective effects are described as follows:

1. Figure 5a shows that the deflection of the reference point increases and the period of the vibration increases, as the value of E reduce.
2. It is seen in Figure 5b that the plastic deflection ($t \cong 7.7$ ms) of the reference point reduces as the value of σ_y increases. However, the changing of σ_y does not affect the cycle of vibration.
3. As shown in Figure 5c, a lower density of buffer provides a better cushion to the model, i.e. a smaller deflection of the bottom plate. Moreover, it results a longer period of vibration.
4. In view of Figure 5d, it is obvious that a heavier transformer will cause a larger deflection and a longer period of vibration.

Table 2 summarizes the effects of the above-mentioned parameters in terms of permanent deflection, acceleration (G), stress and strain of the reference point R.

DISCUSSION AND CONCLUDING REMARKS

Modeling and free drop impact analyses for a mini Hi-Fi audio product have been presented in this work. The impact analysis of the base-drop with buffer hitting the concrete floor is of main concern. The model was created with Pro-E, and the analysis was carried out with PAM-CRASH. The analysis presented is focused on the deformation of the bottom plate that is carrying a transformer, although analyses on other parts in the same generated model could be conducted if required. The deflection measured using a 3D-probe measurement has been presented together with the simulated results. The G-

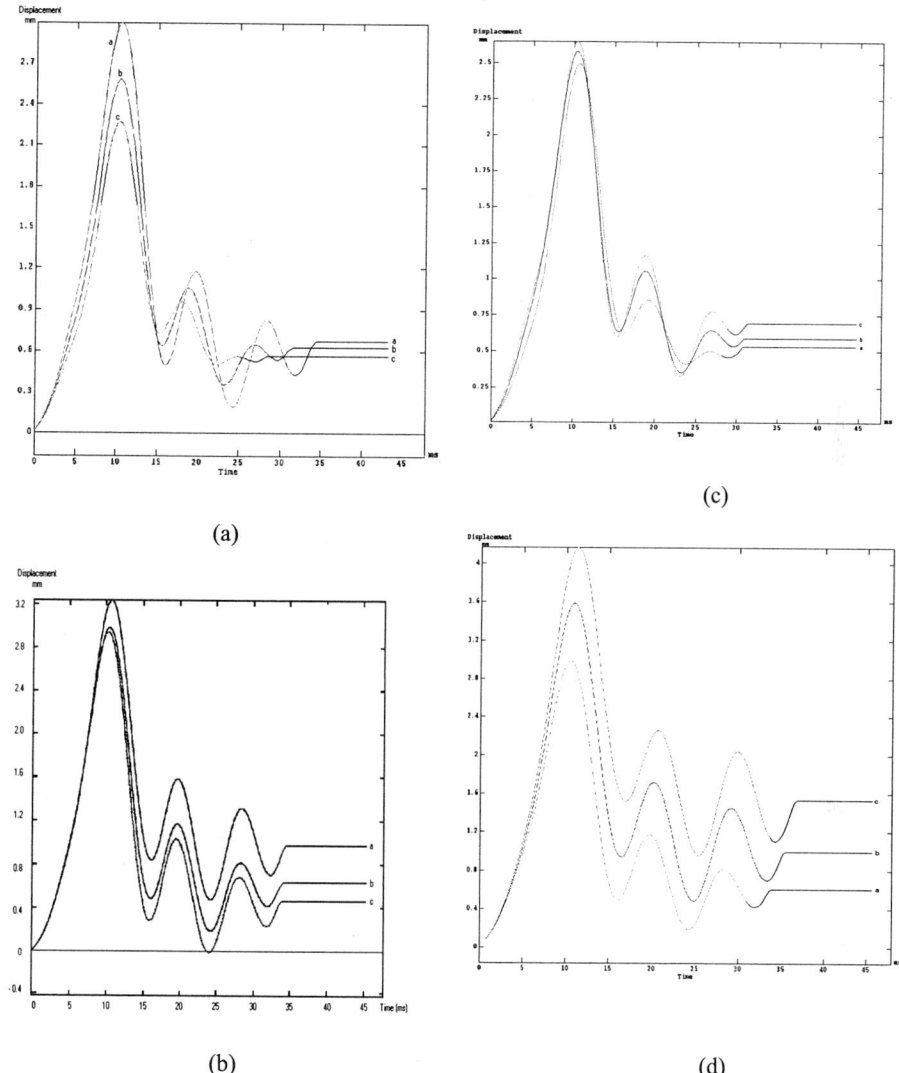

Figure 5a Deflection of Point R on the bottom plate due to changing E: 210 GPa (curve a), 250 GPa (curve b), 295 GPa (curve c)

Figure 5b Deflection of Point R on the bottom plate due to changing σ_y: 0.24 GPa (curve a), 0.27 GPa (curve b), 0.30 GPa (curve c)

Figure 5c Deflection of Point R on the bottom plate due to changing density of buffer: 20 kg/m^3 (curve a), 25 kg/m^3 (curve b), 30 kg/m^3 (curve c)

Figure 5d Deflection of Point R on the bottom plate due to changing mass of transformer: 3.78 kg (curve a), 4.28 kg (curve b), 4.78 kg (curve c)

forces of the bottom plate after drop were also obtained experimentally. Experimental data for the plate and buffer material properties were applied for the drop simulation. The effects of the material properties to the plate deflection under free drop impact have been investigated and discussed. It is found in the parametric study that different values of several parameters can each change the deflection of the bottom plate and the cycle of vibration.

A major difficulty of the above drop simulation and analysis is that some tiny inside substructures, which cannot be simplified as the output is expected, lead to extreme fine meshes. As the step time of explicit finite element analysis used in drop simulation is controlled by the minimum element size; the fine mesh associated to the concerned parts causes too much CPU cost in the whole drop simulation. A case of drop simulation with locally and finely meshed parts will prolong the running time to over many times. For example, the whole simulation for one millisecond of actual time history in the present audio model requires about 5-10 CPU hours. Mass-scaling technique is not suitable for such cases, as the vastly increased mass at the parts to enlarge step time will dramatically change their drop behaviour. It was suggested that a global-and-local method could overcome such a difficulty [18]. The method, coupling global simulation (whole model level drop simulation) and local analysis (only tiny substructures themselves), could be a suitable solution to analyze drop performance of the small parts. Another issue we are concerned with is the material property database availability and reliability, especially for non-linearity, dynamic effect, viscosity and strain rate influence of commonly used materials in the electronic industry [8, 13, 19].

TABLE 2   SUMMARY OF THE EFFECTS STUDY ON THE BOTTOM PLATE

cases compared	E (GPa)	σ_y (GPa)	ρ_{buffer} (kg/m³)	$m_{transfm}$ (kg)	δ_{permn} (mm)	a_{max} (G)	σ_{max} (Pa)	ε_{max} (×10⁻³)
Figure 5a	210	0.27	25	3.783	0.675	95.8	0.3869	9.5846
	250	0.27	25	3.783	0.625	93.8	0.3854	9.1726
	295	0.27	25	3.783	0.563	92.8	0.3792	8.6865
Figure 5b	210	**0.24**	25	3.783	1.000	94.3	0.3458	11.572
	210	0.27	25	3.783	0.675	95.8	0.3869	9.5846
	210	**0.30**	25	3.783	0.450	98.9	0.4189	8.7110
Figure 5c	250	0.27	**20**	3.783	0.531	92.8	0.3869	8.4340
	250	0.27	**25**	3.783	0.625	93.8	0.3854	9.1726
	250	0.27	**30**	3.783	0.700	95.6	0.3874	4.8230
Figure 5d	210	0.27	25	**3.783**	0.675	95.8	0.3869	9.5846
	210	0.27	25	**4.283**	1.000	92.8	0.4031	27.387
	210	0.27	25	**4.783**	1.500	89.7	0.4009	15.388

Note: values in bold refer to the changed variables in the parametric study.

Acknowledgments — The authors would like to thank Dr Juida Guo for his great assistance in conducting the simulation. The support by Mr John Olierook and Mr Lee-Boon Sim of Philips (Singapore) Pte. Ltd. is gratefully acknowledged. The hardware assistance by the Metrology Laboratory and the Robotics Research Centre is also appreciated.

References

1. Choi S. S. (1998). Drop Impact Simulation for Optimum Design of TFT-LCD. *Proc. of the IEEE Intersociety Conference on Thermal Phenomena in Electronic Systems* (ITHERM98), Seattle, Washington, 2-11.

2. Goyal Suresh, Pinson Elliot N., and Sinden Frank W. (1994). Simulation of Dynamics of Interacting Rigid Bodies Including Friction I: General Problem and Contact Model. *Engineering with Computers* **10**, 162-174.

3. Goyal Suresh, Papadopoulos Jim M., and Sullivan Paul A. (1997). Shock Protection of Portable Electronic Products:Shock Response Spectrum, Damage Boundary Approach, & Beyond. *Shock & Vibration* **4**, 169-191.

4. Goyal Suresh, Upasani Sanjay S., and Patel Dhiren M. (1998). Design Best-in-class Impact-Tolerant Cellular Phones and other Portable Products. *Bell Labs Technical Journal* July-September 159-174.

5. Keltie R. F. and Flater K. J. (1993). Guidelines for the Use of Approximations in Shock Response Analysis of Electronic Assemblies. *ASME Trans. Journal of Electronic Packaging.* 115, 124-130.

6. Wu Jason et al. (1998). Drop/Impact Simulation and Test Validation of Telecommunication Products. *Proc. of the IEEE Intersociety Conference on Thermal Phenomena in Electronic Systems* (ITHERM98), Seattle, Washington, May, 330-336.

7. Wu Jason et al. (1998). Drop/Impact Simulation and Test Validation in Motorola. *Proc. of the 5-th Int. LS-DYNA Users Conference,* Southfield, Michigan, Sep. 1-10.

8. Alves M. (2000). Material Constitutive Law for Large Strain and Strain Rates. *ASCE Journal of Engineering Mechanics.* **126**, 215-8.

9. D'Souza A. Frank and Garg Vijay K. (1984). *Advanced Dynamics: Modeling and Analysis.* Prentice-Hall, Inc. Englewood Cliffs, New Jersey.

10. Tornambè A. (1996). Modelling and controlling one-degree-of-freedom impacts under elastic/plastic deformations. *IEE Proc.-Control Theory Appl.* **143**, 470-6.

11. Brogliato Bernard. (1996). *Nonsmooth Impact Mechanics: Models, Dynamics and Control,* Springer-Verlag, London.

12. Abrate Serge. (1998) *Impact on Composite Structures.* Cambridge University Press, New York.

13. Harding J. and Ruiz C. (1998). The mechanical behaviour of composite materials under impact loading. *Key Engineering Materials.* **141**, 403-25.

14. *PAM-CRASH User's Theory Manual.* France: Pam System International SA, April 1988.

15. Low K. H. (1995). Displacement and frequency analyses of vibratory systems. *Computer and Structure* **54:4**, 743-755.

16. Low K. H. (1997). Numerical implementation of structural dynamics analysis. *Computer and Structure* **65:1**, 109-125.

17. *PAM-CRASH Solver Notes Manual.* Pam System International SA, France, May 1998.

18. Wu Jason. (2000). Global and local coupling analysis for small components in drop simulation. *Proc. of the 6-th Int. LS-DYNA Users Conference,* Dearborn, Michigan, Session 11, 17-26.

19. Hurmuzlu Y. (1998). An energy-based coefficient of restitution for planar impacts of slender bars with massive external surfaces. *ASME Trans. Journal of Applied Mechanics* **65**, 952-62.

DEFORMATION AND TEARING OF UNIFORMLY BLAST-LOADED QUADRANGULAR STIFFENED PLATES

S. Chung Kim Yuen and G.N. Nurick

Department of Mechanical Engineering, University of Cape Town,
Private Bag, Rondebosch 7701, South Africa.

Abstract

A series of experimental results and numerical predictions, using the finite element package ABAQUS, on built-in mild steel plates of different stiffener configurations subjected to a uniform blast load are presented. The quadrangular plates are of thickness 1.6mm with stiffener size of (width by height); 3x3mm, 3x7mm, 4x3mm and 4x7mm in configurations of single (S), double (D), cross (C) and double cross (DC) stiffeners. The impulse required to give deformations up to plate tearing is obtained by the explosive centrally laid out in two concentric rectangular annuli.

Observations show that in all the tests where tearing does not occur regardless of the size and configuration of the stiffeners, the plate profiles are characterised by a uniform global dome. Thinning mechanisms at the boundary are observed for all plates in spite of different stiffener sizes and configurations. Thinning, however, is not consistent all around the boundary. Tearing is observed to first occur at the middle of the sides of the plates and then progresses towards the corners as the impulse increases. The numerical predictions which incorporate non-linear geometry, material effects including strain rate sensitivity and temperature effects show an encouraging correlation with experimental data for both the displacement profile and tearing.

Keywords

blast load, stiffened plates, large deflections, tearing, failure.

INTRODUCTION

The investigation into the failure of plates and beams subjected to blast load conditions has been ongoing for the past 50 years. The focus for stiffened plates has been on clamped boundary conditions where the stiffeners and the plate are two separate bodies clamped together by two support plates[1,2]. These studies have shown that the final mid-point displacement of the beam is greater than that of the plate with a gap between the plate and the stiffener. The gap is attributed to the springback effect[3] of the independent components.

In studies on integral plates (DRES panel), Schubak, Olson and Anderson[4,5] observed that for a very high intensity pulse (several times the static collapse pressure) the displacements of the stiffeners and nearby plating were approximately the same. It was suggested that, away from the lateral edges, the one-way stiffened plates behave much like a singly symmetric beam with the plate acting as a large flange. Similarly, a two-way stiffened plate might behave like a grillage of singly symmetric beams.

Recent experimental work on pressure loading of 1m square stiffened plates with and without in-plane restraint was reported by Schleyer, Hsu and White[6]. The pulse was generated by a transient differential pressure, triangular in form with a peak nominal pressure of approximately 0.1MPa, created by the timed blow-down of two pressure loading chambers on either side of the test plate. The test results showed that in-plane restraint was responsible for reducing the maximum transient deflections in the stiffened plates by almost 50% and permanent deformations by more than a factor of 4. There was also no sign of lateral buckling in the stiffeners. On the other hand, the stiffeners appeared to have had little or no effect on the deflections of the plates without in-plane restraint.

Further studies on built-in singly stiffened square plates[7] show that tearing first occurs at the middle of the sides and then progresses towards the corners as the impulse increases.

Predictions[8,9] using computational techniques have also been reported and have shown a good correlation with experiments. However, these predictions did not include the effect of temperature on material properties. Wiehahn, Nurick and Bowles[10] showed how the inclusion of temperature effects can be used to predict tearing. The purpose of this study is to incorporate these effects and hence gain further understanding of the failing mechanism of stiffened built-in plates (where the stiffener and the plate are manufactured as a single unit) when subjected to uniform blast loads.

While an extensive set of experiments and accompanying predictions has been performed[11], only a selected set of results is presented herein.

EXPERIMENTATION

The built-in test plate of hot rolled steel has dimensions of 12mm thickness and approximately 220mm square. Quadrangular sections of dimensions 126 x126mm and of thickness 1.6mm, with rectangular stiffeners 3mm or 4mm wide and 3mm or 7mm in height are machined. Machining effects, such as cutter wear, result in small differences in the thicknesses of the plate and the stiffeners. Hence each plate and stiffener are measured at several locations. The test plate is annealed to remove all residual stress caused by the machining processes.

The plates are uniformly loaded using plastic explosive (PE 4), which has a burn speed of approximately 7500m/s. The explosive is laid out on a 12mm thick polystyrene foam pad in two concentric rectangular annuli made up by rolling the sheet explosive into cylinders and arranged in such a way that there is on average a uniform distribution of explosive mass over the specimen. Two perpendicular strips of explosive at the centre (cross-leaders) interconnect the annuli. A short tail of 1g of explosive holding the detonator is attached to the centre of the cross-leaders. The masses of the explosive tail and the cross-leaders are kept constant for all the tests. Differing masses of the annuli explosive are used, giving different impulses, resulting in large inelastic deformations and complete tearing.

GENERAL EXPERIMENTAL OBSERVATIONS

Plate deformations
In all the tests where tearing does not occur regardless of the size of the stiffener and the configuration of the stiffeners, the plate profiles are characterised by a uniform global dome with the highest displacement occurring at the central part of the plate.

Thinning
For the experiments described herein, thinning mechanisms at the boundary (shown in Figure 1) are observed for all plates regardless of the stiffener sizes and configurations. It is observed that thinning is not consistent all around the boundary. For double stiffened plates, thinning is also observed near the stiffener as shown in Figure 2. However, it is noted that thinning occurs at only the side of the stiffener closest to the boundary.

For cross stiffened plates a reduction in the stiffener width is also observed at the location where two stiffeners cross each other as shown in Figure 3.

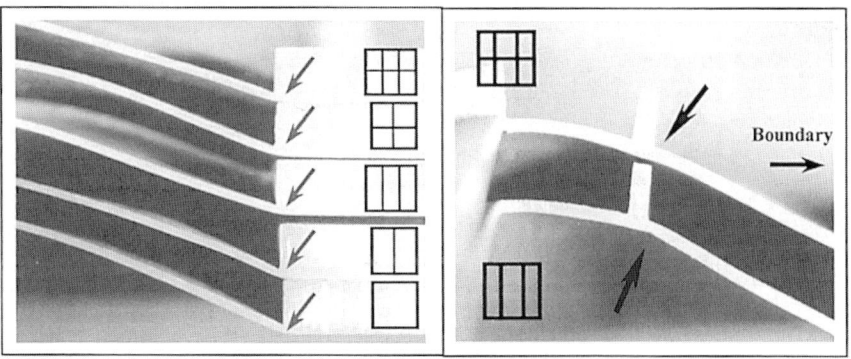

Figure1 Photograph showing boundary thinning for square plates of all stiffener configurations (arrows denote thinning)

From top to bottom:
DC 3x7mm impulse 36.7Ns, C 3x7mm impulse 36Ns, D 3x7mm impulse 37.5Ns, S 4x7mm impulse 40.5Ns, no stiffener impulse 31Ns

Figure 2 Photograph showing plate thinning in the vicinity of the stiffener (arrows denote thinning)

From top to bottom :
DC 3x7mm impulse 36.7Ns,
D 3x7mm at impulse37.5Ns

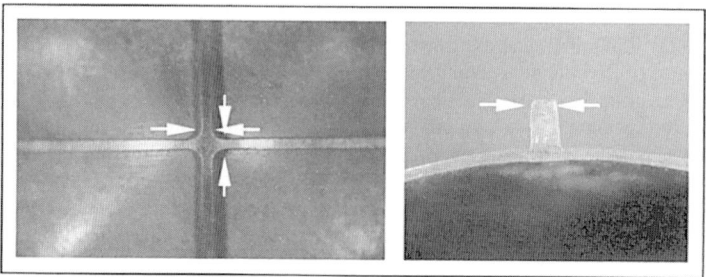

Figure 3 Photograph showing stiffener thinning of cross stiffened plates (C) (denoted by arrow)
From left to right : plan view of stiffener size 3x3mm impulse 36Ns; cross section of stiffener size 3x7mm impulse 36Ns

Tearing

Tearing first occurs at the middle of the sides of the square plates and then progresses towards the corners as the load intensity increases.

For stiffened plates, tearing first occurs at the middle of one of the supported boundaries parallel to the stiffener.

In all cases, as the impulse increases, the number of torn sides increases, progressing from one side of tearing to two sides of tearing to three sides of tearing as also reported by Nurick, Olson, Fagnan and Lewin [7]. It is also observed that when two sides perpendicular to each other tear the plate tears across the corner of the plate. In figure 4, a series of photographs illustrates tearing along different sides of a single stiffened plate.

One side torn   Two sides torn   Three sides with corners torn

Figure 4 Photograph showing tearing of single stiffened plates –
From left to right : stiffener size 4x3 at impulse 39.8Ns; stiffener size 4x7 at impulse 41Ns; stiffener size 3x3 at impulse 40.6Ns

FINITE ELEMENT FORMULATION

The numerical analysis is carried out using the general-purpose finite element code; ABAQUS/Explicit which incorporates non-linear geometry, strain rate sensitivity and temperature effects.

Model geometry

A quarter symmetry model using eight node brick (C3D8R) and six node prism (C3D6) reduced integration and hourglass control continuum elements is used. A diagram of the model illustrating the boundary conditions is shown in Figure 5.

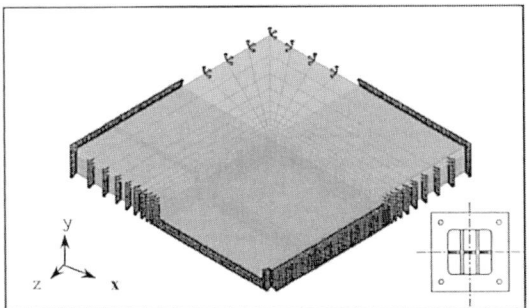

Figure 5 A typical model for a double cross stiffened square plates (DC) with stiffener size 3x7mm

Modelling the blast

The pressure loading from the explosive charge is assumed to be uniformly distributed with a rectangular pulse in time. The pressure magnitude P is calculated to correspond to the experimentally measured impulses according to the relation

$$P = \frac{I}{(At)} \qquad (1)$$

where, I is the total measured impulse, A is the exposed area of the plate and t is the load duration. The duration of the loading is assumed to be 14.5µs, obtained from the explosive configuration and burn speed.

Material Properties

Stress strain curves are obtained from standard uni-axial tensile tests and converted into true stress and logarithmic plastic strain according to the ABAQUS/Explicit code manual[12]. Masui et al.[13] reported that both the Young's Modulus (E) and yield stress (σ_y) are dependent on the material temperature (T) as shown in equations (2) and (3):

$$\begin{aligned} E &= 210 \times 10^9 - 58.34 \times 10^6 \cdot T & \text{for} \quad T \leq 600\,^\circ C \\ E &= 3.1 \times 10^6 \cdot (T-1100)^2 + 97 \times 10^9 & \text{for} \quad 600\,^\circ C < T \geq 1100\,^\circ C \end{aligned} \qquad (2)$$

$$\begin{aligned} \sigma_y &= \sigma_{yo} & \text{for} \quad T \leq 200\,^\circ C \\ \sigma_y &= \sigma_{yo} \cdot [1 - 0.00178 \cdot (T-200)] & \text{for} \quad 200\,^\circ C < T < 700\,^\circ C \\ \sigma_y &= \sigma_{yo} \cdot [0.133 - (T-700) \cdot 3.884 \times 10^{-4}] & \text{for} \quad 700\,^\circ C \leq T \leq 1000\,^\circ C \end{aligned} \qquad (3)$$

where E : Young's Modulus; T : material temperature; σ_y : static yield stress; σ_{yo} : static yield stress at the reference temperature of the tensile test.

The strain rate effects on the material properties are incorporated using the Cowper-Symonds relationship:

$$\frac{\sigma_y^1}{\sigma_y} = 1 + \left(\frac{\dot{\varepsilon}}{\dot{\varepsilon}_0}\right)^{1/\eta} \qquad (4)$$

where σ^1_y : Dynamic yield stress; σ_y : Static yield stress; $\dot{\varepsilon}$: Strain rate; $\dot{\varepsilon}_0$ and η are material constants. $\dot{\varepsilon}_0 = 40$ s^{-1} and $\eta = 5$ are the commonly used values for mild steel[14].

SIMULATION RESULTS

The prediction shows encouraging correlation with experimental data for both large inelastic deformation and tearing of the plate.

Plate deformations

A comparison of the contour plot of the numerical analysis and the experimental data for a double cross stiffened square plate undergoing an impulse of 37Ns is shown in Figure 6. The mid-point displacement measured in the experiment is slightly less than 22mm while the numerical analysis predicts a mid-point displacement of 20.2mm.

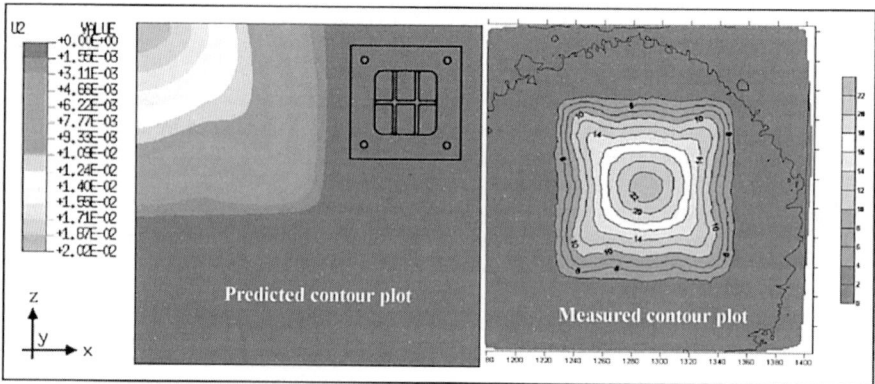

Figure 6 Comparison of predicted contour plot of a double cross stiffened square with stiffener size 3x7mm plate at impulse 37Ns

More detailed comparisons of the results of the numerical analysis are obtained by investigating the deformed quadrangular plate profile along its lines of symmetry. A plate profile plot along the centre lines of a double cross stiffened square plate subjected to an impulse of 37Ns compares the results of the numerical predictions and experimental results in Figure 7.

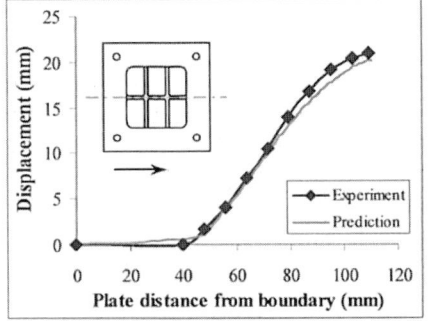 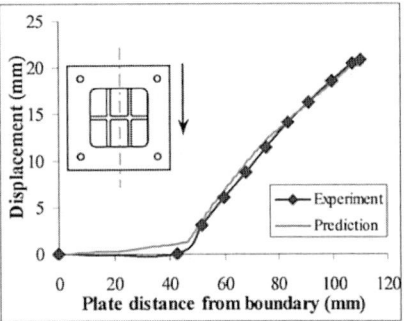

Figure 7 Graphs showing comparisons of plate profiles obtained from experiment and numerical model for a double stiffened square plate with stiffener size 3x7mm at Impulse 37Ns

Tearing and thinning

The finite element models show good correlation with the experimental results for tearing. Tearing is not symmetrical in the experiment due to the complex nature of the loading conditions involving explosives and possible material inconsistencies in the manufactured mild steel. Hence, the prediction of tearing of sequential sides of the plate is not possible.

When tearing occurs the model is subjected to severe temperature rises in the region of fracture, forming a band of high temperatures as a result of greater adiabatic heating due to the large amounts of plastic work and increased strain rate. This results in larger strains until unrealistic element elongation occurs in the direction of displacement (y-direction) exceeding realistic strain (70% - 300%). Hence, this suggests that tearing will occur within the high temperature band.

Figure 8 shows the response of a single stiffened plate which was partially torn at the boundary. A temperature of 896 ^{0}C clearly suggests tearing at the support side parallel to the stiffener. High temperature spots are also formed at the maximum permanent displacement (426 ^{0}C) and at the boundary (493 ^{0}C) where the stiffener joins the plate; indicating locations of thinning.

CONCLUDING REMARKS

This investigation has highlighted the benefits of using the approach to include temperature effects in the modelling of plates undergoing blast loads. The experiments and the predictions demonstrate encouragingly similar results as illustrated by a selection of comparisons – predictions (P) versus experiments (E) are :- no stiffener : 40Ns (P); 43Ns (E), single stiffener : 41Ns (P); 41Ns (E), double stiffener : 40Ns (P); 42Ns (E).

REFERENCES

1. Nurick, G.N., and Conolly, A. G. (1994), Response of clamped single and double stiffened rectangular plates subjected to blast loads. *Structures Under Shock and Impact*, (Ed P Bulson), 207-220.
2. Nurick, G.N. and Lumpp, D.M. (1996), Deflection and tearing of clamped stiffened circular plates subjected to uniform impulsive loads, *Structures Under Shock and Impact*, (Ed N Jones, CA Brebbia, AJ Watson), 393-402.

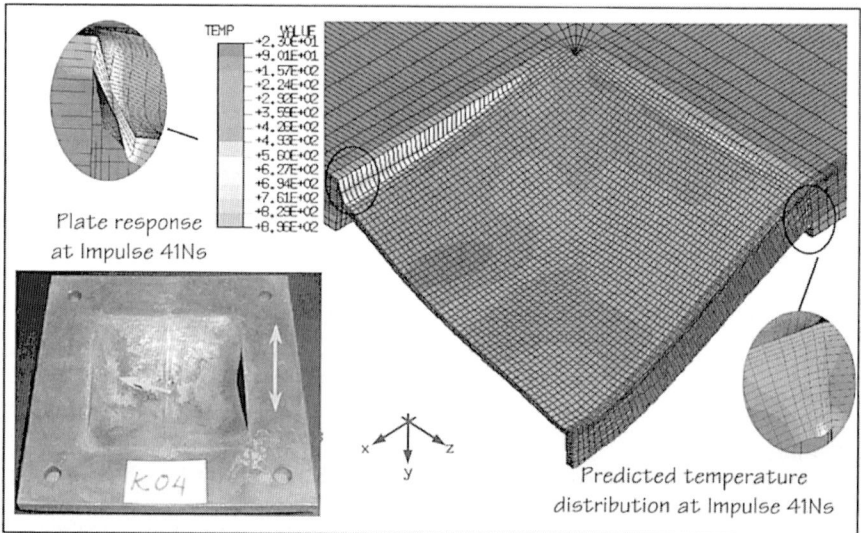

Figure 8 Comparison of predicted partial tearing at the boundary of a single stiffened square plate with stiffener size 4x7mm at impulse 41Ns (arrows show plate tearing)

3. Olson, M. D., Nurick, G. N. and Fagnan, J. R. (1993), Deformation and rupture of blast loaded square plates – predictions and experiments, *International Journal of Impact Engineering*, **13** (2), 279-291.
4. Schubak R. B., Olson, M. D. and Anderson, D. L. (1993), Rigid-plastic modelling of blast loaded stiffened plates – Part I: one way stiffened plates, *International Journal of Mechanical Sciences*, **35** (3/4), 289-306.
5. Schubak, R. B., Olson, M. D. and Anderson, D. L.(1993), Rigid-plastic modelling of blast loaded stiffened plates – Part II: partial end fixity, rate effects and two-way stiffened plates, *International Journal of Mechanical Sciences*, **35** (3/4), 307-324.
6. Schleyer, G. K., Hsu, S. S. and White, M. D. (1998), Blast loading of stiffened plates: Experimental, analytical and numerical investigations, *Structures Under Extreme Loading Conditions*, (Ed by HS Levine), PVP-Vol 361, 237-255.
7. Nurick, G. N., Olson, M. D., Fagnan, J. R. and Levin, A. (1995), Deformation and tearing of blast-loaded stiffened square plates. *International Journal of Impact Engineering*, **16** (2), 273-291.
8. Houlston, R. and DesRochers, C. G. (1987), Non-linear structural response of ship panels subjected to air-blast loading, *Computers and Structures*, 1987, Vol 26, 1-15.
9. Bimha, R.E., Nurick, G. N. and Mitchell, G. P. (1996), Modelling the deformation of blast-loaded stiffened square plates. *Proceedings of the 1st South African Conference on Applies Mechanics (SACAM 96)*, Midrand, South Africa, 260 - 267.
10. Wiehahn, M. A., Nurick, G. N. and Bowles, H .C. (2000), Some insights into the mechanism of the deformation and tearing of thin plates at high strain rates incorporating temperature dependent material properties, *Structures Under Shock and Impact*, (Ed N Jones, CA Brebbia), 207-220.
11. Chung Kim Yuen, S. (2000), Deformation and tearing of uniform blast loaded quadrangular stiffened plates, *MSc thesis*, University of Cape Town, South Africa, 1-181.
12. Hibbit, Karlson and Sorenson (1998), *INC. ABAQUS/Explicit User's Manual*, Volume 1 v5.8.
13. Masui, T., Nunokawa, T. and Hiramatsu, T.(1987), Shape correction of hot rolled steel using an on line leveller, *Journal of Japan Society for Technology of Plasticity*, (**1**).
14. Jones, N. (1989), *Structural Impact*, Cambridge University Press.

NON LINEAR RESPONSE AND ENERGY ABSORPTION OF VEHICLE FRONTAL PROTECTION STRUCTURES

Paul Bignell, David Thambiratnam and Frank Bullen

Physical Infrastructure Centre
School of Civil Engineering
Queensland University of Technology
Brisbane, Australia

ABSTRACT

With the advent of modern safety systems on new passenger vehicles, concern has arisen on the effect that may occur with the installation of frontal protection structures, in particular with the deployment of air bags. This paper discusses the investigation undertaken to assess the performance of frontal protection structures. The investigation includes quasi-static and dynamic experimental testing, as well as finite element analysis, to generate new research information on energy absorption, critical stresses and failure modes. This will facilitate the efficient design of frontal protection structures.

KEYWORDS

Frontal Protection Structures, Static and Dynamic Response, Energy Absorption, Finite Element Analysis

INTRODUCTION

Frontal Protection Structures (FPS) are installed on passenger vehicles to protect the vehicle during a minor impact and to provide impact attenuation during major impacts. Modern vehicles are fitted with safety systems, like crumple zones and air bags, to offer protection to vehicle's occupants during impacts. When an inappropriately designed FPS is fitted to a vehicle it may increase the stiffness of the frontal structure of the vehicle, and possibly affect the deployment of the vehicle's air bags. This concern has resulted in an extensive investigation into the non-linear behavior and energy absorption of the FPS using experimental and analytical techniques.

The objectives of the research project are to:

- Generate research information to enhance our understanding on impact attenuation of FPS, energy absorption, critical stresses and failure modes.
- Develop a simple procedure for evaluating the performance and suitability of a FPS.

- Study the effects of important parameters on FPS design, using finite element technique.

The project has used quasi-static and dynamic testing techniques, as well as extensive non-linear finite element analysis. Results from preliminary investigation, involving static testing of FPS, have been presented elsewhere.

BACKGROUND

FPS have been fitted to vehicles for a number of years. FPS were initially used to protect the vehicle during minor impacts in urban areas, such as animals and roadside objects. With the increasing popularity of recreational vehicles, FPS has become a common accessory on vehicles driven within city areas. This widening of the FPS market has led to changes within the design of the FPS. The FPS has evolved from the aggressive forward protruding angular design, to a more rounded body hugging design that is more cosmetically pleasing. This change has also reduced the risk to pedestrians. Figure 1 shows FPS fitted to the two most popular passenger vehicles sold within Australia.

Figure 1: Vehicles fitted with FPS

With the increased safety within the modern passenger vehicles, concern has arisen on the effect that the FPS may have on these systems. In particular vehicle crumple zones and passenger air bags. Vehicle crumple zones are designed to reduce the forces subjected to the vehicle's occupants during impacts. This is done by absorbing the impacting energy through crumpling of the vehicle's structure. Air bags are used to protect the vehicle's occupants from hard surfaces of the vehicle's interior. Sensors are placed throughout the vehicle to detect when the vehicle is involved in an impact. The air bag system processes this signal to assess whether the impact is severe enough to warrant the necessity of air bag deployment. Air bag deployment characteristics are dependent on the impact properties of the vehicles frontal structure. Thus with the fitment of a FPS to the frontal structure of the vehicle, careful consideration must be conducted on the change that this may make to the stiffness of the vehicle. If the FPS is too stiff premature air bag deployment may occur.

There are two different designs of FPS. These are the additional and replacement types. The additional design is fitted to the vehicle with no modification to the vehicle. Replacement FPS replaces the original bumper, and mounts, of the vehicle. FPS are commonly constructed from aluminium alloy and steel sections. The FPS with mounting system is then bolted to the vehicle.

EXPERIMENTAL TESTING

Experimental testing has been undertaken using both quasi-static and dynamic testing equipment. The results for the testing are correlated and compared, in particular the energy absorbed by the FPS.

Quasi-Static Testing

Quasi-static testing is undertaken using a universal-testing machine. The FPS is mounted to a rigid chassis, to replicate the vehicle's structure, and loaded, as shown in figure 2.

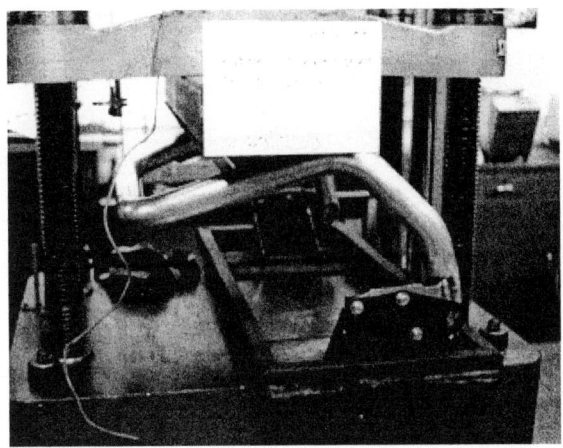

Figure 2: Quasi-static testing of FPS in progress.

The load and deflection of the FPS is recorded and plotted. This is used to calculate the energy absorbed by the FPS up to a certain deflection. This deflection is dictated by the clearance between the top of the FPS and that of the vehicle at a corresponding point. When the FPS has deflected this distance it is assumed that any additional force will be applied directly to the vehicle.

The energy of the traveling vehicle is then calculated and compared to that of the energy absorbed by the FPS. The vehicle energy takes into account the mass of the vehicle plus two 75kg occupants. This comparison is represented as a percentage.

Failure of the FPS often results from yielding of the mounting plates or tubing. It can result from shearing of the mounting bolts.

Dynamic Testing

Dynamic testing is undertaken by using a pendulum test rig. The FPS, and mount, is rigidly fixed to the pendulum test rig, as shown in figure 3. Maximum retardation is recorded through the use of accelerometers, charge amplifiers and memory recorders. The speed of the pendulum is adjusted by varying the height of the pendulum.

It is important in the recording of the peak retardation to take into consideration the deflection during the impacting time period. It has been found that the peak retardation value often occurs after the FPS has traveled sufficient distance to come into contact with the vehicle.

Figure 3: Pendulum test rig.

From the graph of retardation and time history, change in velocity can be calculated. Energy absorbed can also be calculated from a load deflection plot.

Analysis

The energy percentage absorbed in a quasi-static test has been correlated to the observed maximum retardation, 'g', values in dynamic testing, as shown in figure 4. This information together with that obtained during the analytical investigation, described below, will be used to develop efficient designs of FPS and their mounting system.

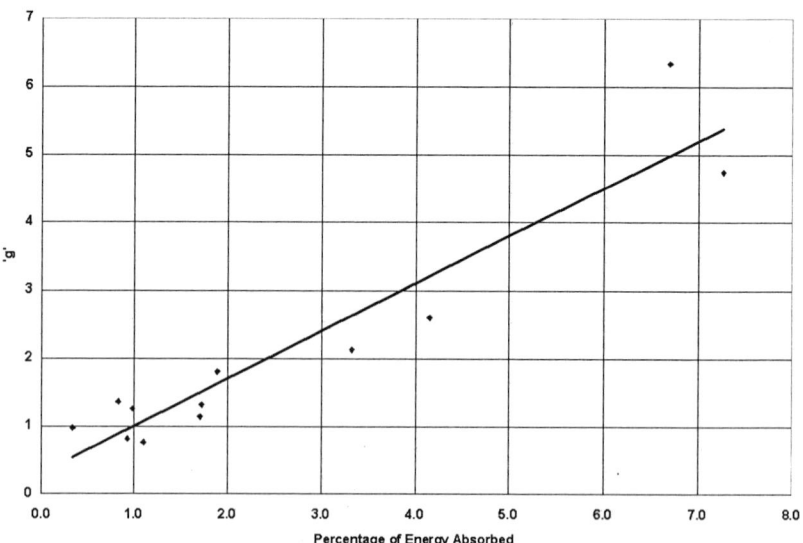

Figure 4: Correlation between quasi-static and dynamic testing

Based on the results of static and dynamic testing it has been possible to develop some preliminary assessment criteria on the suitability of FPS. It has become evident that a FPS which absorbs less than 5% of the kinetic energy of the vehicle, E_v, prior to impact with the vehicle, is satisfactory, ie such a bars will not affect the air bag triggering mechanism. FPS which absorb more than 8% of E_v are unsatisfactory, while evaluation of FPS which absorb between 5% and 8% of E_v should be supplemented with dynamic testing.

ANALYTICAL INVESTIGATION

The analytical investigation involves both dynamic impact analysis and quasi-static analysis using the finite element method. The FPS under investigation is modelled and subjected to quasi-static analysis. Results are compared with those from quasi-static experimental testing and the finite element model is calibrated. Usually, calibration is effected by adjusting the support conditions in the computer model to be better represent the true support conditions. The calibrated computer model is then subjected to dynamic impact analysis using the non-linear explicit finite element program LS/Dyna.

The models use beam and shell elements to represent the FPS and impacting surface, as shown in figure 5. Contact modeling is used to simulate the impact between the FPS and the impacting surface. An initial velocity, or impulse load, is applied to the impacting surface.

Figure 5: Example of FPS model using beam elements.

The finite element analysis is being used to generate a wide range of information, such as dynamic energy absorbed, critical bending moments, axial forces near the supports and failure modes. The dynamic energy absorbed is determined for various velocities of impact, impacting masses and for a variety of FPS section properties. The dynamic impact energies so determined will be compared with the static energy absorbed in order to determine the possible dynamic amplifications. This information will be used to develop improved designs of FPS and their mounting systems, with enhanced levels of safety.

CONCLUSIONS

This project involves quasi-static and dynamic testing, and non-linear finite element analysis of frontal protection structures in passenger vehicles. The key issue is to evaluate the amount of energy absorbed by the FPS before it impacts with the vehicle. Designing the FPS to absorb a predefined amount of energy is critical for the FPS not to have detrimental affects on the safety systems of the intended vehicle. The FPS must both be strong enough to offer protection during minor impacts, and remain flexible enough to avoid influencing the air bag deployment characteristics of the vehicle. All this information will be used in the enhanced understanding of FPS behavior and its design. The research is currently in progress and results will be presented at the conference.

REFERENCES

Bignell P., Thambiratnam D. and Bullen F. (1999). Energy Absorption in Structural Systems Under Random and Unpredictable Loads. *Acta Polytechnica* **39:5**, 147-155

Bignell P., Thambiratnam D. and Bullen F.(2000). Impact Attenuation of Frontal Protection Systems in Passenger Vehicles. *Implast 2000 Structural Failure and Plasticity*, 367-372

Bignell P., Thambiratnam D. and Bullen F.(2000). Behaviour of Frontal Protection Systems in Passenger Vehicles. *ICRASH 2000 Proceedings of the International Crashworthiness Conference.*, 102-112

Livermore Software Technology Corporation. (1999). LS-DYNA Keyword User's Manual, Nonlinear Dynamic Analysis of Structures, Livermore Software Technology Corporation, U.S.A

NON-LINEAR DESIGN OF BLAST / CONTAINMENT REINFORCED GUNITE WALLS FOR COAL MINES IN SA

F. du Toit [1], K. Comninos [1], P. J. Kruger [1]

[1] Civil, Structural and Architectural Department, HATCH Africa (Pty) Ltd,
Woodmead, Johannesburg, South Africa

ABSTRACT

Due to the high levels of methane gas produced in coal mines, the South African Department of Minerals and Energy Affairs (DME) requires the abandoned areas of a mine to be blocked off from the working parts of the mine to prevent methane exchange and/or to contain any possible coal dust explosions. Extensive testing around the world has shown that an equivalent static design pressure of 140 kPa can be used to simulate methane exchange, and 400 kPa for explosion pressures. The large number of walls to be constructed underground requires a cheap, simple and easily constructable design. Rather than using conventional elastic designs of reinforced gunite walls, non-linear yield line analysis techniques and the inclusion of the beneficial effects of membrane action has resulted in a wall with a thickness of only 1/3 of the wall thickness for conventional designs for the same loads. Due to the wide range of dimensions that can possibly occur for each possible underground wall position, design curves for standard wall designs have been produced to facilitate quick and easy construction. Full-scale tests performed underground support the design philosophy and construction method.

KEYWORDS

Containment Walls, Blast Walls, Reinforced Gunite, Non Linear, Yield Line, Membrane Action, Arching Action

INTRODUCTION

The process of developing an economical design specific to a client's requirements is described. The design loads and methodology fall outside the scope of the conventional concrete design codes, as non-linear analysis is required. To satisfy statutory requirements the design and construction methods had to be verified by testing. The results of the test programme, and further development work is also described.

WALL DESIGN REQUIREMENTS

The two main design requirements for the walls were cost and ease of construction. Due to the pillar and bord method of mining used, sealing off unused parts of the mine requires the construction of a considerable number of walls. As an order of magnitude, one mine is looking at constructing several hundred walls per annum. Reducing the cost of constructing the walls is a major factor in the design philosophy. The nature of the mining operations also influences the design philosophy in that the tunnels which have to be blocked off can vary widely in both width and height. Design curves were to be produced to allow for any combination of width and height, with the wall parameters (thickness, reinforcing diameter and spacing etc…) being readily obtained on site from the design curves.

Ingwe Coal Corporation has conducted research, Strydom (1999), on what static design pressure could be representative of the pressures experienced in operation. This topic has been researched in the mining industry extensively, especially in the USA and Australia. This research shows that there are effectively two conditions that need to be considered. The first is a wall in areas where the minimum sealing requirements are used in order to prevent methane exchange between areas, and the second is a wall where the sealed area is subjected to conditions in the "blast range". For these two categories, two different wall types can be developed, each with a different equivalent static pressure. The research showed the required static pressures to be 140 kPa for the minimum sealing requirement, to ensure the wall is air tight preventing methane exchange between sealed and open areas (containment wall), and 400 kPa for the walls in the explosive range (blast wall). These are the statutory parameters specified by the DME, DME (1997).

Wall Geometry

The wall is essentially a reinforced gunited concrete wall, but has been modified to allow ease of construction underground. The wall geometry is shown in Figure 2, with the construction procedure as follows. Timber poles are erected vertically at the blast face of the wall. These are used for lateral support during guniting. PVC sheeting is spanned between the poles to provide a barrier against which guniting can occur. A series of steel wire space meshes with a module size of 50 x 50 x 50 mm are erected against the sheeting. These are provided to prevent the gunite "sliding" down the wall and to control rebound. Roof bolts are drilled and grouted into the rock around the perimeter of the wall. These are intended to provide adequate shear resistance at the wall / rock interface. The wall's main reinforcing is erected on the outside of the space meshes in both directions as tensile steel. A monitor tube for gas sampling and a drain pipe for water flood drainage are built into the wall before guniting. The wall is then gunited to the required thickness.

THEORETICAL DESIGN

Design Approach

Conventional plug walls constructed from reinforced concrete generally range from 600 mm to 1 m thick depending on their particular application. They are designed elastically with limited crack widths to prevent water leakage. However, the blast/containment walls were not required to be water tight, and under the action of the design loads can exhibit significant cracking provided the wall does not rupture. In order to make the blast/containment walls as economical as possible, the ultimate load bearing capacity of the wall was utilised in its design by adopting the yield line method of analysis. In order to estimate the full strength of the wall it was necessary to take into account arching action against the lateral restraint of the side, hanging ad foot walls.

In conventional elastic design, only the bending strength of slabs is considered, and the membrane effects are ignored. A typical load deflection curve for this type of behaviour is shown in Figure 1. However, significant strength is gained from membrane action. If the ends of the slab are restrained from moving laterally, a dome or arching effect can be developed through the thickness of the slab. This arching action can significantly increase the ultimate load carrying capacity of the slab. Unfortunately, as the midspan deflection of the slab increases, the beneficial effect of the arch decays. Generally, arching action occurs when the deflection of the slab is less than half the slab thickness. When the deflection becomes too large, the "arch" snaps through in a stability failure mechanism, and the load carrying capacity drops with the slab experiencing high deflections. As the deflection increases, tensile membrane forces develop. These forces are now dependent on the tensile stresses developed in the slab. In the case of a reinforced concrete slab, this tension is developed in the reinforcing. The load – deflection behaviour of a reinforced slab exhibiting membrane action is also shown in Figure 1.

The effect of membrane action in slabs has been known for many years, and has been well researched, Braestrup & Nielsen (1983). In this case, the lateral support is provided by the surrounding rock, and the design is dominated by ultimate bending and arching action conditions, as well as cost.

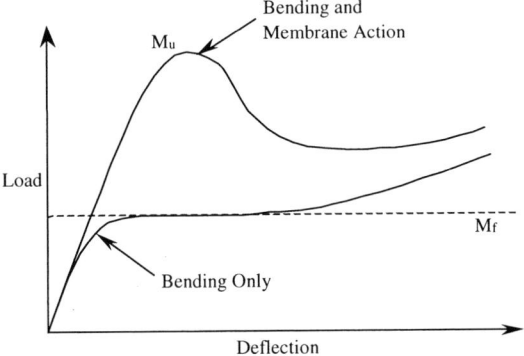

Figure 1: Load-deflection behaviour of slabs

The equations given in Braestrup & Nielson (1983), were used in the theoretical analysis to predict the benefit obtained by including membrane action. The analysis is summarised in following equations:

Symbols:
h = Thickness of concrete section
d_e = Effective depth of concrete section
f^*_{cu} = Modified concrete crushing strength
A_s = Area of tensile reinforcement
f_y = Reinforcing yield stress
Δ = Midspan deflection

The yield moment of a section acting in bending alone is given as:

$$M_f = \phi h^2 f_{cu}^* (d_e/h - 0.5\phi) \tag{1}$$

$$\text{where} \quad \phi = \frac{A_s f_y}{h f_{cu}^*} \tag{2}$$

The yield moment of a section including membrane action is given as:

$$M_u = M_f + \frac{N(N_0 + 0.5N)}{f_{cu}^*} \tag{3}$$

Where N is the membrane force at any misdpan deflection Δ, and N_0 is the membrane force corresponding to zero in-plane extension:

$$N = 0.5 h f_{cu}^* \left(0.5 - \frac{A_s f_y}{h f_{cu}^*} - \frac{\Delta}{h} \right) \quad \text{and} \quad N_0 = 0.5 h f_{cu}^* - A_s f_y \tag{4}$$

f_{cu}^* is a parameter included to account for the increase in the gunite crushing strength due to the presence of the space mesh. The reason for the increase is twofold. Firstly, the compressive strength is increased due to the additional area of steel in compression, and secondly, due to the containing action of the space frame in the transverse direction. The increase was assumed to be 50 % in the theoretical analysis. The above equations give reasonable predictions of the ultimate moment capacity of a concrete section for any given deflection. Unfortunately, they furnish little information about the peak load carrying capacity of a slab, as the deflection is not known at this point. For the blast wall analysis, the ratio of midspan deflection to wall thickness, Δ/h, was assumed to be 0.1. All subsequent calculations used to derive the design curves were based on this assumption. It was expected that the validity of this assumption would be verified during the full-scale tests, which is discussed in more detail later.

The walls were to have dimensions varying between 3 to 9 m in width and 2 to 9 m in height. The exact dimensions of any particular wall to be constructed would only be known once the construction team arrived at the underground wall site. Thus, two sets of design curves (part shown in Figure 2) were developed (one for each of the wall categories), from which the construction team could read the type of wall required, depending on the tunnel dimensions. Eleven different wall "types" were chosen with different thicknesses and reinforcing quantities. The types were selected such that the ultimate moment capacities were spread over the range required to resist either the 140 or 400 kPa loads. As many aspects as possible were kept constant throughout the different wall types so as to minimise the possibility of errors being made during construction. These include the length, size and spacing of the rock bolts, the spacing of reinforcing, lap lengths and gunite strength.

FULL SCALE TESTING

Full-scale tests were conducted on the walls in order to verify the design according to the DME requirements. To date 11 tests have been conducted with varying degrees of success. The main reasons for the test failures have been problems with either the test procedure or construction of the walls. A brief outline of the wall tests is given in Table 1. The first three tests show the development of the wall geometry to its present form.

The tests were conducted underground at Gloria and Matla Mines, thus utilising the intended construction techniques and quality control. Typically a test wall was constructed across a cubby (dead end tunnel), with pipe connections for pressure gauges and a compressor hose cast into the wall. During the initial tests, compressed air was used to apply the pressure behind (the blast face) of the wall. A major problem experienced with using compressed air was, that, as the pressure inside the cubby increased, air was forced through small cracks in the rock adjacent to the wall. This meant that pressures behind the wall could not be raised to, or beyond, the design pressure. For this reason, the sixth and all subsequent tests were conducted using water, supplied from the surface into the cubby by means of a borehole.

Test Results And Problems

Unfortunately, conditions underground were very difficult for testing, being dark, wet and dusty, resulting in very poor extraction of test data. The results from the 11 tests vary considerably, with many of the tests not yielding adequate results. The first two tests were conducted on designs considered before the design curves were produced. At that stage, the required equivalent static pressure was unknown, and very thin walls were used with the result that very low ultimate loads were achieved. The first test wall failed in bending, exhibiting the yield crack pattern expected for a simply supported rectangular slab. The second test wall failed due to construction problems. However, both tests showed that the wall was grossly underdesigned in its present arrangement, and required a more complex, ultimate capacity design. Tests 3, 4 and 5 all experienced problems resulting from the inability to develop pneumatic pressure behind the wall, and no useful results were obtained.

At this stage, the design curves using the yield line analysis and membrane action theory were developed. The method of pressure application was also changed from compressed air to water. The first wall tested using this method (test wall 6) was a Type B wall according to the design curves. During the test, the pressure was increased up to 172 kPa. At this point significant cracking had occurred in the wall, the pattern of which matched the yield line pattern. Unfortunately, the crack widths had become so large, that water was escaping from the cracks faster than it could be pumped into the borehole, and the pressure behind the wall began to drop. Thus, the test wall may still have had additional load carrying capacity over and above the 172 kPa that was used as the "failure load." Unfortunately, no reliable deflection readings could be taken from this test and the assumed deflection-to-thickness ratio of 0.1 used in the theoretical analysis could not be verified. The dimensions of this test wall were substituted into the theoretical analysis, yielding an ultimate load carrying capacity of 181 kPa, compared with 172 kPa in the test.

Following the successful 140 kPa wall test, further tests were conducted on the 400 kPa walls to establish if any additional problems would arise with such high loads and with the thicker walls. Test 11 yielded successful results. The wall was similar to the Type I wall, but with only 2 space meshes as opposed to 5 in the design. This wall sustained an ultimate pressure of 400 kPa before the crack widths again became so large that the water escaped faster than it could be pumped in. During this test, regular deflection measurements were taken using a theodolite to read divisions on a scale bound to a probe stick.

TABLE 1
SUMMARY OF FULL SCALE WALL TESTS

Test #	Wall Type	Failure Pressure	Target Pressure	Failure Mode
1	50mm wall, 1.8 mm mesh reinforcing	Approx. 40 kPa	Undecided	Bending – classic yield line crack pattern developed
2	75mm wall, 1.8 mm mesh reinforcing	Approx. 30 kPa	Undecided	Portion of wall breaking out – construction error
3	150 mm wall, 2 space meshes, 16 diameter roof bolts	Not reached	140 kPa	Test abandoned due to compressor malfunction
4	As for test 3	Not reached	140 kPa	Test abandoned due to compressor malfunction
5	As for test 4	Not reached	140 kPa	Leakage of air at wall to floor junction
6	**Type B**	**172 kPa**	**140 kPa**	**Bending – cracking in assumed yield line pattern**
7	Type H	Approx. 50 kPa	400 kPa	Borehole blocked
8	Type H	Approx. 120 kPa	400 kPa	Portion of wall breaking out – construction error
9	Type H	Approx. 50 kPa	400 kPa	Insufficient water pressure developed through water column
10	Type H	Approx. 180 kPa	400 kPa	Water leakage at large crack at wall to floor junction
11	**Type I with only 2 space meshes**	**400 kPa**	**400 kPa**	**Bending – cracking in assumed yield line pattern**

Comparison Between The Theoretical Design And The Test Results

A direct comparison between the theoretical ultimate strengths and the test ultimate strengths for the two successful test walls is given in Table 2. This comparison can be deceiving, as several factors need to be taken into account. Firstly, the design curves allow a band for each wall type in which any cubby may fall. In the case of the 140 kPa wall, the test wall lies in the lower region of the band, and may thus be expected to resist a pressure greater than the design pressure. On the other hand, the 400 kPa test wall lies in the upper region of the range of that wall type, and was thus expected to fail closer to the design pressure.

TABLE 2
DIRECT COMPARISON BETWEEN THE THEORETICAL AND TEST STRENGTHS

Test No.	Target Strength	Test Strength		Theoretical Strength		% Diff between Test and Theory	Bending Only Strength
		Pressure	% Diff	Pressure	% Diff		
6	140 kPa	172 kPa	22.9	181 kPa	29.2	5.2	75 kPa
11	400 kPa	400 kPa	0.0	564 kPa	41.0	41.0	195 kPa

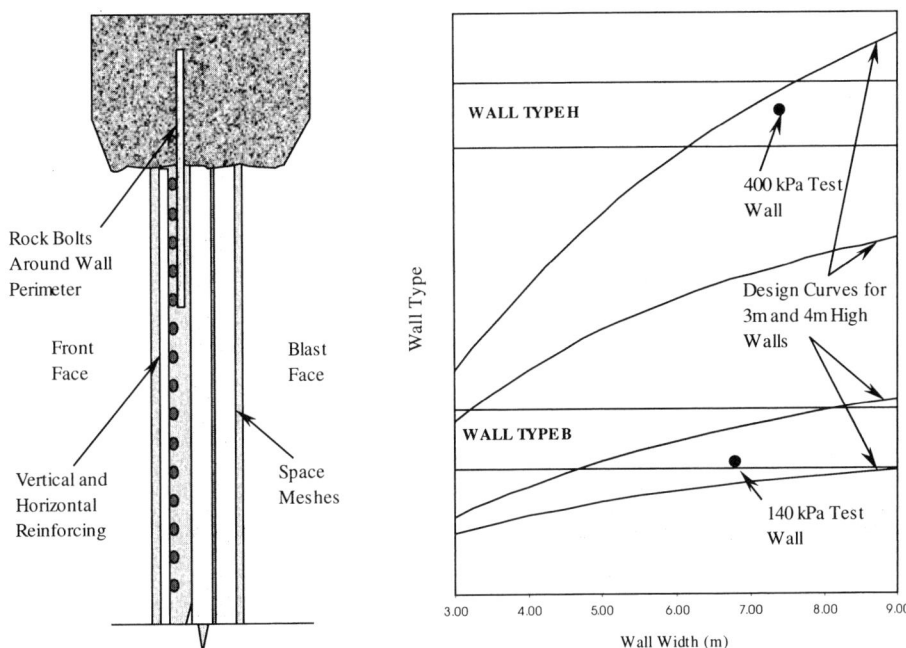

Figure 2: Design curves and test wall positions and geometry

Another problem experienced with the comparison is that the actual ultimate failure strength of the test walls is still unknown. "Failure" was deemed to have occurred when the pressure dropped, but this was due to the breakdown in the testing procedure, rather than collapse of the wall and some additional strength was still available in the walls at the test "failure" load. The exact amount of residual strength however, was not easily determined due to the poor load-deflection results obtained during the testing. The load deflection curve for the 400 kPa test wall is shown in Figure 3. It can be seen that the wall exhibited non-linear behaviour as expected, but had still not conclusively reached it maximum load. The maximum deflection at the test "failure" was only 9 mm, which gives a Δ/h ratio of 0.03. Using this instead of the 0.1 assumed in the theoretical calculations yields an ultimate pressure of 654 kPa. The reason for the increase in strength it that, for a lower deflection at the peak load, the effect of arching action is greater. However, for arching action to be fully developed, a certain amount of rotation is required at the supports to develop a lateral thrust. Thus, from the test data, it cannot be determined exactly how much arching action was developed and how much more could have been developed with further deflection of the wall.

The large difference between the theoretical and test pressures may also be caused by the differences mentioned between the 400 kPa test wall and the Type I wall (2 vs 5 space meshes) earlier. If the effect of the space mesh is ignored, and f^*_{cu} is taken as equal to f_{cu}, then the theoretical pressure calculated is 398 kPa. The reduction is most likely not as great as the neutral axis may well lie within the region of the two space meshes. No real conclusions can be drawn as to the effect of the space frame due to lack of data, but it should rather be noted as a possible variable. Another possible reason for the difference is that the actual cube strength of the gunite may not have been the 30 MPa assumed. From this discussion, it can be seen that the behaviour and basic principles of the theoretical analysis match what is observed in the tests, and that the calculations are numerically in the correct order of magnitude.

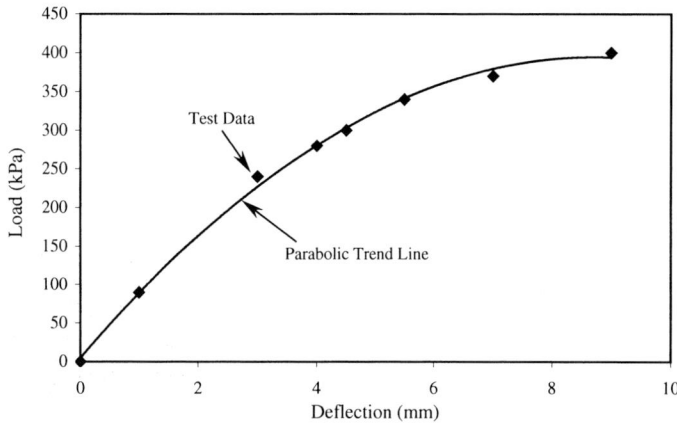

Figure 3: Load-deflection plot for 400 kPa test wall

CONCLUSIONS

The Client's requirements for an economical design for containment and blast walls for underground construction were met. Two wall design charts suitable for underground field use in a coal mine were derived. The construction method for a reinforced gunite wall was proven. A lower bound value for the ultimate strength of a wall slab was derived theoretically, and proven in testing. Unfortunately, due to the lack of data available for the full-scale tests and with all the variables, the assumptions made in the theoretical analysis cannot be properly verified. The next step that has been proposed is to perform an extensive finite element study to correlate the available test data with the theoretical analysis, and to calibrate the theoretical analysis in terms of the assumptions made. With further field and theoretical work the ultimate strength designs of the containment and blast walls can be optimised further.

REFERENCES

1. Braestrup, M. W. & Nielsen, M. P. (1983), *Plastic Methods of Analysis and Design, Chapter 20* Handbook of Structural Concrete, Edited by Kong, F. K., Evans, R. H., Cohen, E. & Roll, F., Pitman Publishing Inc.

2. Strydom, C. (1999), *Explosion Resistant Underground Stoppings,* Ingwe Coal Corporation Limited

3. South African Department of Minerals and Energy Affairs Guideline, GME7/4/118-AC1/1997 (1997), Department of Minerals and Energy Affairs

A PHILOSOPHY FOR BLAST RESISTANT DESIGN

Rolf D. Kratz

Department of Civil Engineering, University of Cape Town,
Rondebosch 7701, Cape Town, South Africa.

ABSTRACT

This paper proposes a formalised procedure for blast resistant design. This involves identification of the various types and magnitudes of blast loading, analytical requirements and design methods. It recommends the grouping of structures into four categories in terms of their vulnerability and sensitivity to blasts in order to have an ordered approach for each category in terms of blast loading, analysis, design and detailing requirements. It is proposed to deal with blast effects only at the Ultimate Limit State. The design and detailing procedures advocated will automatically ensure a high level of serviceability of the structure when subjected to blasts smaller than those designed for at the ULS.

KEYWORDS

Blast, resistant, philosophy, loading, analysis, design, detailing, structural categories.

INTRODUCTION

The need for a formalised approach to blast resistant design should be obvious at a time when bomb blasts have unfortunately become almost a daily occurrence. To be able to design a structure to be blast resistant the design engineer must have guidance on

- philosophy of blast resistant design

- vulnerability and sensitivity of structures to blasts

- the physics and nature of blast loading

- the response of structures to blast loading

- the design and detailing requirements to render a structure sufficiently blast resistant.

There is a great need for such a general formalised design approach in order to assist the design engineer who is completely without guidance at present.

BLAST DESIGN CONCEPTS AND DESIGN PHILOSOPHY

The proposals made in this paper are summarised below and discussed in more detail later:
- Grouping structures into three classes which need blast design considerations and a fourth class for which the "normal" design requirements deem to satisfy any blast design requirements.

- Grouping the members in a structure, for design purposes, into three types, i.e. primary, secondary and tertiary members. A specific design requirement of the tertiary members is for them to have a <u>significantly lower</u> strength than the associated primary and secondary members. This is to safeguard the structural integrity of the main structure, i.e. the tertiary members act as "blow out panels".

- Brittle failure phenomena are to be dealt with even more conservatively than the design for static loading requirements. This is particularly applicable to the design requirements to prevent shear failure.

- The additional strength of materials when subjected to high strain rate loading are to be ignored and in trade-off no ultimate load factors have to be applied to the loads. Since the strain rate factors involved in concrete in compression and reinforcing steel in tension are of the same order, (about 1,3), this amounts to an effective ultimate load factor of 1,3. According to tests performed at the University of Cape Town, shear, bond and other tensile types of failures do not exhibit any significant strain rate related capacity increases. It is therefore proposed to have an additional "load factor" 1,3 on all these effects when compared to standard design requirements resulting from static loading.

- It is further proposed to have a safety feature built into the design of Class 1 and Class 2 structures to make allowance for the complete loss of any primary member without collapse of the structure.

The details of these proposals will be dealt with in the respective sections. The blast resistant design proposal package is summarised in Table 1 below. All following sections refer to the matrix of this table.

TABLE 1
SUMMARY OF PROPOSED BLAST-RESISTANT DESIGN SCHEME

TYPE OF STRUCTURE		CLASS 1			CLASS 2			CLASS 3		
ANALYTICAL REQUIREMENTS		RIGOROUS DYNAMIC ANALYSIS FOR VULNERABLE STRUCTURES OTHERWISE TREATED AS CLASS 2			QUASI STATIC ANALYSIS			QUASI STATIC ANALYSIS		
SPECIFICATION SECTION		LOADING 2.4.1	DESIGN 2.4.2	DETAILING 2.4.3	L	D	D	L	D	D
TYPE OF MEMBER	PRIMARY	P1	P1	P1	P2	P2	P2	P3	P3	P3
	SECONDARY	S1	S1	S1	S2	S2	S2	S3	S3	S3
	TERTIARY	T1	T1	T1	T2	T2	T2	T3	T3	T3

CLASSES OF STRUCTURES

In order to effectively design structures for the various types of blasts it is necessary to identify them in terms of risk, blast consequences and desirability of post disaster serviceability. This can be done by grouping structures into four classes relating to vulnerability and sensitivity. Typical examples are:

Class 1 (high risk): hospitals, government buildings, radio & television stations, etc.
Class 2 (medium risk): arterial road bridges, bank buildings, power transmission towers, any assembly building such as: cinema structures, shopping centres, schools, train stations, etc.
Class 3 (low risk): buildings of four or more stories and other structures falling between the category above and the one below this one.
Class 4 (very low risk): private dwellings, apartment buildings up to three stories, service stations, minor road bridges, etc.

The intention is to require only the first three Classes to be subjected to some form of blast resistant design and that the last Class will deem to satisfy any blast design requirements by the application of the National Loading and Design Codes.

The design of the first three Classes will be subject to different specifications with respect to:

- blast loading (types, magnitude, location, etc.)
- structural configuration (static redundancy requirements after the loss of a primary member)
- analytical requirements (rigorous dynamic, equivalent static, none)
- design requirements (degree of ductility, energy absorption capability, confinement, etc.)
- special detailing requirements (sacrificial protection, shaping of members, spalling plates, confinement of concrete regions, etc.)

LOADING

Any structural design procedure is preceded by an identification of the loading which the structure has to withstand and the response expectations of the structure to the loading, i.e. the analytical requirements. Blast loading can be grouped into two basic categories:

- nuclear explosions
- non-nuclear explosions

In this paper only non-nuclear explosions are dealt with even though the blast phenomenon and the analytical, design and detailing requirements are of a similar nature. The magnitude of nuclear blasts and their global nature, as it affects the structure, would require a different design philosophy to the one proposed in this paper.

Non-nuclear or local explosions can be classified into

- detonation of explosives and stationary bombs
- gas explosions
- explosion of launched bombs (by missiles or dropped by planes)

In this paper only the first two forms of blast effects are dealt with since the last form concerns itself with weaponry whose destructive power goes beyond the scope of blast effects. Only the blast load categories are listed here. The actual blast physics, effective pressures and time scale can be obtained from the appended references. [2 to 14]

- Air blasts in the open, i.e. air shock wave.
- Tamped (or confined) explosions, i.e. shock waves transmitted by soil.
- Contact explosions
- Gas explosions, i.e. unconfined and semi-confined or contained explosions.

All of these blast types can be specified in terms of equivalent weight of TNT and the distance from the structure or the members under consideration.

STRUCTURAL CONFIGURATION

The structural configuration can be grouped into primary, secondary and tertiary member systems in terms of their participation in safeguarding the structural integrity against blast loading. The primary system would consist of the main columns, beams and shear walls forming the backbone of the structure. The secondary system would include secondary beams, slabs and minor supporting members. The tertiary system would comprise facades and cladding elements, infill walls, window units, roofing systems, etc. Protective elements to primary members are also classified as primary members in combination with the members they are protecting or shielding.

ANALYTICAL REQUIREMENTS

For blast sensitive structures it is necessary to perform a rigorous dynamic analysis. Any time stepping analysis program or procedure suitable to rigorous earthquake analysis can be used. Evaluation of natural frequencies is required prior to such an analysis to establish sensitivity. Since only the fundamental frequency is required in most cases, a pseudo static procedure using Rayleigh's method can be used to evaluate it. It may be required to evaluate the fundamental natural frequencies in different orthogonal directions.

Non-vulnerable structures can be analysed by any suitable static analysis procedure using quasi static loading, i.e. the static equivalent of the dynamic blast loading. It is desirable to perform such analyses at collapse, based on plasticity principles, e.g. yield line analysis.

Basic analytical models can be used for local blast effects, e.g. equivalent frame method.

DESIGN CRITERIA

Structures need only be designed at the Ultimate Limit State using the actual working loads and the specified blast load effects in their unfactored form. Only the permanent portion of imposed loading together with the self weight need to be combined with the blast load effects. Since the strain rate material factors are not applied to the concrete and reinforcing steel, the partial material factors used for static design can be kept as they are. This means that the design formulations, charts and tables from the governing design codes can be used without modification.

It must be noted that the strain rate material factors only increase the flexural resistances (including column resistance). In order to ensure ductile type failure and in order to make up for the non-increase in material resistance in tension (which includes bond and shear capacities) an ultimate factor of 1,3 is proposed on all shear, tension and bond design formulations. The design procedure must be done in such a way that those member capacities suffering "brittle failure" are always at least 30% higher than their flexural capacity counterparts. In the case where minimum flexural reinforcement is involved in a member this may increase the actual demand on shear capacity well beyond that required for the loading. It can be noted here that the emphasis is on the total member capacity to resist ultimate loading and not capacity of the sections in a member.

Elastic/plastic forcing functions for the design of beams, columns and slabs can be defined in terms of limiting design parameters. Members (specifically slabs) with axial restraints, which can built up 'membrane action', may require special forcing functions to reflect the suspension action.

In order to better visualise the design requirements a typical design sequence is shown below:

1) Determine the class of the structure to be designed.
2) Determine the loading categories this structure must be subjected to.
3) Choose the structural system or systems, including structural configuration.
4) Use the contact explosion member thickness design requirements where applicable.
5) Determine the relevant natural frequencies of vibration.
6) Establish the sensitivity of the structure to blast effects based on blast impulse durations and natural periods of vibration.
7) Perform the various analyses required to obtain the structural responses and member forces for design purposes.
8) Design the various members according to the design category in Table 1.
9) Detail the reinforcement according to the detailing category in Table 1.

The main differences of blast resistant design relative to 'normal' design can be summarised as follows:

- very high loading levels in local areas
- larger member cross-sectional dimensions (thickness) in regions of contact blasts
- relatively low reinforcement ratios in regions at a distance from member connections.
- relatively small ultimate load factors on gravitational loading

Additional design considerations:

The main blast resistant capacities to be ensured are:
- flexure (with axial compression, tension or pure flexure) and "beam shear"
- flexural and punching shear
- embedment and lap length of reinforcement

Ultimate limit state design (e.g. plastic design, yield line design, etc.) is used throughout. Shear and tension capacities must at all times be at least 1,3 times higher than flexural capacities.

Minimum structural dimensions and minimum reinforcement are largely dependent on contact blast effects.

Design should cover ductility requirements, i.e. minimum ductility ratio of 3,0 for primary and secondary members.

Whenever possible an equivalent static analysis and design approach should be applied.

In order to sufficiently protect exposed and important primary members, consideration must be given to sacrificial member design.

Design for energy absorption, plastic hinges, confinement, etc. will therefore form a major portion of the work. This is especially so for structures belonging to the Class 1 risk category.

The specific loading, analytical, design and detailing requirements as required for the three classes of structures in Table 1 are described in more detail below:

Blast load specifications:

(P1), (S1), (T3), etc. refer to the positions in the matrix of Table 1. The mass of explosive W and the distances X and Y referred to below are values which have to be coded.

Class 1 structures:
Equivalent static blast loading and corresponding quasi static analysis can be used unless otherwise specified or dictated by the blast sensitivity of the structure.

(P1) Primary system of members:
These members must be able to absorb the blast loading transmitted to them by secondary and tertiary systems. The blast loading should consist of some or all of the following:

- air blast from a remote explosion of W1 kg of TNT, Y1 metres above the ground, at X1 metres from the structure or a particular primary member.
- tamped (or confined) explosion of W2 kg of TNT, Y2 metres below the ground, at X2 metres from the structure or a particular primary member. This is needed mainly for the underground portions of a structure and may not be always applicable.
- contact blast of any critical section of unconfined W3 kg of TNT or alternatively contact blast of any critical section of shaped semi-confined W4 kg of TNT
- gas explosion of an equivalent of W5 kg of TNT.

The relevant data like impulse time, impulse value, effective overpressure, reflected pressure, etc. can be read off design charts or can be obtained from design formulae.[2 to 14] The member force response can be evaluated using a rigorous dynamic analysis using the impulse time interval for load application and the natural frequencies of vibration of the primary structural system for the structural response.[1] Only the fundamental natural frequency in each orthogonal direction needs to be considered. A rigorous analysis can only be dispensed with if it can be shown that the impulse interval is longer than the fundamental period or shorter than one quarter of the fundamental natural period. Gas explosions have a duration which is almost always much longer than the fundamental period.[2,9] A rigorous analysis can also be dispensed with if a modal match exists, i.e. if the effective structural system has a simple member mode response.

The type of analysis performed must be based on plastic principles (i.e. collapse type analysis, yield line or equivalent procedure).

(S1) Secondary system of members:
These members must be able to withstand the same blast loading as specified for the primary members.

(T1) Tertiary members:
These members only need to be able to withstand a remote explosion equivalent to half of W1 kg of TNT, Y1 metres above the ground, at X1 metres from the structure. The location is therefore the same as for the primary and secondary members, but the equivalent charge weight is halved. The nature of tertiary members is such that they absorb a major portion of any blast pressure load due to their large exposed areas. It is therefore detrimental to design them for the same loads as those used for the main structural members. They must fail well before their associated primary and secondary members and therefore act as a form of release valve or sacrificial element.

For the same reason these members must only be able to withstand half of the gas explosion pressure, i.e. and equivalent of half of W5 kg of TNT.

Class 2 structures:
Equivalent static blast loading and corresponding quasi static analysis can be used. It is only under exceptional circumstances when a blast sensitive element is detected that rigorous analysis may be required.

(P2) & (S2) Primary and secondary members:
These members have to be designed for 75% of the equivalent charge weights specified for Class 1 structures.

(T2) Tertiary members:
These members must only resist half the specified remote explosion and gas explosion pressures specified for primary and secondary members.

Class 3 structures:
Equivalent static blast loading and corresponding quasi static analysis can be used.

(P3) & (S3) Primary and secondary members:
These members have to be designed for 50% of the equivalent charge weights specified for Class 1 structures. Only remote and gas explosions need to be considered.

(T2) Tertiary members:
These members need not be designed for blast loading at all except that care must be taken that they fail well before their corresponding primary and secondary members.

Design specifications:

Class 1 structures:
(P1) Primary system of members:
1) The primary system of members must have sufficient structural redundancy so that it requires the complete loss of any one primary supporting member plus the formation of at least two more plastic hinges before the structure becomes a mechanism.

2) The primary structural system must be designed for the complete loss of any one of the primary members. This should be a separate loading condition in which only full dead load and the long-term imposed loading is participating. Only under very special circumstances is the response duration such that the blast loading is still acting when the gravitational effect due to the loss of the primary member comes into effect. This may be the case for some of those structures classified as sensitive in Section (P1).

3) Plastic analysis principles should be applied to ensure sufficient capacity against collapse.

4) Contact blasts must be catered for in such a way that spalling of concrete is limited to 'moderate damage', i.e. no shattering or scabbing of concrete should occur. This requires attention to member dimensioning and minimum reinforcement specification. It may require the use of protective measures such as anti-spall plates on the inside of exposed and 'reachable' members.

(S1) Secondary members:
1) The secondary members must be designed in such a way that the loss of any primary member does not in itself cause the collapse of the secondary member. Refer to Section (P1) (2).
2) Plastic analysis principles should be applied to ensure sufficient capacity against collapse. This should in most cases be dealt with by some form of yield line analysis and design.

3) Contact blasts must be catered for in such a way that spalling of concrete is limited to 'heavy damage', i.e. no scabbing should occur. This can in most cases be ensured by member dimensioning (thickness) and minimum reinforcement specification.

(T1) Tertiary members:
These members must not collapse or jump off their supports when subjected to the design loads. Since these members very often transfer substantial portions of the loading to the secondary and primary members (i.e. facades, infill walls, etc.) it is imperative that these members must be designed to fail well before their adjacent supporting members. Of course their supporting members (primary and secondary are still designed for the full blast effect transmitted by the tertiary members.

Class 2 structures:
(P2) Primary system of members:
1) The primary system of members must have sufficient structural redundancy so that it requires the complete loss of one primary member plus the formation of at least one plastic hinge before the structure becomes a mechanism.

2) The primary structural system must be designed for the complete loss of any one of the primary members. Refer to Section (P1) (2).

3) Plastic analysis principles should be applied to ensure sufficient capacity against collapse.

4) Contact blasts must be catered for in such a way that spalling of concrete is limited to 'heavy damage', i.e. no scabbing should occur.

(S2) Secondary members:
1) The secondary members must be designed in such a way that the loss of any primary member does not in itself cause the collapse of the secondary member. Refer to Section (P1) (2).

2) Plastic analysis principles should be applied to ensure sufficient capacity against collapse. This should in most cases be dealt with by some form of yield line analysis and design.

3) Contact blasts must be catered for in such a way that spalling of concrete is limited to 'heavy damage', i.e. no scabbing should occur. This can be ensured by member dimensioning (thickness) and minimum reinforcement specification.

(T2) Tertiary members:
The same design requirements as listed under Class 1 structures apply here.

Class 3 structures:
(P3) Primary members:
Plastic analysis and design principles should be applied to ensure sufficient member capacity against collapse, e.g. three plastic hinges are required on one span of a continuous beam system for local collapse.

(S3) Secondary members:
Plastic analysis and design principles should be applied to ensure sufficient capacity against collapse.

(T3) Tertiary members:
These members always have to fail at lower blast loads than their supporting members.

Detailing specifications:

Some of the structural details which need special attention can be listed as follows:

- shielding of exposed primary members (Class 1,2 and 3 structures)
- detailing of critical primary member geometry in order to improve blast angles and deflection of blast wave. (Class 1 and 2 structures)
- ductility detailing procedures of longitudinal and transverse reinforcement. These would be very similar to the detailing procedures used in earthquake resistant design. They would be mainly applicable to primary and secondary members. Class 1 and 2 structures would need special considerations, especially in the jointing regions of members. Concrete compression block confinement reinforcement may become necessary for normal and reverse flexure.
- ties and/or integrity steel as for earthquake resistant design for Class 1,2 and 3 structures.

CONCLUSIONS AND RECOMMENDATIONS

The additional design requirements for blast resistant design can be substantial for certain vulnerable and blast sensitive structures. Considerations of any measures deemed necessary must therefore be carefully weighed up against the benefits gained from these measures. At the same time it must be pointed out that the above recommendations propose that the majority of structures be subjected to equivalent static loading, analysed by quasi static, relatively simple procedures and designed using standard design charts and formulations. A large group of structures would not have to be designed for blast resistance at all. (Class 4 structures)

It is recommended that these measures be kept as simple as possible but at the same time produce structures with substantially higher blast resistance.

The time has come to do something constructive about practical recommendations to assist the design engineer in order to render structures more blast resistant.

LIST OF REFERENCES

(1) C.H. Norris, R.J. Hansen, M.J. Holley Jr, J.M. Biggs, S. Namyet and J.K. Minami - **'Structural Design for Dynamic Loads'** - Mc Graw-Hill - New York - 1959.

(2) N. Dobbs, E. Cohen and S. Weissman - **'Blast Pressures and Impulse Loads for use in the Design and Analysis of Explosive Storage and Manufacturing Facilities'** - paper presented at the conference on 'Prevention of and Protection against Accidental Explosion of Munition, Fuels and other Hazardous Mixtures' held by the New York Academy of Sciences - October 1966.

(3) **'Structures to Resist the Effects of Accidental Explosions'** - Department of the Army Technical Manual TM5-1300, Department of the Navy Publication NAVRAC P-397, Department of the Air Force Manual AFM88-22 - Department of the Army, the Navy and the Air Force - Washington, D.C. - June 1969, revised March 1971.

(4) W.E. Baker - **'Explosions in Air'** - Southwest Research Institute, San Antonio, Texas - 1973.

(5) W.E. Baker, J.J. Kulesz, P.S. Westine, P.A. Cox and J.S. Wibeck - **'A Manual for the Prediction of Blast and Fragment Loading on Structures'** - Southwest Research Institute, San Antonio, Texas - November 1980, revised August 1981.

(6) W.E. Baker, P.A. Cox, P.S. Westine, J.J. Kulesz and R.A. Strehlow - **'Explosion Hazards and Evaluation'** - Fundamental Studies in Engineering 5 - Elsevier Scientific Publishing Company - New York - 1983.

(7) C.A. Mills **'The Design of Concrete Structures to Resist Explosions and Weapon Effects'** *The Institution of Structural Engineers,* London, FIP notes, 1988/2.

(8) J.M. Biggs - **'Introduction to Structural Dynamics'** - Mc Graw Hill - New York 1964.

(9) W.A. Keenan, J..E. Tancreto, **Blast Environment from fully and partially vented Explosions in Cubicles,** Dept. of the Army, Picatinny Arsenal, TR-828, Nov. 1975.

(10) Forschungsinstitut für MIlitärische Bautechnik, Zürich: Waffenwirkungen und Schutzraumbau, **Ueberdruck und Luftstoss',** 1975.

(11) G.J. Kinney, K.J. Graham, Springer Verlag, **'Explosive Shocks in Air',** 1985.

(12) TWO 1977, Bundesampt für Zivilschutz, Switzerland, **'Technische Weisungen für die Schutzanlagen der Organisation und des Sanitätsdienstes',** 1977.

(13) TWS 1982, Bundesampt für Zivilschutz, Switzerland, **'Technische Weisungen für spezielle Schutzräume',** 1982.

(14) TWP 1984, Bundesampt für Zivilschutz, Switzerland, **'Technische Weisungen für den Pflicht- Schutzraumbau',** 1984.

FIRE SAFETY AND FIRE RESISTANCE

THE EFFECT OF FIRE ON MULTI-STOREY STEEL FRAMES

P.J. Moss[1] and G.C. Clifton[2]

[1]Department of Civil Engineering, University of Canterbury,
Christchurch, New Zealand
[2]Heavy Engineering Research Association, Manukau City,
New Zealand

ABSTRACT

This paper describes advanced analyses being undertaken as part of an international research programme aimed at determining the extent to which unprotected steel beams can be used when the structural fire severity is high. In order to investigate the effect of fire on a multi-storey steel frame building, a particular 17 storey building was studied to determine the behaviour as a fire spread throughout one floor of the building. Resistance to wind or seismic lateral forces acting along the major axis of the building is provided by moment resisting frames on two opposite sides while at right angles the lateral resistance is provided by eccentrically braced frames. A gravity load carrying frame runs along the major axis of the building. In these analyses, the columns and braces were fully fire protected, while the beams (acting compositely with the concrete floor slab) were unprotected. For the analyses carried out, the fire occurred on only the second level and spread from one quarter of the building. The research has shown that this particular multi-storey steel frame building, if constructed using unprotected beams, would possess an adequate level of fire safety. It is expected that this conclusion would apply to a much wider range of multi-storey steel framed buildings with protected columns and unprotected beams and research is planned over the next couple of years to investigate this further.

KEYWORDS

Steel construction, fire behaviour, thermal response, structural response, composite structures.

INTRODUCTION

In New Zealand under the old Building Control System (prior to 1992), design of multi-storey steel framed buildings for fire resistance was undertaken on the incorrect premise that the building will fail to meet the required levels of fire safety unless all the beams and columns are insulated from temperature rise under fully developed fire conditions. The advent, in 1992, of a performance based regulatory system and set of fire safety requirements which foster the use of rational fire engineering design, has led to the now routine use of unprotected steel beams, and sometimes columns, in multi-storey buildings where the structural fire severity is low (such as in car parking buildings) or where it is moderate and the steel members are shielded from immediate exposure to the fire by non-fire-rated linings (such as in hotels and apartment buildings). This paper describes advanced analyses being undertaken as part of an international research programme aimed at determining the extent to which unprotected steel beams can be used when the structural fire severity is high.

BUILDING BEING MODELLED

The building is a 17 storey office building located in the Auckland Central Business District. It was built in 1988 and incorporates a perimeter based seismic-resisting system and an internal gravity load-carrying system.

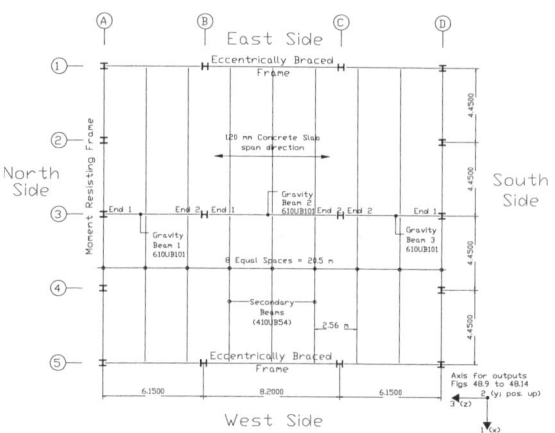

Figure 2 Reflected floor plan of building over the fire floor (second level) showing gravity load-carrying and lateral load-resisting systems

Figure 1 View of building looking towards the north-west corner (the fire floor was on level two)

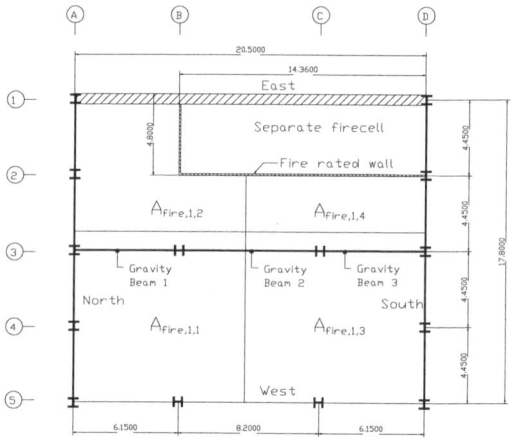

Figure 3 Plan of fire floor (second level) showing extent of firecell and the four areas of fire

Figure 1 shows a view of the building looking towards the north-west corner with the west side being to the right and the north side to the left. Figure 2 shows the reflected floor plan of the third floor level, which forms the top surface of the fire floor. The access for services and stairs is enclosed in a separate firecell in the south-east corner; its shape and extent is shown in Figure 3.

The lateral load-resisting system comprises perimeter moment-resisting frames (MRFs) along the north and south sides, with eccentrically braced frames along the east and west sides, as shown in Figure 2.

The floor system comprises a 120mm thick concrete slab, with $f'_c = 25\text{MPa}$, cast onto a 54 mm deep trapezoidal decking profile and made composite with 410UB54 Grade 250 supporting secondary beams. These span from a central primary beam (610UB101) along gridline 3 out to the east and west sides, as shown in Figure 2.

Because the floors will function as effective fire separations, experience from actual fires shows that a fully developed fire spreads to adjacent floors principally via the windows in external walls, and in the absence of fire service intervention, will take a minimum of 30 minutes. For this reason, the fires were modelled as occurring in one storey only, as the structural effects will occur principally on a floor by floor basis.

Level 2 was chosen, for three reasons. First, it is the lowest typical floor. The ground floor (level 1) has greater ventilation and lower fire load, being intended for access and reception, and hence will generate a significantly lower structural fire severity. The second level is therefore the lowest level on which high structural fire severity could realistically be generated. Putting the fire on the lowest level also means it affects the most heavily loaded columns and has the potentially greatest destabilising effect on the floors above.

ANALYTICAL MODELS USED

Fire model

The fire model used for the majority of the analyses was the large firecell fire model from HERA Report R4-83 (1996).

For the building response modelling required in this project, a fire model is required which accounts for the fact that, in a large firecell, not all the firecell will be subjected to fully developed fire conditions at any one time. Instead, if conditions are favourable for fire growth, the fire will reach full development, at its point of origin, then migrate throughout the remainder of the firecell. Thus, at any one time after full development is reached at the point of fire origin, there will be areas of the firecell not yet subjected to fully developed fire, other areas under full development and other areas which have been burned out.

Heat transfer model

The programme TASEF-2 (Temperature Analyses of Structures Exposed to Fire: Two Dimensional Version) was used to determine the temperature rise in the steel members and the concrete slab (the concrete slab applied to the gravity beams on gridline 3, which have been modelled as fully composite members). Once steel and, where appropriate, concrete temperatures were determined from TASEF-2, they could be used as input for the elements of the cross sections in the structural model.

Structural model

The structural model and analyses used the general finite element analysis program ABAQUS (1998). This program allowed the following aspects to be taken into account:

(1) The effect of elevated temperatures on the mechanical properties of the steel and concrete. The input values for concrete were taken from EC2 (ENV 1992-1-2, 1995). Those for steel, in the first 16 runs, were taken from Stevenson (1993); however, in the later 11 runs, the much more accurate stress-strain-time model of Poh (1996) was used.
(2) The effect of lateral restraint and continuity provided by the floor slab to the columns. The floor slab was modelled as a membrane element.
(3) The effect of composite beam action on the central primary beams along gridline 3. These three beams were modelled as fully composite members.
(4) Column members, which are steel I-sections, were modelled using standard I-beam elements which incorporated temperature dependant material properties.
(5) Connections between perimeter gravity beams A1-B1, C1-D1, A5-B5 and C5-D5 and their supporting columns (see Figure 2) were modelled as continuous.
(6) Connections between primary beams A3-B3, B3-C3 and C3-D3 and their supporting columns were modelled as realistically as practicable as temperature-dependant rotational and axial springs.

SCOPE OF ANALYSES UNDERTAKEN

Thirty-nine different scenarios were analysed, determining the influences of the following parameters:

(1) Two different levels of fire load
   - $e_f = 800$ MJ/m² floor area   (office design fire load)
   - $e_f = 1200$ MJ/m² floor area   (office maximum credible fire load or office library or office storage design fire load)
(2) Three different ventilation conditions, obtained by applying different window heights around the three sides with curtain walling.
(3) The extent of fire spread on building behaviour. This involved runs keeping all loading and ventilation parameters constant and applying the fire over all four areas of fire, then over $A_{fire,1,1}$ and $A_{fire,1,2}$ only, then over $A_{fire,1,1}$ only (see Figure 3).
(4) The presence of a non-fire-rated suspended ceiling forming a radiation barrier between the unprotected beams above and the fire below.
(5) Applied lateral loading (simulated wind loading) acting concurrently with fire; two levels of loading were considered.
(6) Variation in the elevated temperature mechanical properties (stress-strain-time) of the steel, as mentioned above.
(7) Variation in the connection strengths between the primary beams on gridline 3 and the supporting columns.
(8) Variation in the magnitude of the gravity load.
(9) Fire-time-temperature data adapted from the Cardington Demonstration Furniture test.

KEY FEATURES OF THE BUILDING'S BEHAVIOUR

A general overview of the building's behaviour including its performance for both the design fire load and the maximum credible fire load have been presented elsewhere (Moss and Clifton, 1999, 2000). In addition, fire temperatures from the Cardington Demonstration Furniture test were applied to the building and the results compared with those from the maximum credible fire (Moss and Clifton, 2000). Some of those features are covered herein when comparing the behaviour of the basic model.

SENSITIVITY TO THE LENGTH OF TIME THE CEILING STAYS IN PLACE

General

In the basic model, the non-fire-rated suspended ceiling was taken to be ineffective in forming a radiation barrier between the unprotected beams above and the fire below. Several other cases were considered and in this section, the behaviour of the basic model is compared with cases where the ceiling fails 10 or 20 minutes after the 600°C fire temperature is reached, or remains fully effective throughout the fire.

Behaviour of the gravity beams

Figure 4 shows the midspan deflection of the 8.2 m span gravity beam 2 (beam B3-C3) over the two hour analysis time for the basic model. The beam midspan deflection reaches a maximum of 340 mm (span/24), recovering to 255 mm (span/32) at the end of the fire. As the length of time that the ceiling remains in place increases, the peak midspan deflection reduces to 270 mm (span/30) for failure after 10 minutes, 140 mm (span/59) for failure after 20 minutes (Figure 5), but only 20 mm for the case of a fully effective ceiling. These deflections have reduced to 185 mm (span/44), 60 mm (span/137) and 7mm respectively by the end of the fire.

The variations in bending moment at the mid-span of gravity beam 2 are shown in Figures 6 and 7. Also shown is the moment at the quarter span point nearest column C3. This point lies in fire zone 3 (see Figure 3) and consequently shows a different moment time-history response to that of the midspan. The time delay between zones 1 and 3 (or zones 2 and 4 in the amended model) in reaching any particular temperature causes the quarter span moment curves to have two peaks due to the fire effects as they reduce in value. For the case where the ceiling remains fully effective throughout the fire, the moment reduced to 260 kNm after ½ hour before increasing again to 330 kNm after 2 hours.

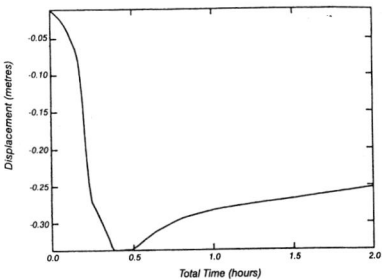

Figure 4   Midspan deflection of gravity beam 2 versus time for the basic model

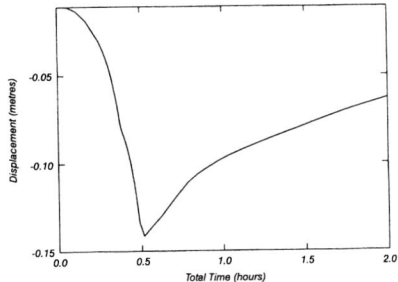

Figure 5   Midspan deflection of gravity beam 2 versus time for the case where the ceiling fails after 20 minutes

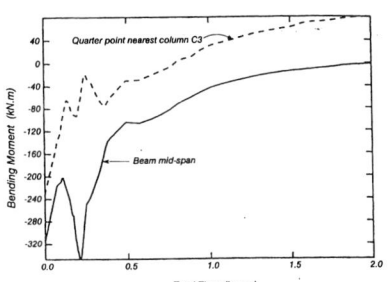

Figure 6   Bending moment in girder 2 versus time for the basic model

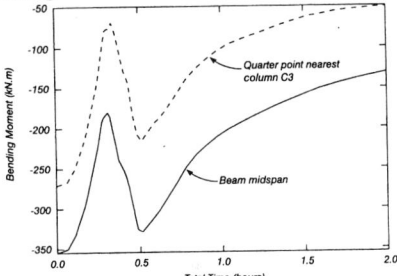

Figure 7   Bending moment in girder 2 versus time for the case where the ceiling fails after 20 minutes

The mid-span deflections of gravity beams 1 and 3 are shown in Figures 8 and 9. The difference between the deflections of the two beams arises from their being in different fire zones, with a time delay as the fire spreads along the building. In Figure 8, the deflection of beam 1, at 170 mm, is about span/36 under the design fire load, while that for beam 3, at 200 mm, is about span/31. For the cases where the ceiling remains in place for a least a short time (Figure 9), both gravity beams undergo much reduced deflections. Where the ceiling remains in place, the initial 5½ mm deflection increases to 10 mm before reducing to 2½ mm.

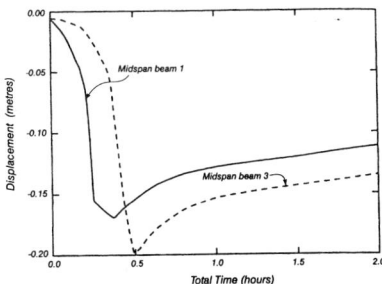

Figure 8   Midspan deflections of gravity beams 1 and 3 versus time for the basic model

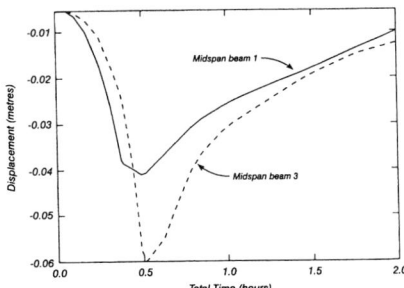

Figure 9   Midspan deflections of gravity beams 1 and 3 versus time for the case where the ceiling fails after 20 minutes

During the heating phase, the beams pushed against the end supports as they underwent a nett expansion. The compression force induced by this reduced during the cooling phase and then became tensile, with a final post-fire residual tensile force remaining through the connection into the supporting columns.

Behaviour of the supporting columns

The maximum temperature reached from TASEF-2 in any of the supporting columns, all of which are protected, was 130°C. This is consistent with the UK test results (Kirby, 1998), except that higher column temperatures were experienced in the more severe tests, with greater effects on fire-induced deflections and actions on the columns. Because of this, the columns remained elastic. Axial compression force in column B3, on the fire floor, for the run involving the basic model, increased from 4,980 kN prior to the fire to 5,250 kN at time of maximum column temperature rise. This corresponded to a growth in column length of 6 mm. Because of the differential heating of the columns at the fire floor, there is a time variation in the load distributed to column B3 by the beams carrying the floor load. The peak column load occurs 1½ hours after the start of the fire while the fire temperatures are cooling. For the case of the ceiling falling after 10 or 20 minutes, the maximum column load increased from 4,980 kN to 5,400 kN over the 2 hour duration of the fire. The load increase is of the order of 5% of the initial load in the column for the basic model but about 8½% in the cases of the falling ceiling. When the ceiling remained intact, the load in column B3 increased to 5280 kN.

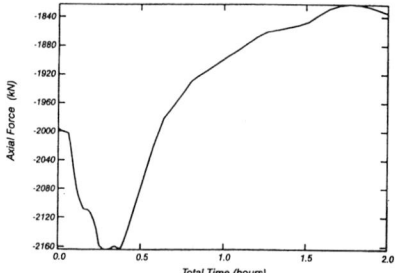

Figure 10 Axial force in column A3 at fire floor versus time for the basic model

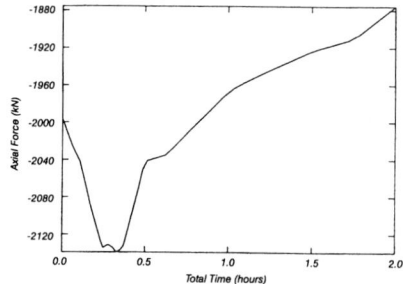

Figure 11 Axial force in column A3 at fire floor versus time for the case where the ceiling fails after 20 minutes

The variation of the axial force in column A3 at the fire floor is shown in Figures 10 and 11. In the case of the design fire load, the growth in column length was 4 mm and the axial force changed by ± 8% from its initial value. When the ceiling falls after 20 minutes, the growth in the column length was 4½ mm and the axial force changed by +6½% and -6%. The moment input into any of the supporting columns from fire-induced actions was also much less than the section moment capacity of the column. The supporting columns therefore remained elastic during each of the analyses.

Lateral stability of the overall building

As previously mentioned, one of the key outputs desired from this research was an indication of the extent to which the building above the fire floor would be destabilised by the fire. The building's high aspect ratio in both plan directions and the exposure of the lateral load-resisting systems on 3 of the 4 sides to the fire meant that the fire would be expected to have more of a destabilising effect on this building than it would on buildings with lower aspect ratios or buildings with a central, lateral load-resisting core (which invariably houses services and access routes and so forms part of a separate fire-cell).

Figure 12 shows the deflection versus time in each direction at the top of the building, location column A3, with no applied lateral load. The vertical deflection (upwards) is due to the lengthening of column A3 on the fire floor. The magnitude of lengthening (3.3mm) reflects the fact that this column is more restrained against lengthening than gravity column B3 and also remains cooler during the fire.

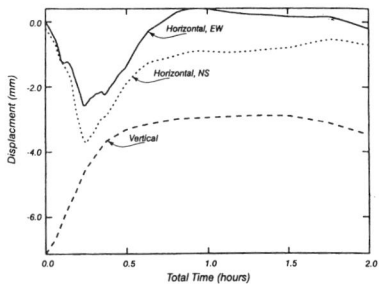 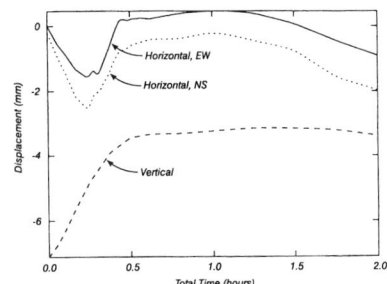

Figure 12 Deflection versus time at top of building, column A3 for the basic model
NS = North-South direction, EW = East-West direction

Figure 13 Deflection versus time at top of building, column A3 for the case where the ceiling fails after 20 minutes

There is a very small horizontal movement in each direction, of less than 4mm, at the peak of the fire, returning to the pre-fire location on cooling. Overall, with no applied lateral loading, the fire at level 2 appears to have no noticeable influence on lateral movement of the overall building above the fire floor. In the case of a ceiling failure, Figure 13 shows that the horizontal deflections are slightly smaller than those in Figure 12 but the vertical expansion is about the same.

THE EFFECT OF DIFFERENT LEVELS OF GRAVITY LOADING

General

In one analysis, the input temperature data for the bare steel beams came from the Cardington Demonstration Furniture test. The column steel temperature data came from the same test with the average temperature being applied to all points in the cross-section, the time variation being the same as that for the maximum credible fire. In this section, the results of using the measured temperatures from the Cardington tests are compared with the results for the same fire data but for cases where the live load component of the total load is increased by factors of 1.5, 2.0 and 2.5, thus reducing the ability of the members to resist the effects of the fire.

Behaviour of the gravity beams

The peak mid-span deflection for the four cases showed a similar time distribution to that of Figure 14, with peak values of 420, 440 and 460 mm as the load factor was increased to 2.5.

The time variation of the mid-span bending moments show similar trends for the four analyses. In the standard case, the moment reduces slightly from its initial value of 350 kNm before increasing to a peak of 400 kNm and then reducing to almost zero. When the live load is increased by 2.5, the initial moment of 530 kNm increases to 610 kNm, then reduces to 110 kNm.

Behaviour of the supporting columns

The variations of the bending moments at the top of column A3 are similar to those shown in Figure 15. For the weak axis moment, M_y, the M_y moment rises rapidly to 430 kNm before, just as rapidly, dropping to half that value then slowly falling over the cooling period to about one quarter of the peak value. As the live load is increased by up to a factor of 2.5, the M_y moment decreases to 360 kNm. The strong axis moment, M_x, in Figure 15 rises in a fluctuating manner to peak at 160 kNm, then falls to two thirds this value. This is the same for all cases as the live load is increased.

In Figure 15, the connection of gravity beam 1 into column A3 is such that the column is bent about its weak axis, while the shorter span beams in the moment resisting frame on line A bend the column A3 about its stronger axis.

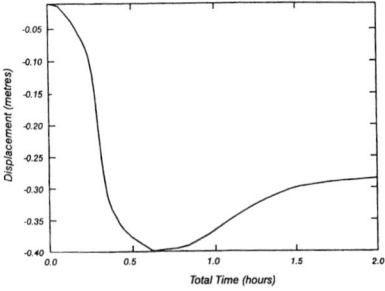 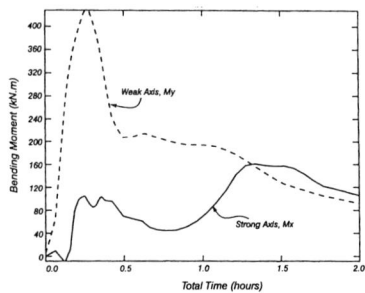

Fig. 14 Midspan deflection of gravity beam 2 versus time for the standard model based on Cardington tests

Figure 15 Bending moments at the top of column A3 at the fire floor for the standard model based on the Cardington tests

CONCLUSIONS

The research has shown that this particular multi-storey steel frame building, if constructed using unprotected beams, would possess an adequate level of fire safety. It is expected that this conclusion would apply to a much wider range of multi-storey steel framed buildings with protected columns and unprotected beams.

ACKNOWLEDGEMENTS

The contribution of all people/organizations involved in HERA's fire research is acknowledged. The principal past and ongoing funding for this project has been provided by the Foundation for Research, Science and Technology, New Zealand

REFERENCES

Clifton G.C. (1996). *Fire Models for Large Firecells,* HERA, Manukau City, HERA Report R4-83.

Clifton, G.C. et. al. (1999). *Behaviour of a Multi-Storey Steel Building in Fully Developed Natural Fires,* HERA, Manukau City, HERA Report R4-95.

ENV 1992-1-2. (1995) Eurocode 2: *Design of Concrete Structures – Part 1-2 General Rules – Structural Fire Design,* CEN, Brussels.

Hibbit, Karlsson and Sorensen, Inc. (1998). *ABAQUS Finite Element Computer Program,* Version 5.8.

Kirby, B.R. (1998) *The Behaviour of a Multi-Storey Steel Framed Building Subject to Fire Attack - Experimental Data,* British Steel Swinden Technology Centre, United Kingdom.

Moss, P.J. and Clifton, G.C. (1999). Behaviour of Multi-Storey Steel Frames in Fires, *Mechanics of Structures and Materials* (M.A. Bradford, R.Q. Bridge and S.J. Foster, eds) being Proc. 16th Australasian Conf. on the Mechanics of Structures and Materials, Sydney, Australia, Dec. 1999, 461-466.

Moss, P.J. and Clifton, G.C. (2000). Behaviour of Multi-Storey Steel Frames in Fires, submitted to J. of Structural Engineering, Inst. Engineers Australia, 10pp

Poh, K.W. (1996). *Modelling Elevated Temperature Properties of Structural Steel,* BHP Research, Melbourne, Australia, Report BHPR/SM/R/055.

Stevenson, P.L. (1993). *Computer Modelling of Structural Steel Frames in Fire,* Master of Engineering Report, University of Canterbury, Christchurch, 1993

FIRE SAFETY DESIGN AND RECENT DEVELOPMENTS IN FIRE ENGINEERING

W. Sha and N.C. Lau

Metals Research Group, School of Civil Engineering, The Queen's University of Belfast, Belfast BT7 1NN, UK

ABSTRACT

The design of a building structure must comply with building regulations. A building must have specified fire resistance, because building materials become weaker with the increase of temperature, which can cause the collapse of the building in event of fire if inadequate fire safety measure is provided. All buildings should have sufficient fire resistance so that there is enough time for the occupants to evacuate, and for fire fighters to perform their duty safely. The measure taken to comply with the building regulations is called fire safety design. This paper introduces the basic concepts and methods of fire safety design. The use of computer software in fire engineering is the main theme. The calculation approach has been made easy with the advance of computation techniques. Computer software was specially designed for calculating moment capacity. In addition, with thermal analysis software, the heat transfer of any new type of steel section in fire can be determined accurately, easily and economically. As an example, the moment resistance in fire of slim floor structures is modelled, using the two separate software packages developed at The Steel Construction Institute, UK.

KEYWORDS

Asymmetric beam, composite beam, fire, fire engineering, fire resistance, fire resistant steel, heat transfer, intumescent coating, moment resistance, slim floor.

INTRODUCTION

The measure taken to comply with the building regulations is called fire safety design [Malhotra (1982), Purkiss (1996)]. In general, fire protection of buildings can be divided into two types of measures, active and passive. The active measure is concerned with the detection and extinction of fire at its early stage, achieved by introducing alarm, control of smoke and other hazardous elements, in-built fire fighting or control and other fire safety management systems. The passive measure of fire safety design is mainly concerned with the structural fire protection and means of escape in case of fire. This can be achieved by means of enhancing structural performance including the use of fire-protected beams and columns, compartmentation, control of flammability of the structural fabric and provision of fixed escape routes [Purkiss (1996)].

Fire engineering is a passive measurement of fire safety for building structures. Traditionally, passive measures such as the provision of protective material to steel columns and beams have been used to achieve fire resistance in steel structures. Fire protection is prescriptive in that the engineer designs the steel structure based on its full strength, and then determines the fire protection required by using pre-defined charts and tables depending on the specified fire resistance. The alternative passive measure, fire engineering, is to analytically design the steel structures by using the properties and behaviour of steel at high temperature. This type of passive measure is more economical in that the needs for fire protection may be minimised or eliminated.

The recent development of fire engineering based on the analytical approach has resulted in steel structures that have built-in fire resistance such as slim floor beam construction [Mullett & Lawson (1993), Newman (1995), Lawson et al. (1997), Sha (1998a)]. Two types of slim floor have been developed by Corus Group (formerly British Steel) and The Steel Construction Institute (SCI), UK, the SLIMFLOR and the SLIMDEK. The SLIMFLOR system consists of a SLIMFLOR beam and either pre-cast floor slab with in-fill concrete or composite construction with deep decking. The composite SLIMDEK system is constructed using an asymmetric SLIMFLOR beam (ASB) with deep decking. Another option of fire engineering in steel structures is to use fire resistant steels, steels that have better high temperature strength than conventional steels [Sha (1998a), Sha (1998b), Sha & Kelly (1998), Sha et al. (1998), Kelly & Sha (1997)].

This paper is concerned with the development and use of fire engineering computer software. It uses two types of software originally developed by SCI to model respectively the moment capacity and temperature development in fire of floor beams. The first type of program has been used to calculate the moment capacity at elevated temperatures for slim floor beams and conventional composite I-beam floor with fire protection, which can then be used to calculate the fire resistance. The second type of program, TFIRE, is used to model the heat transfer and the temperature development in SLIMFLOR beams with intumescent coating protection.

METHODS OF APPROACH

There are two fire-engineering approaches for steel structures. One is the prescriptive approach, by assessing the performance of structures in fire from tabulated or graphical data derived from standard or fire tests by manufacturers. In this approach, the structure is designed based on ultimate Limit State and then the fire protection requirement is calculated from pre-defined data tables and charts in various codes of practice. The other is the calculation approach, by calculating the capacity of members in fire to assess the performance of the structure. In this approach, the limiting temperature and the load ratio, i.e. the ratio between moment capacities of structural members in fire and normal conditions, are determined. Both approaches are centred on limiting the steel temperature of structural members, because steel begins to lose its strength at 400°C and it decreases rapidly with increasing temperatures.

DESIGN CONCEPT OF THE CALCULATION APPROACH

The focus of this paper is based on the calculation approach for steel structures. A design concept has been introduced that incorporates fire safety into the structural members themselves instead of designing the structural members and then providing protection to them. There are two fire safety design methods, the limiting temperature/load ratio method and moment capacity method. The latter is used in the present work.

If the temperature distribution through a structural member section is known, the reduced strength of all the elements in the cross-section can be calculated. The plastic neutral axis and, hence, the moment capacity in fire conditions can then be determined directly [Newman (1990)]. In the moment capacity method, if the member's moment capacity at the required period of fire is not smaller than the applied moment at the fire Limit State, the member may be considered to have adequate fire resistance without protection. Otherwise, the member will require fire protection.

The state-of-the-art in fire engineering is to build fire resistance into the structures themselves. The greatest advantage of this is that it does not require any traditional fire protection (board or spray) or only require minimum protection, that can lead to construction that is considerably more economical. This approach has become more acceptable in Western Europe from the early 1980s, which has resulted in the development of slim floor construction.

SLIM FLOOR CONSTRUCTION

Slim floor is a relatively new form of floor construction using steel as the structural member. The development of slim floor construction is an excellent example of achieving built-in fire resistance in building structures. The beam used in slim floors is contained almost totally within the depth of the concrete floor. The floor has a flat appearance similar to a reinforced concrete floor, and has excellent fire resistance, as only the bottom face of the steel section would be exposed to the heat in event of a fire. As the concrete floor protects the steel beam, the structure can withstand fire for longer time without any traditional fire protection material. Originated in Scandinavia, the principles of the slim floor construction were adapted by Corus Group and The Steel Construction Institute (SCI), UK, to develop two types of new slim floor beams, the SLIMFLOR and the ASB SLIMDEK [Mullett & Lawson (1993), Newman (1995), Lawson et al. (1997), Sha (1998a)].

A SLIMFLOR beam is a universal column with a steel plate welded to its bottom flange. There are two types of SLIMFLOR beam construction system, the more popular and desirable in-situ composite decking system and the pre-cast unit. In the latter system, the slab rests on the steel plate and the remaining space is filled with in-situ concrete. ASB is the transformed steel section of the SLIMFLOR beam, in which the plate and bottom-flange are replaced by a larger bottom flange. The ASB system uses composite construction.

FIRE RESISTANT STEELS

The development of fire resistant steels is another achievement in fire engineering. This type of steel can retain a higher strength in fire condition, and thus increase the fire resistance. This is achieved by developing new compositions and rolling processes. A main effort in developing such steels has been from Nippon Steel of Japan where some fire resistant steels were developed [Sha (1998a), Kelly & Sha (1997), Sha et al. (1998), Sha & Kelly (1998)]. Several new experimental fire resistant steels have been developed at The Queen's University of Belfast (QUB) [Sha (1998b)].

SOFTWARE

There are three computer programs developed at SCI to calculate respectively the moment capacity of I-beam, SLIMFLOR and ASB floors in fire conditions. The principal theory is based on plastic analysis which are permitted by EC4 Part 1.2 [BS ENV (1994)]. The model is in essence a finite element model. The plastic moment capacity of a beam at elevated temperatures is calculated by dividing the beam into several elements such as bottom flange, lower web, upper web, and top flange.

In the calculation, the software considers the steels to have reached 2% strain at failure, and the strength reduction factors at elevated temperatures are applied to the normal strength to obtain the elevated temperature strength. For each element, the area, position of centroid and reduced strength are calculated. The total resistance of all elements is then found. In pure bending the plastic neutral axis must be found which divides the total resistance into equal tension and compression. After finding the plastic neutral axis, moments are taken about any convenient axis and moment capacity at the temperature is thus obtained.

The elevated temperature in a steel beam section due to the heat transfer in fire determines its moment capacity. The temperatures of each element for each type of beam can be measured in a standard fire test. Although such test give good and representative results of the temperature development in beams and their structural behaviour, it is expensive and therefore there are limited test data on newly developed steel beams such as SLIMFLOR and ASB. Thermal analysis software can be used to supplement fire tests. A new steel beam section can be designed more economically by using the software to model the heat transfer in the section, and hence determine its feasibility before manufacturing the section and conducting fire tests. The thermal analysis software used in this paper is TFIRE. It is a two-dimensional finite difference heat transfer program developed by SCI and has been verified against fire test data. It was used to model the temperature development in the SLIMFLOR and ASB during the course of developing these sections [Newman (1995)].

Based on thermal conduction and heat transfer models, TFIRE can be used for any form of structural sections. The program can calculate the heat flow in a steel section and changes of temperature distribution in the section in fire condition. With the temperature data obtained by the software, moment capacity of the steel section can be determined by using the moment capacity software.

TFIRE uses the following basic heat input expression to model the heat transfer [Purkiss (1996), Newman (1995)]:

$$\frac{dq}{dt} = SV(a\varepsilon_f T_f^4 - \varepsilon_s T_s^4) \quad (1)$$

where $\frac{dq}{dt}$ = rate of heat transfer per unit area
S = Stephen-Boltzman constant
V = View factor of the element
a = Surface absorbivity
ε_f = Flame emissivity
ε_s = Surface emissivity
T_f = Furnace or gas temperature
T_s = Element temperature.

The consideration of the interface resistance between nominally touching surfaces is important in the modelling. For normal conduction between elements in perfect contact, the following expression applies:

$$\frac{dq}{dt} = \frac{T_2 - T_1}{\frac{1}{K_1} + \frac{1}{K_2}}. \quad (2)$$

If an interface resistance exists, this expression becomes:

$$\frac{dq}{dt} = \frac{T_2 - T_1}{\frac{1}{K_1} + \frac{1}{K_2} + \frac{1}{K_i}} \tag{3}$$

where K_1, K_2 = Thermal conduction terms
K_i = Interface resistance coefficient.

All software used in this work was written in Visual BASIC. All have been previously tested and verified [Newman (1995)]. Figures 1 and 2 show examples of approximate cross-section for a slim floor beam used in TFIRE and the corresponding output screen.

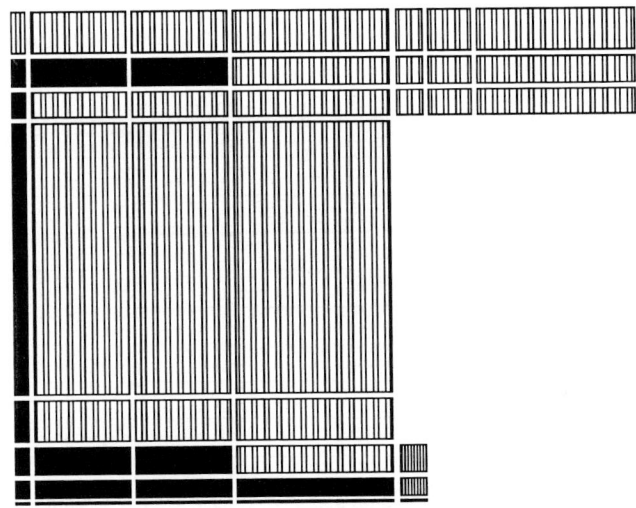

Figure 1: Approximated cross-section based on user input data

```
Nullifire  10:19:39  11-03-1998
 Time    20.0  Rtime   0.3  Ave inc  1.59  Inc   1.98  ratio   57.7   Fire    781




          21   21   21   26   55   62   64
          23   22   21   35  108  137  143
          25   21   22   50  370  411  419
          38   25   31  122
         135  100  100  138
         174  189  201  211
         356  358  383  484



       Grp   1   420   Grp   2   194
```

Figure 2: Output screen showing a calculated result by TFIRE

MOMENT RESISTANCE OF UNPROTECTED SLIM FLOORS IN FIRE

Firstly, TFIRE was used to calculate the temperature distribution in a floor section when the compartment underneath is in fire. Secondly, its moment resistance was calculated. For the fabricated SLIMFLOR beam, it was divided into six rectangular elements, plate, bottom flange, lower half of the web (lower web), upper half of the web (upper web), top flange and concrete compression flange [Newman (1995)]. For asymmetric beams, the division was more detailed [Sha (1998a)] and, for the beams designed for British market (280 ASB 100 and 300 ASB 153), consideration was given for the fillet at the junctions of the web and the bottom flange. The width and height of each element are given in Table 1, for three beam sizes.

TABLE 1
WIDTH AND DEPTH OF ASB ELEMENTS (FROM BOTTOM TO TOP, ALL DIMENSIONS IN mm)

Beam designation	280 ASB 100		300 ASB 153		European	
Element	Width	Depth	Width	Depth	Width	Depth
1	300	16	300	24	375	20
2	44.7	6.4	55.9	7.2	10	10
3	31.8	6.4	41.4	7.2	10	30
4	19	7.2	27	5.6	10	85
5	19	20	27	20	10	85
6	19	20	27	20	10	20
7	19	184	27	202	200	20
8	190	16	190	24		

For fabricated beam, the calculation was based on a 254×254×89 Universal Column (UC) section, with flange plate thickness 15 mm and width 455 mm. The concrete depth above the top flange was 30 mm. Decking was made with ComFlor CF210 deck [Mullett & Lawson (1993)]. S355 steel grade was used, and the concrete strength was taken as 20.1 MPa. Temperature distribution at 60 minutes in fire was calculated using TFIRE, with fire temperature taken from the standard fire curve. The calculation gave a flange plate temperature of 822°C, bottom flange 611°C and lower web 433°C. All other parts of the steel section were below 400°C, therefore full strength being preserved. With this temperature distribution, the fire moment resistance was found 352 kNm, compared with 523 kNm for moment resistance at room temperature,. These give a load ratio of 0.67.

For asymmetric beams, based on experimental evidence [Lawson et al. (1997)], the bond strength limits between the steel sections and concrete when hot were taken as 0.9 and 0.4 MPa, for British (280 ASB 100 and 300 ASB 153) and European versions, respectively. The beam configurations and calculation results are detailed in Table 2. The temperatures of those elements not included in the table are below 400°C and therefore they are at full strength. The moment resistance data at room temperature are included for comparison. The moment resistance in fire was based on fire temperatures at 60 minutes.

TABLE 2
BEAM PARAMETERS, TEMPERATURES IN FIRE AT 60 MIN AND MOMENT RESISTANCE OF ASB

Beam designation		280ASB100	300ASB153	300ASB153	European
Top flange width (mm)		190	190	190	200
Bottom flange width (mm)		300	300	300	375
Beam depth (mm)		276	310	310	270
Flange thickness (mm)		16	24	24	20
Web thickness (mm)		19	27	27	10
Decking		CF210	CF210	CF225	CF210
Concrete depth above beam flange (mm)		30	30	0	40
Steel temperature (°C): Element 1		786	747	747	804
	2	684	635	635	612
	3	643	597	597	485
	4	599	564	564	≤400
	5	532	509	509	≤400
	6	422	410	410	≤400
Moment resistance Fire		253	461	429	190
(kNm)	Room temperature	554	889	739	418
Load ratio		0.46	0.52	0.58	0.45

The values of the load ratio, i.e., the ratio between the load resistance at the fire Limit State and that under the normal "cold" condition, were centred on 0.5, which is the normal design fire load ratio. This confirms that the slim floors have up to one-hour fire resistance without traditional fire protection. Although the calculations were based on five case parameters, the general conclusion should be applicable to other beam sizes and floor configurations.

CONCLUSIONS

A series of modelling work has been carried out using computer software for calculating the steel beam section temperatures and moment resistance in fire. It has been found that, without any applied fire protection, SLIMFLOR and asymmetric beam (ASB) slim floors can support around 50% of the designed load in fire at 60 min. This is due to the inherently good fire resistance in these structures offered by the protection of concrete surrounding steel beams.

REFERENCES

BS ENV (1994). *Eurocode 4: Design of Composite Steel and Concrete Structures, Part 1.2: Structural Fire Design (including UK National Application Document)*, British Standards Institution, London, UK.

Kelly, F.S. and Sha, W. (1997). Mechanical Properties of Fire-resistant Steels for Construction. *Euromat 97 – Proceedings of the 5th European Conference on Advanced Materials and Processes and Applications: Materials, Functionality & Design*, vol. 1, Metals and Composites, ed: L.A.J.L. Sarton and H.B. Zeedijk, Netherlands Society for Materials Science, Zwijndrecht, The Netherlands, pp. 35-40.

Lawson, R.M, Mullett, D.L. and Rackham, J.W. (1997). *Design of Asymmetric Slimflor Beams Using Deep Composite Decking*, Steel Construction Institute, Ascot, UK.

Malhotra, H.L. (1982). *Design of Fire-resisting Structures*, Surrey University Press, Guildford, UK.

Mullett D.L. and Lawson R.M. (1993). *Slim Floor Construction Using Deep Decking*, Steel Construction Institute, Ascot, UK.

Newman, G.M. (1990). *The Fire Resistance of Composite Beams with Unfilled Voids*, Report Number 135, Steel Construction Institute, Ascot, UK.

Newman, G.M. (1995). Fire Resistance of Slim Floor Beam. *Journal of Constructional Steel Research*, **33:1-2**, 87-100.

Purkiss, J.A. (1996). *Fire Safety Engineering Design of Structures*, Butterworth-Heinemann, Oxford, UK.

Sha, W. (1998a). Fire Resistance of Floors Constructed with Fire-resistant Steels. *Journal of Structural Engineering* **124:6**, 664-670.

Sha, W. (1998b). Fire Resistant Steels for Construction: Design, Properties and Microchemistry. *Proceedings of the Third China Association for Science and Technology Conference of Young Scientists – Materials Science and Technology*, China Science and Technology Publishers, Beijing, China, pp. 233–237.

Sha, W. Blair, P.J. and Kelly, F.S. (1998). Tensile Properties of Mo and Mo-Nb Microalloyed Fire Resistant Steels. *Proceedings of the Third Pacific Rim International Conference on Advanced Materials and Processing*, vol. 1, ed: M.A. Imam, R. DeNale, S. Hanada, Z. Zhong and D.N. Lee, TMS, Warrendale, PA, pp. 247-252.

Sha, W. and Kelly, F.S. (1998). Mechanical Property and Microstructure of Structural and Fire Resistant Steels. *Proceedings of the Third Pacific Rim International Conference on Advanced Materials and Processing*, vol. 1, ed: M.A. Imam, R. DeNale, S. Hanada, Z. Zhong and D.N. Lee, TMS, Warrendale, PA, pp. 199-204.

FIRE RESISTANCE OF STEEL FLOORS CONSTRUCTED WITH EXPERIMENTAL FIRE RESISTANT STEELS

W. Sha[1], N.C. Lau[1] and T.L. Ngu[2]

[1]Metals Research Group, School of Civil Engineering, The Queen's University of Belfast, Belfast BT7 1NN, UK
[2]Subur Tiasa Holdings Berhad, 167681-D, No. 8-14, 1st-3rd Floor, Jln. Chengal, 96000 Sibu, Sarawak, Malaysia

ABSTRACT

This paper reports a continuation of research on the modelling of the fire resistance of steel floor structures, using available computer software. Modelling of the heat transfer in steel sections and the moment resistance for floor structures has been made. The moment capacity software has been used to calculate the fire resistance of I-beam, Slimflor beam and Asymmetric Slimflor beam (ASB) floor structures, constructed with three types of steel with different strength characteristics. The first a conventional structural steel, S355, and the other two are experimental fire resistant steels developed in Belfast. The compositions of the experimental steels are 0.08C-0.38Si-1.32Mn-0.54Mo-0.26Nb and 0.02C-0.36Si-0.87Mn-0.16Mo-0.63Nb, respectively. Strength reduction factors at 2% strain were used in calculating beam moment capacity at fire condition. This work follows a previous investigation by one of the authors who used the software to obtain the moment capacity of beams constructed with Nippon Steel fire resistant steels, and then calculated the fire resistance. It has been found that the use of fire resistant steels can increase the fire resistance of conventional floor structures by nearly half an hour, but the improvement in slim floors is limited to under 20 minutes. None of the fire resistant steels could achieve thirty minutes fire resistance in conventional non-composite and composite floors without protection.

KEYWORDS

Fire resistance, fire engineering, steel structures, fire resistant steel, moment resistance, micro-alloyed steel.

INTRODUCTION

Fire resistance of steel building structures has always been a major design criterion, as steel has high conductivity and low strength at elevated temperatures. Thus, preventive and protective measures are needed to minimise loss of life and properties from fire hazard. Steel strength, member section factor and types of protective measure taken govern the fire resistance of a steel structure. These are passive measures in term of fire safety engineering terminology.

An interpretative document published by European Commission [Lawson & Newman (1996)] has set out the main objectives of fire safety:
(1) the load bearing capacity of the construction can be assumed adequate for a specified period;
(2) the generation and spread of fire within the works are limited;
(3) the spread of fire to neighbouring construction works is limited;
(4) occupants can leave the works, or other means can be used to rescue them;
(5) the safety of rescue teams is taken into account.

Structural fire design is different from fire safety design. Within the discipline of fire safety engineering, elements can be readily identified which relate separately to the safety of life and property. The two areas are not mutually exclusive as an action that increases life safety also increases property safety. Fire protective measures include alarm, smoke control, in-built fire-fighting or fire control system, limitation of hazardous contents, access for external fire fighting and fire safety management system. These measures are known as "active" measures because they seek to reduce the severity of a fire. Providing adequate compartmentation, control of flammability of the structural fabric, provision of fixed escape route and provision of adequate structural performance are passive measures. The requirements lie in building design and material selection in the construction. However, in the aspect of structural design, only the former is directly relevant to the structural designer in providing adequate fire resistance. Most of the structural designers are more concerned with the so-called "passive" protection, by achieving adequate load ratio to prevent the collapse of the steel structure due to excessive heating and loss of strength in elevated temperatures.

Fire resistance design of steel structures is based on Eurocodes [BS ENV (1993), BS ENV (1994)] and fire test criteria are established by ISO 834 (1985). Modelling of steel floor structures using computer software is a significant advancement in fire resistance design. The present work has used thermal modelling and moment capacity software developed by The Steel Construction Institute (SCI) to analyse different types of novel steels and protection.

UNPROTECTED CONVENTIONAL STEEL FLOOR STRUCTURES

The Steel Construction Institute (SCI), UK, has developed reliable software for modelling the fire resistance of steel floor structures. The present work is concerned with the modelling of the fire resistance of unprotected floor structures constructed with conventional and fire resistant steels. As there are no data available, Tfire was used to obtain the temperature distribution first, which was then used in moment resistance calculations. Details of the approach used were given in an earlier publication [Sha (1998a)]. Standard fire curve was used to represent the fire temperatures in all cases.

In the moment resistance calculations, different steels are characterised by their strength reduction factors, the ratio between stress at 2% strain at elevated temperatures and room temperature yield strength. Standard values were used for the conventional steel, S355 [Lawson & Newman (1996)]. Data for fire resistant steels manufactured by Nippon Steel, FR1 and FR2, were given in a previous publication [Sha (1998a)]. Two experimental fire resistant steels, P8123 and P8124, designed by the Queen's University of Belfast, were also studied. Their strength reduction factors were measured in laboratory using tensile tests (Figure 1). The room temperature yield strength values for P8123 and P8124 are 594 and 411 MPa, respectively. The I-beam floor configuration was given in Sha (1998a). The load ratio is defined as the ratio between the moment resistance at elevated temperature to that at room temperature, a factor always no greater than unity.

The modelling work concentrates on two types of universal beams, 305×102×33 kg/m and 457×191×98 kg/m. Two configurations as far as thermal behaviour is concerned were modelled, with the top flange of the steel beam in contact or away from the concrete floor above. In the former case, the upper half of the steel beam would have a lower temperature than the lower half, due to heat transferring from steel into the concrete. Figures 2 and 3 give temperature calculation results using Tfire. In the latter case, the steel beam is symmetrical with regard to its centroid, thus the upper and lower halves having symmetrical temperature, too. In other words, the upper web would have the same temperature as the lower web, and the top-flange same as the bottom flange. In this case, the temperatures follow those of the bottom flange and lower web in Figures 2 and 3. For an examination of the effect of beam size, the temperature distributions in larger beams with top flange in contact with concrete were calculated and are given in Table 1.

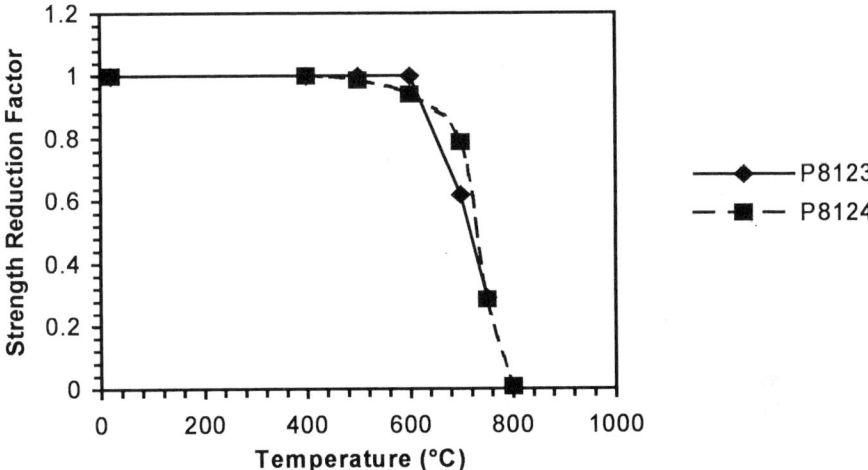

Figure 1: Strength reduction factors for the experimental fire resistant steels

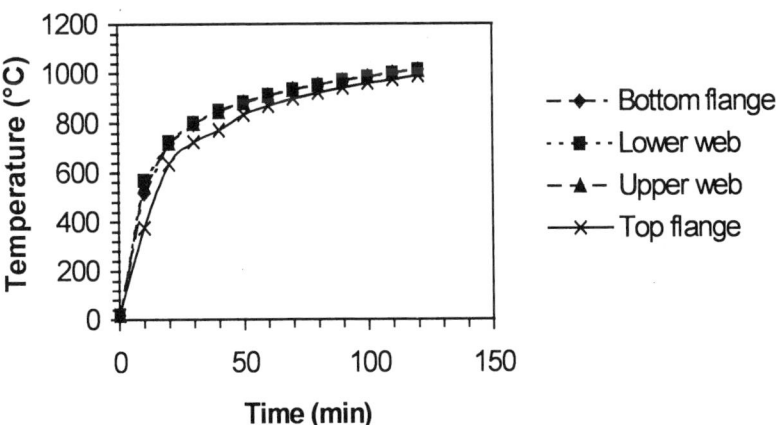

Figure 2: Temperature-time curves of 305×102×33 beam. Top flange in contact with concrete

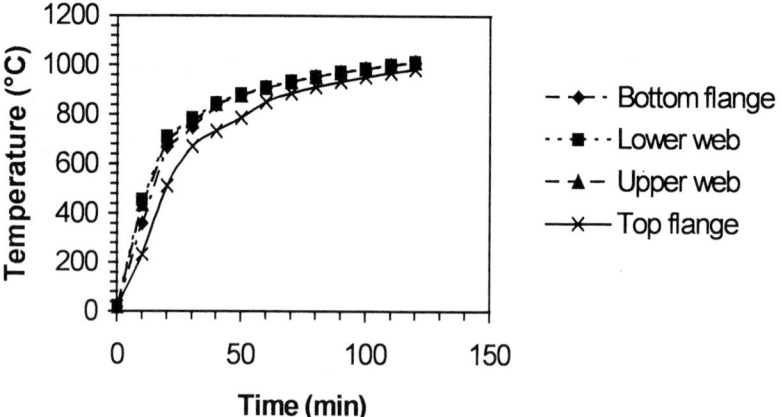

Figure 3: Temperature-time curves of 457×191×98 beam. Top flange in contact with concrete

TABLE 1
BEAM TEMPERATURES (°C) CALCULATED USING TFIRE

Beam size	Time (min)	Bottom flange	Lower web	Upper web	Top flange
533×210×122	20	641	700	684	448
610×305×179	30	735	764	755	617

Based on the temperature results given in Figures 2 and 3, the moment resistance was calculated. The results are given in Figures 4–12. No steels achieved 30-minute fire resistance for a load ratio of 0.5. As the temperature in the bottom flange reached 750°C, the load ratio fell rapidly for as much as 65%. Although steel sections in contact with concrete would allow heat transmission into the concrete, the temperature in the top half is not lowered enough to make effective.

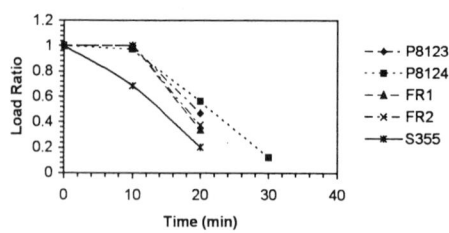

Figure 4: Load ratio versus time in fire for non-composite 305×102×33 beam not in contact with concrete floor

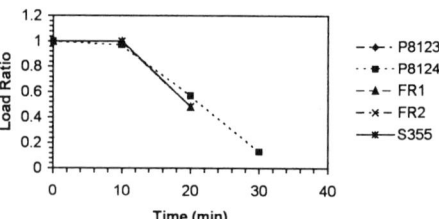

Figure 5: Load ratio versus time in fire for composite 305×102×33 beam not in contact with concrete floor

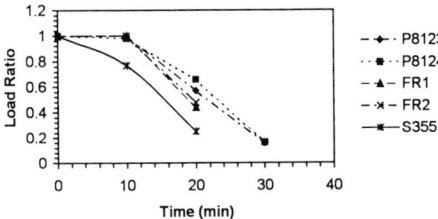

Figure 6: Load ratio versus time in fire for non-composite 305×102×33 beam in contact with concrete floor

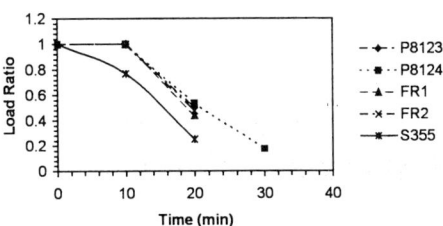

Figure 7: Load ratio versus time in fire for composite 305×102×33 beam in contact with concrete floor

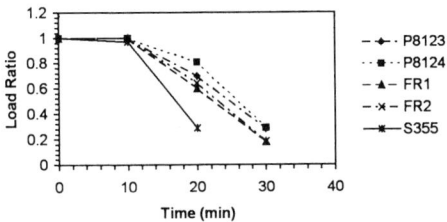

Figure 8: Load ratio versus time in fire for non-composite 457×191×98 beam not in contact with concrete floor

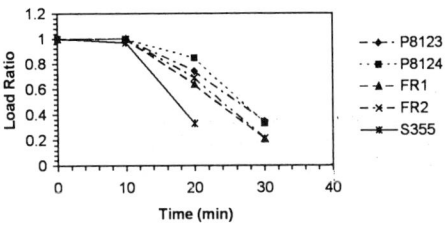

Figure 9: Load ratio versus time in fire for composite 457×191×98 beam not in contact with concrete floor

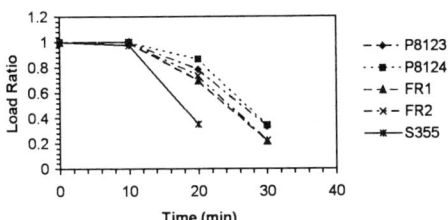

Figure 10: Load ratio versus time in fire for non-composite 457×191×98 beam in contact with concrete floor

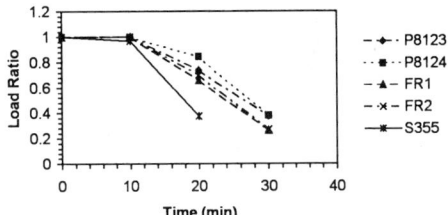

Figure 11: Load ratio versus time in fire for composite 457×191×98 beam in contact with concrete floor

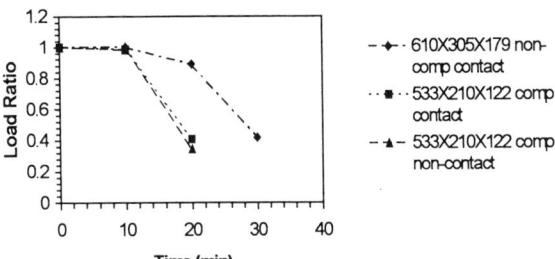

Figure 12: Load ratio as a function of time in fire for larger beams constructed with conventional structural steel, S275

The room temperature yield strength does not have any significant influence on the fire resistance because it is measured according to load ratio. The fire resistance is determined by the strength reduction factor.

FIRE RESISTANCE OF VARIOUS TYPES OF FLOORS

The moment capacity software has been used to calculate the fire resistance of I-beam, SLIMFLOR beam and ASB floor structures, constructed with three types of steel with different strength characteristics. The first is a conventional structural steel (S355) and the other two are experimental fire resistant steels developed in Belfast, P8123 and P8124. Compositions and strengths of the novel fire resistant steels were given in an earlier publication [Sha (1998b)] where P8123 was referred to as Steel A and P8124 Steel B. The fire resistant steels were tested up to 750°C, and the strength reduction factors at higher temperatures were assumed to follow those of conventional steels. The present investigation is based on the same method and software used previously [Sha (1998a)].

All three types of beams (I-beam, SLIMFLOR beam and ASB) used in the calculation are based on those used in standard fire tests carried out by Warrington Fire Research Centre (WFRC), UK. Details of the test and data analysis including temperatures of member elements at different times were described in earlier publications [Sha (1998a) and the references therein].

The moment capacity calculation was based on temperatures of elements and steel reduction factors experimentally determined. Figure 13 shows the split elements of the composite I-beam. The load ratio was determined by dividing moment capacity at elevated temperatures by that at room temperature. The calculated room temperature moment capacities for the specific beam configurations used in fire tests [Sha (1998a)] are given in Table 2. The calculated fire resistance as a function of load ratio is shown in Figure 14.

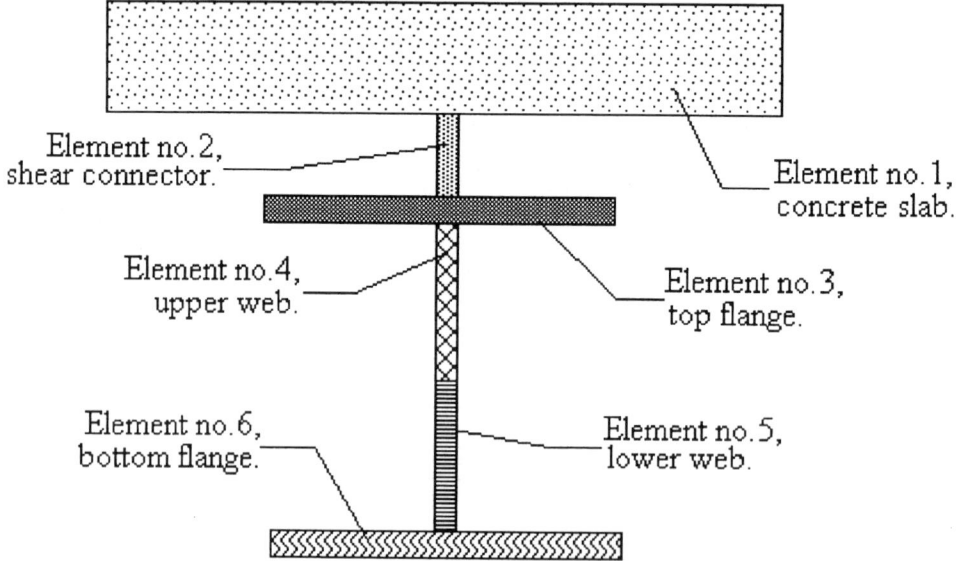

Figure 13: Split elements of the composite I-beam

TABLE 2
ROOM TEMPERATURE MOMENT CAPACITY (kNm) OF I-BEAM, SLIMFLOR AND ASB CONSTRUCTED WITH DIFFERENT STEELS

Floor type	S355	P8123	P8124
I-Beam	375	551	427
SLIMFLOR	429	718	497
ASB	555	869	624

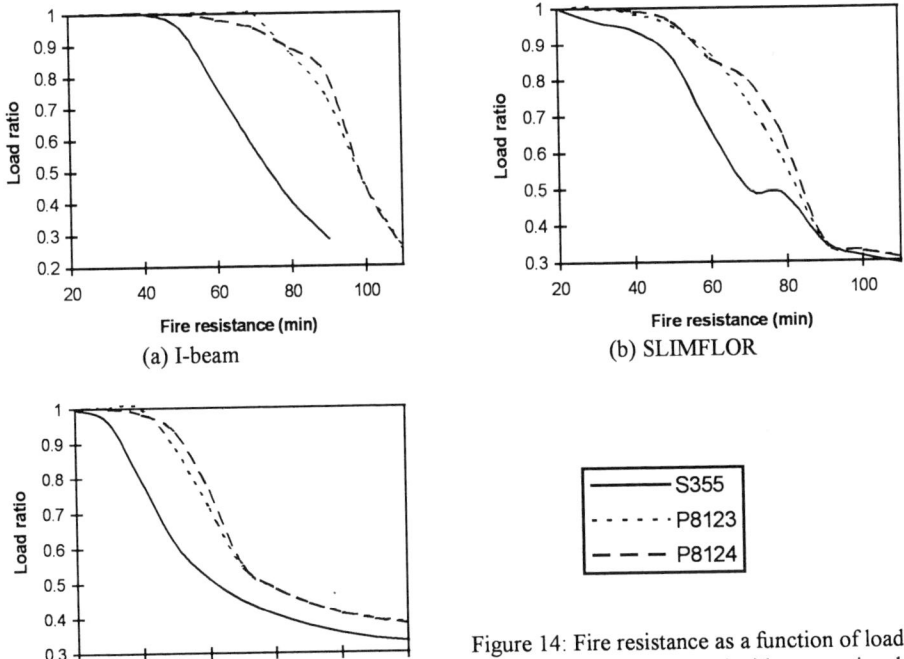

(a) I-beam

(b) SLIMFLOR

(c) ASB

Figure 14: Fire resistance as a function of load ratio for beams constructed with conventional (S355) and fire resistant (P8123 and P8124) steels

Table 3 summarises the fire resistance for the load ratio of 0.6 of model beams using S355 and experimental P8123 and P8124 steels, obtained from graphs in Figure 14. The two fire resistant steels perform similarly. P8123 has lower fire resistance than P8124 although its room temperature yield strength is 30% higher [Sha (1998b)]. This is because P8124 has a higher strength reduction factor at high temperatures, around 700°C. These steels have slightly higher strength reduction factors than Nippon fire-resistant steels [Sha (1998a), Sha (1998b)], resulting in a marginal improvement in fire resistance.

TABLE 3

FIRE RESISTANCE (MIN) USING CONVENTIONAL AND FIRE RESISTANT STEELS AT LOAD RATIO OF $0.6^{\dagger}$

Floor type	S355	P8123	P8124
I-Beam (protected)	68	95 (+27)	96 (+28)
SLIMFLOR	63	78 (+15)	82 (+19)
ASB	51	67 (+16)	68 (+17)

†Numbers in parentheses are increased fire resistance due to the use of the fire resistance steels.

In conclusion, for a load ratio of 0.6, conventional floor structures constructed with the experimental fire resistant steels can have 27-28 minutes extra fire resistance. However, in the novel and more fire resistant slim floor structures these steels can only achieve 15-19 minutes extra fire resistance. Combined use of fire resistant steels and fire engineering structural design does not produce the maximum yield.

REFERENCES

BS ENV (1993). *Eurocode 3: Design of Steel Structures, Part 1.2: Structural Fire Design (including UK National Application Document)*, British Standards Institution, London, UK.

BS ENV (1994). *Eurocode 4: Design of Composite Steel and Concrete Structures, Part 1.2: Structural Fire Design (including UK National Application Document)*, British Standards Institution, London, UK.

ISO 834 (1985). *Fire Resistance Tests: Elements of Construction ISO*, International Standards Organisation.

Lawson, R.M. and Newman, G.M. (1996). *Structural Fire Design to EC3 and EC4, and Comparison with BS 5950*, Steel Construction Institute, Ascot, UK, p. 22.

Sha, W. (1998a). Fire Resistance of Floors Constructed with Fire-resistant Steels. *Journal of Structural Engineering* **124:6**, 664-670.

Sha, W. (1998b). Fire Resistant Steels for Construction: Design, Properties and Microchemistry. *Proceedings of the Third China Association for Science and Technology Conference of Young Scientists – Materials Science and Technology*, China Science and Technology Publishers, Beijing, China, pp. 233–237.

THE INFLUENCE OF CONNECTION CHARACTERISTICS ON THE BEHAVIOUR OF BEAMS IN FIRE

K. S. Al-Jabri[1], I. W. Burgess[2], and R. J. Plank[3]

[1]Department of Civil Engineering, Sultan Qaboos University, PO Box 33, Muscat 123, Sultanate of Oman
[2]Department of Civil and Structural Engineering, University of Sheffield, Sheffield, S1 3JD, UK
[3]Department of Architectural Studies, University of Sheffield, Sheffield, S10 2TN, UK

ABSTRACT

The analytical models developed to study the behaviour of steel beams within steel-framed structures treated the connections as either fully rigid or pinned without taking into considerations the significance of connections in fire when they have beneficial effects on the survival time of the structure. This paper presents results from a series of parametric studies carried out to investigate the behaviour of steel-framed beams with different connection characteristics. A finite element program VULCAN developed at Sheffield University is used to conduct the analysis. The connection characteristics incorporated to simulate the connection response are obtained from a series of beam-to-column connection tests. These tests were conducted in a portable junction furnace to study the moment-rotation characteristics of typical connections. The results showed that the structural connections can have considerable influence on the behaviour of beams at elevated temperatures.

KEYWORDS

Connections, Fire, Frame, Beam, Temperature, Flush endplate, Flexible end-plate, Load ratio.

INTRODUCTION

Steel is seriously affected by fire, losing strength and stiffness leading to large deformations and often collapse. Using applied fire protection remains the most common way of satisfying structural fire resistance requirements, despite its cost. Alternative approaches which account for steel's inherent fire resistance are beginning to develop, based on experimental and analytical studies of structural behaviour. Utilisation of such approaches in the design demands a better understanding of the factors which govern the behaviour of both individual steel members and the structure as a whole in fire conditions.

Traditionally it is well known that, at ambient-temperature but even more at elevated temperatures, a complete structure behaves better than its individual members in isolation. This is due to the considerable influence of structural continuity provided by the connections and also influence by the restraint forces due to the adjacent cool structures. However, most of previous fire work were concentrated on isolated members and current design codes reflect this which limit the effective use of numerical models. The analytical models developed to study the behaviour of steel beams within steel-framed structures treated the connections as either fully rigid or perfectly pinned without taking into consideration the significance of connections in fire when they have beneficial effects on the survival time of the structure. This paper presents results from a series of parametric studies carried out to investigate the behaviour of steel-framed beams with different connection characteristics. A finite element program, VULCAN developed at Sheffield University was used to conduct the analyses. The connection characteristics incorporated to simulate the connection response are obtained from a series of beam-to-column connection tests. The results showed that some connections can have considerable influence on the behaviour of beams at elevated temperatures.

THE "VULCAN" FINITE ELEMENT PROGRAM

A finite element code for the analysis of the structural performance of two- and three-dimensional steel and composite framed buildings under fire loading has been developed at the University of Sheffield, and is known as VULCAN [Bailey,(1995)]. The formulation retains higher order terms, resulting in a very accurate treatment of geometric non-linearities.

To stimulate the connection characteristics within the model, two-noded spring element with zero length and eight degrees-of-freedom in local co-ordinates is adopted [Bailey,(1995)]. The independent modelling of the eight degrees-of-freedom allows each to be modelled by a separately specified moment-rotation or force-displacement relationship. These may be subjected to elevated temperatures by following a family of Ramberg-Osgood curves for the reduction in both strength and stiffness of the connection with increasing temperature.

REPRESENTATION OF SEMI-RIGID CONNECTIONS AT ELEVATED TEMPERATURES

To model the behaviour of semi-rigid frames in fire, the moment-rotation characteristics of the connections at various temperatures need to be known. Although a vast of experimental data exists for connection characteristics at ambient temperatures, very little has been published for high temperature response.

The moment-rotation characteristics of the connections adopted in the analysis are based on an experimental test results conducted by Al-Jabri [Al-Jabri (2000)]. A series of elevated temperature connection tests was conducted to study the influence of parameters such as member size, end-plate type and thickness, and composite slab on the connection response in fire. Five different configurations were considered. The connection types included two flush end-plate and one flexible end-plate bare steel connections and two flexible end-plate composite connections. For each configuration a series of tests were conducted, each at a different load level.

The Ramberg-Osgood type of curve fitting was adopted in order to enable the experimental data to be incorporated in the model. The Ramberg-Osgood moment-rotation function may be expressed in the following form:

$$\phi = \frac{M}{A} + 0.01\left(\frac{M}{B}\right)^n \qquad (1)$$

where, ϕ, M = connection rotation and the corresponding level of moment respectively;
A, B and n = temperature dependent parameters.

The stiffness and strength of the connection may be degraded with increasing temperatures by modifying the temperature dependent parameters A and B respectively. From each group of tests, a family of moment-rotation-temperature curves was derived, two of which are shown in Fig. 1.

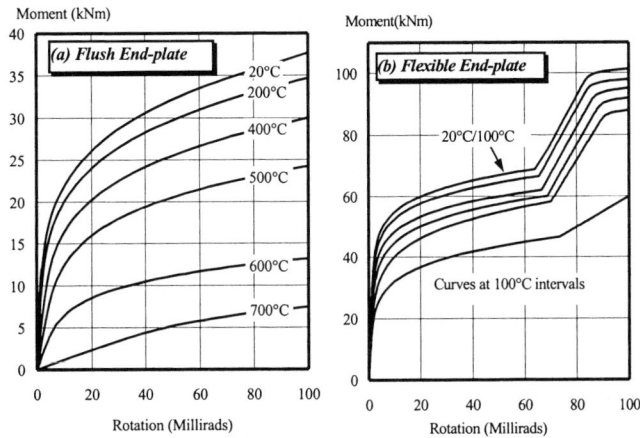

Figure 1: Moment-Rotation-Temperature Curves for Typical Connections

SUB-FRAME LAYOUT AND TEMPERATURE DISTRIBUTION

When conducting parametric studies it is very helpful to consider the analysis of a 'representative' sub-frame rather than a complete structure. BS5950: Part 1 [British Standard Institute (1990)] adopted a number of sub-frame arrangements, which were suggested to be suitable for representing beam behaviour in rigidly connected frames. Analytical studies [El-Rimawi (1993)] have demonstrated that these sub-frames are capable of producing results that are reasonably comparable with those from the full-frame under fire conditions using a fraction of the computing effort required. The same sub-frame layout has been adopted in the present studies, along with the isolated member arrangement for comparison. A similar arrangement was used corresponding with the member sizes utilised in the connection fire tests [Al-Jabri (2000)]. The general sub-frame arrangement and temperature distribution adopted are illustrated in Fig. 2.

Analyses were performed using the recorded material properties of the members obtained from connection tests [Al-Jabri (2000)]. Restraint conditions at the end beams were modified to allow lateral movement, while columns were allowed to expand vertically. The beams and columns were loaded with a load ratios of 0.6 and 0.4 respectively (load ratio is as defined in BS5950: Part 1). The lengths of the primary and secondary beams were 4.5m and 2.25m respectively, whilst the column length was 6.0m.

Heating regimes adopted in the analysis are illustrated in Fig. 2. The influence of the composite slab is investigated by considering a 130mm deep profiled slab consistent with the test arrangement. The effective width of the slab adopted was one sixth of the span. The properties of the concrete, and reinforcement used in the analysis are those obtained from material tests [Al-Jabri. (2000)]. A thermal expansion coefficient of 14.0x10^{-6} /°C and Poisson's ratio of 0.2 is used. The temperature of the slab is assumed to be 20% of the hottest part of the steel.

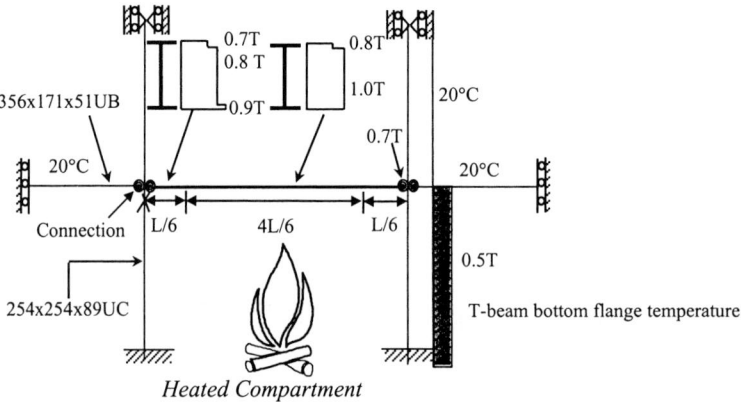

Figure 2: Sub-Frame Arrangement and Temperature profile used in the Fire Zone

The term 'failure temperature' used in the parametric studies is assumed to be the temperature at which the central deflection of the beam reaches a limiting deflection of span/20. It should be noted that a limiting temperature of 950°C is imposed by the program due to the lack of material data beyond this.

PARAMETRIC STUDIES ON SUB-FRAME ARRANGEMENT

To study the effect of connections on the behaviour of the connected structural members, it is necessary to conduct systematic parametric studies. This is only practically possible using numerical modelling, and the finite element program described above has been used for this. Various parameters affecting the beam behaviour at elevated temperature are considered.

General Beam Response at Elevated Temperatures

The beam response is shown in Figs. 3*(a)*, 3*(b)*, and 3*(c)* for flush and flexible bare-steel connections and flexible end-plate composite connections respectively, along with the idealisation of the connection characteristics as 'fully' rigid and 'perfectly' pinned. The beam response is also compared with that obtained for an isolated beam with no axial restraint. It can be seen that the central deflections for the sub-frame in most cases are consistently less than for the isolated beam due to the structural continuity which provides some degree of restraint. This difference is small for temperatures up to 650°C and 750°C for pinned and semi-rigid connections respectively. At higher temperatures the isolated beam continues to deflect rapidly, whereas the rate of deflection of the beam within the sub-frame arrangement eases a little.

Influence of Connection Type

In order to study the influence of connection type, the beam response with both flush and flexible connections as shown in Figs. 3*(a)* and 3*(b)* respectively is compared, together with the 'failure temperature' of the beam corresponding to the type of connection used. This comparison is illustrated in Fig. 4.

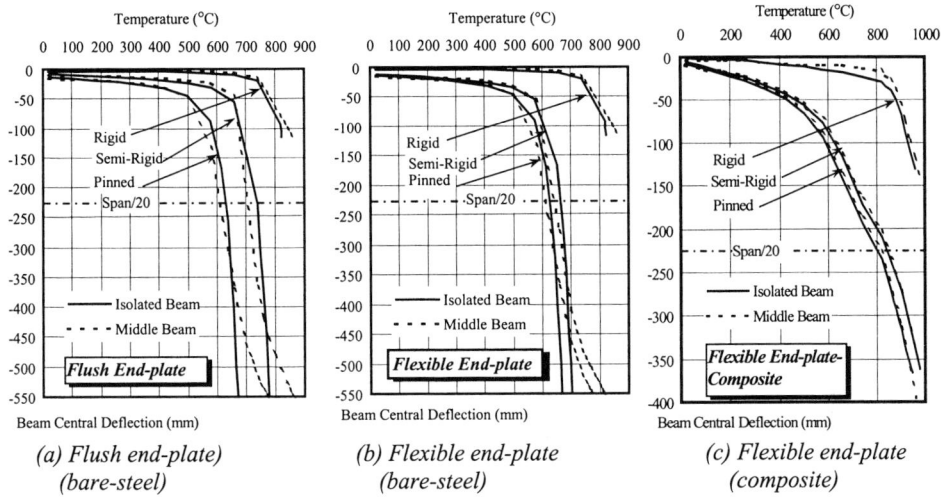

(a) Flush end-plate) (bare-steel) *(b) Flexible end-plate (bare-steel)* *(c) Flexible end-plate (composite)*

Figure 3. Beam Response Resulted from Incorporating Different connections Characteristics

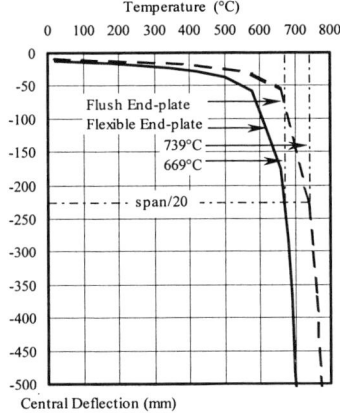

Figure 4: Influence of Connection Type on Beam Response with Increasing Temperatures

It may be seen from Figs. 3*(a)* and 3*(b)* that the central deflections of the beam gradually increase up to temperatures of approximately 720°C and 450°C for both rigid and pinned connections respectively. Further increase in temperature causes a progressive increase in the deflection rate until the onset of failure at temperatures of 850°C and 650°C respectively. The incorporation of semi-rigid connection characteristics representing a flush end-plate, results in an intermediate response between these extremes with a significant reduction in the beam deflections compared with pinned ends. However, the incorporation of flexible end-plate characteristics results in frame response comparable with that for a 'pinned' connection (as shown in Figs 3*(b)* and 3*(c)*). Also it may be seen from Fig. 4 that as expected the central deflection of the beam with flexible end-plates is consistently higher than for the beam with flush end-plates. This is because flexible end-plates are possessing negligible degree of end-restraint allowing little redistribution of forces at high temperatures. In contrast flush end-plates provide a high degree of restraint. This considerably reduces the central deflection and hence enhances the performance of the beam in fire. As can be seen from Fig. 4 that there is an increase in the beam failure temperature of about 70°C when using flush end-plate. It is apparent that as the rigidity of the connection increases, the survival time of the structure increases and the beam mid-span deflections decrease.

To study the effect of end-plate thickness on the beam behaviour, two analyses were conducted with end-plate thickness of 8mm and 12mm (results are not shown). Despite the variation in the end-plate thickness, there was negligible variation in the central deflection of the beam for temperatures up to 730°C, beyond which the deflection of the beam connected with 8mm end-plate thickness is slightly greater.

Influence of Connection Temperature

The effect of the relative temperatures of the beam and connection is shown in Fig. 5 for various bare-steel and composite connection details. It may be seen that for the arrangement with flush end-plates, there is negligible variation in the beam response for a connection temperature-ratio up to 0.5, beyond which there is a gradual decrease in the failure temperature with increasing temperature-ratio. Thus the temperature of the connection has a significant effect on the beam response for flush end-plate connections. Therefore in order to improve the performance of the beam in fire it is necessary to keep the connection temperature at a relatively low (i.e. not exceeding half of the connected beam). It seems that for both bare-steel and composite flexible end-plate connections the relative temperature of the connection has little influence on the failure temperature. This is probably due the inherent nature of flexible end-plates, with relatively large rotations and low moment capacity.

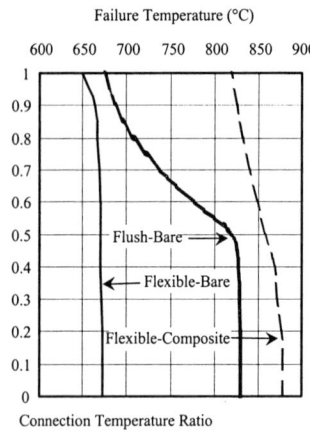

Fig. 5: Influence of Connection Temperature

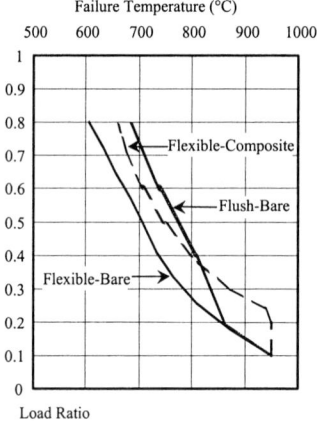

Figure 6: Influence of Applied Load-Ratio

Influence of The Applied Load-Ratio

For the parametric studies described above, a load ratio of 0.6 was generally used representing typical full service loading. Fig. 6 shows failure temperatures for load ratios up to 0.8, for bare-steel and composite sections with flexible connection characteristics. There is a significant decrease in the failure temperature as the load ratio increases for both bare-steel and composite cases. At a very low load ratios of 0.1 and 0.25 for bare-steel and composite connections repectively, failure temperatures exceed 950°C. For the bare-steel connections, the applied load-ratio has almost a linear effect on the overall failure-temperature. This decrease in failure temperature with increasing load ratio is due to the fact that the connections undergo considerable rotations resulting in a significant increase in deformation of the beam and consequently reducing the failure temperature.

It is apparent from Fig. 6 that the load ratio can have a significant influence on the performance of the member in fire irrespective of connection type adopted. It should be remembered that adopting a load ratio of 0.6 forms a realistic maximum bound normally used in fire design, since it corresponds to the full design condition (at ambient temperatures).

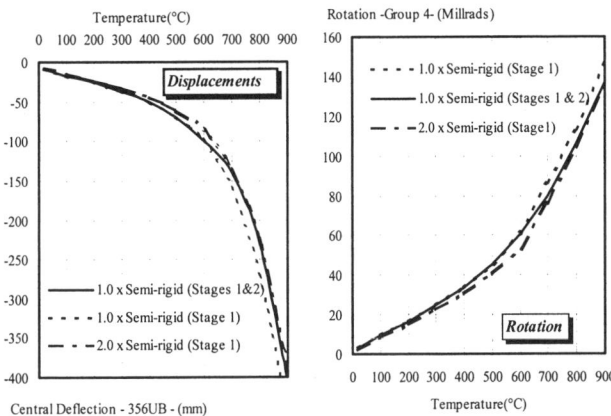

Figure 7: Deflections and Corresponding Rotations for the beam Incorporating Flexible End-Plate Composite Connection Characteristics

Influence of Connection Response Stage

Some flexible end-plate connections have two stages of response:
1. the unobstructed rotation of the end-plate, until;
2. the beam bottom flange comes into contact with the column, resulting in significant increase in stiffness with further rotation (as shown in Fig. 1*(b)*).

Despite the fact that large levels of rotation are generated, Nethercot et al (Nethercot (1986)] reported that connection behaviour at rotations beyond 50 millrads had little practical significance (i.e. after contact). In fire conditions however the structure experiences deflections significantly greater than normal conditions and the entire rotation range may therefore be relevant. In order to study this the composite sub-frame was analysed incorporating the moment-rotation-temperature characteristics for the flexible end-plate assuming the entire connection response. Fig. 7 shows the central deflection of the beam and the corresponding rotation of the connection assuming the entire response, along with those obtained ignoring the influence of the second stage.

For temperatures up to approximately 625°C beam deflections resulted from incorporating the entire connection response remained unchanged whereas the corresponding rotation at this temperature is approximately 65 millirads. This suggests that stage 1 rotation (before the beam and the column comes into contact) is still governing the connection response. With increasing temperatures beyond 625°C it may be seen that the connection gradually enters stage 2 (after the contact) of the response causing a small decrease in both the deflection of the beam and the rotation of the connection. This occurred as a result of enhancement in the connection response due to the contact between the beam lower flange and the face of the column. This corresponds to a beam deflection of about 145mm which is obviously less than the specified 'limiting deflection' (i.e. 225mm) corresponding to the nominal failure temperature of the beam. Despite the fact that the use of the entire connection response has resulted in a reduction in the beam deflection, this enhancement in the beam behaviour is not significant.

CONCLUSIONS

Parametric studies have been conducted to investigate the effect of connection characteristics on beam response in fire. This effect is considered by incorporating the moment-rotation characteristics for flush and flexible end-plate connections within both bare-steel and composite sub-frame analyses. The Ramberg-Osgood expression was adopted to represent the connection characteristics at elevated temperature. The analyses were conducted using the finite element code, VULCAN.

The results obtained for the bare-steel sub-frame demonstrated that the failure temperature for the beam with flush end-plate connection was almost 70°C higher than for the beam with flexible end-plates and the response was approximately mid-way between that for rigid and pinned characteristics. This suggests that the connection type has considerable influence on the performance of the structural member in fire. However, there was only a slight enhancement in the beam response with flexible end-plate for both bare-steel and composite arrangements compared with pinned condition, suggesting that the connected beams may be treated as simply supported. From the investigation it was found the thickness of the end-plate has little influence on the response of the beam. For bare-steel flush end-plates it was concluded that the beam failure temperature is sensitive to the connection temperature, while it had little effect in the case of flexible end-plate connections. A number of applied load ratios was considered from which it was found that increasing load ratio has a considerable effect in reducing the failure temperature of the beam irrespective of the connection type. In the case flexible end-plates it was found that the enhanced response resulting from the contact between the beam and the column had only a small influence on the survival time of the beam in fire.

REFERENCES

Al-Jabri, K.S. (2000). The Behaviour of Steel and Composite Beam-to-Column Connections in Fire. *PhD Thesis*. University of Sheffield, UK.

Bailey, C. G. (1995). Simulation of The Structural Behaviour of Steel-Framed Buildings in Fire, *PhD Thesis*. University of Sheffield, UK

BS 5950 Structural Use of Steelwork in Building: Part 1: Code of Practice for Design in Simple and Continuous Construction (1990*), British Standard Institution*, London.

El-Rimawi, J. A., Burgess, I. W., and Plank, R. J. (1993). Modelling the Behaviour of Steel Frames and Sub-frames with Semi-Rigid Connections in Fire. *Research Report DCSE/93/S/02,* Department of Civil and Structural Engineering, University of Sheffield, UK.

Nethercot, D. A., Davison, J. B., and Kirby, P. A.(1986). Connection Flexibility and Beam Design in Non-sway Frames*, American Society of Civil Engineers, Structural Convention*, New Orleans.

HEAT TRANSFER IN STEEL STRUCTURES AND THEIR FIRE RESISTANCE

W. Sha[1] and T.L. Ngu[2]

[1]Metals Research Group, School of Civil Engineering, The Queen's University of Belfast, Belfast BT7 1NN, UK
[2]Subur Tiasa Holdings Berhad, 167681-D, No. 8-14, 1st-3rd Floor, Jln. Chengal, 96000 Sibu, Sarawak, Malaysia

ABSTRACT

This paper is concerned with the fire resistance of composite beams with unfilled voids and temperature modelling with fire flux data and intumescent coating. The objective of the first part of work is to quantitatively predict the amount of additional protection required when the voids between the top flange and the underside of the steel deck are unfilled. Two fire tests were carried out on a conventional I-beam floor structure with the same 18-mm board protection, but one with filled voids and the other with unfilled voids. Moment resistance calculations involved dividing the beam into four elements, the top flange, the upper half of the web, the lower half of the web, and the bottom flange, and considering the slab one single element. The strengths of these elements depend on their temperatures at different times during fire tests. In the second part, temperature development in a steel section during fire has been calculated with a heat transfer and thermal conduction model, TFIRE, developed at The Steel Construction Institute (SCI). The steel section temperatures have been calculated with the data of heat flux into the steel measured in a fire test, instead of the fire temperature. In the third and final part of the paper, the behaviour of intumescent coating under fire, in particular the swelling thickness has been investigated. The heat conductivity value of intumescent coatings was modelled, based on fire test data.

KEYWORDS

Composite beam, fire, fire protection, fire resistance, heat flux, heat transfer, slim floor, steel structures, intumescent coating, thermal conduction, thermal conductivity.

INTRODUCTION

The first part of the work investigated composite floor structures, as a continuation of the theme of the programme of research on fire engineering modelling of steel building structures. Modern multi-storey steel framed buildings widely use composite construction for floor slabs and beams. A concrete topping is designed to act in a composite way with a steel deck to form the floor slab. The floor slab is then designed to act compositely with the steel beams to form composite beams [Johnson (1994)].

Two types of deck profile are used, the re-entrant dovetail deck and the open trapezoidal deck (Figure 1). For this type of structures, steel beams require fire protection. For dovetail decks, the voids formed by the deck profile are small, and it is normally not necessary to fill them to maintain the required fire resistance. However, for trapezoidal decks, the voids can result in a significantly increased rate of heating of the steel section effectively unprotected above its top flange. To fill these voids has severe economic disadvantages because it is very labour intensive. This serves against the use of trapezoidal decks with board-type fire protection. The purpose of the present research is to quantitatively evaluate the effect of not filling the voids formed with open trapezoidal decks on the fire resistance of the structure.

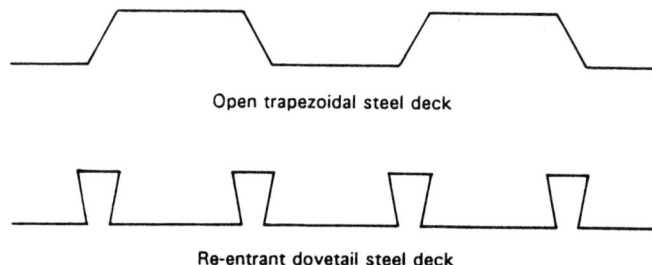

Open trapezoidal steel deck

Re-entrant dovetail steel deck

Figure 1: Types of profiled steel decks (courtesy Newman and Lawson)

The second part of this paper involves computer modelling of temperature development in steel sections during fire using heat flux data measured in fire tests.

Tfire is a program that can consider the behaviour of intumescent coating in the thermal analysis. The intumescent swelling thickness, thermal conductivity, emissivity and absorbivity are needed for constructing analysis software for calculating the temperature development. There has been no published information on these parameters. These are the subject of the third part of the paper.

FIRE RESISTANCE OF COMPOSITE BEAMS WITH UNFILLED VOIDS

The essential parameter is the load ratio, the ratio between the load capacities at the fire Limit State and under normal "cold" condition. For most applications, a load ratio of 0.6 is adequate. The approach taken is as follows:

(i)   Obtain the bottom flange temperature, T_1, that will give a load ratio of 0.6. Temperature values of various parts of the beam were required for this calculation, which were obtained from real temperature distributions from fire test data [Newman (1990)].

(ii) The thickness required of fire protection board is proportional to $(T_{bottom\ flange}-140)^{-1.3}$ [Newman (1989), Newman & Lawson (1991), Fire (1992)]. Therefore, the thickness to achieve a load ratio of 0.6 (d_1) can be calculated using:

$$\frac{d_1}{d_0} = \left(\frac{T_0 - 140}{T_1 - 140}\right)^{1.3} \tag{1}$$

where T_0 is the bottom flange temperature measured in the tests with protection thickness d_0 and T_1 is the bottom flange temperature to give the load ratio of 0.6.

(iii) Repeat (i) and (ii) for the situation of voids being unfilled and obtain the protection thickness (d_{1u}) to achieve the load ratio of 0.6. Temperatures corresponding to unfilled voids are indicated by a subscript u.

(iv) Obtain the percentage increase in protection thickness (d_{1u}/d_1) for the same load ratio (0.6).

The calculations were based on the actual parameters used in the fire tests (Table 1).

TABLE 1
BEAM PARAMETERS FOR COMPOSITE BEAMS WITH UNFILLED VOIDS

Beam	305×102×33 UB S275
Slab depth	125 mm
Deck depth	60 mm, open trapezoidal
Concrete	Normal weight, Grade 30

The bottom flange temperatures and the calculated corresponding load ratios are given in Table 2. For 60 minutes fire resistance, the ratio of fire protection thickness required to achieve a same load ratio of 0.6 without and with the voids being filled can be obtained:

$$\frac{d_{1u}}{d_1} = \frac{\frac{d_{1u}}{d_0}}{\frac{d_1}{d_0}} = \frac{\left(\frac{T_{0u}-140}{T_{1u}-140}\right)^{1.3}}{\left(\frac{T_0-140}{T_1-140}\right)^{1.3}} = \frac{\left(\frac{575-140}{555-140}\right)^{1.3}}{\left(\frac{528-140}{584-140}\right)^{1.3}} = \frac{1.063}{0.839} = 1.27 \quad (2)$$

Therefore, the thickness of the protection need be increased by 27%.

Table 2
BOTTOM FLANGE TEMPERATURE AND CORRESPONDING LOAD RATIO

Time (min)	Voids	$T_{bottom\ flange}$ (°C)	Load ratio
60	Filled	528 (T_0)	0.71
	Unfilled	575 (T_{0u})	0.54
90	Filled	706 (T_0)	0.28
	Unfilled	775 (T_{0u})	0.14
-	Filled	584 (T_1)	0.6
	Unfilled	555 (T_{1u})	0.6

For 90 minutes fire resistance, similar calculation can show that the thickness of the protection board also need be increased by 27%, although the temperature values (T_0 and T_{0u}) are very different from those for 60 minutes. It is therefore expected that the same level of increased protection will be necessary for other fire resistance periods such as 120 minutes. Further, although the above calculations are based on a load ratio of 0.6, it is reasonable to assume that similar amount of extra protection would be required for other load ratios.

The present study gave a quantitative answer regarding the amounts of extra fire protection needed when the voids between the top flange and the underside of the steel deck in a composite beam structure are unfilled. The result essentially provides the design engineer with an option of either filling the voids, or increasing the thickness of the board protection (by 27%) and leaving the voids unfilled. Given the labour and materials consumed in filling the voids, the latter is perhaps more economical and definitely offers a more speedy construction.

TEMPERATURE MODELLING WITH FIRE FLUX DATA

In TFIRE, a model developed at SCI [Newman (1995)], the standard consideration is to calculate the rate of heat transfer from fire to steel section using the temperature difference between the two elements. In a standard fire test, the fire temperature, measured by thermocouples in the test furnace, is made to follow a standard fire curve [Lawson & Newman (1996)]. An alternative is to measure directly the heat flux into the steel.

Heat flux was measured by Ulster University Fire Safety Engineering Research and Testing (FireSERT) Centre during a standard fire test at Warrington Fire Research Centre (WFRC), UK (Figure 2). Thermal calculations were then made with TFIRE program to obtain the bottom flange temperature of an asymmetric slim floor beam (Figure 3). Good agreement is demonstrated, showing the accuracy of both the flux measurement and the model. This work opens the possibility of modelling section temperatures under real fire conditions, with any kind of (uneven) fire temperature distribution, so long as the local heat flux is measured.

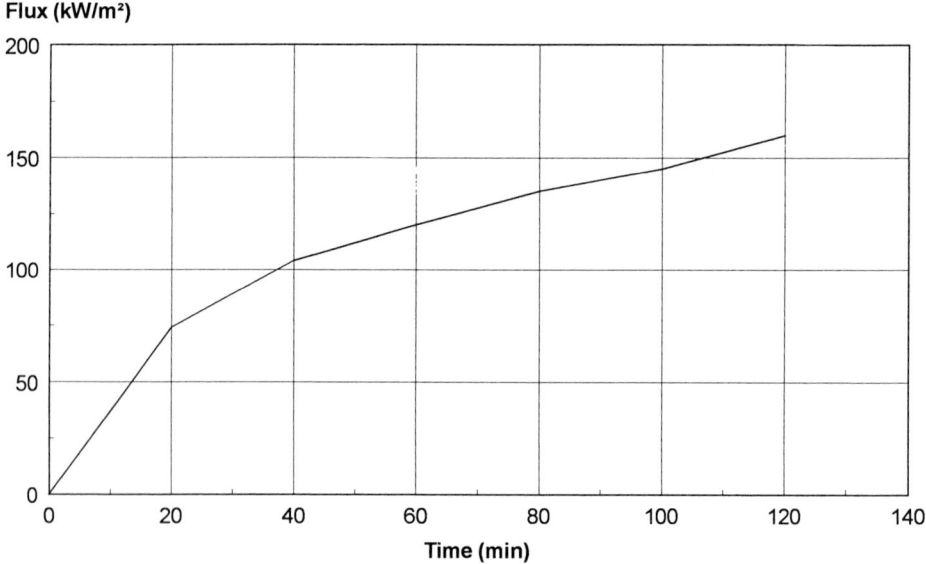

Figure 2: Measured heat flux as a function of time in standard fire test

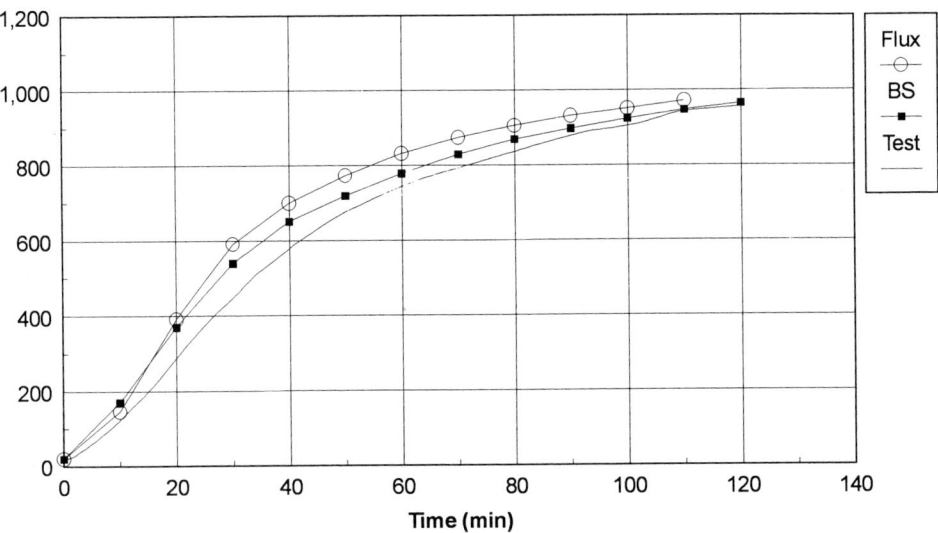

Figure 3: Bottom flange temperatures of asymmetric beam in fire. Temperatures were calculated using measured heat flux data ("Flux") and standard fire temperatures ("BS"). Actual test temperatures ("Test") are included for comparison

INTUMESCENT COATING

Intumescent material is extensively used in steel frame buildings to protect the steel surface from exposure to heat by providing a physical barrier. It responds to heat with endothermic chemical reactions generating multiple tiny bubbles causing the material to swell into a thickness of 5 to 100 times that of the original material [Butler (1997), Butler et al. (1994)]. The low thermal conductive bubbles solidify into a thick multicellular char layer.

Nullifire System S has a long record in both laboratory tests and real fire situations. Its fire resistance can last up to 120 minutes depending on the thickness applied. System S is differentiated for external or internal use. In general, sections with their section factor H_p/A greater than 90 m^{-1} will not achieve fire resistance longer than 30 minutes. Once applied with 0.3 mm of intumescent material, 30 minutes can be achieved easily.

S605 is a single-pack aromatic solvent based intumescent coating for fire protection of both internal and external structural steelwork, particularly for the protection of exposed structural steel. S607 is a water borne thin film intumescent coating for internal structural steel work. It is complimented by water borne acrylic top seal.

Fire tests have been carried out to measure the performance of intumescent coatings in fire. Temperature-time curves given in Figure 4 of slim floor flange plate for different thickness of coating show the significant fire protection effect of these coatings.

In the present work, laboratory tests were carried out with thin steel plate of 200×100×3 mm painted with intumescent coating all round with varying thickness. The furnace temperature was made to follow that in a standard fire curve. One series of tests was made with Nullifire S607 system with a dry thickness of 0.17-mm (300 g/m²). It was observed that at 300°C, the coating started to emit smoke and fume. Significant expansion started at 375°C, and continues until 625°C. Figure 5 shows the measured coating thickness as a function of furnace temperature. The expanded coating remains black until 750°C when it started to turn white through combustion and to crack. At 900°C the coating was totally white and started to drop off from the steel plate. At 1050°C, the coating almost completely dropped off and the heat separation effect was lost. White powder resulted.

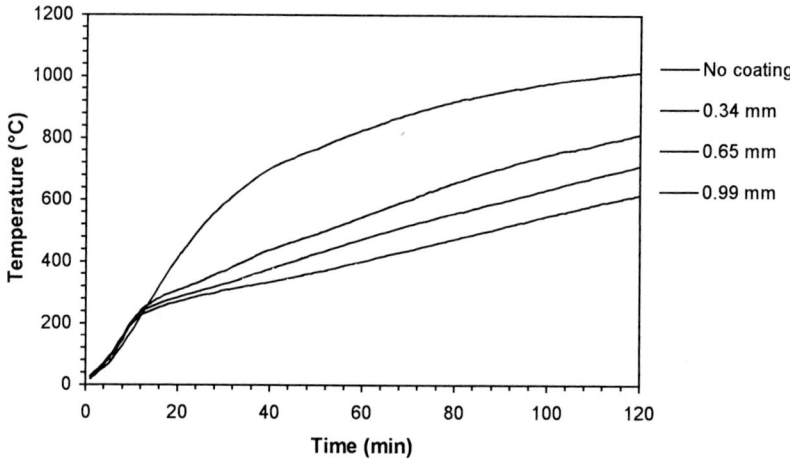

Figure 4: Temperature-time curves from fire tests of bottom steel plate in a Slimflor structure with S605 coating at different thickness applied

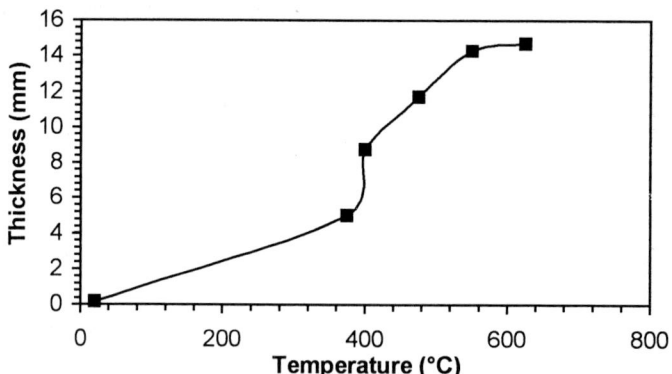

Figure 5: Coating thickness as a function of temperature. The furnace temperature follows that in a standard fire curve

Thicker coating can expand more. When the coating thickness was increased to 0.5 mm (900g/m²), it expands to approximately 55 mm after being heated up to 840°C within 30 minutes time. At 1005°C it was almost fully combusted.

In the following the temperature development of steel section in a slim floor structure protected with intumescent coating is modelled using the thermal analysis software, Tfire. Temperatures were measured at three locations in the bottom plate [Sha (1998a)]. Only the bottom plate temperatures were modelled. The section size is 203×203×60. Slimflor structure was used, except in one case where a "Top hat" slim floor structure is used (Figure 6).

In the model, a swelling thickness of intumescent coating of 40 times the initial thickness was used. In the heat transfer model, however, the thickness of any element is always used in conjunction with its thermal conductivity (λ, in W/m°C), appearing in the form of the ratio of the two. Therefore the variation of the two can be considered together, by only varying the thermal conductivity value. Emissivity and absorbivity were assumed 0.9. Figure 7 gives the λ value required in the model in order to achieve an average bottom plate temperature same as obtained from fire tests. Results are erratic, although the λ values at below 800°C are consistent. As mentioned above, in the heat transfer model, λ and the coating thickness are always used in the form of their ratio, so the modelling can also be regarded as using a fixed λ value with varying coating thickness. However, the heat emissivity and absorbivity of the intumescent coatings may not be constant as assumed in the model, which may significantly affect the modelling results.

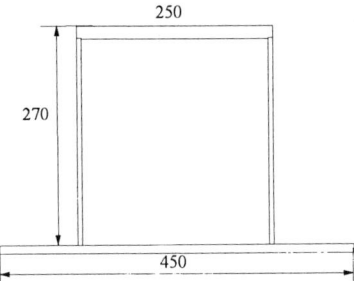

Figure 6: "Top hat" slim floor beam cross-section for Nullifire tests. Thickness of top flange, web and bottom flange is 16, 5 and 10 mm, respectively

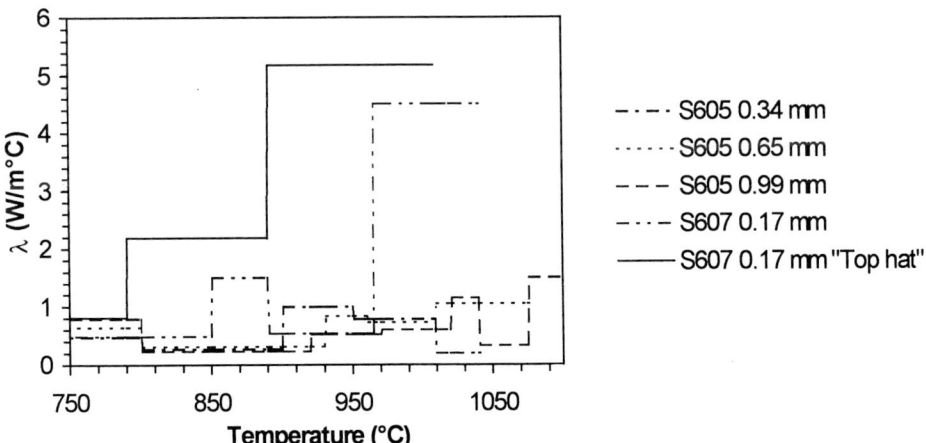

Figure 7: λ for obtaining same bottom plate temperatures between modelling and fire tests, for Slimflor except in one occasion for "Top hat". The coating type and thickness are given

CONCLUSIONS

The following conclusions can be made from this study.

1. For composite floor using I-beam, when the voids between the top flange and the underside of the steel deck are not filled, an extra 27% protection is required. This percentage of extra protection is the same for different fire resistance periods.

2. Thermal modelling calculation using measured heat flux data gives reasonable section temperature, showing the accuracy of flux measurement and the model.

ACKNOWLEDGEMENTS

This work was carried out during one of the authors, WS's secondment to The Steel Construction Institute, UK. The author is grateful to Mr. G.M. Newman for useful discussion.

REFERENCES

Butler, K.M. (1997). Physical Modelling of Intumescent Fire Retardant Polymers. *ACS Symposium Series*, vol. 669, ch. 15, Science and Technology, ed: K.C. Khemani, American Chemical Society, Washington, pp. 214-230.

Butler, K.M., Baum, H.R., Kashiwagi, T. (1994). A Three-dimensional Kinetic Model for the Swelling of Intumescent Materials. *Proceedings of Annual Conference on Fire Research*, National Institute of Standards and Technology, Gaithersburg, pp. 109-110.

Fire Protection for Structural Steels in Buildings (1992). 2nd edn., Association of Specialist Fire Protection Contractors and Manufacturers (ASFPCM), Aldershot, UK.

Johnson, R.P. (1994). *Composite Structures of Steel and Concrete*, vol. 1, Beams, Slabs, Columns, and Frames for Building, 2nd edn., Blackwell Science, Oxford, UK.

Lawson, R.M. and Newman, G.M. (1996). *Structural Fire Design to EC3 and EC4, and Comparison with BS 5950*, Steel Construction Institute, Ascot, UK, p. 22.

Newman, G.M. (1989). *Reports to British Steel on the Fire Resistance of Composite Beams*, Document SCI-RT-070/9, Steel Construction Institute, Ascot, UK.

Newman, G.M. (1990). *The Fire Resistance of Composite Beams with Unfilled Voids*, Report Number 135, Steel Construction Institute, Ascot, UK.

Newman, G.M. (1995). Fire Resistance of Slim Floor Beam. *Journal of Constructional Steel Research*, **33:1-2**, 87-100.

Newman, G.M. and Lawson, R.M. (1991). *Fire Resistance of Composite Beams*, Steel Construction Institute, Ascot, UK.

TEMPERATURE DEVELOPMENT DURING FIRE IN SLIM FLOOR BEAMS PROTECTED WITH INTUMESCENT COATING

W. Sha and N.C. Lau

Metals Research Group, School of Civil Engineering, The Queen's University of Belfast, Belfast BT7 1NN, UK

ABSTRACT

The TFIRE program was revised to model the heat transfer in SLIMFLOR beam with intumescent fire protection. The intumescent material is a relatively new type of fire protection material, consisting of mastics based on epoxy resins that soften and release carbon dioxide when the temperature exceeds 150°C. This leads to swelling and giving a low-density honeycomb with good insulating characteristics. The original version of TFIRE does not have the facility to model intumescent coatings. However, the program has a sub-routine that can be modified to cater for new protection material. Because of the lack of information on the properties of intumescent coating in fire, assumptions had to be made. These properties are swelled coating thickness, thermal conductivity, surface absorbivity and emissivity, all some functions of temperature and time in fire. In the first step of the modelling, the intumescent coating was assumed to act as a normal fire-protection board without temperature/time dependent characteristics. The coating was assumed to swell 40 times to its original thickness from the start of fire. The values of surface absorbivity and surface emissivity were assumed the same, 0.8. The thermal conductivity value to give the best fit was found 0.5 W/m°C.

KEYWORDS

Asymmetric beam, composite beam, fire, fire engineering, fire resistance, fire resistant steel, heat transfer, intumescent coating, slim floor.

LABORATORY TESTS

Seven fire tests [Sha (1997a)] were carried out at Nullifire Ltd., UK, on unloaded slim floor beams protected using S607 and S605 intumescent coatings [Malhotra (1982), Fire (1992)]. The test furnace was approximately 1-m cube. The furnace temperature was made to follow the standard fire curve. In all SLIMFLOR tests, UC 203×203×60 beams were used, and the bottom plates were 15 mm thick and approximately 200 mm wider than the flange of the UC section. The first test carried out was on an unprotected slim floor beam. In the tests with protection, the intumescent protective coatings were applied on the lower surface of the bottom plate. Only the bottom plate was directly exposed to fire, and the UC section welded to the plate was buried in sand (Figure 1). This made the experiments considerably easier to conduct as casting and drying of concrete were avoided, although the beam temperature development would be somewhat different compared with situations that are more realistic where the beam is covered by concrete.

Temperatures were measured along the beams at two sections, each approximately one third of the exposed beam length from the exposed end. The positions of thermocouples in each section are shown in Figure 2. Four thermocouples were used in each experiment to monitor and control the furnace temperature.

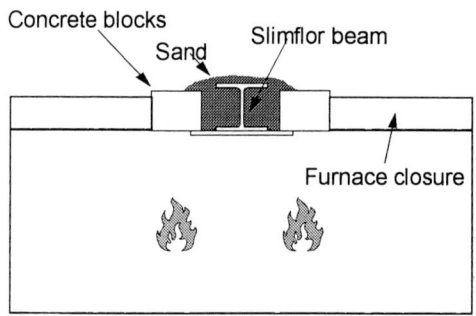

Figure 1: Nullifire furnace configuration for fire testing of fabricated SLIMFLOR beam (cross-section)

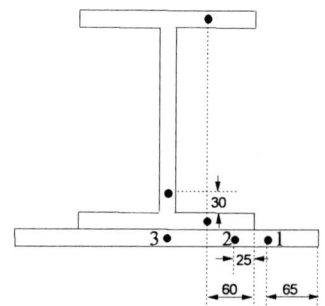

Figure 2: Fabricated SLIMFLOR beam cross-section with thermocouple positions for Nullifire tests

Temperature-time profiles for all twelve thermocouples on each test sample were obtained. Data useful for modelling and fire performance assessment were temperatures of plate, bottom flange and the lower part of the web (referred to as lower web throughout this paper). These are given in Figure 3, in which each temperature was averaged from two readings at two sections. The top flange temperature was always below 400°C and therefore the steel was at full strength.

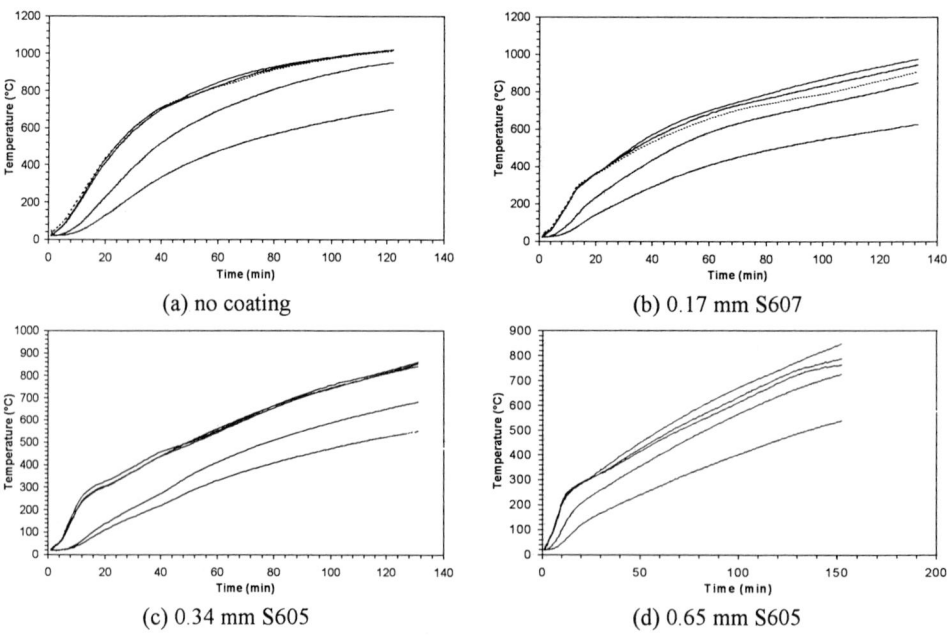

(a) no coating

(b) 0.17 mm S607

(c) 0.34 mm S605

(d) 0.65 mm S605

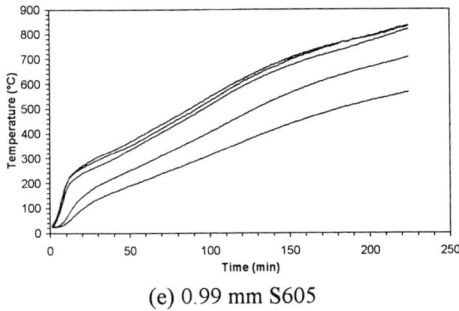

Figure 3: Temperature-time profiles obtained from fire test of SLIMFLOR beam. The five curves are, respectively from the top of the graph, for plate thermocouples 1, 2 and 3, bottom flange and lower web

(e) 0.99 mm S605

In addition to the tests on SLIMFLOR beams two tests were conducted on a "Top Hat" slim floor beam (Figure 4). The results of the tests are given in Figure 5. In these figures, thermocouples 1 and 2 on bottom flange are 20 and 60 mm from the edge, respectively, and, thermocouples 1 and 2 on web are 30 and 60 mm from bottom flange, respectively.

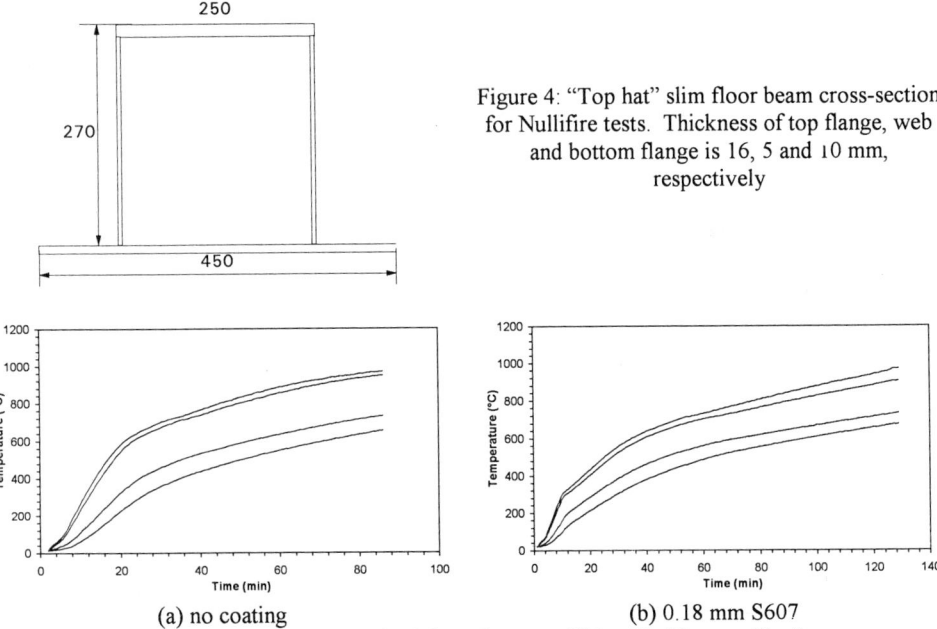

Figure 4: "Top hat" slim floor beam cross-section for Nullifire tests. Thickness of top flange, web and bottom flange is 16, 5 and 10 mm, respectively

(a) no coating   (b) 0.18 mm S607

Figure 5: Temperature-time profiles obtained from fire test of "Top Hat" beam. The four curves are, respectively from the top of the graph, for thermocouples 1 and 2 on the bottom flange and thermocouples 1 and 2 on the web

TFIRE MODELLING OF SLIMFLOR BEAMS

In modelling the SLIMFLOR beams, the cross-section of the beams was split into small elements by rectangular grid (Figure 6). The rectangular grid was set up in such a way that the division of elements corresponded to the positions of thermocouples in the fire tests (Figure 2). Properties of materials, such as steel, concrete and fire protection, and boundary conditions (fire) were then given to each element. It was found that the concrete humidity and the use of sand in place of concrete do not significantly influence the result. The coating parameters used in the model are listed in Table 1.

TABLE 1
COATING PARAMETERS

Coating thickness	40 times original thickness from the start of fire
Surface absorbivity	0.8
Surface emissivity	0.8
Thermal conductivity	0.5 W/m°C

The computation took place in small time-increments of about 6 seconds. At each increment, the heat flow between adjacent elements was calculated and the total heat flow into each element was used to calculate the elevated temperature over the time increment.

The results are shown in Table 2 together with the test data. Note that only the results of bottom plate and bottom flange and lower web of UC are presented.

TABLE 2
PREDICTED AND TESTED TEMPERATURES (°C) FOR SLIMFLOR BEAM

Protection	Time (min)	Plate		Bottom flange		Lower web	
		Test	Predicted†	Test	Predicted†	Test	Predicted†
0.17 mm S607	30	471	489 (+18)	349	299 (−50)	227	171 (−56)
	60	697	684 (−13)	588	496 (−92)	414	317 (−97)
	90	802	809 (+7)	708	623 (−85)	513	422 (−91)
	120	903	893 (−10)	806	715 (−91)	598	507 (−91)
0.34 mm S605	30	383	402 (+19)	215	245 (+30)	173	144 (−29)
	60	559	606 (+47)	418	419 (+1)	336	270 (−66)
	90	714	717 (+3)	557	550 (−7)	446	376 (−70)
	120	818	813 (−5)	655	642 (−13)	526	462 (−64)
0.65 mm S605	30	335	306 (−29)	265	187 (−78)	172	113 (−59)
	60	483	499 (+16)	406	333 (−73)	279	217 (−62)
	90	605	631 (+26)	532	466 (−66)	377	323 (−54)
	120	718	706 (−12)	641	558 (−83)	465	408 (−57)
	150	798	792 (−6)	724	635 (−89)	536	475 (−61)
0.98 mm S605	30	292	246 (−46)	194	150 (−44)	141	93 (−48)
	60	388	427 (+39)	282	279 (−3)	216	182 (−34)
	90	498	560 (+62)	377	405 (+28)	286	285 (−1)
	120	608	651 (+43)	477	503 (+26)	368	370 (+2)
	150	694	713 (+19)	565	575 (+10)	456	437 (−19)
	180	753	785 (+32)	634	639 (+5)	501	493 (−8)
	210	804	824 (+20)	685	683 (−2)	549	543 (−6)

†Numbers in parentheses are differences in temperatures between predicted results and fire test data.

In the heat transfer model, the thermoconductivity (λ) and thickness (d) of the coating appear in the combined form λ/d. The plate and bottom flange temperatures as a function of λ/d at 120 minutes in fire were calculated (Figure 7). The relationships between plate and bottom flange temperatures and λ/d were found to be:

$$T_{plate} = 1310 + 160\ln\frac{\lambda}{d} \ (r^2 = 0.9986) \text{ or } (T_{plate} - 140)^{-1.3} = 3.51 \times 10^{-6}\frac{d}{\lambda} + 2.20 \times 10^{-4} \ (r^2 = 0.9997) \quad (1)$$

$$T_{flange} = 1087 + 137\ln\frac{\lambda}{d} \ (r^2 = 0.996) \text{ or } (T_{flange} - 140)^{-1.3} = 1.88 \times 10^{-6}\frac{d}{\lambda} + 1.76 \times 10^{-4} \ (r^2 = 0.995) \quad (2)$$

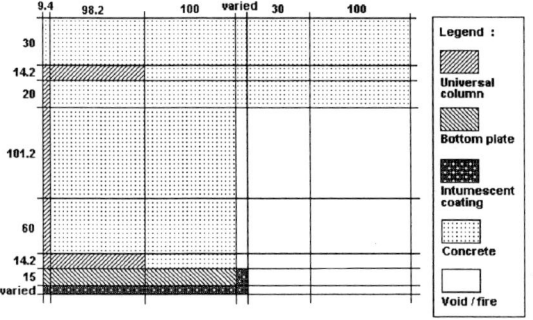

Figure 6: Division of SLIMFLOR section into elements for TFIRE analysis. All dimensions in mm (not to scale)

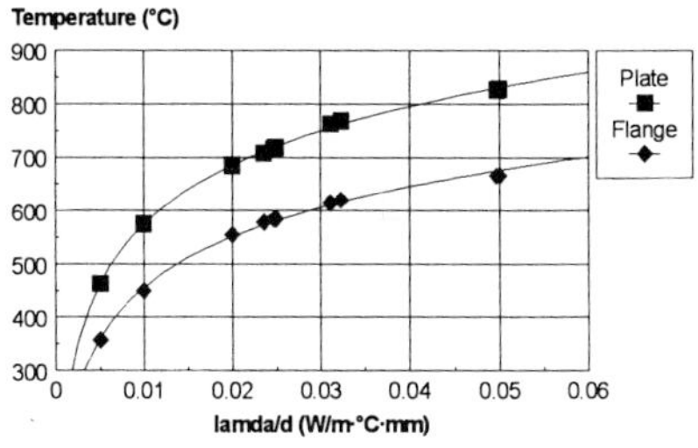

Figure 7: Plate and bottom flange temperatures of the SLIMFLOR as a function of λ/d at 120 minutes, modelled with TFIRE

Figure 8 shows the plate and bottom flange temperatures at 120 minutes in beams protected with a coating with λ/d of 0.032 W/(m°C•mm). As expected, larger beams have lower temperatures.

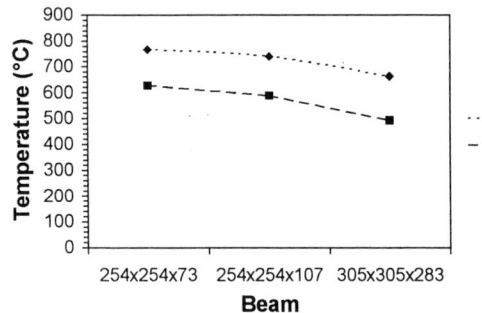

Figure 8: Plate and bottom flange temperatures of the SLIMFLOR as a function of beam size at 120 minutes, modelled with TFIRE. λ/d = 0.032 W/(m°C•mm)

Section temperature as the function of time was also obtained. Figure 9 shows a graph of plate temperatures for two conditions, 0.34 and 0.65 mm S605. For these calculations, a thermoconductivity of 0.38 W/m°C was used, instead of 0.5 W/m°C for Table 2. Figure 10 shows plate and flange temperatures for different λ/d values.

It can be seen that a thermal conductivity of 0.5 W/m°C with the assumption that the coating swells 40 times its original thickness from the beginning of fire is reasonably correct, especially for bottom plate temperatures. The differences of predicted and fire test temperatures are within –49 to 62°C. The predicted temperatures of UC bottom flange and lower web are significantly lower than the test temperatures for beams with coating thickness up to 0.65 mm. However, the differences were not varied drastically between the time at 60 to 120 minutes in all beams. The predicted temperatures at UC bottom flange and lower web are more accurate for beam with 0.98 mm coating. The effects of S605 and S607 appear similar which is largely as expected since the two types of coating have similar protection properties [Fire (1992)].

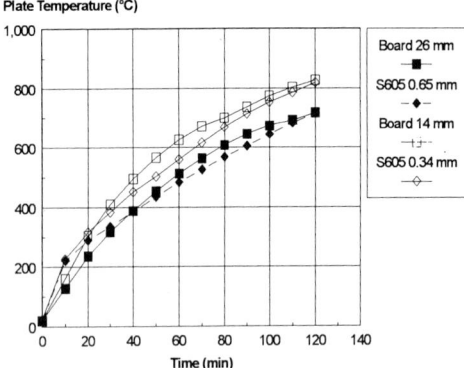 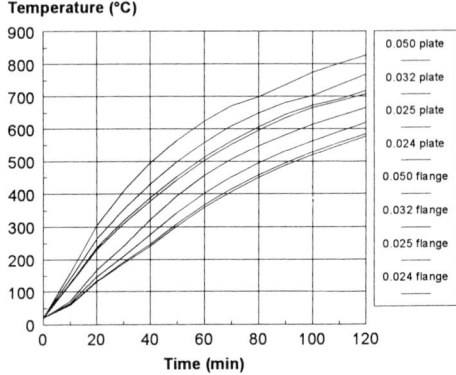

Figure 9: Calculated (■ and □) and test (♦ and ◊) plate temperatures in SLIMFLOR protected with S605. Coating assumed to swell 40 times

Figure 10: SLIMFLOR plate and bottom flange temperatures for different λ/d values. Line identification same order as in legend

The above modelling used fixed thermal conductivity and thickness of the intumescent coatings throughout the fire temperature range. In reality, these coatings have varied thermal conduction and thickness as they respond to high temperature. At room temperature and in the beginning of fire before swelling occurs, the coating is thin and thermal conductivity is high. When the temperature increases, the coating swells progressively and the thermal conduction reduces, achieving the effect of fire protection. At prolonged fire, the insulation reduces as the coating cracks and loses adhesion with the steel. Effectively, this corresponds to an increase in conductivity, or, equivalently, a decrease in thickness. In the present work, modelling using temperature-dependant thermal conductivity was attempted. The calculations were concentrated on the plate temperature.

Different combinations of λ/d for different temperature ranges are investigated (Figure 11). So is the effect of temperature at which the change of coating thermal properties (λ/d) occurs (Figure 12). Figure 13 gives further results with different combinations of λ/d for different temperature ranges, including a case where λ/d changes gradually with increasing temperature.

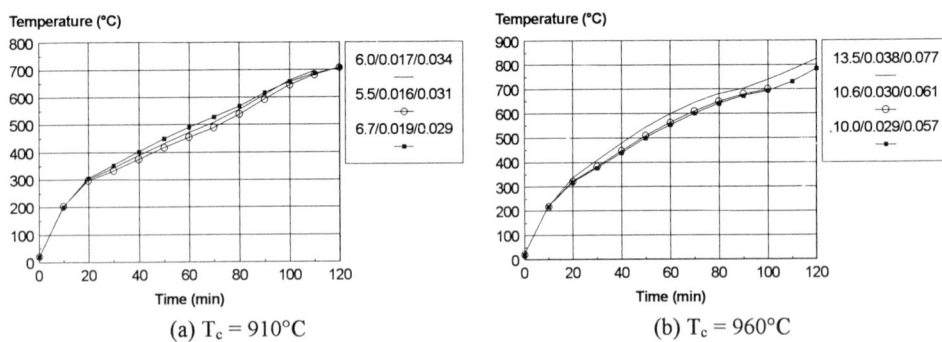

(a) $T_c = 910°C$

(b) $T_c = 960°C$

Figure 11: Intumescent protected SLIMFLOR plate temperatures as a function of time in fire as calculated using TFIRE. λ/d values (in W/(m°C•mm)) are given in legend. The first values are for below 200°C, the second between 200°C and T_c and the third above T_c

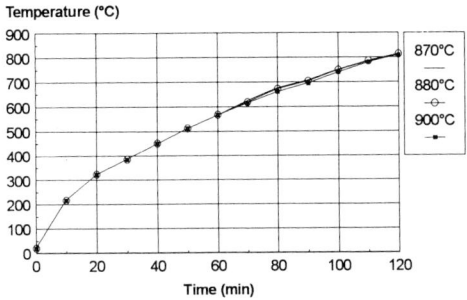 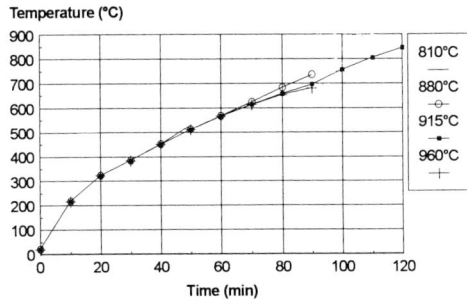

(a) 0.047 W/(m°C•mm) above T_c (b) 0.078 W/(m°C•mm) above T_c

Figure 12: Intumescent protected SLIMFLOR plate temperatures as a function of time in fire as calculated using TFIRE. λ/d = 10.9 W/(m°C•mm) below 200°C, 0.031 W/(m°C•mm) between 200°C and T_c. T_c is indicated in legend

The best fit for 0.34 mm S605 was achieved with the thermoconductivity starting with 150 W/m°C, dropping to 0.43 W/m°C at 300°C and increasing to 0.64 W/m°C at 870°C. For 0.65 mm coating, this needs to start with 175 W/m°C, dropping to 0.5 W/m°C at 300°C and increasing to 0.75 W/m°C at 950°C. The coating thickness was assumed to be 40 times of dry film thickness in both cases. The error is kept within 10°C at all times (Figure 14). The variation of thermoconductivity values with temperature is different for the two coatings, so the effect of film thickness is not linear. In particular, the modelling results suggest that the thinner coating is more efficient per unit coating thickness. In addition to considering the thickness and conductivity as a function of temperature, further work should investigate the effect of interface resistance between steel and the coatings and the gradual dropping-off and loss of effectiveness of the coatings. Further experimental work is also needed as uncertainties in fire test repeatability and reproducibility may well contribute to the difficulties unresolved here.

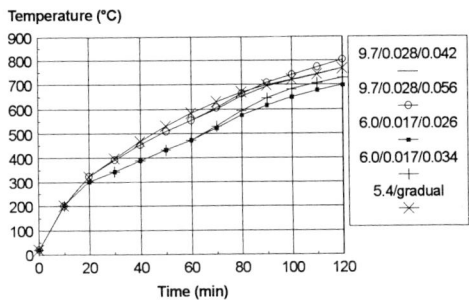 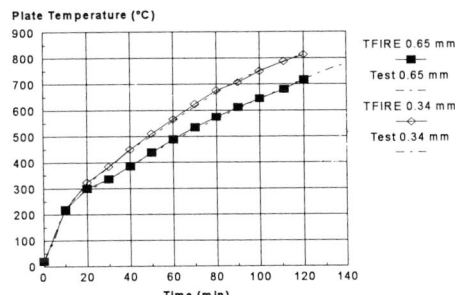

Figure 13: Plate temperatures, λ/d given for <300/300-880/>880°C. "5.4/gradual", λ/d 5.4 <300°C and follows 0.0038 (1+0.01T) >300°C

Figure 14: Best fit between calculated and test results of intumescent protected SLIMFLOR plate temperatures

Finally, Figure 15 gives a modelling result for an asymmetric slim floor beam (ASB) protected with intumescent coating. In this test, the coating thickness was 500 g/m², and other beam and test parameters were given in previous publications [Lawson et al. (1997), Sha (1997b), Bailey (1996)].

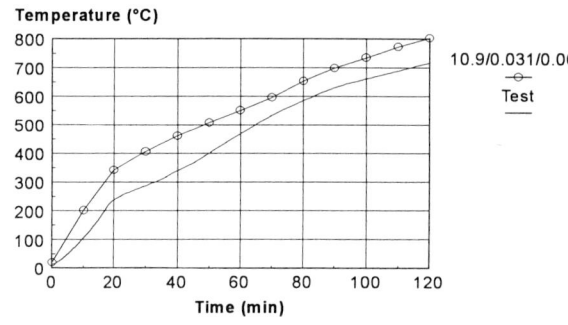

Figure 15: Intumescent protected ASB bottom flange temperatures as a function of time in fire as calculated using TFIRE. λ/d values (in W/(m°C·mm)) are given in legend. The first value is for below 300°C, the second between 300°C and 880°C and the third above 880°C. Temperatures obtained in fire test were given for comparison

CONCLUSIONS

1. Intumescent coatings can significantly reduce the steel temperature of slim floor beams (SLIMFLOR, ASB and "Top Hat") when only applied to the exposed bottom section, effectively improving the fire resistance.

2. Section temperatures in fire may be modelled for SLIMFLOR protected with intumescent coatings. Calculations assuming constant material parameters give reasonable estimations. Assuming varying material parameters with temperature may reduce deviation from test data. However, further work is needed to fully understand the thermal and perhaps stickability behaviour of the intumescent coatings.

ACKNOWLEDGEMENTS

Part of the work was carried out during WS's secondment to the Steel Construction Institute, UK, who is grateful to Mr. G. M. Newman and Dr. R.M. Lawson for direction and discussion.

REFERENCES

Bailey, C.G. (1996). *Fire Resistance of Asymmetric Slim Floor Beams*, Report to British Steel, Document RT607, Steel Construction Institute, Ascot, UK.

Fire Protection for Structural Steels in Buildings (1992). 2nd edn., Association of Specialist Fire Protection Contractors and Manufacturers (ASFPCM), Aldershot, UK.

Lawson, R.M, Mullett, D.L. and Rackham, J.W. (1997). *Design of Asymmetric Slimflor Beams Using Deep Composite Decking*, Steel Construction Institute, Ascot, UK.

Malhotra, H.L. (1982). *Design of Fire-resisting Structures*, Surrey University Press, Guildford, UK.

Sha, W. (1997a). *The Fire Resistance of Slimflor Beams Protected Using Intumescent Coatings S605 and S607*, Report to Nullifire Ltd., Document RT629, Steel Construction Institute, Ascot, UK.

Sha, W. (1997b). *The Fire Resistance of Asymmetric Slimflor Beams Protected Using Intumescent Coatings S605 and S607*, Report to Nullifire Ltd., Document RT655, Steel Construction Institute, Ascot, UK.

STRUCTURAL SAFETY AND RELIABILITY

ON THE NON STATIONARY SPECTRAL MOMENTS AND THEIR ROLE IN STRUCTURAL SAFETY AND RELIABILITY

S. Caddemi[1], P.Colajanni[2] and G.Muscolino[2]
[1]Dipartimento di Ingegneria Civile ed Ambientale, Università degli Studi di Catania, Catania, Italy
[2]Dipartimento di Costruzioni e Tecnologie Avanzate,Università degli Studi di Messina, Messina, Italy

ABSTRACT

The reliability of structures subjected to non stationary stochastic loads is addressed by studying the role of spectral quantities called *spectral moments*. They have been introduced for the stationary case, and extended to the non stationary case, according to a frequency domain *geometric* formulation. However, the exact formulation (called *non geometric*) has to be conducted in the time domain by means of the study of a complex process called pre-envelope. In this study, the non geometric approach is followed and the non geometric spectral moments are obtained by means of formulation and solution of their governing differential equations, hence comments on the approximations of the probability of survival and comparisons with the geometrical approach are presented.

KEYWORDS

Mean upcrossing rate, Envelope process, Pre-envelope covariances, Spectral moments, Time domain reliability, Probability of failure, Probability of survival.

INTRODUCTION

The assessment of time domain reliability of structures implies the evaluation of the probability of survival under the action of external dynamic stochastic forces. The probability of survival requires the definition of the hazard function. In the literature various approximations of the hazard function have been proposed. All of them require the definition of spectral moments first introduced by Vanmarcke (1972) as geometric moments in the frequency domain of the stationary power spectral density function and consequently extented to the non stationary case. Then, the spectral moments were defined by Yang (1972), Di Paola (1985) and Muscolino (1988) according to a non geometric time domain formulation, applied to the non stationary case also, as cross-covariances of the response process and its Hilbert transform.

In this work the role of the non geometric time domain non stationary spectral moments in the approximation of the hazard function is explored. In particular a well established approximation of the hazard function for stationary responses is examined in a non stationary context in order to show the difference of the performance of the non geometric spectral moments with respect to the geometric ones. Restrictions of the approximated expression of the hazard function are also highlighted. The non geometric time domain spectral moments are determined by means of a novel approach based on the formulation and solution of their governing differential equations alternative to the cumbersome integral formulation.

TIME DEPENDENT RELIABILITY OF A SDOF OSCILLATOR

The equation of motion of a linear single degree of freedom (SDOF) oscillator subjected to a non stationary white process can be written as follows

$$\ddot{X}(t)+2\varsigma_0\omega_0\dot{X}(t)+\omega_0^2 X(t)=\varphi(t)W(t) \quad \text{s.t.} \quad X(0)=0 \text{ with probability 1} \quad (1)$$

where $W(t)$ is a zero mean stationary white noise, with power content S_0, modulated by a deterministic modulating function $\varphi(t)$ of time t. The oscillator governed by Eqn.1 is characterised by the natural frequency ω_0 and the damping ratio ς_0. The response of the oscillator to the external noise is represented by the process $X(t)$. The response process $X(t)$ can be considered satisfactory if pre-established requirements, called limit state, are fulfilled. However, since the applied load possesses stochastic properties, the fulfilment or violation of the limit state has to be defined by means of a probability function which is the object of structural reliability. Since, more precisely, the applied load is a stochastic process, a time dependent reliability problem will be addressed to in what follows. If violation of the limit state implies structural collapse, the time dependent reliability aims at evaluating the *probability of failure* of the structural system during its lifetime, or the complementary function, called *probability of survival*. If we consider the limit state as a pre-established barrier η not to be violated by the response process $X(t)$, the probability of survival is the probability that the maximum response value, within a prescribed time interval, does not exceed the given barrier.

Let us define the problem by considering the stochastic process $X_{max}(t)$, which is the extreme value of $|X(t)|$ in a defined time interval $[0,T]$, as $X_{max}(T)=\max_{0\le t\le T}|X(t)|$. The cumulative probability distribution of $X_{max}(t)$ is defined as follows $F_{X_{max}(T)}(\eta)=Prob[X_{max}(T)\le\eta]=L_X(\eta,0)L_X(\eta,T)$ in which $L_X(\eta,T)$ defines the probability that the process $|X(t)|$ does not cross the barrier η and $L_X(\eta,0)$ the probability that the process $|X(t)|$ starts below the barrier η. The cumulative distribution $F_{X_{max}(T)}(\eta)$ is hence the probability of survival, since η is here considered as a critical value for $|X(t)|$ corresponding to some failure mode of the system.

Since for the case under study $L_X(\eta,0)=1$, the probability function $L_X(\eta,T)$ coincides with the probability of survival and can be given the following form

$$L_X(\eta,T)=\exp\left[-\int_0^T \alpha_\eta(\tau)d\tau\right] \quad (2)$$

where $\alpha_\eta(\tau)$ is known as *hazard function*. It can be proved that the hazard function $\alpha_\eta(\tau)$ can be regarded as the conditional rate of upcrossings of the barrier η by the process $|X(t)|$ given the initial condition and the circumstance that there has been no prior upcrossing. Since the evaluation of the hazard function is not an easy task no exact solution is available in the literature, however approximations are adopted for solving practical problems.

The roughest approximation consists in replacing the hazard function with the unconditional mean upcrossing rate $v^+_{|X(t)|}(t)$ of the barrier η by the process $|X(t)|$ given as follows

$$\alpha_\eta(t) \approx v^+_{|X(t)|}(\eta,t) = 2v^+_{X(t)}(\eta,t) = \sqrt{\frac{\sigma_{\dot{X}}(t)}{\sigma_X(t)}} \frac{\sqrt{1-\rho^2_{X\dot{X}}(t)}}{\pi} \exp\left[-\frac{\eta^2}{2\sigma_X^2(t)(1-\rho^2_{X\dot{X}}(t))}\right]$$

$$\left\{1+\frac{\eta\rho_{X\dot{X}}(t)}{\sigma_X(t)}\sqrt{\frac{\pi}{2(1-\rho^2_{X\dot{X}}(t))}}\exp\left[\frac{\eta^2\rho^2_{X\dot{X}}(t)}{2\sigma_X^2(t)(1-\rho^2_{X\dot{X}}(t))}\right]\right\}\left\{1-\Phi\left[-\frac{\eta\rho_{X\dot{X}}(t)}{\sigma_X(t)\sqrt{2(1-\rho^2_{X\dot{X}}(t))}}\right]\right\} \quad (3)$$

where $\rho_{X\dot{X}}(t)$ is the correlation coefficient given as follows $\rho_{X\dot{X}}(t) = \sigma_{X\dot{X}}(t)/\sigma_X(t)\sigma_{\dot{X}}(t)$. The approximation $\alpha_\eta(t) \approx v^+_{|X(t)|}(\eta,t)$ provided by Eqn.3 implies the conditioning event for the hazard function, that there as been no prior upcrossing, being neglected. Hence, according to this approximation, the mean upcrossing rate is considered independent of the past history of the process and the time intervals elapsed between upcrossings will be independent and, as a result, the integer valued process counting the number of upcrossings is a Poisson process. The hypothesis so far discussed (Poisson approximation) is accurate when the process is very broadband and the barrier η is very large, otherwise the results may be significantly in error. In fact, for narrow band processes or small barrier η a single upcrossing is likely to be associated with several dependent upcrossings, and this is inconsistent with the Poisson approximation requiring the time intervals between upcrossings being independent. However, in view of the conditioning event for the hazard function, it can be assumed in most of the cases that $\alpha_\eta(t) \leq v^+_{|X(t)|}(\eta,t)$, hence the Poisson approximation leads to an underestimation of the probability of survival, at least for the cases of primary importance.

A further approximation based on the Poisson assumption requires the definition of the envelope $a_X(t)$ of the response process which is a smooth process joining the peaks of the response process. In fact, it can be assumed that the extreme value process is the same as the envelope process. In this case the Poisson approximation for the occurrences of the upcrossings by the envelope process seems more appropriate, hence the hazard function can be approximated by the mean upcrossing rate $v^+_{a_x}(\eta,t)$ by the envelope process provided for zero mean non stationary Gaussian input process as follows

$$\alpha_\eta(t) \approx v^+_{a_x}(\eta,t) \quad (4)$$

However for large values of the barrier η, it is generally found that many of the upcrossings of η by the envelope process are not accompanied by upcrossings of the process $|X(t)|$ (empty envelope upcrossings). This circumstance leads to $v^+_{a_x}(\eta,t)$ being much larger that $v^+_{|X(t)|}(\eta,t)$. On the other hand, the hazard function $\alpha_\eta(t)$ (coincident with the conditional mean upcrossing rate) tends to $v^+_{|X(t)|}(\eta,t)$ as η becomes very large, hence there must be some limits to the adoption of the last discussed approximation provided by Eqn.4.

An improved approximation of the extreme value distribution can be obtained by estimating the fraction of the upcrossings by the envelope process which are accompanied by upcrossings by the process $|X(t)|$ (qualified envelope upcrossings). Hence, the hazard function can be taken as this subset of upcrossings by the envelope process and is provided as follows [Vanmarcke (1975)]

$$\alpha_\eta(t) = v^+_{|X(t)|}(\eta,t)\left\{\frac{1-\exp\left[-v^+_{a_x}(\eta,t)/v^+_{|X(t)|}(\eta,t)\right]}{1-\left(v^+_{X(t)}(\eta,t)/v^+_{X(t)}(0,t)\right)}\right\} \quad (5)$$

Different expressions for the hazard function have been provided in the literature, however Eqn.5, which has been proposed for the stationary case, has been claimed to be one of the most accurate, and, for this reason, is considered in this study.

THE NON STATIONARY SPECTRAL MOMENTS

Equation 5, in the stationary case, requires the evaluation of the following expressions.

$$v^+_{X(t)}(0) = \frac{1}{2\pi}\sqrt{\frac{\lambda_{2,X}}{\lambda_{0,X}}} \quad ; \quad v^+_{|X(t)|}(\eta) = 2 v^+_{X(t)}(\eta) = v^+_{X(t)}(0) exp\left[-\frac{\eta^2}{2\sqrt{\lambda_{0,X}}}\right] \quad (6)$$

$$v^+_{a_X}(\eta) = \sqrt{2\pi} \, q_X \, v^+_{X(t)}(\eta)\frac{\eta}{\sqrt{\lambda_{0,X}}} \quad ; \quad q_X = \sqrt{1 - \frac{\lambda^2_{1,X}}{\lambda_{0,X}\lambda_{2,X}}}$$

Equations 6 rely on the definition and evaluation of the quantities $\lambda_{0,X}, \lambda_{1,X}, \lambda_{2,X}$ named *spectral moments* by Vanmarcke (1972), in view of their definition given by the following expression $\lambda_{i,X} = 2\int_0^\infty \omega^i S_{XX}(\omega) d\omega$, $i = 0,1,2$. The definition of spectral moments for the quantities $\lambda_{0,X}, \lambda_{1,X}, \lambda_{2,X}$ is due to the meaning of moments of the power spectral density (PSD) function $S_{XX}(\omega)$ of the response with respect to the origin. Equations 6 have been subsequently extended by Corotis et al. (1972) to the non stationary case by simply generalising the definition of spectral moments as moments of the evolutive PSD function $S_{XX}(\omega,t)$ of the response as follows

$$\lambda_{i,X}(t) = 2\int_0^\infty \omega^i S_{XX}(\omega,t) d\omega \quad , \quad i = 0,1,2.$$ This approach has been referred to as *geometric approach* by Michaelov at al. (1999). Corotis et al. (1972) applied initially this approach to the case of unit step modulating function (i.e. the case of transient of a stationary response of a quiescent oscillator) and noticed divergences of the first and second order spectral moments $\lambda_{1,X}(t), \lambda_{2,X}(t)$ leading to the adoption of the their convergent part only. The question concerning the performance of the geometric approach for the case of general modulating function remained open until a discussion of Bucher and Lin (1988) evidenced the low accuracy of the mean upcrossing rate in some cases. The geometric approach has been also adopted by Irschik & Ziegler (1991) who claimed the simplicity of the expressions and proposed approximated formulae for Eqns.6 to be applied to general modulating functions.

It has to be pointed out that, according to Eqn.5, the quantities involved in the evaluation of the hazard function require the study of the envelope response process and the evaluation of its mean upcrossing rate. In line with this comment the study of the envelope process has been conducted by Yang (1972), Di Paola (1985) and Muscolino (1988) on the basis of the definition of a complex pre-envelope process $Y(t)$ whose imaginary part is the Hilbert transform of the given real response process. The modulus of the pre-envelope process $Y(t)$ is the envelope or amplitude process $a_X(t)$. The mean upcrossing rate of a given barrier η by the envelope response process for a non stationary input can be obtained as follows:

$$v^+_{a_X}(\eta,t) = \int_0^\infty p_{a\dot{a}}(\eta,\dot{a};t) \, \dot{a} \, d\dot{a} \quad (7)$$

where $p_{a\dot{a}}(a,\dot{a};t)$ is the joint probability density (JPD) function of the envelope process and its time derivative. The JPD function $p_{a\dot{a}}(a,\dot{a};t)$ is strictly related to the evaluation of the JPD function $p_{Y\dot{Y}}(y,\dot{y};t)$ of the pre-envelope process $Y(t)$ and its time derivative $\dot{Y}(t)$ which requires the so-called pre-envelope cross-covariances $E[Y^* Y], E[Y^* \dot{Y}], E[\dot{Y}^* Y], E[\dot{Y}^* \dot{Y}]$ collected by the cross covariance matrix $R_{2,Y}(t)$ as follows $R_{2,Y}(t) = \begin{bmatrix} E[Y^* Y] & E[Y^* \dot{Y}] \\ E[\dot{Y}^* Y] & E[\dot{Y}^* \dot{Y}] \end{bmatrix} = \begin{bmatrix} \lambda_{0,Y}(t) & i\lambda_{1,Y}(t) \\ -i\lambda_{1,Y}(t) & \lambda_{2,Y}(t) \end{bmatrix} = \Lambda_{2,Y}(t)$

where the quantities $\lambda_{0,Y}(t), \lambda_{1,Y}(t), \lambda_{2,Y}(t)$, related to the pre-envelope cross-covariances, fulfil the following definitions and relations, in view of the properties of the Hilbert transform:

$$\lambda_{0,Y}(t) = E[Y^*(t)Y(t)] = E[X^2(t)] = \sigma_X^2(t) \quad , \quad \lambda_{1,Y}(t) = -i\, E[Y^*(t)\dot{Y}(t)]$$
$$\lambda_{2,Y}(t) = E[\dot{Y}^*(t)\dot{Y}(t)] = E[\dot{X}^2(t)] = \sigma_{\dot{X}}^2(t) \tag{8}$$

The quantities $\lambda_{0,Y}(t), \lambda_{1,Y}(t), \lambda_{2,Y}(t)$ are different from the spectral moments defined by Vanmarcke (1972), however they can be proved to be coincident for the stationary case only.

Once the cross-covariance matrix $R_{2,Y}(t)$ is known the JPD function $p_{Y\dot{Y}}(y,\dot{y};t)$ and consequently $p_{a\dot{a}}(a,\dot{a};t)$ can be evaluated and replaced into Eqn.7 leading to the following expression for $v_{a_x}^+(t)$

$$v_{a_x}^+(t) = \frac{\eta q_Y(t)}{\lambda_{0,Y}(t)} \sqrt{\frac{\lambda_{2,Y}(t)}{2\pi}} \exp\left[-\eta^2\left(\frac{1}{2\lambda_{0,Y}(t)} + \psi_{1,Y}^2(t)\right)\right]\left\{1 + \psi_{1,Y}(t)\eta\sqrt{\pi}\exp\left[(\psi_{1,Y}(t)\eta)^2\right]\left(1 + \Phi[\psi_{1,Y}(t)\eta]\right)\right\}$$

$$\psi_{1,Y}(t) = \frac{\mathrm{Im}[\lambda_{1,Y}(t)]}{q_Y(t)\lambda_{0,Y}(t)\sqrt{2\lambda_{2,Y}(t)}} \quad , \quad q_Y(t) = \sqrt{1 - \frac{\lambda_{1,Y}(t)\lambda_{1,Y}^*(t)}{q_Y \lambda_{0,Y}(t)\lambda_{2,Y}(t)}} \quad , \quad \Phi(b) = \frac{2}{\sqrt{\pi}} \int_0^b \exp\left[-\frac{\eta^2}{2\lambda_{0,Y}(t)}\right] \tag{9}$$

Since the latter approach, based on the definition of the quantities $\lambda_{0,Y}(t), \lambda_{1,Y}(t), \lambda_{2,Y}(t)$, does not correspond to any moment of the PSD function $S_{XX}(\omega,t)$ of the response, it has been addressed to by Michaelov et al. (1999) as *non geometric approach* which represents the exact approach.

PRE-ENVELOPE CROSS-COVARIANCE DIFFERENTIAL EQUATIONS

Aim of this section is the evaluation of the pre-envelope cross-covariances of the response process by means of a procedure, alternative to the integral formulation, based on the solution of their governing differential equation. The given oscillator subjected to the pre-envelope of the input process $F_W(t)$ is governed by the following differential equation:

$$\ddot{Y}(t) + 2\zeta_0\omega_0\dot{Y}(t) + \omega_0^2 Y(t) = F_W(t) \quad ; \quad F_W(t) = \frac{1}{\sqrt{2}}\varphi(t)\left[W(t) + i\hat{W}(t)\right] \tag{10}$$

where $\hat{W}(t)$ is the Hilbert transform of the zero mean stationary white noise $W(t)$. The response process $Y(t)$ of Eqn.10 represents the complex pre-envelope process of the real response process $X(t)$. Equation 10 can also be written by means of a state variable approach and converted into an Ito type stochastic differential equation as follows

$$d\mathbf{Z} = \mathbf{D}\,\mathbf{Z}\,dt + \mathbf{V}\,dF_B \quad , \quad \mathbf{Z}(t) = \begin{bmatrix} Y(t) \\ \dot{Y}(t) \end{bmatrix} \quad ; \quad \mathbf{D} = \begin{bmatrix} 0 & 1 \\ -\omega_0^2 & -2\zeta_0\omega_0 \end{bmatrix} \quad ; \quad \mathbf{V} = \begin{bmatrix} 0 \\ 1 \end{bmatrix} \tag{11}$$

where $dF_B(t) = F_W(t)dt$ is the differential of the pre-envelope Wiener process for which it can be proved that $dF_B^*(t)dF_B(t) = 2\pi S_0\, \varphi^2(t)dt$, hence $dF_B(t)$ is a differential of order $dt^{1/2}$ according to the extension of the stochastic differential calculus to the complex field. In order to evaluate the differential equations governing the covariances of $Y(t)$, the increment of $\mathbf{Z}^* \otimes \mathbf{Z}$, by applying the Kronecker product $\otimes$ to the $\mathbf{Z}$ vector must be evaluated as follows

$$\Delta(\mathbf{Z}^* \otimes \mathbf{Z}) = \mathbf{Z}^* \otimes d\mathbf{Z} + d\mathbf{Z}^* \otimes \mathbf{Z} + d\mathbf{Z}^* \otimes d\mathbf{Z} + O(dt^2)$$
$$= \mathbf{D}_2(\mathbf{Z}^* \otimes \mathbf{Z})dt + (\mathbf{V} \otimes \mathbf{I}_2)dF_B^*\mathbf{Z} + (\mathbf{I}_2 \otimes \mathbf{V})\mathbf{Z}^* dF_B + \mathbf{V}^{[2]} dF_B^* dF_B + O(dt^2) \tag{12}$$

where $D_2 = (I_2 \otimes D) + (D \otimes I_2)$, I_2 is the identity matrix of order 2 and the exponent in square brackets means Kroneker power $V^{[2]} = V \otimes V$. In view of the properties of $dF_B(t)$, the term $dF_B^*(t) dF_B(t)$ is an infinitesimal of order dt; furthermore it can be shown that the following relationships between the pre-envelope white input process and the output process Z hold $Re\{E[dF_B^* Z]\} = 0$; $Im\{E[dF_B^* Z]\} = Im\{E[F_W^* Z]\}dt \neq 0$ hence the imaginary part of $E[dF_B^* Z]$ is an infinitesimal of order dt. Making the stochastic average of both sides of Eqn.12, accounting for the properties of dF_B, dividing by dt and neglecting infinitesimals of higher order than dt lead to the differential equations governing the evolution of cross-covariances of the pre-envelope response process $Y(t)$ in the form

$$\dot{E}[Z^* \otimes Z] = D_2 E[Z^* \otimes Z] + V^{[2]} \varphi(t)^2 2\pi S_0 + i Im\{W_2\} \tag{13}$$

$$Im\{W_2\} = (V \otimes I_2) Im\{E[F_W^*(t) Z]\} + (I_2 \otimes V) Im\{E[Z^* F_W(t)]\} = 2 S_0 \varphi(t) \begin{bmatrix} 0 \\ \int_0^t \frac{h(\tau)}{\tau} \varphi(t-\tau) d\tau \\ -\int_0^t \frac{h(\tau)}{\tau} \varphi(t-\tau) d\tau \\ 0 \end{bmatrix} \tag{14}$$

where $h(t) = 1/\omega_D \exp[-\zeta_0 \omega_0 t] \sin(\omega_D t)$, $\omega_D = \omega_0 \sqrt{1-\zeta_0^2}$, and $\dot{h}(t)$ is the time derivative of $h(t)$. According to the procedure outlined so far, the evolution of the pre-envelope covariances is reduced to the solution of the deterministic first order differential equation called *pre-envelope covariance differential equation* (Eqn.13). The term $E[Z^* \otimes Z]$ obtained by means of a step by step solution of Eqn.13 contains the quantities $\lambda_{0,Y}(t), \lambda_{1,Y}(t), \lambda_{2,Y}(t)$ according to the following relationship:

$$E[Z^* \otimes Z] = [E[Y^* Y] \ E[Y^* \dot{Y}] \ E[\dot{Y}^* Y] \ E[\dot{Y}^* \dot{Y}]]^T = [\lambda_{0,Y}(t) \ i\lambda_{1,Y}(t) \ -i\lambda_{1,Y}(t) \ \lambda_{2,Y}(t)]^T \tag{15}$$

In view of the properties of the pre-envelope process, the following relationships between the cross-covariances of the real response process $X(t)$ and its time derivative $\dot{X}(t)$ hold:

$$\sigma_X^2(t) = E[X^2(t)] = \lambda_{0,Y}(t) \ , \ \sigma_{X\dot{X}}(t) = E[X(t)\dot{X}(t)] = Re\{E[Y^*(t) \dot{Y}(t)]\} = Im\{\lambda_{1,Y}(t)\}$$
$$\sigma_{\dot{X}}^2(t) = E[\dot{X}^2(t)] = \lambda_{2,Y}(t) \tag{16}$$

The mean upcrossing rates $v_{X(t)}^+(\eta,t), v_{|X(t)|}^+(\eta,t), v_{X(t)}^+(0,t), v_{a_X}^+(\eta,t)$ required by the hazard function $\alpha_\eta(t)$ (Eqn.5), can now be evaluated according to the non geometric approach in view of Eqns.3 and 9.

NUMERICAL RESULTS

Let us consider a linear oscillator with $\omega_0 = 1 \ rad/sec$, $\zeta_0 = 0.1$, subjected to a zero mean stationary Gaussian white noise with $S_0 = 1/\pi$ modulated by the deterministic function of the type $\varphi(t) = t \exp[-\alpha t]$. Two cases have been considered: 1) $\alpha = 0.2$; 2) $\alpha = 2$, corresponding to the two different modulating functions plotted in Fig.1. In the case 1) the modulating function reaches its peak at a time instant close to the natural period $T_0 = 2\pi/\omega_0$, while in the case 2) the peak is reached more rapidly. In the case 1) the spectral moments of the geometric $\lambda_{0,X}(t), \lambda_{1,X}(t), \lambda_{2,X}(t)$ and non geometric $\lambda_{0,Y}(t), \lambda_{1,Y}(t), \lambda_{2,Y}(t)$ approaches are plotted in Fig.2a and show only a slight difference. On the contrary the quantities $\lambda_{1,X}(t), \lambda_{2,X}(t)$ and $\lambda_{1,Y}(t), \lambda_{2,Y}(t)$ plotted in Fig.2b for the case 2) show a substantial difference while $\lambda_{0,X}(t)$ and $\lambda_{0,Y}(t)$ coincide as expected. It must be noted that in the case 2) there is an oscillatory behaviour consequent to the rapid occurrence of the peak of the modulating function $\varphi(t)$.

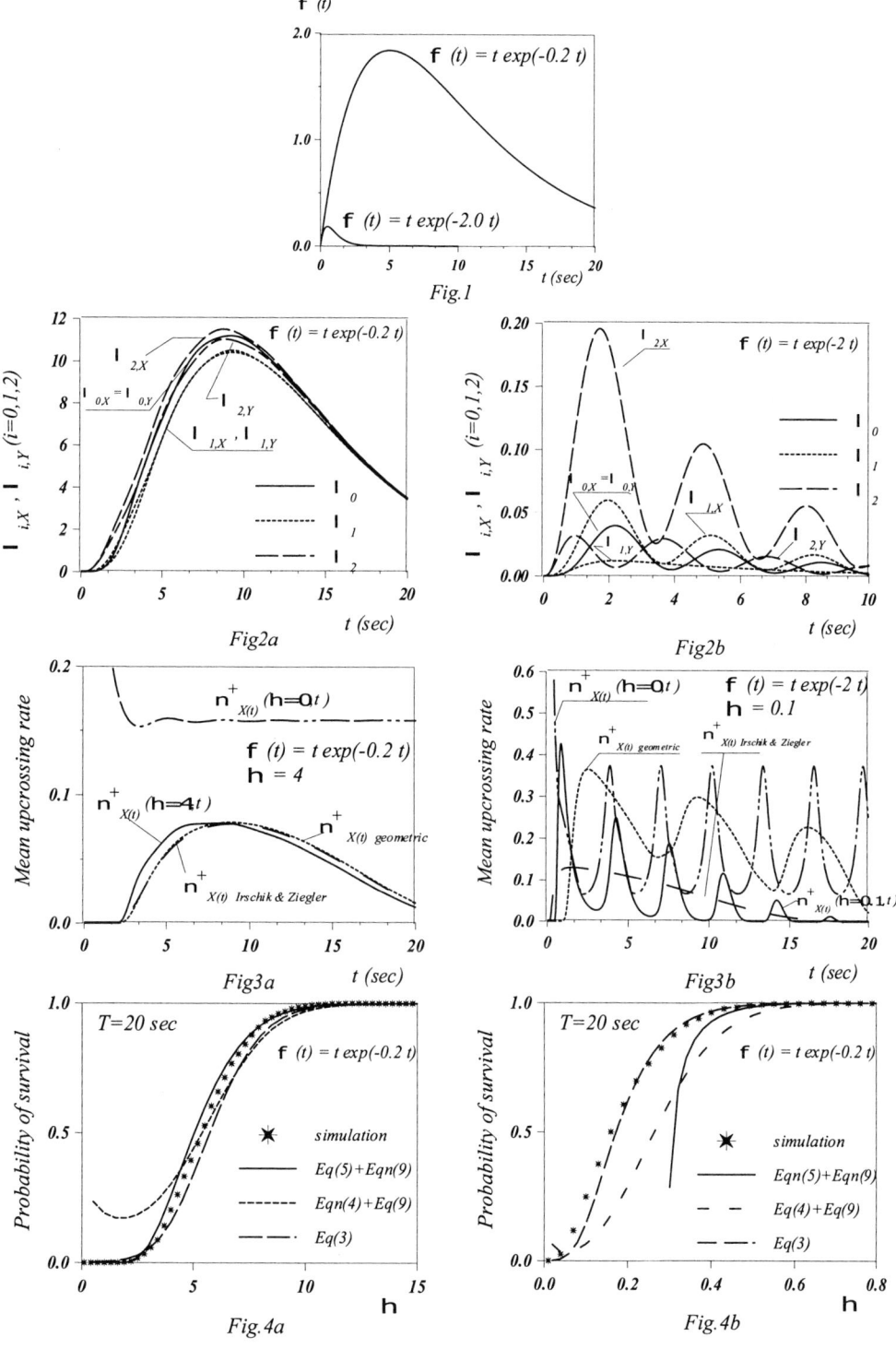

Fig.1

Fig.2a

Fig.2b

Fig.3a

Fig.3b

Fig.4a

Fig.4b

The results reported in Figs.3 in terms of mean upcrossing rate $v^+_{X(t)}(\eta=\eta_0,t)$ for pre-established barriers $\eta_0 = 4, 0.1$ reflects the differences highlighted in Figs.2. It has to be pointed out that, for the case 2) the curve of $v^+_{X(t)}(\eta=0,t)$ is exceeded by $v^+_{X(t)}(\eta=\eta_0,t)$ in certain time intervals, as a consequence the expression for the hazard function given by Eqn.5 looses accuracy since provides indeterminate or negative values which cause the inappropriateness of the expression of the probability of survival given by Eqn.2. In Figs.3 is also plotted the expression of $v^+_{X(t)}(\eta=\eta_0,t)$ provided by Irschik and Ziegler (1991) who claimed that the elimination of the oscillatory term provides a better approximation of the mean upcrossing rate in the geometric approach. However, here it can be stated that for cases such as 1) no oscillatory terms are present and the geometric formulation provides solutions reasonably accurate. While for cases such as 2) (including the case $\varphi(t)$ as the unit step function, i.e. transient of a stationary response) the exact response provided by the non geometric formulation presents oscillatory terms which are not captured by the geometric one. The correction of $v^+_{X(t)}(\eta,t)$ proposed by Irschik and Ziegler is not able to correct the inappropriateness of Eqn.5 for the hazard function. Finally, in Figs.4 the probability of survival for each case is plotted. For case 1) Eqn.5 coupled with the non geometric approach provide an accurate solution. While for case 2), Eqn.5 cannot provide a meaningful solution in correspondence of a wide range of barriers and the approximation provided by the Poisson assumption seems to provide better results.

References

Bucher C.G. and Lin Y.K. (1988) A note on spectral moments in non stationary random vibration. *Prob.Eng.Mech.*, **3**, 53-55.

Corotis R.B., Vanmarcke E.H. and Cornell C.A. (1972) First passage of non stationary random processes. *J. Engrg. Mech. Div. (ASCE)*, **98**(2), 401-414.

Di Paola M. (1985) Transient spectral moments of linear systems. *S.M.Archives*, **10**, 225-243.

Irschik H. and Ziegler F. (1991) Non stationary random vibrations: approximate spectral formulation versus exact theory. *Prob.Eng.Mech.*, **6**(2), 83-91.

Michaelov G., Sarkani S. and Lutes L.D. (1999) Spectral characteristics of non stationary random processes – A critical review. *Structural Safety*, **21**, 223-244.

Muscolino G. (1988) Non stationary envelope in random vibration theory. *J. Engrg. Mech. Div. (ASCE)*, **114** (8), 1396-1413.

Vanmarcke E. H. (1972) Properties of spectral moments with application to random vibration. *J. Engrg. Mech. Div. (ASCE)*, **98** (2), 425-446.

Vanmarcke E. H. (1975) On the distribution of the first passage time for normal stationary random processes. *J. Appl. Mech. (ASME)*, **42**, 215-220.

Yang J. N. (1972) Non stationary envelope process and first excursion probability. *J. Struct. Mech.* **1**, 231-248.

DEVELOPMENTS IN STRUCTURAL RELIABILITY BOUNDS

Kanagasabai Ramachandran

Department of Civil and Environmental Engineering,
Imperial College of Science, Technology and Medicine, London, UK

ABSTRACT

Until early nineties, probabilistic concepts were used only in the analysis of civil engineering structures and systems. Recently more researchers and practitioners have started to use probabilistic concepts in the assessment of existing structures and systems. Better estimation of system bounds thus becomes more important. Since the publication of the author's earlier paper, a few papers have only been published on system bounds. This paper presents the progress made on reliability bounds during the last fifteen years.

KEYWORDS

Reliability bounds, Structural reliability, Probability, Structural safety, System bounds, Structural assessment, Probabilistic assessment.

INTRODUCTION

Recently many researchers and practitioners have started to use probabilistic concepts in the assessment of civil engineering structures and systems. Furthermore, some researchers are also using better models to represent the statistical variability of important parameters in order to get better reliability estimates, Karimi and Ramachandran (1999). Some of these models lead to the evaluation of the probability of union of a number of highly correlated events. The probability associated with these events are, unfortunately, rather high and therefore improvements on the estimation of system bounds demands high priority.

In sixties and seventies Cornell (1967), Moses and Kinser (1967), Kounias (1968), Vanmarcke (1973), Hunter (1973) and Ditlevsen (1979) made valuable contributions in the field of probability bounds. The author published comprehensive reviews on system bounds in 1984 and 1992, and produced better reliability bounds using graph theory. The next notable publication was made by Greig in 1992. Greig, while appreciating the importance of the ordering of events, has limited his contribution to extend the bounds to fourth order bounds. In this paper, author's research at Imperial College during the last decade is described, and reference is made to other published works to highlight the differences between the approaches.

PROBLEM DEFINITION

A structure may have a number of limit states (ultimate and serviceable) and the structure is said to have failed if any one of the limit states is reached. Each limit state may be represented by a continuous function $g_i(\mathbf{x})$ (failure function or safety margin M).The event $M \leq 0$ is referred to as the failure event F_i and its probability of occurrence is denoted by $P(F_i)$ or simply P_i. Since the probability of failure of a structure is given by the probability of occurrence of any one of the possible failure events (limit states, failure modes), the probability of failure of the structure can be obtained by evaluating the probability of union of all the failure events $P(\cup F_i)$. This paper deals with the evaluation of $P(\cup F_i)$ and assumes that P_i, P_{ij} -intersection probability $P(F_i \cap F_j)$, P_{ijk} – intersection probability $P(F_i \cap F_j \cap F_k)$, P_{ijkl} etc are available.

SECOND ORDER BOUNDS

These bounds are now widely used in civil, structural and geo-technical engineering fields. The gap between the upper and lower bounds is unfortunately too wide for systems with more than six events (failure modes) and for high correlations between the events. The need for closer bounds became more apparent when assessment of corrosion of reinforcement in reinforced concrete bridges is undertaken with random field modeling for the variables, Karimi and Ramachandran (1999).

It has been now accepted that the second-order bounds may be conveniently expressed as

$$P(UF_i) \leq \sum_{i=1}^{n} (P_i - \underset{j \prec i}{Max}\, P_{ij}) = P_{2u}$$

$$P(UF_i) \geq \sum_{i=1}^{n} \left\{ P_i - \sum_{j \prec i} P_{ij} \right\} = P_{2L}$$

The terms inside the parenthesis {} should be considered zero if the value is less than zero. The values of the second-order upper bound P_{2u} and lower bound P_{2L} obviously depend on the ordering of the events P_i.

ORDERING OF EVENTS

Our aim is to get the smallest value of P_{2u} and the largest value of P_{2L}. In the following sections, methods will be presented for ordering events which would give optimal bounds in most cases.

Lower Bound

The procedure for reordering events to get the lowest P_{2L} is as follows:

Steps:
1. Select the largest P_i.
2. Select the largest $\{P_i - P_{ij}\}$, $j \neq i$ and $j \in (1,2,...n)$.
3. Select the largest $\left\{P_j - \sum\limits^{m} P_{jk}\right\}$, $k \neq j$ and $k \in (1,2,....n)$
4. Repeat Step 3 for m = 2,3,4 ..n-1 or until the value of the term inside the parenthesis { becomes negative.

Upper Bound

The same procedure can be used to get the largest P_{2u}.

Steps:
1. Select the smallest P_i.
2. Select the smallest $(P_i - P_{ij})$ for all possible values of j (i $\Box$ j).
3. Repeat Step 2 until all the events are used. Repeat if both i and j have already been used.

This method of ordering for the upper bound P_{2u} is computationally less efficient than the method described in the author's earlier paper. But the advantage is that it follows the same logic as for the lower bound P_{2L}.

The upper bound can be rewritten as follows:

$$P_{2u} = \sum_{i=1}^{n} P_i - \mathrm{Max} \sum_{\substack{i \neq j \\ i,j \prec n}} P_{ij}$$

This formulation allows a faster computational procedure to select the optimal bound as described below.

Steps:
1. Choose the largest P_{ij}.
2. Choose the next largest P_{ij}.
3. Repeat Step 2 until (n - 1) bisections P_{ij} are selected. Repeat if both i and j have already been used.

THIRD ORDER BOUNDS

Third-order bounds were first produced and published by the author in an earlier paper, Ramachandran (1984). They are reproduced below for easy reference.

$$P(\cup F_i) \geq P_1 + P_2 - P_{12} + \sum_{i=3}^{n} \left\{ P_i - \sum_{j=1}^{i-1} P_{ij} + \text{Max } P_{ilk} \right\} = P_{3L}$$

$$P(\cup F_i) \leq P_1 + P_2 - P_{12} + \sum_{i=3}^{n} \left(P_i - \text{Max}[P_{ij} - P_{ik} + P_{ijk}] \right) = P_{3U}$$

To get optimal lower bound, use of graph theory has been suggested.

Furthermore it was suggested that P_{3u} may be extended to include more trisections P_{ijl} and bisections P_{il} and P_{jl} provided they improve the bound. Author has modified his earlier upper bound to incorporate these suggestions in his internal report and computer program (Ramachandran 1992). Greig, in his paper, extended these bounds to include the probability of intersections of four events P_{ijkl} - $P(F_i \cap F_j \cap F_k \cap F_l)$, to produce fourth-order bounds in a systematic way.

He used the expression

$$P(\bigcup_{i=1}^{n} F_i) = \sum_{i=1}^{n} P\left[F_i \cap (\overline{\bigcup_{j \prec i} F_j})\right] = \sum \left[P_i - P\left(\bigcup_{j \prec i} (F_i \cap F_j)\right)\right]$$

to generate bounds of any order for P_f. This formulation also brings out the similarity between the different order bounds clearly as shown below.

$$P_{3L} = \sum_{i=1}^{n} \left\{ P_i - \sum_{j \prec i} \left(P_{ij} - \text{Max}_{k \prec j} P_{ijk} \right) \right\}$$

$$P_{3u} = \sum_{i=1}^{n} \left(P_i - \left\{ P_{ij} - \sum_{j \prec k} P_{ijk} \right\} \right)$$

$$P_{4L} = \sum_{i=1}^{n} \left\{ P_i - \sum_{j \prec i} \left(P_{ij} - \sum_{k \prec j} \{P_{ijk} - \sum P_{ijkl}\} \right) \right\}$$

$$P_{4u} = \sum_{i=1}^{n} \left(P_i - \sum_{j \prec i} \left\{ P_{ij} - \sum_{k \prec j} \left(P_{ijk} - \text{Max}_{l \prec k} P_{ijkl} \right) \right\} \right)$$

Greig's formulation, although gives bounds similar to existing ones, may be considered a good contribution in system bounds development, due to the ease with which bounds of any order can be produced.

THIRD AND HIGHER ORDER BOUNDS

In this section a summary of the author's work on the development of higher order bounds will be described. Full details are available in his report (Ramachandran, 1992).

It is well known that

$$P\left(\bigcup_{j=1}^{n} F_j\right) = P(F_1) + P(F_2 \cap S_1) + P(F_3 \cap S_2 \cap S_1) +$$
$$P(F_i \cap S_{i-1} \dots \cap S_2 \cap S_1) \dots + P(F_n \cap S_{n-1} \cap S_{n-2} \dots S_2 \cap S_1)$$

In order to get bounds for $P\left(\bigcup_{j=1}^{n} F_j\right)$, it is sufficient to get bounds for $P(F_i \cap S_{i-1} \cap S_{i-2} \dots S_l \dots S_2 \cap S_1)$.

Third Order Bounds

Denote $P(F_i \cap S_{i-1} \cap S_{i-2} \dots S_l \dots S_2 \cap S_1)$ by $P(F_i \cap S_j, j=1, i-1)$ and the lower and upper bounds of $P(F_i \cap S_j, j=1, i-1)$ are denoted by P_i^L and P_i^U.

It can be shown that P_i^L and P_i^U are given by

$$P_i^L = P_i - \sum_{j=1}^{i-1} P_{ij} + \text{Max} \sum P_{ijk}$$

$$P_i^U = P_i - \sum_{j=1}^{i-1}\left(P_{ij} - \sum_{k=1}^{j-1} P_{ijk}\right)$$

where Max $\sum P_{ijk}$ indicates that the summation is taken over the branches of a tree spanning over the vertices 1 to i-1.

The upper bound P_i^U can also be written as

$$P_i^U = P_i - \sum_{\substack{j=1 \\ j \neq l}}^{i-1}\left(P_{ij} - \sum_{k=1}^{j-1} P_{ijk}\right) - \left(P_{il} - \sum_{k=1}^{l-1} P_{ilk}\right)$$

Also $\qquad P\bigl((F_i \cap S_j, j=1, i-1) j \neq l\bigr) < P_i - \sum_{\substack{j=1 \\ j \neq l}}^{i-1}\left(P_{ij} - \sum_{k=1}^{j-1} P_{ijk}\right)$

This gives us the result $$P_i^U = P_i - \sum_{j=1}^{i-1}\left\{P_{ij} - \sum_{k=1}^{j-1} P_{ijk}\right\}$$

But $$P(F_i \cap S_j, j=1, i-1) \leq P(F_i \cap S_j, j=1, i-1), j \neq l)$$

Hence $$P_i^U = P_i - \sum_{\substack{j=1\\j\neq l}}^{i-1}\left\{P_{ij} - \sum_{k=1}^{j-1} P_{ijk}\right\}, \text{ if } \left\{P_{ij} - \sum_{k=1}^{l-1} P_{ijk}\right\} \text{ is negative.}$$

Hence $$P_{3L} = \sum_{i=1}^{n}\left\{P_i - \sum_{j=1}^{i-1}\left(P_{ij} + \text{Max}\,\Sigma\, P_{ijk}\right)\right\}$$

$$P_{3u} = \sum_{j=1}^{i-1}\left[P_i - \sum_{j=1}^{i-1}\left\{P_{ij} - \sum_{k=1}^{j-1} P_{ijk}\right\}\right]$$

Higher Order Bounds

These bounds can be extended further to include higher order intersections, say m - event intersections, as shown below.

$$P_i^L = P_i - \sum_{j=1}^{i-1}\left(P_{ij} - \sum_{k=1}^{j-1}\left\{P_{ijk} - \sum_{l=1}^{k-1} P_{ijkl}\right\}\right) \qquad \text{for } m=4$$

$$P_i^U = P_i - \sum_{j=1}^{i-1}\left\{P_{ij} - \sum_{k=1}^{j-1}\left(P_{ijk} - \text{Max}\,\Sigma\, P_{ijkl}\right)\right\} \qquad \text{for } m=4$$

$$P_i^U = P_i - \sum_{j=1}^{i-1}\left[P_{ij} - \sum_{k=1}^{j-1}\left(P_{ijk} - \sum_{l=1}^{k-1}\left\{P_{ijkl} - \sum_{p=1}^{l-1} P_{ijklp}\right\}\right)\right] \qquad \text{for } m=5$$

$$P_i^L = P_i - \sum_{j=1}^{i-1}\left\{P_{ij} - \sum_{k=1}^{j-1}\left(P_{ijk} + \sum_{l=1}^{k-1}\left(P_{ijkl} - \text{Max}\,\Sigma P_{ijklp}\right)\right)\right\} \qquad \text{for } m=5$$

EXAMPLE

Different bounds obtained for a simple example is given below. The reliability indices are (0.5, 1.2, 1.5, 1.1, 1.3, 1.1) and the correlation matrix ρ is given by

$$\rho = \begin{vmatrix} 1.00 & 0.64 & 0.39 & 0.29 & 0.25 & 0.21 \\ 0.64 & 1.00 & 0.64 & 0.39 & 0.29 & 0.25 \\ 0.39 & 0.64 & 1.00 & 0.64 & 0.39 & 0.29 \\ 0.29 & 0.39 & 0.64 & 1.00 & 0.64 & 0.39 \\ 0.25 & 0.29 & 0.39 & 0.64 & 1.00 & 0.64 \\ 0.21 & 0.25 & 0.29 & 0.39 & 0.64 & 1.00 \end{vmatrix}$$

Second-order bounds (unordered) = 0.57   0.42
Second-order bounds (ordered)   = 0.56   0.43
Third-order bounds   = 0.49   0.45

CONCLUSIONS

This paper gives an up-to-date summary of the progress made on system bounds. The problem of ordering the events to get optimal bounds has also been addressed. The ordering problem has been solved for second-order bounds and research is still progressing at Imperial College to get an optimal reordering for third-order bounds. These optimal bounds are useful in both reliability analysis and assessment work.

REFERENCES

Cornell, A.C. (1967). Bounds on the reliability of structural systems. *J. Struct. Div. ASCE*, **93**, 171-200.

Ditlevsen, O. (1979). Narrow reliability bounds for structural systems. *J. Struct. Mech.*, **4**, 453-472.

Greig, G.L. (1992). An assessment of higher-order bounds for structural reliability. *Structural Safety*, **5**, 213-225.

Hunter, D. (1973). An upper bound for the probability of a union. *J. Applied Prob.* **13**, 597-603.

Karimi, A. (2001). Probabilistic assessment of concrete bridges, *Ph.D Thesis* (to be submitted)

Karimi, A and Ramachandran, K. (1999). Probabilistic estimation of corrosion in bridges due to chlorination. *ICASP 8, Civil Engineering Reliability and Risk Analysis, Balkema, Rotterdam.*

Kounias, E. (1968). Bounds for the probability of a union with applications *SIAM J.Appl. Maths*, **39**, 2154-2158.

Moses, F and Kinser D.E. (1967). Analysis of structural reliability. *J. Stru. Div. ASCE*, **83**, 147-164.

Ramachandran, K. (1984). System bounds: a critical study. *Civil Engineering Systems*, **1**, 123-129.

Ramachandran, K. (1992). New developments in system bounds. *Internal report, Imperial College.*

Vanmarcke, E.H. (1973) Matrix formulation of reliability analysis and reliability based design. *Computers and Structures.* **3**, 757-770

ISOSAFETY PARAMETERS FOR FINK-TYPE STEEL ROOF TRUSSES

J.O. Afolayan and A. Ocholi

Civil Engineering Department, Ahmadu Bello University,
Zaria, NIGERIA

ABSTRACT

Isosafety parameters for a structural system with uncertain properties and acted upon by random loads are combinations of loads and geometrical properties leading to a prescribed reliability level against the occurrence of a specified limit state. Based on the design requirements of BS5950 (1985) and the First-Order Reliability Method (FORM), the safety indices for the individual members, joints which are assumed filet-welded and the entire truss system for varying inputs are determined. It is noted that the design criteria are grossly inadequate as the system failure level is extremely high. Consequently, a computer program capable of iteratively selecting design parameters for both members and joints that may be combined to ascertain an overall constant reliability level for the entire truss system is developed. The program is illustrated using a practical example.

KEYWORDS

Isosafety, reliability index, design requirements, sectional profiles, limit states, probability of failure, computer program

INTRODUCTION

For more than a decade, structural codes and standards for design of buildings in Europe and North America have moved toward a unified probability-based limit states approach with common load factors and load combinations for all construction materials (Ang and Cornell, 1974; Corotis, 1985). One important aspect of the approach is that a target reliability index is identified as a quantitative measure of acceptable performance in respect of a given limit state (Frangopol and Nakib, 1986). Although the development is quite elegant, its focus has been limited to the safety of individual members. Many studies (e.g., Ellingwood and Galambos, 1982; Chou et al, 1983; Ocholi, 2000) have shown that the reliability of a structure, considered as a system, differs from the reliability of its individual components.

In this paper, the implied reliability levels in the BS5950 (1985) criteria for individual members and joints of a Fink-type truss system are determined. Then using the system reliability procedure, the entire system safety level is estimated. Due to the inconsistencies noted in the reliability levels for the members, joints and the entire truss system for varying inputs, Fortran modules, based on FORM and the BS5950 (1985) design format are developed to iteratively select member profiles and joint details which are elements of the sets of limit state functions for the individual members and joints of the truss at a specified overall system reliability index. The combination of these parameters lead to the maintenance of a uniform target safety level.

FIRST-ORDER RELIABILITY METHOD (FORM)

The determination of the probability of failure of structural components and system has generated a great deal of interest. The basic procedures involved are briefly reviewed.

Component Analysis

The component analysis is an essential step in performing a system reliability analysis. FORM has been developed for calculating the necessary levels of probability of failure. Theoretical details of the method are presented in a number of publications with varying applications (e.g., Rackwitz & Fiessler, 1978; Madsen et al, 1986; Jang et al, 1994; Sitar et al, 1987).

In the analysis of a truss system, failure may be defined as the occurrence of the load effect being equal to or greater than the specified resistance of a member or strength of a joint. Therefore a reliability problem may then be formulated in terms of a limit state function g(**X**), where **X** is a vector of random variables, and g(**X**) ≤ 0 connotes a region in which the resistance is met or exceeded. A boundary called the limit state surface exists between the region in which the resistance is not exceeded and that in which it is exceeded. The probability of g(**X**) ≤ 0 can be obtained from

$$P[g(X) \leq 0] = \int_{g(X) \leq 0} f_X(x)dx \quad (1)$$

in which $f_X(x)$ is the joint probability density function of **X** and the n-fold integral is over the unsafe region. In practice, a direct numerical evaluation of Eqn. 1 is prohibitively cumbersome if not virtually impossible. Therefore an approximate method coded in Fortran modules (Gollwitzer et al, 1988) is preferred in this work. The procedure generates design points in a transformed space of standard normal variates. The shortest distance from the origin to the limiting surface in this space becomes the safety index, β, from which the probability of failure, P_f, is derived using

$$P_f = \Phi(-\beta) \quad (2)$$

so that $\Phi(\ )$ is the standard cumulative normal probability.

System Analysis

In reliability classification, a determinate pin-jointed structure is a series system. Thus, for a given series system which has a set of random variables **X** and a set of failure modes with limit state functions $g_i(X)$; i=1, ..., m, the failure probability is defined by

$$P_{SYS} = P[\cup_i g_i(X) \leq 0] \quad (3)$$

Assuming two modes of failure,

$$P_{SYS} = P[g_1(X) \leq 0 \cup g_2(X) \leq 0] \qquad (4)$$
$$= p_1 + p_2 + p_{12}$$

in which $p_j = P[g_j(X) \leq 0]$, $j = 1, 2$, are the individual probabilities of failure, and $p_{12} = P[g_1(X) \leq 0 \cap g_2(X) \leq 0]$ represents the joint probability of failure of modes 1 and 2. It is shown that P_{SYS} is bounded and details of this can be found in Ditlevsen (1979b) while the first-order approximation of the joint probability for any modes, say, i and j can be obtained from an integral expressed in Madsen et al (1986). Finally the generalised safety index for the system is deduced from (Ditlevsen, 1979a)

$$\beta_{SYS} = \Phi^{-1}(1 - P_{SYS}) \qquad (5)$$

DESIGN REQUIREMENTS AND SAFETY MARGINS

The designs of the members and joints of a Fink truss are to meet the requirements of BS5950 (1985). These requirements lead to the generation of performance or limit state functions from which the reliability indices are computed. As needed for this purpose, the safety margins, G(X), as a function of the basic variables **X** based on the design criteria are as follows:

For top chord members acting under axial and lateral loads.

$$G(X) = 1 - \frac{F}{A_g P_y} + \frac{M_x}{M_{cx}} \qquad (6)$$

For members in tension,

$$G(X) = P_t - F \qquad (7)$$

while for members in compression

$$G(X) = P_c - F \qquad (8)$$

Since welded joints are assumed, for the joints,

$$G(X) = P_{weld} - F \qquad (9)$$

In Eqns. 6 to 9, F = member forces, P_t = tensile capacity, A_g = gross cross-sectional area, P_c = compressive capacity, P_y = design strength, P_{weld} = weld capacity, M_x = applied moment in x-x axis and M_{cx} = moment capacity of the section about x-x axis.

ISOSAFETY DESIGN PROCEDURE

It is assumed that before the members and joints can be designed the complete truss has been analysed under the given load to determine the bar forces. The subsequent procedure is summarised as follows:

1. Perform a preliminary design for members and joints for assumed sections.

2. Conduct a reliability analysis to obtain the safety index, β

3. Check if β in Step 2 is approximately equal to a target value β_T. If $\beta \approx \beta_T$, move to Step 4, otherwise, return to Step 1.

4. List the design variables as the isosafety functions for members and joints.

ILLUSTRATIVE EXAMPLE

The proposed design procedure will be illustrated using a typical Fink Truss shown in Figure 1.

The Basic Structure and Loading

The example truss forms a part of a roof system over a building of 60x20m² in plan. The assumed dead and live loads are 0.50kN/m² and 0.75kN/m² (no access to the roof) respectively. The wind speed for the environment where the structure is recommended is 44m/s.

The load analysis showed that dead + imposed loads combination was more critical (Ocholi, 2000). Profiles commensurate with this loading were selected and tabulated in Table 1. For assumed distributions of the decision variables, the associated safety indices for varying failure modes for members and joints are calculated.

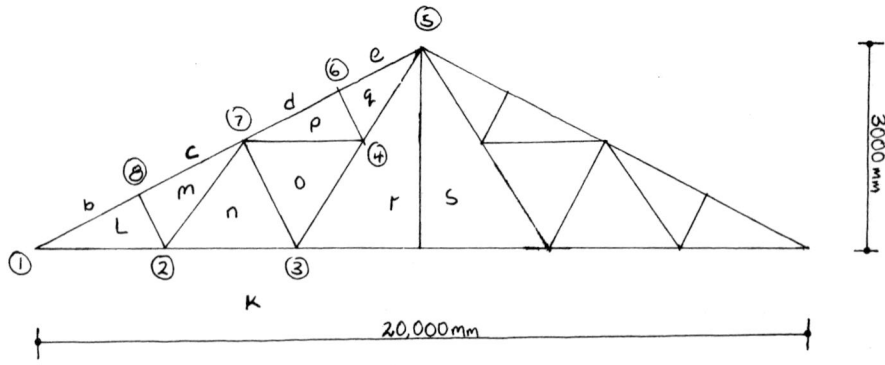

Figure 1: Example Fink Truss

TABLE 1
SECTIONS ADOPTED AFTER DETERMINISTIC DESIGN

Member	Section adopted	Serial size
Top chord member	T-section	146 x 127 x 22 kg/m
Bottom chord member	T-section	133 x 102 x 15 kg/m
Internal members (STRUTS)		
lm	L-section	45 x 45 x 5 kg/m
no	L-section	50 x 50 x 5 kg/m
pq	L-section	45 x 45 x 5 kg/m
Internal members (TIES)		
mn	L-section	45 x 45 x 5 kg/m
op	L-section	45 x 45 x 5 kg/m
qr	L-section	50 x 50 x 6 kg/m
or	L-section	50 x 50 x 6 kg/m

Component Analysis Results

The component analysis is an integral part of the system reliability analysis if FORM is used. Using the adopted profiles in Table 1, the numerical results of the component reliability analysis for some selected members and joints are shown in Figures 2 and 3 respectively. It is shown that reliability indices vary from member to member and joint to joint. As the ratio of dead to live load increases, the safety indices drop for the members and more significantly for the joints. Among the members, member qr appears to be the weakest with a dramatic loss of integrity as the load ratio augments. It is also clear that the joints are more critical than the individual members as their safety indices are generally much lower.

System Analysis Results

Table 2 gives the results of the system reliability analysis for varying load ratio. The β or P_f values correspond to any member or joint reaching its limit state and the consequence of that on the entire truss failure. The system safety indices catastrophically drop from 0.847 at $\alpha = 0.2$ to a negative value indicating a total collapse. These results signify that the BS5950 (1985) design criteria fall short of the established categorisation (NKB, 1978).

ISOSAFETY PARAMETERS

The calibration of the BS5950 (1985) design requirements specially applied to Fink-truss system cannot yield a reasonable target system safety index. As earlier recommended (NKB, 1978) a safety index range of 3.1 to 5.12 will be adequate for an ultimate limit state design while a range of 1 to 2 is sufficient for serviceability limit state. These ranges are excluded in the value of β_{SYS} in Table 2.

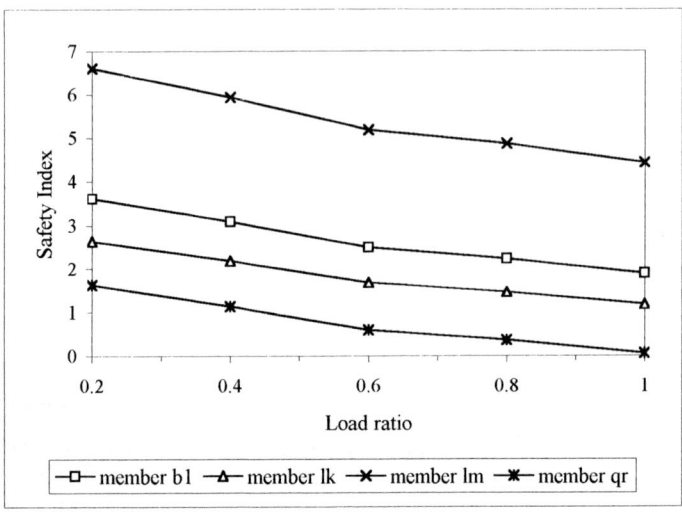

Figure 2: Safety indices against load ratio for selected members

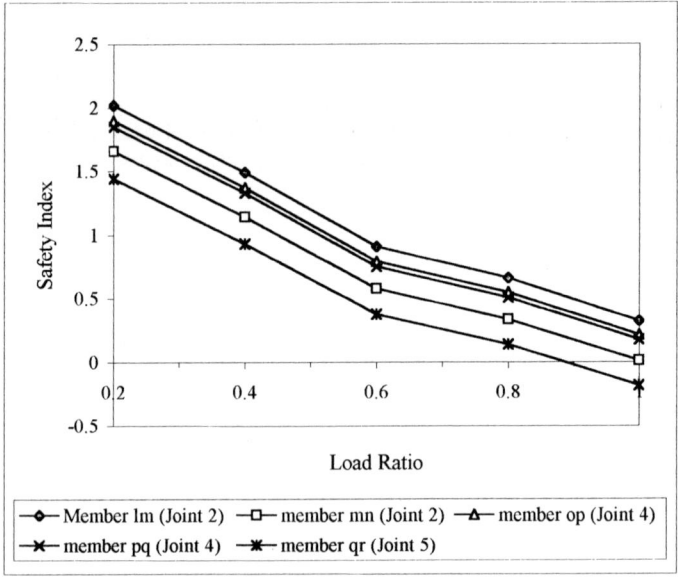

Figure 3 Safety indices against load ratio for selected joints

TABLE 2
SAFETY INDICES FOR THE ENTIRE SYSTEM

Load ratio	Members		Joints		System	
	β	P_f	β	P_f	β_{SYS}	P_f
0.2	1.567	0.0586	1.080	0.1400	0.847	0.1986
0.4	1.022	0.1534	0.488	0.3127	0.085	0.4662
0.667	0.324	0.373	-0.090	0.5358	-1.334	0.9080
0.8	-0.067	0.5226	-	-	-	-
1.0	-0.904	0.8170	-	-	-	-

For the illustrative example, a target safety index of 3.0 is assumed for the entire truss system. With the developed program modules, the parameters for a top chord member appropriate for the realisation of this target on a uniform basis are tabulated in Table 3 at a load ratio of 0.64. More details pertaining to other members and joints are in Ocholi (2000).

TABLE 3
COMPARISON OF PROGRAM AND ADOPTED STEEL PROFILE VALUES FOR A TOP CHORD MEMBER

Section parameter	Design values based on the program	Steel profile selected	% Difference
Gross area, A_g (mm^2)	2716.4	2670.0	1.71
Flange thickness, T (mm)	11.19	12.1	7.52
Section depth, d (mm)	161.15	153.3	4.87
Web thickness, t (mm)	8.0	8.0	0.0
Root radius, r (mm)	9.55	8.9	6.81
Flange breadth, B (mm)	140.55	124.3	11.56

CONCLUSION

A probabilistic approach has been followed to probe the adequacy of the BS5950 (1985) design requirements for members and joints of a Fink truss system. It is established from the results that the criteria for joints make them more prone to failure than the individual members of the truss. Thus, the overall safety level of the truss is grossly inadequate. In order to correct this seemingly alarming occurrence, program modules capable of selecting member and joint details for a constant overall system reliability level are developed. A demonstration example showed that the developed program is practically reliable.

REFERENCES

Ang A.H-S. and Cornell C.A. (1974). Reliability Bases of Structural Safety and Design, *Journal of Structural Division, ASCE,* **100**, 1755-1769.

BS5950 (1985). *Code of Practice for Design in Simple and Continuous Construction: Hot Rolled Sections, Part 1*, The Majesty's Stationery Office, London

Chou K.C., McIntosh C. and Corotis R.B. (1983). Observations on Structural System Reliability and the Role of Model Correlations, *Structural Safety*, **1**, 189-198.

Corotis R.B. (1985). Probability-Based Design Codes, *Concrete International*, **7**, 42-49.

Ditlevsen O. (1979a). Generalized Second Moment Reliability Index, *Journal of Structural Mechanics*, **7:4**, 435-451.

Ditlevsen O. (1979b). Narrow Reliability Bounds for Structural Systems, *Journal of Structural Mechanics*, **7:4**, 453-472.

Ellingwood B. and Galambos T.V (1982). Probability-Based Criteria for Structural Design, *Structural Safety*, **1**, 15-26.

Frangopol D.M. and Nakib R. (1986). Isosafety Loading Functions in System Reliability Analysis, *Computers and Structures*, **24:3**, 425-436.

Jang Y-S., Sitar N. and Der Kiureghian A. (1994). Reliability Analysis of Contaminant Transport in Saturated Porous Media, *Water Resources Research*, **30:8**, 2435-2448.

Madsen H.O., Krenk S. and Lind N.C. (1986). *Methods of Structural Safety*, Prentice-Hall, Englewood Cliffs, NJ

Melchers R. (1987). *Structural Reliability Analysis and Prediction*, Ellis Horwood Limited, Chichester, West Sussex, England

NKB (1978). *Recommendation for Loading and Safety Regulations for Structural Design*, Nordic Committee on Building Regulation, Report No. 36.

Ocholi, A. (2000). Reliability-Based Design of a Steel Roof Truss System, *Unpublished M.Sc. Thesis*, Ahmadu Bello University, Zaria, Nigeria

Rackwitz R. and Fiessler B. (1978). Structural Reliability under Combined Load Sequences, *Computers and Structures*, **9**, 489-494.

Sitar N., Crawlfield J.D. and Der Kiureghian A. (1987). First-Order Reliability Approach to Stochastic Analysis of Subsurfaces Flow and Contaminant Transport, *Water Resources Research*, **23:5**, 794-804.

COST-MODELING FOR THE ECONOMIC APPRAISAL OF JOINT DETAILS IN STEEL TRUSSES

J.O. Afolayan

Civil Engineering Department, Ahmadu Bello University,
Zaria, NIGERIA

ABSTRACT

Codes of practice are expected to provide or indicate tolerable levels of risk that yield equitable balance in benefits and costs. However, no non-probability-based design method accedes this expectation explicitly. Thus, the potential losses associated with the design requirements of BS5950 (1985) for joints of truss systems are examined. Variables describing loadings, material and geometrical properties are treated as random. Modules based on value analysis and the first-order reliability method are developed to highlight the trends of the potential losses as critical parameters are varied.

KEYWORDS

Potential loss, welded joints, probability of failure, truss system, random variables, cost analysis.

INTRODUCTION

The purpose of a design, is to ensure a low probability of getting action values higher than the resistances. It is not easy to obtain a specific value of this low probability because there is often a conflict between simple but sometimes conservative models and the more complicated models that may better reflect the behaviour of a system though with a higher risk of making errors and overlooking failure modes (Larsen, 1995). Generally joints require large areas of contact which may eventually give rise to local eccentricities not designed for.

Hicks (1987) considered the effects of welding on the shape of a component or structure along with the setting up of residual stress patterns. It is remarkable that both distortions and residual stresses have a significant bearing on the subsequent response of a component or structure to the design loadings. Hence, the proneness to failure may be higher than expected. This is more serious during the periods of high economic activities when attention may be directed at the construction of new physical structure which are brought into use as rapidly as possible (Lind, 1983).

Structural quality is dependent on human intervention at every stage of a building process. Thus, the performance of risk assessment from the design stage rather than sitting back till failure occurs is necessary (Villemeur, 1992).

In this paper, it is intended to economically appraise the design requirements of BS5950 (1985) for steel connections with specific reference to truss systems. Engineering phenomena are often associated with uncertainties of varying magnitudes, therefore variables affecting loading, material and geometry are considered random. With value analysis and reliability concept, the trends of potential losses inherent in the design criteria of BS5950 (1985) for joint details in two different truss layouts are highlighted.

DESIGN REQUIREMENTS FOR JOINTS

The requirements for the design of joints are well documented in BS5950 (1985). This investigation concerns joints that are filet-welded and the relevant condition is as follows.

For a welded connection, the condition for a satisfactory performance is

$$0.7\, s.L.P_w > F \tag{1}$$

where s = leg length, L = effective length, P_w = Design strength of weld and F = the member force to be transferred which is a function of the dead and live loads.

ECONOMIC APPRAISAL MODULE

The economic appraisal module consists of a cost model and a First-Order Reliability Procedure (FORP). In order to determine an expected potential loss, FORP estimates the probability of failure which is fed into the cost model.

Potential Loss Model

This model is developed to show trends rather than numerical details. In view of this limitation, the material and labour costs are not separated as done by Li et at (1997). They are rather built into a dimensionless expected initial cost parameter.

It is easily accepted to assert that the initial cost of any structure is directly proportional to its estimated resistance, while the risk cost decreases with the increasing design resistance. Mathematically, we state that:

$$T_c = I_c + P_c \tag{2}$$

in which T_c = the expected total cost, I_c = initial expected cost and P_c = risk cost. If we assume that the life-time failures are related to acceptable economic fluctuations, then Eqn. 2 can be rewritten in an expanded form as (Afolayan, 1999):

$$T_c = I_c + V_c r_f \sum_{i=1}^{N} P_{ci} P_{fi} \tag{3}$$

with

$$V_c = \frac{1 - e^{-(I-F)T_L}}{T_L(I-F)}$$

being a function which accounts for economic fluctuations and it depends on the interest rate I, inflation rate F, and the design life T_L. In Eqn. 3, r_f is the life-time failure probability related to the probability P_{fi} associated with the event due to cause i out of N resulting into failure and P_{ci} denotes the potential loss in respect of failure cause i.

Dover and Bea (1979) showed that

$$r_f = 1 - e^{-P_f T_L} \tag{4}$$

where P_f is the probability of a resistance being exceeded by the load effect. Therefore in a dimensionless form Eqn. 3 becomes

$$T_c^* = 1 + \frac{P_c}{I_c} r_f P_f \left(\frac{1 - e^{-(I-F)T_L}}{T_L(I-F)} \right) \tag{5}$$

so that $T_c^* = T_c/I_c$. The second term of Eqn. 5 represents the risk cost index and can be expressed as

$$R_I = \frac{1 - e^{-(I-F)T_L}}{T_L(I-F)} r_f P_f r_o \tag{6}$$

where $r_o = P_c/I_c$. Hence, Eqn. 6 yields the basis for appraising the potential loss associated with a joint failing when all quantifiable uncertainties in material, loading and geometrical properties are considered.

Failure Probability Estimation Model

It is reasonable to employ a probabilistic analysis in the assessment of safety of a structure when the actual scatter of each variable involved in determining its response to an uncertain loading is recognised. One of the probabilistic methods suitable for many such engineering problems is the well documented FORP (Madsen et al, 1986; Gollwitzer et al, 1988). In this method, if a function, Z, with the basic design variables z_i, i = 1, 2, ..., N represents the performance of a structure, the probability of failure can be estimated from

$$P_f = P(Z < 0) \tag{7}$$

Due to computational demands should Z have more than two variables, FORP employs some approximation which leads to the determination of a safety index

$$\beta = \min_{Z \in F_s} \left(\sum_{i=1}^{N} z_i^2 \right)^{1/2} \tag{8}$$

where F_s is the failure surface, Z = 0, in the z-coordinate system. Thus, Eqns. 7 and 8 are related in

1140

$$P_f = \Phi(-\beta) \tag{9}$$

with $\Phi(\ )$ being the standard cumulative function.

DEMONSTRATION PROBLEMS

Two different truss layouts (Fink and Pratt) as shown in Figures 1 and 2 respectively were used for illustration.

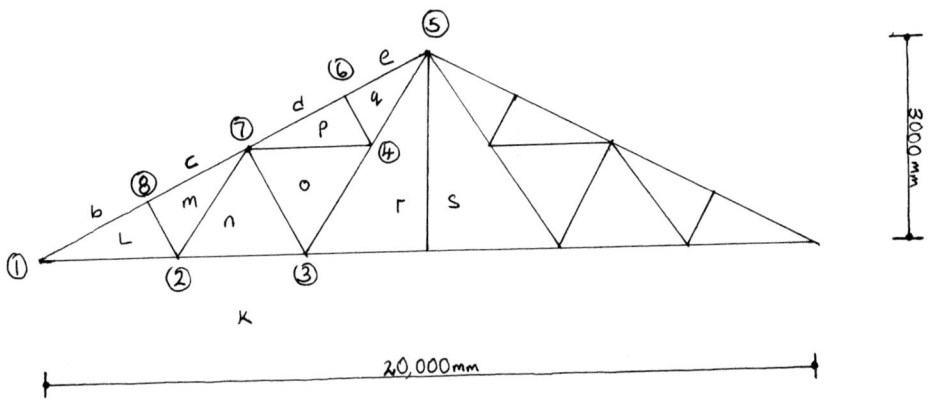

Figure 1: Example Fink truss

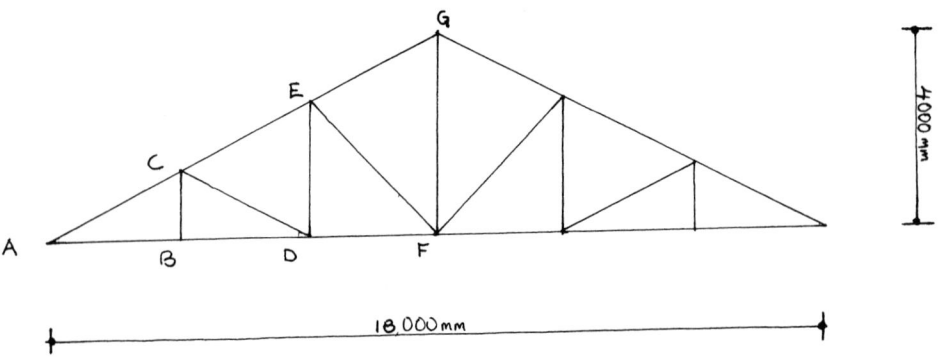

Figure 2: Example Pratt truss

The assumed loads are: dead load = 0.5kN/m² imposed load = 0.75kN/m² and wind speed = 44m/s. Load analysis shows that, the live plus dead load combination is the most critical. The adopted profiles after design are listed in Tables 1 and 2 for the Fink and Pratt trusses respectively.

TABLE 1
SELECTED PROFILES FOR THE FINK BARS

Member	Section adopted	Serial size
Top chord member	T-section	146 x 127 x 22 kg/m
Bottom chord member	T-section	133 x 102 x 15 kg/m
Internal members (STRUTS)		
lm	L-section	45 x 45 x 5 kg/m
no	L-section	50 x 50 x 5 kg/m
pq	L-section	45 x 45 x 5 kg/m
Internal members (TIES)		
mn	L-section	45 x 45 x 5 kg/m
op	L-section	45 x 45 x 5 kg/m
qr	L-section	50 x 50 x 6 kg/m
or	L-section	50 x 50 x 6 kg/m

TABLE 2
ADOPTED PROFILES FOR THE PRATT BARS

Member	Section adopted	Serial size
Top	Angles	2L 100 x 75 x 10 kg/m
Bottom	Angles	2L 80 x 60 x 8 kg/m
Tie and Strut	Angles	2L 60 x 60 x 10 kg/m

Starting with the various sectional properties listed in Tables 1 and 2 and employing well established probability distributions for loading, material and geometrical properties, the P_f implicit in BS5950 (1985) criteria for filet-welded joints were estimated. Consequently, the associated potential losses were calculated. It is assumed that I - F = 10%, T_L = 50 years and r_o = 50. Other values of these parameters may be chosen but due to space limitation the relevant details in that direction are excluded in this work.

The results in Tables 3 and 4 show clearly that the potential losses associated with the BS5950 (1985) requirements for joints are dependent on truss configuration. It is difficult to do a direct comparison between the two truss systems selected because the profiles adopted after design slightly differ. However, it is observed that the degradation of the safety indices with increasing load ratio for the Fink truss is much higher than for the Pratt. As reiterated by Racher (1995), the joint detail for a Pratt truss is simpler than for a Fink and as such the result so far may be considered realistic.

TABLE 3
POTENTIAL LOSSES FOR SELECTED JOINTS IN FINK TRUSS

Joints	Load ratio	Member			
		lm		mn	
		β	R_I	β	R_I
2	0.2	2.02	0.14	1.44	0.44
	0.4	1.49	0.65	0.58	1.26
	0.6	0.91	1.81	0.34	2.80
	0.8	0.66	2.52	0.01	3.67
	1.0	0.32	3.70	-	-
4		op		pq	
		β	R_I	β	R_I
	0.2	1.90	0.22	1.85	0.25
	0.4	1.37	0.84	1.33	0.90
	0.6	0.79	2.13	0.75	2.24
	0.8	0.55	2.92	0.51	3.04
	1.0	0.22	4.12	0.18	4.27

TABLE 4
POTENTIAL LOSSES FOR SELECTED JOINTS IN PRATT TRUSS

Joint	Load ratio	Member			
		1		2	
		β	P_L	β	P_L
A	0.2	3.64	0.000009	3.91	0.000001
	0.4	3.22	0.0002	3.49	0.00003
	0.6	2.86	0.0022	3.13	0.00038
	0.8	2.53	0.014	2.80	0.0031
	1.0	2.24	0.059	2.51	0.016

The results on Tables 3 and 4 are quite revealing. The rate of degradation of structural integrity is much higher for the Fink truss than the Pratt truss should any joint fail. It is indicative that the rate of increasing the potential loss associated with joint failure is alarmingly higher for the Fink truss than the Pratt truss. For example, in the neighbourhood of load ratio being equal to unity, a 20% increase in load ratio increases the potential loss by 120% for the Fink truss while it is 4.5% for the Pratt truss. Meanwhile it appears more attractive to solicit for welded joints in Pratt trusses than Fink Trusses.

CONCLUDING REMARKS

The cost implication of BS5950 (1985) requirements for welded joints in structural steel truss systems has been investigated using the concepts of value analysis and FORP. Two different truss layouts are considered and the results highlight the trend of potential losses associated with loading intensity. It is also established that the simpler the joints of a truss the safer the truss system and the less the potential loss. In fact, the rate of degeneracy of the reliability of a truss is dependent on the complexity built into its joints. In order to minimise loss, simpler joint details are recommended for the application of BS5950 (1985). Otherwise, the current criteria may require a re-appraisal.

REFERENCES

Afolayan J.O. (1999). Economic Efficiency of Glued Joints in Timber Truss System, *Building and Environment*, **34:2**, 101-107.

BS5950 (1985). *Part 1: Code of Practice for Design in Simple and Continuous Construction: Hot Rolled Sections*, Her Majesty's Stationery Office, London

Dover A.R. and Bea R.G. (1979). Application of Reliability Methods to the Design of Coastal Structures, *Paper presented at the Coastal Structures '79 Conference*, Alexandria, Virgina, March 14-16, 6-17.

Gollwitzer S., Abdo T. and Rackwitz R. (1988). *First-Order Reliability Method (FORM5)*, RCP GmbH, Munich

Hicks J.G. (1987). *Welded Joint-Design*, 2nd Edition, BSP Professional Books, Oxford

Larsen, H.J. (1995). Limit State Design and Safety Format , in *Timber Engineering (STEP1)*, Ed. Blass H.J. et al, Centrum Hont, the Netherlands, A2/1 - A2/8.

Lind N.C. (1983). Control of Structural Quality, in *Reliability Theory and its Application in Structural and Soil Mechanics,* Ed. Thoft-Christensen P., Martinus Nijhoff Publishers, The Hague, 225-235.

Li T.Q., Nethercot D.A. and Tizani W.M.K. (1997). Integrated Design System for Semi-rigidly Connected Steel Frames, *Advances in Structural Engineering*, **1:1**, 47-61.

Madsen H.O., Krenk S. and Lind N.C. (1986). *Methods of Structural Safety*, Prentice-Hall, Englewood Cliffs, NJ

Racher P. (1995). Mechanical Timber Joints General, in *Timber Engineering (STEP1)*, Ed. Blass H.J. et al, Centrum Hout, The Netherlands, C1/1-C1/10.

Villemeur A. (1992). *Reliability, Maintainability and Safety Assessment - Vol 2*, John Wiley and Sons, Chichester

STRUCTURAL OPTIMISATION

A GA APPROACH FOR SIMULTANEOUS STRUCTURAL OPTIMIZATION

M. Kahraman and F. Erbatur

Department of Civil Engineering, Middle East Technical University,
Ankara, 06531, Turkey

ABSTRACT

This paper deals with simultaneous layout structural optimization of trusses using genetic algorithms (GAs). The inclusion of shape and topology variables in addition to size variables makes the optimization problem highly complex. This is because, firstly, the number of design variables is increased and secondly, a simultaneous treatment of design variables of different types requires working on quite complicated design spaces. In view of the computational efficiency of the GA, an adaptive 3-phase search approach is proposed to use in conjunction with the GA. Several examples from the literature are solved and compared with the solution algorithm proposed in the paper. Besides, optimum designs with respect to only size and both size and shape design variables are also obtained for the example problems to examine the effects of considering different variables on the optimum solution.

KEYWORDS

Optimisation, Structural layout optimisation, Genetic Algorithms, Simultaneous optimum design, Adaptive three phase search approach.

INTRODUCTION

Layout structural optimum design of skeletal structures concerns simultaneous optimization with respect to size, shape and topology variables. Such a consideration results in a very large design space and requires simultaneous treatment of the design variables of different natures, i.e., discrete variables for size optimization, continuous variables for shape optimization and integer variables for topology optimization. For the last few decades several artificial intelligence methods, in particular genetic algorithms (GAs) and simulated annealing (SA) have attracted a lot of interest of researchers in structural optimization, Hajela (1999). A main reason of suitability of GAs as an optimization method in layout optimization lies under its capability to treat all the above stated variables, and also it has been shown that GAs are flexible and versatile methods with good problem solving capacities, Adeli & Cheng (1994), Shrestha & Ghaboussi (1998).

This paper is concerned with the use of GAs in simultaneous optimum layout design of trusses, where the objective is to minimize the weight imposed by a set of constraints controlling the structural behaviour. Besides, an increased computational efficiency of the GAs in this course is emphasised. Accordingly, the so-called adaptive 3-phase search approach is proposed and incorporated into the solution algorithm. The main idea behind this approach is to separate the whole process into three phases, where the experience gained over the development of the best design in a former phase is utilized to adaptively restrict the design space for the following search phases. This is achieved by reducing the number of topology variables at the end of the first phase and by narrowing the design sets of shape variables at the end of the second phase. By this way, it is aimed at evolving a more rapid and favourable solution under restricted number of generations. Two numerical problems taken from the literature are used (i) to compare the GA with other techniques, (ii) to test the success of the adaptive 3-phase search in comparison to classical approach, and (iii) to analyze the influence of considering different design variables (only size variables, size and topology variables size and shape variables, size and shape variables plus topology variables) on the optimum solution.

SIMULTANEOUS SIZE, SHAPE AND TOPOLOGY OPTIMIZATION PROBLEM

The simultaneous size, shape and topology structural design optimization problem is posed as follows:

Find the set of size (A), geometry (R) and topology (T) variables

to minimise $W(A,R,T)$

subject to $G_j^L \leq G_j(A, R, T) \leq G_j^U \quad j = 1, 2, .., q$

where, A and R the vectors of member cross-sectional areas and joint coordinates, respectively. The vector T represents the set of existing or non-existing members. W is the objective function, which is taken as the weight or volume of the structure. The superscripts L and U denote lower and upper limits, respectively. G_j's are the constraints that limit the pertinent design variable sizes, structural responses and selected design variable domains. Accordingly, the following constraint functions can be identified: member cross-sectional areas, joint coordinates, member topological variables domain, member allowable stresses, member buckling strengths and joint allowable displacements.

GENETIC ALGORITHMS

Generation of an initial population of individuals (designs) is the first step in the implementation of any GA. This is done randomly by the algorithm itself as starting solutions of optimization process. After the creation of the initial population, each individual is assigned a numerical fitness value reflecting how good it is compared to other individuals of the population, Michalewicz (1996). Afterwards, the selection-reproduction scheme of a GA is implemented, where individuals of higher fitnesses are selected to undergo a series of genetic operations with generally more than one copy (reproduction), Goldberg (1989). The selected and reproduced individuals are then manipulated to provide the next generation. The basic genetic operators applied here are crossover and mutation. Crossover is the mechanism by which design characteristics between any paired individuals are exchanged to form two new (child) individuals. Mutation is applied on the genes of child individuals after crossover, altering a gene of '1' to '0' or vice-versa for a random investigation of the design space. Crossover and mutation are activated according to chosen probability values. Probability of crossover is typically taken around 0.90. However, it is kept low for mutation (around 0.01) to avoid a severe damage on the chromosomal structures of child individuals. A GA cycle is completed with the generation of a new population. The algorithm repeats itself until a specified stopping criterion is met.

Adaptive Three Phase Search Approach

GAs aim at achieving a wide search over the entire design space to avoid getting stuck in a local optimum. However, there is no direct mechanism in GAs to avoid revisiting those regions which have been discovered and are found unfavourable. In other words, every solution -no matter good or not- is a potential search point from the start to the end of the optimization process. This consideration obviously reduces the efficiency of the algorithm, allowing also the search to be uncontrollably shifted to the inappropriate regions. In this study a proposed adaptive 3-phase search approach is incorporated into the solution algorithm to eliminate this drawback. The main idea underlying this approach is to reduce the design space in a systematic way by making use of the experience gained through the search process.

In this approach, firstly the optimization process is divided into three equal phases over the generation number. In the first phase, the search is carried out with the initial design sets assigned for all the design variables, complying with the original design space. Upon the completion of the first phase, a design space reduction is performed over the topology set of the problem. For this purpose, the best design is taken and examined for topological variables and those structural members appearing in the topological model of this design are no more considered as topological variables in the following phases. Here, the goal is to identify those structural members the existences of which are necessary for the stability of the truss and to impel the subsequent designs to include those members so as to have stable topology configurations. Thus, for the second phase while the design space is reduced by eliminating those topology variables, the search is prevented to move to those regions incorporating unstable topology configurations. At the end of the second phase, the best design is taken but this time is examined for the shape variables. Here, a further reduction of the design space is performed by changing the ranges for the shape variables. For a shape variable this is done by assigning new design interval which is one fourth length of its initial interval with the centre moved to the value of the variable in the best design. Once again, the search ability is improved by narrowing the joint coordinate ranges for the shape variables. And, in the third phase the emphasis is dominantly focused on sizing of structural members for the truss, the geometric model of which is more or less shaped up in the previous two phases.

ILLUSTRATIVE PROBLEMS

Two example problems taken from the literature are solved both (i) to carry out a comparative study with other techniques, and (ii) to test the efficiency of the adaptive search approach as compared to the classical treatment. Besides, in the examples the influence of the choice of design variables on the optimum solution is investigated by processing the algorithm with combination of different types of design variables. The following control parameters are used for both problems: population size = 50, generation number = 400, mutation probability = 0.005 and crossover probability = 0.90.

10-member plane truss

Figure 1 shows the initial geometry of the 10-member truss with the joint and member numbering details as well as loading. The truss is subjected to a stress limitation of 25ksi both in tension and compression, and the displacements of the joints are restricted to be less than 2 in. The following design variables are considered: cross-sectional areas of all truss members for size variables, the coordinates of the joints 1,3,5 in y direction for the shape variables and the presence/absence of the members 2, 5, 6, 7, 8, 9 and 10 for the topology variables. The size variables are chosen from a discrete set of 42 sections changing from 1.62 in^2 to 33.5 in^2. A complete and detailed list of design data is presented in Table 1.

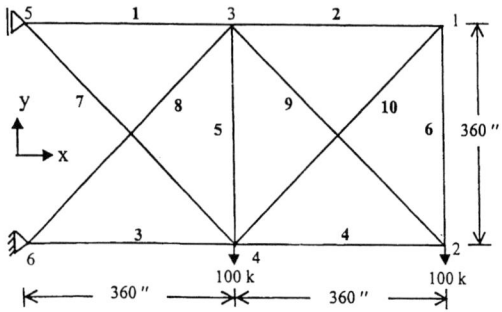

Figure 1: Initial geometry of 10-member plane truss

The following optimization cases are performed: (i) size optimization (SO), (ii) size and shape optimization (SSO), (iii) size and topology optimization (STO) and (iv) size, shape and topology optimization (SSTO). The optimum designs obtained in the four optimization cases and the comparison with the previous studies for the cases STO and SSTO are listed in Table 2. Also, for the cases STO and SSTO, the geometric models of the optimum designs obtained in this study are shown in Figure 2.

TABLE 1
DESIGN DATA FOR 10-MEMBER PLANE TRUSS

Design variables	
Size variables	$A_1; .. A_i ..; A_{10}$, $i = 1,..,10$
Shape variables	$Y_1; Y_3; Y_5$
Topology variables	$T_2, T_5, T_6, T_7, T_8, T_9, T_{10}$
Design set for size variables	
$A_i \in$ [1.62,1.80,1.99,2.13,2.38,2.62,2.63,2.88,2.93,3.09,3.13, 3.38,3.47,3.55,3.63,3.84,3.87,3.88,4.18,4.22,4.49,4.59,4.80, 4.97,5.12,5.74,7.22,7.97,11.5,13.5,13.9,14.2,15.5,16.0,16.9, 18.8,19.9,22.0,22.9,26.5,30.0,33.5], $i = 1,..,10$;	
Constraint data	
Allowable stress in tension	25 ksi
Allowable stress in comp.	25 ksi
Max. x-displacement	2 in
Max. y-displacement	2 in
Loading data	
Case Joints	F_x (kips) F_y (kips)
1 2, 4	-100 -100
Material properties	
Modulus of elasticity, E	1.0E4 ksi
Density, ρ	0.1 lb/in^3

Table 2 indicates that the improvement of the optimum design in relation to the choice of the types of design variables is as follows: SO < STO < SSO < SSTO. Concerning the comparative study for the case STO, the optimum design weight obtained in this study is 4976.43 lb, which is very close to those given in the literature. Amongst the other studies, while Schmit & Feleury (1980) and Rajeev & Krishnamoorty (1997), respectively, report better values of 4962.42 lb and 4925.80 lb for the same problem, a poorer optimum design weight of 4981.70 lb is reported in Rajan (1995). For the optimization case SSTO, the efficiency of the adaptive search approach is tested by manipulating the solution algorithm both in the absence and presence of this approach. Here, an optimum design weight of 2708.20 lb is found with the adaptive search algorithm, whereas the algorithm yielded an optimum design weight of 2810.36 in the absence of adaptive search. These two values are better as compared to Rajan (1995), where an optimum design weight of 3177.90 is reported for the problem.

TABLE 2
THE OPTIMUM DESIGNS AND COMPARISON TABLE FOR 10-MEMBER PLANE TRUSS

Design variables	Optimization case								
	SO	SSO	STO				SSTO		
	This study	This study	This study	Schmit (1980)	Rajeev (1997)	Rajan (1995)	This study	This study[†]	Rajan (1995)
Size and topology variables, in^2									
A_1	33.50	13.90	30.00	30.00	30.00	30.00	11.50	14.20	11.50
A_2	1.62	11.50	removed	removed	removed	removed	removed	removed	12.00
A_3	22.90	11.50	22.00	19.90	18.80	22.00	11.50	11.50	13.60
A_4	13.50	1.80	15.50	15.00	15.50	15.50	7.22	7.22	1.00
A_5	1.62	1.62	removed	removed	removed	removed	removed	1.62	removed
A_6	1.99	3.63	removed	removed	removed	removed	removed	removed	11.00
A_7	5.74	3.87	7.22	7.22	7.22	7.22	5.74	3.84	9.40
A_8	26.50	3.38	22.90	22.00	22.90	22.90	3.09	2.38	2.00
A_9	22.00	3.13	19.90	22.00	22.90	19.90	13.50	13.5	removed
A_{10}	1.62	5.74	removed	removed	removed	removed	removed	removed	9.40
Shape variables, in.									
Y_1	fixed	2.0	fixed	fixed	fixed	fixed	removed	removed	198.5
Y_3	fixed	470.0	fixed	fixed	fixed	fixed	485.0	475.0	545.0
Y_5	fixed	784.0	fixed	fixed	fixed	fixed	783.0	865.0	790.0
Weight, lb	5448.34	2824.10	4976.43	4962.42	4925.80	4981.70	2708.20	2810.36	3177.90

[†] GA algorithm without adaptive three-phase search approach.

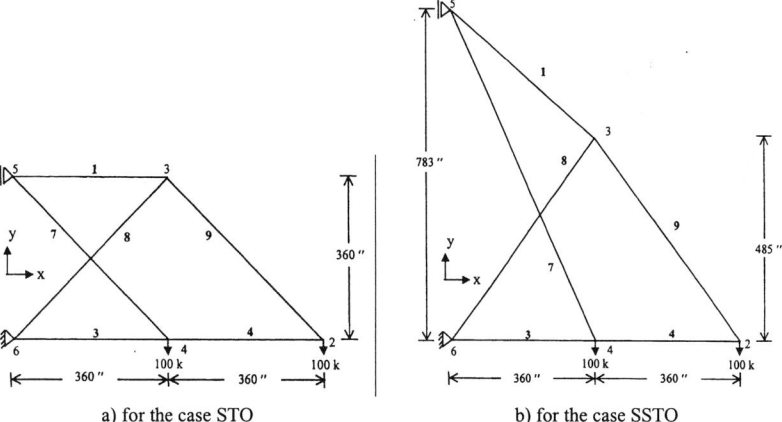

Figure 2: Geometric models of the optimum designs of 10-member truss a) for the case STO, and b) for the case SSTO.

25-Member Plane Truss

The second example problem is the 25-member space truss, initial geometry of which is shown in Figure 3. The independent size and shape design variables are assigned by grouping of structural members and linking of nodes in such a manner that symmetry of the structure along x-z and y-z planes is also retained in the optimum design geometry, and no topology variable is considered. Accordingly, 25 truss members are grouped to yield eight independent size variables, as shown in Table 3 and five independent shape variables are chosen as X_4, Y_4, Z_4, X_8 and Y_8. Side constraints for shape variables are set as follows: 20 in $\leq X_4 \leq$ 60 in, 40 in $\leq Y_4 \leq$ 80 in, 90 in $\leq Z_4 \leq$ 130 in, 35 in $\leq X_8 \leq$ 80 in, 100 in $\leq Y_8 \leq$ 140 in.

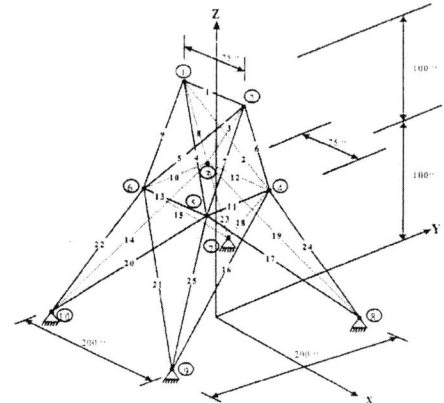

Figure 3: Initial geometry of 25-member space truss

TABLE 3
MEMBER GROUPING DETAILS FOR 25-MEMBER SPACE TRUSS

Independent size variable	Members
A_1	1
A_2	2 - 5
A_3	6 - 9
A_4	10, 11
A_5	12, 13
A_6	14 - 17
A_7	18 - 21
A_8	22 - 25

TABLE 4
DESIGN DATA FOR 25-MEMBER SPACE TRUSS

Design variables			
Independent size variables	$A_1; .. A_i ..; A_8$, $i = 1,..,8$		
Independent shape variables	$X_4; Y_4; Z_4; X_8; Y_8$		
Design set for size variables			
$A_i \in$ [0.1, 0.2, 0.3, 0.4, 0.5, 0.6, 0.7, 0.8, 0.9, 1.0, 1.1, 1.2, 1.3, 1.4, 1.5, 1.6, 1.7, 1.8, 1.9, 2.0, 2.1, 2.2, 2.3, 2.4, 2.5, 2.6, 2.8, 3.0, 3.2], $i = 1,..,8$;			
Constraint data			
Allowable stress in tension	40 ksi		
Allowable stress in comp.	40 ksi		
Max. displacement in all dir.	0.35 in		
Euler buckling constraints			
Loading data			
Case	Joints	F_x (kips)	F_y (kips)	F_z (kips)
1	1	1.0	-10.0	-10.0
	2	0.0	-10.0	-10.0
	3	0.5	0.0	0.0
	6	0.6	0.0	0.0
Material properties			
Modulus of elasticity, E	1.0E4 ksi		
Density, ρ	0.1 lb/in^3		
Buckling coefficient	12.5		

Two design cases are studied regarding the choice of behavioural constraints imposed on the problem. In the first design case, the truss is subjected to a stress limitation of 40ksi both tension and compression, and the displacements of the joints in any direction is restricted to be less than 0.35 in. In the second design case, in addition to the constraints taken in the first case, Euler buckling constraint is involved. The size variables are chosen from a discrete set of 29 sections changing from 0.1 in^2 to 3.2 in^2. A complete and detailed list of design data is presented in Table 4.

Both design cases are solved with respect to both SO and SSO optimization cases with respect to the choice of the design variables. The optimum designs produced for these cases and comparison with the previous studies are listed in Table 5. As is seen here, in all these cases the proposed algorithm gives better results in comparison to the literature results. For the second design case, the x-z and y-z plane projections of the initial and optimum shapes of the truss obtained in SSO is shown in Figure 4.

TABLE 5
COMPARISON TABLE FOR 25-MEMBER SPACE TRUSS

Design variable	Case 1: Stress and displacement constraints					Case 2: Case1 + Euler buckling const.			
	SO			SSO		SO		SSO	
	This study	Rajeev (1997)	Wu (1995)	This study	Wu (1995)	This study	Wu (1995)	This study	Wu (1995)
Independent size variables, in^2									
A_1	0.1	0.2	0.1	0.1	0.1	0.1	0.3	0.1	0.9
A_2	0.2	1.8	0.6	0.1	0.2	0.7	0.9	0.8	0.8
A_3	3.4	2.3	3.2	0.9	1.1	3.2	3.0	1.0	1.3
A_4	0.1	0.2	0.2	0.1	0.2	0.1	0.4	0.1	0.5
A_5	2.0	0.1	1.5	0.1	0.3	0.7	1.0	0.2	0.3
A_6	1.0	0.8	1.0	0.1	0.1	1.1	1.1	0.3	0.6
A_7	0.6	1.8	0.6	0.2	0.2	1.2	1.2	1.1	1.2
A_8	3.4	3.0	3.4	1.0	0.9	3.0	3.0	1.6	1.6
Independent Shape variables, in									
X_4	fixed	fixed	fixed	40.0	41.07	fixed	fixed	20.0	22.22
Y_4	fixed	fixed	fixed	55.0	53.47	fixed	fixed	50.0	49.01
Z_4	fixed	fixed	fixed	125.0	124.60	fixed	fixed	117.0	106.98
X_8	fixed	fixed	fixed	50.0	50.80	fixed	fixed	39.0	44.60
Y_8	fixed	fixed	fixed	134.0	131.48	fixed	fixed	120.0	102.44
Weight, lb	485.38	546.76	491.72	125.09	136.20	512.80	525.20	258.13	304.78

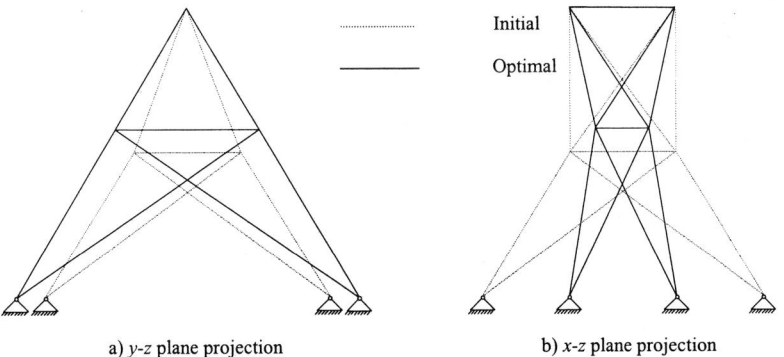

a) y-z plane projection   b) x-z plane projection

Figure 4: Comparison of the initial and optimum shape for 25-member space truss under stress, displacement and Euler buckling constraints a) on y-z plane projection, b) on x-z plane projection.

CONCLUSION

The paper is concerned with simultaneous layout structural optimization of trusses using GAs. A so-called 3-phase search algorithm where the number of topology variables are reduced at the end of the first phase and the design sets of shape variables are narrowed at the end of the second phase for increased computational efficiency is introduced. Two numerical examples taken from the literature are presented for illustrative purposes. Comparison of results only for size variables, size and topology variables, size and shape variables, and finally size, shape and topology variables reveals the efficiency of the proposed algorithm.

References

Adeli H. and Cheng N.-T. (1994). Concurrent Genetic Algorithms for Optimization of Large Structures. *Journal of Aerospace Engineering, ASCE* **7:3,** 276-296.

Goldberg D.E. (1989). *Genetic Algorithms in Search, Optimization and Machine Learning.* Addition-Wesley.

Hajela P. (1999). Nongradient Methods in Multidisciplinary Design Optimization – Status and Potential. *Journal of Aircraft* **36: 1,** 255-265.

Michalewicz Z. (1996). *Genetic Algorithms + Data Structures = Evolution Programs.* Springer-Verlag Berlin Heidelberg, New York.

Rajan D. (1995). Sizing, Shape and Topology Design Optimization of Trusses Using Genetic Algorithm. *Journal of Structural Engineering ASCE* **121:10,** 1480-1487.

Rajeev S. and Krishnamoorty C.S. (1997). Genetic Algorithms-Based Methodologies for Design Optimization of Trusses. *Journal of Structural Engineering ASCE* **123:3,** 350-358.

Schmit L.A. and Feleury C. (1980). Discrete-Continuous Variable Structural Synthesis Using Dual Methods. *AIAA Journal* **18,** 1515-1524.

Shrestha S.M. and Ghaboussi J. (1998). Evolution of Optimum Structural Shapes Using Genetic Algorithm. *Journal of Structural Engineering* **124:1,** 1331-1338.

Wu S.J. and Chow P.T. (1995). Integrated Discrete and Configuration Optimization of Trusses Using Genetic Algorithms. *Computers and Structures* **55,** 695-702.

GENETIC ALGORITHM OPERATORS FOR OPTIMISATION OF PIN JOINTED STRUCTURES

J. Mahachi[1] and M. Dundu[2]

[1]Division of Building & Construction Technology, CSIR
PO Box 395, Pretoria 0001, South Africa
[2]Department of Civil Engineering, University of the Witwatersrand
P Bag 3, Wits 2050, South Africa

ABSTRACT

Traditional Genetic Algorithms are based on binary-string structures and genetic structures such as selection, cross-over and mutation. For complex problems, however, the use of simple Genetic Algorithms (GA) is not efficient. In order to speed up convergence, hybrid solutions and more efficient genetic operators are required. This paper is concerned with the structural mechanics of pin-jointed structures. An investigation is carried out in order to determine the GA operators for the optimisation of pin-jointed structures. Numerical examples are illustrated to show the efficiency of the parameters compared to simple Genetic Algorithms and traditional mathematical programming techniques.

KEYWORDS

Genetic Algorithm, Optimisation, Pin-Jointed

INTRODUCTION

Structural optimisation of pin-jointed structures involves minimising the weight (or cost) of the structure subject to the members strength not exceeding the allowable stresses, and the deflection not exceeding the allowable limits. This problem involves choosing a sub-set of areas from a discrete set of available areas. As soon as the number of members increase to more than 5, the traditional gradient-based mathematical optimisation algorithms become more complex and cumbersome.

Several attempts to explore this problem has been presented by Yeh (1999), Adeli and Cheng (1993, 1994) using Genetic Algorithms (GA). Unlike gradient optimisation algorithms, GAs are a class of probabilistic algorithms that start with a population of randomly generated strings (chromosomes). The chromosomes represent a solution at hand, and evolve toward better solutions by applying genetic operators. The normal genetic operators used are the crossover, mutation and inversion. These operators will be described in the next sections.

One of the key issues in the implementation of a GA is the representation or coding of the chromosome. The most common coding method that has been reported in the literature (Jenkins (1991), Rajeev and Krishnamoorthy (1992)) is to transform the variables to a binary string of a specific length. The binary code technique, however, may lead to very large chromosomes, leading to high computational time. In this paper, we explore a different coding mechanism, which is applicable to any number of members in a pin-jointed structure. In addition, we investigate the capabilities and efficiency of different cross-over techniques to solve the optimisation problem.

PROBLEM FORMULATION

The Genetic Algorithm (GA) will be used to optimize a pin-jointed structure, with the objective of minimizing the weight of the structure subject to stress and displacement constraints.

If there are n available cross-sectional areas, and the truss has m members, then the optimisation problem can be written as:

$$\text{Minimise} \quad f = \sum_{i=1}^{m} \rho_i L_i A_i \quad (1)$$

$$\text{Subject to} \quad |\sigma_i| \leq \sigma_a \quad ; \quad i = 1, 2, \cdots m \quad (2)$$

$$\delta_j \leq \delta_a \quad ; \quad j = 1, 2, \cdots k \quad (3)$$

where
- ρ weight per unit volume
- σ_i stress in member i
- σ_a allowable stress in member. For this problem the allowable stress in compression and tension were assumed to be the same.
- δ_j displacement of joint j
- δ_a allowable displacement of joint
- A_i cross-sectional area of member i
- L_i length of member i

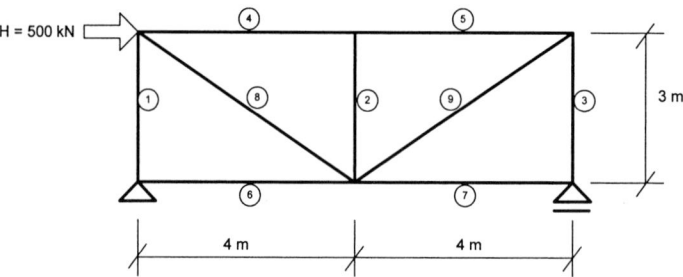

Figure 1: Truss configuration

The method of the GA will be illustrated by an example. A pin-jointed truss that was studied is shown in Figure 1, where the number of members m = 9.

It is assumed that 10 angle sections (ie n =10) are available, with the following cross-sectional areas (taken from the SAISC handbook (2000)).

$$\overline{A} = \{658, 719, 811, 901, 1020, 1270, 1550, 2160, 2320, 3390\} \quad (4)$$

This set of available areas is then coded as:

$$\overline{A} = \{1, 2, 3, 4, 5, 6, 7, 8, 9, 10\} \quad (5)$$

Other parameters:

$$\sigma_a = \pm 300 \text{ MPa}; \qquad E = 200 \times 10^3 \text{ N/mm}^2; \qquad \rho = 7850 \text{ kg/m}^3$$

The horizontal force, H = 500 kN, and the allowable horizontal displacement at the point of load application, δ_a = 15 mm.

The vector for the length of members is:

$$L = \{3, 3, 3, 4, 4, 4, 4, 5, 5\} \text{ metres} \quad (6)$$

GENETIC ALGORITHM TECHNIQUE

Coding Technique

Step 1: Chromosome representation

The first step in a GA is the coding scheme of the design variables. For the truss optimisation problem, it was decided to represent all the members as one chromosome (string), with n (=9) genes. Each gene represents the cross-sectional area of a member. This is illustrated as follows:

Chromosome number 1

5	9	10	2	4	1	6	7	3

(7)

The above chromosome has 9 genes. Each box represents a gene. For each gene, a random number is generated between 1 and $n(=10)$, where n is the number of available cross sectional areas. The set of available cross-sectional areas is given by Eqn. 4.

When chromosome number 1 is decoded, the following areas are obtained:

1020	2320	3390	719	901	658	1270	1550	811

(8)

The first gene (corresponding to member number 1) with a code of 5, has a cross-sectional area A_1 of 1020 mm^2 (See Eqns. 4 and 5). Similarly, the second member has a code of 9, and hence the cross-sectional area A_2 is 2320 mm^2.

The vector of cross-sectional areas for chromosome number 1 is therefore;

$$A^1 = \{A_1^1, A_2^1, \cdots A_m^1\}$$

i.e. $A^1 = \{1020, 2320, \cdots, 811\}$ \hfill (9)

where the superscript 1 refers to chromosome number 1.

Genetic Algorithms consider a group of chromosomes in the search space in every iteration, called a population p of chromosomes. The population normally is problem dependent and also depends on the speed of the processor used. For our problem, the population p was set at 50, i.e. 50 chromosomes were randomly generated.

Step 2: Evaluate the objective function

For each chromosome (i), the objective function f^i is evaluated as:

$$f^i = \sum_{j=1}^{m} \rho L_j A_j \quad ; \quad i = 1, 2 \cdots p \tag{10}$$

For chromosome number 1 given by Eqn. 8, the objective function is:

$$f^1 = 7850 \cdot [3(1020 + 2320 + 3390) + 4(719 + 901 + 658 + 1270) + 5(1550 + 811)] \times 10^{-6}$$
$$\therefore f^1 = 362.6 \text{ kg}$$

Step 3: Constraint violation

The force, and hence the stress in each member is calculated using simple equilibrium equations or by interfacing with a structural analysis software. The constraint Eqns. 2 and 3 can be re-written as:

$$g_i = \frac{|\sigma_i|}{\sigma_a} - 1 \tag{11}$$

$$h_j = \frac{\delta_j}{\delta_a} - 1 \tag{12}$$

In order to convert the constraint problem to the unconstrained problem, a penalty-based transformation method similar to that proposed by Rajeev and Krishnamoorthy (1992) and Yeh (1999) was used. However, in our case, we decided to use two penalty factors K_1 and K_2. K_1 is for the violation of stress constraint, and K_2 for the displacement violation. The stress violation occurs when $g_i > 0$, and displacement violation when $h_i > 0$. More details of the stress/displacement violation coefficients are given in the appendix.

Depending on the degree of violation, the objective function is then modified to give the modified objective function φ^i for each chromosome as follows:

$$\varphi^i = f^i(1 + K_1 C_1^i + K_2 C_2^i) \quad ; \quad i = 1, \cdots p \tag{13}$$

where C_1^i and C_2^i are stress violation and displacement violation coefficients (see appendix).

Step 4: Evaluate the fitness of each chromosome

The selection criteria for the next generation was based on the normalization technique, originally proposed by Cheng and Gen (1997). The fitness of each chromosome F^i was evaluated as:

$$F^i = \frac{\varphi_{max} - \varphi_i + \gamma}{\varphi_{max} - \varphi_{min} + \gamma} \quad ; \quad i = 1, \cdots p \tag{14}$$

where φ_{max} and φ_{min} are the best and worst raw fitness in current population, respectively and γ is a small positive number which prevents division by zero.

Once the fitness for all chromosomes have been defined, the Roulette wheel approach is adopted as the selection procedure for a new population of p chromosomes. Strings or chromosomes having larger fitness are more likely to be reproduced.

Step 5: Crossover

After reproduction, crossover of chromosomes proceeds. Crossover operator is the mating operator that allows production of new chromosomes through partial information exchange between pairs of chromosomes. In this analysis, an investigation of the performance of different operators was carried out, and these will be discussed in some detail.

a. Simple crossover (SX) technique

In this technique, a one cut off point is randomly selected for 2 parents. The left parts of the two parents are then exchanged to generate the off-spring.

Example: If the random cut off point for 2 parents is say 3, then the following 2 parents will generate the offspring as follows:

Parent 1 — Cut off point ↓

| 4 | 8 | 1 | 6 | 3 | 7 | 9 | 2 | 5 |

+

Parent 2

| 3 | 2 | 9 | 7 | 4 | 10 | 1 | 6 | 8 |

=

Offspring 1

| 3 | 2 | 9 | 6 | 3 | 7 | 9 | 2 | 5 |

+

Offspring 2

| 4 | 8 | 1 | 7 | 4 | 10 | 1 | 6 | 8 |

Note that, the genes 4, 8, 1 of parent 1 are interchanged with the genes 3, 2, 9 of parent 2 on a one to one basis.

b. Partially matched crossover (PMX)

For this technique, two cut off points are selected at random. The genes bounded by the cut off points are exchanged to produce two offspring. This is illustrated below, where the two cut off points are at position 3 and 6.

Parent 1 — Cut off point (3), Cut off point (6)

| 4 | 8 | 1 | 6 | 3 | 7 | 9 | 2 | 5 |

+

Parent 2

| 3 | 2 | 9 | 7 | 4 | 10 | 1 | 6 | 8 |

=

Offspring 1

| 4 | 8 | 1 | 7 | 4 | 10 | 9 | 2 | 5 |

+

Offspring 2

| 3 | 2 | 9 | 6 | 3 | 7 | 1 | 6 | 8 |

Genes 6, 3, 7 of parent 1 are interchanged with genes 7, 4, 10 of parent 2 on a one to one basis to produce the two offspring.

COMPUTATIONAL RESULTS

Case 1

A first analysis was carried out assuming that each member can take any of the available areas A_1, ... A_{10}. With a horizontal force of 500 kN and a deflection limit of 15 mm, the following areas were obtained using both the Simple Cross-over and the Partially Matched Cross-over techniques.

$$\overline{A} = \{A_1, A_2, A_3, A_4, A_5, A_6, A_7, A_8, A_9\}$$
$$\overline{A} = \{1020, 658, 719, 1270, 1550, 2160, 658, 2320, 2160\} \text{ mm}^2$$

The corresponding objective function (weight of the structure) is 409.3 kg. With these areas, all the stresses in the members are less than the allowable stress of 300 MPa, and the horizontal deflection is 15 mm. The following parameters were used in the GA.

$K_1 = 5$ $K_2 = 2$ $\gamma = 1$ Population of chromosomes = 50.

The SX technique converged after 30 generations, while the PMX converged after only 20 generations.

Case 2

The next study was performed assuming that all vertical members $\{A_1, A_2 \text{ and } A_3\}$ have the same cross-sectional area, and the rest of the members $\{A_4, A_5, A_6, A_7, A_8 \text{ and } A_9\}$ have the same area.

The following areas were obtained using both the SX and PMX techniques:

$$\{A_1, A_2, A_3\} = 719 \text{ mm}^2$$
$$\{A_4, A_5, A_6, A_7, A_8, A_9\} = 2160 \text{ mm}^2$$

For these area combinations, the objective function was 457.7 kg, the deflection was 15 mm, and all stresses were less than the allowable stress of 300 MPa. SX converged after 50 generations, and PMX after 30 generations.

To benchmark the solution, a manual calculation was performed using geometric mathematical programming technique. The results obtained are the same as those obtained using a GA.

CONCLUSIONS

The GA technique offers an easy methodology to optimize pin-jointed structures. The coding mechanism suggested in this paper allows any number of members to be analysed without making the chromosome length very long. Further, the PMX approach reduces the computational time compared to the SX technique. In this algorithm, mutation of genes has not been considered. The mutation operator alters the value of a chromosome to create a completely new chromosome. This operator is occasionally introduced so as to increase the variability of the population.

REFERENCES

Adeli H. and Cheng N.T. (1993). An augmented Langragian genetic algorithm for structural optimization. Aerospace Engineering, ASCE, **7:1**, 104-118.

Adeli H. and Cheng N.T. (1994). Concurrent genetic algorithm for optimization of large structures. Aerospace Engineering, ASCE, **7:3**, 276-296.

Gen M. and Cheng R. (1997). Genetic Algorithms and Engineering Design, John Wiley and Sons, New York.

Jenkins W.W. (1991). Towards structural optimization via the genetic algorithm. Computers and Structures, **40:5**, 1321-1327.

Rajeev S. and Krishnamoorthy C.S. (1992). Discrete optimization of structures using genetic algorithms. Structural Engineering, ASCE, **118:5**, 1233-1250.

SAISC handbook (2000). Southern African Steel Construction Handbook. Southern African Institute of Steel Construction, Johannesburg, SA.

Yeh I.C. (1999). Hybrid genetic algorithms for optimization of truss structures. Computer-Aided Civil and Infrastructure Engineering. **14**, 199-206.

APPENDIX

Determination of stress violation penalty factor

For i = 1 to p (population of chromosomes)
    For j = 1 to m (number of members)

$$\sigma_j = F_j/A_j \qquad (F_j \text{ is the force in member j})$$

$$g_j = \frac{|\sigma_j|}{\sigma_a} - 1$$

If $g_j > 0$ then $c_{ij} = g_j$, Else $c_{ij} = 0$

Next j

$$C_i^1 = \sum_{j=1}^{m} c_{ij}$$

Next i

Determination of displacement violation penalty factor

For i = 1 to p

$$\delta_i = \frac{H}{64E}\left[\frac{9L_1}{A_1} + \frac{0L_2}{A_2} + \frac{9L_3}{A_3} + \frac{16L_4}{A_4} + \frac{16L_5}{A_5} + \frac{64L_6}{A_6} + \frac{0L_7}{A_7} + \frac{25L_8}{A_8} + \frac{25L_9}{A_9}\right]$$

$$h_i = \frac{\delta_i}{\delta_a} - 1$$

If $h_i > 0$ then

$$C_i^2 = h_i$$

Else

$$C_i^2 = 0$$

Next i

USE OF A STOPPER FOR THE STRESS REDUCTION IN BEAM-BLOCK SWITCHING SYSTEMS OF AUDIO PRODUCTS

K. H. Low[1] and H. P. Sin[2]

[1]School of Mechanical and Production Engineering
Nanyang Technological University, Singapore 639798, Republic of Singapore
[2]Philips Singapore Pte. Ltd. 620A Lorong 1 Toa Payoh, Singapore 319762,
Republic of Singapore

ABSTRACT

A set of buttons is designed and attached to the cabinet of mini Hi-Fi audio products. The press buttons provide a series of operating functions of the audio product. The button system is modeled by a series of beam-blocks. Each beam-block model can be represented by a rigid block attaching to two identical stepped beams. The stepped height of the beam is to provide the spring-back effect. The spring-back effect is to enable the rigid block going back to its original position after the pressing force is released. It will however cause a beam breaking near narrow neck due to the stress concentration, if a higher force is continually applied to the block. A stopper is therefore suggested to reduce the stress and thus lower the chance of beam breaking. The stopper is placed below the rigid block with an initial gap from the ground. In fact, the model is a statically indeterminate structure owing to the presence of the stopper and a switch at the end of the block.

In this work, symbolic expressions of the beam-block deflection and stress distribution are obtained by solving different sets of simultaneous equations. Experimental beam deflection is compared with the numerical results. The effect of different loading parameters on stress distribution, with and without stopper, is studied and presented. Results of different locations of the stopper and the pressing force are given for discussion. The results show that installing a stopper can generally reduce the beam stress. The stopper height is another parameter to be studied, as it may reduce the spring-back effect if not properly designed.

KEYWORDS

Stress reduction, stopper, indeterminate structure, beam, deflection, buttons, parametric study.

INTRODUCTION

The press buttons of an audio Hi-Fi product is quite crucial to the operating function of the set. The button system can be modeled by a series of beam-blocks. Each beam-block model is a rigid block carried by two identical stepped beams. The stepped height of the beam is to provide the spring-back effect. The effect is to enable the rigid block going back to its original position after the pressing force is released. It will however cause a beam breaking near narrow neck due to the stress concentration, if a high force is continually applied to the block. A stopper is therefore suggested to reduce the stress and thus lower the chance of beam breaking. In this work, symbolic expressions of the beam-block deflection and stress distribution are obtained by solving different sets of simultaneous equations. Experimental beam deflection is compared with the numerical results. The effect of different loading parameters on stress distribution, with and without stopper, is studied and presented. Results of different locations of the stopper and the pressing force are given for discussion.

MODEL CONSIDERED

Figure 1 shows six buttons in series that are attached to the cabinet of mini Hi-Fi audio products. Each button system is of the similar design with different dimensions of button and the two identical hinges (i.e. dual-hinge). Figure 2 shows a single button system that consists of button, end-switch and dual-hinge. Note that the end-switch is to provide tactile sensing upon the pressing. As mentioned, the dimensions in the figure are design parameters that can be different for each button system (see Figure 1).

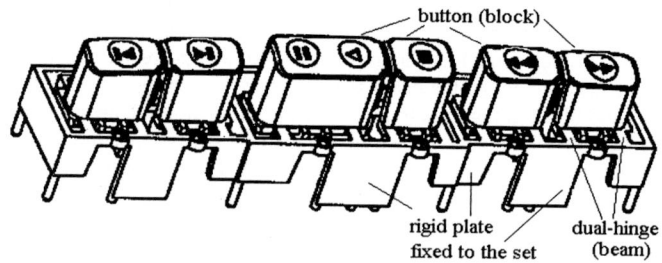

Figure 1   Series of buttons

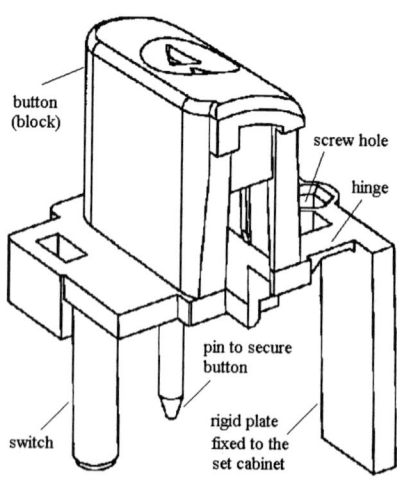

Figure 2   Details of a single button

Simplified model

A simplified model is used in the present analysis, as shown in Figures 3 and 4. Each stepped cantilever beam that is rectangular in cross section and of length L carries a button at its free end. The button is considered as a block of rigid mass. Pressing force F is applying at an arbitrary position of the button. A spring-like switch is connected at the free end of the block. Furthermore, a stopper is located below the block at a desired position. Dimension parameters and location of forces are designated in Figure 3. Note that $b_{AB} = 2b_c$. It is obvious that $a_r = l_r - L$, $a_f = l_f - L$, and $a_s = l_s - L$. Also, δ_s and δ_r are the gap (or height) of the stopper and switch from the ground, respectively. Beams A and B are identical. The beam heights h_i ($i = 1$ to 4) are varied along the longitudinal axis x in different span lengths s_i.

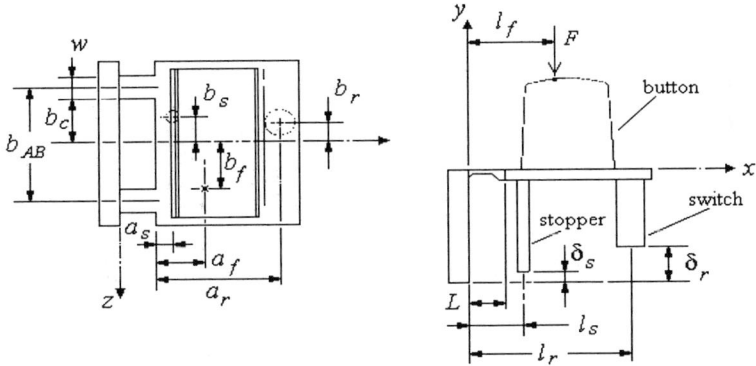

Figure 3   Dimensions and variables of the beam/block system

PROBLEM FORMULATION

The model shown in Figure 4 is in fact a statically indeterminate structure owing to the presence of the stopper and the end-switch. The complete problem solving will then require both displacement and force analyses.

Displacement analysis

Beam deflection
The changing beam-height along the x-axis is defined by a height ratio
$$n_i = h_{i-1} / h_i, \quad i = 1 \text{ to } 4. \tag{1}$$

The expression for the beam height along the x-axis can thus be obtained [1, 2],
$$h(x) = h_{i-1} + (h_i - h_{i-1})x/s_i = h_{i-1}e(x) = h_i g(x), \quad i = 1 \text{ to } 4, \tag{2}$$

in which
$$g(x) = n_i + \frac{x}{s_i}(1 - n_i) \quad \text{and} \quad e(x) = 1 + \frac{x}{s_i}\left(\frac{1}{n_i} - 1\right) = g(x)/n_i, \quad i = 1 \text{ to } 4. \tag{3}$$

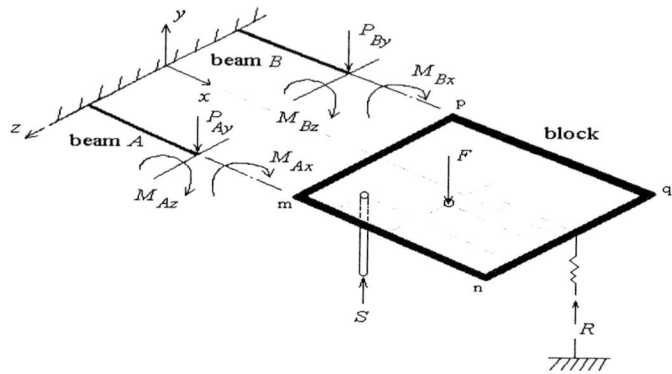

Figure 4   Free-body diagram of the beam/block system

The deflection of the cantilever beam in different span length is given by [3-7],

$$y_{xi} = \frac{1}{EI_z} \iint M_{ei}\, dxdx + C_{1i} \int dx + C_{2i}, \quad l_{i-1} < x < l_i, \quad i = 1 \text{ to } 4, \tag{4}$$

in which M_{ei} is the total moment owing to the force/moment (P_y/M_z) applied at the free end [4, 5]:

$$M_{ei} = -P_y(L-x) - M_z, \quad i = 1 \text{ to } 4. \tag{5}$$

Note that I_z is the second-area moment inertia of the beam's cross section relative to the z-axis. Also, $l_i = \sum_{j=1}^{i} s_j$ ($i = 1$ to 4) in Eq. (4) and $l_0 = 0$.

Block deflection
The deflection of the block edge (line mn or pq in Figure 4) extending from the beam is taken as

$$\begin{aligned} y_{5s1} &= C_{15s1}x + C_{25s1}, \quad l_4 < x < l_s, \\ y_{5s2} &= C_{15s2}x + C_{25s2}, \quad l_s < x < l_r. \end{aligned} \tag{6}$$

Beam/block deflection
After solving different sets of equations involving the unknown constants C, the deflection of the beam/block from the fixed end to the block free-end can then be written in a general form as

$$y_{bb} = y_{x1} + y_{x2} + y_{x3} + y_{x4} + y_{5s1} + y_{5s2}, \quad 0 < x < l_r. \tag{7}$$

The results of the respective deflections in terms of P_y and M_z are obtained as

$$y_{x1} = y_{x2} = y_{x3} = y_{x4} = \frac{P_y(x^3 - 3Lx^2) - 3M_z x^2}{6EI_z};$$

$$y_{5s1} = y_{5s2} = \frac{L(2l_s P_y L^2 + 3l_s M_z L + 6EI_z \delta_{bks}) - x(2P_y L^3 + 3M_z L^2 + 6EI_z \delta_{bks})}{6EI_z(L - l_s)}, \tag{8}$$

where δ_{bks} is the corresponding block deflection at $x = l_s$ due to the stopper at $\{l_s, b_s\}$. Note that the term I_z is varied with respect to x. Furthermore, the generalized forces P_y and M_z associated to the beam can be found in a force analysis, as will be presented next.

Force Analysis

Figure 4 is the free-body diagram of the two beams with the block. For clarity, only the internal forces and moment at the beam-ends are shown. For example, P_{By} is the reaction from the block to beam B in y-direction. The applying force (F) to the block, reaction forces from the ground to the stopper (S) and the end-switch (R) are also shown in Figure 4. By virtue of the free-body diagram of the block and taking into account the specified directions, seven equations have been used to solve the reaction forces/moments at the beam-ends and the switch, as listed below:

$$\begin{aligned}
&\text{eq1 } (\Sigma F_y = 0): \quad P_{Ay} + P_{Ay} - F + R + S = 0 \\
&\text{eq2 } (\Sigma M_{x\_mn} = 0): \quad -F(b_c - b_f) + R(b_c - b_r) + S(b_c - b_s) + M_{Ax} + M_{Bx} + P_{By} b_{AB} = 0 \\
&\text{eq3 } (\Sigma M_{z\_mp} = 0): \quad M_{Az} + M_{Bz} - F a_f + R a_r + S a_s = 0 \\
&\text{eq4 (constraint 1): } \quad M_{Az} = M_{Bz} \\
&\text{eq5 (constraint 2): } \quad M_{Ax} = M_{Bx} \\
&\text{eq6 (constraint 3): } \quad \delta_{SAe} - \delta_{SBe} = b_{AB} M_{Bx} /(GJ_p) \\
&\text{eq7 (constraint 4): } \quad \frac{\delta_{SAe} - \delta_{SBe}}{b_{AB}} = \frac{\delta_r - \delta_{SBe}}{b_c + b_r}
\end{aligned} \qquad (9)$$

gf_sols $\Rightarrow$ solve {eq1, eq2, eq3, eq4, eq5, eq6, eq7} for $\{P_{Ay}, P_{By}, M_{Ax}, M_{Bx}, M_{Az}, M_{Bz}, R\}$

Note that the reaction force (S) to the stopper is taken as [5, 6]

$$S = \frac{3EI_r \delta_s}{l_s^3}, \qquad (10)$$

in which I_r is the second-area moment inertia of the cross section, relative to the z-axis, for the switch.

It is worth mentioning that constraints 3 and 4 in eqn. (9) are obtained by assuming the linear displacement of rigid block due to the switch height δ_r. The rigid-block deflection (d) along the z-axis can be related to the end deflections (d_A and d_B) according to the assumption of linear displacement.

By using the same concept, we can relate the displacements, δ_s, δ_{bksA} and δ_{bksB}:

$$\begin{aligned}
&\text{eq1bks}: \quad \frac{\delta_{bksA} - \delta_{bksB}}{b_{AB}} = \frac{\delta_s - \delta_{bksB}}{b_c + b_s} \\
&\text{eq2bks}: \quad \delta_{bksA} - \delta_{bksB} = \frac{b_{AB} M_{Bx}}{GJ_p}
\end{aligned} \qquad (11)$$

delta_bks_ sols $\Rightarrow$ solve {eq1bks, eq2bks} for $\{\delta_{bksA}, \delta_{bksB}\}$

where δ_{bksA} and δ_{bksB} are the block deflection at the intersecting point of lines mn and pq with $x = l_s$ associated to the stopper height δ_s, respectively.

The following relationships [5] were applied to eqns. (4), (9) and (11):

$$G = \frac{E}{2(1+\upsilon)}, \quad I_z(x) = \frac{wh(x)}{12} \quad \text{and} \quad J_p = \frac{wh(w^2 + h^2)}{12}. \qquad (12)$$

By combining the displacement and force analyses as modeled by eqns. (7), (9) and (11), the deflection and reaction forces due to F and δ_r can then be obtained. Note that the following relationships have been substituted into eqn. (8) for Beams A and B:

Beam A: $M_z = M_{Az}$, $P_y = P_{Ay}$, $I_z = I_A$, $\delta_{bks} = \delta_{bksA}$;

Beam B: $M_z = M_{Bz}$, $P_y = P_{By}$, $I_z = I_B$, $\delta_{bks} = \delta_{bksB}$.   (13)

Maximum Stress

A simple expression is used here to obtain the maximum stress along the beam [4]:

$$\sigma_{max}(x) = \frac{M(x)c(x)}{I_z(x)} = \frac{M}{S_z}. \qquad (14)$$

Note that $c(x) = y_{max}(x) = h(x)/2$ and the term S_z ($= I_z/c$) is called the section modulus. Note that σ_{max} for each section of the beam, s_i ($i = 1$ to 4), is taken as σ_{maxsi}. Again, different I_z and M should be accounted for different beams similar to eqn. (13).

SYMBOLIC EXPRESSIONS

The present button design analysis involves four steps after the problem formulation. It is worth mentioning that the first step is to define the fixed hinge parameters, while the second step allows engineers to input design parameters for the button systems. By virtue of the solving scheme presented, symbolic expressions in different forms can be generated in each of the two steps for forces, deflections, and stresses. Only those results directly relevant to the analysis will be listed in this paper, as the expressions are tedious and lengthy.

Expressions Without Stopper

Without the stopper ($S = 0$), the expressions for the stopper force and deflections have been derived in terms of loading parameters (F, l_f, b_f) are listed below:

$$R_0 = 0.0597433(l_f - 4.75)F + 0.148649F - 0.162161 - 0.160131 \times 10^{-9} b_f F$$

Beam A

$y_{x1A0} = y_{x2A0} = y_{x3A0} = y_{x4A0} = 0.317019 \times 10^{-6} b_f x^3 F - 0.116138 \times 10^{-4} x^3 l_f F$
    $- 0.451748 \times 10^{-5} b_f x^2 F + 0.220661 \times 10^{-3} x^3 F + 0.78807 \times 10^{-4} x^2 l_f F$
    $+ 0.315232 \times 10^{-4} x^3 - 0.149734 \times 10^{-2} x^2 F - 0.179681 \times 10^{-2} x^2$

$y_{x5A0} = - 0.679507 \times 10^{-4} b_f F + 0.53341 \times 10^{-3} (l_f - 4.75) F - 0.76012 \times 10^{-2} F$
    $- 0.371625 \times 10^{-1} + 0.23327 \times 10^{-2} (x - 4.75)[- 0.919885 \times 10^{-2} b_f F$
    $- 0.16049 \times 10^{-1} (l_f - 4.75) F + 0.22865 F - 6.40289]$

Beam B

$y_{x1B0} = y_{x2B0} = y_{x3B0} = y_{x4B0} = - 0.317019 \times 10^{-6} b_f x^3 F - 0.116138 \times 10^{-4} x^3 l_f F$
    $+ 0.451748 \times 10^{-5} b_f x^2 F + 0.220661 \times 10^{-3} x^3 F + 0.78807 \times 10^{-4} x^2 l_f F$
    $+ 0.315232 \times 10^{-4} x^3 - 0.149734 \times 10^{-2} x^2 F - 0.179681 \times 10^{-2} x$

$y_{x5B0} = 0.679507 \times 10^{-4} b_f F + 0.53341 \times 10^{-3} (l_f - 4.75) F - 0.76012 \times 10^{-2} F$
    $- 0.371625 \times 10^{-1} + 0.23327 \times 10^{-2} (x - 4.75)[0.919885 \times 10^{-2} b_f F$
    $- 0.16049 \times 10^{-1} (l_f - 4.75) F + 0.22865 F - 6.40289]$ (15)

Note that the deflection expressions for Beams A and B are the same, except for the opposite sign of the terms involving b_f. This is one of the advantages by attaching the reference frame xyz at the mid-line between the hinges.

Expressions With Stopper

The symbolic expressions, in terms of loading parameters (F, l_f, b_f; δ_s, l_s, b_s), for the cases involving stopper are listed as follows:

$R = N_{R\_lp}/D_{R\_lp}$
$y_{x1A} = y_{x2A} = y_{x3A} = y_{x4A} = N_{bA\_lp}/D_{bA\_lp}$
$y_{x1B} = y_{x2B} = y_{x3B} = y_{x4B} = N_{bB\_lp}/D_{bB\_lp}$
$y_{x5A} = N_{bkA\_lp}/D_{bkA\_lp}$
$y_{x5B} = N_{bkB\_lp}/D_{bkB\_lp}$ (16)

Details of the terms N_{R_lp}, D_{R_lp}, N_{bkB_lp}, etc. are not shown here due to the space limit.

The symbolic forms in terms of design and loading parameters are more useful to the engineers. However, the expressions are in general quite lengthy. Note that the block and the beams are of the same materials (same E and G) but different cross-section (different I_A, I_B and I_r). By further assuming $I_A = I_B = I_{beam}$, the symbolic expressions can be simplified significantly, for example,

$$D_{bA} = I_{beam}^2 l_s^3 E^2 \times 10^6 (2.708 b_r^2 - 2.708 b_r b_s - 0.57 l_s b_r^2 + 0.57 l_r b_r b_s + 0.57 l_r b_c^2 - 0.57 l_s b_c^2$$
$$-0.18 l_s b_c^2 a_r + 0.18 l_r b_c^2 a_r) + I_{beam} l_s^3 EGJ_r \times 10^7 (l_s - l_r)(2.03626 + 0.643029 a_r). \quad (17)$$

RESULTS AND COMMENTS

The button with the hinges is made of the plastic material ABS. The measured stress-strain curve showed that $E = 1200$ MPa, $\sigma_{fatigue} = 33$ MPa, $\sigma_u = 35$ MPa; $F_{max} = 1200$ N. The measured data and the following values are used in the present numerical evaluation: $w = 5$ mm; $h_0 = 0.95$ mm, $h_1 = h_2 = 0.55$ mm, $h_4 = 2.00$ mm; $s_0 = 0.4$ mm, $s_1 = 2.5$ mm, $s_2 = 1.45$ mm, $s_4 = 0.4$ mm; $L = 4.75$ mm, $l_r = 19$ mm, $b_c = 6.5$ mm, $b_r = 0$ mm, $\delta_s = -0.15$ mm, $\delta_r = -0.25$ mm; $\upsilon = 0.3$, $E = 1200$ MPa.

Numerical/Experimental Comparisons

Figure 5 shows the comparison between the experimental deflections and those obtained based on the derived expressions without stopper. The results are associated to the centre-loading with $F = 27.47$ N and 37.28 N acting at position $l_f = 11.875$ mm and $b_f = 0$ mm.

Parametric Study

For the parametric study, a key value is assigned to each of the following loading parameters: $F=27.47$ N, $\delta_s = -0.15$ mm; $l_f = L+1 = 5.75$ mm, $b_f = 4$ mm; $l_s = L+1 = 5.75$ mm and $b_s = 0$ mm. In each effect study, only one of these parameters is taken as the variable while the rest are kept constant with its key value. Six effect study cases were considered with seven values as listed below:

(i) Effect of F: $F = 5, 10, 15, 20, 25, 30, 35$ N;
(ii) Effect of δ_s: $\delta_s = -0.1, -0.2, -0.3, -0.4, -0.5, -0.6, -0.7$ mm;
(iii) Effect of l_f: $l_f = 5.75, 7.75, 9.75, 11.75, 13.75, 15.75, 17.75$ mm;
(iv) Effect of b_f: $b_f = 0, 1, 2, 3, 4, 5, 6$ mm;
(v) Effect of l_s: $l_s = 5.75, 7.75, 9.75, 11.75, 13.75, 15.75, 17.75$ mm;
(vi) Effect of b_s: $b_s = 0, 1, 2, 3, 4, 5, 6$ mm.

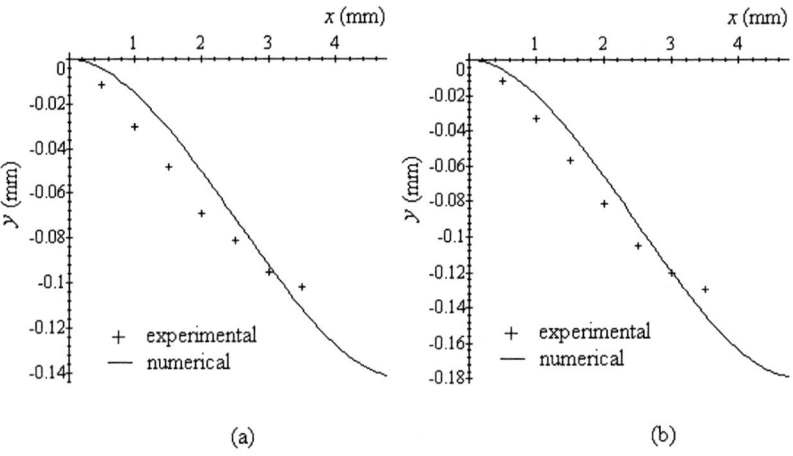

Figure 5   Comparison of beam deflections without stopper. (a) $F = 27.47$ N; (b) $F = 37.28$ N

Results of all these six cases with different values of the respective parameter were obtained and compared. Only Case (ii) on the stress reduction due to changing δ_s is presented here with Figure 6.

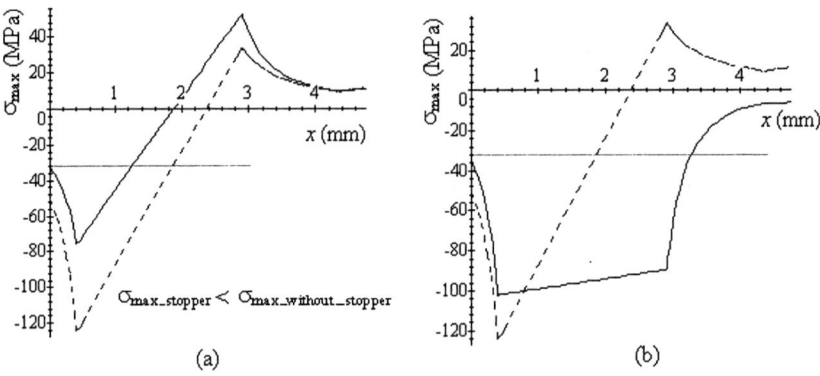

Figure 6   Effect of stopper with changing δ_s on the stress distribution of beam A
(dash line: without stopper; solid line: with stopper). (a) $\delta_s = -0.1$ mm;(b) $\delta_s = -0.6$ mm

The following trends can be concluded in views of results obtained:

(1)   A higher applied force (F) at a specific position will produce higher deflection. For the same magnitude of force F, the beam deflection is smaller if it is further away from the fixed end of the hinges. Furthermore, the deflection increases for the beam/block A if the same force applied nearer to the beam.

(2) A higher displacement (δ_s) at a specific position yields smaller deflection. For the same magnitude of displacement δ_s, the deflection is smaller if the stopper is located nearer to the fixed end of the hinges. Similar to the case with b_s, the deflection increases for the beam/block A if the stopper is installed nearer to the beam.

(3) The stresses will all be within the safety limit if the applied force is small or the acting point is nearer to the end-switch.

(4) In most of the cases, the maximum stress can be reduced by installing a stopper, i.e. $\sigma_{max\_stopper} < \sigma_{max\_without\_stopper}$.

(5) No significant different for the stress distribution by changing b_f and b_s.

CONCLUDING REMARKS

A structural analysis for a button system of an audio Hi-Fi product has been presented in this work. Symbolic expressions of the beam-block deflection and stress distribution were derived by solving different sets of simultaneous equations. Experimental measurement of the beam deflection has been compared with the numerical results. The effects of individual loading parameters on stress distribution, with and without stopper, are studied and presented. Results of different locations of the stopper and the pressing force are given for discussion. The results have shown that the beam stress can, in general, be reduced by installing a stopper. The stopper height could be another parameter to be studied, as it may reduce the spring-back effect if not properly designed.

Acknowledgments – The authors would like to thank Ms Jing Bai for her help on FEM analysis on the button design. Thanks are also due to Miss Judy Lim for her help in obtaining the experimental data of the button material. The support of Philips (Singapore) Pte. Ltd. on the work is greatly appreciated.

References

1. Demeter G. Fertis. (1996). *Advanced Mechanics of Structures*, M. Dekker, New York.

2. William A. Nash. (1994). *Schaum's Outline of Theory and Problems of Strength of Materials*, 3rd ed., McGraw-Hill, New York.

3. P.P. Benham, R.J. Crawford and C.G. Armstrong. (1996). *Mechanics of Engineering Materials*, 2nd ed., Longman, Harlow, Essex.

4. Richard G. Budynas. (1999). *Advanced Strength and Applied Stress Analysis*, 2nd ed., WCB/McGraw-Hill, Boston.

5. Roy R. Craig, Jr. (2000). *Mechanics of Materials*, 2nd ed., John Wiley, New York.

6. James M. Gere and Stephen P. Timoshenko. (1984). *Mechanics of Materials*, 2nd ed., Brooks/Cole Engineering Division, Monterey, CA.

7. Egor P. Popov. (1999). *Engineering Mechanics of Solids*, 2nd ed., Prentice Hall Upper Saddler River, New Jesey.

DAMAGE PREDICTION AND DAMAGE ASSESSMENT

FIBER OPTIC ASSESSMENT OF DAMAGE IN FRP STRENGTHENED STRUCTURES

M. Zhao[1], Y. Zhao[1], and F. Ansari[1]

Department of Civil and Materials Engineering, University of Illinois at Chicago, Chicago, Illinois 60607-7023, USA

ABSTRACT

A distributed fiber optic sensor is developed for condition monitoring of civil infrastructure systems. The fiber optic sensor is especially useful in applications involving structures strengthened by fiber reinforced polymer composites (FRP). The sensor principles are simple and therefore, practical for detection of cracks, debonding and deformation measurements. Structural monitoring capability of the sensor was evaluated through experiments with cracked reinforced concrete beams strengthened by way of CFRP fabrics. Sensor principle and experimental results are presented and discussed.

KEYWORDS

Bridges, Concrete, Condition Monitoring, Concrete, CFRP, FRP, Fiber Reinforced Polymer Composites, Fiber Optic Sensors, Optical Fibers, Repair, Reinforced Concrete, Structures

INTRODUCTION

Fiber-reinforced plastics (FRP) have been increasingly used in conjunction with civil structural materials, either as reinforcement for concrete structures or for repair and strengthening of existing structural elements (Dolan, et al, 1998). In comparison with steel reinforcement, the primary attributes of FRP bars are of reduced mass, high tensile strength, corrosion resistance, lower relaxation, high strength-to-weight ratio, non-magnetic properties ease of handling and cutting. Important applications of FRP composites pertain to rehabilitation of cracked concrete members. In such applications, it is important to ensure adequate composite action between the FRP reinforcement and the existing structures. Five types of failure modes can be expected when reinforced concrete beams are strengthened with CFRP fabrics. These failure modes include: (1) laminate rupture; (2) crushing of concrete in compression; (3) yielding of steel rebar; (4) shear failure of the beam; (5) peeling of CFRP fabric.

The first three failure modes are typical in cases where the CFRP debonding and shear collapse do not occur prematurely. These failure types are accurately predicted based on the existing knowledge in reinforced concrete design. Shear failure, is due to mismatch in shear capacity of the strengthened member and the increased load capacity due to FRP. This type of failure can be avoided through appropriate shear design considerations. The peeling of FRP plate is a unique failure phenomenon, and such failure modes is caused by the complex normal, and interface shear stresses that exist at the plate cut-off section.

Review of literature reveals that many researchers have investigated the development of models to predict the debonding failure mode. Roberts (1989) developed an approximate procedure to calculate the normal and shear stresses at the FRP cut-off point. Triantafillou (1991) developed an analysis technique for reinforced concrete beams prestressed with external FRP sheets. Täljsten (1997) derived relationships for computation of shear and normal stress distributions in beams subjected to four-point-bending. While it is important to develop predictive models for failure analysis of FRP bonded beams, there is also an urgency to develop techniques for condition monitoring of FRP strengthened structures. Fiber optic sensors provide unique capabilities for monitoring of strains and deformations in such structures. Strain gauge techniques do not provide adequate information in these applications. Determination of bond and strain over the entire interface between the FRP and concrete beam requires sensors with capability to discern the deformations over long gauge lengths. Fiber optic sensors will provide long gauge length strain measurement capability. The objective for the research presented here was to develop a long-gauge length quasi-distributed sensor for the particular application in FRP repaired structures. A test program was designed to use the distributed optical fiber sensor system in conjunction with reinforced concrete beams strengthened with CFRP fabrics. The experimental program included testing of CFRP strengthened beams in four point bending and examining the strain sensing capability of the sensor along the length of the interface.

FIBER OPTIC DISTRIBUTED SENSORS

Most of the fiber optic sensor research activities have involved development of localized sensors. However, survey of recent literature reveals a number of activities pertaining to the development of distributed and or multiplexed sensors (Ansari, 1997). Brillouin scattering, optical switching, wavelength division multiplexing, and optical time domain reflectometry (OTDR) have been used in the development of distributed and multiplexed sensors. Gu, et al (1999) developed a distributed sensor by connecting a number of individual optical fibers of desired gauge lengths in series. The strain transduction mechanism was intensity based by using a high resolution OTDR through measurement of intensity losses at the spliced joints.

Other investigators developed long-gauge length distributed sensors using Brillouin scattering based on the Doppler shift in frequency (Lechochea, 2000). Wavelength division multiplexing by using Bragg gratings is another technique. In using this technique, a broadband light source is employed for scanning a number of Bragg gratings in series and or in parallel. In using this technique, a broadband light source was employed for scanning a number of Bragg gratings in series and or in parallel. The reflected wavelength of each Bragg is slightly different from the other. Other techniques employ optical switches for the development of a white light interferometric sensor. These types of sensors are based on a Michaelson interferometer. Other multiplexed interferometric sensors are based in time domain. The research presented here describes the development of a quasi-distributed fiber optic sensor that provides distributed strain measurement capability for structures. The optical fiber sensor operates based on white light interferometry. This system will be described next and the experimental investigations with FRP bonded reinforced concrete members will be presented.

DISTRIBUTED SENSING METHODOLOGY

Schematic representation of the quasi-distributed sensor consisting of *m*-segments in series is shown in Fig. 1. Each segment represents one gauge length, and distributed sensing and measurement is accomplished by splicing of several sensors in series. The theoretical basis pertaining to the optical demodulation of signals from the individual sensors, the strain-optic relationships leading to quantification of strains sensed by the individual sensors as well as the interferometer are described in this section.

The system consists of two parts: sensing interferometer module, and the receiving interferometer. As described earlier, the sensor is comprised of a number of individual single mode fibers of desired gauge lengths. The individual fibers are mechanically connected through ferrules and a portion of beam is reflected when the light wave passes through them. The receiving interferometer consists of a Michelson white light interferometer with a scanning translation stage, signal processing and the system control unit.

White light from a wide-band LED is fed into the sensor by way of a 2X1 coupler. The power is transmitted through the entire sensor system and returned to the interferometer by way of a coupler. The returned optical field, E, from the individual sensor segments is expressed in the following form:

$$E = \sum_{i=1}^{m} E_i \exp\{-j[2\pi \bar{v} t - 2\bar{k}(L_0 + \sum_{h=1}^{h=i} L_h \cdot n)]\} \qquad (1)$$

Where E_i is the amplitude of the reflected beam, $\bar{v}$ is the mean frequency of the signal, $\bar{k}$ is the mean wave number in vacuum, L_0 is initial optical path length, L_j is gauge length of the strain sensor, n is the effective index of refraction of the sensing fiber, m is the number of the sensors, $i=1,2,3,...m$ and $h=1,2,3,...i$. The individual reflected signals contain the information pertaining to the measurand.

The set of reflected signals is then interrogated by the white light receiving interferometer. The interferometer consists of the reference arm and the sensing arm, which is composed of a GRIN lens and a scanning mirror mounted on a motorized translation stage. The reflected optical fields from the reference (E_r) and the sensing arms of the interferometer (E_m) can be expressed as:

$$E_r = \sum_{i=1}^{n} E_i' \exp\{-j[2\pi \bar{v} t - 2\bar{k}(L_0 + \sum_{j=1}^{j=i} n \cdot L_j + L_r)]\} \qquad (2)$$

$$E_m = \sum_{i=1}^{n} E_i' \exp\{-j[2\pi \bar{v} t - 2\bar{k}(L_0 + \sum_{j=1}^{j=i} n \cdot L_j + L_m)]\} \qquad (3)$$

Where E_r' and E_m' refer to the amplitude of the signals, L_r, L_m are optical path lengths for the reference and sensing arms in the interferometer, and $L_m = L_r + x$, where, x, is the scanning displacement of the translation stage in the interferometer. The output intensity I of the interference can be obtained by taking the time average of the production of the overall output electric field from the interferometer and its complex conjugate:

$$I_c = (\langle E_r + E_m \rangle \cdot \langle E_r + E_m \rangle^* \\ = \sum_{i=1}^{n} E_i'^2 + 2\sum_{i=1}^{n} E_i' E_{i+1}' \cos 2k[n \cdot L_i - x] \cdot \gamma_{11}(n \cdot L_i - x) + \cdots \cdots \qquad (4)$$

Where $\gamma_{11}(\tau)$ is visibility of the fringe. Since the spectral distribution of LED used in these experiments is Gaussian then, the visibility is given by:

$$\gamma_{11}(L_i - x) = \exp\left\{-\left[\frac{1.66(L_i - x)}{L_c}\right]^2\right\} \quad (5)$$

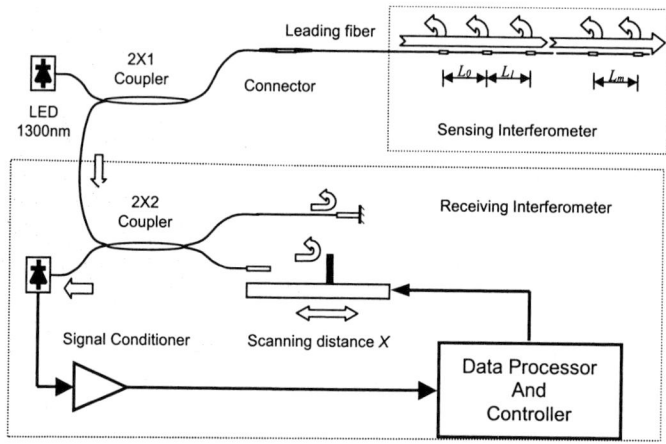

Figure 1: Fiber optic white light distributed sensor and the scanning interferometer system

White light fringes develop every time when the optical path difference in the interferometer matches L_j ($j=1,2,...n$), at which, $\gamma_{11}(\tau)$ becomes non-zero. The change of scan length, x, for individual segments of the optical fiber along the length of the sensor represents the gauge length change due to measurand, i.e. strain for instance.

The generalized strain – optic relationship between the optical length change (OLC) and the axial strain, ε_{xx}, induced over the gauge length of a single segment of the optical fiber is given below:

$$OLC = \frac{\Delta x_i}{2} = nL\{n - \frac{1}{2}n^2[P_{12} - v_f(P_{11} + P_{12})]\}\varepsilon_{xx} \quad (6)$$

Where P_{11} and P_{12} are Pockels constants, v_f is the Poisson's ratio, n is the refractive index of the fiber, and L is the length (gauge length) of the optical fiber.

For single mode silica optical fibers at a wavelength of 1300 nm, the refractive index, n=1.46. Furthermore, $P_{11}\approx 0.12$, $P_{12}\approx 0.27$, $v_f =0.25$, and therefore Eqn .6 can be re-written as:

$$\varepsilon_{xx} = \frac{\Delta x_i}{2.38L} \qquad (7)$$

Eqn. 7 corresponds to the amount of scanning displacement, Δx, required by the interferometer for the sensing of the induced strain, ε_{xx}, by the sensor. Therefore, the interferometer measures scanning displacements from various segments along the length of the sensor and these scanning displacements are converted to strains by way of Eqn. 7.

EXPERIMENTAL PROGRAM

The experimental program consisted of testing six reinforced concrete beams subjected to four-point bending. All the beams were 10-ft in length, had a nominal cross sectional area of 6×12 inch. The beams were reinforced with two grade 60 number 6 bars in the longitudinal direction, and the shear reinforcement consisted of grade 40, number 3 bars spaced 4-inches apart. These beams were loaded in an MTS closed-loop testing machine in displacement control up to the cracking load of the beam. At this point, the loading was kept at the cracking load of approximately 6-kips, and CFRP sheets were adhered to the beam, mimicking repair under service load conditions. There were approximately 4 to 6 tension cracks were observed at this stage. The load points and reaction supports were designed to load the beam in reverse (Fig.2). This arrangement provided easy access to the tension face of the beam for application of the CFRP sheets from the top of the beam. The specimens were kept under the service load through the duration of curing period for the adhesives involved in bonding the CFRP sheets to the beam.

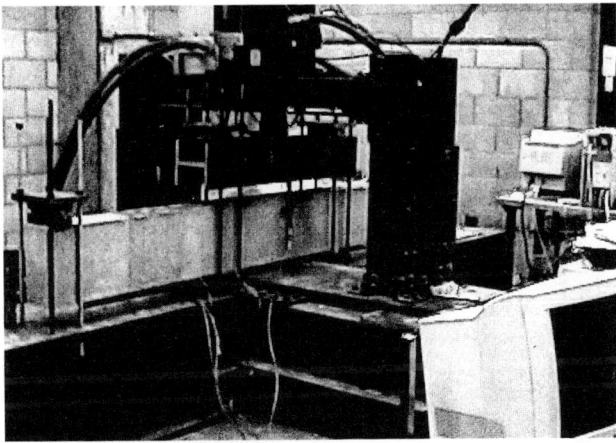

Figure 2: Beam subjected to reverse four-point loading arrangement.

Following the curing period the static load was applied to failure. As shown in Fig.2, a Linear Variable Differential Transformer (LVDT) was employed for measurement of center-span deflection of the beams. The feedback signal from the same LVDT was employed for control of the loading program. Due to symmetry of loading, only half of the span was instrumented with the fiber optic sensor system. The distributed sensor was comprised of four segments, each at a gauge-length of 12-inches. This system is shown in Fig.3, and the sensors segments are designated by numerals 1 through 4. Strain gauges were also used for comparison with the strain measurements by the fiber optic sensors. Four strain gauges were adhered to the surface of the CFRP and centered within the four 12-inch fiber optic sensor gauges lengths. One additional strain gauge was applied at the edge of the CFRP sheet near the support in order to detect the initiation of CFRP debonding.

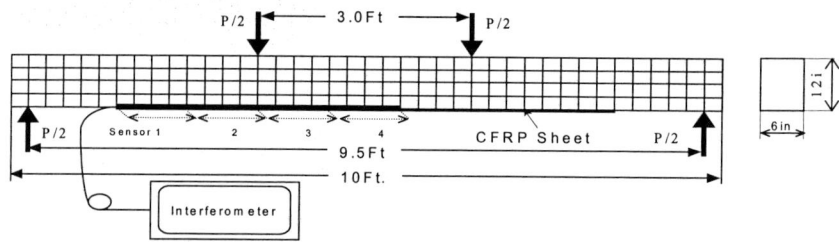

Figure 3: Sensor arrangement and beam dimensions.

EXPERIMENTAL RESULTS

All the beams were monotonically loaded to failure. Acquisition of load, deflection, and sensor data was accomplished through a computer GPIB interface. Figs. 4 through 7 correspond to strain measurements along the symmetrical half of the beam length as measured by sensor segments 1 through 4. In these figures, fiber optic measured strains are compared with the strain gauge and strains evaluated by elastic beam theory.

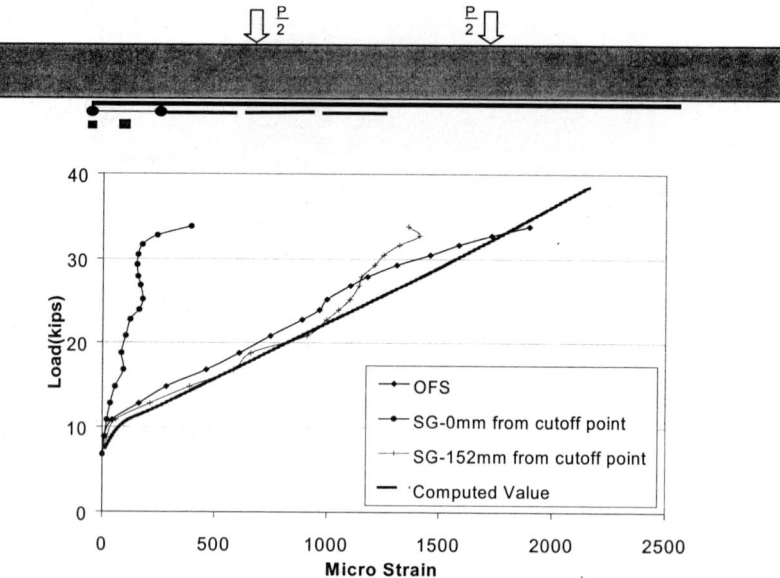

Figure 4: Typical load – strain data within segment-1 of the beam

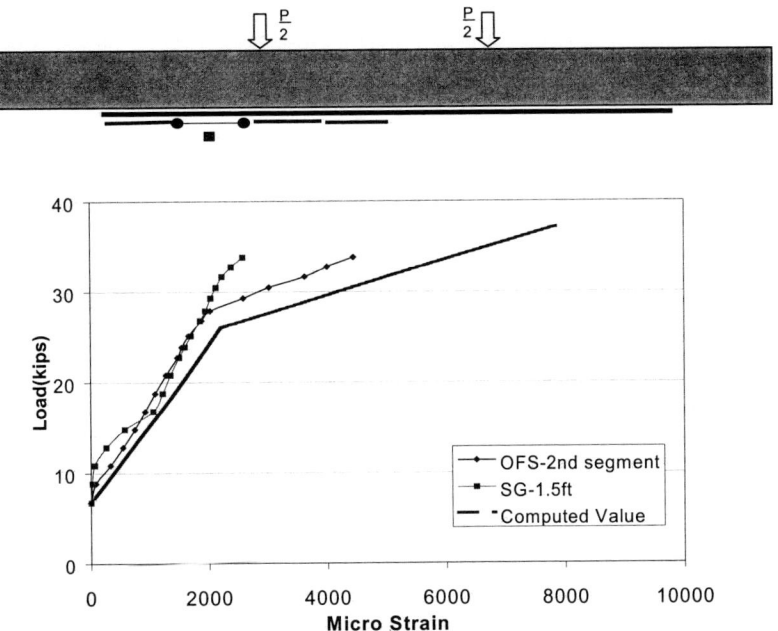

Figure 5: Typical load – strain data within segment-2 of the beam.

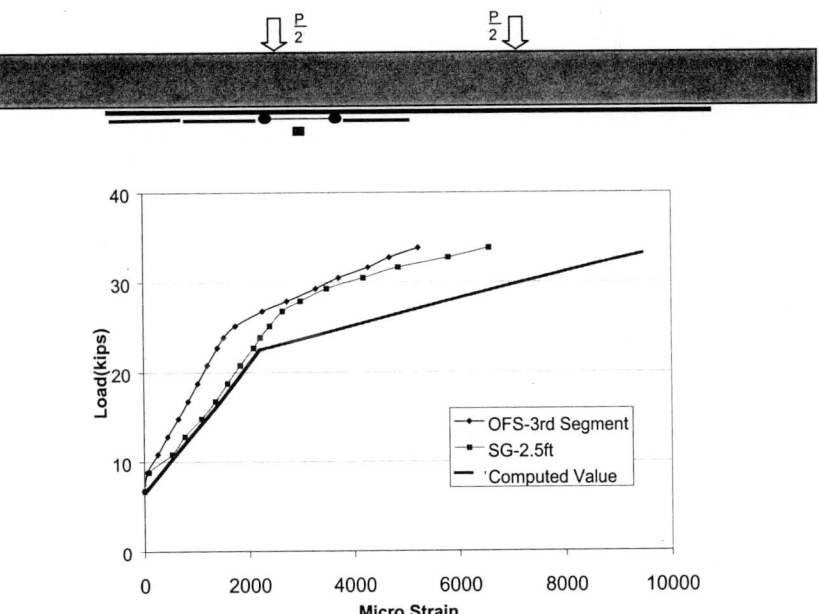

Figure 6: Typical load – strain data within segment-3 of the beam.

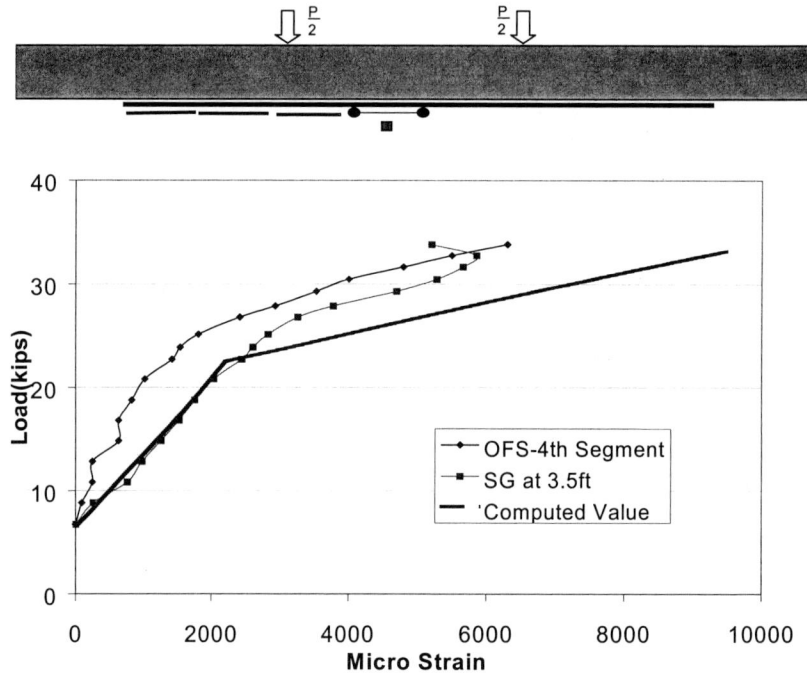

Figure 7: Typical load – strain data within segment-4 of the beam.

In general, fiber optic measured strains are larger than the ones obtained by the strain gauges. This is due to the fact that fiber optic measurements correspond to deformations over large gauge lengths, i.e. 12-inces. Within the same segments, strain gauge measurements pertained to half-inch gauge lengths. Experimental Observations of beam failure at various stages of loading indicates that the peeling force is initially formed at the concrete cover and in between the flexural cracks. These cover cracks propagate into the beam until total brittle failure occurs through full peeling of CFRP at the end section (Fig. 8). In this case, ultimate strength is not governed by general flexural theory, and it is dependent on the normal and shearing stresses generated at the interface zones

CONCLUSIONS

A distributed fiber optic sensor system was developed for application in measurement of strains over the interfacial zones of CFRP bonded reinforced concrete beams. Feasibility of the sensor system was examined through experiments with beams subjected to four-point bending. This is a preliminary study to investigate appropriate usage of such sensors in rehabilitated bridges. Further experiments are under way to fully exploit this technology in practice.

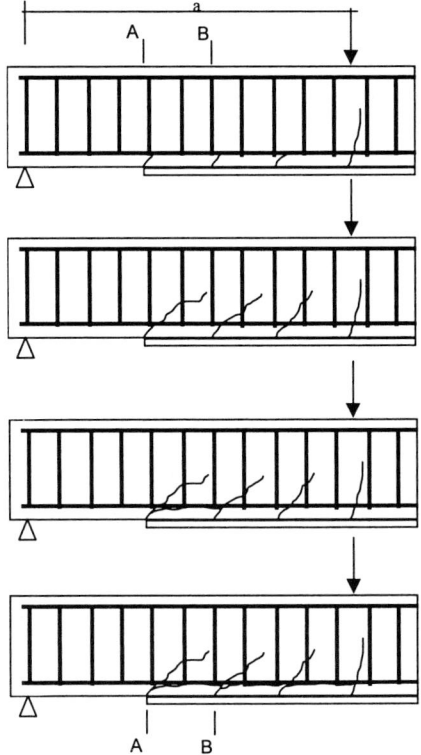

Fig.8 Propagation of peeling cracks at the concrete cover.

References

Dolan, Brendan E.; Hamilton, H.R. (1988), "Strengthening with bonded FRP laminate, Concrete International, Vol.20, No. 6.
Roberts, T.M., (1989), "Approximate Analysis of shear and normal stress concentrations in the adhesive layer of plated RC beams,", Structural Engineer, Vol. 67, No.12, June,1989
Triantafillou, T.C., Deskovic, N., "Innovative Prestressing with FRP sheets: Mechanics of Short-Term Behavior," Journal of Engineering Mechanics, vol. 117, 14, 1991
Täljsten, B., "Strengthening of Beams by Plate Bonding, Journal of Materials in Civil Engineering," Vol. 9, No. 4, 1997.
Ansari, F., (1997) " Theory and Applications of Integrated Fiber Optic Sensors in Structures," *in Intelligent Civil Engineering Materials and Structures*, Ed. Ansari F., ASCE Press, Reston, VA, pp.2-28.
Gu, X., Chen, Z., and Ansari, F. (1999), "Method and Theory for Multi-Gauge Distributed Fiber Optic Crack Sensor," *Journal of Intelligent Material Systems and Structures*, V.10, N.4, PP. 266-273.
Lecoeuche,V., Webb,D.J., Pannel,C.N., and Jackson, D.A.(2000), "Distributed Sensor for Detection of impending Structural Failure Along a 25-Km Optical Fiber with 2 Meters Spatial Resolution," *Journal of Structural Control*, V.7, N.1, PP. 23-34.

STRUCTURAL HEALTH MONITORING

J. L. Humar and M. S. Amin

Department of Civil and Environmental Engineering, Carleton University,
Ottawa, On. K1S 5B6, Canada

ABSTRACT

Early detection of damage is of special concern for civil engineering structures. If not identified in time, damage may have serious consequences, both safety related and economic. The traditional methods of damage detection include visual inspection or instrumental evaluation. A comparatively recent development in the health monitoring of civil engineering structures is vibration-based damage detection. Vibration characteristics of a structure, that is, its frequencies, mode shapes, and damping are directly affected by the physical characteristics of the structure including its mass and stiffness. Damage reduces the stiffness of the structure and alters its vibration characteristics, which are, in fact, global properties. Therefore, measurement and monitoring of vibration characteristics should theoretically permit the detection of both the location and severity of damage. The advantages and limitations of vibration-based damage detection are discussed. A computer simulation study of the identification of damage in a three-dimensional aluminum truss structure is presented. The results obtained from two different algorithms for damage detection are reviewed.

KEYWORDS

Vibration-based damage detection, structural health monitoring, modal residual vector, matrix update method, quadratic nonlinear optimisation.

INTRODUCTION

Early detection of damage to civil engineering structures has assumed a special importance because of the aging infrastructure, increased demand, complexity and size of some of the modern structures, and lack of long-term experience with innovative materials and structural shapes that may be incorporated in a structure.

Damage may be detected by visual inspection or instrumental evaluation. These methods require that all portions of the structure are accessible. This may be impractical, particularly when the structure is complex and/or large in size. Certain type of damage, for example, internal delamination and fibre fracture in a composite, and fracture of prestressing strands in a prestressed concrete girder cannot be detected by visual inspection. Several non-destructive damage detection techniques have been developed to detect damage that may not be visible to the naked eye. Such techniques include, for example, acoustic methods, magnetic field methods, radiography, and eddy current methods. Application of such methods requires *a-priori* knowledge of the possible damage sites and access to such sites. Also, the results of instrumental evaluation are often inconclusive or difficult to evaluate. Instrumental evaluation may involve measurement of local stress and strains and accelerations through sensors installed on critical elements and components. Changes in stress and strain states or accelerations can provide warning of possible damage. However, it may be too expensive or impractical to instrument all elements and components that are possibly critical.

A comparatively recent development in the health monitoring of civil engineering structures is vibration-based damage detection. Vibration characteristics of a structure, that is, its frequencies, mode shapes, and damping are directly affected by the physical characteristics of the structure including its mass and stiffness. Damage reduces the stiffness of the structure and alters its vibration characteristics, which are, in fact, global properties. Therefore, measurement and monitoring of vibration characteristics should theoretically permit the detection of both the location and severity of damage.

Vibration based assessment offers several advantages; one of which is that the location of damage need not be known *a-priori*. Often, the sensors required to measure the vibration characteristics need not be located in the vicinity of the damage. In addition, a limited number of sensors can provide sufficient information to locate the damage and assess its severity, even in a large and complex structure. Vibration measurements do not require use of bulky equipment, except when forced vibration tests are carried out. However, in practice there are a number of limitations associated with vibration-based damage assessment. They can be summarized as follows.

Low Sensitivity to Damage

Vibration characteristics are global properties of the structure, and although they are affected by local damage, they may not be very sensitive to such damage. As a result the change in the global properties may be difficult to identify unless the damage is very severe or the measurements are very accurate and are made with extra care. Accurate vibration measurements on civil engineering structures are difficult because such structures are large, so that controlled force excitation is often impractical. As a result reliance must be placed on measurement of vibrations induced by ambient forces.

Complexity of the Damage Identification Algorithms

The identification of a possible damage site and severity of damage on the basis of a change in global properties derived from measurements at a limited number of sensor locations is a problem that has a non-unique solution. Sophisticated and complex mathematical techniques including non-linear programming need to be employed to obtain the most probable solution. In fact, this is currently an area of ongoing research. The methods that are currently available cannot deal with situations where the damage introduces nonlinearity in the structure. Such nonlinearity may, for example result from the presence of cracks. Closing and opening of the cracks alters the stiffness of the structure introducing nonlinearity in its behaviour. Nonlinearity may also result from loose connections that slip under load.

Effect of Factors Other than Damage

Global vibration characteristics are often affected by phenomena other than damage, including environmental effects, such as change of mass caused by water waves and snow accumulation, and thermal effects caused by temperature variation. Whenever the structural system is constrained or indeterminate, thermal effects introduce axial stresses in the structural elements. The presence of such axial stresses changes the stiffness of the structure and may alter its vibration characteristics. The boundary conditions in a structure can have a significant effect on its stiffness, and if these boundary conditions are prone to change with the age of the structure, they may lead to a change in the vibration characteristics even when there is no damage in the structure.

A large number of research studies are currently being carried out to address the difficulties associated with the practical application of vibration-based damage detection. The present study focuses on the evaluation of some of the vibration-based damage detection procedures from tests on a 3D aluminium truss. The truss is first tested to obtain the baseline vibration characteristics of the undamaged structure. A finite element model of the structure is constructed and refined so that the analytical properties derived from the model match the measured properties. Damage of specified severity is now introduced in the specimen at predetermined locations. Next, the altered vibration characteristics are measured. Various damage detection algorithms are then used to predict the location and severity of damage and the predicted values are assessed for their accuracy. The present paper reports on the first part of the study, which is related to a computer simulation of the damage and its detection. Details of the test specimen are provided. A 3D truss model is then used to obtain the first few non-rigid-body modes and frequencies of the structure. Specified damage is now introduced in the model and the mode shapes and frequencies of the damaged model are computed. The new vibration characteristics are then used to predict the damage. Prediction is based on two different algorithms.

DESCRIPTION OF THE TEST SPECIMEN

The test specimen used in this study is an erectable aluminium space frame made from commercially available hardware (Meroform M12). The hardware consists of standardized aluminium nodes and aluminium tube struts of several different sizes. Figure 1 shows details of the nodes and struts. The design of the joint node allows the frame to be assembled into numerous configurations in any of three orthogonal directions, thereby providing structures with varying complexity. The struts have threaded solid steel end connectors, which when tightened into the node also clamp the tube by means of an internal compression fitting. This feature allows any of the frame struts to be replaced by another one of a different (smaller) size without disassembly of the entire unit, which is very useful in simulating damage in any of the frame struts.

A finite element model of the space frame is shown in Fig. 2. The frame consists of eight bays, each of which is a cube with 707 long mm sides. Since the modal tests are to be conducted in a free-free condition, no supports are identified in the model. The nominal physical properties of the frame components are listed in Table 1. All tubes in the vertical (x-y) planes are 30 mm in diameter and have a wall thickness of 1.5 mm. Tubes in the horizontal (x-z) planes, other than those already included in the vertical planes, are 22 mm by 1.5 mm. Lumped masses, each of 1.75 kg, are added to nodes 4, 9, 25, 28, and 36. Masses of 2.75 kg are added at nodes 6, 17, and 30. These design features give well-separated modes and are adopted so as to facilitate the evaluation of damage detection methods. Finite element studies showed that the frame behaviour was very similar to that of a 3D truss; hence a truss model has been used in all the results presented here.

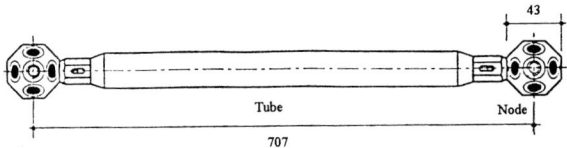

Fig. 1: Meroform Tube and Node

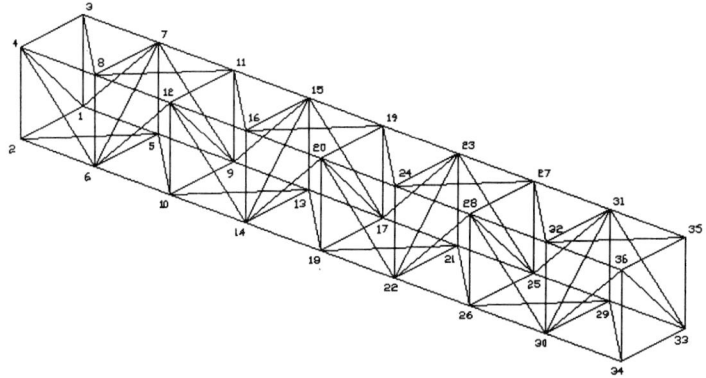

Figure 2: Finite Element Model of Eight-bay Space Truss

Table 1: Properties of the Frame Components

Component	Dimensions (mm)	Cross-sectional area (mm^2)	Inertia (mm^4)	Weight (gm)
Aluminium node	46 diameter	-	-	80
Tubes in (x-y) planes	707×30×1.5	134.3	13673.7	350
Tubes in (x-z) planes	707×22×1.0	65.97	3645.0	230
Diagonal tubes in (x-y) planes	1000×30×1.5	134.3	13673.7	450
Diagonal tubes in (x-z) planes	1000×30×1.0	65.97	3645.0	275

ALGORITHMS FOR VIBRATION-BASED DAMAGE DETECTION

A great deal of research has been carried out in the past decade or so on the development of analytical techniques of vibration-based damage detection. Since vibration-based SHM found its early application in aerospace and mechanical engineering much of the research has been carried out in these fields. Research studies on the application on vibration-based assessment in the health monitoring of civil engineering structures are more recent. Doebling et al (1996) provide an extensive bibliography related to this subject. Farrar et al (1994) have reviewed the literature on vibration testing and damage detection in bridges.

A number of different analytical techniques have been developed for the identification of damage from detected changes in vibration properties. The following two techniques are employed in the present study.

1. Method based on modal residual vector
2. Matrix update methods

Vibration-based damage detection algorithms use the basic eigenvalue equation, which for the healthy structure is given by:

$$\mathbf{K}\phi_i = \lambda_i \mathbf{M}\phi_i \qquad (1)$$

where $\mathbf{K}$ is the stiffness matrix, ϕ_i is the ith mass-orthonormal mode shape, λ_i is the associated eigenvalue (squared frequency), and $\mathbf{M}$ is the mass matrix. Damage in the structure may change both the stiffness and the mass matrices, altering the frequencies as well as the mode shapes. The eigenvalue equation for the damaged structure is given by

$$(\mathbf{K} + \delta \mathbf{K})(\phi_i + \delta \phi_i) = (\lambda_i + \delta \lambda_i)(\mathbf{M} + \delta \mathbf{M})(\phi_i + \delta \phi_i) \qquad (2)$$

Assuming that damage does not alter the mass matrix, $\delta\mathbf{K}$ is small so that the higher order terms can be neglected, premultiplying both sides of Eqn. 2 by ϕ_i^T and using Eqn. 1, we get:

$$\delta\lambda_i = \phi_i^T \delta\mathbf{K} \phi_i \qquad (3)$$

Method Based on Modal Residual Vector

This method relies on the measurement of both the frequencies and mode shapes of the damaged structure. The eigenvalue equation for the damaged structure can be written as:

$$(\mathbf{K} + \delta\mathbf{K})\phi_{di} - \lambda_{di}\mathbf{M}\phi_{di} = 0 \qquad (4)$$

where ϕ_{di} is the ith mode shape of the damaged structure, λ_{di} is the associated eigenvalue, and it is assumed that damage does not cause a change in the mass.

In Eqn. 4 all quantities other than $\delta\mathbf{K}$ are known or have been determined through modal test. Equation 4 can be written in the following alternative form:

$$\mathbf{K}\phi_{di} - \lambda_{di}\mathbf{M}\phi_{di} = \mathbf{R}_i = -\delta\mathbf{K}\phi_{di} \qquad (5)$$

Evaluation of the left hand side of Eqn. 5 provides the modal residual vector $\mathbf{R}_i$ for the ith mode. Matrix $\delta\mathbf{K}$ will have non-zero terms corresponding to only those degrees of freedom (d.o.f.) that are connected to the damaged elements. Correspondingly the non-zero terms in $\mathbf{R}_i$ will also lie along the same d.o.f. Knowledge of the affected d.o.f. and the connectivity relation between the elements and the d.o.f. will allow the determination of damage location. When more than one mode is determined an absolute sum of the modal residuals can be used to determine the damage location. When using the absolute sum, the mode shape should be appropriately normalized so that individual elements of the various mode shapes have the same relative order of magnitude. This can be achieved by mass-orthonormalization of the mode shapes.

Having identified the damaged elements it is possible to express $\delta\mathbf{K}$ as the weighted sum of the stiffness matrices of the damaged elements. The weighting factors, which are the unknowns in the problem, define the severity of damage in the affected elements. For example, if the reduction in the stiffness of element j is expressed as $\beta_j \mathbf{k}_j$, we have

$$\delta\mathbf{K} = \sum_j \beta_j \mathbf{k}_j \qquad (6)$$

where the summation is carried out over all the damaged elements. Equation 5 can now be expressed as

$$\left(\sum_j \beta_j \mathbf{k}_j\right)\phi_{di} = -\mathbf{R}_i \tag{7}$$

When more than one mode is determined, the following equation may be used in place of Eqn. 7.

$$\sum_i \left|\left(\sum_j \beta_j \mathbf{k}_j\right)\phi_{di}\right| = \sum_i |\mathbf{R}_i| \tag{8}$$

Equation 7 or 8 can be solved to obtain the values of factors β_j. In general, more than one value will be obtained for each of the factors β_j. The different values can be averaged to obtain the best estimate.

Matrix Update Methods

The matrix update methods constitute the largest class of methods developed for the identification of damage from measured vibration properties. The methods are based on the determination of perturbations in the property matrices, such as $\delta \mathbf{K}$, that will satisfy the eigenvalue Equation 4. The identified perturbation provides both the location and severity of the damage. In general the number of unknown parameters in $\delta \mathbf{K}$ is significantly greater than the number of measured frequencies and mode shapes. The problem is therefore underdetermined and has an infinite number of solutions. A unique solution can be obtained through the minimization of an objective function subject to some specified constraints. The matrix update methods can thus be classified on the basis of the objective function selected and the constraints used.

Many different objective functions may be used. They include the norm of the estimated solution, rank of the perturbation matrix $\delta \mathbf{K}$, norm of the perturbation matrix, norm of the modal residual vector $\mathbf{R}_i$ and others. The constraints may include satisfaction of the eigenvalue equation, and preservation of the sparsity and connectivity of the property matrices. In addition β_j should be greater than or equal to zero but no greater than 1.

Because of the many different ways in which the objective function and constraints can be defined and the different numerical techniques that can be used in the solution of the resulting optimisation problem, a large number of methods have been developed. Here we present only one such method for the purpose of illustration.

Using Eqn. 6, Eqn. 3 can be expressed as

$$\delta \lambda_i = -\sum_{j=1}^n \phi_i^T \mathbf{k}_j \phi_i \beta_j \tag{9}$$

or

$$\mathbf{D}\beta = -\delta\lambda \tag{10}$$

where n is the number of elements, $\mathbf{D}$ is a m by n matrix whose elements are $d_{ij} = \phi_i^T \mathbf{k}_j \phi_i$, β is the n-vector of the unknown changes in element stiffness matrices, and $\delta\lambda$ is the m-vector of measured eigenvalue changes. If $m = n$, Eqn. 10 can be directly solved for β. In general however m will be less than n so that the problem defined by Eqn. 10 is underdetermined and has an infinite number of solutions. In order to obtain a unique solution we may minimize the quadratic norm of the stiffness changes given by

$$J = \beta^T \beta \tag{11}$$

with the constraint that Eqn. 10 must be satisfied. If these are the only constraints the problem has a closed-form solution given by

$$\beta = -\mathbf{D}^T (\mathbf{D}\mathbf{D}^T)^{-1} \delta\lambda \tag{12}$$

This solution is also useful in refining the analytical model of the undamaged structure so as to ensure that the analytically determined frequencies and mode shapes match those obtained by modal testing. The refined analytical model then serves as the baseline model in the subsequent determination of damage.

For a damaged structure the following additional constraint must be placed on the stiffness changes.

$$0 \le \beta \le 1 \tag{13}$$

The problem now becomes a nonlinear optimisation problem with the objective function defined by Eqn. 12, equality constraints given by Eqn. 10 and inequality constraints given by Eqn. 13. The problem can be solved by using any one of the several algorithms available for the solution of constrained quadratic nonlinear optimisation problems, for example, those in the computer software Matlab (1999).

RESULTS OF COMPUTER SIMULATION

Modal Residual Method

A simulated damage is introduced in member No. 20, which connects nodes 15 and 19, and in member No. 78, which connects nodes 12 and 14. In both cases it is assumed that for the purpose of testing the 30 mm tubes used in these members will be replaced by 22 mm tubes, reducing the cross-sectional area by 50.9%, so that in each case $\beta = 0.509$. The modal residual vector is now calculated by evaluating the left-hand side of Eqn. 5. The 1st mode residuals along the x and y coordinates of each of the 36 nodes are shown in Fig. 3. The only non-zero residuals are at nodes 12, 14, 15, and 19. It is evident from the connectivity that damage exists in members 20 and 78. The severity of the damage is calculated from Eqn. 7, using just the first mode. Because the damaged mode shapes and frequencies used in this case are exact, and because complete mode shapes are available, the damage factors obtained exactly match the simulated damage. In real life, however, there will be errors in the measurements. Also it is unlikely that modal deflection would be measured at all of the d.o.f., as a result complete mode shapes would have to be obtained by some sort of extrapolation. The effect of these factors will be studied in a subsequent phase of this work.

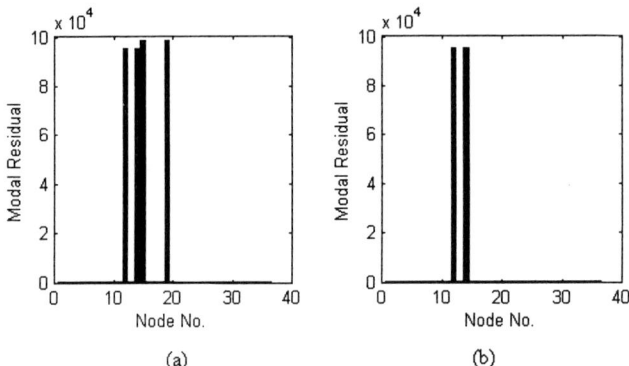

Fig. 3: Modal Residuals, (a) x Coordinates, (b) y Coordinates

Matrix Update Method

Analytical results are obtained for the case of simulated damage in member Nos. 20 (node 15 to 19), 21 (node 19 to 23), 28 (node 16 to 20), and 29 (node 20 to 24). In each case the damage factor is 0.509. Damage is estimated by using Eqn. 12. The estimates are shown in Fig 4. The results show that the constraint represented by Eqn. 13 is violated in many cases and negative values are obtained for β. This would be acceptable in the refinement of a finite element model, but not in a damage detection model. There is also a smearing of the damage, so that although damage has been located in the elements that were in fact damaged, many other non-damaged members have also been identified as being damaged. This is always a likely result in the solution of an underdetermined optimisation problem. It may be noted that if the equality constraint of Eqn. 10 is satisfied by a solution that identifies a certain damage in a particular member, it is also likely to be satisfied if the damage is divided between that member and others that are symmetrically placed in the structure. The minimization process tends to prefer the latter solution. The Matlab routines on constrained minimization were unable to find a solution when the constrained given by Eqn. 13 was specified. If the constrained was relaxed a solution identical to that given by Eqn. 12 was obtained. An advantage of the matrix update method presented here is that it relies on the frequency changes alone and mode shapes of the damaged structure need not be measured.

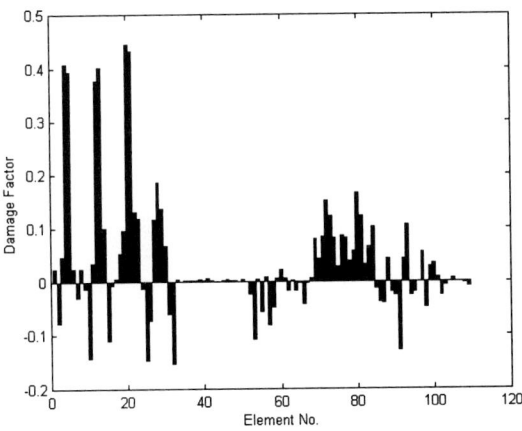

Figure 4. Identification of Damage by Matrix Update Method

REFERENCES

Coleman, T., Branch, M.A., Grace, A. (1999). *Optimization Toolbook, for use with MATLAB*, The MathWorks Inc., Natick, MA.

Doebling, S.W., Farrar, C.R., Prime, M.B., Shevitz, D.W. (1996). "*Damage Identification and Health Monitoring of Structural and Mechanical Systems from Changes in Their Vibration Characteristics: A Literature Review*," Los Alamos National Laboratory, Los Alamos, New Mexico, Report No. LA 13070-MS.

Farrar, C.R, Baker, W.E., Bell, T.M., Cone, K.M., Darling, T.W., Duffey, T.A., Eklund, A., Migliori, A. (1994). "*Dynamic Characterization and Damage Detection in the I-40 Bridge Over the Rio Grande*," Los Alamos National Laboratory, Los Alamos, New Mexico, Report No. LA 12767-MS.

3D SEISMIC DAMAGE SIMULATIONS FOR AN EXISTING BRIDGE SUBSTRUCTURE USING NONLINEAR FEM CALIBRATED WITH MODAL NDT

C. L. Mullen, P. Tuladhar, B. LeBlanc, S. Shrestha

Department of Civil Engineering, University of Mississippi,
203 Carrier Hall, University, MS-38677, USA

ABSTRACT

A seismic damage evaluation procedure is discussed and multi-level computational simulations are presented for an existing three-span, nine lanes, Interstate Highway Bridge in north Mississippi, situated on the rapidly growing southern perimeter of metropolitan Memphis, Tennessee. Nonlinear damage response and displacement ductility performance predictions are presented for the reinforced concrete bents and their foundations using nonlinear, static and dynamic, finite element models. Both 2D and 3D models have been developed using a variety of features of the commercial software, ABAQUS. Of particular relevance here are the beam-column elements which accepts the user-specified damage points on the moment versus curvature relations in each direction, estimated here using section fiber models in the software, BIAX. A performance-based evaluation is presented including: 1) a pushover analysis of a 2D model of an individual bent to estimate the shear and drift capacity, 2) eigenvalue analysis of 3D models having different representations of the foundation boundary conditions, including a) fixed base, b) 6 DOF linear springs for soil beneath the footings, and 3) 3D computational simulations, both spectral and time history, of the bridge system including substructures and superstructure, under earthquake motions applied at the foundation level, at an intensity of M=7.5. Prior to embarking on the damage simulation, field vibration tests were conducted using a portable 4-channel accelerometer system placed in two arrays targeting a characteristic lateral and a longitudinal deformation mode, respectively. Frequency response functions and mode shape animations produced using commercial software, SigLab and StarModal provided confirmation of the eigenvalue analyses of the 3D ABAQUS models. The application of multilevel simulations to seismic vulnerability evaluation have been demonstrated previously in a recent computational study at the University of Mississippi, involving an aging highway bridge located on the University campus. The simulations prove invaluable in quantifying, using performance-based design concepts, benefits of proposed seismic retrofits schemes, prior to embarking on expensive procedures.

KEYWORDS

Nonlinear FE simulations, seismic response, RC bridge, field test, modal analysis.

INTRODUCTION

The bridge is located near Memphis in Desoto County, Mississippi, on Highway No. 302 (Goodman Road) over Interstate I-55, which functions as an important lifeline to the rapidly growing community of Southaven. The bridge also functions as an important lifeline to the Baptist Memorial Hospital – DeSoto, a hospital, which regional planners view as a critical backup facility for hospital in Memphis in the event of a catastrophic seismic event along the southern portion of the New Madrid fault system.

According to the bridge plans, the selected bridge was designed in 1988 to satisfy AASHTO 1983 criteria for safety under operating and environment loading. Significant changes have occurred since 1983 in AASHTO criteria as well as in the profession's knowledge of the characteristics of likely earthquake motions in the New Madrid seismic zone and likely damage response of reinforced concrete highway bridges subject to ground shaking.

OBJECTIVES

The primary objective of the work is to establish seismic vulnerabilities for the selected bridge. Vulnerabilities are defined here as the failure to meet performance expectations that may vary over a range of likely earthquake intensities. The focus of this assessment is on the substructure elements including the three bents, the two abutments, and each of the bent footings.

BRIDGE STRUCTURE

The bridge plans indicate that the structure consists of two adjacent skewed (9° RF) four span concrete underpasses, built using phase construction, supporting a total of nine 12-ft wide lanes of traffic on MS 302 over six 12-ft wide lanes of traffic on I-55. The southern portion (Phase I) was constructed first alongside the original MS 302 Bridge and carries five 12-ft wide lanes. The northern portion (Phase II) was then constructed after demolition of the original MS 302 Bridge and carries four 12-ft wide lanes. In the current operation, the northern portion carries westbound traffic while the southern portion carries eastbound traffic.

In each phase, the substructure consists of three reinforced concrete (RC) two-column intermediate bents supported on shallow RC footings, except for the central bent which is supported on an array of precast prestressed concrete (PC) piles under the footing. The columns are trapezoidal in plan with longitudinal reinforcement bars that extend into the cap at the top and are lap-spliced to dowels that extend into the footing at the base. The base of the column has construction joints with shear keys. The RC caps at each intermediate bent are separated by an inch gap at the median line except where short dowels extend into the cap ends.

The end bents consist of two lines of PC batter piles with a continuous rectangular RC cap. A RC end wall sits atop the cap with vertical reinforcement bars lap-spliced at the base to a U-shaped reinforcement bar cast inside the cap. The end wall tapers down the exterior end of the abutment. The deck superstructure consists of nineteen PC I-girders of two lengths tied together transversely with various spaced RC diaphragms and joint beams. The girders support a poured-in-place PC slab that is poured continuous at the intermediate bents, and includes a RC diaphragm at intermediate bent. The deck transmits self-weight dead load and live load from vehicular traffic, wind and seismic inertial live loads to the substructure by means of neoprene pads atop the caps. There is a two-inch (depending on the temperature) gap between the deck slab and the end wall at both end bents.

ASSESSMENT PROCEDURE

The procedure for assessing seismic vulnerabilities of the bridge substructure consists of a synthesis of the following multi level investigations.

Level 1. Uses two-dimensional (2D) nonlinear static analyses of lateral load versus deformation to determine the stiffness, capacity, and the intermediate damage states of the bent under quasi-static monotonic loading.

Level 2. Linear elastic 3D multi-mode system eigenvalue and response spectrum analyses to establish compliance with current AASHTO LRFD Bridge Design Specifications for a regular multi-span critical bridge in Seismic Zone 2 under the Extreme Event 1 load combination.

Level 3. Nonlinear dynamic, 3D finite element, time-history, computational simulations including soil-structure interaction, to develop damage response indices needed for a performance-based engineering evaluation. This level is considered the most realistic.

LEVEL 1: (NONLINEAR PUSHOVER) ANALYSIS:

The primary quantities of interest in the Level 1 analysis are the key points on the load-deformation curve associated with damage limit states such as first concrete cracking, first steel yielding, and formation of either a plastic hinge in the case of flexure or of brittle failure in the case of shear.

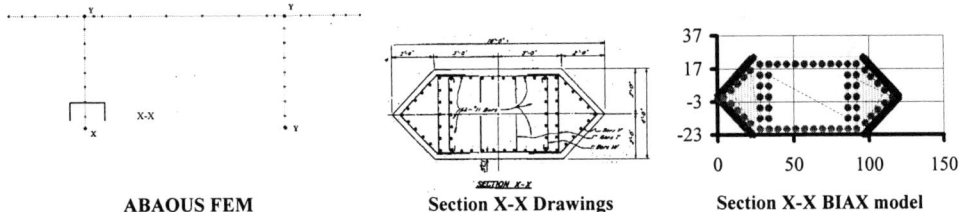

X- Input at base of column level
Y- Output at various positions on the bent

Figure 1: An intermediate bent model

This analysis was carried out using nonlinear finite element analysis. An intermediate bent was modeled in ABAQUS using beam elements (B21). The mesh for the model has 35 elements. The model is as shown in the Figure 1.

The base of the bent is fixed in all 3 degrees of freedom. A vertical load of 45 Kips (slab dead load) is applied at the nodes where the girders rest on the bent. In this analysis a lateral load is applied at one end of the beam. It is gradually increased until the bent fails.

MATERIAL PROPERTIES

The material used are, concrete having the strength of 4,500 psi and Steel having yield strength of 60,000 psi. Figure 2 and 3 show the stress strain curves of concrete and steel respectively, used by BIAX (a computer program for the Analysis of Reinforced Concrete and Reinforced Masonry Sections) to get the nonlinear properties for each of the sections. Figure 4 shows a typical BIAX result for the column section.

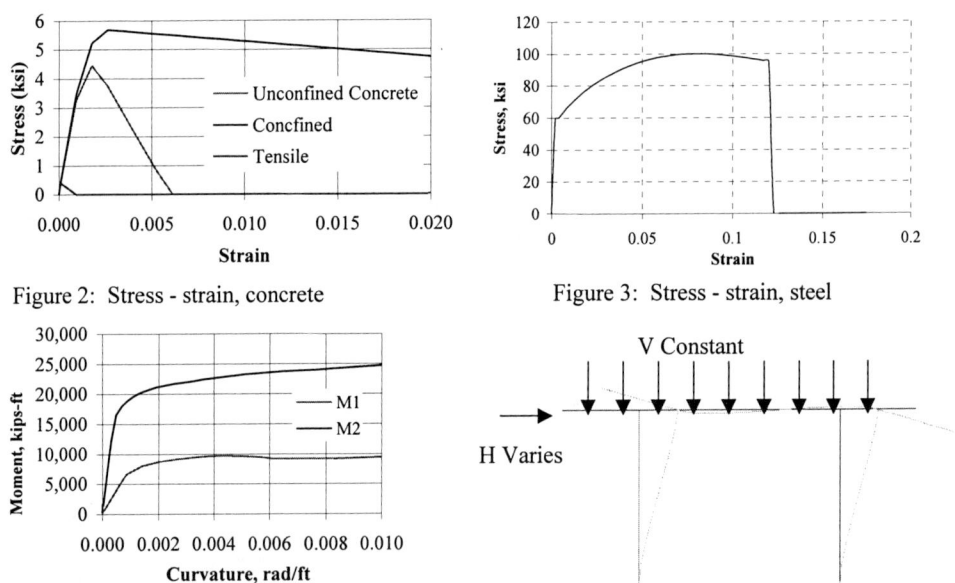

Figure 2: Stress - strain, concrete

Figure 3: Stress - strain, steel

Figure 4: M-K curve for the column

Figure 5: Displaced shape of the intermediate bent

ABAQUS RESULT

Figure 5 shows the displaced shape of the intermediate bent after the application of lateral load. From the load-drift curve (Figure 6) and moment curvature (Figure 7), it is seen that the bent yields at 1491Kips. A collapse mechanism may be predicted using plastic methods at a load of 2119 Kips This push over analysis is carried out since it is one of the primary mode that causes the failure of the substructure. The eigenvalue analysis (shown later) shows that this mode shape occurs at a frequency of 3.83 Hz.

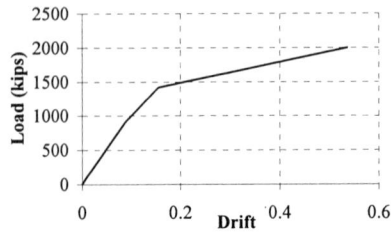

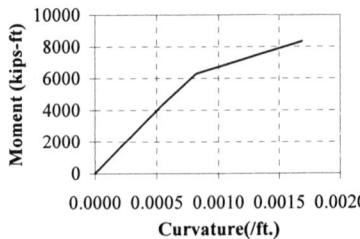

Figure 6: Load-Drift between top and base of column

Figure 7: M-K at the base of column

Level 2: (Eigenvalue and Response Spectrum) Analysis:

This analysis establishes whether the bridge would be considered safe if designed as a new bridge under current design specifications. An eigenvalue analysis provides important information on the frequencies and mode shapes of the existing system prior to the inducement of any damage. Such information characterizes the dynamic tendencies of the undamaged system without regard to any loading from the specific earthquake event. For an event of any given intensity, this analysis can only predict whether the system will respond dynamically in a way that will exceed the first cracking limit state, after which the properties of the existing system change and the assumptions of linear elasticity become invalid.

Figure 3 shows a 3D model of the bridge created for analysis in ABAQUS. The bent beams and columns were modeled as nonlinear beam elements, the deck beams as linear beam elements, and the deck slabs as linear shell elements. Two boundary conditions were considered: **fixed** base case and **spring** base case. In the fixed base case, the base of the column at the footing level is fixed in all 6 degrees of freedom for intermediate bents. There is a 2 inch gap between the end wall and the deck slab. The steel rocker bearing allows frictional sliding in 1 and 2 axes and rotation in the 2 axis. In both, the fixed base case and spring base case, frictional effect has been neglected in the result presented in this paper. Vertical translation is fixed at the end bents for the Fixed base case. In the Spring base case, linear springs replaced all fixed degrees of freedom in the Fixed base model. The stiffness of the springs was determined as per Federal Guidelines (FEMA, 1997). The soil properties for the Southaven area used in calculating the spring stiffness are as shown in Table 1.

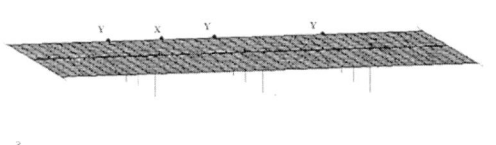

Accelerometer positions for ambient vibration measurements
X- Input at roadway level
Y- Output at bent cap level

Figure 8: 3-D model of the bridge

TABLE 1
SOUTHAVEN SUBSURFACE GEOLOGICAL PROPERTIES

Layer	Z (ft.)	H (ft.)	Weight Density (k/ft^3)	Shear Wave Velocity (ft/s)	Poisson's Ratio	Mass Density	Shear Modulus (k/ft^2)	Young's Modulus (k/ft^2)
1	0	33	0.116	554	0.33	0.0036	1106	2941
2	33	13	0.114	987	0.32	0.00354	3449	9105
3	46	62	0.118	1233	0.35	0.00367	5571	15042
4	108	49	0.106	1937	0.4	0.00329	12351	34583

From the eigenvalue analysis of the fixed-base model, the fundamental frequency of the structure is seen to be 1.29 Hz. This corresponds to the vertical translation of the deck. The vertical translation of the deck does not contribute much to the bending of the columns, which are of primary interest. The first mode that causes bending in the columns occurs at 3.20 Hz (Mode 17). It is caused by the rotation of the deck. The first mode that causes bending in the columns in the lateral direction occurs at 3.83 Hz (Mode 18). It is caused by the lateral translation of the deck. The first mode that causes bending in the columns in the longitudinal direction occurs at 5.25 Hz (Mode 33). It is caused by the longitudinal translation of the deck. Figure 9 through Figure 12 shows the comparison between the fixed base model and spring base model. It is seen that the springs do not behave very well to capture the behavior of the soil at lower frequencies. At higher modes the frequencies are reduced. This is explained by the reduction of stiffness at the boundary. In further study, with the addition of geology, there should be some significant change in the frequency.

Figure 9: Lateral direction (Fixed) - 3.20 Hz

Figure 10: Lateral direction (Spring) – 2.9 Hz

Figure 11: Longitudinal direction (fixed) - 5.25 Hz

Figure 12: Longitudinal Direction (spring) – 4.85 Hz

VALIDATION OF THE MODEL

In order to confirm the modeling assumptions, ambient vibration measurements were performed using portable accelerometers. This technique offers no risk of damage to the bridge. The Figure 13 shows the basic layout of the field measurement.

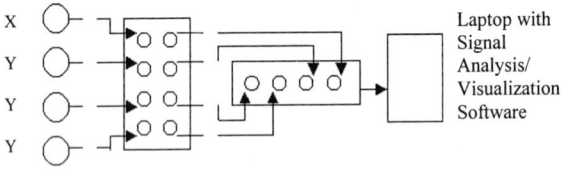

Figure 13: Ambient vibration measurement scenario

The data received at the laptop were stored onsite. These stored data were later filtered and processed in a modal analysis software called StarModal. In StarModal software, classical mode models of the bridge elements where measurements were taken was created. Nodes were provided at the measurement points. The data from the Siglab were imported to Star Modal and processed to obtain the mode shapes. These mode shapes and frequencies were then compared with those obtained from ABAQUS. Figure 14 shows one of the Star Modal results, which is compared with the numerical results from ABAQUS. From the figures, we see that the frequencies and the mode shapes are similar. Hence we conclude that the modeling of the bridge in ABAQUS is acceptable.

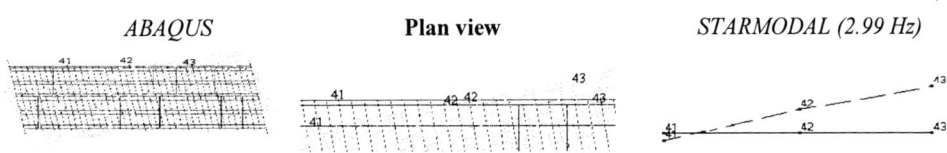

Figure 14: Comparison of experimental and analytical results (see Figures 9 and 10)

RESPONSE SPECTRUM ANALYSIS

Response Spectrum analysis was carried out on the linear fixed model in ABAQUS. The acceleration spectrum used for the analysis is shown in Figure 15 and Table 2 shows the peak responses obtained from the analysis.

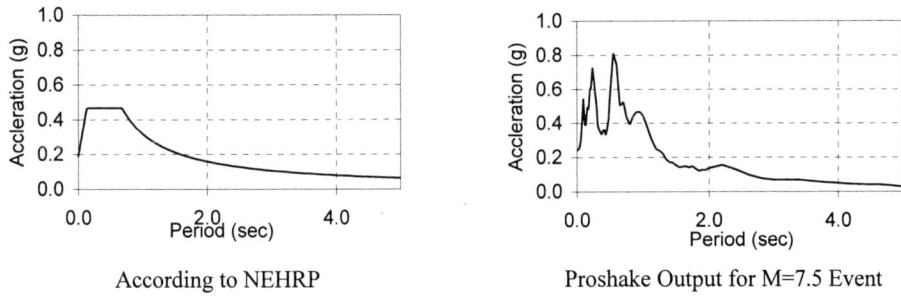

Figure15: Acceleration spectrum

TABLE 2
PEAK RESPONSES FOR THE FIXED MODEL WITH NEHRP DESIGN ACCELERATION SPECTRUM

Location		Response
Top of bent 4 of Phase I	0.1398	Peak Acceleration (g) in 1-axis
Top of bent 2 of Phase I	0.4680	Peak Acceleration (g) in 2-axis
Top of bent 2 of Phase I	0.0451	Peak Acceleration (g) in 3-axis
Top north column bent 2 Phase I	0.0070	Peak Displacement (ft) in 1-axis
Top north column bent 2 Phase II	0.0260	Peak Displacement (ft) in 2-axis
North column of bent 4 Phase I	0.0003	Peak Drift Ratio in 1-axis
North column of bent 4 Phase II	0.0012	Peak Drift Ratio in 2-axis

LEVEL 3: (NONLINEAR DYNAMIC) ANALYSIS:

A dynamic analysis was carried out for an input motion of M=7.5. It was generated using the SMISM program provided by David Boore of the USGS, which generates motion at the bedrock. Proshake, a commercial program by EduPro, Inc., for 1D equivalent linear ground response analysis was used to propagate the acceleration time history at the bedrock level to the required level of soil at the footings and further to the abutments. From the analysis, from the moment-curvature criteria of failure, there seems to be considerable damage to the columns in the longitudinal directions. Figure 16 and 17 shows the acceleration time history and moment curvature hysteresis respectively. Final conclusions regarding vulnerability await the completion of the final Level III model, which will include soil-structure interaction and consider three SMSIM seismic events, each at varying levels of intensity.

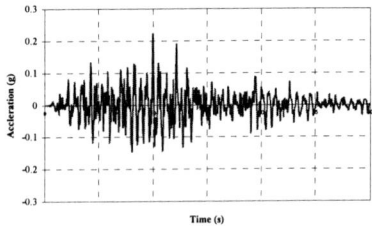
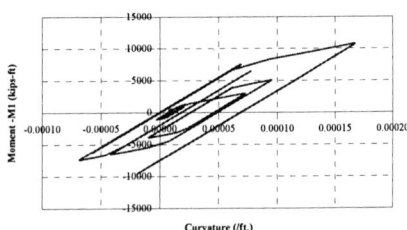

Figure 16: X- acc. response at top of bent 2 of phase I

Figure 17: Moment (M1) curvature hysteresis at bottom of south column of bent 2 phase I

CONCLUSION

A multilevel performance based engineering analysis procedure has been proposed for seismic vulnerability assessment of a reinforced concrete highway bridge in northern Mississippi. The nonlinear response during damage has been predicted using a 2D static FE model of a typical bent. Mode shapes and frequencies of 3D models have been computed and give insight to deformations and sensitivity to assumptions regarding boundary conditions. Reliability of the model predictions is enhanced by ambient vibration measurements keyed to the 3D model. Such a calibrated FE model is used in computational simulations of damage during seismic events. Response spectrum analysis gave an idea of the peak responses of the bridge. From the nonlinear dynamic analysis, it was possible to observe the damage response in more detail, as needed in a performance-based evaluation.

REFERENCES

AASHTO 1994, '*LRFD Bridge Design Specification 1st ed.*', American Association for State Highway and Transportation Officials, Washington D.C.

Hibbit, Karlsson and Sorensen, '*ABAQUS/Standard User's Manual, Version 5.8*', Hibbitt, Providence, Rhode Island, 1996.

Wallace, J.W, '*BIAX: Revision 1, A Computer Program for the Analysis of Reinforced Concrete and Reinforced Sections*', Report No. CU/CEE-92/4 Structural Engineering Mechanics and Materials, Department of Civil Engineering, Clarkson University, Postdam, New York, Feb 1992.

Federal Emergency Management Agency, '*NEHRP Guidelines for the Seismic Rehabilitation of Buildings, (FEMA Publication 273)*', October 1997

DAMAGE ASSESSMENT OF CORRODED REINFORCED CONCRETE BEAMS USING MODAL TESTING

H. Abdul Razak [1] and F.C. Choi [1]

[1] Department of Civil Engineering, University of Malaya
50603 Kuala Lumpur, Malaysia

ABSTRACT

A laboratory investigation to study the effect of general corrosion on the modal parameters of reinforced concrete beams was conducted. Full scale beams were subjected to reinforcement corrosion until an appreciable amount of steel corrosion damage was introduced using an accelerated technique namely galvanostatic method. The states of damage in the test beams were assessed through measurement of crack width, loss of area and half-cell potential. Modal tests were performed on the test beams after corrosion damage and the modal parameters extracted were compared against that from control beams. The results obtained showed changes in the modal parameters especially natural frequencies and mode shapes. The intensity of the drop in the measured natural frequencies was sensitive to the deterioration state of the beams. Hence, changes in natural frequencies could be established as a damage indicator, which relate to the loss in flexural stiffness of the corroded beams. Using modal assurance techniques to perform orthogonality checks on sets of mode shape from the control and corroded beams, changes could be visualised which were dependant on the degree of corrosion damage induced. Finally, the load carrying capacity of the beams were determined through static load test and the results were correlated with the state of corrosion damage and changes in the modal parameters. This investigation provided further insight on the use of modal parameters to detect and assess damage in structural concrete elements, which can be useful for structural appraisal purposes when applied to full scale structures.

KEYWORDS

Corrosion, reinforced concrete, modal test, natural frequency, mode shapes, load carrying capacity.

INTRODUCTION

For the past three decades, the construction of structures and infrastructures was heavily relied on the use of reinforced concrete. It was based on the presumption that protection of the passive film layer on the steel reinforcement was permanent. Unfortunately, experiences have proved otherwise in that the reinforcement can be depassivated due to carbonation of the concrete or contamination by chloride ions and setting the scene for corrosion. Therefore, it is important to assess the health condition of reinforced concrete structures especially that constantly exposed to weathering agent and one of the resulted deteriorations is reinforcement corrosion damage. Past investigations and surveys have revealed that deterioration caused by reinforcement corrosion causes the design life span of structures to be shortened. As reported by Wallbank (1989), of 200 concrete highway bridges examined, 25 had minor blemishes, 114 classed as fair and 61 were categorised as being in poor condition. The primary reason for this deterioration was caused by corrosion of reinforcement.

This phenomenon has been confirmed by researches worldwide. Transport and Road Research Laboratory, UK (1991) studied on the effects of corrosion deterioration on the assessment of concrete bridges. The results showed that severe general corrosion could cause a complete breakdown of the bond between concrete and the reinforcement. The result Yoshihiro et al. (1990) found that the reduction in stiffness and load carrying capacity occurred in the corrosion damaged beams. Pritpal and Elgarf (1999) had present an experimental study on corroded reinforced concrete beams to determine its residual flexural capacity. The results show marked reductions in flexural strength due to reinforcement corrosion and nomograms were established to predict the long-term residual strength of corroding flexural members.

Besides quantitative assessment of the state of art of the test structures mentioned above, qualitative assessment is also popular as an appraisal tool. Several assessment standards have been established such as Baker et al. (1977) and Bridge Unit, Department of Road Work (JKR), Malaysia conditioning rating (1995). Peterman et al. (1999) had also carried out durability assessment of bridges with full span prestressed concrete form panels exposed to chloride ions. One of the qualitative techniques used by the authors was visual inspection. The result revealed that the steel reinforcement most extensively corroded is the longitudinal bars. Significant loss of area of the steel reinforcement due to corrosion was not observed.

RESEARCH SIGNIFICANCE

This paper is concerned with the strength of deteriorated beams caused by corrosion of reinforcement. The primary objective of this experimental study is to establish a relationship between the strength of defect beams and the changes of its natural frequencies and modal damping. Visual inspection, half-cell potential and load carrying capacity test were adopted to assess the degree of damage due to reinforcement corrosion. Modal tests on the beams were conducted to determine the modal parameters. By comparing the data obtained, the effect of corrosion damage on the modal parameters such as natural frequencies and mode shapes can be investigated. Furthermore, the modal experimental investigation presented in this paper would provide an alternative corrosion assessment tool besides the conventional techniques.

EXPERIMENTAL PROGRAMME

In this investigation, reinforced concrete beams were cast as test specimens. Two of the beams were induced with different state of reinforcement corrosion while the remaining two acted as controls.

Preparation of test beam

The concrete for the test beams were designed in accordance to DOE method for Grade 30 with 150mm slump. The mix proportion is shown in Table 1. The beam with cross section dimensions of 150mm in breadth and 250mm in height had an overall length of 2200mm. It was reinforced with two 16mm diameter high yield bars and no stirrups were provided. In order to ensure that corrosion of the bars occurred under accelerated conditions, minimal concrete cover of 20mm was provided. In addition, 5% of sodium chloride by weight of cement was introduced into the concrete mix to assist in accelerating the corrosion process.

TABLE 1
MIX PROPORTION FOR GRADE 30 CONCRETE

Mix Properties (per m^3)	Weight (kg)	Percent (%)
Ordinary Portland Cement (OPC)	350	15.5
Fine Aggregate	750	33.2
Coarse Aggregate (20mm)	950	42.0
Water	210	9.3
Total	2260	100.0
W/C Ratio	0.6	
Sodium chloride (NaCl)	5.0% by weight of cement	

Accelerated corrosion

Corrosion in the beam was induced and accelerated through a process known as the galvanostatic method (1990). The method was modified to suit the requirements of a full scale beam in this investigation. The set-up of the accelerated test and the electrical circuit are shown in Figure 1 and Figure 2, respectively. The copper plate acted as the anode while the main reinforcement bars in the beam acted as the cathode. The ponding of sodium chloride solution provided the electrolyte to complete the electrical circuit.

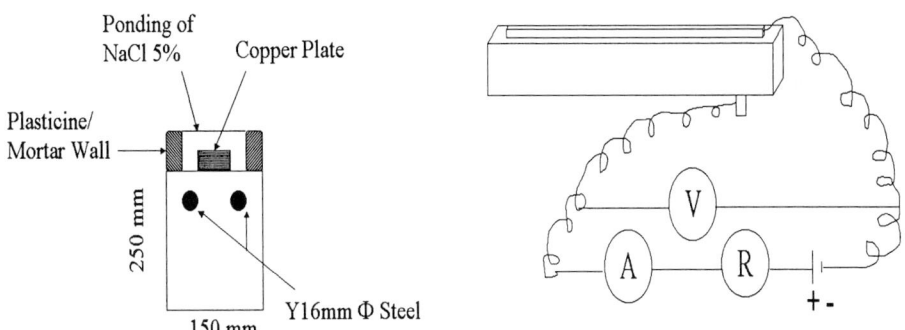

Figure 1: Galvanostatic method   Figure 2: Proposed electrical circuit

Prior to applying accelerated corrosion on the beam, trial mock up tests on small scale specimens were performed to establish the parameters and beam details necessary to induce the worst possible damage effect in the structural element. The damage considered included rust staining, cracking, spalling, delamination of concrete cover, loss in member stiffness and structural capacity. It was finalised that the beam should be subjected to the condition of 1.0A of current for about 6 to 7 hours a day by the galvanic action.

Corrosion assessment

Signs of corrosion, cracking and crack pattern were recorded through visual inspection. Longitudinal crack widths on the side of the beam were measured using a crack microscope. The state of corrosion for the defect beams were classified according to ASTM STP 629 (1977). The visible defects on the test specimen can be categorised from very slight cracking to severe cracking. The loss of area is determined by direct measurement of remaining steel reinforcement. Modal tests were performed on all test beams prior to load testing the beams to failure to ascertain the load carrying capacity. The extent and severity of the damage on one of the defect beams is illustrated in Figure 3.

Figure 3: Overview of crack on side of the test beam

Modal test

Modal testing was adopted to obtain the modal parameters namely the natural frequencies and mode shapes, using the transfer function method as illustrated in Figure 4. The transfer function method is based on the use of digital signal processing techniques and the fast Fourier transform (FFT) algorithm to measure transfer function between different points on the test structure. In this investigation, white noise was used as the excitation signal for the vibration shaker, which was permanently positioned at the reference point number 72 as shown in Figure 5. The input force was measured by means of a force transducer mounted on to the soffit of the beam and connected to the shaker by means of a flexible rod. The sine wave rated force of the vibration shaker is 294N.

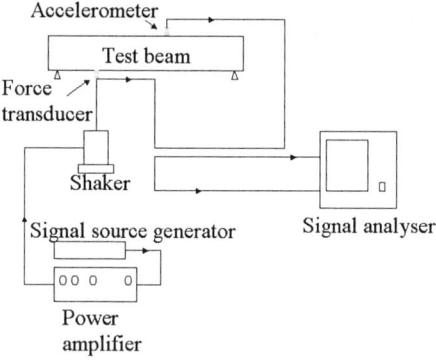

Figure 4: Experimental setup of modal testing

The response signal was picked up using a single general-purpose low impedance accelerometer having a sensitivity of 100mV/g. The accelerometer was moved from point to point until all the measurement locations were covered. The measurement points, which were located on the top surface of the beam, are also shown in Figure 5. It can be seen that points 1, 56, 57, 112, 113 and 168 were located at the supports. The distance between each point was 40mm and 50mm in the longitudinal and transverse directions of the beam respectively.

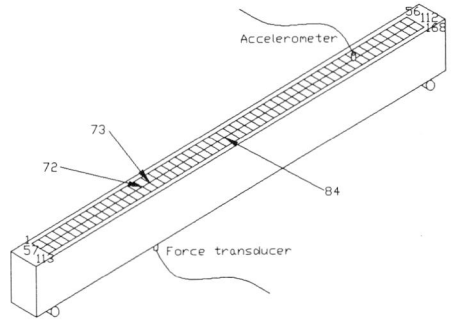

Figure 5: Measurement location

Initially, the transfer function spectrum within a 4kHz frequency span was obtained in order to locate roughly the resonant frequency peaks of all the flexural modes within the band. Subsequently, zooming within a 100Hz span of the resonant frequency peak of a particular mode was carried out. The measurements were made using a block size of 400 lines thus giving a resolution of 0.25Hz per spectral line. By using modal analysis software packages namely SMS-STAR and ICATS, the curve fitting process was performed on the transfer function spectrums obtained to extract the natural frequencies and mode shapes for the first seven modes. Subsequently, the mode shape data was extracted and analysed using ICATS software to perform modal assurance techniques such as modal assurance criterion (MAC) and co-ordinate modal assurance criterion (COMAC).

Load test

The setup for the load test, which was conducted after modal testing, is shown in Figure 6. A concentrated line load was applied across the beam at mid span using a load actuator controlled by a servo hydraulic pump. The loading rate was controlled and maintained at a constant rate until the beam failed. Displacement transducers positioned at the beam soffit at mid span and a load cell placed directly above the load actuator measured displacements and applied loads respectively. These measurements were then acquired using a high-speed multichannel digitizer thus enabling the load-deflection plot to be displayed while the test was being conducted. Finally, the failure mode, crack width and crack pattern were recorded.

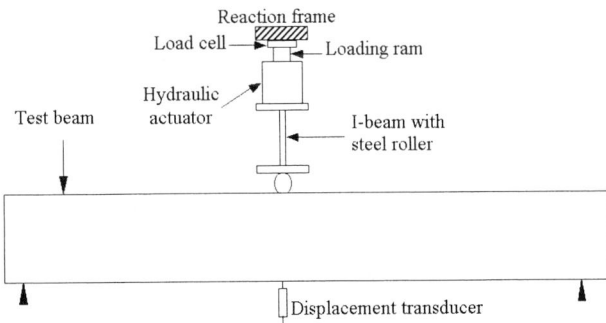

Figure 6: Load test setup

RESULTS AND DISCUSSIONS

The results obtained from the test beams were compared. The control beams provided the reference readings, which form the basis of the comparison of the modal parameters as described below. Results from the other two beams, which were subjected to corrosion, was considered as the defect readings.

State of corrosion

Visual examination revealed that both the beams subjected to corrosion had cracks running longitudinally along the sides and soffit, prior to the loading tests. The level of severity of the corrosion damage was determined visually according to Baker et al. (1977). Load test results based on the applied load at mid-span are also tabulated together with the failure mode in Table 1.

Defect Beam D1 and defect Beam D3 were inspected visually in accordance with ASTM STP 629 (1977). Based on the classification, the degree of corrosion damage for Beam D1 and Beam D3 are very slight spalling and moderate cracking, respectively.

From Table 2, it is apparent that the average crack width of Beam D1 is wider compared to Beam D3 which results in a lower failure load from the load test. The average crack width and failure load for Beam D1 are 5.0mm and 28.59kN while that of Beam D3 are 1.0mm and 41.85kN. Based on these measured values, the severity of the corrosion damage experienced by two beams are notably different and as such it is anticipated that the modal parameters being studied should also reflect these observations. Cracking and spalling of concrete in the beams were caused by swelling pressures exerted by the formation of corrosion products on the reinforcement bars. This leads to development of tensile hoop stresses within the concrete adjacent to the reinforcement thus causing a progressive loss of bond between concrete and the reinforcement. Subsequently when load is applied, there is inadequacy in the transmission of the stresses across the concrete-steel interface resulting in a reduction of load carrying capacity. As the corrosion damage becomes more severe, as in the case of excessive spalling and loss of concrete cover, the ability to transfer the stresses will be diminished. Consequently, the load carrying capacity drops immensely and premature failure is only prevented by the sole action of the reinforcement acting as a tie.

After the beams were load tested, it was possible to expose the reinforcement bars and to closely examined and obtained measurements for loss in steel area in the corroded test beams. As indicated in Table 2, both beams suffered a loss of area of less than 8% and the difference between the two beams was insignificant. Thus for this particular investigation, the loss of steel area is not an indicator of the severity of damage as reflected by the measured crack widths and failure loads. This could be explained by the fact that in general corrosion, the loss of steel area experienced is small even though the corrosion has reached an advanced stage. Yoshihiro et al. (1990) has also suggested that calculation of loss of area may be a useful estimation of the integrity of corrosion damaged beams only if the beams fail in flexure, implying that this is probably realised in the case of localised corrosion. However, in general corrosion the tendency is that loss of bond between concrete and the reinforcement has a significant contribution to the failure mode of the beam in general.

TABLE 2
CORROSION DAMAGE AND STRUCTURAL CAPACITY

Beam	Crack width* (mm)	Degree of concrete cracking/ spalling**	Loss of area*** (%)	Load carrying capacity (kN)	Failure mode
Beam D1	5.0	Very slight spalling	7.77	28.59	Bond-Flexure
Beam D3	1.0	Moderate cracking	7.65	41.85	Bond-Shear
Control	--	--	--	66.64	Flexure-Shear

* *Average major crack width due to corrosion.*
** *Evaluation of degree of concrete cracking/spalling by using ASTM STP 629[6].*
*** *As percentage of material loss by direct measurement of the remained sound material to the sound member before corroded.*

Referring to Table 3, it shows readings of half-cell potential of Beam D1 and Beam D3. The analysis of these data shows that the average value of half-cell potentials in Beam D1 and Beam D3 were –360mV and –610mV, respectively. It was obvious that the potential of Beam D1, in the range of –300mV and –420mV, is less negative relative to Beam D3, in the range of –570mV and –660mV. The potential readings were quite consistent throughout the whole specimen for both cases with standard deviations of –33mV for Beam D1 and –20mV for Beam D3. The highest negative values of half-cell potential readings were obtained from Beam D3, where less corrosion deterioration occurred. Arup (1983) described general corrosion as a general loss of passivity, resulting from either carbonation or the presence of excessive amounts of chloride and the potentials are typically within –450mV to –600mV. Judging from similarity of the results obtained and the conclusions made by Arup (1983), it is possible to categorise the induced defect in Beam D1 and Beam D3 as general reinforcement corrosion.

TABLE 3
RESULTS OF HALF-CELL POTENTIAL

Statistics	Beam D1 (-mV)	Beam D3 (-mV)
Minimum	302	572
Maximum	423	660
Average	362.83	610.27
Standard Deviation	33.06	19.85

Modal test results

When a system is subjected to a certain degree of damage or deterioration, it experiences a change in stiffness. Subsequently it causes the dynamic parameter namely natural frequency and mode shape to chage. The relationship for a simply supported beam as suggested by Demeter (1973) is as follows:

$$f_n = \frac{n^2 \pi}{2} \sqrt{\frac{EI}{mL^4}} \qquad (1)$$

where f_n = natural frequency
n = mode number

E	=	modulus of elasticity
I	=	second moment of area
M	=	mass per unit length
L	=	span length

Since the test beams have similar dimensions and simply supported over the same span,

Therefore
$$f_n \propto \sqrt{EI} \qquad (2)$$

From Eqn. 2, it is apparent that the natural frequency of the beam is proportional to the stiffness EI of the beam. Consequently, a reduction in stiffness leads to a reduction in the natural frequencies.

Drop in natural frequency

Two sets of modal data were compared namely the natural frequencies, obtained before and after the dynamic system experiences a change. In other words, if there is a drop in natural frequency in a system, it can be inferred that the system has undergone changes. The magnitude of the changes is also an indicator of the severity or state of the change experienced. In this investigation the changes are related to the damage induced by corrosion of the reinforcement. This is apparent in the changes in the natural frequencies of the defect beams as compared to the control beam, for the first seven modes, as shown in Table 4.

There is a higher decrement in natural frequencies for Beam D1 compared to Beam D3. The drop for Beam D1 ranges from 4% to about 11% while that of Beam D3 the range is between 1% to 4%. The drop in load carrying capacity is also consistent whereby Beam D1 recorded a higher drop of 57% compared to 34% for Beam D3. As the effect of corrosion was more severe in Beam D1 as compared to Beam D3, it experienced a greater loss in bending stiffness, lower load carrying capacity and bigger drop in natural frequencies. It can also be observed that in general the trend of the magnitude of the percentage change in natural frequencies increases with higher modes. Similar trends were also reported in the investigations conducted by M. Kato (1986) and O.S. Salawu (1995).

TABLE 4
NATURAL FREQUENCY AND LOAD CARRYING CAPACITY

Mode	Natural frequencies (Hz)		Drop in natural freq. (Hz)	Drop in natural freq. (%)	Drop in load carrying capacity[#] (%)	Natural frequencies (Hz)		Drop in natural freq. (Hz)	Drop in natural freq. (%)	Drop in load carrying capacity[#] (%)
	D1	C4				D3	C4*			
1	70.82	73.815	3.00	4.06	58.62	63.01	63.67	0.66	1.04	33.77
2	461.42	482.45	21.03	4.36		449.53	457.21	7.68	1.68	
3	774.98	805.06	30.08	3.74		783.58	795.65	12.07	1.52	
4	1153.98	1222.35	68.37	5.59		1173.40	1225.29	51.89	4.23	
5	1539.87	1666.94	127.07	7.62		1629.71	1658.98	29.27	1.76	
6	1967.98	2137.71	169.73	7.94		2089.61	2130.49	40.88	1.92	
7	2345.70	2630.02	284.32	10.81		2558.97	2614.48	55.51	2.12	

D1, D3, C4 and C5 are the modal data for the defect Beam D1, Beam D3 and control Beam C4 and Beam C5, respectively.
[#] As a percentage compared to the control Beam C4.
* Control readings taken again later to eliminate the concrete maturity effect.

Mode shape data

Fundamentally, mode shapes express the oscillation of an underdamped single degree of freedom (SDOF) system when it is subjected to an initial perturbation. Subsequently when it is left to move freely, the system will oscillate about the static equilibrium position. This is referred to as its natural modes of vibration. Mathematically speaking, mode shape of a dynamic system is characterized through an unique property described by its free vibration natural frequency.

The first seven flexural vibration modes were recorded as shown in Figure 7 for purpose of comparison. The first flexural vibration mode as exhibited in Figure 7(a) is basically a representation of the amplitudes in the form of a half sinusoidal curve. The second flexural vibration mode shape would be a full sinusoidal curve, while the third flexural vibration mode is a one and half sinusoidal curve and so on as the mode increase. It is also apparent that as the mode increase, the mode shape becomes more complex. The fourth flexural mode shape was not considered here due to the quality of the measured FRF and the reason being the excitation point is too close to the node of the considered mode.

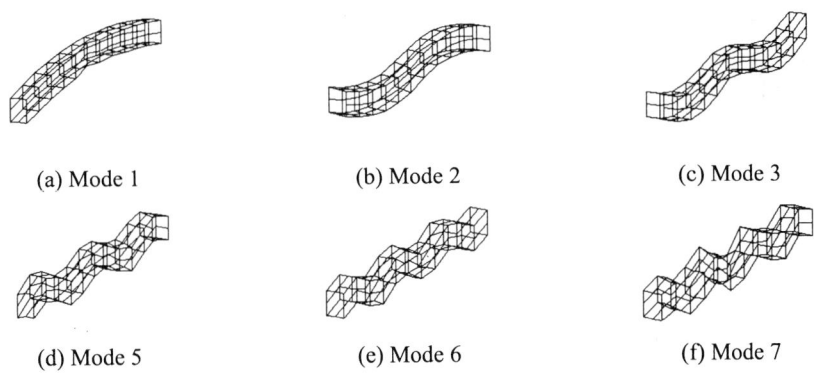

(a) Mode 1   (b) Mode 2   (c) Mode 3

(d) Mode 5   (e) Mode 6   (f) Mode 7

Figure 7: Flexural vibration mode shapes

Modal assurance criterion (MAC)

Modal assurance criterion (MAC) is a technique to compare two sets of mode shapes by performing an orthogonality check on the eigenvectors. The orthogonality check is carried out to ascertain the angle between the two sets of eigenvectors is perpendicular or 90 degrees. If changes occur in a test structure, there would be a modification to the mode shapes. Thus, the mode shapes before and after the system experience changes will cause incompatibility between these mode shapes and its eigenvectors will not be perfectly orthogonal. In this investigation, the changes are related to the damaged induced by the corrosion of the reinforcement.

Based on the MAC values in this study for the control beams as shown in Table 5, all the modes demonstrate MAC value higher than 95 percent, ranging from 96.6 to 99.2 percent. The fourth mode was ignored in the threshold consideration due to the quality of the measured FRF. The reason for this is that the excitation point is too close to the node of the considered mode. Therefore, a value of 90 percent for the MAC was suggested as an appropriate threshold to indicate changes to the mode shape which are caused by damage induced in the system. This suggested threshold value has taken into consideration human errors and noise during data acquisition as well as other inaccuracies during the analysis process.

The correlation values for all the first seven modes are shown in Table 5. These modes have been taken as real part of the eigenvectors and were calculated by using the post processing software namely ICATS (1997) with the exception of mode 4 for the reason mentioned earlier. The ICATS programme is able to perform post processing analysis to produce eigenparameters such as eigenvalues and eigenvectors. The produced eigenvectors are compiled in a cmp file. The cmp files are then used to compute the MAC. The MAC results are shown in diagrammatic form and the actual values are also given.

From Table 5, the higher modes in the more damaged Beam D1 that is modes 5, 6 and 7, registered values less than 90%. While Beam D3 with less corrosion damage, it only registered a value less than 90% for mode 7. It is apparent that for a more deteriorated beam, a MAC value below the threshold would first occur at a lower natural frequency mode as compared to a less deteriorated beam.

The reason being the mechanical properties, such as bending stiffness, of the beam changes tremendously with the corroded reinforcement. In addition, the rust products also caused the deterioration of the concrete-steel interface. Subsequently, for a higher mode which has a more complex mode shape, the modification on the mode shape becomes more severe. This may due to a relatively higher non-synchronous movement of concrete and steel during the excitation for higher modes in the deteriorated beam compared to a control beam. As the beam deteriorated further, the MAC value registered lower than 90 percent would occur on much earlier mode compared to the less deteriorated beam. This trend has confirmed the earlier estimation in corrosion assessment and shift of frequency that Beam D1 suffers more damage than Beam D3.

TABLE 5
MAC DATA FROM ICATS

Mode	MAC (%)		
	D1-C4	D3-C4*	C4-C5
1	99.7	99.4	99.2
2	98.7	99.1	97.1
3	98.3	99.5	98.5
4	59.7	46.6	62.8
5	81.4	93.0	99.1
6	77.7	97.6	99.1
7	52.2	59.2	96.6

D1, D3, C4 and C5 are the modal data for the defect Beam D1, Beam D3 and control Beam C4 and Beam C5, respectively.
* Control readings taken again later to eliminate the concrete maturity effect.

Co-ordinate modal assurance criterion (COMAC)

Co-ordinate modal assurance criterion (COMAC) is a technique to indicate the correlation between mode shapes that is the eigenvectors at pre-determined measurement points on the test structure. If local changes occur in a test structure, there would be a modification to the mode shapes. Thus, the eigenvectors before and after the system experience changes will cause incompatibility between mode shapes.

By using COMAC values, it is possible to infer damage locations on the test structure. A COMAC value close to 100 percent indicates good correlation at the pre-determined locations between two sets of mode shape data. In order to compute the COMAC values, correlated modes from the two sets data are paired, for example, eigenvector for mode 1 of control beam compared to eigenvector for mode 1 of defect beam.

The COMAC values obtained are illustrated in Figure 8. COMAC values were calculated by using the post processing software namely ICATS (1997). The modes have been assumed to be real part of the eigenvectors. Eigenvectors of all modes were used in calculating the COMAC values except for mode 4 as for the reasons mentioned earlier. The COMAC values are shown in diagrammatic form only without giving the actual values for all the three rows of measurement points.

Based on the COMAC values in this study for the control beams as shown in Figure 8(a), combination of all the modes demonstrate COMAC close to 100 percent. Therefore, the COMAC values in Figure 8(a) suggested an appropriate reference to indicate, generally, no changes to the local mode shape data in the system. A qualitative assessment is drawn from Figures 8 (b) and (c) where Beam D1 shows, generally, a lower value of COMAC as compared to Beam D3. Furthermore, the COMAC values near the support for Beam D1, such as measurement locations number 1, 56, 57, 112, 113 and 168, demonstrate a much lower value compared to Beam D3. Generally, the COMAC values for Beam D1 and Beam D3 show a significant drop for most of the measurement locations. Hence, it is possible to infer that Beams D1 and Beam D3 had experienced general corrosion damage. The severity of damage in Beam D1 is higher compared to Beam D3 with most of its COMAC values dropping more than Beam D3 as weighted against the reference from the control beams as shown in Figure 8(a).

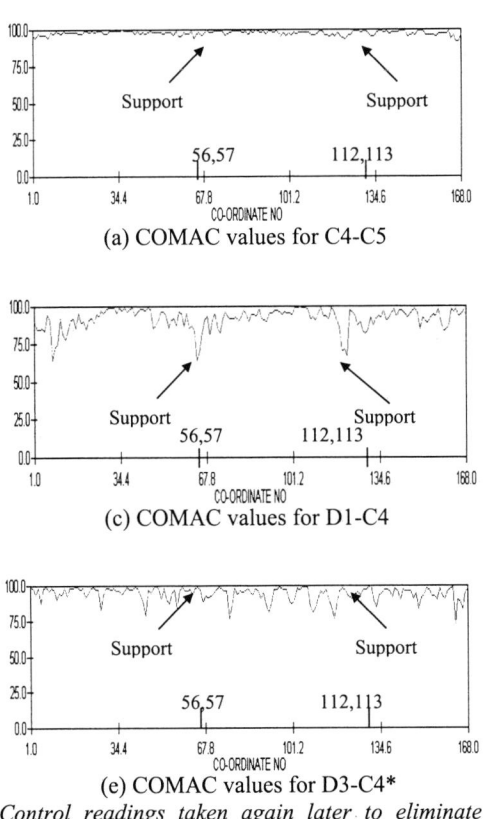

* Control readings taken again later to eliminate the concrete maturity effect.

Figure 8: Illustration of COMAC values from ICATS

CONCLUSIONS

The conclusions of the test results presented herein, show that corrosion of reinforcement in the test beams caused significant changes to the structural capacity and to the modal parameters namely natural frequency and mode shape. The changes in natural frequencies and mode shapes are also consistent with the severity of the damage induced and the load carrying capacity obtained from static load tests. Furthermore, the changes in natural frequencies for all the seven modes investigated can be established as a damage indicator.

Mode shape can also be used as a damage indicator and simultaneously as a damage assessment tool qualitatively. By employing the modal assurance techniques, there is a significant trend of modification shown on the mode shapes of the damage beams. The MAC values show a higher drop as the severity of corrosion damage increases. This consistent trend is more distinct for the higher modes. While the COMAC values could also be utilised to assess qualitatively the state of damage, whereby values significantly different from a reference threshold indicates presence of damage. The more severe the damage the greater will be the difference. It is also possible to use the values to indicate the damage location although in this study this is not obvious due to the global nature of the damage caused by general corrosion.

ACKNOWLEDGEMENT

The authors gratefully acknowledge the financial assistance provided by the National Council for Scientific Research and Development through a research grant under the Intensification of Research in Priority Areas (IRPA) programme under project 03-02-03-0610.

REFERENCES

Albert F. Daly (1991). Effects of deterioration on the assessment of concrete bridges. *Technical Paper BD/TP/89/91*, Department of Transport, London.

Arup, H. (1983). The Mechanisms of the protection of Steel by Concrete in Corrosion of Reinforcement in Concrete Construction. *A.P. Crane*, Ed., Ellis Horwood, Ltd., 151.

Baker E.A., Money, K.L. and Sanborn, C.B. (1977). Marine corrosion behaviour of bare and metallic-coated reinforcing rods in concrete. *Chloride corrosion of steel in concrete, ASTM STP 629*, Tonini D.E. and Dean S.W.Jr.: Eds. American Society for Testing and Materials, 30-50.

Bridge Unit, Department of Road Work (JKR) (1995). JKR annual bridge inspection manual. *Bridge Unit*, Department of Road Work (JKR), Kuala Lumpur, Malaysia.

Demeter G. Fertis (1973). Free vibration of two-degree spring-mass systems. *Dynamic and vibration of structures*, A Wiley-Interscience Publication, 31-35.

Imperial College Analysis, Testing and Software (ICATS) (1997). MODENT, MODESH, MODACQ and MESHGEN Reference Manual. *Mechanical Engineering Department*, Imperial College of Science, Technology and Medicine, London.

Kato M. and Shimida S. (1986). Vibration of bridge during failure process. *Journal of Structural Engineering*, ASCE, **Vol. 112, No. 7**, July, 1692-1703.

Mangat P.S. and Elgarf M.S. (1999). Flexural strength of concrete beams with corroding reinforcement. *ACI Structural Journal*, **V. 96, No. 1**, January-February, 149-158.

Peterman R.J., Ramirez J.A. and Poston R.W. (1999). Durability assessment of bridges with full-span prestressed concrete form panels. *ACI Materials Journal*, **V.96, No. 1**, January-February, 11-19.

Salawu O.S. and Williams C. (1995). Bridge assessment using forced-vibration testing. *Journal of Structural Engineering*, ASCE, **Vol. 121, No. 2**, February, 161-173.

Wallbank R.J. (1989) The performance of concrete in bridges: a survey of 200 highway bridges. *Department of Transport*, HMSO, London.

Yoshihiro T., Ken-Ichi M., Yasuo K. and Mitsunori K. (1990). Mechanical behaviour of RC beams damaged by corrosion of reinforcement. *Corrosion of reinforcement in concrete*: Edited by Page, C.L., Treadaway, K.W.J. and Bamforth, P.B. Elsevier Applied Science, 178-187.

EVALUATION OF THE STRUCTURAL INTEGRITY OF AN AGING MINE SHAFT

M.M. Khan and G.J. Krige

Anglo American
PO Box 61587, Marshalltown, 2107, South Africa
Tel: +27 (11) 638-4221, Fax: +27 (11) 638-4840
e-mail: mkhan@anglotechnical.co.za

ABSTRACT

Mineshafts are amongst the most important components of the infrastructure in an underground mine, as they are used for transporting men and material to and from the ore body, as well as hoisting the ore to the surface. They are, however, susceptible to corrosion and mechanical damage. This paper addresses one shaft as a case study, a sub-vertical shaft running between 1800 m and 2900 m below surface in a deep gold mine. The shaft was over 20 years old, with a remaining mining life expectancy of 5 years. The shaft steelwork had extensive corrosion problems, which raised concern about the integrity of the shaft steelwork structure.

An analysis of the shaft steelwork was used to define the lateral loads developed in the steelwork as a result of the dynamic interaction between the conveyance and its guiding system. The dynamic interaction between conveyances and the shaft steelwork is an interesting phenomenon, because it depends on the stiffness of the steelwork. As the steel corrodes, its strength reduces, but the applied forces also reduce. This analysis took cognizance of the effects of corrosion by successively reducing the thickness of the steel members, and calculating the resulting stresses. The extent of corrosion also made the possibility of supports failing a real concern, so certain supports were omitted in the analysis, to investigate the likely result of a support failure. The limiting criteria in this analysis were the local buckling stress of the thin residual material, and the guide deflections. Based on this analysis, acceptance criteria for the extent of corrosion were developed.

KEYWORDS

Shaft Steelwork, Corrosion, Buntons

INTRODUCTION

The South African gold mining industry operates many mines which are now over 20 years old and have an aging infrastructure. Mineshafts are amongst the most important components of the infrastructure in an underground mine, as they are used for transporting men and material to and from the ore body, as well as hoisting the ore to the surface. Access is gained by means of conveyances that are located in the shaft on guides that run continuously from top to the bottom of the shaft. These guides are typically attached to a horizontal grid of beams, commonly referred to as buntons. These form the lifeline of the mineshaft, but are susceptible to corrosion and mechanical damage. A structural collapse within a shaft, in addition to possible injury or loss of life, typically leads to significant production loss. It is, therefore, pertinent that the structural integrity of shafts is maintained to provide continued safe and reliable operation of production. As a result, it has become necessary to carefully evaluate the condition of the shaft steelwork, to ensure that it remains functional.

This paper addresses one shaft as a case study and sketches the procedure used in establishing the functional integrity of the steelwork.

MECHANICS OF SHAFT STEELWORK

It is paramount that the dynamic interaction of the conveyance and its guiding system is well understood in order to define the loads and displacements experienced by the shaft steelwork. The dynamic interaction between the conveyance travelling and the shaft steelwork providing the guidance system has been simplified into what is commonly referred to as the "slamming" phenomenon. Ideally the shaft steelwork should be aligned such that the conveyance travels straight up. However, in practice it has to negotiate inevitable misalignment in the steelwork, and this results in lateral impact forces, referred to as slam loads, being developed at the slipper plates of the conveyance. Research done by Thomas on behalf of COMRO (1990) showed that the lateral slam forces can be defined as a general function:

$$F = f(k_b, k_g, k_c, m_e, v, L, e) \qquad (1)$$

Where:

- k_b is the bunton stiffness. The slipper plate load increases as the bunton stiffness increases.
- k_g is the guide stiffness. The slipper plate load is strongly dependent on the ratio of the bunton stiffness to guide stiffness, k_b/k_g.
- k_c is the conveyance stiffness.
- m_e is the effective mass of the conveyance.
- v is the hoisting speed of the conveyance.
- L is the bunton spacing.
- e is the misalignment of the guide from one bunton to the next. The slipper plate loads are directly proportional to the misalignment.

If the buntons are stiff, the slam loads are high as the conveyance negotiates the misalignment. Hence, in this case corrosion can have positive results, as corrosion leads to the loss in bunton stiffness and thereby a reduction in the slipper plate loads. However, the converse of this is the increase in the stresses due to the reduced material thickness, and the reduced local buckling stress. Reduction in stiffness also leads to increased guide displacement and hence a greater risk that the conveyance may be derailed in the shaft.

CASE STUDY

A sub-vertical shaft running between 1800 m and 2900 m below surface in a deep gold mine is presented here as a case study. The shaft was over 20 years old, with a remaining mining life expectancy of 5 years. The shaft steelwork had extensive corrosion problems, which raised concern about the integrity of the shaft steelwork structure. In some places bunton supports were corroded to the extent that they no longer fulfilled their intended structural function, and only a completely different load carrying action was preventing collapse. From many visual inspections it was clear that corrosion in shafts affects the buntons to a greater extent than the guides. Corrosion of buntons leads to a general reduction in wall thickness, as well as more severe localised corrosion, particularly at the shaft wall connections. A typical corroded bunton is shown in Figure 1.

Figure 1: Typical corroded bunton

Given the short remaining life of the mineshaft, fatigue was not of concern, so the structural considerations in assessing the deteriorated condition of the shaft were the strength, the guide displacement, and the effect of a possible bunton support failure.

Strength of steelwork

The outside buntons shown in Figure 3 were the most seriously corroded buntons. Hence only part of the shaft steelwork was modeled on the computer, comprising of beam elements as shown in Figure 2. The analysis took cognizance of the effects of corrosion by successively reducing the thickness of the bunton from the initial installed thickness of 8 mm to 5mm and then to 1mm. The guide and bunton stiffness for the various thicknesses was determined. The slam loads were calculated for varying conveyance velocities corresponding to the successive reduction in bunton thickness, based on the following assumptions:

- Mass of conveyance   15 336 kg
- Mass of payload   20 000 kg
- Maximum misalignment   20 mm
- Varying hoisting velocities of 5 m/s, 10 m/s and 18 m/s
- The existing buntons were made of Gr 43 steel with yield stress 250 MPa.
- The buntons experienced a uniform reduction in thickness due to corrosion.

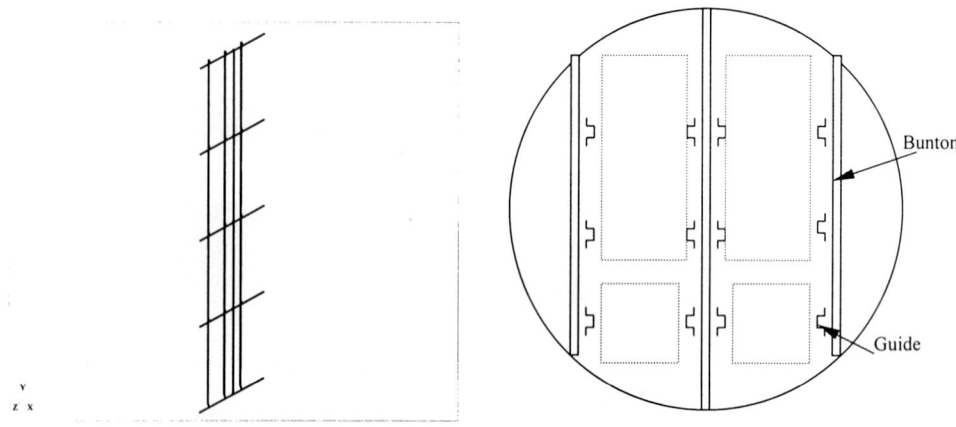

Figure 2: Structural model of the shaft steelwork   Figure 3: Shaft Layout

Figure 4 shows the calculated lateral slam loads as a function of the extent of corrosion and for varying hoisting velocities. The reduction in slam loads due to the increased flexibility of corroded buntons can be clearly seen. The bunton stresses obtained from the analysis for varying bunton wall thickness and three different conveyance hoisting velocities are presented in Figure 5.

When the bunton walls have become very thin due to corrosion the maximum stress that can be applied is reduced by the onset of local buckling. The limiting stress shown in Figure 5 is the allowable local buckling stress calculated from the equations given in SABS 0162-II (1993) as:

$$\sigma_{allowable} = \frac{\sigma_{cr}}{1.5} = \frac{btf_y}{200t} = 67.1t(1-0.059t) \qquad (2)$$

Where: f_y is yield stress
         t is the residual thickness

$$b = Bt = 0.95t\sqrt{\left(\frac{KE}{f_y}\right)}\left[1 - \frac{0,208}{W}\sqrt{\frac{KE}{f_y}}\right] \quad (3)$$

Assuming yield stress $f_y = 250$ MPa and the web depth as 200 mm, manipulation of equation 3 yields:

$$b = 53.7t(1 - 0,059t) \quad (4)$$

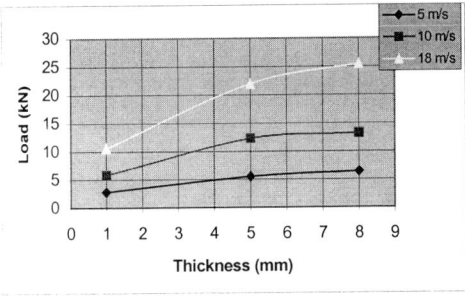
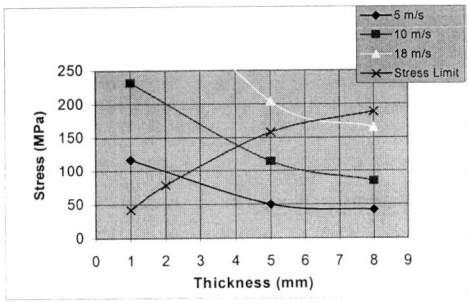

Figure 4: Slam load Vs Thickness   Figure 5: Stress Vs Thickness

The stress information presented in Figures 5 suggests the limiting bunton wall thicknesses for various hoisting velocities. The limiting thicknesses for the conveyance are given in Table 1.

TABLE 1
LIMITING BUNTON WALL THICKNESS

Hoisting velocity (m/s)	Limiting bunton wall thickness (mm)
18	6.4
10	4.1
5	2.4

Displacement of guides

The dynamic conveyance displacement, which is the conveyance displacement relative to the steelwork, was also investigated to ensure the conveyance does not disengage from its guiding system. The relationship between the dynamic displacement and the bunton thickness for varying hoisting velocities is presented in figure 6. The guide shoe on conveyances typically overlap the guides by about 65 mm. Construction tolerances, wear of the slipper plates, and the design clearance between the slipper plates and the guides, together account for the guide shoe moving up to about 45 mm away from the guide. It is generally desirable to keep the conveyance dynamic displacement to less than 15 mm.

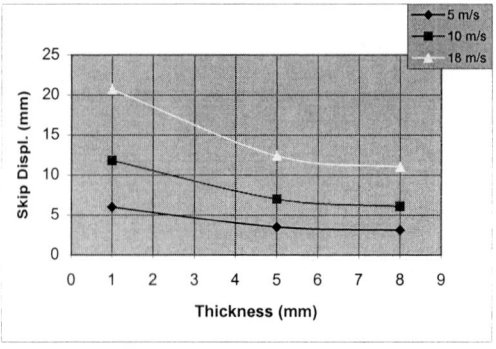

Figure 6: Displacement Vs Thickness

Failure of a bunton support

In order to assess the risk associated with possible failure of a bunton support due to localised corrosion, the structural analysis was repeated with one bunton freed from the walls at both ends, to simulate failure of that bunton. A residual wall thickness of 1 mm was used for the other buntons. This analysis showed results that were very similar to the analysis with the bunton fully attached to the walls, as shown in Table 2.

TABLE 2
COMPARISON OF ANALYSIS WITH FIXED AND FREED BUNTONS

Speed m/s	Fixed End		Free End	
	Slam Load (kN)	Displacement (mm)	Slam Load (kN)	Displacement (mm)
5	2.8	6	2.1	7.7
10	5.8	11.8	4.1	15
18	10.6	21	6.9	25

This meant that the operation of conveyances at a hoisting velocity of not more than 10 m/s would be safe even if a bunton support were to fail.

Repair recommendations

The outcome of the analysis was to enable rational prioritization in the maintenance schedule. The philosophy adopted in making the recommendations for repair to steelwork in the shaft was to:

(a) Ensure that the shaft should at all times be safe to operate.
(b) Reduce hoisting speeds if necessary within the constraint of the production requirements.

(c) Repair members where possible, and only replace them if absolutely necessary, or if replacement is an easier option then repair.

Thus it was recommended that a schedule for bunton repair or replacement should be drawn up, based on the following criteria:

- Hoisting speed should not exceed 10 m/s.
- The aim is to ensure that at least every second bunton supporting each guide is sound, i.e. that its remaining thickness exceeds that limiting thickness indicated in Table 1. Thus buntons must be repaired or replaced where more than two consecutive buntons are unsound.
- Replacement is required where an entire bunton is generally badly corroded, or where the end connections have deteriorated to the extent that they are inadequate. Buntons may be repaired by welding on patches of 5mm to 8 mm thick plate where corrosion has led to localised thin areas.
- Where more than two consecutive buntons require replacement, it is recommended that every fourth bunton should be replaced initially, and then the intermediate buntons can be replaced at a later stage.

CONCLUSION

This paper gives a rational method of evaluating the condition of an existing shaft and providing guidance to a maintenance plan without disruption to the production, whilst ensuring that the functional integrity of the steelwork is not compromised.

REFERANCES

Krige, G.J. (1999). *Subvertical Shaft Structural Condition*, Anglo American Internal report 99-S-11.

Krige, G.J. and van Schalkwyk, W. (2000). *State-of-the-art shaft steelwork in vertical mine shafts*, Mine Hoisting 2000, SAIMM Symposium series S25, Johannesburg.

SABS0208:4 (2000). *Design of Structures for the Mining Industry. Part 4: Shaft System Structures,* South African Bureau of Standards, Pretoria.

SABS0162:2 (1993). *The structural use of steel. Part 2: Limit-states design of cold-formed steelwork,* South African Bureau of Standards, Pretoria.

Thomas, G.R. (1990). COMRO, *Design for the Dynamic Performance of Shaft Steelwork and Conveyances,* COMRO Guidelines 21, Johannesburg..

Thomas, G.R., Greenway, M.E. and Backeberg, R.A. (1991). *A New Approach to the Design of Shaft Steelwork,* Trends in Steel Structures for Mining and Building, SAISC, Johannesburg.

PROBLEMS IN MEASURING STRAINS IN THE REMAINING PARTS OF PARTIALLY DEMOLISHED BRIDGE

R. Masih

Department of Civil and Environmental Engineering
The University of Connecticut, Groton, CT06340, USA

ABSTRACT

This paper outlines the difficulties and problems encountered, in measuring the strains in the remaining concrete, when the concrete bridge is partially demolished for renovation or expansion. This work was part of a $100,000 research project awarded to this author by New England Transportation Consortium (NETC) to find analytically as well as experimentally how badly the adjacent remaining concrete can be affected by the demolition process. This paper also suggests ways to avoid such problems and how to get around them, if avoiding them is not possible. Encountered problems are the difficulty in complying with the construction schedule, if the strain measurements project is not part of the construction contract, the shortcoming of the measurement equipment and how to minimize such effect, the local conditions of the project, the preparation and installation of the delicate strain gages and the effect of its accuracy on the gage performance, organizing the work in order to avoid the mix up between the gages and the output of the measurement equipment and protecting the gages against the weather and vandalism. Observing each one of these components is extremely important for the successful conclusion of the work.

KEYWORDS

Concrete bridge, Demolition, Renovation, Heavy construction equipment, Damage assessment, Strain measurement, Measurement problems

INTRODUCTION

The US Federal government is embarked on an ambitious program of renovating the infrastructure in USA. That includes the decaying roads and bridges, or modifying them to fit the modern transportation means. Billions of dollars were allocated for such projects. While the work on such projects continues, NETC found that other benefits could be extracted from those projects, thus it was decided to award a research project of one of those bridges called West Hartford bridge, Vermont, shown in Figure.1, located on interstate Highway 89, to investigate the effect of the heavy demolition equipment on the remaining concrete parts, and how far such damaging effect can spread through the remaining parts of the bridge. The research project was to

Figure.1: West Hartford Bridge in Vermont, USA

monitor the concrete strains in the neighborhood of the demolition points, where a 4 inch (0.10 m) Hoe-Ram is to be used in the demolition, then to conclude how far such damage could spread in the adjacent concrete. The project included also a theoretical analysis and comparison between the experimental and the theoretical analysis results, then writing a set of guidelines to help NETC in future projects. The construction project involved the removal of the pavement, the demolition of the deck, raising the steel girders to install shock absorbers to minimize the effect of any possible future earthquakes on the bridge, modifying the abutments to accommodate the installation of the shock absorbers, then casting a new deck. Several locations were chosen on the bridge to install strain gages. Depending on the construction schedule and its sequence, a suitable area of the deck was chosen to install the strain gages. Strain gages were also installed in several locations on the abutments and on one of the piers. Because the research project was not part of the construction project, it was necessary to submit the drawings of the gages installation, shown in Fiure.2 and Figure.3, to the construction team for approval, and to avoid any conflict between the two projects.

Because the research project was so large in its scope, its mathematical and computer model, and its problems were extensive, therefore covering all that in one article would have required numerous pages. This article is one of six articles written for presentation at different national and international conferences.

PROBLEMS OF CONSTRUCTION SCHEDULE

Because the research project was not part of the demolition and construction project, it was necessary not to have any conflict between the two projects. Any delay in the construction schedule caused by the research project will entitle the construction contractor to file claims for time extension and monetary damage due to the idling of the labor and the equipment. Such problem caused many revisions in the installation plans of the

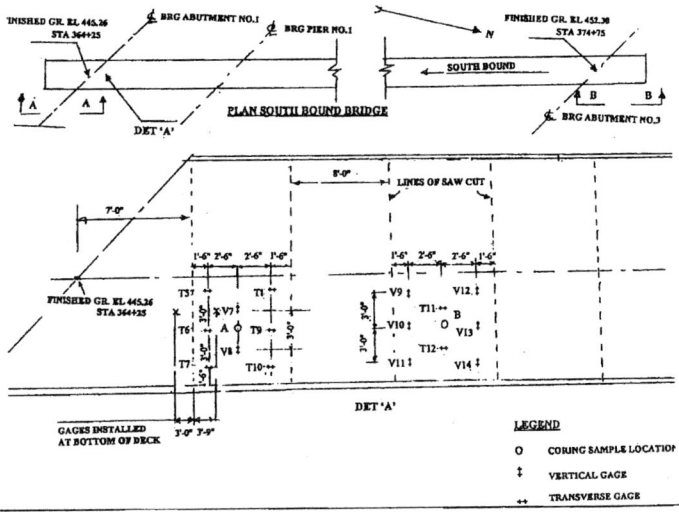

Figure.2: Showing the plan of the bridge and location of the gages on the deck

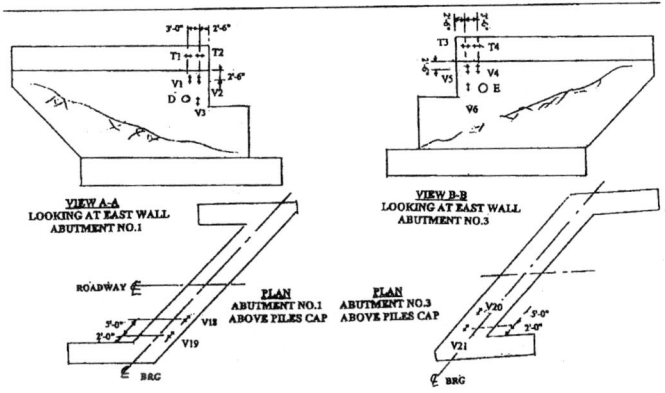

Figure.3: Showing the location of the gages on the abutments of the bridge

gages, which belonged to the research project. It also caused hardship for the research team, since the research team and the bridge were located in different states. The research team had no choice in picking up the time of installation of the gages and measuring equipment, has no choice in picking up the time of observation and monitoring of the demolition process. All of that has to follow the construction schedule, which was changing due to the weather and other considerations. Thus the research team has to rush to the site under a short notice to comply with the construction schedule. Nevertheless the cooperation with the construction team and compliance with the changing construction schedule went very smooth to the end of the monitoring task and the experimental stage of the project.

SHORTCOMING OF THE MEASUREMENT EQUIPMENT

The measurement equipment, as shown in Figure.4, consisted of the four modules SCX1-1321, each one of them has sixteen channels, each channel capable of monitoring one strain gage. The modules are mounted on the chassis of the monitoring equipment SCX1-1121, which was capable of converting the electrical signals into digital signals, through special software installed on what is called a daq card. The SCX1-1120 is linked to a lap top computer, which records the strain gage readings and plot them instantaneously on the screen, enabling the researcher to monitor the behavior of the sixteen gages of anyone of the four modules at the same time.

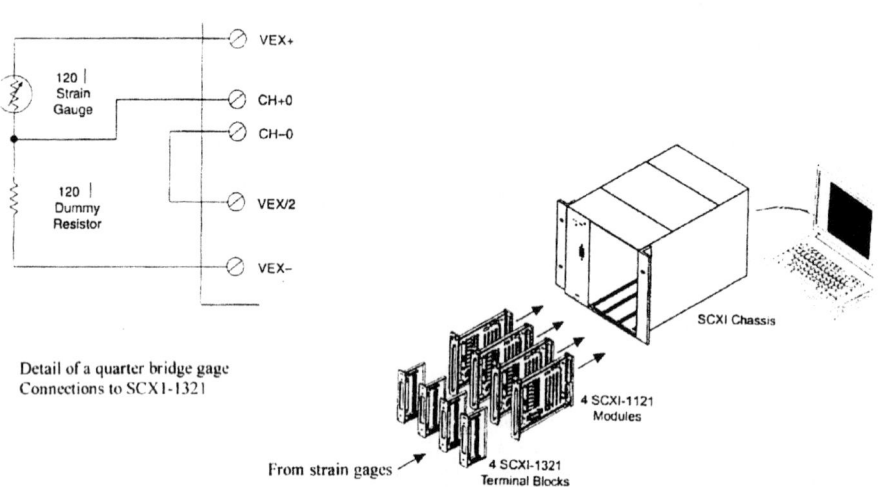

Figure.4: Showing details of quarter bridge and equipment connection

The problem with the daq card was its slot, which enables it to be inserted in the lap top computer. That slot was slightly short in length, making the contacts between the pins of the card and the computer part not so perfect, any small movement caused its separation and malfunction, every time such separation took place caused the computer not recognize the card and its software. Such difficulty caused the crash of the computer several times, even though the computer was brand new, bought exclusively for this job. The manufacturer was made aware of the problem, but such a problem could not be fixed in the proper time and the research team has to live with the problem. The only available solution was to try several times until the contact is made, and the computer recognized the card and its software. Thus the card was left inside the computer throughout the process of the experimental part, and the ribbon cable connecting the SCX1-1121 with the computer was always pulled carefully out of the card, in order not to disturb the connection.

Another problem with the monitoring equipment was its sensitivity to the electronic noise. The equipment was working perfectly if it was used in the lab, but it recorded a lot of electronic noise, when it was used in the field at the bridge. Such electronic disturbance gave readings for the gages equivalent to 6000 strains, while the concrete ultimate strain is about 3000 strains. This meant that the gage could show that the concrete has reached the ultimate strain, while it is hardly under any strain. The manufacturer was made aware of the problem. However, after investigation, it was found that the equipment was at no fault and the reason was the existence of a broadcasting station in the nearby area of the bridge causing the equipment to record such electronic noise. Since it was not possible to stop the broadcasting station during the demolition, therefore the data recorded by the equipment was added together along the time scale, causing the positive noise to cancel the negative noise and combined readings showed a clear strain in the gages, whenever the demolition equipment worked in the neighborhood of the gage. Furthermore, when the gage showed suddenly a jump to a 50,000 strains, it was a clear indication that the strain at the gage has exceeded the ultimate strain.

LOCAL CONDITIONS

Since it was not possible to cause any delay in the construction schedule, and because the installation of the strain gages on the deck must be completed in a short period of time. Such short time was falling between the stripping of the pavement and the demolition of the deck. Therefore it was necessary to install all the other gages before the arrival of the cold season, when all the construction work must be shut down due to the heavy rain and snow, Furthermore, some of the gages, which were supposed to be installed on the pier, could be flooded by the rising water level in the river, thus all the gages installed prior to the cold season must be winterized in order to protect them from the rain, snow and the rising water level of the river. All the connecting wires between the gages and the monitoring equipment must be installed and tagged.

The gages must be installed on a layer of smooth epoxy surface, placed on the old concrete. It was found that the epoxy layer do not dry up in the cold and humid conditions of the season in that area. Therefore a hair dryer was used to hasten the drying of the epoxy layer, before the installation of the gages. This procedure was badly needed in the case of the installation of the gages in the deck area, otherwise a delay in the construction schedule could have been caused because of that problem.

HANDLING THE DELICATE GAGES

The strain gages are delicate thin plastic filaments ¼ inch wide and 2 inch long (6 mm x 50 mm). It has a thin wire electric resistance embedded in it. The resistance is a U shape each end has a terminal to be soldered to a short thin wire of gage 22-24. The soldering must be done very carefully, under temperature control, not to exceed 250° C. Any deviation or error in the process could melt the plastic or destroy the terminals. All that should be done in the lab. Then the gages with the short wires must be stored carefully in preparation for its shipment to the site. At the site, the concrete surface must be cleaned from dirt and grease, using alcohol, then sanded to a clean surface, before applying a thin layer of epoxy. After the epoxy dries, it should be sanded again using grade 300 and 400 sand pad to make it extremely smooth, so that the gage will stick to the concrete properly, which will make any strain in the concrete transmitted exactly to the gage. Two adhesive chemicals must be brushed on the epoxy layer and the gage, then the gage must be pressed against the epoxy layer within few seconds, before the chemical adhesive dries out. Any careless handling could split the wire from the gage, or destroy the gage and makes it useless. After that, the task of winterizing the gage starts, by securing the wires against any unforeseen pull, which could detach the wire from the gage terminal, then adding layers of protective Butane rubber, then covering the whole thing with aluminum adhesive tape to prevent any moisture from filtering through to the gage. Such moisture could short circuit the resistance of the gage.

ORGANIZING THE WORK

This part of the work is the most monstrous part, with so many gages installed and so many long wires extending along the bridge (twelve wires in a cable, each cable serves four gages, because each gage requires three wires), it is hard to tell which wire belongs to which gage and which channel of the equipment will measure the strain of which gage. Bearing in mind that there are three wires coming from each gage, one of them is to be connected to the dummy resistance, another one is to be connected to the equipment channels, numbered 1 to 16 of one of the four modules, and the third wire is to be connected to the VEX+ terminal in the equipment. Furthermore, there are some internal connections to be made in the equipment to complete the circuits for the gages. Such situation could make a big mess out of the project if the work was not clearly organized and the homework done prior to the installation of the gages and their wires in the field. The fact that the installation work has to be done in a short period of time, so that it will not have any adverse effect on the main construction contract adds another reason for doing a good organizational work, in the office as well as in the field. It was decided that any work can possibly be done in the office must be completed in the office. To avoid opening the modules repeatedly to do the wiring connections from the gages to the channels, it was decided that one black wire and one red wire of reasonably bigger diameter (smaller gage number) were to be connected to the VEX- and VEX+ respectively, which were extended outside the module. Those wires were partially exposed in order to connect to them the dummy resistances and soldering short pieces of wires in preparation to insert the short wire and the gage wire in a connector, crimp it by a crimper, which will do the connection in a very short period of time, rather than soldering the wires, which can take a lot longer time to do.

It was needed to have a system capable of avoiding the confusion of which gage is connected to which wire, and which wire goes to which channel, which ultimately meant which channel is reading the strain of which gage and recording it in the data, stored in the computer. Such system was accomplished by tagging the two ends of each wire showing the gage number, then logging the gage numbers and the channels reading those gages on a logging sheet.

PROTECTION AGAINST WEATHER AND VANDALISM

Because the work was done in several stages, it was necessary to protect all the gages against the weather by sealing the cover on top of the gages against the concrete surface. Special adhesive was used for that purpose and extra layer was added to perfectly secure the system against any leaks. The connecting wires were secured by attaching them firmly to the concrete surface, just before the wires enter into the winterized system of the gage. This was done to avoid the transmission of any pull to the wire from being transmitted to the gage, which could destroy the gage or its terminals. Furthermore, each individual gage was covered by the red plastic cone, which is used in the construction work to route the traffic away from running over the gages. All gages were tested to make sure that every one of them was alive by indicating that its resistance was 120 Ω less the few Ω resistance of the long wires. Nevertheless, when the other ends of the wires were connected to the measuring equipment later on, it was found that some of the gages were dead, due to the carelessness of the construction workers and equipment operators, trafficking along the bridge back and forth during the demolition and construction process. There was no time to replace such dead gages, at the time of recording the readings, therefore they were crossed out from the logging sheets as dead gages.

COMPARISON BETWEEN ANALYSIS AND TESTING

To be able to make a decent comparison between the theoretical analysis results (not shown here) and the experimental results, there should be longitudinal and transverse strain readings, thus it was decided to place the gages in longitudinal and transverse directions. Although the theoretical analysis is not yet finished, it requires static analysis and dynamic transient analysis for each loading case of so many cases, using finite elements program called ANSYS. Using the energy method to predict the effect of the Hoe-Ram on the parts of the structure to remain was needed too. Nevertheless the preliminary results, obtained from the theoretical analysis show consistency with the results of the experimental analysis. Such consistency does not mean thaT the results are the same in both cases, but they are consistent in creating stress influence in both directions at the same distance from the center of the demolition. Furthermore, the stresses are not too divergent.

CONCLUSION

It was conceived that using the strain gages to measure the stresses in a concrete structure is a simple process, stick the gages on the concrete and connect them to the reading equipment. However, the reality was totally different from such perception. There were all kind of problems with the gages and the reading equipment, problems of the field conditions, which needed to be resolved, problems of the conflict with the demolition and construction schedule and problems of organizing the work to make it successful and to avoid any mess and confusion created by the magnitude and diversity of the work. Failure to resolve all those problems could have made the research project a failure.

References

1- Masih, R., Wang, T and Zahid, G. The effect of the powerful demolition equipment on the remaining parts of the concrete bridge. Proceedings of the second international conference on computational methods for smart structures. PP 233-234. Madrid, Spain, June 2000.
2- Masih, R. Effect of demolition on remaining parts of concrete bridge-numerical analysis vs. experimental results. Proceedings of the IKM 2000 promise and reality, PP 24-29, Weimar, Germany, June 2000.
3- Masih, R. and Hambertsumian, V, Predicting stability of lift slab structures by energy method, ASCE J. of Performance of Constructed Facilities, Aug 1997, Vol. II, No. 3, PP 141-144.

REPAIR, REHABILITATION AND STRENGTHENING

SEISMIC REHABILITATION OF BEAM-COLUMN JOINTS

A. Ghobarah

Department of Civil Engineering, McMaster University
Hamilton, Ontario L8S 4L7, Canada

ABSTRACT

During recent earthquakes such as the 1999 Kocaeli (Turkey) earthquake, failure of beam-column joints was identified as the cause of many failures of the 5 to 7 storey moment resisting frame buildings. Inadequate transverse reinforcement in the joint and weak column-strong beam design were the principal reasons for the observed joint shear failures. It is important to develop effective and economic rehabilitation techniques for the upgrade of the joint shear resistance capacity in existing structures. The objective of this research is to develop effective selective rehabilitation schemes for reinforced concrete beam-column joint using advanced composite materials.

The study is mainly experimental in nature. Several reinforced concrete beam-column joints were constructed. The joints were designed to simulate nonductile detailing characteristic of pre-seismic code construction. No transverse reinforcement was installed in the joint. The beam and column above and below the joint represent the lengths to the point of contraflexure in the frame beams and columns. The control specimen showed joint shear failure when subjected to cyclic load applied at the beam tip. Different fibre wrap rehabilitation schemes were applied to the joint panel with the objective of upgrading the shear strength of the joint. Cyclic load tests were conducted on the rehabilitated joints with increasing loads to failure.

The tested rehabilitation techniques for beam-column joints with the objective of upgrading the joint shear capacity were successful in eliminating or delaying the shear mode of failure. Instead, the failure mode was transferred to flexural hinging of the beam which is a ductile mode of failure. The rehabilitation technique was found to be effective, simple to install and non disruptive to the function of the building.

KEYWORDS

Beam-column joint, rehabilitation, reinforced concrete, moment resisting frame, shear, seismic upgrade, experimental, FRP, repair

INTRODUCTION

There are many existing reinforced concrete (RC) moment resisting frame structures that were designed before the development of current seismic codes or to earlier codes before ductile reinforcement detailing was required. These structures may have nonductile reinforcement detailing in the beam-column joint area in the form of inadequate or even no shear reinforcement, and short anchor length of the longitudinal beam bottom steel bars. A brittle joint shear failure greatly reduces the ductility of the RC moment resisting frame. Evidence from recent earthquakes such as the 1999 Kocaeli (Turkey) earthquakes shows that in many cases, a brittle failure in the frame joints was the principal cause of the total collapse of many structures. Due to the large extent of the problem, it is necessary to develop economic methods to upgrade the joint's capacity in order to prevent a brittle failure and instead shift the frame failure towards a beam flexural hinging mechanism, which is a more ductile type of failure.

The focus of available research is on the modeling and behavior of typical interior and exterior joints in newly designed structures under cyclic loading. However, few studies were concerned with testing joint strengthening techniques in existing vulnerable beam-column joints (Migliacci et al., 1983; Corazao and Beres et al., 1992). Alcocer and Jirsa (1993) tested a few strengthened beam-column joints. Prion and Baraka (1995) proposed a method that involves encasing the beam-column joint with a grouted steel jacket. The casing, which is of a circular and rectangular geometry, increases the moment capacity of the section and forces most of the specimens to fail outside the jacket. The circular casing was proven to be more effective than rectangular casing, although the latter also met the general expectations. Ghobarah et al. (1997a) proposed the use of mechanical anchors to prevent the bulging problems associated with flat steel jackets. The procedure enables the effective use of flat steel jackets in the rehabilitation of columns and joints. Ghobarah et al. (1997b) investigated a retrofitting technique in which corrugated steel jackets were used to encase deficient RC beam-column joints. The proposed method was found to be quite efficient in upgrading the shear strength of the joint.

The use of fibre reinforced polymers (FRP) for retrofitting piers of overpasses was investigated by Gergely et al. (1998) and Pantelides et al. (1999). Nine 1/3-scale T-shaped specimens were fabricated and tested. Several wrapping configurations using advanced composite materials were applied to improve the shear capacity of the joints. Improvements in the overall characteristics of the specimens such as peak load, ductility and energy dissipation capacity were observed. The retrofit method is shown to be fast, non-intrusive and corrosion resistant.

The use of the fibre reinforced polymer materials in the rehabilitation of RC beam-column joint offers several advantages such as fast application, light weight, resistant to corrosion, economic, simple and effective. The objective of this study is to introduce a joint strengthening technique using glass fibre reinforced polymer (GFRP) laminates for seismic rehabilitation of RC beam-column joints. The effectiveness of the rehabilitation technique in increasing the shear strength of the joint is investigated.

EXPERIMENTAL PROGRAM

Specimen Description

Reinforced concrete beam-column joints were constructed. The height of the column and the length of the beam represent the distance to the points of contraflexure in the frame. The column is 3,000 mm high with cross section dimensions of 250x400 mm The beam's length is 1,750 mm from the face of the column to the free end with cross section of 250x400 mm. The longitudinal reinforcement used in the column is 6 M20 bars (equivalent to 19.5 mm diameter bar) in addition to 2 M15 bars (equivalent to 16.0 mm diameter bar) without splicing. The transverse reinforcement in the column is M10 rectangular ties with a single M10 supplementary central leg. Following the practice before the seismic design codes were available, no transverse reinforcement was installed in the beam–column joint. The top and bottom longitudinal reinforcement of the beam are 4 M20 bars each. The transverse reinforcement of the beam is M10 rectangular ties starting 75 mm from the face of the column. The ties are spaced at 150 mm for 600 mm and then spaced at 200 mm for 1000 mm and ending at 75 mm from the free end of the beam. The test setup is shown in Figure 1. The beam-column joint is expected to fail by joint shear before a plastic hinge is formed in the beam due to the lack of transverse reinforcement in the joint. The beam-column joint specimen designated as T1 shown in Figure 2, was tested as control specimen. Three rehabilitated joints T1R, T2R and T9 were tested.

Figure 1 Test set-up and rehabilitation specimen T1R

Figure 2 Joint shear failure of control specimen T1

Rehabilitation schemes

Two different rehabilitation techniques were tested. One system consists of wrapping the reinforced concrete joint area with one layer of GFRP laminates in the form of a "U". The free ends of the "U" are tied together using threaded steel rods driven through the joint section and a steel plate, as shown in Figures 1, 3 and 5. In specimen T1R, the height of the FRP laminates was limited to the depth of the beam at the joint. It is assumed that the potential presence of a slab will prevent extending the fibre laminate application above the joint region. The fibre is extended above and below the joint and wrapped around the column in specimen T2R. This system may be applied to strengthen the joint area as well as the column. The FRP laminates did not extend onto the beam to avoid unnecessary flexural strength enhancement which may adversely modify the relative strength ratios of the connected beam and column. In specimen T1R, one layer of GFRP laminate was applied while in specimen T2R two layers of GFRP laminate were used. The proposed rehabilitation scheme is expected to provide lateral confinement and shear resistance to the joint area, hence adding strength and ductility to the joint. The presence of the bolted steel plate allowed the laminate to develop its full capacity and prevented premature delamination of the fibre wrap. In the second technique of specimen T9 shown in Figure 4, three layers of GFRP were wrapped in the direction of diagonal tension forces in the joint at 45° with the vertical. By strengthening the shear resistance of the joint, the possibility of shear failure can be eliminated or delayed which will create the opportunity for a ductile plastic flexural hinging in the beam.

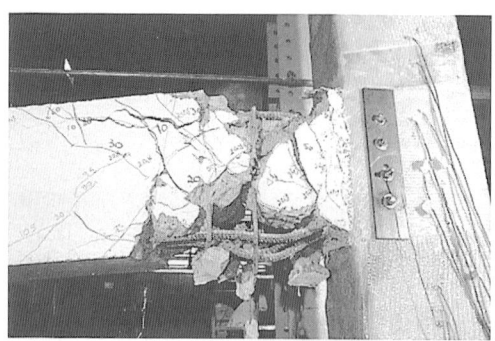

Figure 3 Rehabilitation to eliminate joint shear failure specimen T2R

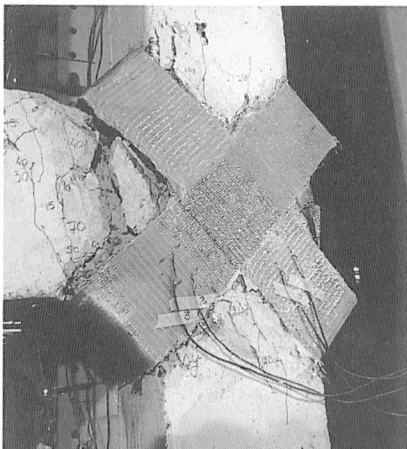

Figure 4 Specimen T9

RESULTS

Behavior of the specimen

For the control specimen, T1, the first crack was recorded at the column face. Before first yield of longitudinal beam steel, a diagonal shear crack was noted in joint area in each loading direction forming an X-pattern. At failure, these cracks extended to the back of the column. A considerable degradation in strength occurred at a ductility factor of 2, which caused the termination of the test at ductility factor of 2.5 as the load sustained by the specimen dropped to 30 % of the maximum load. The specimen failed in classical joint shear failure pattern as shown in Figure 2.

For specimen T1R, partial delamination was observed at a ductility factor of 2.5 near the middle of the fibre laminate as slight fingertip tapping on the laminate revealed a hollow sound. This delaminated area increased until the whole laminate separated from the column sides. The presence of the bolted plate kept the jacket in place despite the delamination and prevented the premature failure of the joint. Meanwhile, a flexural plastic hinge in the beam of length that is approximately equal to the depth of the beam developed starting from the face of the column. Due to the intentional underdesign of the fibre tensile strength capacity, tension failure in the fibre composite material started at a ductility factor of 4. The final crack pattern of the specimen T1R is shown in Figure 5.

In specimen T9, the fibre wrap was successful in delaying the joint shear failure until a flexural hinge was developed in the beam at the face of the column. The fibre wrap delaminated as shear failure started in the joint. The failure pattern of specimen T9 is shown in Figure 4. The two GFRP layers used in specimen T9 were successful in preventing the joint shear failure and instead a ductile flexural plastic hinge formed in the beam as shown in Figure 3.

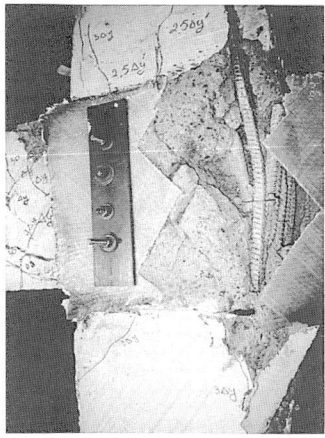

Figure 5 FRP Failure of specimen T1R

Hysteretic behavior

The beam tip load-displacement relationships for specimens T1, T1R and T9 are shown in Figures 6, 7 and 8, respectively. The hysteretic behavior was examined in terms of strength as well as ductility. The hysteretic loops of specimen T1 representing the existing structure as shown in Figure 6, showed considerable pinching and severe stiffness degradation, especially after displacement ductility factor of 2. The loss of stiffness was primarily attributed to concrete deterioration in the beam-column joint region. The pinching of the hysteretic loops for cycles after initial yielding of the beam's longitudinal reinforcement indicates that the stiffness is fairly low near the zero displacement point in these cycles because the cracks did not close. Although specimen T1 almost reached the theoretical ultimate flexural resistance at the first cycle of a displacement ductility factor of 2, a rapid strength deterioration and stiffness degradation were noted due to the joint shear failure which was caused by the lack of joint transverse reinforcement.

The hysteretic loops of specimen T1R, representing the repaired structure, showed an improved ductile behavior as compared to specimen T1 due to the joint FRP rehabilitation which provided confinement to the concrete and increased the joint shear strength and ductility. The limited ductility of T1R is attributed to the failure in the FRP laminate due to the underdesign of the fibre strength. The specimen reached ductility factor of 4 and sustained the maximum load for several cycles as shown in Figure 7. The tensile failure of the fibre is indicated by the drop in the beam-tip load at a tip displacement of about 85 mm.

The behaviour of specimen T9 is shown in Figure 8. The specimen reached ductility factor of 4. Delamination of the fibre wrap occurred at beam tip displacement of about 90 mm. Following delamination, the confinement of the joint was reduced which caused the shear failure of the joint to be initiated. The rehabilitation scheme was successful in delaying the shear failure sufficiently for the ductile flexural hinge in the beam to occur.

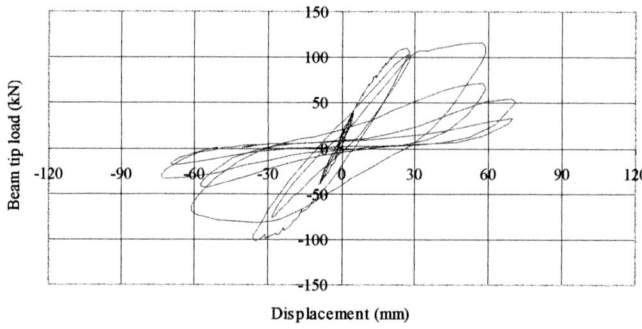

Figure 6 Beam tip load-displacement plot for specimen T1

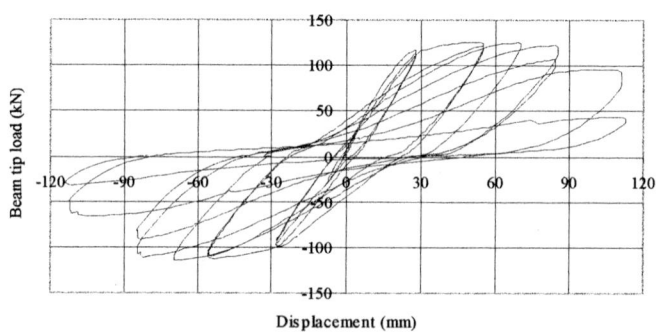

Figure 7 Beam tip load-displacement plot for specimen T1R

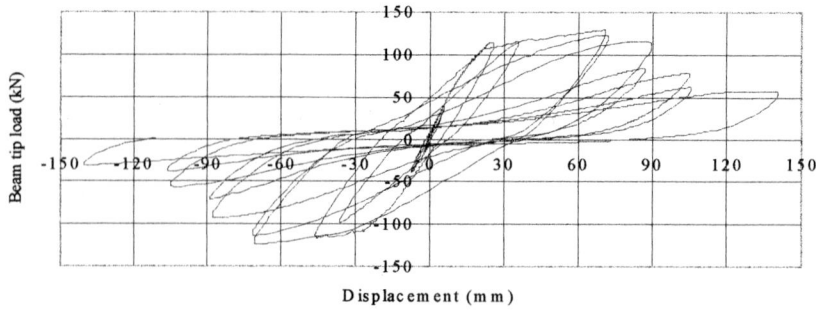

Figure 8 Beam tip load-displacement plot for specimen T9

DISCUSSION

The comparison between the load-deflection plots in Figures 6 to 8 shows a significant difference between the performance of specimen T1 representing a deficient joint, T1R and T9 representing the rehabilitated joints. Specimen T1R had a slightly higher yield load of 117 kN as compared to 109 kN for the control specimen T1. The increase in the yield and ultimate loads are small and might be attributed to the use of higher strength concrete in the repair of the joint (38 MPa as compared to 30.8 MPa for specimen T1). For specimen T1R, a slight drop in strength is observed in both directions on the cycle corresponding to a ductility factor of 4 concurring with the point where the fiber laminate failed in tension in each direction at a time. The envelopes of the beam tip load-displacement curves indicate that a lower rate of deterioration of strength and a higher ductility up to failure (increased by 60%) in the repaired specimen T1R and a slightly higher initial stiffness for the repaired specimen T1R.

Specimen T2R was designed with two layers of fibre wrap to strengthen the joint area as compared with one layer of GFRP used in specimen T1R. The joint sustained a maximum beam tip load of 129.9 kN and reached ductility factor of 5. Throughout the test, no GFRP delamination was observed and no shear cracking developed in the joint area. Instead, the failure mode was transferred to flexural hinging of the beam which is a ductile mode of failure. This rehabilitation system was successful in preventing the beam-column joint shear failure with ductile flexural hinging failure occurred in the beam as shown in Figure 3.

The beam tip load-displacement plot for specimen T9 shown in Figure 8, indicates that the hysteretic loops represent an improvement in performance, strength and ductility when compared to the control specimen T1. The specimen had a high ultimate load of 129.5 kN as the steel angles used under the GFRP wraping, moved the beam critical section away from the face of the column. Although the jacket was adequately designed, it did not provide proper confinement to the joint due to the bulging of the fibre material near the middle of the joint. Similar situation was observed in steel jacketing of rectangular columns (Ghobarah et al., 1997a). The less than adequate confinement delayed the shear failure but did not fully prevent it. The theoretical capacity of the beam section was reached taking into consideration the shortened moment arm due to the steel angle effect.

The hysteretic behaviour of specimens T1R, T2R and T9 indicated much higher energy dissipation capability as compared to the control specimen T1 which is a desirable property when the structure is subjected to a severe seismic event.

CONCLUSIONS

Based on the presented experimental observations and test results the following conclusions are drawn:
1. Two effective method for rehabilitating existing deficient beam-column joints are presented. A comparison between the performance of the original specimen and the rehabilitated ones shows that the GFRP jacket was capable of increasing the shear resistance of the joint and enhancing the performance of the connection from ductility point of view.
2. The rehabilitated specimens, T1R, T2R and T9, exhibited energy dissipation characteristics that are superior to that of the original specimen, T1.
3. The tested rehabilitation techniques for beam-column joints with the objective of upgrading the joint shear capacity were successful in eliminating or delaying the shear mode of failure. Instead, the failure mode was transferred to flexural hinging of the beam which is a ductile mode of failure. The FRP can be designed to prevent the brittle joint failure due to shear and allow a plastic hinge to develop in the beam. The rehabilitation technique was found to be effective, simple to install and non disruptive to the function of the building.

ACKNOWLEDGMENTS

The author wishes to acknowledge the support of the Intelligent Sensing for Innovative Structures network, ISIS Canada. The support of the Fyfe Co. and the R.J. Watson who supplied the materials used in this study, is gratefully acknowledged.

REFERENCES

Alcocer, S.M., and Jirsa, J.O. (1993). Strength of reinforced concrete frame connections rehabilitated by jacketing. *ACI Structural Journal*, **90:3**, 249-261.

Beres, A., El-Borgi, S., White, R.N., and Gergely, P. (1992). Experimental Results of Repaired and Retrofitted Beam-Column Joint Tests in Lightly RC Frame Buildings. Technical report NCEER-92-0025, National Center for Earthquake Engineering Research, State University of New York at Buffalo, NY.

Corazao, M., and Durrani, A.J. (1989). Repair and strengthening of beam-to-column connections subjected to earthquake loading. NCEER technical report No. 89-0013, National Center for Earthquake Engineering Research, State University of New York at Buffalo, N.Y.

Gergely, I., Pantelides, C.P., and Reaveley, L.D. (1998). Shear Strengthening of Bridge Joints with Carbon Fiber Composites. Proceedings of the Sixth U.S. National Conference on Earthquake Engineering, EERI, Seattle, WA.

Ghobarah, A., Biddah, A., and Mahgoub, M. (1997a). Seismic retrofit of reinforced concrete columns using steel jackets. *European Earthquake Engineering*, **11:2**, 21-31.

Ghobarah, A., Aziz, T.S., and Biddah, A. (1997b). Rehabilitation of Reinforced Concrete Frame Connections Using Corrugated Steel Jacketing. *ACI Structural Journal*, **94:3**, 283-294.

Migliacci, A., Antonucci, R., Maio, N.A., Napoli, P., Ferretti, A.S., and Via, G. (1983). Repair techniques of reinforced concrete beam-column joints. Final report, IABSE Symposium on Strengthening of Building Structures - diagnosis and therapy, Venice, Italy, 355-362.

Pantelides, C.P., Gergely, J., Reaveley, L.D., and Volnyy, V.A. (1999). Retrofit of RC Bridge Pier with CFRP Advanced Composites. *Journal of Structural Engineering*, ASCE, **125:10**, 1094-1099.

Prion, H.G.L., and Baraka, M. (1995). Grouted Steel Tubes as Seismic Retrofit for Beam to Column Joints. Proceedings of The Seventh Canadian Conference on Earthquake Engineering, Montreal, Canada, 871-878.

TR 55: DESIGN GUIDANCE FOR STRENGTHENING CONCRETE STRUCTURES USING FIBRE COMPOSITE MATERIALS – A REVIEW

C. Arya[1], J.L. Clarke[2], E.A. Kay[3] and P.D. O'Regan[3]
[1]Department of Civil and Environmental Engineering, University College, London WC1E 6BT, UK
[2]The Concrete Society, Crowthorne, Berkshire RG45 6YS, UK
[3]Halcrow Group, 44 Brook Green, London W6 7BY, UK

ABSTRACT

This paper reviews the contents of a recent Concrete Society publication (Technical Report No. 55) on the use of externally bonded FRP materials for strengthening concrete structures. The paper describes the background, nature and mechanics of producing the Report. Set in the context of limit state philosophy, the Report provides design guidance on:
- flexural strengthening of beams and slabs;
- shear strengthening of beams and columns;
- flexural and compressive strengthening of columns.

These procedures are briefly explained. Other aspects of the Report, which deal with material properties and manufacture, field applications, workmanship and installation, and long-term monitoring, are summarised. The paper concludes by outlining areas of needed research.

KEYWORDS

Fibre composites, strengthening, concrete, externally bonded FRP, flexural strengthening, shear strengthening.

INTRODUCTION

There are a number of situations where it may become necessary to increase the load carrying capacity of a structure in service. These include change of loading or use and the cases of structures that have been damaged as a result of impact or material deterioration. In the past, the increase in strength has been provided by casting additional reinforced concrete, dowelling in additional reinforcement or by externally post-tensioning structures. More recently, attaching steel plates to the surface of the tension zone by use of adhesives and bolts has been used to strengthen concrete structures. Even more recently, the use of fibre reinforced plastic (FRP) plates, generally using carbon fibres, has been developed using the same basic techniques as for steel plate bonding.

FRP plates have many advantages over steel plates in this application and their use can be extended to situations where it would be impossible or impractical to use steel. For example, FRP plates are lighter than steel plates of equivalent strength, which eliminates the need for temporary support for the plates while the adhesive gains strength. Also, since FRP plates used for external bonding are relatively thin, neither the weight of the structure nor its dimensions are significantly increased. The latter may be important for bridges and tunnels with limited headroom or when strengthening in two directions. In addition, FRP plates can be easily cut to length on site. These various factors in combination make installation much simpler and quicker than when using steel plates. This is particularly advantageous for bridges because of the high costs of lane closures and possession times on major highways and railway lines.

Equally important is the fact that the materials used to manufacture FRP plates i.e. fibres and resin, are durable if correctly specified, and hence requirements for maintenance are low. If the materials are damaged in service, it is relatively simple to repair them, by bonding an additional layer.

In addition to plates, various types of fibres are available in the form of fabrics, which can be bonded to the concrete surface. The chief advantage of fabrics over plates is that they can be wrapped round curved surfaces, for example around columns and chimneys or completely surrounding the sides and soffits of beams.

Over recent years an appreciable number of structures around the world have been strengthened using fibre reinforced polymer materials and the rate at which the technique is being used is increasing rapidly. However, there is still little independent guidance on how the design of strengthening works should be carried out. Some FRP suppliers provide limited advice but this tends to be system specific. Also, this advice may not be compatible with national design codes. Because of this lack of design guidance, the UK Concrete Society set out in 1999 to produce a report on strengthening structures with externally bonded FRP, which would be applicable to all types of FRP and all strengthening systems, and set in the context of British codes and standards. The resulting design guidance, Technical Report No. 55[1], is due for publication late 2000. The aim of this paper is to briefly describe the mechanics of producing the report and summarise its scope and salient features.

DRAFTING OF TR55

Responsibility for drafting of TR55 was undertaken jointly by the Halcrow Group, The Concrete Society and University College London. A Steering Group consisting of representatives of client organisations, composite material suppliers, design consultants, contractors and research organisations reviewed the work.

SCOPE OF REPORT

The overall aim of the Report is to provide design and construction guidelines on strengthening concrete structures strengthened with fibre composite materials, set in the context of British design codes and standards. Specifically, the following subjects are covered in the Report:

Chapter 1   Introduction
Chapter 2   Background
Chapter 3   Material types and properties
Chapter 4   Review and applications
Chapter 5   Structural design of strengthened members
Chapter 6   Design of members in flexure
Chapter 7   Shear strengthening
Chapter 8   Column strengthening
Chapter 9   Workmanship
Chapter 10   Long term inspection and monitoring

In addition, an appendix contains some information on the properties of commercially available FRP materials, along with a list of composite suppliers with contact addresses. A specification was felt to be inappropriate but a list of some significant points on quality control of the FRP and the method of installation was considered to be useful and has been included in an appendix.

Prestressed FRP strengthening systems are not covered in detail in the Report. Also the discussion on retrofitting of concrete structures for enhanced seismic resistance is limited since this is not a major loading case for most UK structures. The emphasis in writing this Report, particularly those sections dealing with detailed design, was on providing guidance on design and construction and not necessarily a state-of-the-art report on strengthening.

MATERIAL PROPERTIES AND MANUFACTURE

The properties of FRP materials and manufacturing methods are discussed in so far as they have a direct effect on the design and performance of the strengthened structure.

Currently, the most suitable fibres are glass, carbon or aramid. Each is a family of fibre types and not a particular one. Typical values for the properties of the materials are given in Table 1. The fibres all have a linear elastic response up to ultimate load, with no significant yielding. The fibres are combined with suitable resins to form a composite. In each case the strength and stiffness will be lower than the value for the pure fibre given in Table 1; for a composite they will be in the region of 65%.

TABLE 1
TYPICAL FIBRE PROPERTIES

Fibre	Tensile strength N/mm^2	Modulus of elasticity kN/mm^2	Elongation %	Density
Carbon	3430-4900	230-240	1.5-2.1	1.8
	2940-4600	390-640	0.45-1.2	1.8-2.1
Aramid	3200-3600	124-130	2.4	1.44
Glass	3500	75	4.7	2.6

Fibre composite materials used for strengthening works are available as fabrics, plates and preformed shells. Composite fabric is formed by weaving fibres and may be supplied dry or be pre-impregnated with a resin. The properties of the sheet materials will depend on the amount, type and arrangement of the fibre used. The most widely used process for manufacturing composite plates is pultrusion. As pultrusion is a continuous process, very long lengths of materials are available, which may eliminate the need for laps and joints. Preformed shells have been used to strengthen columns on a number of structures in North America. The first UK application is currently going ahead. For a circular column, the most appropriate manufacturing process is probably filament winding, although the hand-lay process is also widely used.

The selection of the appropriate type of fibre to be used in a particular application will depend on many factors, including the type of structure, the likely loading and environmental conditions. Some indications of appropriate materials to be used in various applications are given in TR 55.

The properties of typical adhesives used to bond FRP composites to concrete surfaces are also discussed in the Report. Common methods of application and ways of preventing degradation are described. To ensure compatibility, it is recommended that the composite materials and the adhesive, along with any paint or additional protective materials, should be obtained from a single supplier.

FIELD APPLICATIONS

To help convince clients of the merits of this technique, brief summaries of projects where externally bonded FRP has been used to strengthen existing structures has been included. The list is by no means exhaustive and mainly covers examples of strengthening works to structures, mainly in the UK. The typical applications are grouped according to:

- Flexural strengthening of beams and slabs
- type of structure i.e. building, bridge, others
- type of element i.e. beam, slab, wall, column
- reason for strengthening i.e. requirement for additional load capacity, structural alteration, insufficient or incorrectly located reinforcement, repair.

Where possible, references to published articles are included. It was felt that these might provide a source of further information and also help the reader to identify individuals and organisations with experience of this technique.

DESIGN GUIDANCE

In common with most British Codes of practice for structural design, the design guidance contained in the Report is based on limit state philosophy. This ensures that a structure will not become unfit for its intended use, that is, it will not reach a limit state during its design life. In assessing the effect of a particular limit state on the structure, the designer will need to assume certain values for the loading and the strength of the materials.

The Report identifies and discusses limit states relevant to the use of FRP for strengthening. In general, these are similar to those encountered in conventional reinforced concrete design but they also include checks for plate separation failure, which is peculiar to this method of construction. In limit state design, the variability of material properties is taken into account by assuming a characteristic strength. Usually, characteristic strength is taken as the value below which not more that 5% of test results lie. A similar approach is used to define characteristic strength of FRP in the Report but the acceptable failure rate has been reduced to 1%.

The design strength is obtained by dividing the characteristic strength by a partial safety factor, which is a function of the type of fibre and the manufacturing route. In addition, a partial safety factor is also applied to the Young's modulus of the FRP, which is a departure from conventional concrete design.

The Report contains procedures for the following:

- Flexural strengthening of beams and slabs
- Flexural strengthening of beams and slabs
- Shear strengthening of beams and columns.
- Flexural and compressive strengthening of columns.

The following sub-sections briefly review the recommended design procedures.

Flexural strengthening

Table 2 lists ultimate and serviceability limit states relevant to flexural strengthening of beams and slabs. The design of FRP strengthening systems is mainly concentrated on the ultimate limit state of strength but checks on the serviceability limit states of deflections and crack widths will also be necessary.

Bonding FRP to the tension face increases the flexural strength of concrete elements. Failure of the element may then occur as a result of either the concrete reaching its ultimate compressive strain or the FRP reaching its ultimate tensile strain. Laboratory tests have consistently shown, however, that the latter rarely occurs in practice. The element generally fails prematurely as a result of plate separation. This is undesirable since the failure load is difficult to predict. In design, anchoring the FRP and limiting the design strain in the FRP below its ultimate value normally avoids this condition. Strengthening against bending may lead to shear failure and it is vital that this is considered in the design process.

TABLE 2
LIMIT STATES

Ultimate	Serviceability

Strength
    bending
    shear
    plate separation
	Deflection
	Cracking

For buildings, fire should also be included in the above limit states as it will influence the properties of both the FRP and the adhesive used to attach the FRP to the concrete. Other relevant limit states are fatigue, creep rupture and durability. Structures that cannot resist working loads without collapsing should not normally be considered suitable for strengthening.

The area of FRP required for strengthening can be determined iteratively using a computer program by finding the internal equilibrium of forces acting on the concrete cross-section. Alternatively, hand calculations can be used. The actual steps involved are discussed in the Report as well as a paper[2], which also includes a worked example to illustrate the design procedure.

Shear strengthening

Externally bonded FRP laminates and fabrics can be used to increase the shear strength of reinforced concrete beams and columns. Figure 1 shows examples of possible configurations.

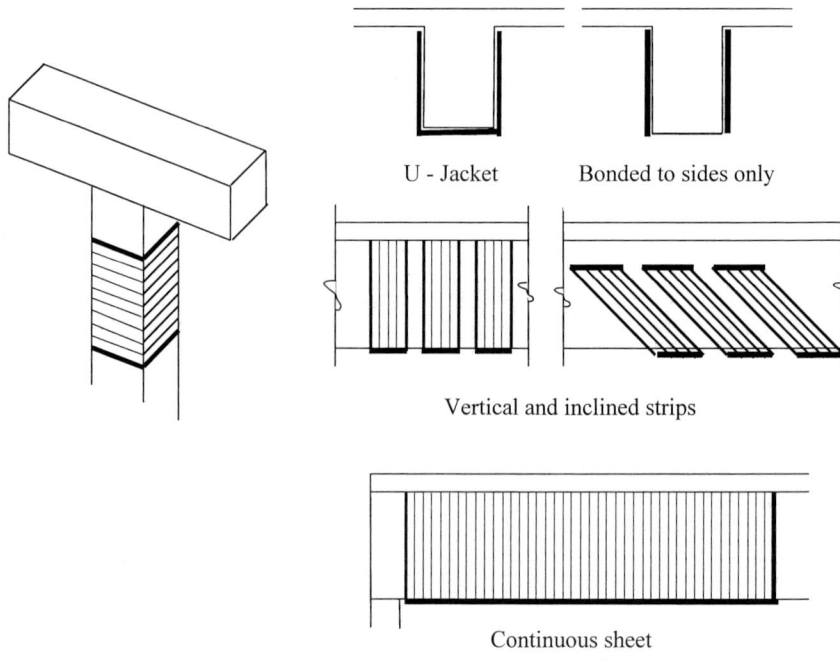

Figure 1: Beam and column shear reinforcement configurations

It can be seen that the shear strength of columns can be easily improved by wrapping with a continuous sheet of FRP to form a complete ring around the member. Shear strengthening of beams, however, is likely to be more problematic since they are normally cast monolithically with slabs. This increases the difficulty of anchoring the FRP at the beam/slab junction and exacerbates the risk of debonding failure.

The design of shear strengthening systems made from FRP follows the same procedure used in conventional reinforced concrete. Basically it involves calculating the design shear force due to ultimate loads and comparing with the shear capacity of the concrete and any steel shear reinforcement present. Where a deficiency exists, FRP shear reinforcement will be needed. The amount of FRP required can be calculated using the same principles adopted in conventional reinforced concrete design, that is assuming a crack pattern and multiplying the area of the FRP reinforcement intersecting the potential crack by the failure stress.

Strengthening of columns

Concrete columns may require upgrading to enhance some or all of the following properties:
- flexural capacity

- axial load capacity
- shear capacity.

Generally, bonding axial FRP to the column surface enhances the flexural strength of the member. Hoop wrapping will also be necessary and should be placed over the axial FRP. The hoop wrapping increases the concrete strain capacity, which can significantly improve the efficiency of the strengthening design. It also improves buckling of the axial fibres, potentially enabling them to contribute in compression. However, this contribution will be small since FRP materials are weaker in compression than tension.

To calculate the thickness of axial FRP, the effect of hoop wrapping on compressive strength and strain to failure of the concrete must be known. Tests by Cuninghame et al[3] on wrapped circular columns have shown that provided the effective hoop stiffness of the FRP is greater than 320 N/mm^2, irrespective of the type of fibre and resin system used, the compressive strength of concrete may be assumed to vary linearly with strain from f_{cu} at a strain of 0.0035 to $1.5f_{cu}$ at a strain of 0.01. This relationship enables the confined concrete stress block to be modeled, which is fundamental to flexural design of wrapped columns.

Bonding hoop FRP to the column surface enhances axial load capacity, shear capacity and ductility of columns. The hoop FRP resists lateral deformations due to the axial loading, resulting in a confining stress to the concrete core. This confinement delays rupture of the concrete, thereby enhancing both the ultimate compressive strength and the ultimate compressive strain of the concrete. This process is reported to be significantly more efficient with circular rather than square or rectangular columns: with the latter, the confinement action is mostly concentrated at the corners. Current estimates suggest that the confinement efficiency of square columns may be around 30-70% of circular columns and is believed to decrease still further with columns of rectangular cross-section and/or large side dimensions.

Critical to compressive strength enhancement of columns is the behaviour of the strengthened member at service loads. The maximum compressive strain in the concrete should not be excessive otherwise loss of confining pressure due to accidental damage, fire, vandalism, etc may result in brittle collapse because the concrete is fissured. To prevent the possibility of this problem arising, it is recommended that the axial compressive strain in the concrete should not exceed 0.0035 under working loads.

WORKMANSHIP AND INSTALLATION

The installation of fibre composite materials must be carried out correctly, to ensure good long-term performance. All installations should comply with the requirements of the relevant Health and Safety at Work regulations. In addition, all materials must be used in accordance with the manufacturer's requirements. It is crucial to the success of the installation that an experienced contractor is appointed. The contractor should have Quality Assurance procedures in place and should have a demonstrated proven track record in the installation of composites. The contractor must be able to demonstrate competency and be approved for the application of the system. This approval may be obtained by providing evidence of the manufacturer's training of the operatives or by documentary evidence of experience on similar projects.

TR55 gives detailed guidance on the installation of plate and fabric materials. The key areas are:

- Adequate preparation of the concrete surface, to remove laitance and irregularities.
- Application of the composite, taking care to ensure a uniform adhesive layer and the exclusion of air bubbles.

- Correct curing of the adhesive.

LONG-TERM INSPECTION AND MONITORING

As with all structural elements there will be a need to check the fibre composite strengthening system as part of the regular inspection and monitoring of the structure. This may be difficult if the fibre composite material has been covered with a protective coating. It is possible that this may have deteriorated in which case it will need to be replaced. The surface of the fibre composite should also be inspected visually for signs of crazing, cracking or delamination. De-bonding of the material from the concrete may be determined by tapping or thermography. Any identification/warning labels should also be checked. Where repairs are required, it is important that the material has similar characteristics to the material in place.

AREAS OF NEEDED RESEARCH

The Steering Group has identified that long-term inspection and monitoring is a key concern to the owners of structures. It is hoped that a subsequent project will look at the long-term performance of the adhesive and the fibres used and develop suitable inspection techniques. In addition, as experience of strengthening grows, it may be possible to modify some of the conservative assumptions made in the design recommendations.

ACKNOWLEDGEMENTS

The work of preparing this Report was supported by research establishments, leading specialist designers, contactors, and the following oraganisations, whose help is gratefully acknowledged: London Underground Ltd, Railtrack, The UK Highways Agency, DML Composites, Du Pont de Nemours International, Exchem Mining and Construction, Feb MBT, SBD Weber & Broutin, Sika, Sumitomo Corporation (Europe) Ltd., Toray Europe Ltd. and the ICE Enabling Fund.

REFERENCES

1. Arya, C., Clarke, J.L., Kay, E.A. and O'Regan, P.D. (2000) *Technical Report No. 55, Design guidance for strengthening concrete structures using fibre composite materials*, The Concrete Society, Crowthorne, UK.
2. Arya, C. and Farmer, N.S., *Design guidelines for flexural strengthening of concrete members using FRP composites*. Submitted for publication.
3. Cuninghame, J.R., Jordan, R.W. and Assejev, A. (1999). *Fibre reinforced plastic strengthening of bridge supports to resist vehicle impact*, Transport Research Laboratory, Crowthorne, Berkshire, U.K., Unpublished Project Report.

FORCE TRANSMITTING FILLING OF WET AND WATER FILLED CRACKS IN CONCRETE STRUCTURES BY MEANS OF CRACK INJECTION WITH NEWLY DEVELOPED EPOXY RESINS

K. P. Grosskurth[1] and W. Perbix[2]

[1] Institute for Building Materials, Constructions and Fire Resistance
in Civil Engineering (iBMB), Technical University of Braunschweig,
Hopfengarten 20, D-38102 Braunschweig, Germany

[2] DYWO Dyckerhoff Wopfinger Umweltbaustoffe GmbH,
Wopfing 156, A-2754 Waldegg

ABSTRACT

As shown in several guidelines in different countries ductile as well as force transmitting filling of cracks in reinforced and prestressed concrete constructions, f. e. bridges, by use of the injection of polymeric resins in the meantime has been advanced to ordinary repair methods. For the force transmitting type epoxy resin systems mainly are used. Epoxides usually are successful if the internal crack surface is dry or has only a low content of humidity. Then the stiffness of an uncracked construction and the steel reinforcement corrosion protection nearly will be restored.

However, in wet or water filled cracks until now polymeric resins are only suitable for water stopping ductile filling of cracks. In small scale tests as well as in on-site tests it could be shown that newly developed epoxy resin systems are able to realize similar success in force transmitting filling of wet or water filled cracks. Then the bond strength between resin and concrete measured by adhesion tests exceeds the tensile strength of the concrete itself. By means of additional scanning electron microscopic studies of the adhesion zone the micromorphological details and reasons for the remarkable quality of the bond (adhesion) and the cohesion of the resin could be obtained.

KEYWORDS

Concrete, wet cracks, repair, crack injection, epoxy resins, interface, morphology, bond strength

INTRODUCTION

Mainly in the last 25 years a lot of prestressed concrete constructions, mainly in highway bridges, but also ordinary reinforced concrete buildings show more or less intensive cracking phenomena. Depending of the crack width and the surrounding climatic and chemical conditions in cases of endangered durability and/or stability the closing of those cracks can be necessary. In general two kinds of crack injection methods exist: Ductile filling of cracks in order to get water tightness of active cracks with movement in the crack size, induced for example by differential settlement, temperature changes or traffic load, and force transmitting filling of so called passive cracks nearly without any movement.

For ductile filling of cracks polyurethanes and acrylic gels are used, for force transmitting filling mostly reactive resins based on epoxides but also some cement lime. Although with cement lime the filling of cracks with high concrete humidity is possible, this material in several cases can not be used, specially when there is still some small movement in the crack. Then mostly epoxy resins are preferred for crack injection, because this polymeric material allows some strain in the range of 1 or 2 % and its reaction shrinkage is rather small compared with other reactive resins and cement lime.

The results of injection with epoxy resins worldwide are excellent, if the crack is clean and the concrete dry enough. The last point is not exactly to define, but successful injections can be expected for cracks in concrete exposed to tropic climates up to air humidities of 80 to 90 %.

If the pores in the cement matrix are nearly completely filled with water or if the cracks are already wet or water filled, normally the bond strength between the epoxy resin and the crack surface becomes – contrary to the bond behaviour in dry cracks - much lower than the tensile strength of the concrete. Therefore in cooperation with the chemical industry modified hydrophilic epoxy resin/hardener components were developed and tested in the view of their mechanical and morphological behaviour.

EXPERIMENTAL

Crack Injection and Mechanical Tests

To study the injectability of the resins and their adhesion to the concrete surface cracks in reinforced concrete beams (maximum aggregate diameter 8 mm) with an average width of 0.3 to 0.5 mm has been injected under the following conditions:

- ➢ "dry": concrete humidity in climate 23 °C / 50 % r.h.
- ➢ "water-filled": crack filled with demineralized water for 4 h; directly before injection the water was blown out by air-pressure
- ➢ "under water-pressure": crack filled with demineralized water under pressure of 10^4 Pa

After 7 days under normal laboratory temperature (23 °C) and humidity conditioning (50 % r. h.) the injected beams were sawed into prisms (dimensions: 150 x 40 x 40 mm³). The bond strength of the injected resins was measured by means of adhesion tests for tensile strength and bending tests for tensile bending strength. The test arrangements are schematically shown in Fig. 1.

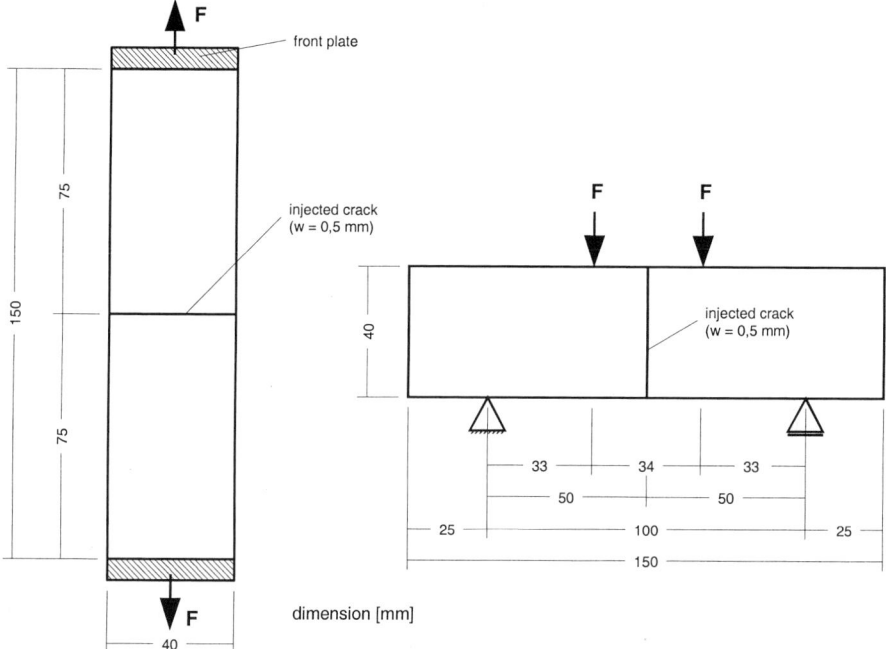

Figure 1: Scheme of experimental arrangements for testing tensile strength (arrangement on the left) and tensile bending strength (arrangement on the right)

The results of the adhesion tests are shown in Fig. 2 (relative tensile strength) and Fig. 3 (relative tensile bending strength).

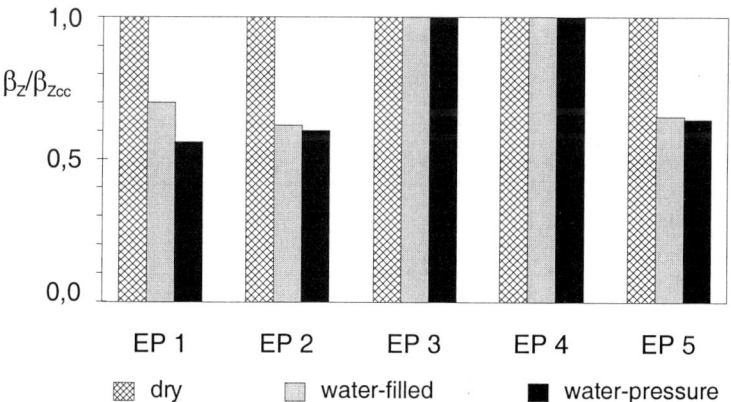

Figure 2: Relative tensile strength β_Z/β_{Zcc} of different injected epoxy resins. β_Z adhesion tensile strength; β_{Zcc} tensile strength of concrete. $\beta_{Zcc} = 3{,}5$ N/mm²

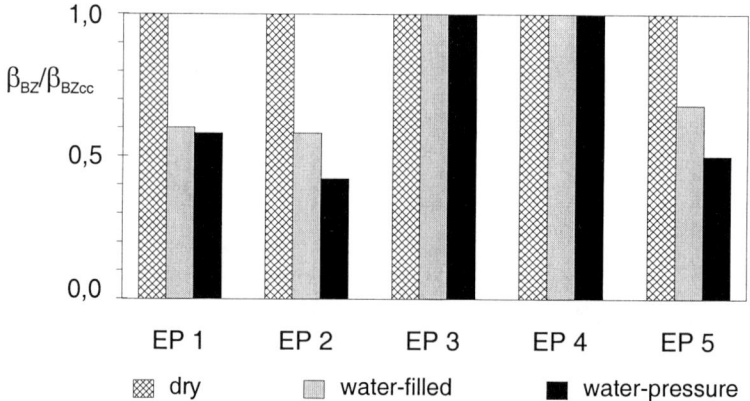

Figure 3: Tensile bending strength β_{BZ}/β_{BZcc} of different injected epoxy resins. $\beta_{BZcc} = 6{,}7$ N/mm². β_{BZ} adhesion tensile bending strength; β_{BZcc} tensile bending strength of concrete

After overloading the prisms in the bending test the adhesion areas of the specimen were visually investigated. The type of the failure zone is described in Tab. 1.

Resin	Crack Conditioning	Zone of Failure
Resin 1	dry	concrete
	water-filled	adhesion zone
	under water-pressure	adhesion zone
Resin 2	dry	concrete
	water-filled	adhesion zone
	under water-pressure	adhesion zone
Resin 3	dry	concrete
	water-filled	concrete
	under water-pressure	concrete
Resin 4	dry	concrete
	water-filled	concrete
	under water-pressure	concrete
Resin 5	dry	concrete
	water-filled	adhesion zone
	under water-pressure	adhesion zone

Table 1: Zone of failure in overloading bending test of the injected concrete specimen as a function of the crack conditioning

The mechanical test results indicate that the presence of water has extremely negative influence on the bond behaviour, if commercial injection resins – here the by the german traffic ministry approved resins 1, 2 and 5 - are used for force transmitting filling of cracks. However both of the modified resins 3 and 4 succeeded in exceeding the tensile strength of the concrete and therefore restore the uncracked state of the concrete.

Morphological Investigations

The interface between epoxy resin and concrete was studied by means of scanning electron microscopy. The shape of a crack in dry concrete injected under static load with the commercial and approved epoxy resin 1 is shown in the upper third of Fig. 4. At higher magnification it is visible that the crack surface wetting and the penetration of microcracks and pores in the crack near concrete area is without significant failure, what is shown in Fig. 5 for the bond of resin and hydrated cement and in Fig. 6 for the bond of resin and aggregate, surrounded on the crack side by a thin cement layer.

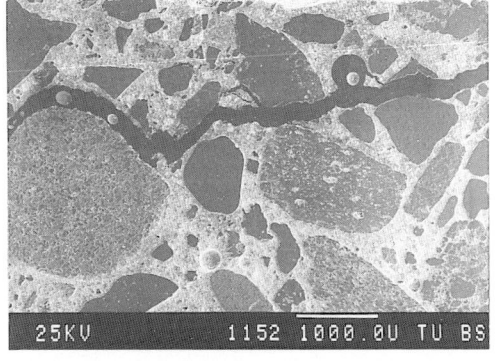

← concrete

← injected crack

← concrete

Figure 4:
Under static load injected crack in dry concrete (resin 1)

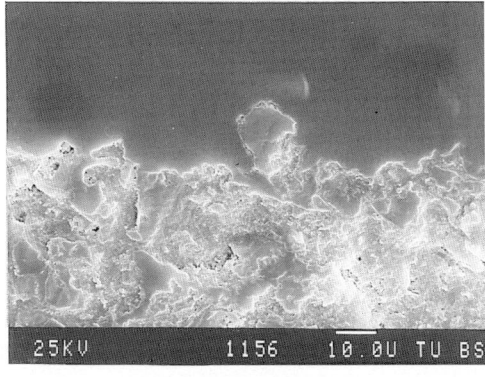

← injected crack

← hydrated cement

Figure 5:
Under static load injected crack in dry concrete; interface resin / cement (resin 1)

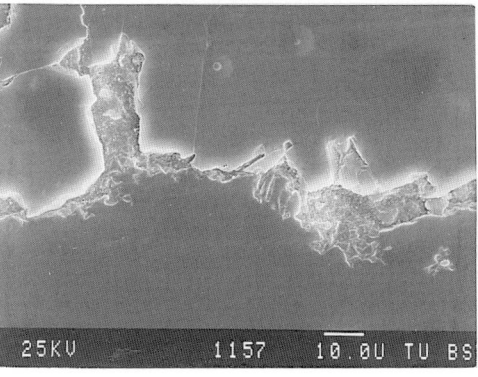

← aggregate

← cement layer

← injected crack

Figure 6:
Under static load injected crack in dry concrete; interface resin / aggregate (resin 1)

If a crack was water-filled for 4 h and then injected under static load after blown out the water the morphological results in Fig. 7 occur: Obviously at the surface of the aggregate at the lower side of the micrograph a thin water layer harms the adhesion. In the epoxy resin a lot of small bubbles and voids exist as a result of gaseous reaction between the components of resin and water.

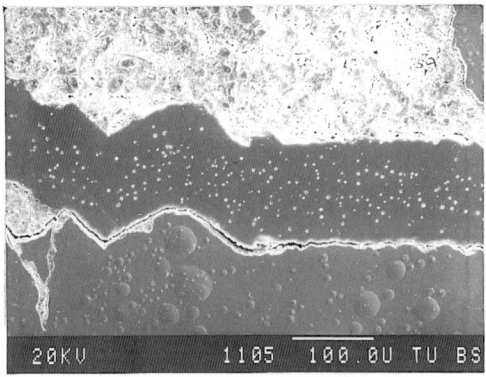

← hydrated cement

← injected crack

← aggregate

Figure 7:
Under static load injected water-filled crack (resin 1)

The view onto the microstructure of the hydrated cement demonstrates in Fig. 8 that the water-filled microcracks and the pores are not penetrated by epoxy resin. Although the surface wetting seems to be rather good the bond is not sufficient; several failures and microcracks exist between the resin and the cementitious surface of the crack.

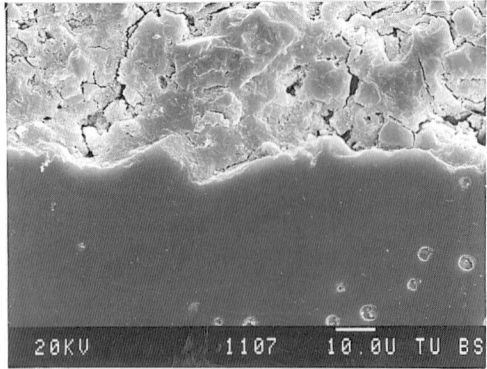

← hydrated cement

← injected crack

Figure 8:
Under static load injected water-filled crack; interface resin / cement (resin 1)

Under the same water-filling conditions, but crack injection under dynamic load of the concrete bar a completely destroyed morphology of the epoxy resin could be observed. As shown in Fig. 9 the cohesion of the resin itself is quite desolate; the crack is only partially filled with resin and no adhesion could be noticed. Obviously the dynamic load during injection in consequence has a vibration effect to the resin and harms any bond to the aggregates and to the hydrated cement. Furthermore the resin under these conditions is not able to cure to a homogeneous matrix.

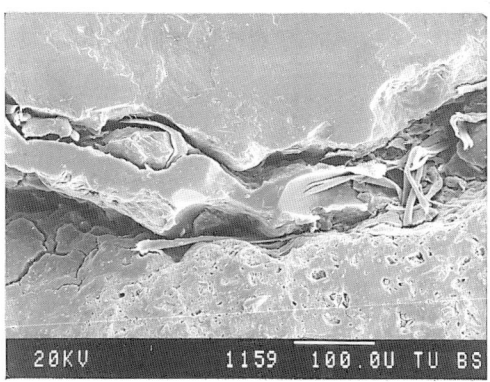

← aggregate

← injected crack

← hydrated cement

Figure 9:
Under dynamic load injected water-filled crack (resin 1)

In the last step of scanning electron microscopy the morphological behaviour in the case of injection with the hydrophilic modified resin-hardener-components – resin 3 and resin 4 – has been investigated.

According to Fig. 10 the adhesion zones in the case of resin 3, injected under dynamic load in a water-filled crack, are characterized by a chain of bubbles. Only the bubble walls seems to connect resin and crack surface. Obviously highly intensive gaseous reactions happen between water and resin. Therefore it is remarkable that the bond strength under such critical micromechanical conditions is higher than the tensile strength of the concrete.

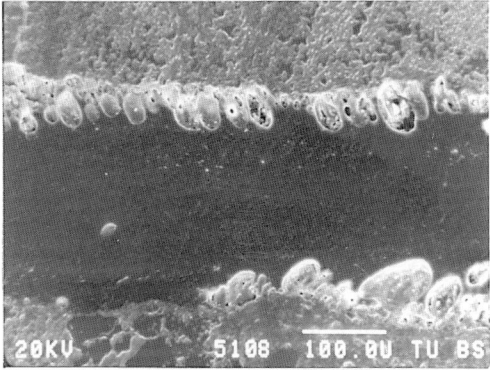

← hydrated cement

← injected crack

Figure 10:
Under dynamic load injected water-filled crack (resin 3)

Much better results are obtained, if water-filled cracks are injected under dynamic load with resin 4. As shown in Fig. 11 only a few and small gas bubbles occur within the epoxy resin and in the adhesion zone. The wetting of the crack surface in relation to the injection conditions and all other results is excellent. Therefore it is understandable that in this case the bond strength is higher than the tensile strength of the concrete. It could be expected that here the bond is suitable to bear long term dynamic loads and therefore to restore the durability of the concrete construction.

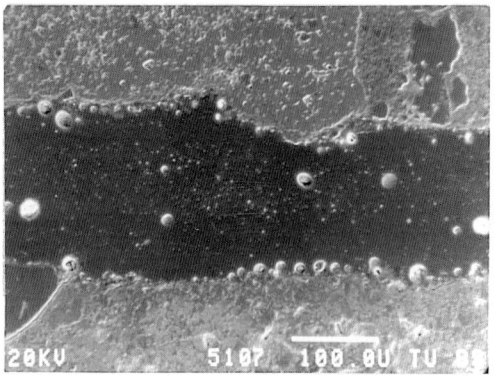

← aggregate

← injected crack

Figure 11:
Under dynamic load injected water-filled crack (resin 4)

CONCLUSIONS

Morphological investigations by means of scanning electron microscopy show that the main reason for a weakening and premature bond failure must be sought in the interface between resin and crack surface. By using commercial and approved epoxy resins for force transmitting injection of cracks their bond behaviour extremely depends on the moisture state of the crack surfaces, because such resins in general are hydrophobic materials like most of the epoxides. With specially developed epoxy resins possessing hydrophilic properties successful force transmitting injection of water-filled cracks without water-pressure as well as under water-pressure conditions are possible. The morphological behaviour of the interfacial zone is in good correlation to the results of mechanical tests. The bond and the tensile bending strength of the injection in all those cases where a combination of good wetting, efficient penetration of the crack surface, specially the hydrated cement, and a low rate of structural failures could be observed was higher than the tensile strength of the concrete itself.

REFERENCES

Chen, S.-W., Perbix, W. and Grosskurth, K.P. (1992). The interface and adhesive strength between reactive resin and concrete. *International RILEM Conference "Interfaces in cementitious composites, Toulouse, France, 21.-23.10.1992. Proceedings*, 99-106.

Engelke, P. and Iványi, G. (1985). Kraftschlüssiges Verpressenvon Rissen in Überbauten von Massivbrücken mit Epoxidharzen. *Beton- und Stahlbetonbau* **80:2**, 29-35, ibid. **80:3**, 79-82.

Grosskurth, K.P. and Perbix, W. (1986). Improvement in the durability of cracked concrete elements injected with synthetic resin by optimizing the bond behaviour. *International RILEM Symposium "Adhesion between Polymers and Concrete", Aix en Provence, France, 16.-19.09.1986, Proceedings*, 403-409.

Grosskurth, K.P. and Perbix, W. (1990). Untersuchungen an Reaktionsharzen zur Bauwerksinstandsetzung. *Adhäsion* **34:3**, 33-37.

Perbix, W. (1993). *Feuchteabhängiges Tragverhalten von Epoxidharzen für das kraftschlüssige Füllen von Rissen in Betonbauteilen*. Dissertation, Technical University of Braunschweig, Shaker Verlag, Aachen, Germany.

NUMERICAL SIMULATION OF MOISTURE DIFFUSION IN A CONCRETE PATCH REPAIR

Muhammad Kalimur Rahman

Research Engineer, Research Institute
King Fahd University of Petroleum and Minerals, Dhahran-31261

ABSTRACT

Durability of a patch repair has become a major concern in the recent times because of cracking and failures of repair layers over a short span of time. A phenomenological model which is capable of modeling a chain of *critical processes* that is set off in a patch repair system and which results in the generation and dissipation of stresses in the system has been developed. Moisture diffusion, a primary critical process in a repair system is addressed in this paper. A non-linear 2-D finite element program **DIANA-2D** (**DI**ffusion **ANA**lysis) has been developed for repair systems. A phenomenological approach in which various transport mechanisms at microlevel are lumped together by defining "diffusivity" of the material at macroscopic level is adopted. A miniature patch repair system consisting of a thin layer of repair material placed over a hardened concrete substrate is analyzed. The spatial and temporal moisture content is obtained and the moisture loss history over the specified time intervals is computed.

KEYWORDS

Patch Repair, Critical processes, Nonlinear finite element, Diffusion Analysis, Diffusivity, Moisture profile, Cementitious repair material

INTRODUCTION

When a cementitious repair layer is cast over a hardened concrete substrate, a chain of processes is initiated in the composite repair system. These processes are set off due to physico-chemical phenomenon and mass-energy transfer in the repair layer. Some of these processes are coupled and they have a profound influence on the deformational response in the system without any external mechanical action. It results in engendering of stresses and development of cracks in the repair layer. In a simple phenomenological model for prediction of response in a repair system under environmental loading it is desirable to neglect the coupling effects of various processes, and consider only the critical processes in the system as independent processes. The processes which are critical in the quantification of stresses in a repair system includes expansion due to time-state activated expansive agents, evolution of strength and stiffness, moisture transport within the repair layer, drying shrinkage evolving as a consequence of moisture desorption, drying tensile creep under evolving stress states due to shrinkage, and cracking and propagation of cracks at zones of high stresses (Rahman 1999, Rahman et.al 1999).

The prediction of spatial distribution and time history of moisture content in a cementitious structural/non-structural repair is of paramount importance. It governs the development of shrinkage, creep, thermal dilation, strength, and durability of repair layer. Moisture transport involves a complex interaction between a numbers of different transport processes. The moisture transport in a 3-D solid can be segregated as: (i) moisture transport at the boundaries of the system and (ii) moisture transport within the domain of the system. The mechanism of domain moisture transport in cementitious material is extremely complicated. In spite of some in-depth studies revealing the moisture state and transport mechanism at microlevel (Queenard and Sallee 1992, Garboczi and Bentz 1992), it is difficult to assess quantitatively the contribution of each of these mechanisms to global moisture transport in the system. A phenomenological approach in which various transport mechanisms are lumped together by defining "diffusivity" of the material at macroscopic level is generally adopted. Drying in a cementitious system at macroscopic level is strongly dependent on the moisture concentration in the system. Hence, the diffusivity is considered to be dependent on moisture concentration in spatial domain of the system and this renders the moisture diffusion equation nonlinear. The physics of boundary layer moisture transport is in itself quite involved. A simplified treatment considers the transport in the form of a constant flux boundary condition.

Nonlinear moisture diffusion theory in terms of moisture content has been used by Sakata (1983), Penev and Kawamura (1991), Asad et al. (1997), and Rahman et al. (1999). Bazant and Najjar (1972) have considered moisture diffusion in terms of relative humidity rather than moisture content. They assert that due to porosity becoming non-uniform with time, the usage of evaporable water content as a basic variable in diffusion equation is erroneous. Bazant's formulation has been used in several studies for predicting moisture loss and shrinkage in concrete. Jonnason (1978), Garbozyskwi and Shah (1989), and Alvaredo (1994) are some examples. Granger, Torrenti and Acker (1997) have analyzed the two classical hypotheses treating moisture diffusion equation in terms of relative humidity and water content. They have shown that in many cases, the good linearity of the desorption isotherm between 50% and 100% relative humidity allows for both types of models. Sadouki and Van Mier (1997) have used moisture diffusion to predict crack growth in a bi-layer repair system.

A two-dimensional finite element program DIANA-2D (DIffusion ANAlysis) has been developed in this study for numerical simulation of moisture transport in repair materials using nonlinear moisture diffusion equation. This model requires an empirical moisture dependent diffusivity law. Non-linear finite element method combined with nonlinear least square fit method was used for the formulation of diffusivity law for selected repair materials (Rahman et al., 2000). A miniature patch repair system consisting of a thin layer of repair mortar placed over a hardened concrete substrate is analyzed for 2-D moisture diffusion. The spatial and temporal moisture content is obtained and the moisture loss history at specified time interval is computed. The effect of temperature on moisture diffusion of a selected repair material is also presented.

GOVERNING EQUATION FOR 2-D MOISTURE DIFFUSION

In a thin 3-D solid of thickness h, moisture in the solid diffuses from the exposed surfaces, the predominant flow being from the top surface (Figure 1). The flow in a 2-D slice of the 3-D solid is principally in the x_1–x_2 direction (Figure 2). The governing differential equation for 2-D moisture diffusion is given as:

$$\frac{\partial C(x_k,t)}{\partial t} = \frac{\partial}{\partial x_i}\left[D(C)\frac{\partial C(x_k,t)}{\partial x_i}\right] \quad \begin{array}{l} k=1,2 \\ i=1,2 \end{array} \tag{1}$$

$C = C(x_i, t)$, $i=1,2$ is the moisture content varying in domain and with time
$D(C)$ = Isotropic moisture diffusivity coefficient which is function of C.

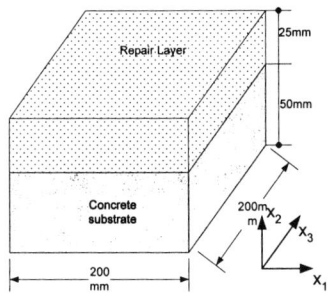
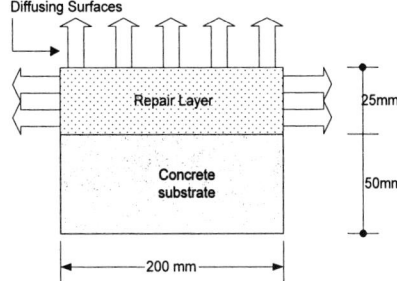

Figure 1: A Typical Repair System   Figure 2: A 2-D Model of the Repair System

Initial And Boundary Conditions

Initial Conditions: $C(x_k, 0) = C_o$ @ $t = 0$ $k=1,2$ for 2-D   (2)

Prescribed Surface Moisture: $C(x_k, t) = C_1(x_k, t)$ $x_k \in \Gamma_c$ (3)

Prescribed Moisture Flow at the Surface $D(C)\dfrac{\partial C(x_k, t)}{\partial x_i} n_i = q_i(x_k, t)$ $x_k \in \Gamma_q$ (4)

Convection Boundary Condition $D(C)\dfrac{\partial C(x_i, t)}{\partial x_i} n_i - h_f(C_e - C_s) = 0$ $x_i \in \Gamma_f$ (5)

Where n_i is the outward normal at the boundary, q_i is the specified amount of moisture flow @ x_k on Γ_q, h_f is the surface moisture transfer coefficient, C_e is the moisture of external environment, C_s is the surface moisture content of the solid, and $\dfrac{\partial C}{\partial \underset{\sim}{n}}$ is the moisture flux normal to the exposed surface.

SALIENT FEATURES OF THE PROGRAM *DIANA-2D*

Moisture diffusion analysis program DIANA-2D is a specific purpose computer program for the computation of moisture diffusion in a bi-layer two-dimensional domain of a repair system. Based on finite element method, it computes temporal variation of moisture field in a repair system. The program comprises of pre- and post-processing modules in MATLAB and computational modules in fortran. DIANA-2D carries out computations for the evolution of moisture field in the repair system under various boundary conditions. A semi-empirical model, with moisture content evolution based on Fick's second law and moisture content dependent diffusivity law obtained empirically for selected repair materials, is used in this module.

DIFPRE, developed in MATLAB, is a preprocessor for moisture diffusion analysis and has the capability of automatically generating the bulk of input data including mesh generation and load definition for simple 2-D rectangular domains. The input data is generated in vectorized/matrix form. A sub-function of this module, FEGRID generates the 2-D mesh over the regular domain, the connectivity matrix and displays the finite element mesh. A sub-function BOUNDC generates the bounding surface elements and assigns the appropriate specified boundary conditions. DIFPRE then generates a data file as an input to the solution of moisture diffusion equation in the module DIANA-2D. DIPOST is a postprocessor developed in MATLAB and has the capability to plot moisture surfaces in the repair system at various time intervals, the moisture contours and moisture loss history at various nodes.

COMPUTATIONAL ALGORITHM

The program is designed to compute the moisture content history in a cementitious mortar due to moisture diffusion in the system. Finite element method is used to solve the nonlinear moisture diffusion equation. The matrix differential equation governing moisture diffusion in a finite element system containing 'n' nodes, derived from Eqn. 1 and 5 is given as

$$\underline{M}(C)\underline{C}(t) + \underline{V}\underline{\dot{C}}(t) = \underline{F} \qquad (6)$$

where $\underline{M}(C)$ = Moisture diffusivity matrix, $\underline{V}$ = Moisture velocity matrix, $\underline{C}(t)$ = Vector of nodal moisture at time t, $\underline{\dot{C}}(t)$ = Vector of time rate of change of nodal moisture, and $\underline{F}$ = Vector of external moisture load. Eqn. 6 represents a set of nonlinear equations, which is to be solved over the finite element domain to obtain the nodal moisture content as a function of time. The transient nature of this equation is accounted for by a step-by-step time integration scheme. The moisture content at time $t = 0$, $\underline{C}(0)$ is known apriori. In this scheme, the time is incremented by Δt and the equation is solved to obtain the moisture content C_1 at time t_1. The process is repeated, gradually generating a sequence of moisture states in the domain $\{\underline{C}_o, \underline{C}_1, \underline{C}_2, \ldots \underline{C}_n\}$, approximating the moisture content history $\underline{C}(t)$ in the domain. The computational algorithm is shown in Table 1 and Figure 3 shows the flowchart of the computational steps and the fortran subroutines activated in the program DIANA-2D.

TABLE-1
TIME STEP COMPUTATIONS FOR MOISTURE DIFFUSION

Initial Computations

1. Form the moisture velocity matrix $\underline{V}$, $\underline{V} = \int_{\Omega_e} \underline{N}^T \underline{N} d\Omega_e$

3. Form element moisture diffusivity matrix, $\underline{M}_S = h_f \int_{\Gamma_{qe}} \underline{N}^T \underline{N} d\Gamma_{qe}$

4. Form the moisture load vector $\underline{F} = h_f C_e \int_{\Gamma_{qe}} \underline{N} d\Gamma_{qe}$

5. Set initial conditions $t=0, \underline{C}_o = \underline{C}(0) = C_{init}$. For first time interval $t_i = t_1$

Time Step Computations

Nonlinear analysis is carried out using time step integration in temporal domain and an iterative solution procedure using a sequence of linear analysis in the geometric domain.

6. Moisture at current time step t_i for first iteration $\underline{C}_i^o = \underline{C}_{i-1}$

7. Form diffusivity matrix $\underline{M}_D(\underline{C}_i^r)$ for r^{th} iteration $(\underline{M}_D)_i^r = \int_{\Omega_e} D(C_i^r) \left[\frac{\partial \underline{N}^T}{\partial x} \frac{\partial \underline{N}}{\partial x}\right]_i^r d\Omega_e$

8. Add surface diffusion contributions to domain diffusivity matrices $\underline{M}_i^r = \underline{M}_S + (\underline{M}_D)_i^r$

9. Compute effective element system matrices for the current iteration (r^{th}) and current time step (i^{th})

    $\underline{A}_i^r = \underline{V} + \theta \underline{M}_i^r$ and $\underline{P}_i^r = \underline{V} + (1-\theta) \underline{M}_i^r \Delta t_i$

10. Compute the effective moisture load vector $\overline{\underline{F}}_i^r = \underline{P}_i^r \underline{C}_{i-1} + \underline{F}$, $\underline{C}_{i-1}$ is the previously converged moisture content. It does not change during the iterations.

11. Solve for nodal moisture $\underline{C}_i^r$ for r^{th} iteration of current time step $\underline{A}_i^r \underline{C}_i^r = \overline{\underline{F}}_i^r$

12. Update element diffusivity matrix $(M_D)_i^{r+1}$ based on current moisture content C_i^r as defined in step 7 and add the contribution from surface diffusion.
13. Form the element system matrices using updated moisture diffusivity matrix M_i^{r+1}

$$\underline{A}_i^{r+1} = \underline{V} + \theta \underline{M}_i^{r+1} \quad \underline{P}_i^{r+1} = \underline{V} + (1-\theta) \underline{M}_i^{r+1} \Delta t_i$$

14. Compute the updated effective load vector $\overline{F}_i^{r+1} = \underline{P}_i^{r+1} \underline{C}_{i-1} + \underline{F}$
15. Check for convergence of solution.
    (i) Find the internal force $\underline{R}_i^r = \underline{A}_i^{r+1} \underline{C}_i^r$ (ii) Compute residual force vector $\delta \underline{R}_i^r = \underline{R}_i^r - \overline{F}_i^{r+1}$

    (iii) Check if certain norm of $\delta \underline{R}_i^r$ $\|\delta \underline{R}_i^r\| < specified \ tolerance$

16. If convergence tolerance is satisfied, then $\underline{C}_i = \underline{C}_i^j \quad j = n,$ n is the iteration number at convergence

Update the time step to get a new time increment t_{i+1} where $i = 1,2,...N$, where N is the maximum number of time increments. Go to step 6.

17. If convergence criteria is not satisfied, increase iteration counter $r = r + 1$ and proceed as follows:
18. Using out of balance force, solve for increment in moisture content $\Delta \underline{C}_i^{r+1}$, $\Delta \underline{C}_i^{r+1} = \left[\underline{A}_i^{r+1}\right]^{-1} \delta \underline{R}_i^r$

19. Find the updated moisture content. $\underline{C}_i^{r+1} = \underline{C}_i^r + \Delta \underline{C}_i^{r+1}$
20. With updated moisture content, repeat steps 13 through 15. If the convergence tolerance is satisfied in step 15, proceed to step 6 for a new time increment. If the solution does not converge, increment the iteration counter to $r+2$ and process the steps 16 through 18. Repeat the process till the convergence is satisfied or the divergence flag is up.

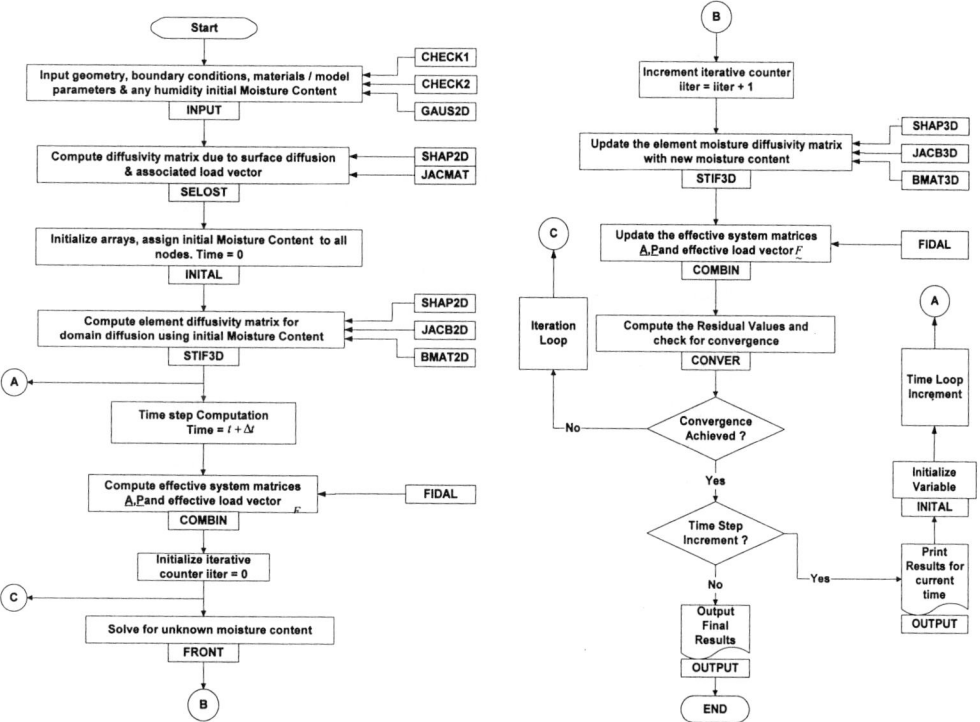

Figure 3: Flow Chart for the Computer Program DIANA-2D

2-D NUMERICAL MODELING OF A MINIATURE PATCH REPAIR SYSTEM

A miniature patch repair system consisting of a thin layer of repair material placed over a hardened concrete substrate was analyzed using the computer program DIANA-2D. The repair system consists of a 25 mm thick repair layer placed over 50 mm thick dried concrete host layer. The repair layer is 200 x 200 mm in plan (Figure 1). The substrate for the analysis is considered to be either fully saturated or mechanically isolated from the repair layer by sealing the interface by an adhesive agent. This means that the movement of the moisture from the repair layer into the substrate does not take place. In view of above finite element simulation of only the repair layer is required for diffusion analysis.

Strictly speaking this is a 3-D problem with moisture diffusing from the top and the sides. Predominant moisture movement is however, from the top surface of the repair layer. Towards the corner of the repair layer the moisture diffusion is essentially three-dimensional. A 2-D model of the miniature repair system is shown in Figure 2. The moisture diffusion in this model occurs from the top edge and from the sides of the repair. In view of symmetry one-half of the system can be modeled. Finite element discretization of the repair layer using 9-noded Lagrangian elements is shown in Figure 4.

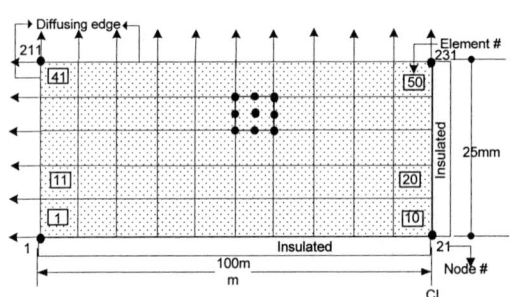
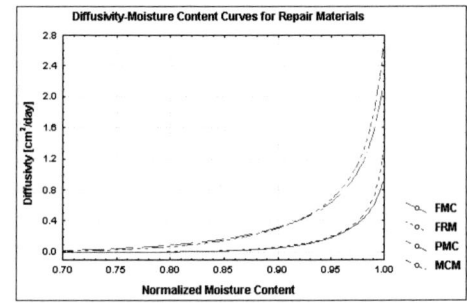

Figure 4: F.E Model for Repair Layer   Figure 5: Diffusivity-Moisture Content Curves

Although a 4-noded linear element would have given a sufficient accuracy in a diffusion analysis, for the convenience of mapping the moisture loss field to the free shrinkage strain field, 9-noded elements are used. For stress analysis, higher order elements are needed for accuracy. The finite element mesh has a total of 231 nodes and 50 elements. A non-linear finite element analysis for moisture diffusion in the system is carried out for the following three cases. The spatial and temporal moisture content is obtained and the moisture loss history over the specified time intervals is computed.

Case-I: 2-D moisture diffusion analysis for the material FMC (Flowing Microconcrete)
Case-II: 2-D moisture diffusion analysis for the material PMC (Polymer Modified Concrete)
Case-III: 2-D moisture diffusion analysis in PMC under extreme environmental temperature

A diffusivity law for repair materials using trigonometric function of the form $D(C) = a \tan (bC^n)$ with three unknown parameters (a, b and n) was developed and found to give good results. This functional form was incorporated in the program DIANA-2D. The variation of diffusivity of repair materials as a function of the moisture content is shown in figure 5. Drying tests were conducted on thin specimens of various sizes with unidirectional and bi-directional moisture movement at 22±2 °C temperature and 65±3% R.H. These tests were carried out for computing diffusivity of the repair materials. Figure 6 and 7 shows the experimental and predicted mean moisture loss in 25x25x285mm prism of repair material FMC and 40x40x300mm prisms of repair material PMC.

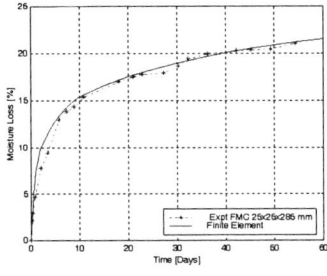
Figure 6: Experimental and Predicted Mean Moisture Loss in FMC

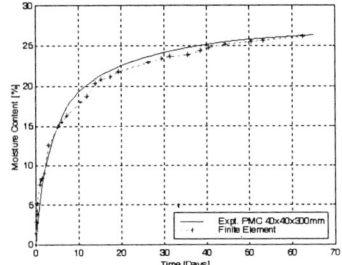
Figure 7: Experimental and Predicted Mean Moisture Loss in PMC

TABLE-2
PARAMETERS USED IN DIFFUSION ANALYSIS

Parameters	Case-I	Case-II	Case-III
Repair materials	FMC	PMC	PMC
Env. Moisture C_e	65%	65%	22%
Initial Moisture C_i	99%	99%	99%
Surface Coeff h_f (cm/day)	0.3	0.3	0.5
Diffusivity Law $D(C) = a\tan(bC^n)$	a=0.5 b=1.1 n=22	a=0.65 b=1.3 n=10	a=0.65 b=1.2 n=3

Case-I: 2-D Moisture Diffusion Analysis -Repair Mortar FMC

Moisture diffusion analysis of the 2-D model of miniature repair system was carried out for the repair mortar FMC. The parameters used in the analysis are shown in Table-2. Figure 8 shows the moisture content surfaces at selected times during the time history of the diffusion process. It can be seen from the figure that there is steep decrease in moisture content at the corner at the onset of the diffusion process. This sharp decrease at the corner is due to the bi-directional moisture diffusion at the corner. The moisture content in the elements adjacent to the diffusing surface also shows a sharp decrease reaching a value of 85% in 0.5 days. As the process of diffusion continues the outer layers lose moisture continuously, whereas the core shows a slow and steady decrease in the moisture content. The repair material FMC has a very low diffusivity (Figure 5) as the moisture content drops below 90%. The moisture is therefore retained in the system. After 28-days of exposure the moisture in the core was above 90% and after 60 days of exposure the moisture content was still about 90%.

Figure 9 shows the moisture content history at various nodes in the repair domain. It can be seen from this figure that moisture content at the top left corner node 211, where a bi-directional moisture movement takes place the moisture content decreases very sharply at the onset of the diffusion process and is in equilibrium at 0.5 days. At node 21 at the center of the core the moisture content decreases very slowly, being 95% at 14 days, 92% at 28 days and 90% at 60 days. The prediction from moisture diffusion model was also observed in drying experiments conducted on this material. After an initial rapid loss the sample weight-loss time curve became asymptotic after a few days and the moisture was lost from the system at an extremely slow rate. A large quantity of moisture was however, lost on heating at high temperature (Rahman 1999).

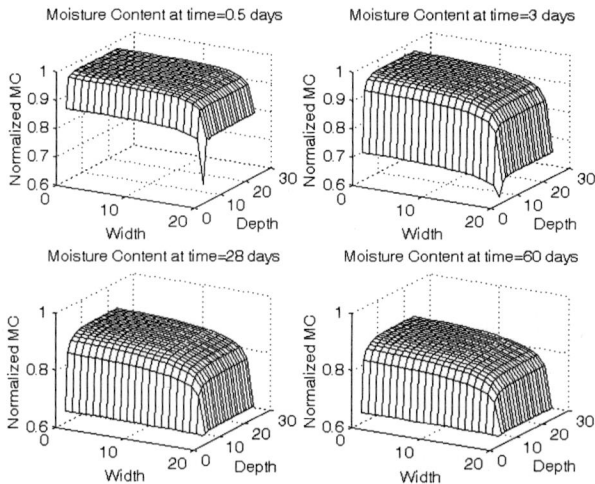

Figure 8: Moisture Content Surfaces – Repair Mortar FMC

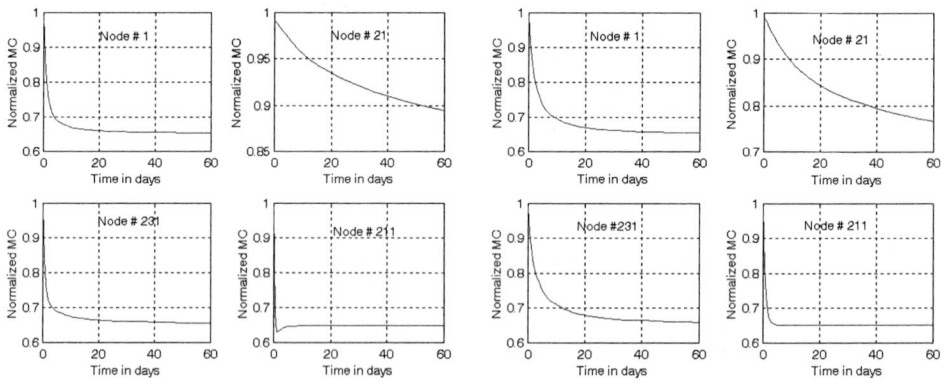

Figure 9: Moisture Content at Nodes for FMC   Figure 10: Moisture Content at Nodes for PMC

Case-II: 2-D Moisture Diffusion Analysis -Repair Mortar PMC

Moisture diffusion analysis of the 2-D model of miniature repair system was carried out for the repair mortar PMC. The parameters used in the analysis are shown in Table 2. Figure 10 shows the moisture history at various nodes in the repair domain and Figure 11 shows the moisture content surfaces at selected times during the time history of the diffusion process. A steep decrease in moisture content at the corner similar to that in FMC at the onset of the diffusion process can be seen. The moisture content in the elements adjacent to the diffusing surface also shows a sharp decrease reaching a value of 85% in 0.5 days. As the process of diffusion continues the outer layers lose moisture continuously. The core for this material however, shows a much higher decrease in the moisture content as compared to the material FMC. The repair material PMC has much higher diffusivity as compared to FMC (Figure 5). After 28-days of exposure the moisture at the core is about 80% and after 60 days of exposure the moisture content is less than 75%. The prediction from moisture diffusion model was also observed in drying experiments conducted on this material. After an initial rapid loss the sample continued to lose moisture at much higher rate as compared to the material FMC.

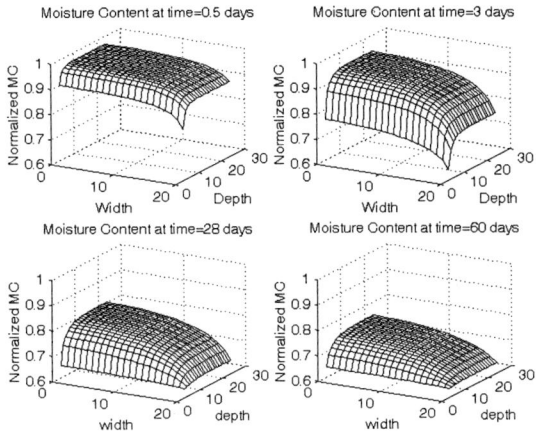

Figure 11: Moisture Content Surfaces-Repair Material PMC

Case-III: 2-D Moisture Diffusion Analysis – High Environmental Temperature

Moisture diffusion analysis of the 2-D model of miniature repair system was carried out for assessing moisture diffusion in repair material PMC at high environmental temperature and a low humidity. It has been found that the diffusivity of repair material is substantially enhanced at an environmental temperature of 50°C together with a higher surface transfer coefficient (Rahman 1999). The parameters used in the analysis are shown in Table 2. Figure 12 shows the moisture content surfaces at selected times during the time history of the diffusion process. From the moisture content surfaces it is evident that the moisture diffusion process occurs at a highly accelerated rate in this case. The moisture loss in several elements adjacent to the diffusing surface occurs rapidly and the core also loses moisture at an accelerated pace. The outer layer is in equilibrium with the environmental humidity in 7 days, whereas, the core attains a moisture content of 70% at this age. At 28 days the average moisture content over the domain is below 40%. The higher temperature increases the energy in the system and dilates the pore structure to facilitate the removal of moisture.

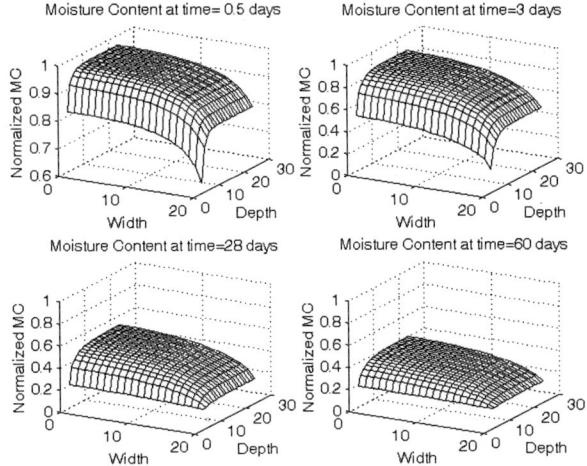

Figure 12: Moisture Content Surfaces-High Environmental Temperature

CONCLUSION

Moisture diffusion from repair domain in a patch repair system is a key critical process stimulating several mechanisms at macro level (shrinkage and tensile creep) that is of paramount importance for repair systems integrity and durability. A 2-D finite element program DIANA-2D developed using non-linear diffusion equation for predicting moisture transport in a repair mortar is presented. An empirical trigonometric form of diffusivity law derived for commercially available repair mortars is used in the model. The proposed model gives a good replication of drying tests conducted on 1-D and 2-D moisture diffusion prisms made of various repair mortars. The model has been used to predict temporal variation of moisture loss over the spatial domain of a miniature patch repair system. High environmental temperature is found to have a significant effect on moisture diffusion in cement based repair mortars. This may be attributed to an increase in the energy of the water vapor molecules and dilation of the pores due to thermal changes.

ACKOWLEDGEMENTS

The support of the Department of Civil Engineering and Research Institute at the King Fahd University of Petroleum & Minerals is acknowledged with thanks; also, the support of repair companies FOSAM, MBT and SIKA in the form of complimentary supply of repair materials for laboratory component of research work is gratefully acknowledged.

REFERENCES

Alvaredo, M.A. (1994). Drying Shrinkage and Crack Formation. *Ph.D Dissertation*, Swiss Federal Institute of Technology, Zurich.

Asad, M., Baluch, M.H. and Al-Gadhib, A.H. (1997). Drying Shrinkage Stresses in Restrained Concrete Systems. *Magazine of Concrete Research* **49:181**, 283-293.

Bazant, Z.P. and L.J Najjar. (1972). Nonlinear water diffusion in nonsaturated concrete. *Materials & Structures* **5:25**, 3-20

Garboczi, E.J. and Bentz, D.P. (1992). Computer Simulation of the Diffusivity of Cement-Based Materials. *Journal of Materials Science* **27**, 2083-2092.

Granger, L., Torrenti, J.M. and Acker, P. (1997). Thoughts about Drying Shrinkage: Scale Effects and Modelling. *Materials and Structures* **30:3**, 96-105.

Grzybowski, M. and Shah, S.P. (1989). Model to Predict Cracking in Fiber Reinforced Concrete due to Restrained Shrinkage. *Magazine of Concrete Research* **41:148**, 125-135

Jonasson, J.E. (1978). Analysis of Creep & Shrinkage in Concrete & Its Application to Concrete Top Layers *Cement & Concrete Research* **8:4**, 441-454.

Penev, D. and Kawamura, M. (1991). Moisture Diffusion in Soil-Cement Mixtures and Compacted Lean Concrete. *Cement & Concrete Research* **21:1**, 137-146

Quenard, D. and Sallee, H. (1992). Water Vapour Adsorption and Transfer in Cement-Based Materials: A Network Simulation. *Materials and Structures* **25**, 515-522

Rahman, M.K. (1999). Simulation & Assessment of Concrete Repair System. *Ph.D. Dissertation*, King Fahd University of Petroleum & Minerals, Dhahran, Saudi Arabia, 530 p.

Rahman, M.K. and Baluch, M.H., Al-Gadhib, A.H. and Alfarabi, S. (2000). A Finite Element Based Approach for determination of Diffusivity Law of Repair Materials. Submitted to Second International Conference on Engineering Materials, 2001, Aug 16-19, San Jose, California, USA.

Rahman, M.K. and Baluch, M.H., and Al-Gadhib, A.H. (1999). Modeling of Shrinkage and Creep Stresses in Concrete Repair. *ACI Materials Journal* **96:5**, 542-550.

Sadouki, H. and Van Mier, J.G.M. (1997). Simulation of Hygral Crack growth in Concrete Repair Systems. *Materials and Structures* **30:11**, 518-526.

Sakata, K. (1983). A Study on Moisture Diffusion in Drying and Drying Shrinkage of Concrete. *Cement & Concrete Research* **13:2**, 216-22

ANALYTICAL MODELLING OF HISTORICAL MASONRY STRUCTURES FOR THE EVALUATION OF STRENGTH CAPACITY OF THEIR VULNERABLE ELEMENTS

A. I. Unay

Department of Architecture, Building Science
Middle East Technical University, Ankara, TR-06531, Turkey

ABSTRACT

The analytical studies for the structural preservation of historical structures mainly have two aspects. Firstly, structural demand should be determined by a decisive structural analysis. Then, the strength capacity of the structure must be evaluated. A comparison of these two concluding stages results in the ultimate safety. In this paper the following steps of analytical study is proposed to evaluate the safety level of historical structures. Structural demand of any given structure can be determined by using the Finite Element Method. Then, vulnerable parts of the structure could be singled out. The strength capacity of these structural elements could be achieved by an analytical method derived. In this method, using the material properties of masonry, the interaction relationship between bending moment and the axial force is generated.

KEYWORDS

Historical Structures, Analytical Modeling, Finite Element Analysis, Masonry Structures, Masonry Strength, Structural Safety.

INTRODUCTION

The main principle of structural conservation is to protect the structure against environmental and man-caused hazards. Repair and restoration are needed in previously damaged structures. Historical structures are strengthened because of deterioration of their structural material, due to the effects of time. Most historical structures are, unfortunately, in poor condition. Natural disasters, foundation settlements and fires have taken their toll. Centuries of neglect have eroded away many treasures.

Historical structures, at the present, show serious cracking of masonry and, deterioration of structural integrity, with accompanying partial or almost full collapse. Primitive and unskilled attempts have been made to seal cracks and repair roofs and walls, but what is overlooked is that such repair efforts cannot be attempted before understanding why these faults have occurred. The original and presently defective load transfer mechanism of the structure must be well studied before deciding on the rehabilitation, restoration or strengthening techniques. If this is not done, then the preservation efforts may even unduly harm the historical structure, let alone help it.

It is thus evident that, methods of structural analysis for historical structures must be developed to understand how loads are transferred. Analytical models of the historical structure under study can be developed, with satisfactory closeness to reality. On such a model, various load effects (natural and artificially induced) can be applied and their actions are analyzed. Through such analyses, the load carrying mechanism can be understood. The architect and engineer are then in a position to attempt a method of repair. They are aware of the forces that are in equilibrium. They know the vulnerable parts of the structure. They can assess the effects of the repair methods, which are to be applied to the structure which will change and modify and hopefully improve the present forces that are in "some kind" of equilibrium.

SAFETY OF HISTORICAL STRUCTURES

Structural safety and factor of safety can be expressed by considering probability of failure of a structure. Therefore, to determine the safety factor of a structure, first a structural analysis must be performed to determine the distribution of load effects (**F**) on the total structure. Then, a strength (**R**) capacity determination of the critical elements must be evaluated. A comparison of **F** and **R** will yield the safety factor.

$$SF = \frac{R}{F} \qquad (1)$$

In Eqn. 1 the **SF** is called the safety factor, **R** is the resistance of the cross-section of the element and **F** is the load effect on the element.

In determining the safety factor of historical masonry structures, various criteria should be considered. Consequently, it is not so easy to reach an exact solution by conventional engineering analysis. In order to estimate the safety of a historical masonry structure, behavior of the structure must be determined well. Principally, geometrical form of structure, materials of construction, the loading and the support conditions must be studied with as much precision as possible, Ersoy (1989).

If the failure of an individual element causes overall collapse of the structure, it becomes important to detect, at what time and for what reason this element fails. Masonry structural elements frequently have nonlinear material properties and the uniformity of material characteristics is uncertain throughout the whole length of the element. For such cases, the exact solution for the evaluation of safety is to determine the ultimate strength of the element. Therefore, if stress-strain relationship is known, the ultimate strength of any masonry cross-section can be determined.

Axial force, bending moment, shear force and torsional moment are the main internal forces that influence the masonry elements. Torsion and shear forces may be more characteristic under special conditions. However, because of inherent material properties, cross-sectional properties and overall structural form, historical masonry structural elements are subject to axial compressive forces and bending moments.

When axial force (**N**) and bending moment (**M**) are governing load effects to cause failure, ultimate load carrying capacity of the element can be represented by **N-M** interaction diagram. Due to determination of ultimate capacity of structural elements under axial force and bending moment, the generated **N-M** interaction diagram can only be meaningful, if the element is assumed to have linearly elastic material properties.

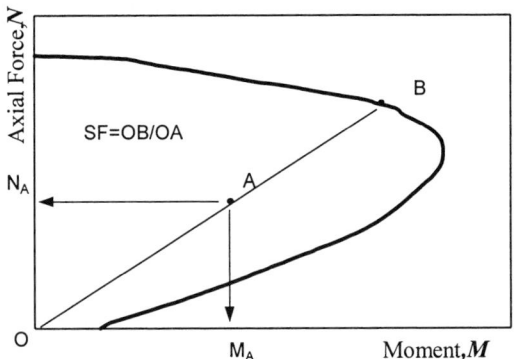

Figure 1: Evaluation of safety factor using N-M interaction diagram

The **N-M** interaction diagram shown in Figure 1 illustrates the evaluation of the safety factor of a masonry element. As a result of structural analysis, axial force N_A and bending moment M_A is determined at a point **A** on the interaction diagram. Since structure is analyzed by assuming linear elastic material characteristics, gradually increasing loads lead to linear increase in the internal forces. Therefore a line connecting the origin and the point **A** intersect the interaction diagram at point **B**. At point **B** the element is assumed to have reached its load carrying capacity. Consequently, the safety factor can be defied by simply the ratio of line **OB** to **OA**.

$$SF = \frac{OB}{OA} \qquad (2)$$

Indeed, in addition to the mathematical meaning of the Eqn. 2, graphical significance of this ratio is more meaningful for the concept of safety evaluation.

This approach facilitates the analysis of historical masonry structures for safety investigation and exploration of structural load carrying capacity under various disturbances. Safety of historical masonry structures is dependent on several factors and interpretation of these factors is very difficult. Therefore, a systematic study is needed for the evaluation of the safety level.

DETERMINING THE STRENGTH CAPACITY OF MASONRY

The best way to define the strength capacity of masonry is to plot axial load (**N**) – bending moment (**M**) interaction diagram for the element section. In determining the strength capacity of masonry structural elements analytically, some basic assumptions should be considered. In this analytical study, both moment and axial force load carrying capacities are determined simultaneously. Basic assumptions in the determination of combined flexural strength of masonry structural elements are as follows:

- Plane sections before bending remain plane after bending.
- There is a perfect bond between distinctive materials such as masonry unit and mortar.
- The strain in the masonry units is equal to the strain in the mortar at the same level.
- The stress-strain curve of the material is known.
- The stress in the masonry unit and mortar can be computed from strain using the stress-strain curves for masonry materials.
- Tensile strength of material is also to be considered in section analysis.
- The strain distribution in the section is uniform under pure axial force.
- The strain distribution in the section is linear under flexure.
- Masonry is assumed to fail when the compressive strain reaches a limit value.

There are different material characteristics of masonry structural elements and different stress-strain relationship of distinct materials. This should be considered in the determination of load carrying capacity of elements due to flexure. In order to determine the ultimate strength of masonry structural elements, stress-strain relationship of materials must be known. This can be achieved by testing material samples of the structural element.

The use of stress-strain (σ-ϵ) relation, equilibrium and compatibility conditions in the section are a quick and simple way to evaluate the load carrying capacity of masonry structural elements. In this study a method is proposed which is based on the estimation of the load carrying capacity of the section for any combination of axial force (**N**) and bending moment (**M**) in the structural element, namely N-M interaction diagram of the section, Unay (1997).

Strength of the masonry element subjected to pure axial force and strength of a general masonry element subjected to flexure can be determined by considering true strain and stress distribution throughout the element cross-section. At any level, the stress integrated over the section must add up to the required force effects M and N as shown in Figure 2.

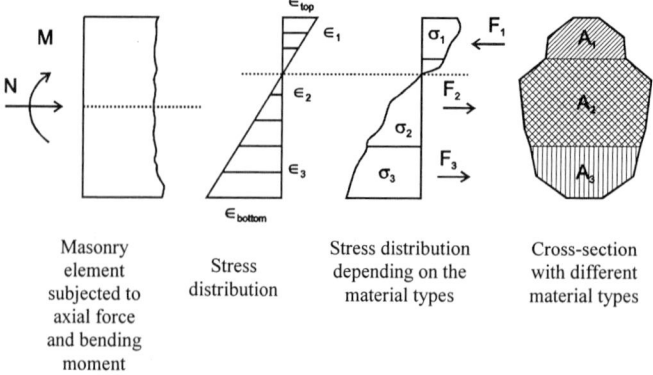

Masonry element subjected to axial force and bending moment

Stress distribution

Stress distribution depending on the material types

Cross-section with different material types

Figure 2: Equilibrium of internal forces in a masonry element subjected to axial force and moment

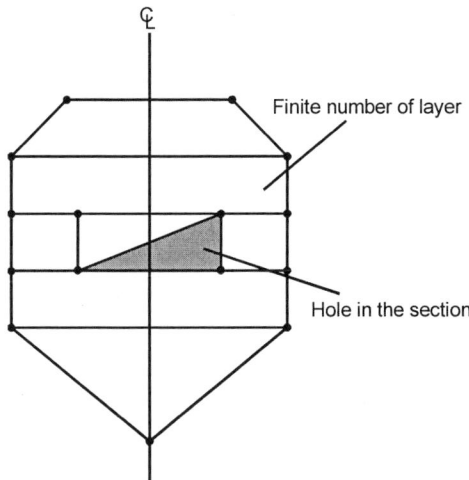

Figure 3: Definition of element cross-section by finite number of layers

In order to determine the strength capacity of a masonry section, axial force (**N**) – bending moment (**M**) interaction diagram is plotted for the section. As shown in Figure 3, element cross-section is defined by a finite number of layers parallel to main bending axis. The number of layers that is going to be used is related to the dimensions and shape of the structural element. Furthermore, while modeling elements, distinct kind of materials can be used for every different layer. The voids in the cross-section can be considered also. Material characteristics of the cross-section are defined in terms of stress-strain (σ-ϵ) curve. Since axial force (**N**) and bending moment (**M**) carrying capacity of masonry structural elements with arbitrary geometry can be analyzed easily, safety factors of these structures can also be determined in an approximate manner.

Generation of N-M Interaction Diagram

In association with **N-M** interaction diagram, load carrying capacity of all kind of masonry element cross-sections can be evaluated by satisfying the equilibrium conditions in the cross-section. Masonry structures have unusual cross-sectional shapes because of their architectural form and construction techniques. By dividing the overall section into small rectangular or triangular parts, all kind of irregular element cross-sections and different material characteristics in cross-section of certain masonry elements could be designated by finite number of regions. Differential areas should be taken into account to satisfy the equilibrium and compatibility condition in the section.

Employing adequate number of sub-areas according to the shape and material characteristics of the section, the force, equilibrium condition can be satisfied. As shown in Figure 2, equilibrium equation can be written as according to location of neutral axis

$$N = A_1 \times \sigma_1 + A_2 \times \sigma_2 + \ldots\ldots\ldots\ldots = A_i \times \sigma_i \qquad (3)$$

In this equation σ can be evaluated in two steps. Initially the strain (ϵ_i) corresponding to differential area (A_i) at this level is determined from strain distribution in the section, and then, σ is determined by using the σ-ϵ curve of the identical material.

Therefore, equilibrium state of computed internal forces is ensured in the cross-section from Eqn. 3, and then internal moment equation can be written as

$$M = \sum F_i \times y_i \qquad (4)$$

Applying this procedure, axial force **N** and bending moment **M** values in the section can be determined. Moreover, tensile and compressive crushing strain $\epsilon_{crushing}$ should not exceed in the strain variation in the section at uppermost and lowermost fiber of the section. Exceeding the crushing strain $\epsilon_{crushing}$ at the section means failure of the masonry element. Calculated axial force **N** and bending moment **M** can be considered as the maximum load carrying capacity, Unay&Atimtay (1997).

APPLICATION OF THE METHOD ON A CASE STUDY

In order to show the versatility of the method, the most unusual cross-sectioned example is selected as shown in Figure 4. A case study is done on Yivli Minaret in Antalya, which is built by Seljuks. Internal forces due to gravity and seismic forces are determined by the Finite Element Method. The mathematical model, deformed shape of the structure and force distribution is shown in Figure 5. The structure is analyzed first under its own weight. Then, it is analyzed under the maximum credible earthquake by the Response Spectrum method using SAP2000 Structural Analysis Program, Wilson & Habibullah (1998).

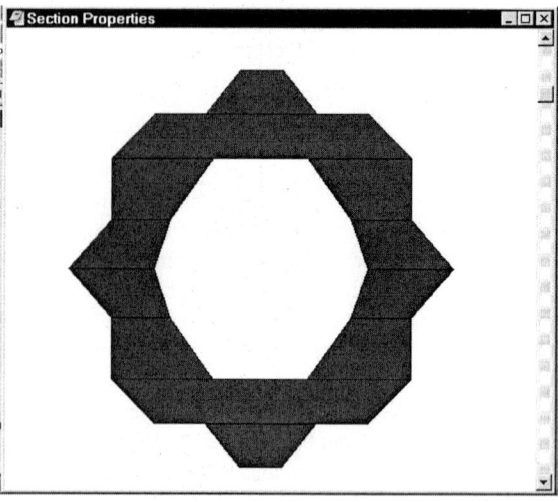

Figure 4: Section properties of Yivli Minaret

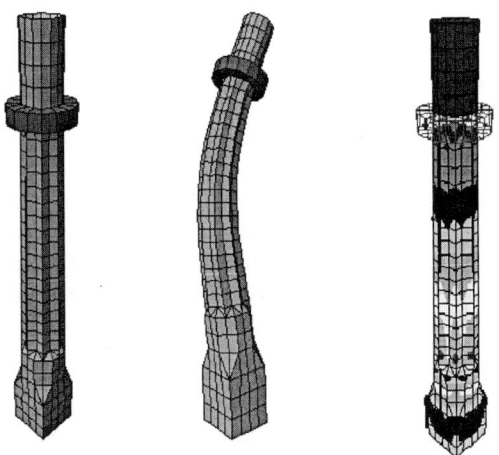

Figure 5: Finite Element Analysis of Yivli Minaret

By using the results of the first analysis, (N_1,M_1), a point is plotted in the interaction diagram which shows the equilibrium position under its own weight. By considering this point of equilibrium as the origin, a second point is plotted by using the results of the earthquake analysis, (N_2,M_2). This second point should fall within the interaction diagram for the element to stay safe under the attack of the earthquake. If this second point is extended to cut the interaction diagram, the failure point is determined, (N_f,M_f). The ratios of the distance between (N_1,M_1) to the distance between (N_2,M_2) and (N_f,M_f) will give the factor of safety under the considered earthquake. Figure 6 shows the determination of factors of safety for the Yivli Minaret. It is seen that, under an earthquake of base rock acceleration of 0.4 g, the minaret is unsafe.

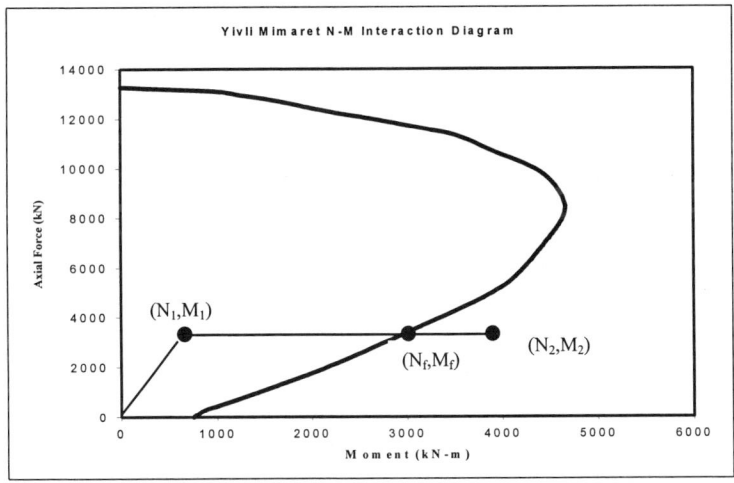

Figure 6: Determination of safety level

CONCLUSIONS

The safety factor can be defined as the ratios of the strength capacity of the historical structure to the load effects imposed. Therefore, in order to define the safety level rationally, a well performed structural analysis is needed. The most available method of doing this is the Finite Element Method.

The strength capacity of the historical structure is determined by developing the axial force – bending moment interaction diagram of the cross-section that are selected as vulnerable by the extensive analyses already performed. Using realistic stress-strain relationships for masonry, the interaction diagram between axial force and bending moment can be determined. The combination of **(N,M)** due to gravity loads and seismic loads can be plotted on the interaction diagram to determine the safety factor.

Repair and rehabilitation process can be done with much greater confidence, once the vulnerable parts of the structure are known and the factor of safety quantified. The methodology to define the inherent safety level of historical structures thus becomes an invaluable tool for the engineer and architect who attempt to repair or rehabilitate historical structures.

REFERENCES

Ersoy U. (1989). Diagnosis, Assessment and Emergency Intervention for Historic Masonry Structures, *Proceedings of Structural Conservation of Stone Masonry, Proceedings of International Technical Conference,* pp13-19, ICCROM, Rome.

Unay A. I. (1997). *A method for the Evaluation of The Ultimate Safety of Historical Structures*, Ph.D Thesis Dissertation Presented to the Faculty of Architecture of the M.E.T.U, Ankara.

Unay A. I.and Atimtay E. (1997). Determining the Ultimate Safety of Historical Structures, *Computers in the Practice of Buildings and Civil Engineering, Worldwide ECCE Symposium*, pp 305-309, Lahti, Finland.

Wilson L. E. and Habibullah E. (1998). *SAP 2000 – Integrated Structural Analysis and Design Software*, Computers & Structures Inc., Berkeley, California.

LOADINGS AND CODE DEVELOPMENTS

ACTUAL PROBLEMS OF STEEL-DESIGN – FUTURE OF THE CODES

F. Werner[1] and P. Osterrieder[2]

[1]Department of Civil Engineering, Bauhaus University Weimar, Germany
[2]Department of Civil Engineering, University of Brandenburg, Cottbus, Germany

ABSTRACT

The development of the planning job is characterised by factors like growing project sizes, very short processing times, widely use of computers, automation of the design processes a.s.o.. The development of computer technology in conjunction with the development of structural mechanics allows for qualitatively new ways in the organisation of analysis and design of steel-, steel-glass, and steel-concrete constructions. The progress in the area of the structural design does not only depend on the technical possibilities but also demands robust software products which can be effectively and safely used in practice and a normative bases which offers advise and protection to the engineer in practical work. Examples from different practical fields of activity show sustainable possibilities for design methods. Advanced techniques in analyses and assessment of steel structures call for a new philosophy of design codes. The currently observed trends particularly in the Eurocodes, which extensively portray recipe books character do not meet the demands of the future.

KEYWORDS

Analysis, design, steel construction, code, flexural torsional buckling, plastic design

GENERAL

The development of the planning job is characterised by several factors:

- Growing project sizes;
- Growing complexity (volume and technical level) of the projects;
- Very short processing times;
- Effectiveness in the usage of the building-materials;
- Automation of the design processes;
- Cooperative engineering based on CAE and CAAD.

This leads to new demands on structural design. On the other hand the development of computer technology in conjunction with the development of structural mechanics allows for qualitatively new

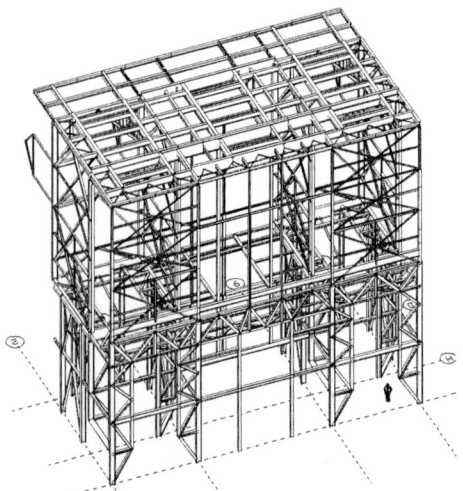

Figure 1: Steel structure with about 2000 elements; calculation time: 30s - second order analysis for 10 load combinations

ways in the organisation of analysis and design of steel-, steel-glass, and steel-concrete constructions. This brings to an end an epoch characterised by:

- Analysis through breakdown of structures into single elements;
- Linear calculations in reference to material and geometry;
- Application of extensive hand formulas with very restricted scopes of application and very idealistic boundary conditions;
- Purely intuitive variant construction without real optimization procedures;
- Global safety considerations.

It is not easy to forecast when these mentioned developments will become operative and this varies regionally. To achieve these objectives however we have to formulate new initiatives in research, teaching and in the development of the necessary design tools!

The progress in the area of the structural design does not only depend on the technical possibilities but also demands solutions on the following areas:

- Highly trained engineering personal;
- Robust software products which can be effectively and safely used in practice;
- A normative bases which offers advise and protection to the engineer in practical work.

The difference between scientific analysis carried out by theoretical mechanics and the practical analysis or design, even if this may be very demanding, is substantial. In practice only methods and procedures are used whose reliability has been confirmed. The probability that the demanded reliability of the constructions is warranted must be very high (Werner). For this reason the standards acquire great significance on one hand and their form and contents must on the other hand be brought into line with the new qualities. The currently observed trends particularly in the Eurocodes, which extensively portray recipe books character, depict a dim future, Werner (2000)!

ANALYSIS AND DESIGN OF BEAM STRUCTURES

The number of the details in steel constructions is very high. In the previous codes the proofs of details played only a subordinate roll. References have only been given to general questions like connecting elements, one-sided angle connections, cable connections or design of thick flanges. Within the last few years extensive investigations on beam connection behaviour have been performed. At present the results of this work are found in textbook like form in the codes. If one examines these carefully, it is not difficult to find several contradictions to the other parts of and to the principle requirements of the codes (Eurocode, 1993):

- The description of sway frame is done without a consistent and exact theoretical exact basis (Eurocode 3, part 5.2). But the given moment characteristics for connections however then require non-linear analyses.
- The influence of semi-rigid connections on internal forces is e.g. negligible for practically important frame systems. The influence from stabilisation forces perpendicularly to the frame plane which could be of importance in many cases is not mentioned.
- Any use of the calculation algorithms is meaningful and effective only with computer programs (Eurocode 3, app. J).
- As a matter of principle, a code should not take textbook character → the size of the descriptions in current documents is too big (Eurocode 3, app. J).

The consideration of single structural planes or structural elements is at present only justified for simple systems. The model of complex and in general spatial structures, very often reveals interactions which lead to remarkable rearrangements of internal forces. This is particularly the case in bracing systems, walls and floors where simply defined loads are assigned to two-dimensional systems. This can be accepted for the preplanning phase but otherwise should contain more exact analyses. The application of a simplified geometrically nonlinear calculation, e.g. as second order theory is the current state of technology (Fig. 1). For the purpose of the effectiveness real geometrically nonlinear analyses should only be used for systems with big deformations, like cable structures.

The important problem of definition of imperfection forms will be discussed later. Material savings for structural steel frames in the background of reliable design procedure are of high relevance for competitive steel design. An essential development is the creation of automated systems for the design of such special types of structure. Besides the problems of the optimisation, which create the basis of such synthesis systems, the reliability of the analysis software has to be checked under new considerations. In comparison to practical approaches, the system design based on stochastic methods can take completely nonsensical forms. In these cases simplified analysis methods can lead to completely incorrect conclusions. For frames with non-uniform members a procedure for practical application was developed using stochastically optimisation to find a cross section distribution for optimal weight design. Compared to traditional design methods weight savings of up to 15 % were achieved (Fig. 3). Design requirements are based on specifications in the national German Code DIN 18800 1 and the EC 3. The used beam element is defined by the cross section at the left and right element end, by its material constants and by the number of integration segments along the beam axis (Fig. 2). The stiffness within the element is computed in the integration points of the actual cross section. As a result, the changeable element stiffness is registered exactly in terms of mechanical calculation. For the description of deformation behaviour perpendicular to the beam axis an approximation approach of 5th order and along the beam axis an approach of 2nd order is applied. These form functions are very flexible, so that one element is sufficient to model one tapered member, Müller, Osterrieder, Werner (1999).

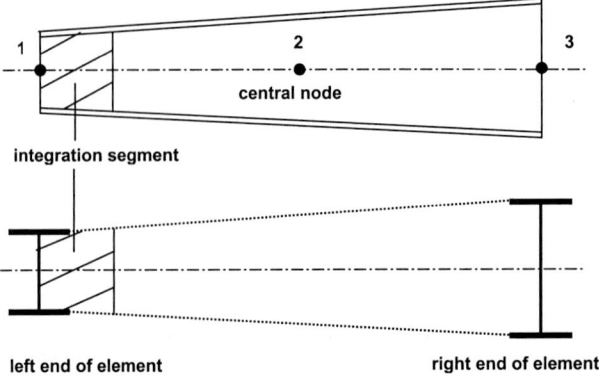

Figure 2: Tapered beam element with additional central node

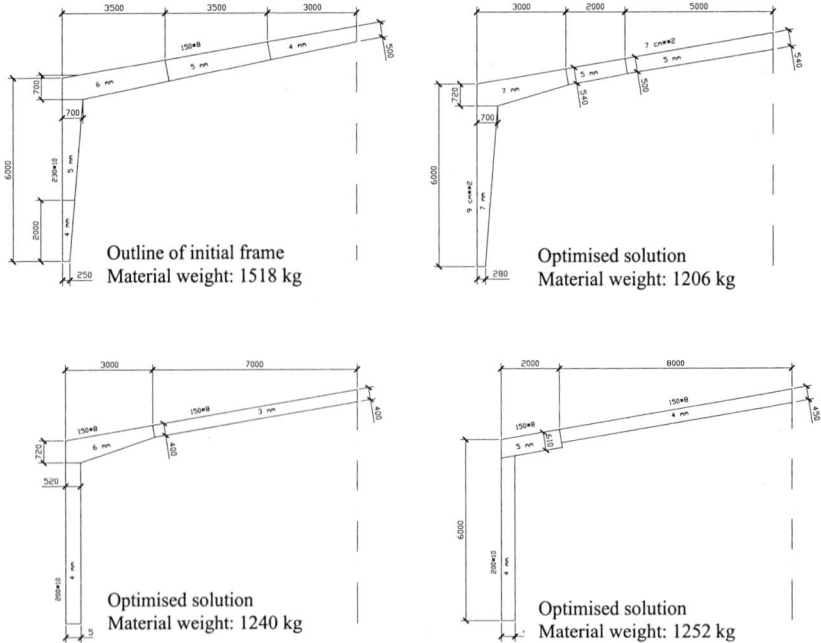

Figure 3: Shapes of an optimised frame structure

The next step in the development is the use of second order analysis on the basis of flexural-torsional theory. To achieve this all necessary checks –strength and stability – will be performed in one step (see part STABILITY CHECKS).

Real plastic calculations, i. e. application of the geometrically nonlinear yield area theory, play a subordinate role in the steel structures. Besides the problems of the required software tools the fields of practical application where any advantages are evident are not defined. The determination of the plastic limit load capacity of cross-sections is through the codes, where they rather don't belong (and generally existing methods) and is limited to simple symmetrical I- sections. The use of calculation methods based on optimization algorithms creates a larger application field for cross sectional forms (Fig. 4) as well as for internal force combinations. For example warping moments can be taken into account without major problems, Osterrieder, Werner, Kretzschmar (1997).

For any set of internal forces, in a defined section, implied in the analyses of strength or buckling problems and given by the reference vector:

$$\overline{F} = \begin{bmatrix} \overline{N} & \overline{M_y} & \overline{M_z} & \overline{M_\omega} \end{bmatrix}$$

the multiplication factor ∀, which extends the force vector **F** towards the yield surface has to be determined. The yield limit state is defined by:

Equilibrium conditions   yield conditions

$\Sigma \sigma_i A_i \quad - \quad \alpha \overline{N} \quad = \quad 0$

$\Sigma \sigma_i z_i A_i \quad - \quad \alpha \overline{M}_y \quad = \quad 0 \qquad\qquad \sigma_i \leq +f_y$

$\Sigma \sigma_i y_i A_i \quad - \quad \alpha \overline{M}_z \quad = \quad 0 \qquad\qquad \sigma_i \geq -f_y$

$\Sigma \sigma_i \omega_i A_i \quad - \quad \alpha \overline{M}_\omega \quad = \quad 0$

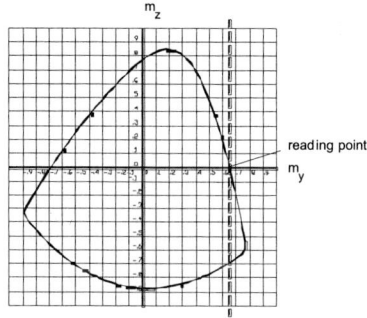

Figure 4: Yield surface for L-profile

The linear objective function has the form: ∀ ⇒ **Maximum**
By solving this numerical optimization problem, f.i. by Revised Simplex, load carrying capacities for any type of profile are to be calculated in an easy way..

Yield surface for L-profile:

Section: 100x75x9   n = 0.4

Internal forces: $M_{y,d} = 7{,}0$ kNm
$N_d = 143{,}44$ kN
$M_z = 0$

steel grade St37: $f_{y,k} = 24$ kN/cm^2

first yield design leads to:

$$\max \sigma_{S,d} = \frac{700}{181} * 6{,}91 + \frac{143{,}4}{15{,}1} = 36{,}22 \frac{kN}{cm^2} > \sigma_{R,d} = 21{,}8 \frac{kN}{cm^2}$$

first hinge design leads to:
fully plastic properties : $N_P = 358.6$ kN $M_{y,P} = 10.87$ kNm
$n = 143{,}44/358{,}60 = \mathbf{0{,}4}$, $m_y = 7{,}0/10{,}87 = \mathbf{0{,}644}$

The given load combination satisfies the yield condition. Thus allows for an load increase of **67%**.

STABILITY CHECKS

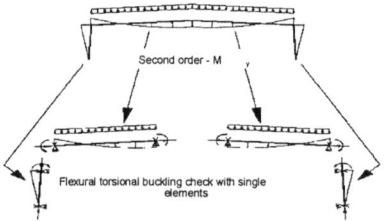

Figure 5: Concept for Lateral Torsional Buckling - DIN 18800 and EC3

An important and critical area where up to now no satisfactory solutions are available is stability checks. It is only for plane problems where reasonable solutions are available. The concepts for lateral torsional buckling and plate buckling in the codes have only been improved in detail during the last years (Fig. 5). The modern codes (DIN 18800, EC3) contain necessary details on imperfections. But research is still required in this area in order to come up with more certain predefinitions for spatial systems. Furthermore the boundary conditions must be specified for imperfections. In current documents they are treated like safety factors, but cannot be easily and systematically assigned to them. One of the problems is shown in figure 6. According to the code the first eigenmode is to be the imperfection shape. At the same time the imperfection has to meet the most unfavourable strain situation. For a column with a slenderness 8 in the middle range the first eigenmode does not meet these requirements. In this case the internal moments at the supports can be higher for an imperfection form according to the deflections. In its current form the EC3 expresses the uncertainty of a practical use of this method of calculation. In a long and winded way simplified formulas are mixed with particular instructions and theoretical references (EC3, part 5.2). Currently developments are in progress in this field and that will lead to more effective and reliable results than the presently available methods according to the codes. The problem here, as already described, is that the standards still show essential gaps in defining boundary conditions.

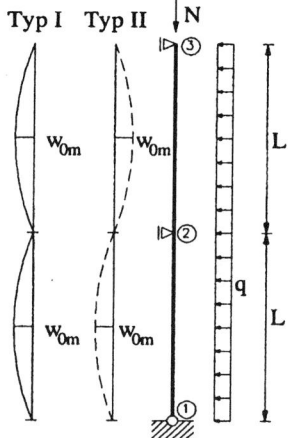

Figure 6: Imperfection forms for a column:
I - similar to the deflection
II - similar to the first eigenmode

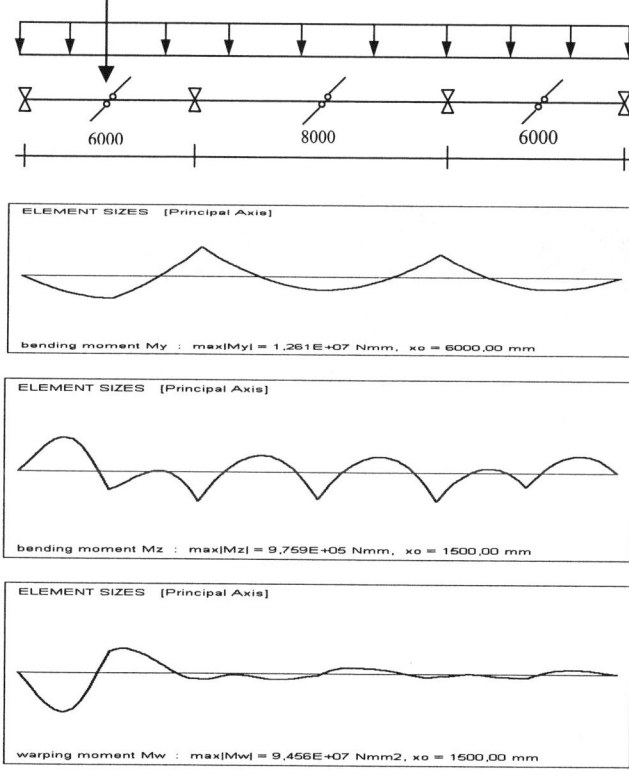

Figure 7: Purlin system analysed by beam elements with an additional torsional degree of freedom v`

And there are very few prospects for change! Improving the handling of lateral torsional buckling problems is of great importance. The available formula apparatuses with confusing coefficients can only seldom meet practical requirements, such as:
- Multispan beams with unequal spans and arbitrary loads both in magnitude and point of application;
- Semi-rigid support conditions at the ends and in the spans;
- Coupling of components in spatial structures etc.

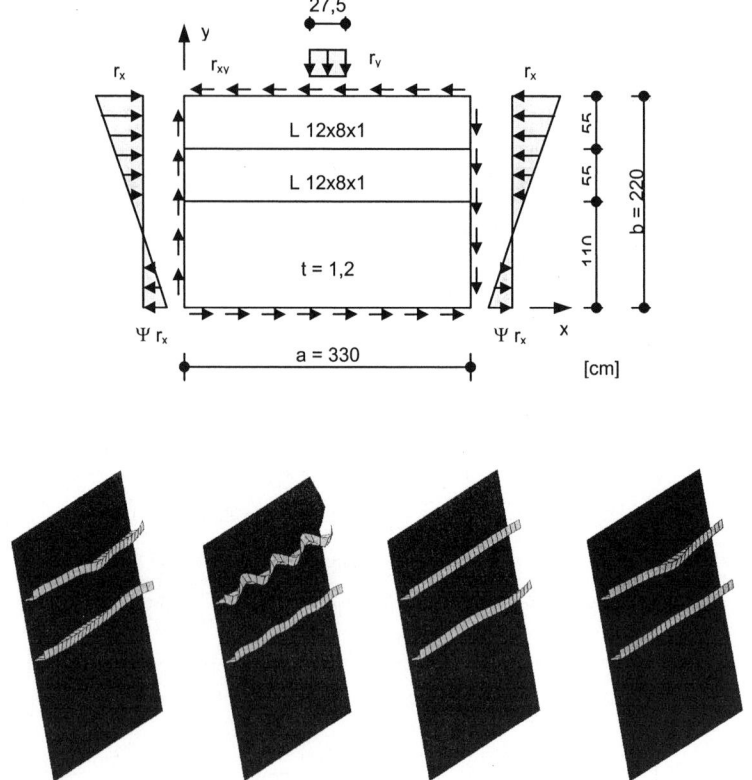

Figure 8: Buckling problem with eigenmodes for: $\Phi_x \rightarrow 8_{cr,x} = 3{,}03$ $\Phi_y \rightarrow 8_{cr,x} = 4{,}37$
$\vartheta \rightarrow 8_{cr,x,y} = 1{,}96$ overall stress: $\rightarrow 8_{cr,o} = 2{,}03$

The use of beam elements with an additional torsional degree of freedom v` allows the creation of robust analysis systems which solve very effectively a large class of problems (Fig. 7). Development work still needs to be done, particularly in defining transition conditions between profiles with different shapes and frame splices and connections. The changes in conditions for section-deformation, especially warping conditions can not be easily described. Simplified presumptions lead in general to uneconomic solutions. Particularly for crane support beams and purlin systems, very economic and realistic design with the necessary safety level are possible. It is also possible to generate practical design tools which can replace the restricted valid concepts of the codes for thin-walled, asymmetrical cross sections, like purlins (Fig. 7). Thus it is possible to eliminate a big portion of the differences in load carrying capacities normally observed between results from experiments and those from calculations. Especially boundary conditions like elastic restraints from claddings or special internal supports can be modelled in a practical way. The load carrying capacity is very sensitive to elastic restraints, so with small spring-stiffness a noticeable increase of the critical load can be achieved. Research work is still necessary on connecting conditions for beams with different shapes, splices, frame corners etc. structures. The description of the warping restraints or transformation of warping parameters in this area requires a new quality of models. One of problems here is also the definition of boundary conditions, like:

- Imperfection forms
- Support conditions
- Limitation of deflections.

These issues are important for congruence between theoretical model and reliability of calculation results as shown by Werner, Osterrieder & Lehmkuhl (1999). The large field of the plate buckling particularly in cases where horizontal and vertical stiffeners and torsion-stiff flanges exist is difficult to handle on the basis of the code. Advanced software tools are coming up allowing fast checks and design even for sophisticated problems. Due to missing normative conditions for real nonlinear analyses, it is reasonable to use the way prescribed by the codes. Modern FE-analysis tools are applied to determine the eigenvalues and then the normal buckling coefficients of the codes are used. According to the codes the plate buckling problems are separated in different load cases [Fig. 8]. The design is then realised through interaction formulas. This is a cumbersome method but which the practising engineer however trusts. For FE-methods which can analyse complex failure forms assumptions are still missing for corresponding buckling factors. Suggestions are available to handle this problem in a practical way, Ortlepp, Werner, Osterrieder (2000).

SUMMARY

The historically major objective of the codes i.e. the creation of a safety standard, must be shifted into the center of focus once again. For the representation and storage of knowledge, methods and catalogues very effective media systems are today at our disposal. They could complement the new generation of codes in a very effective way.

REFERENCES

[1] DIN 18800 - Stahlbauten (Steel structures; design and construction) 11/1990
[2] Eurocode 3 - Design of steel structures, ENV 1993
[3] Eurocode 1 – Basis of design and actions on structures, ENV 1991
[4] Müller, A.;Osterrieder, P.; Werner, F.: Optimal Design of Steel Frames with Non-uniform Members. International Conference on Steel Structures, Hongkong 1999
[5] Ortlepp, O.; Werner; F.; Osterrieder, P.: Practical Buckling Capacity Curve for Coupled Instabilities of Pates. CIMS 2000, Lissabon Portugal
[6] Osterrieder, P.; Werner, F.; Friedrich, M.; Ortlepp, O.: Advanced Finite Element Buckling Analysis in Engineering Practise. SDSS 99, Timishuara 1999
[7] Werner, F.: Probleme der Normengestaltung (Problems of code-configuration); http://www.uni-weimar.de/Bauing/stahlbau/norm_ref.pdf, 2000
[8] Werner, F;.Osterrieder, P.; Lehmkuhl, H.: Stability Design of Thin Walled Memebers including Local Buckling. 2nd International Conference on Thin-Walled Structures, Singapore 1998
[9] Werner, F.; Lehmkuhl, H.: Use of Numerical Methods for Practical Analyses of Structural Elements. CIMS 2000, Lissabon Portugal 2000
[10] Osterrieder, P.; Werner, F.; Kretzschmar, J.: Plastic Flexural-Torsional Buckling Design of Beams with Open Thin-Walled Cross SDSS 97, Nagoya 1997

CALIBRATION OF LOAD FACTORS FOR THE SOUTH AFRICAN LOADING CODE

T.R. Ter Haar[1], J.V. Retief[2] and A.R. Kemp[3]

[1]BKS (Pty) Ltd Engineering and Management, Durban, South Africa
[2]Department of Civil Engineering, University of Stellenbosch,
South Africa
[3]Department of Civil Engineering, University of the Witwatersrand,
Johannesburg, South Africa

ABSTRACT

The South African Loading Code SABS 0160:1989 is presently under review to rectify deficiencies and as a preamble to the revision of materials design codes. As part of this review programme the load combination rules and partial load factors are investigated and evaluated. A code calibration methodology has been developed at the University of Stellenbosch for the rational, systematic and efficient calibration of structural codes. The implementation of the first stage of this methodology for the calibration of the SABS 0160:1989 is presented in this paper. The first stage involves the derivation of partial load factors as well as target values for the derivation of resistance factors in the second calibration stage. The results obtained from this first calibration stage are presented for different degrees of uncertainty in resistance models that are representative of most failure modes in common structural materials. Target partial safety factors are derived for load statistics representative of South African conditions. The results are also compared to that obtained for the Eurocode and ASCE 7. Based on the results obtained recommendations are made regarding proposed changes to the current South African Loading Code.

KEYWORDS

Code calibration, structural reliability, load factors, load combinations, safety targets

CURRENT SABS 0160 PROVISIONS

The derivation of partial load factors for SABS 0160 (1989) was the result of the implementation of the first stage of a two-stage calibration procedure. This first stage involves the identification of partial load factors and load combination factors from available statistical data for the common types of loading (refer to Table 1 for the load statistics used for the calibration of SABS 0160). The load factors pertaining to ultimate limit states were selected to give design load effects in accordance with a target probability of 1% of exceeding the design load value for a 50-year life. At that stage the uncertainty in the resistance was not explicitly included in the calibration model. Instead it was provisionally accounted for in the target exceedance probability of 1% with respect to the design loads. This method is also called the load index approach. The actual exceedance probability over a practical range of load combinations was allowed to deviate from the target value in order to obtain results comparable with the existing practice at the time. The load statistics for the dead, live and wind loads that were used for the calibration of the current SABS 0160 (1989) are given in Table 1. The corresponding load statistics for the European (Eurocode 1) and American (ASCE 7) loading codes are also listed in Table 1 as well as the load factors adopted for each of these codes (Kemp, 1999).

TABLE 1
COMPARISON OF LOAD STATISTICS AND LOAD FACTORS FOR DIFFERENT CODES

Type of Load	Code	Distribution	Mean/Nominal	c.o.v.	Load Factor
Dead Load (Permanent)	SABS 0160	Lognormal	1.05	0.10	1.2
	EUROCODE	Normal	1.05	0.10	1.35
	ASCE 7	Normal	1.05	0.10	1.2
Lifetime Maximum Live load (office)	SABS 0160	Gumbel (Type I)	0.96	0.25	1.6
	EUROCODE	Gumbel (Type I)	0.80	0.25	1.5
	ASCE 7	Gumbel (Type I)	1.00	0.25	1.6
Lifetime Maximum Wind Load	SABS 0160	Gumbel (Type I)	0.41	0.52	1.3
	EUROCODE	Gumbel (Type I)	0.71	0.30	1.5
	ASCE 7	Gumbel (Type I)	0.78	0.37	1.3

The second code calibration stage as proposed by SABS 0160 is concerned with the identification of partial material and resistance factors for each of the different structural codes. The responsibility has been assigned to the individual code committees to derive material and resistance factors. General guidance was given to the material codes that for materials with a coefficient of variation between 0.10 and 0.30, a safety index of $\beta=3.0$ could be achieved if the partial material and resistance factors are selected on the basis of achieving a 1% resistance fractile.

Material code development since publication of SABS 0160 essentially consisted of the adoption of reference codes from elsewhere. One such example is SABS 0162 (Structural Steel) that is based on the Canadian Code with the Canadian loading code provisions that are very similar to that of SABS 0160. Since stage two has not been implemented, extensive calibration results are not available to evaluated the consistency in the reliability that can be achieved through appropriate choices of resistance factors in combination with the utilisation of the load factors derived in stage one. In the following section a code calibration methodology will be presented that addresses the above mentioned shortcomings of the load index approach.

CODE CALIBRATION METHODOLOGY

Ideally it would be possible to calibrate the South African Loading Code and all the associated material codes at the same time. It is considered more practical to calibrate the loading code and material codes in two separate stages. For this purpose a code calibration methodology has been established (Ter Haar & Retief, 2000) for the rational, systematic and efficient calibration of structural codes. The main features of the methodology are the following:

1. Generic resistance model
2. Transforming target reliabilities into target safety factors
3. Basic load factors derived from single load cases
4. Target resistance factors for chosen load factors

Generic Resistance Model
The calibration results using the load index approach is equivalent to assuming a deterministic resistance model for the calibration of the load factors. An important feature of the calibration procedure developed at the University of Stellenbosch is that it incorporates a generic probabilistic model for the resistance. This not only allows parametric studies with respect to the degree of uncertainty in the resistance model, but also establishes an important link between the first and second stages of code calibration.

Resistance models in general are not only subject to various sources of inherent variability, but also to resistance model uncertainty. Depending on the complexity and type of resistance function, different partial factors are used in codified resistance specifications. Material factors may be used to factor material strength properties, model factors for specific targeted model uncertainties (e.g. rectangular stress block for concrete bending model) and resistance factors may be applied to the overall resistance model. For the purpose of obtaining partial safety factors to achieve consistent reliability as described in the calibration procedure below it is deemed sufficient to adopt a simple generic resistance model. This generic resistance model is represented by a single stochastic variable. A lognormal distribution has been assumed as an appropriate and most representative probability distribution for resistance. The mean value of the resistance is determined in accordance with the requirement that the resultant reliability is equal to the target reliability. The coefficient of variation of the resistance, R_{cov} is varied to represent varying degrees of uncertainty in the different modes of failure of the various materials.It should be noted that for the calibration of the loading code the different load effects are also modelled as single variables, whereas they are in fact a function of a number of variables, for example the density of material, geometrical properties and natural phenomena.

Transforming Target Reliabilities into Target Safety Factors

The general format for limit states design specifications is represented by equation 1.

$$\sum_i \gamma_i S_i \leq \phi R_n \tag{1}$$

where S_i = nominal design value for i^{th} load effect
γ_i = load factor for i^{th} load effect
R_n = nominal resistance
ϕ = resistance factor

For a particular set of load statistics and any specified target β value, the values of the partial safety factors γ_i and ϕ corresponding to the most likely failure point can be calculated through FORM analysis. These values are called the ideal partial safety factors. Once the target β values for the range of load ratios and R_{cov} parameter values have been transformed to sets of ideal partial safety factors, all further calculations are performed utilising these target safety factors. The most powerful expression in this regard is the ratio of nominal resistance to total load that is expressed in terms of the partial safety factors. This ratio can also be regarded as a global safety factor. For a chosen set of load factors a resistance factor can be calculated that will result in the same ratio of nominal resistance to total load value corresponding to the required level of safety.

Basic Load Factors Derived from Single Load Cases

Ter Haar and Retief (2001) has shown that load factors obtained from calibration analysis for single load cases (defined here as basic load factors) results in the actual reliability consistently being greater than or equal to the target value over the entire range of load ratios. Note that single load cases define the extreme points of any load ratio where all load values except one are zero.

For a particular set of load statistics the basic load factors are a function of the coefficient of variation of the resistance. The variation of the basic load factors for different degrees of uncertainty of the resistance have been calculated for the South African load statistics as well as those assumed for Eurocode and ASCE7. The results are presented elsewhere in this paper.

Target Resistance Factors for Chosen Load Factors

For each single load case the unique ratio γ_i/ϕ is defined by the ideal partial factors corresponding to the target reliability index β. One of the most fundamental loading code requirements is that the load factors, γ_i are chosen such that the same resistance factors are required for all load cases to achieve the designated target reliability. For this purpose it is required that one convenient safety factor value be chosen and that all the other partial safety factors are then calculated to achieve the target load ratios for each of the single load cases. This can best be achieved through the following procedure:

1. Choose a convenient dead load factor
2. For each R_{cov} value calculate the required resistance factors from the target load ratio obtained from the single dead load case
3. Use these calculated resistance factors to obtain the target load factors for all the other loads
4. If the resistance factor and partial load factor values seems to be too high or too low a new value can be chosen as dead load factor. Steps 1-3 are then repeated to obtain factors that are adjusted pro-rata to the adjustment in the dead load factor.

It is important to note that for a given target β value and load statistics there is a unique relationship between all the partial safety factors. This unique relationship is defined by the target load ratios, which in turn are defined by the ideal partial factors obtained for the single load cases. Once one partial safety factor has been fixed all the other safety factors are defined in terms of these unique relationships. The decision to start with the choice of a fixed dead load factor is not only a convenient starting point, but can also be regarded as the most rational choice for the following reasons:

1. The dead or permanent load of any structure is the most common load present in most load combinations. The single dead load case therefore represents the one extreme load ratio value for all load cases that involves dead load.
2. From an evaluation of the load statistics in Table 1 it can also be seen that of all the common loads the most widespread agreement between various code writing authorities exists for the load statistics of the dead load. The only difference is that the South African code assumes a lognormal distribution, whereas a normal distribution is assumed for both the European and American loading code. From the data in Table 2 it can be seen that the calibration results are not sensitive to the choice of lognormal or normal distribution for the dead load.

CALIBRATION RESULTS

All the calibration results presented in this paper were obtained using the load statistics listed in Table 1. All the results are further presented for a chosen dead load factor of 1.2. Both SABS 0160 and ASCE 7 currently prescribe 1.2 as the dead load factor. However, it should be noted that the results for any other chosen dead load factor is similar with only an appropriate pro rata adjustment in the partial safety factors.

Single Load Cases

The results of the procedure described in the previous section for the determination of target resistance factors are given in Table 2. With $\gamma_D = 1.2$ the target resistance factors corresponding to a beta value of 3.0 was calculated from the reliability results of the single dead load cases. With the resistance factors now fixed, the target γ_L and γ_W factors were calculated for different R_{cov} values. The variation of the target live load factors over a range of R_{cov} values for the different codes is graphically represented in Figure 1.

TABLE 2
TARGET PARTIAL SAFETY FACTORS FOR $\beta = 3.0$ AND $\gamma_D = 1.2$

R_{cov}	Target Resistance Factor, ϕ			Target Live Load Factor, γ_L			Target Wind Load Factor, γ_W		
	SABS	EURO	ASCE7	SABS	EURO	ASCE7	SABS	EURO	ASCE7
10%	0.75	0.75	0.75	1.65	1.39	1.74	1.10	1.36	1.70
15%	0.66	0.67	0.67	1.56	1.30	1.63	1.03	1.28	1.58
25%	0.50	0.50	0.50	1.42	1.18	1.48	0.90	1.14	1.40
40%	0.32	0.32	0.32	1.29	1.07	1.34	0.78	1.02	1.23

Figure 1 shows that for a target β value of 3.0, $\gamma_D = 1.2$ and common resistance factors, the required live load factors for the Eurocode are significantly lower than the corresponding values for SABS 0160. These differences are due to the different mean to nominal value ratios that are assumed for the live load distribution (see Table 1). Note that the assumed distributions as well as the coefficient of variation for the live load are the same for all three codes. This difference in the mean to nominal live load ratio can probably be justified since Retief et al (2000) has shown that the nominal specified live loads for the European code is consistently higher than those specified in SABS 0160.

From Table 2 it can be seen that for $\beta = 3.0$ and $\gamma_D = 1.2$ significant differences exist in the target wind load factors for the three codes. This is due to the differences in the load statistics that are used for the calibration of the different codes. It can therefore be concluded that, in general, different load factors should be used in combination with the nominal load values specified by the different codes to achieve the same level of safety.

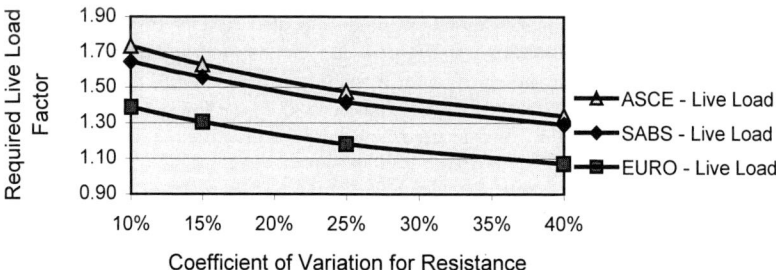

Figure 1: Variation in Target Live Load Factors for Different Codes

The target resistance factors in Table 2 are called unbiased factors, as they apply to the mean resistance values. These factors should be used in the calibration of the material codes in the second calibration stage. With the nominal resistance, material property values and model uncertainty identified for a particular failure mode, the unbiased resistance factors could be used to determine the required partial material and resistance factors for nominal property values applicable to the particular failure mechanism.

Target safety factors for $\beta = 4.0$ and $\gamma_D = 1.2$ have also been calculated. The results for both $\beta = 3.0$ and 4.0 for the South African load statistics are graphically shown in Figure 2. Similar results have been obtained for the European and North American load statistics. It should be noted that the R_{cov} values for most structural materials are between 10 and 30%, with 10% to 20% being typical values for reinforced concrete and structural steel failure mechanisms. From Figure 2 the following observations can be made for a R_{cov} value of 15%:

1. An unbiased resistance factor of 0.66 is required for $\gamma_D = 1.2$ and $\beta = 3.0$.

2. With $\gamma_D = 1.2$ and $\phi = 0.66$ the target γ_L value is 1.56. The current specified γ_L value of 1.6 will therefore result in reliability slightly higher than $\beta = 3.0$ for live load dominated load cases.

3. Similarly the target γ_W value is 1.03 for $R_{cov} = 15\%$. Therefore, the current specified value of 1.3 will lead to an achieved β value of approximately 4 for wind dominated load cases.

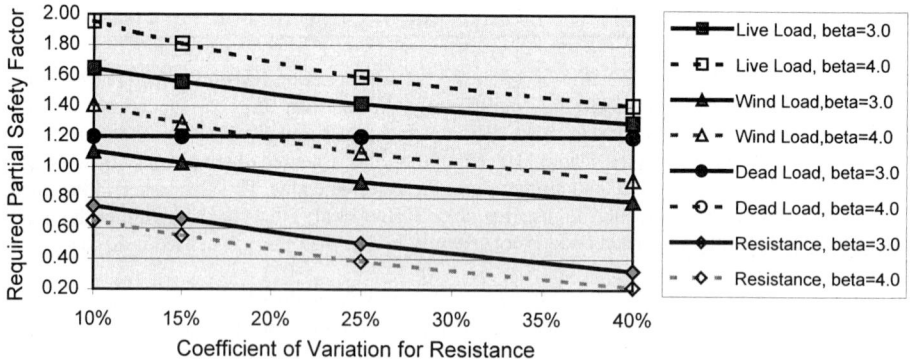

Figure 2: Target Safety Factors for SABS0160

Similar observations can be made for other R_{cov} values. In general the target live and wind load factors for $\beta = 3.0$ will be larger for R_{cov} values smaller than 15%, and smaller for R_{cov} values larger 15%. Results such as those for $\beta = 4.0$ as shown in Figure 2 can also be used to derive recommendations for the adjustment in partial safety factors derived for $\beta = 3.0$ to achieve higher levels of safety (Ter Haar and Retief, 2000). The above results are for single load cases. In the following section the application of these results for load combinations is evaluated.

Load Combinations

As discussed earlier the application of load factors derived from single load cases will always lead to conservative results over the range of load ratios for any load combination. First consider the dead plus live load combination. The codified performance function in terms of the partial safety factors for this load case is given by:

$$\gamma_D D_n + \gamma_L L_n \leq \phi_i R_{ni} \qquad (2)$$

For each ratio of nominal live to dead load a different set of ideal partial safety factors is obtained through FORM analysis. Comparing the value of the nominal resistance to total load ratio calculated from the ideal partial factors (the target value) for each load ratio with the values obtained from the basic load factors, the resultant percentage difference between the actual and target values can be obtained over the range of load ratios. The resultant % conservatism using $\phi = 0.66$, $\gamma_D = 1.2$ and $\gamma_L = 1.6$ for all load ratios are shown for $R_{cov} = 15\%$ in Figure 3.

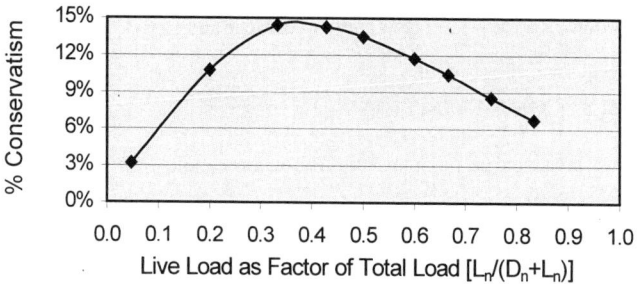

Figure 3: Conservatism using Basic Safety Factors for D + L Load Combination

Similar results are obtained for other R_{cov} values. These results as well as the maximum resultant % conservatism for the dead plus wind load case with $\gamma_D = 1.2$ are summarised in Table 3. The following observations can be made from the results displayed in Figure 3 and Table 3:

1. With $\gamma_D = 1.2$ and using a live load factor of 1.6 the maximum conservatism over the full range of live to dead load ratios varies between 12% and 18%, depending on the R_{cov} value. This is mainly due to the fact that it is unlikely that extreme unfavourable values for the dead load will occur simultaneously with extreme unfavourable live load values (Ter Haar and Retief, 2001).

2. With $\gamma_D = 1.2$ and $\gamma_W = 1.3$ the maximum conservatism over the range of wind to dead load ratios varies from 20% to 31%. This is significantly higher than the corresponding percentages for the dead plus live load case. The increased conservatism can be attributed to the fact that the wind load factor of 1.3 is significantly larger than the basic load factors derived for the single wind load case. It can be shown that should a wind load factor of 1.0 be used the resultant maximum conservatism is between 11 % and 18% which is of the same magnitude than for the dead plus live load case.

TABLE 3
RESULTANT MAXIMUM % CONSERVATISM USING CHOSEN SAFETY FACTORS

R_{cov}	Chosen ϕ value	Maximum % Conservatism	
		Dead plus Live, $\gamma_L = 1.6$	Dead Plus Wind, $\gamma_W = 1.3$
10%	0.75	12%	20%
15%	0.66	14%	24%
25%	0.50	18%	31%

Should it be deemed necessary to reduce the degree of conservatism it is possible to derive two or more sets of load factors that apply to different portions of the load range (NKB, 1999). It is further recommended that the resistance factors for different failure mechanisms are chosen such that the resultant reliability is larger than the target value for all practical load ratios applicable to the particular failure mechanism (and not simply the entire range of load ratios). For example, consider the case where the practical load ratio for a particular failure mechanism with $R_{cov} = 15\%$ is defined by the live load being between 20% and 70% of the total load. Using Figure 3 it can be shown that the resistance factor given in Table 2 ($\phi = 0.66$) could be increased by 10% without compromising the resultant minimum required safety over the practical load ratio.

The results obtained for the single load cases can also be used for the derivation of target partial factors when the dead load acts as a counteracting load. In such a case the dead load acts as a resistance and therefore the results obtained for $R_{cov} = 10\%$ apply (Ter Haar and Retief, 2000).

CONCLUSIONS & RECOMMENDATIONS

Based on the findings presented in this paper the following conclusions and recommendations can be made:
- The application of the code calibration methodology that has been developed at the University of Stellenbosch leads to a rational, systematic and efficient two-stage calibration procedure for structural codes.
- It can be concluded that the 1.2 and 1.6 load factors for dead and live load respectively have been found to be compatible for a beta value of 3.0. As a result there is no basis for changing these current load factors.
- However, the calibration analysis indicated that a wind load factor of 1.3 in combination with a dead load factor of 1.2 do not lead to optimum consistency in reliability. The results obtained with the load statistics as listed in Table 1 show that a wind load factor of 1.3 lead to conservative designs. It is therefore recommended that further research be done to investigate the wind load statistics, particularly since the wind load statistics used differ significantly from that assumed for Eurocode and ASCE 7.
- Due to different nominal values and load statistics adopted for different codes, the load factors for these different codes can be expected to be different in order to achieve the same levels of safety.
- Target resistance factors have been obtained from the first stage calibration work that could be used for the second stage of code calibration, that is the calibration of the material codes

References

KEMP A. R. (1999). *Position Paper on Lack of Harmonisation of Load Factors and Load Combination Factors between North American (including South African) and European Loading Codes*. Report to SAICE Working Group on SA Loading Code. Document D6.2

NORDIC COMMITTEE ON BUILDING REGULATIONS, NKB (1999), *Basis of Design of Structures: Proposals for Modification of Partial Safety Factors in Eurocodes*, NKB Committee and Work Reports 1999:01 E.

RETIEF J.V., DUNAISKI P.E., DE VILLIERS P.J. (October 2000). *Evaluation of SABS0160:1989 Minimum Imposed Loads: Comparison with other codes*. Report to SAICE Working Group on SA Loading Code. Document H 5.2

SABS (SOUTH AFRICAN BUREAU OF STANDARDS).(1990) *Code of Practice for the General Procedures and Loadings to be Adopted in the Design of Buildings, SABS 0160 - 1989 (as amended 1990)*, Pretoria.

TER HAAR, T. R. AND RETIEF, J. V. (2000) *Development of a Methodology for Structural Code Calibration. ISI Report no. R00/1*, University of Stellenbosch, South Africa., 2000.

TER HAAR, T.R. AND RETIEF, J.V. (2001). *A Methodology for Structural Code Calibration*. International Conference on Safety and Risk in Engineering, Malta 2001.

AN EVALUATION OF IMPOSED LOADS
FOR APPLICATION TO CODIFIED STRUCTURAL DESIGN

JV Retief, PE Dunaiski, PJ de Villiers

Department of Civil Engineering, University of Stellenbosch
Private Bag X1, Matieland 7602, South Africa

ABSTRACT

A critical evaluation of provisions for imposed loads in the South African Loading Code for design of structures, SABS 0160:1989, by comparison with other codes is reported. A broad and extensive comparison where all aspects of imposed loads are considered is followed by an intensive evaluation of critical aspects identified in the broad survey. The results are then measured against the two potential reference codes ANSI/ASCE 7-95 and Eurocode ENV1991-2-1:1995.

The survey shows that SABS 0160:1989 essentially provides for all aspect of imposed loads found in other codes. Minimum imposed load values for the South African code are significantly lower as compared to other codes. Values are on average 18% lower then those of the other codes. The differences occur across the range of general, special, auxiliary and industrial occupancy types. It is concluded that the imposed load values need to be adjusted to be satisfactorily compatible with a selected reference code. No clear preference for a reference code could be established.

KEYWORDS

Structural design, loading code, design lode, imposed load, code comparison

INTRODUCTION

Revision of the South African Code of Practice for the General Procedures and Loadings to be Adopted in the Design of Buildings (SABS 0160-1989) provides the opportunity to reassess provisions made for minimum imposed loads prescribed in the code. There is no evidence that the present provisions are deficient, nor are there substantial recent research results to warrant substantial revision of the relevant prescriptions. This review is therefore limited to a thorough comparison with other structural design loading codes. The scope of type of loading that must be considered such as floors, roofs, area effects, walls, distributed and concentrated loads, etc., is fairly well developed and therefore does not vary significantly between the various codes. The degree of detail, to which occupancy types are provided for, varies significantly amongst the codes. Since this factor has various implications, it is evaluated more carefully. Most attention was given to the quantitative values for the minimum imposed loads.

CODIFIED PROVISION FOR IMPOSED LOADS

Basis for Imposed Load Values

Simplified models of loads such as uniformly distributed loads, which are a function of area, are used to reflect the load effects of the various activities that could occur during the life of the building structure. Conservative assumptions are made to obtain design values. The floor of a living room can be designed for instance by providing for a grand piano with nine strong men to carry it, all together multiplied by three! Realistically conservative values can be obtained through the judgement of experienced designers and the performance of structures designed to previous codes. Load surveys are sources of objective information on actual loads. The statistic nature of load surveys together with incompleteness, require stochastic treatment of load survey data. Reliability modelling provides significant improvement in the rationality of imposed load values. The prudent approach is only to reduce sufficiently conservative loads to the extent that the necessary evidence can substantiate such reduction.

Codes Selected for Comparison

A representative set of four codes was selected for use as a basis in the evaluation of the SABS Loading Code. Motivation for the selection is provided in TABLE 1. The primary objective of the study is to identify deficiencies in SABS to be rectified in code revision. The formal adoption of an international suite of structural design codes is an option to be considered for code development for South Africa (South African National Conference on Loading (1998)). Particular emphasis was therefore placed on Eurocode and ASCE 7 as primary contenders as a reference code. A truly international code will be the ideal solution; ISO 2394 is therefore included as a reference, although it is not suitable for current practice.

TABLE 1. CODES USED FOR EVALUATION OF SABS 0160:1989 IMPOSED LOADS

Code Prescribing Imposed Loads	Short Title	Motivation
SABS 0160: 1989 (Revised 1990) *General Procedures and Loadings to be Adopted in the Design of Buildings*	SABS Code	Evaluation of prescription of imposed loads.
BS 6399 Part 1:1996 *Dead and Imposed Loads*; Part 3:1988 *Imposed Roof Loads*	BS Code	Served historically as reference to code development for South Africa.
ENV 1991-2-1:1995 Part 2-1 *Actions on Structures*	Eurocode	Results from comprehensive code development process; to replace BS Important international reference.
AS 1170.1-1989 Part 1: *Dead and live loads and load combinations*	AS Code	Similarities in conditions between the two countries.
ANSI/ASCE 7-95 *Minimum Design Loads for Buildings and Other Structures*	ASCE Code	Important source of reference for structural design internationally, and for South Africa.
ISO 2394: 1998 *General principles on reliability of structures*	ISO Code	An international code. Not directly applicable due to different nature

Basis for Comparison of Code Provisions

The comparison of provision for imposed loads in various structural design codes is complicated by the fact that these codes vary significantly in philosophy and approach, layout and structure, the manner in which safety is partiallised throughout the code and even the level of reliability. A common basis was therefore required, to which the properties of the various codes could be mapped for comparison.

The approach followed was first to use the most extensive range of factors such as load and occupancy type, as the basis for a comprehensive comparison. This ensured that all the factors related to imposed loads considered in all the codes were substantially reflected in the comparison. The factors considered in this study were uniformly and concentrated floor and roof loads and area load reductions, imposed loads on walls and balustrades, all considered against an extensive complement of occupancies. Critical aspects in need for more detailed evaluation were then addressed in an intensive survey, where SABS provisions for imposed loads were measured against the other codes. Particular attention was given to provisions of the two potential codes of reference, ASCE 7 and ENV 1991-2-1.

EXTENSIVE COMPARISON OF ALL CODES

All aspects of imposed loads that need to be considered for structural design were taken into account in the extensive comparison of the various codes. In spite of the substantial differences in approach and detail of treatment of relevant aspects, a reasonable match could be compiled for comparison. A rather detailed exercise was required, which is fully reported by DE VILLIERS (2000). Aspects that were considered are summarised in TABLE 2.

TABLE 2. SCOPE OF EXTENSIVE COMPARISON OF CODES

General aspects of code provisions that were compared	<ul><li>Pattern loading, permanency of live loads, dynamic forces, impact, movable partitions for floor loads.</li><li>Imposed floor load reduction</li><li>Minimum concentrated floor load intensities</li><li>Minimum imposed roof loads: classification and load intensities; load reduction; curved roofs; additional loads on roof trusses.</li><li>Forces on parapet walls, balustrades and railings</li></ul>
Minimum imposed floor loads as a function of occupancy	<ul><li>Building and floor occupancy classification used as basis for imposed load values</li><li>Comparison of imposed load values prescribed by various codes, for all possible occupancy types</li><li>Treatment of special components of facilities such as stairs, landings, corridors, hallways, cantilever balconies</li><li>Treatment of special occupancies such as industrial and storage areas</li></ul>

General Properties of Imposed Floor Loads

A systematic comparison was made of the methods and values in which the properties of imposed loads, as summarised in TABLE 2, are specified in the various codes. A summary of the most important aspects is presented in TABLE 3.

Minimum Concentrated Imposed Floor Load Intensities

Provision for minimum concentrated imposed floor loads in the SABS code generally compared well with that of the other codes, and no deficiencies could be identified.

Minimum Uniformly Distributed Imposed Floor Load Intensities

Minimum values for uniformly distributed imposed floor loads for the various occupancy types form a vital component of a loading code. A proper comparison between the various codes requires a comprehensive list of occupancy types as basis. This was done by using the AS code, with its eight building types, with an average of 15 floor uses each, as reference. Where necessary the list was extended to include all occupancies provided for in all the codes. Thus, the imposed load values of the various codes could be compared over the full spectrum of floor occupancies.

TABLE 3. GENERAL PROPERTIES OF IMPOSED FLOOR LOADS

SABS 0160-1989	ASCE 7-95	ENV 1991-2-1	BS6399-1-1996	AS 1170.1-1989
Pattern Loading				
Applied to produce most severe effects	Applied to a portion of structure if effects are more severe	Applied to most un-favourable tributary zone per storey	No provisions	
Movable Partitions				
Load intensities are prescribed. Distinction between different weights of partitions.	Provision where specified live load < 3.8 kN/m². No prescribed load intensities.	No provisions. Partitions are taken as part of the nominal dead load.	Same as SABS Less conservative prescribed minimum. No distinction is made for different weights of partitions	Same as SABS Less conservative prescribed minimum. No distinction is made for different weights of partitions
Reduction of Prescribed Minimum Imposed Load Intensities				
Tributary area Member type - no Assembly, storage, manufacturing, garaging: Area > 80m²; 30% max Other: > 20 m²; 50%	Influence area Member type Number of floors Not to assembly areas One floor - 50% max More than one floor - 60% max	Tributary area Member type Number of floors Not to storage, heavy industrial, garaging Area > 20 m² Light storage 30% Assembly, shopping 40%; other 50% max.	Tributary area. Member type Number of floors Not to storage, industrial, garaging Vertical member - 50% max. Horizontal member - 25% max.	Tributary area Not to storage, industrial, garaging areas, one-way slabs. For > 5 kN/m² - 20% < 5 kN/m² - 50% max.

The comparison gave persuasive evidence that imposed load values prescribed by SABS are systematically lower than those of other codes are. This is true for some parts of residential areas and areas such as offices, parking areas, storage and others. In a number of instances the differences were substantial, for example stairs and corridors, classrooms, retail areas. However it was also clear that the extensive comparison magnified this point significantly.

General Properties of Imposed Roof Loads

Aspects of imposed roof loads generally provided for in the various codes are the classification of roofs, load intensities and their reduction for different types of roof, curved roofs, additional loads on roof trusses, accidental loads during maintenance and provision for snow load if not provided for separately. Load intensity is the most important component of the provisions. The comparison is summarised in TABLE 4.

Forces on Parapet Walls, Balustrades and Railings

All of the loading codes classify parapet walls, balustrades and railings into a number of categories according to the type of area they serve. Imposed load values are then assigned to the various categories. The total number of categories for the various loading codes provides an indication of the level of detail to which loads are specified: four to six categories are generally used, except for the BS code which use 14 categories. There is substantial agreement between SABS and the other codes.

TABLE 4. COMPARISON OF IMPOSED ROOF LOADS INTENSITIES

SABS 0160-1989	ASCE 7-95	ENV 1991-2-1	BS6399-1-1996	AS 1170.1-1989
Accessible roofs: Load values are prescribed. If the roof is to be used as a floor use the prescribed floor load intensities. *Inaccessible roofs*: Load values are prescribed. A reduction with increase in tributary area is allowed.	*Special-Purpose roofs*: Load values are prescribed. The values are more conservative than those of the SABS. *Ordinary, flat or curved roofs*: Load values are prescribed. Reduction based on tributary area, roof slope. Values substantially more conservative than SABS.	*Accessible roofs*: Use floor load values. *Special-Purpose roofs*: Load values to be determined for the particular case. *Inaccessible roofs*: Load values dependant on roof slope only are prescribed For roof slopes < 20°, values substantially more conservative than SABS.	*Accessible roofs*: Values prescribed. Provision for snow loads. Values less conservative than SABS *Inaccessible roofs*: General buildings. Small buildings: Load values (dependant on roof slope) are prescribed. Values more conservative than SABS	*Accessible roofs*: Values prescribed. Reduction based on tributary area. Differentiate between houses and other buildings. More conservative than SABS. *Inaccessible roofs*: Load values are prescribed. Reduction with tributary area Values close to SABS over the range of tributary areas.

INTENSIVE COMPARISON OF OTHER LOADING CODES TO SABS

The broad and extensive comparison of the various codes was followed by an intensive evaluation of the SABS code, considering areas of discrepancies and potential deficiencies. Attention was primarily given to imposed load values and the corresponding definition and classification of occupancies. The occupancy classes of building or floor zones used in SABS 0160 was used as the basis for the comparison; a few occupancies not specified in the SABS code were added to the set. This comparison not only provides focus on the critical aspects of the SABS code that were identified through the extensive evaluation; it also provides a more balanced evaluation of the relative rating of the SABS code. The complete intensive comparison is reported in RETIEF (2000).

Comparison of Load Intensities Using the SABS 0160:1989 Table 4 as Basis

The load categories used in SABS 0160 Table 4 to group all occupancy classes with a common imposed load intensity were subdivided to provide compatibility with occupancies and imposed load values of other codes. The approach applied was to maintain the overall structure of the extended SABS Table 4, but to separate occupancy classes identified from differences with the other codes, whilst maintaining a minimum set of sub-categories. A summary of the comparison of the distributed imposed load values of the SABS code to that of the other codes is given in TABLE 5. A representative and synoptic description of occupancies, grouped together in SABS Load Categories, is provided. Average values of the ratio of value of each of the other codes to that of the SABS code is also tabulated. Average ratios are given per occupancy class and per code.

It is quite apparent that SABS imposed load values are, excepting a few instances, significantly and systematically lower than that of the other codes. Although weighted average ratios are required to provide a proper reflection of the comparative imposed load values, inspection shows that values are of the order of 16% to 18% lower for SABS. The exception is the BS code, which prescribes an average only marginally higher imposed loads.

TABLE 5. COMPARISON OF SABS IMPOSED LOAD VALUES TO OTHER CODES

OCCUPANCY	SABS	ASCE	ENV	BS	AS	Ratio
Dwelling house/unit	1.5	1.4	2	1.5	1.5	1.07
Bedrooms, wards, dormitories, etc in hospitals, hotels		1.9		2	2	1.32
Corridors, lobbies, landings to dwelling house				1.5	3	1.42
Stairs to dwelling house			3			2.00
Classrooms, lecture theatres	2	1.9	3	3	3	1.36
Operating theatres, x-ray rooms		2.9	3	2	3	1.36
Reading rooms in libraries		2.9	3	2.5	2.5	1.36
Garages, parking areas: < 25 kN gross weight	2	2.4	2	2.5	3	1.24
Offices for general use	2.5	2.4	3	2.5	3	1.09
Offices with data processing equipment	3			3.5		1.06
Cafes, restaurants, dining rooms lounges	3	4.8	3		2	1.09
Kitchens, laundries in hotels, offices, educational etc			2	3	4	1.00
Communal bathrooms, toilets in hotels, offices, etc		1.9	2	2	2	0.66
Entertainment areas		3.6		3		1.10
Light industrial		6		2.5	4	1.39
Assembly areas; fixed seating in residential buildings	4	2.9	4	4	3	0.87
Assembly halls, theatres, sport complex; fixed seats		2.9	4	4	4	0.93
Grandstands with fixed seating		4.8		5	5	1.23
Retail shops, department stores: sales and display	4	4.8	5	4	5	1.18
- upper floors		3.8				1.10
Light laboratories, banking halls				3	3	0.75
Assembly halls, sport complex; without fixed seats; stairs, corridors, landings of grandstands; public assembly areas, cantilever balconies.	5	4.8	5	5	5	1.00
Stages to assembly halls, theatres		7.2		7.5	7.5	1.36
Filing and storage: offices, hotels, institutions	5		6	5	5	1.07
Stack rooms: books, stationary				2.4/m	4/m	
Shelved areas to libraries		7.2	6	4	3.3/m	1.15
Exhibition halls			5	4		0.90
(1) Ratio of values relative to SABS - Average		1.18	1.17	1.05	1.18	1.16
- Standard deviation		0.34	0.30	0.21	0.33	0.26

Evaluation of SABS 0160 Imposed Uniformly Distributed Load Intensities

An evaluation of the distributed load intensities of SABS relative to the values of the other codes is summarised in TABLE 6 for the critical occupancy classes. The comparison is primarily made against ASCE 7 and Eurocode as reference codes. The availability of reliability based load models as provided in the Probabilistic Model Code (JCSS 2000) is also listed. Load models can be applied in the adjudication process.

Important occupancy classes for which uniform imposed floor loads need to be reconsidered for SABS are those which cover large areas in certain buildings: bedrooms, wards and such for residential buildings; offices; retail areas; garages and parking areas. Although increased loads for general occupancies will influence structural costs over large areas, these loads are also related to wide public exposure to structural performance. Satisfactory motivation will have to be provided for not being compatible with at least one of the reference codes.

Specialist areas to be reconsidered are lecture and operating theatres, reading rooms, restaurants, entertainment areas, stages, commercial storage and shelved areas. Auxiliary areas such as stairs, corridors and lobbies need to be carefully evaluated to provide for the effects of crowding and vehicles.

TABLE 6. COMPARISON OF SELECTED IMPOSED LOAD INTENSITIES

SABS Cat	Value kN/m²	Occupancy class	Evaluation and Recommendations	Load model available
1	1.5	(a) All rooms in a dwelling unit and dwelling house		Residence
		(b) Bedrooms, wards, dormitories, private bathrooms and toilets in educational buildings, hospitals, hotels and other institutional occupancies	Category 1(b) represents large areas with exposure to the public. An increase to 2 kN/m² will achieve compatibility with reference codes.	Hotel guest room, Patient room, Lobby
		(c) Corridors, stairs, lobbies, landings to a dwelling house		
		(d) Stairs	Compatibility with Eurocode - 3 kN/m²	
2	2	(a) Classrooms, lecture theatres	Use load model to adjudicate between ASCE (2 kN/m²) & Eurocode (3 kN/m²)	School room
		(b) Operating theatres, x-ray rooms.	An increase to 3 kN/m² will achieve consensus with other codes.	-
		(c) Reading rooms in libraries		Libraries
3	2	Garages and parking areas for vehicles of gross weight < 25 kN	Reference to ASCE requires increase.	
4A	2.5	Offices for general use	Reference to Eurocode requires increase	Office
5	3	Entertainment	Reference to ASCE requires increase	
		Light industrial	Reference to ASCE requires substantial increase. Refine definition. Use model.	Light industrial
6	4	Grandstands with fixed individual seating	Confirm merit of reduced load for seating by SABS as compared to other codes.	
7	4	Sales and display areas in retail shops and department stores	Lower than both ASCE & Eurocode. Use load model to evaluate change	Merchant / Retail
9	5	Shelved areas to libraries	Substantially lower than Eurocode and particularly ASCE values.	Storage
10	3	Corridors, stairs, lobbies, aisles, hallways, landings to buildings other than Category 1	SABS: same as adj. Room, but ≥ 3 kN/m² Other codes consider crowd loads and wheeled vehicles, with much higher loads	Concentration of people

A discrepancy exists in the way in which the SABS and the other codes assign imposed load values to storage and industrial areas. The SABS and the Eurocode assigns specific minimum load intensities to these areas and stresses that the load should be determined for the specific case in practise. The AS, BS and ASCE codes presents specific examples of such floor occupancies to which imposed load values are assigned (exceeding the absolute minimum given by the SABS) and in the case of storage areas the assigned load values are dependant on the height of the stacked materials. For relatively low storage heights of stacked materials the SABS is already non-conservative in comparison with the BS and AS.

Although differences between the SABS imposed load values and those of the ASCE and Eurocode respectively did not match, no clear preference for a reference code could be determined. This is confirmed by the statistics of load ratios listed in TABLE 5. Compatibility with either ASCE or Eurocode would require an increase for seven occupancy classes each, and ten increases to satisfy both codes.

Occupancy Classification System of the Loading Codes

The occupancy classification system of SABS was compared to the reference codes with respect to comprehensiveness, clarity and convenience. The ASCE code has the most comprehensive list of 63 occupancies arranged alphabetically; Eurocode has a sparse set of 21 occupancies, arranged according to intensity values, somewhat similar to the arrangement of the 35 SABS occupancy classes. The layout of SABS 0160 Table 4 can be retained even if values are adjusted and the number of classes is increased.

CONCLUSIONS

Minimum imposed load values prescribed by SABS 0160:1989 are lower than did those of a representative set of codes or as compared to the two possible reference codes, ASCE-7 and Eurocode. This is the case for a range of general and specialist occupancy classes. The compatibility of the SABS code to at least one of the potential reference codes should be improved. Deviations should be justified sufficiently.

REFERENCES

ANSI/ASCE 7-95: AMERICAN SOCIETY OF CIVIL ENGINEERS *Minimum Design Loads for buildings and Other Structures.*

AS 1170.1-1989: AUSTRALIAN STANDARD, Part 1: *Dead and live loads and load combinations.*

BS 6399: Part 1: 1996: BRITISH STANDARD *Loading for buildings Part 1. Code of practice for dead and imposed loads.*

BS 6399: Part 3: 1988: BRITISH STANDARD *Loading for buildings part 3. Code of practice for imposed roof loads.*

ENV 1991-2-1:1995: RATIFIED EUROPEAN TEXT, *Part 2-1: Actions on structures - Densities, Self-weight and Imposed Loads.*

DE VILLIERS J, RETIEF JV and DUNAISKI PE (2000). *Imposed Loading: Comparison of imposed loading as prescribed by various codes.* Report to SAICE Working Group on SA Loading Code. Document F5.2. January 2000.

INTERNATIONAL ORGANIZATION FOR STANDARDIZATION (1998). *General principles on reliability of structures.* International Standard ISO 2394: 1998

JCSS (2000) *Joint Committee on Structural Safety Probabilistic Model Code. Part 2 Load Models.* 11th Draft.

RETIEF JV, DUNAISKI PE and DE VILLIERS PJ. (2000). *Evaluation of SABS0160:1989 Minimum Imposed Loads: Comparison with other codes.* Report to SAICE Working Group on SA Loading Code. Document H 5.2. October 2000.

SABS 0160-1989: SOUTH AFRICAN STANDARD. *Code of Practice. The general procedures and loadings to be adopted in the design of buildings.* South African Bureau of Standards, Pretoria.

SOUTH AFRICAN NATIONAL CONFERENCE ON LOADING (1998) *Towards the development of a unified approach to design loading on civil and industrial structures for South Africa.*

SOUTH AFRICAN WIND LOADINGS: WHERE TO GO

A. M. Goliger, R.V. Milford and J. Mahachi
Division of Building and Construction Technology, CSIR,
P O Box 395, Pretoria 0001, RSA

ABSTRACT

This paper summarises several investigations into the proposed revision of wind-loading stipulations of the South African structural design code SABS 0160-1989. Initially, a brief evaluation of selected aspects of the current code is made, by comparing them with modern trends and overseas loading codes. The feasibility of adopting the European and American codes is discussed. Finally information on the proposed principles of the wind loading stipulations is presented. This work follows on from the CSIR's several years of involvement in the development of the SABS 0160, including the preparation of its 1989 version (RV Milford and JAP Laurie), input to the 1998 SAICE Loading Conference and current involvement and inputs to the SAICE/SABS Loading Committee.

KEYWORDS

Wind loading, codification, structural design.

BACKGROUND

The current South African loading code (SABS 0160-1989) was developed between 1970 and 1985. Its wind loading stipulations, which form a substantial portion of the code, are based on the old British Code (CP3 1952) and on the research of the second co-author of this paper (Milford, 1985, 1987a and 1987b). At that stage our code was fairly in line with the state of research, scientific trends and design practices.

In the past 10 to 15 years a tremendous amount of wind engineering research and effort throughout the world went into the modernising of the existing codes. As a result of these activities, several overseas codes, including the CP3, went through a process of significant (and in many cases radical) changes aimed at
- optimising,
- simplifying, and
- standardising.

Another trend is also apparent - in which the constant pursuit for more economic design has increased the complexity (i.e. decreased the clarity) of some of the codes or their selected aspects. This is in particularly relevant in respect to the British and Eurocode.

The modernising process is still in place in particular in respect to the European, ISO and Australasian codes.

This paper presents a proposal regarding the new wind-loading stipulations of SABS 0160. A brief evaluation of the most relevant aspects of the code is made, by comparing them with modern trends and overseas loading codes. This is followed by an examination of the applicability of European and American codes. The paper includes the information on the principles, structure and content of the proposed wind-loading stipulations.

COMPARISON WITH OTHER CODES

A more detailed evaluation of the loading stipulations of the SABS 0160 against selected international codes of practice was presented at the SAICE Loading Conference (Goliger et al, 1998) and also in Goliger and Niemann (1998). The present paper highlights only the selected issues and comparisons so as to set the scene for the proposed set of principles of the new wind-loading stipulations.

Basic design wind speed, vertical profile and terrain categories

The use of relevant and statistically sound wind data is of utmost importance to the design (Melbourne, 1975). Various codes of practice use different definitions of wind speeds, based on different averaging periods. Like several other codes, the current SA loading code is based on the 3-second gust wind speed (having a 50-year return period) derived from the analysis of 15 Weather Bureau recording stations (Milford, 1987b).

In the past 10 years or so a philosophy of adopting the mean wind speed has emerged due to its relevance and applicability to the dynamic design of structures. Codes which implemented this philosophy reverted then to hidden conversions to the peak wind speeds for the purpose of compatibility with the data on the pressure coefficients. For the vast majority of structures (which are not dynamically sensitive) this principle appears to be artificial and some of the leading modern codes (AS 1170.2 and ASCE 7-95) did not follow the above philosophy and adopted the peak gust wind speed. Even the Eurocode, in which the initial versions were based on the 10-min mean wind speed, has been recently modified to use the peak gust wind speed. A comparison of averaging periods used in various codes of practice is presented in Table 1.

TABLE 1
COMPARISON OF AVERAGING PERIODS AND WIND PROFILES

Standard / Code	Averaging period	Wind profile	No. of terrain cat.
SABS 0160-1989	3 sec	Power-law	4
BS 6399; 1997	1hr [1]	Logarithmic [4]	3
ISO 4354 (1990)	10 min [2]	Power-law and logarithmic	4
ASCE 7-95: 1996	3 sec	Power-law	4
AS 1170.2-1989	2-3 sec	Logarithmic	4
ENV 1991-2-4; 1994	10 min (1 sec) [3]	Logarithmic,	4

[1] subsequently converted to peak
[2] allows for other
[3] recently modified to 1 sec (verbal communication)
[4] incorporated in the 'terrain and building' factor

Two models of wind speed profile are used for codification purposes, namely the logarithmic and power-law. Although the logarithmic formulae are slightly more accurate for large heights, they are more difficult to use (more complicated integration) and produce unreal negative speeds at elevations lower than effective ground roughness. For these reasons the power-law approach is popular amongst practising engineers (Liu, 1991). A comparison of the profile stipulations included in various international codes, including the SABS0160, is given in Table 1.

The opposite transpires from two of the leading overseas codes. The Australian code adopts a complex logarithmic formulae based on meteorological science and extensive theoretical and full-scale research (Deaves and Harris, 1978), while the US code is based on a simple power-law profile, derived from the previous code and evaluated against the Australian profile.

In view of the above conflicting trends a detailed comparative study of the differences between the profiles stipulated in the SABS, ANSI and ASCE codes has been undertaken (Goliger and Milford, 2000). The analysis has shown that, in general, the differences between profile multipliers stipulated in those codes are relatively small, typically of up to 5% in terms of wind speeds and 10% in terms of pressure. (Larger differences are apparent for terrain category 4 and low elevations.) Figures 1a to 1c present the differences for terrain category 1. In Figure 1a, all three wind profiles stipulated in SABS 0160 are compared with the Australian and US profiles. Figures 1b and 1c present the percentage differences between the profiles in terms of wind speed and pressures. (Note that in Figures 1b and 1c, the Australian profile has been adopted as a reference and for the SABS code the 3-sec profile (class A), has been used.)

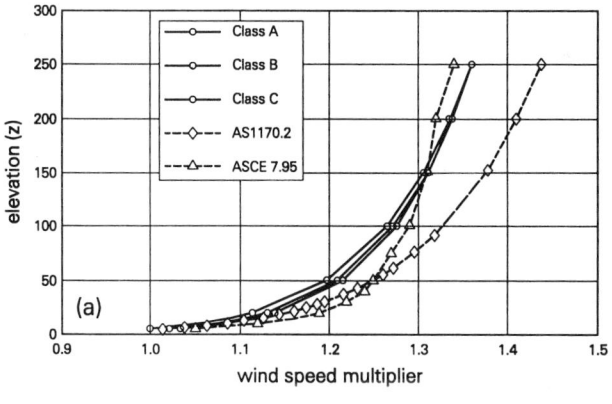

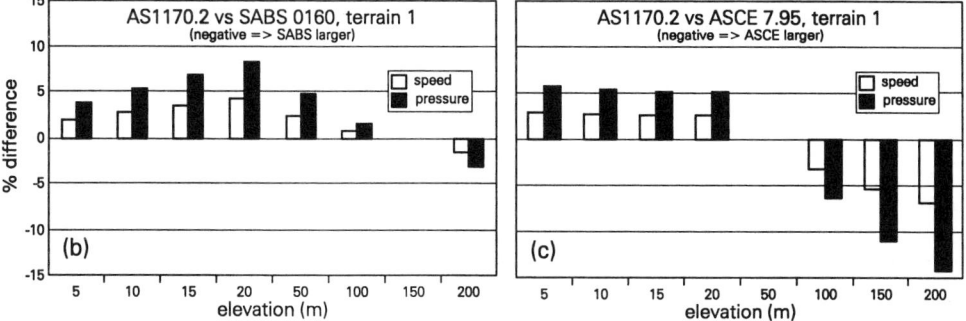

Figure 1a to 1c: Comparison of wind speed profiles; terrain cat 1

At low elevations, and depending on terrain category, it is assumed that the wind is strongly influenced by local obstructions (which may introduce acceleration of speed) and the wind speed multipliers are kept constant. This assumption is of a conservative nature and it often raises concerns / criticism of the designers of low-rise structures in South Africa. It has to be noted that most of the overseas codes of practice include similar considerations although the cut-off elevations differ.

Correlations of pressures

In SABS 0160 the effect of size of the structure is included in a division into classes of structures with different gust wind speed profiles, averaged over 3, 5 and 10 seconds (Classes A, B and C).

This principle seems to be outdated and modern codes of practice stipulate 3-second profiles only but introduce other 'correlation-related' provisions in the form of either an interpolation of pressure coefficients in terms of the extent of the loaded area (Eurocode) or a 'size effect factor' (British, Australian and US codes). A comparative interpretation of various codes visualising the differences between various approaches has been presented in Goliger et al (1998) and is reproduced in Figure 2.

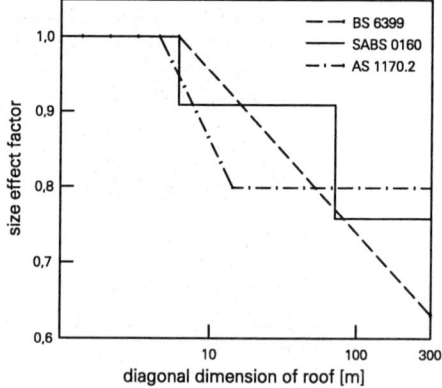

Figure 2: Comparison of pressure correlation stipulations (SABS 0160, BS 6399 and AS 1170.2)

Of particular interest is the simplicity of the approach adopted by the Australian code in which an area reduction factor (applied in terms of pressures) is introduced as a function of the tributary area, i.e. the area which contributes to the force being considered. It has to be noted that this factor is independent of the terrain category and elevation i.e. the turbulence of flow, which seems to be a significant simplification of the principles of the boundary-layer theory.

External pressure coefficients

The external pressure coefficients included in the South African code follow the old British code (CP3-1952) which, in turn, like many other national codes, is based on the old Swiss code (reference unknown). Most of the coefficients quoted in the old codes of practice are based on wind-tunnel measurements which were carried out without due consideration to boundary-layer modelling.

The recent standards still follow the principle of the equivalent gust model (ie. the pressure coefficients are mean values).

> *Although this approach does not take into account the real-time fluctuations of pressure coefficients, it has been adopted for its simplicity and continuity with previous codification.*

The main disadvantages of this approach are that it is not suitable for large structures and for cases in the which mean pressure coefficients are close to zero (Holmes et al, 1990).

The values adopted in the modern codes (e.g. British, Australian, US) have, however, been updated to include information from extensive research programmes (often involving comparative studies at several wind-tunnel facilities) and full-scale observations. The wind-tunnel studies include an adequate modelling of turbulence, surface averaging and the filtering process to reduce the amplitude and frequency of pressure fluctuations. (The British code includes even pressure coefficients for a broader range of wind orientations - i.e. not only the 0° and 90°.)

It has to be noted that a large scatter in the data on pressure coefficients appears in the international literature. This is illustrated in Figure 3 in which the envelope of normalised integrated uplifts derived from ten international codes, for a duo-pitch roof of a low-rise house and for range of roof pitches, are given (Moritz, 1999).

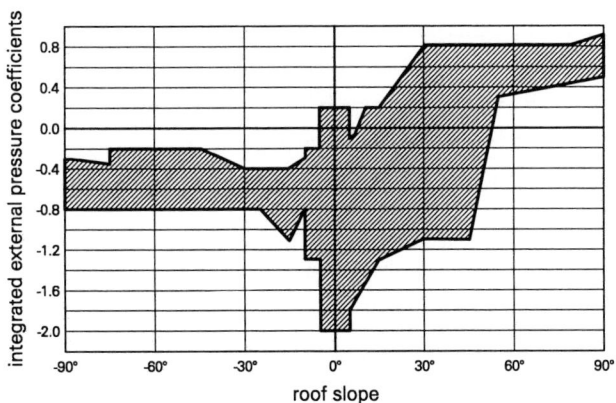

Figure 3: Envelope of differences between codes of practice (pressure coefficients)

It is also apparent that sub-division into more loading zones with more representative values of pressure coefficients adopted by modern overseas codes will, in most cases, lead to a 'leaner' design than that according to the current SABS code. This is presented in Figure 4, in which a comparison of loading zones stipulated by the new British code and the SABS code for a typical roof of a small house is presented.

ADOPTING THE EUROCODE OR AMERICAN CODE

At certain stage the SA Loading Committee did consider the feasibility of adopting one of the overseas loading codes as opposed to developing and modernising the existing SABS document. Two international codes have been identified for a brief survey, namely the proposed Eurocode and the recent ASCE code (Goliger, 2000 and Forbes and Goliger, 2000). The most relevant comments regarding the feasibility of adopting these documents are listed below.

Eurocode: ENV 1991-2-4; versions 1994, 1996

- The Eurocode document on wind loading is broader and more comprehensive than SA's. Furthermore, in a sense it represents a state-of-the-art document on recent wind engineering scientific thinking.

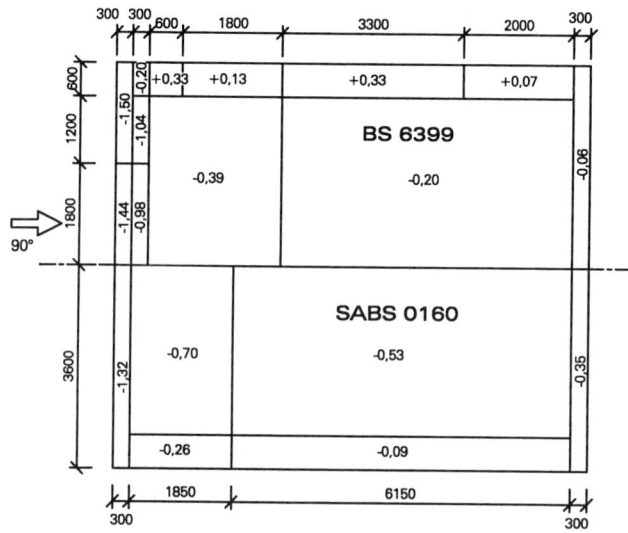

Figure 4: Comparison of roof pressure coefficients (SABS 0160 and BS 6399)

- Despite nearly ten years of work the code is still not complete and the versions of the code available to us contain several changes, both in principle and structurally. (For example, the 94 and 96 versions were based on 10 minutes mean wind speed while the recent version returned to peak wind speeds.) Because of the inputs of a large number of people / organisations, the code is: complicated, user unfriendly and contains inconsistencies (especially its 1996 draft). Currently the member countries are in the process of developing so-called: National Application Documents, which in many cases will differ substantially from the original document.

American National Code ASCE (7-95)

- The recent edition of the ASCE code constitutes a drastic departure from the US codification tradition in a sense that it introduces the SI notation (as opposed to Imperial) and furthermore it stipulates the basic design wind speed in terms of the 3-sec and gust and no longer the 'fastest mile'.

- The general format and logic of the ASCE code is fairly simple (in comparison with the Eurocode or British code) although substantially different from that of the SABS 0160. This is especially evident in respect of the format in which the actual loading information is presented (e.g. graphical as opposed to tabularised and in combination with the 'gust effect factor'). Some of the differences are not readily noticeable, and in case of adopting the ASCE code, may become a source of misinterpretation and therefore error. As an example, there is the issue of sequencing the terrain categories (i.e. roughness), which is opposite to that stipulated in the SA code.

- Initial comparative calculations (Forbes and Goliger, 2000) indicate that use of the ASCE code will increase the amount of design work as well as the wind loads.

PRINCIPLES FOR REVISING SABS 0160

The review of various codes of practice, other scientific information, feedback from the design practitioners as well as discussions with overseas wind experts involved in the development of codes have led us to believe that it would be of benefit to follow certain principles in order to optimise our future wind loading stipulations.

- In view of the conflicting trends and discrepancies which transpire from comparing the modern codes of practice we propose to follow a 'path of simplicity' and continuity with the current SABS 0160 version, where possible.

- In our opinion it would be of merit to utilise / follow several of the principles and stipulations included in two of the modern codes i.e. Australian and ASCE (US). In particular they refer to the overall logic of the Australian code and its information on pressure coefficients, and the simplicity of certain procedures / stipulations included of the US code.

- We propose to elaborate more (and under a separate section) on the scope and definitions, that are specific to wind loading stipulations.

- The choice of design methods should be spelt out explicitly at the beginning of wind loading stipulations and, furthermore, the simplified procedure should be included before others.

- It is proposed to follow the logic of the Australian code in respect to the division into design methods, namely Simplified, Detailed Static and Dynamic. For vast the majority of structures the resonant dynamic effects are small and can be ignored and, therefore, the proposed approach provides simplicity and allows the designer of most structures to ignore the dynamic issues.

- We propose to maintain the principle of the design peak wind speed, and four terrain categories. Furthermore, in view of the variances in pressure coefficients quoted in various codes, we propose to adopt a simple power-law profile, similar to that included in the US code (i.e. without the zero-plane displacement).

- The principle of the classes of structures (e.g 3, 5 and 10-sec profiles) appears to be outdated and should be replaced with a procedure on correlation of pressures. This procedure would also cater for the local pressure coefficients.

- Because of that we propose to maintain the principle of using two different maps of design wind speed (3-sec and hourly mean), one included in the section dedicated to detailed static design and the other to dynamic design.

- We propose the correction factor for the mean return period be replaced with an importance factor.

- We propose to maintain, where possible, the order / logic of the current SABS 0160 (i.e. in respect of the 'detailed procedure') but minimise the amount of cross-referencing to various places in the document - even if it may mean a certain amount of duplication. (Such logic has been introduced in the Australian code.)

- We propose to stipulate a simplified procedure to account for changes in terrain category and adopt the detailed procedure on the influence of dominant topography, which follows other modern codes of practice.

ACKNOWLEDGEMENTS

We would like to acknowledge the encouragement and interactions with the members of the SAICE Loading Committee under the chairmanship of Prof A Kemp.

REFERENCES

ANSI/ASCE 7-95: 1996. *ASCE Standard, Minimum design loads for buildings and other structures*, American Society of Civil Engineers, 1996.

AS 1170.2-1989. *Australian Standard, SAA Loading Code, Part 2: Wind loads*, Standards Australia, 1989.

BS 6399: Part 2: 1997. *British Standard, Part 2, Code of practice for wind loads*, BSI, 1997.

CP3: Chapter V: Part 2: 1952. *Code of basic data for the design of buildings.* BSI, 1952.

Deaves D M and Harris R I (1978). A mathematical model of the structure of strong winds. *CIRIA*, Report 76.

ENV 1991-2-4: 1994. *European Prestandard, Eurocode 1: Basis of design and actions on structures, Part 2-4: Wind actions*, European Committee for Standardisation, 1994.

Forbes R A and Goliger A M (2000). *A brief overview and design comments on the US wind loading code (ASCE 7-95)*, Internal Report BOU/I151, Division of Building Technology, March 2000.

Goliger A M, Niemann H-J (1998). *Assessment of the South African lading code in comparison to other loading standards; Part III: Wind Loads*, Internal Report BOU/I95, Division of Building Technology, 1988.

Goliger A M, Niemann H-J and Milford R V (1998). Assessment of wind-load specifications of SABS 0160-1989. *SAICE Loading Conference, Johannesburg*, September 1998.

Goliger A M (2000). *An assessment of the wind-loading stipulations of the Eurocode (ENV 1991-2-4: 1994 and 1996)*, Internal Report BOU/I158, Division of Building Technology, January 2000.

Goliger A M and Milford R V (2000). *Selected aspects of the proposed revision of loadings (SABS 0160)*, Internal Report, draft in preparation.

ISO/DIS 4354: (1990). *Draft International Standard; Wind actions on structures*, International Organisation for Standarisation, 1990.

Liu H (1991). *Wind engineering: A handbook for structural engineers.* Universiry of Missouri-Columbia, Prentice-Hall. Inc., 1991.

Melbourne W H (1975). The relevance of codification to design. *Proceedings of the 4th International Conference on Wind Effects on Building Structures, Heathrow, UK 1975.*

Milford R V (1985). *Extreme value analysis of South African gust speed data*, Internal report 85/4, Structural and Geotechnical Engineering Division, NBRI, CSIR, May 1985.

Milford R V (1987a). Annual maximum wind speeds from parent distribution functions. *Journal of Wind Engineering and Industrial Aerodynamics*, **Vol 25,** (1987), pp 163-178.

Milford R V (1987b). Annual maximum wind speeds for South Africa. *The Civil Engineer in South Africa, January 1987*, pp 5-15.

Moritz G (1999). *Influence of roof pitch on the wind loads of low-rise buildings*, MSc thesis, Aerodynamik im Bauwesen, , Ruhr-Universität Bochum, Germany.

SABS 0160-1989. *South African standard code of practice for the general procedures and loadings to be adopted in the design of buildings*, The South African Bureau of Standards, reprint 1994.

SEISMIC DESIGN PROVISIONS BASED ON UNIFORM HAZARD SPECTRUM

J. L. Humar[1] and M. A. Mahgoub[1]

[1] Department of Civil and Environmental Engineering, Carleton University,
Ottawa, On. K1S 5B6, Canada

ABSTRACT

In the equivalent static load procedures specified in many seismic codes, including the National Building Code of Canada (NBCC), the elastic base shear is obtained from a design response spectrum using the first mode period of the structure. The elastic response spectrum generally is obtained by applying appropriate amplification factors to the peak ground motion bounds. In recent years methodologies have been developed that allow the direct determination of maximum spectral accelerations for specified values of the period of a single-degree-of-freedom system and for a given probability of exceedance. A plot of such spectral acceleration values is referred to as a uniform hazard spectrum (UHS). Several codes propose to use UHS as the basis for determining the seismic design base shear. However, the base shear obtained from a UHS corresponding to the first mode period needs to be adjusted for higher mode effects. Base shear adjustment factors are derived here for several different structural types including moment-resisting frames, braced frames, and flexural walls. The corrected base shear can be distributed across the height according to the NBCC procedures. However, because the NBCC distribution is primarily in the form of the first mode, the resulting overturning moments generally overestimate the true moments, which arise from a combination of various modes. Adjustment factors to be applied to the overturning moments are also derived in this paper for the various structural types.

KEYWORDS

Seismic design, uniform hazard spectrum, base shear, overturning moment, higher mode effects, shear adjustment factor, moment adjustment factor.

INTRODUCTION

The objective in seismic design of a building can be expressed in terms of the performance level the building should achieve when subjected to the design earthquakes. In reality there is a whole spectrum of earthquakes with different severity and different probability of occurrence to which the building may be exposed. For practical design only a small number of earthquake events with different probability of occurrence, ranging from frequent to very rare, are selected. Such earthquake events are referred to as the design earthquakes. A frequent earthquake event is likely to be the least severe while a very rare earthquake event will be very severe. The performance level that the building must achieve is specified corresponding to each level of design earthquake. In fact, most building codes further simplify this procedure by defining only a single design earthquake level and the associated performance objective. For example, the National Building Code of Canada (NBCC) (Canadian Commission, 1995) requires simply that the buildings be life-safe (should not collapse) in an earthquake event that has 10% probability of exceedance in 50 years, and that the storey drifts under such an earthquake do not exceed specified limits.

The NBCC requirements are satisfied by designing the building to resist without collapse a design base shear obtained from an idealised response spectrum for a single-degree-of-freedom system with 5% damping. In the 1995 NBCC this spectrum was derived from two ground motion parameters: the peak ground acceleration and peak ground velocity for the seismic region in which the building was located. A number of other seismic codes throughout the world have used a similar approach. However, it has been pointed out that this method of obtaining the design response spectra may involve significant errors (Atkinson, 1991); more rational procedures for determining the design response spectra are therefore desired.

Since the mid-1970s, methodologies have become available for deriving linear elastic spectra for a given site and a given hazard level. Such spectra are called uniform hazard spectra (UHS). For a given site, a uniform hazard spectrum represents the envelop of spectral acceleration responses, at specified values of the period, of an elastic single-degree-of-freedom (SDOF) system subjected to the earthquakes that contribute to the seismicity of the site. The UHS acceleration responses are produced by different earthquakes with varying magnitudes and source distances. Thus the maximum short-period spectral values may result from short-distance earthquakes, while the long-period values may be contributed by more distant earthquakes. Because UHS provide a response parameter that is directly related to the design earthquake forces, they are preferable to the spectra derived indirectly from peak ground motion bounds. Many seismic codes are considering the adoption of UHS as the basis for obtaining the design forces. In the revisions to the NBCC currently under consideration it is proposed that the idealized response spectra of NBCC 1995 be replaced by UHS. The UHS that will form the basis of the next NBCC have been produced by the Geological Survey of Canada (Adams et al, 1999). The spectra correspond to earthquakes with a return period of 2500 years. They have been specified in terms of the median values of elastic spectral acceleration of a 5% damped single-degree-of-freedom system for selected values of the period.

In using a UHS to obtain the seismic forces for which a building must be designed, a number of issues need to be addressed. One of these issues is related to the fact that a UHS is an envelope of the maximum spectral accelerations experienced by a SDOF system. Appropriate adjustments are therefore required when the seismic forces in a multi-storey building are obtained from a UHS. For the next revision of NBCC it is being proposed that the effect of higher modes on the seismic base shear and overturning moments be accounted for explicitly by including appropriate adjustment factors in the base shear and overturning moment formulas. This paper presents (1) a methodology for deriving the adjustment factor to be applied to the base shear obtained from the UHS as well as representative values of such factors, (2) procedures for distributing the shear across the height, and (3) values of adjustment factors to be applied to the overturning moments obtained from the lateral forces obtained by distributing the base shear as in item 2.

FORMULATIONS IN NBCC FOR THE BASE SHEAR AND ITS DISTRIBUTION

It is being proposed for the next version of NBCC that the design earthquake base shear, V, in a building be obtained from the following formula:

$$V = \frac{S_a(T_1)M_v IFW}{R_o R_d} \quad (1)$$

where $S_a(T_1)$ is the spectral acceleration corresponding to the first mode period T_1, M_v accounts for the higher mode effects, I is a factor reflecting the importance of the building, F accounts for the effect of foundation soil on the spectral acceleration, W is the weight of the building during the earthquake, R_o accounts for the dependable overstrength in the structure, and R_d is a factor that specifies how much the design base shear could be reduced when the structure may be strained into the inelastic range without compromising its performance. This last factor is dependent on the ductility capacity of the structure.

The use of the first mode spectral acceleration in Eqn. 1 signifies that the base shear determined from it is associated with a SDOF system having a period T_1. However, in a multi-storey building the base shear is affected by the higher elastic modes. Factor M_v has been included in Eqn. 1 to account for the contribution of such higher modes.

In the 1995 NBCC provision, the base shear is distributed across the height of the structure according to a shape that is representative of the first mode of the structure. The force at floor level i is thus given by

$$F_i = \frac{w_i h_i}{\sum w_i h_i} V \quad (2)$$

where w_i is the weight assigned to the ith storey and h_i is the height of the ith storey above the base.

The distribution shape given by Eqn. 2 is an inverted triangle, which is a reasonably good approximation of the first mode. The code recognizes that the first mode distribution fails to account for the effect of higher modes, which tend to increase the shear in the upper storeys. Also, the higher mode effect becomes more significant as the fundamental period increases. These factors are taken into account in the code by specifying that a portion of the base shear, F_t, be assigned to top floor level and the remaining shear distributed according to Eqn. 2 with V replaced by $V-F_t$. The top force is given by

$$F_t = \begin{cases} 0 & T_1 \leq 0.7 \\ 0.07 T_1 V & 0.7 < T_1 < 3.6 \\ 0.25 V & T_1 \geq 3.6 \end{cases} \quad (3)$$

The provisions related to the distribution of base shear are expected to remain unchanged in the next version of NBCC. It may be noted that the overturning moments produced by seismic forces depend both on the magnitude of the base shear and how it is distributed across the height of the structure. For a given base shear the largest overturning moments are produced when that shear is distributed according to the first mode. The moment becomes proportionally smaller when higher mode contributions are accounted for in the distribution of the base shear. Since code-prescribed distribution of shear is based predominantly on the first mode, the overturning moments calculated from such forces overestimate the true moments. The 1995 NBCC specifies correction factors to account for this phenomenon. A factor J is applied to the base overturning moments while factor J_x is applied to the overturning moment at level x. The higher mode effects are more predominant in flexural wall structures than in shear frame structures. Also, the relative contribution of higher modes is strongly affected by the shape of the response spectrum. Obviously a single formula for distribution of base shear across the height together with a single expression for J or J_x cannot capture the variations caused by different structural characteristics and different spectral shapes. These factors need to be accounted for in the development of new provisions.

METHODOLOGY FOR ESTIMATING M_v AND J FACTORS

Humar and Rahgozar (1998) have studied the variation of M_v and J factors for frame and wall models for two locations, Vancouver and Montreal. In their study they used the UHS for these two cities and carried out response spectrum analyses for determining M_v and J factors. In the present work Humar and Rahgozar's study has been extended in the following manner. First, several new configurations have been included. These include braced frames, coupled wall systems and hybrid systems comprising moment frame and a wall. Second, studies have been carried out for 22 different cities, 10 located in the western region of Canada and 12 located in the eastern region. It is assumed for the purpose of these studies that the structure remains elastic. Based on this assumption the adjustment factor to be applied to the base shear obtained by assuming that the entire response is in the first mode is given by

$$M_V = \frac{\sqrt{\sum \{S_a(T_i)W_i\}^2}}{S_a(T_i)W} \tag{4}$$

where $S_a(T_i)$ is the spectral acceleration corresponding to the ith mode having a period T_i, W_i is the modal weight in the ith mode, and W is the total weight of the structure. Base shear obtained by taking the square root of the sum of squares of modal base shears is assumed to be a reasonably close estimate of the true value. It is evident from Eqn. 4 that for a given spectrum M_v factor depends only on the modal periods and modal weights. The base shear obtained by modal analysis is distributed according to the provisions of NBCC to yield storey level forces. The overturning moments calculated from these forces are compared with their values obtained directly from the response spectrum analyses to produce the overturning moment adjustment factors, J related to the base and J_x related to storey level x.

DESCRIPTION OF STRUCTURAL MODELS

A simplified symmetric multi-storey building model is selected for the present study. The building floors are assumed to be infinitely rigid in their own planes. The torsional effects are neglected so that the response of the building can be studied by analysing a single planar frame. The entire mass of the structure is assumed to be uniformly distributed at the floor levels. The storey height and floor mass are assumed to be uniform across the height of the building. The lateral force resisting planes are selected to be regular and simple. The resisting planes may comprise columns and beams, shear walls, or combination of these elements. For the sake of brevity results for only the moment-resisting frames, concentrically braced frames, and flexural walls are presented here.

Moment-resisting Frame (MRF)

In this case each of the lateral force-resisting planes is a single-bay 10-storey frame comprised of rigidly connected beams and columns. The frame width is taken to be 8.0 m. The ratio of the beam stiffness to the sum of the column stiffnesses in each storey is taken as 1/4. This is representative of a strong-column-weak-beam system which is the system preferred by most seismic codes. The relative values of the column and beam stiffnesses across the height are adjusted such that under a set of storey forces distributed in an inverted triangular shape the interstorey displacements are approximately the same. The displaced shape under the selected forces is thus linear. The fundamental period of the frame is now matched to a specific value by selecting an appropriate value for the modulus of elasticity. In other words the same frame configuration is used for the entire range of periods studied, but with different values of the modulus of elasticity. The mass tributary to each level of the frame is taken as 55.18 Mg. It should be noted that in the study of an elastic system such as this the important consideration is to produce the relative modal periods and modal weights that are representative of the structural type. The shape of the individual element, the absolute values of the element stiffness, and the number of storeys are irrelevant.

Concentrically Braced Frame

Two different single-bay braced frame structures are used for the study. Each frame is modelled as a simple vertical truss. The first braced frame consists of 2-storey cross bracing. It is 8 storeys high, has a width of 8.0 m and a storey height of 3.6 m. The mass tributary to each level of the frame is taken as 55.18 Mg. The second braced frame employs Chevron bracing. It is 12 storeys high, has a bay width of 8.0 m and a storey height of 3.8 m. The storey tributary mass is taken as 55.18 Mg. For each of the two frames the desired value of the first mode period is achieved by adjusting the value of the modulus of elasticity.

Flexural Wall

A 10-storey flexural wall system is used. The wall has a uniform width of 8.0 m and a uniform thickness of 0.4 m across the height. Again, the first mode period of the structure is matched to a specified value by adjusting the modulus of elasticity. The mass tributary to each storey level is taken as 55.18 Mg.

ANALYTICAL RESULTS

The selected buildings are assumed to be located in twenty-two cities in Canada, ten in the western region and twelve in the eastern region. Models with 7 different values of the fundamental period, T_1 = 0.5, 0.7, 1.0, 1.5, 2.0, 2.5, and 3.0 s, are studied. In calculating the base shear it is assumed that $I = 1$ and $F = 1$. The spectral shapes for the various regions of Canada can be divided into two broad categories. In both regions the spectral acceleration drops with the period, however in the eastern region the drop is much sharper than in the west. For this reason the results presented her are grouped by regions, one set for the west and the other for the east.

Analytical results for base shear and base overturning moment for moment-resisting frames, concentrically braced frames with cross braces, and flexural walls are presented for 22 locations, including Montreal and Vancouver, in the form of graphs in Figures 1 through 4. The curves related to Montreal and Vancouver are shown in bold lines. The following observations can be made on the basis of results presented.

Base shear and overturning moment adjustment factors M_v and J are both period dependent. Factor M_v increases with period while factor J decreases with period. The rate of increase of M_v and the rate of decrease of J with period are both higher for the eastern region than for the western region. For example, in the west M_v factor varies between 0.687 and 1.197, while in the east it ranges from 0.748 to 3.701. Factor J varies from 1.030 to 0.601 in the west, while it varies from 0.950 to 0.304 in the east.

Both M_v and J are strongly dependent on the type of ground motion. For the same fundamental period, structures subjected to records in the east usually have larger M_v factor and lower J factor values than their counterparts in the west. This can be attributed to the difference in the spectral shapes for the two regions. Thus, in the east the spectrum drops more rapidly with period in comparison to that in the west. A consequence of this is that the higher mode contribution is more predominant in the east than in the west.

Based on the analytical results obtained for the 22 cities, empirical expressions are derived for M_v and J values for use in design. These values are shown in Table 1 for three structural types: moment-resisting frames, braced frames, and walls. The suggested values are also shown in Figures 1 through 4 in dashed lines. The values vary with period and differ for the western and eastern regions. In Table 1 the two regions are identified by the spectral ratio $S_a\ (0.2)/S_a\ (2.0)$. This ratio is less than 8.0 for the eastern region and more than 8.0 for the western region where the spectrum drops more rapidly with the period. It will be noted that M_v and J are assumed to be constant for $T_1 \geq 2.0$ s. This is justified by the uncertainty associated with the determination of spectral accelerations for periods larger than 2 s.

REFERENCES

Adams, J., Weichert, D. H., and Halchuck, S. (1999). *Trial seismic hazard maps of Canada - 1999: 2% in 50 year values for selected Canadian cities*, Geological survey of Canada open file 3724.

Atkinson, G. M. (1991). *Use of the uniform hazard spectrum in characterizing expected levels of seismic ground shaking*. 6th Canadian Conference on Earthquake Engineering, Toronto, ON. 469-476.

Canadian Commission on Building and Fire Codes (1995). *National Building Code of Canada 1995*, National Research Council, Ottawa, ON.

Humar, J. L. and Rahgozar, M. A. (2000). *Application of uniform hazard spectra in seismic design of multistorey buildings*. Canadian Journal of Civil Engineering, **27**:1-18.

TABLE 1
Proposed base shear and overturning moment adjustment Factors, M_v and J

$S_a(0.2)/S_a(2.0)$	Type of lateral force resisting system	M_v for $T \leq 1.0$	M_v for $T \geq 2.0$	J for $T \leq 0.5$	J for $T \geq 2.0$
> 1.0 but < 8.0	Moment-resisting frame	1.0	1.0	1.0	1.0
	Braced frame	1.0	1.0	1.0	0.8
	Wall	1.0	1.2	1.0	0.7
> 8.0	Moment-resisting frame	1.0	1.2	1.0	0.7
	Braced frame	1.0	1.5	1.0	0.5
	Wall	1.0	2.5	1.0	0.4

Note: Values of M_v and J for intermediate periods are obtained by interpolation.

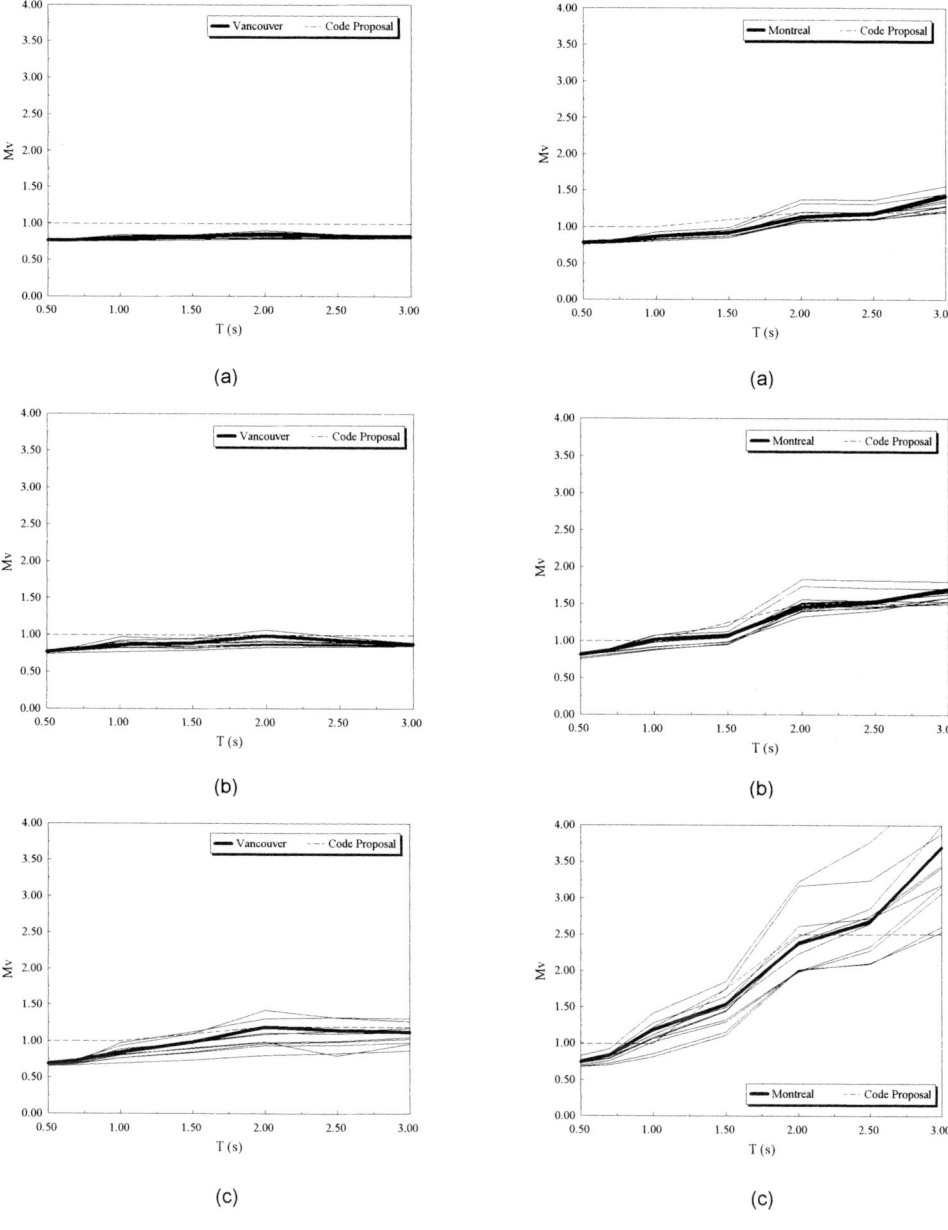

Figure 1: Mv factors for buildings in western regions of Canada (a) moment resisting frames (b) braced frames (c) walls.

Figure 2: Mv factors for buildings in eastern regions of Canada (a) moment resisting frames (b) braced frames (c) walls.

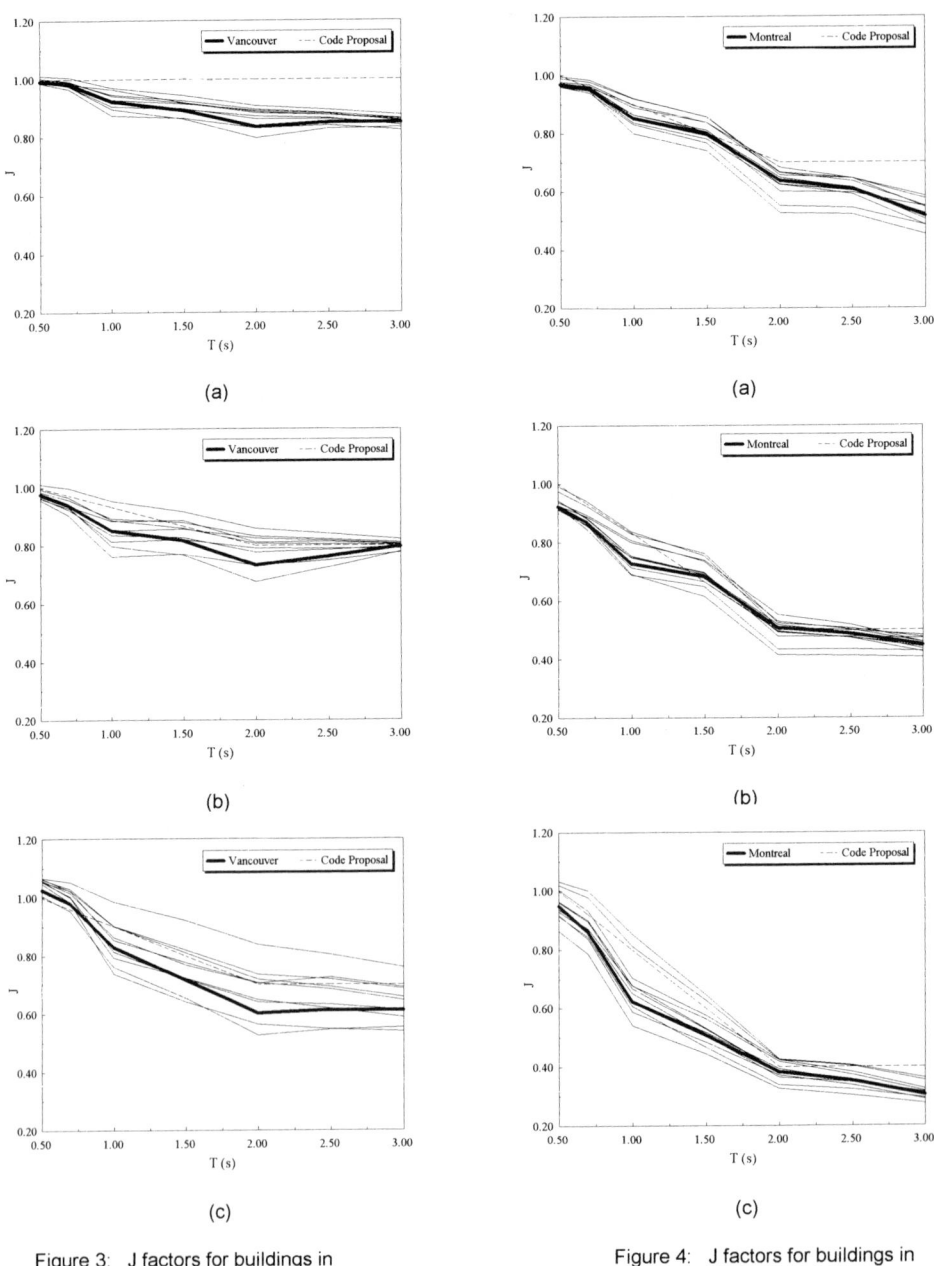

Figure 3: J factors for buildings in western regions of Canada (a) moment resisting frames (b) braced frames (c) walls.

Figure 4: J factors for buildings in eastern regions of Canada (a) moment resisting frames (b) braced frames (c) walls.

REVIEW OF PROVISION OF LOADS TO STRUCTURES SUPPORTING OVERHEAD TRAVELLING CRANES
With reference to SABS 0160 (1989)

P E Dunaiski [1], H Barnard [1], G Krige [2], R Mackenzie [3]

[1] Department of Civil Engineering, University of Stellenbosch, Private Bag X1, Matieland, 7602, South Africa.
[2] Anglo American, Anglo Technical Division, P O Box 61587, Marshalltown, 2107, South Africa.
[3] South African Institute of Steel Construction (SAISC), P O Box 1338, Johannesburg, 2000, South Africa.

ABSTRACT

Operational difficulties are experienced in South Africa with overhead travelling cranes and their support structures. With industry support, a programme was launched to research the basis of design of crane support structures systematically, in order to improve future designs. This paper reflects the insights developed from a combination of theoretical studies from the research programme, with considerations from practical experience with the design and maintenance of such facilities. The material was developed as input to the review of the South African Loading Code SABS 0160 (1989), which is presently in progress. An initial evaluation of SABS 0160 (1989) shows that crane loading provisions are considerably simplified in relation to the provisions of BS 2573 (1983), from which it was derived. Further comparison with other design codes such as Eurocode prENV 1991-5 (1997) and DIN 15018 (1984) shows that SABS 0160 (1989) does not possess the necessary conservatism that should be expected relative to the simplistic nature of the current load specifications. There is no evidence of deficiencies with regard to the safety of structures designed according to SABS 0160 (1989). Failures regarding serviceability limit state criteria do however occur. The most notable of these are fatigue failure of the support structure, crane wheel and rail wear, crane performance failure due to structures that are too stiff or too flexible. These failure occurrences suggest that provisions are adequate for a limited range of cranes, depending on the operational frequency of the crane and the loading state relative to the maximum capacity of the crane. These observations led to a more extensive comparison between the South African loading code and a range of codes and standards applied elsewhere, notably, ISO 8686-1 (1989), DIN 15018 (1984), NEN 2018 (1983), AISE TR 13 (1979) and BS 2573 (1983). The comparison shows marked differences in loads considered in the various design documents, and their respective magnitudes. The necessary adjustments to the present code are considered. An important component of the codified design process, is the crane classification. A classification system specifically for the support structure, reflecting the relevant design considerations, is proposed.

KEYWORDS

Overhead, Cranes, Classification, Design Codes, Serviceability, Dynamic loads.

INTRODUCTION

A research project at the University of Stellenbosch is investigating Overhead Travelling Crane (OHTC) behaviour and Support Structure response. The aim of the project is to study load cases and load magnitudes that are imposed onto Support Structures by OHT Cranes. A review of Codes of Practice from abroad was performed to assess the load cases and load magnitudes specified in these sources. A comparison is made with SABS 0160 (1989), the general procedures and loading to be adopted in the design of buildings.

The results from the Codes of Practice review and comparison with SABS 0160 (1989) are presented in the next section. The results of this review show that there exists a classification system for OHT Cranes only. The need for a classification system for OHTC Support Structures was identified and this classification system is under development. The proposed classification system is presented in the following section. Also, it was inferred from serviceability limit state failures of OHTC Support Structures that many serviceability failures occur due to lack of experience regarding the design and construction detailing of such structures. A list of design and construction considerations was therefore composed. These considerations are related to the proposed service classification for OHTC Support Structures and the purpose of these considerations is discussed in the section titled "OHTC Support Structure Design and Construction Considerations". The concluding section comments on the further use of the proposed Support Structure service classification.

CODE OF PRACTICE REVIEW AND COMPARISON

Table 1 shows a comparison of load cases by SABS 0160 (1989), ISO 8686-1 (1989), DIN 15018 (1984), NEN 2018 (1983), AISE TR 13 (1979), and BS 2573 (1983). The load cases considered are those which describe forces of a regular occurrence due to everyday operational activities of the crane. Other load cases e.g. loads due to mechanical failure, tilting forces, forces on end stop, and emergency cutout are not included in this paper. The load cases are divided into (1) forces in the vertical direction and (2) forces in the horizontal plane.

TABLE 1
LOAD CASE COMPARISON

TYPE OF LOAD LOAD EFFECT	SABS 0160	ISO 8686	DIN 15018	NEN 2018	AISE TR 13	BS 2573
Vertical Load						
Static wheel loads due to weight of crane.	√	√	√	√	√	√
Static wheel loads due to weight of payload.	√	√	√	√	√	√
Hoisting an unrestrained grounded load.	√	√	√	√	√	√
Sudden release of part of the payload.		√	√	√		
Travelling on an uneven surface.		√	√			
Horizontal Loads						
Loads due to movement of the crab.	√	√	√	√	√	√
Loads due to movement of the crane bridge.	√	√	√	√	√	√
Loads due to crane wheel / rail misalignment	√				√	
Loads due to oblique travelling (skewing).	√	√	√	√		√

Forces acting on the crane girder in the vertical direction

Static loads due to the weight of the crane and the payload: The static loads refer to the forces that depend on the weight of the crane, load lifting attachments, and the load to be lifted. These forces are imposed onto the crane girder at the crane wheel positions. The depth of explicit description of how to determine the static crane wheel forces varies from code to code.

Hoisting an unrestrained grounded load: This case is treated by all codes. There are differences in the magnitude of the wheel forces prescribed by the codes. OHTC properties taken from Krige (1998) are used as an example to illustrate this variation in force magnitude by means of numerical values calculated for a random choice of a sample crane. The curves in Figure 1 represent the difference in maximum wheel force magnitude calculated by means of SABS 0160 (1989) versus DIN 15018 (1984), and SABS 0160 (1989) versus ISO 8686-1 (1989). The load impact factors used for determining the maximum wheel forces were taken as that for crane hoist class 2. The curves in Figure 1 clearly show a positive difference in maximum wheel force magnitude at load lifting speeds below 0,165m/s for the comparison with DIN 15018 (1984), and below 0,68m/s for the comparison with ISO 8686-1 (1989). At these "break-even" speeds there is a change of sign in the difference of maximum wheel force. SABS 0160 (1989) therefore overestimates the maximum wheel force for a particular range of load lifting speed where-after it underestimates the maximum wheel force when compared to DIN 15018 (1984) and ISO 8686-1 (1989). Another concern is also the large difference in "break-even" speed. This shows that there is a significant difference between DIN 15018 (1984) and ISO 8686-1 (1989).

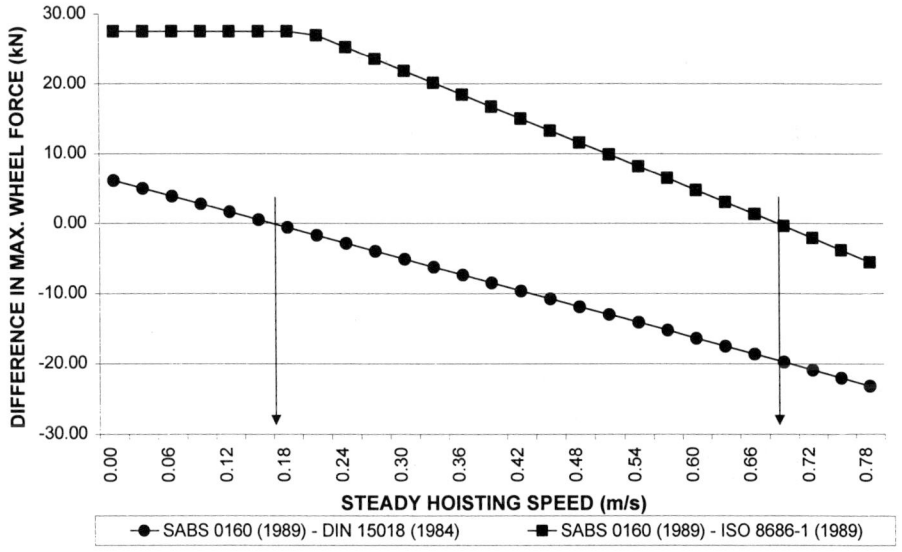

Figure 1: Difference of Vertical Wheel Force - Code Comparison

Effects of sudden release of part of the payload: Only ISO 8686-1 (1989), DIN 15018 (1984), and NEN 2018 (1983) describe this load case explicitly. This case is a shortcoming of SABS 0160 (1989) because inexperienced design engineers may disregard the dynamic effects due to the release of part of the payload as a critical load case.

Loads caused by travelling on an uneven surface: Only ISO 8686-1 (1989) and DIN 15018 (1984) provide for this event explicitly. This is a shortcoming of SABS 0160 (1989), however it is practice in South Africa to weld together the ends of rails for structures with frequent or continuous working cranes. This eliminates gaps or steps at rail joints and therefore also the sources of rail surface discontinuities causing an uneven surface.

Forces acting in the horizontal plane

Loads due to acceleration / braking of the crab or crane bridge: Forces due to the acceleration and braking of the crab or crane bridge are treated by all Codes of Practice. The depth of description for this load case varies according to code. The European codes usually specify that loads should be determined due to acceleration of deceleration of the crane or crane mechanisms. ISO 8686-1 (1989) provides a detailed rigid body kinetic model of the crane that may be used to determine the forces caused by acceleration. SABS 0160 (1989) provides for these loads by empirical means through specifying the forces as a percentage of the weight of the crane and/or payload.

Loads due to crane wheel / rail misalignment: Only SABS 0160 (1989) and AISE TR 13 (1979) provide for this particular load case. This case provides for forces of equal magnitude at crane wheel positions. The forces act in the horizontal plane and in the same direction (transverse to the longitudinal axis of the crane girder).

Loads due to oblique travelling (skewing): This case is treated by all Codes of Practice tabulated in Table 1. For this load case there is also a difference in force magnitude depending on the code used for design purposes, and different load models that are used for design calculation. Figure 2 (i) shows the load model from ISO 8686-1 (1989), DIN 15018 (1984), and NEN 2018 (1983) for a crane with a total of four wheels. Figure 2 (ii) shows the load model from SABS 0160 (1989) and BS 2573 (1984) for a crane with a total of four wheels. The internal force distribution (shear, moment, and torsion) for the crane girder and the reaction forces of the crane girder transferred to the support structure will differ completely depending on which Code of Practice is used for design.

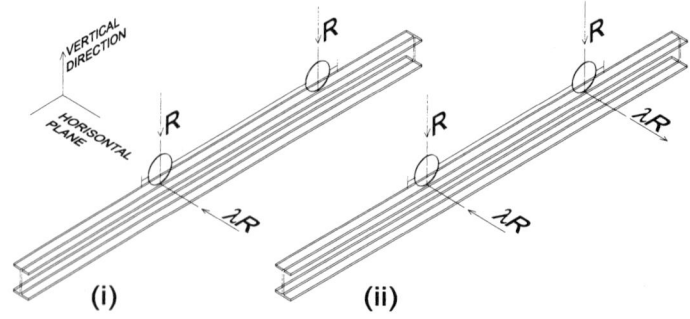

Figure 2: Skewing Load Case Comparison

OHTC SUPPORT STRUCTURE CLASSIFICATION

The proposed OHTC Support Structure service classification system is encapsulated in Table 2 and Table 3. Table 2 shows the criteria that are used in the process of determining the service class of the Support Structure. The criteria are as follows: (1) the frequency of utilisation of the OHTC within the Support Structure or parts thereof, (2) the loading state of the OHTC and therefore the Support Structure. Table 3 shows a 2-dimensional matrix where criterion (1) and criterion (2) are used to classify the Support Structure as a service class A, B, C, or D.

The purpose of this proposed service classification system for OHTC Support Structures is to provide the design engineer with the means to assess the level of reliability that is required when considering the choice of structural system and structural components, conservatism of the design, and the choice of construction details. For example: An OHTC Support Structure with a service classification D has an crane that is an integral part of the operation activities within the building, the crane is working continuously. The crane is almost always loaded near to its maximum capacity and it may also operate at high speeds. These considerations that are derived from data for the operational activities translates to severe dynamic loading and wear on the crane and the Support Structure. A shutdown for long time period in order to perform maintenance or repairs to Support Structure components/members e.g. crane rails, crane girders damaged due to fatigue, etc. is not possible and this must be considered at the design and constructional phase of the structure.

TABLE 2
DESCRIPTION OF DEGREE OF UTILISATION AND LOADING STATE

DEGREE OF UTILISATION	
CATEGORY	**REMARKS**
Infrequent	Cranes in this category are used for initial installation and infrequent maintenance of equipment and plant.
Intermittent	Cranes in this category are an integral part of the operational requirements of the plant during a normal working day. Sufficient time for the maintenance of the supporting structure is available.
Frequent	Cranes in this category are an integral part of the day and night operational requirements of the plant where limited time is available for the maintenance of the supporting structure.
Continuous	Cranes in this category are an integral part of continuous operational requirements of the plant. Maintenance of the supporting structure is only possible during scheduled maintenance shutdowns. Therefore the highest level of reliability with special attention to ease-of-maintenance is required.
LOADING STATE	
CATEGORY	**REMARKS**
Light	The crane is very rarely subjected to the safe working load, and usually subjected to light loads only.
Moderate	The crane is fairly frequently subjected to the safe working load, but usually subjected to rather moderate loads only.
Heavy	The crane is frequently subjected to the safe working load, and usually subjected to loads of heavy magnitude.
Very heavy	The crane is regularly subjected to the safe working load.

TABLE 3
MATRIX OF SERVICE CLASSIFICATION OF OHTC SUPPORT STRUCTURES

LOADING STATE	DEGREE OF UTILISATION			
	Infrequent	Intermittent	Frequent	Continuous
Light	A	A	B	C
Moderate	A	B	C	C
Heavy	A	B	C	D
Very heavy	B	C	D	D
SERVICE CLASS	REMARKS			
A	<ul><li>Low operational requirements.</li><li>Generally slow vertical lifting speed and precise handling of load.</li><li>This class normally includes manually operated cranes.</li></ul>			
B	<ul><li>Moderate operational requirements.</li><li>Generally slow to moderate vertical lifting speed.</li></ul>			
C	<ul><li>High operational requirements.</li><li>Generally moderate to high vertical lifting speed.</li></ul>			
D	<ul><li>Critical operational requirements.</li><li>Generally high vertical lifting speed.</li></ul>			

OHTC SUPPORT STRUCTURE DESIGN AND CONSTRUCTION CONSIDERATIONS

Discussions with experienced design engineers and experienced fabricators and erectors of OHTC Support Structures shows that many serviceability failures can be avoided with more careful attention to the design of such structures and the choice of constructional details. A comprehensive list of design and construction considerations was composed. This list will provide engineers in practice with guidance to aspects that must be considered carefully in the design calculations and details used for construction. Table 4 shows examples of these design and construction considerations. For each design or construction consideration it is shown if that strict compliance with that particular consideration is "Mandatory" (M), "Recommended" (R), or "Not Required" (N). The options "Mandatory", "Recommended", and "Not Required" depend on the service classification of the Support Structure and therefore the level or reliability required for structural components/members.

CONCLUSION

In this paper it is shown that there are significant differences between load cases, load case models, and load magnitudes when looking at Codes of Practice for OHTC Support Structures. This justifies the research into OHTC behaviour and Support Structure response. A service classification system for OHTC Support Structures is presented. It is shown that this proposed service classification for OHTC Support Structures could be used as a means to address design and construction considerations. This will eliminate serviceability limit state failures at future OHTC Support Structures due to design oversights and poor construction details. It is proposed that the service classification system for OHTC Support Structures must be included in the current revision of SABS 0160, with the complete list of design and construction considerations added as an addendum to SABS 0160. This service classification system also provides a good framework for the specification of load magnitudes and limit state load factors. The use of the classification system for this purpose will be assessed with the results from the ongoing research on OHTC Support Structures at the Department of Civil Engineering, University of Stellenbosch.

TABLE 4
EXAMPLES OF DESIGN AND CONSTRUCTION CONSIDERATIONS

	DESCRIPTION	SERVICE CLASSIFICATION			
		A	B	C	D
Denomination: N = Not Required; R = Recommended; M = Mandatory, DC = Design Consideration, CC = Construction Consideration					
1 DC	Design drawings for the building shall show crane load criteria for all cranes in the building including: safe working load, self weight of bridge, crab and lifting attachment(s), maximum and minimum static wheel loads, long travel speed, crab cross travel speed, payload vertical lifting speed, buffer impact loads at the ends of the crane runway, and fatigue loading criteria for vertical and horizontal crane-induced loads.	M	M	M	M
2 DC	The relative vertical deflection of the crane girder between the supports under static wheel loads shall not exceed the indicated ratios of the span.	R $\frac{1}{600}$	R $\frac{1}{600}$	R $\frac{1}{800}$	R $\frac{1}{1000}$
. . .					
6 CC	Camber shall be considered for crane girders with a span exceeding 12m.	R	R	R	R
. . .					
9 CC	The accumulated fabrication and erection tolerances for crane supporting structure members shall satisfy the requirements of SABS1200H.	M	M	M	M
10 DC CC	Simply supported crane girders are preferred to continuous crane girders.	N	R	R	R
. .					
13 DC	Transverse loads from cranes shall be distributed with due regard for the lateral stiffness of the structures supporting the rail.	M	M	M	M
14 DC	A comprehensive fatigue analysis shall be conducted.	N	R	M	M
15 CC	Ends of simply supported crane girders shall be free of restraint to rotation in the plane of the web and free from prying action on bolts in the girder to column connection.	R	M	M	M
. .					

REFERENCES

Association of Iron and Steel Engineers – Technical Report (TR) 6 (1996). *Specification for Electric Overhead Travelling Cranes for Steel Mill Service.*

Association of Iron and Steel Engineers – Technical Report (TR) 13 (1979). *Guide for the Design and Construction of Mill Buildings.*

British Standard – BS 2573 (1983). *Rules for the Design of Cranes. Part 1: Specification for Classification, Stress Calculations and Design Criteria for Structures*, British Standards Institution.

Deutsche Norm – DIN 15018 (1984). *Cranes. Part 1: Steel Structures. Verification and Analysis.*

European Pre-Standard – prENV 1991-5 (1997). *Actions Induced by Cranes and Other Machinery (CEN/TC250/SC1 Document).*

International Standard – ISO 8686-1 (1989). *Cranes. Design principles for loads and load combinations. Part 1: General.* International Standards Organisation.

International Standard – ISO 8686-5 (1992). *Cranes. Design principles for loads and load combinations. Part 5: Overhead Travelling and Portal Bridge Cranes.* International Standards Organisation.

Krige G.J. (1998). *Overhead Crane and Machinery Loads.* Proceedings of the South African National Conference on Loading.

Nederlandse Norm – NEN 2018 (1983). *Hijskranen. Belastingen en belastingscombinaties.* Nederlandse Normalisatie-instituut.

South African Bureau of Standards – SABS 0160 (1989). *The General Procedures and Loading to be adopted in the Design of Buildings.*

BACKGROUND TO WIND DAMAGE MODEL FOR DISASTER MANAGEMENT IN SOUTH AFRICA

A. M. Goliger[1] and J.V. Retief[2]

[1]Division of Building and Construction Technology, CSIR,
P O Box 395, Pretoria 0001, RSA

[2]Department Civil Engineering, University of Stellenbosch
Private Bag X1, Matieland, RSA

ABSTRACT

This paper presents the background to a proposed wind-damage and disaster-management support model for South Africa. The relevance of wind disaster is initially highlighted on the basis of selected statistics, both local and international. The information transpiring from international research is summarised, including the role of codification and construction guidelines/building regulations. Selected aspects of South African research are presented. A summary of factors that influence the extent of damage and the philosophy of the proposed model are given.

KEYWORDS

Wind damage, wind disaster, risk model.

WIND DISASTERS

Throughout the world, as well as in South Africa, wind forms one the dominating environmental loadings affecting structural design. It constitutes a large percentage of disastrous events to the built-up environment, as presented in Figure 1 (which also compares the economic implications of various disasters). The issue of wind damage becomes even more relevant in light of the constant pursuit for more economic design of structures and the development of materials with higher strength, which in many cases leads to development and use of lighter components and also the introduction of "leaner" design techniques. In addition, buildings and facilities that do not comply with building regulations pose particular challenges to wind disaster management in South Africa.

In view of these technological trends, a significant amount of theoretical and experimental research has been carried out throughout the world into the effects of wind on structures and ways to increase their wind resistance, and therefore to minimise the amount of damage.

Nevertheless, insurance statistics suggest that losses due to windstorm disasters continue to rise (Berz, 1991). This could partly be attributed to an increase in the extent of the built-up environment, but it also posses questions as to the possibility of global environmental change leading to the increase in

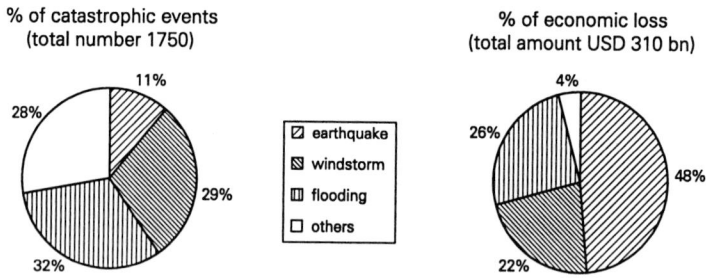

Figure 1: World statistics of natural disasters for the period 1994-1996 (derived from Munich Re)

severe windstorms. The situation also suggests that scientifically derived knowledge on wind-resistant structural design has not yet been implemented across the entire construction sector. Some attempts to bridge the gap between scientific work, structural design and the actual construction have been made in various countries (mainly the USA, Australia and certain parts of Europe e.g. Minor and Mehta, 1979; Reardon, 1980; Gandemer and Moreau, 1999) but not in South Africa.

There are no statistics on windstorm disasters in South Africa and only recently a database on wind disasters has been set up at the CSIR (Goliger, 2000). Its initial statistics indicate a trend in which the number of damage events reported is increasing, as presented in Figure 2. Following the setting up of the database, a wind damage and disaster management support model for South Africa is currently under development and the summary of this model presented in the present paper.

RESEARCH

Over the last 30 years or so a significant amount of research on wind damage to structures has been carried out throughout the world. The underlying aim of that research has been to analyse and gain an understanding of past events in order to minimise and predict devastation from future events. The research can broadly be grouped into the types summarised below.

Site investigations and analysis of damage

Most significant windstorm devastation throughout the world is usually followed by site inspection and evaluation of the damage, undertaken by local building authorities supported by structural researchers/consultants (e.g. Thompson, 1994; NIST, 1998). In South Africa in the last 20 years or so the CSIR has been actively involved in wind engineering research, which have included numerous windstorm site investigations, some in cooperation with the SA Weather Bureau. These investigations have included major disasters like that in Welkom in 1990 (CSIR/Eskom, 1992) or, more recently, that in Umtata in 1998 (Mdlekeza et al, 1999) and Cape Flats 1999 (Goliger and De Coning, 1999).

In our experience there are several factors affecting the reliability of investigations, including their timing (immediately after the event), the inclusion of visual documentation of damage with structural detail, the consideration of weather/climatic aspects, the determination of spatial information on the damage and inputs of reliable witnesses regarding the duration, directional characteristics, etc.

The analysis of wind damage should be aimed at determining the type of extreme wind event, its geographical extent, the likely magnitude of devastating wind speeds, the extent of damage and structural aspects of the damage, including the non-engineered structures.

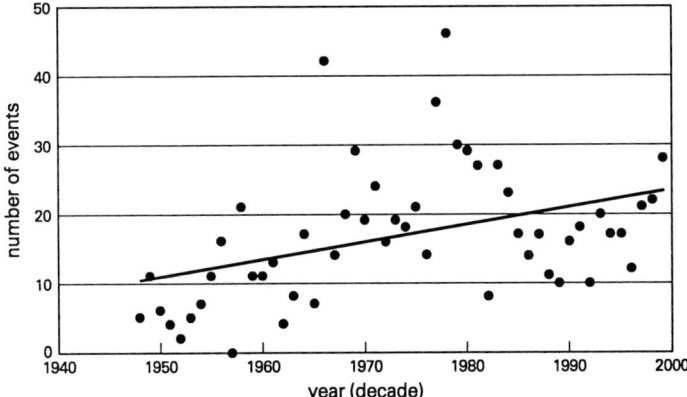

Figure 2: Yearly distribution of wind damage events in South Africa

Statistics of damage

The statistics of damage are usually derived from the number and type of structures which have been affected and the extent of the damage. These depend largely on the spatial extent (i.e. the size) and severity of the extreme events. For large-scale events the damage is often expressed in terms of the number of housing units which have been affected. For example, Hurricane Andrew damaged the roofs of 135 000 houses and totally destroyed 28 000 houses.

In South Africa no statistics of wind damage are available apart from the initial analysis on the CSIR's database. This database considers broad types of structures. Unfortunately, the information on damage to houses is inconsistent (and not suitable for further processing) because in most cases descriptors like *large number* or *hundreds* are used in the reports, instead of the actual number of units that have been damaged. Figure 3a presents an analysis of damage reports in respect of various types of structures which have been affected. It can be seen that in more than 70% of cases damage to buildings and houses has been reported. A number of those reports refer to damage to several hundreds (even thousands) of houses and informal housing units. Information on the number of people affected by wind damage (which could also serve as an indication of the amount of damaged dwellings) is even more sketchy. The limited information on the number of people rendered homeless due to wind damage is plotted in Figure 3b. Of interest, nearly 15 000 people were left homeless in 1999 as a result of several unrelated wind events.

Economic consequences

Numerous studies have been carried out, mainly in Europe and America, on the economic impact of wind disasters, both from the national and the insurance industry point of view. Statistics by Munich Re suggest that between 1994 and 1996 wind damage constituted more than 20% of the world's total economic loss due to disasters (Figure 1).

The effects on the insurance industry depend on the level of cover, which is typically related to the wealth of the society. In some cases it can be devastating. For example, the Atlantic coast of the US, which is exposed to hurricanes, involves $13,5 trillion of insured property. Hurricane Andrew, which affected only a small portion of that coast, caused the bankruptcy of about 30% of insurance companies in the US and had a serious impact on Lloyds (INEL, 1997).

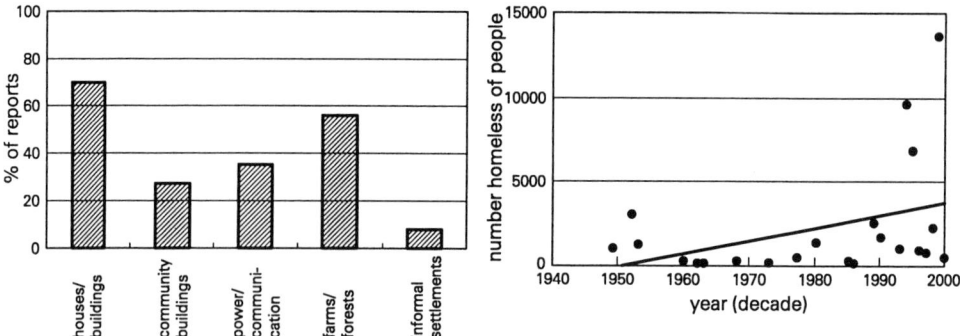

Figure 3: (a) Damage reports to various types of structures; (b) number of homeless people

In South Africa no records and statistics of the economic consequences of wind damage are kept. The costs of significant events, like the Welkom or Cape Town tornadoes, have been estimated at R400m and R170m respectively (at current values).

Scenarios of damage and risk models

The scenes of wind disasters form an important tool for predicting typical and/or worst implications on national and regional scale, as well as the impact on the insurance industry. For example, in the period between 1989 and 1994, total wind damage in the USA was about $85 billion, but it is predicted that a future loss from a single event in Miami could be $75 billion (INEL, 1997). In comparison the scenario of a likely tornado event crossing the Johannesburg metropolitan area (Goliger, 1996) predicts a loss in excess of R0,5 billion.

Information on the risk of damage forms the fundamental contribution to disaster management activities. In a nutshell, the risk model of wind damage should include:
- the probability of occurrence of extreme events per unit area (or administrative region) in terms of their intensities;
- the vulnerability of the particular area (i.e. exposure of population and built-up environment);
- the vulnerability of the built-up environment (e.g. strength of structures); and
- potential consequence (i.e. damage per time unit).

Several risk models of tornado strikes have been developed (e.g. McDonald, 1983; Twisdale, 1988). Other models refer to hurricanes and cyclones (Leicester et al, 1979; Chiu et al, 1995) and thunderstorms (Holmes, 1999). Figure 4 presents a risk map of tornadic strike in South Africa (Milford and Goliger 1997).

Construction guidelines

As an outcome of ongoing site investigations and analysis of damage, several construction, repair-work and architectural guidelines have been developed (Melaragno, 1982). These refer to various measures to mitigate the extreme uplift of roofs and lightweight wall systems. For example, damage surveys in the USA revealed a fairly consistent pattern in which small interior compartments within houses and buildings escaped unscathed. This has led to research and development of architectural and construction guidelines on 'protective' areas and shelters (FEMA, 1999). No information of a similar nature for South Africa has been developed.

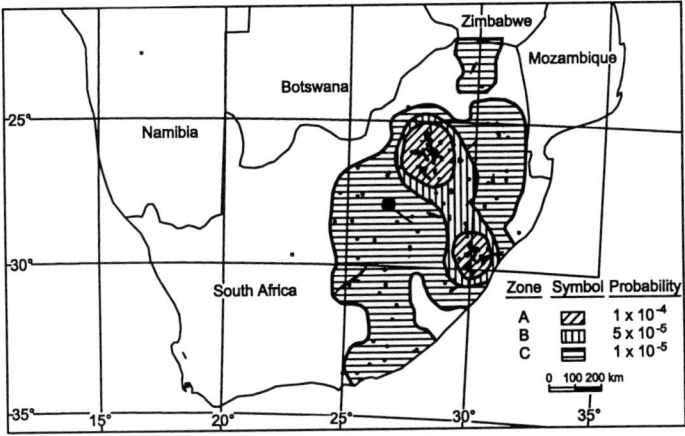

Figure 4: Mean rate of occurrence of tornadoes in South Africa (per year per km^2)

Policies

All natural disasters have a tremendous socio-economic impact resulting from human loss and damage to property. A trend can be noted, in which major windstorm disasters affect the poorest countries and communities in the world. This could be attributed to a couple of factors like the geographical distribution of extreme events coinciding with distribution of underdeveloped countries, insufficient financial resources leading to poor construction practices and lack of insurance cover.

Another trend is also apparent across the field of natural hazards in which the emphasis is placed upon mitigating measures and not on the reliance on relief and reconstruction in the wake of disasters (Davenport, 1995; Cohen, 1999). This philosophy has been adopted by the International Decade for Natural Disaster Reduction (IDNDR) under the auspices of the United Nations.

ROLE OF CODIFICATION

The role of structural loading codes is to ensure an acceptable level of risk of damage to structures by taking into account the climatic situation (i.e. the probability of winds with damaging magnitude and structural aspects of damage). This statement implies a direct and inseparable relationship between wind damage and codification.

The only way to investigate the reliability (i.e. the performance of structural design codes) is to monitor, evaluate and develop the statistics of the full-scale damage. Unfortunately no formal procedure and activities for site evaluation of wind damage from a structural point of view have been carried out in South Africa.

A fairly consistent message in that respect transpires from a review of several analyses of wind disasters undertaken overseas. Walker et al (1987) claim that strict compliance with modern regulations and design codes in Australia have ensured that practically no damage to recently built structures subjected to cyclonic conditions have been observed. The opposite has been reported by Thompson (1994) in Florida (and echoed by McDonalds), where the pursuit for affordable housing forced the relaxation of building regulations, leading to disastrous effects under Hurricane Andrew.

Cook (1984) reports that a survey of wind damage in the UK suggests that most occurs at wind speeds far below the design speed stipulated in the structural design code.

All the above findings emphasise the critical role of codification and building regulations within the disaster-management chain of activities, and suggest that potential wind damage can to a large extent be minimised or avoided by adherence to modern, scientifically and statistically sound structural codes and regulations.

INFLUENCING FACTORS

There are several factors that influence the extent of damage caused by extreme wind events. These can broadly be grouped into types.

- Type and characteristics of wind event, including the type of the storm generating excessive winds, the magnitude of the devastating winds, their duration, geographical distribution and probability of occurrence, the extent of the event, and changes in wind characteristics with elevation, directional and seasonal characteristics.

- Local factors affecting the wind field, which include exposure (i.e. terrain roughness, the influence of topography, density and shielding due to other structures). The effect of local factors can be significant and an example of such a situation is presented in Figure 5, in which wind-profile characteristics at three locations at front, on top and on the lee side of Cape Town's Table Mountain are plotted (Goliger et al, 1992).

- The distribution of the built-up environment and assets; this is related to the accessibility of resources which are determined to a large extent by the distribution of the population and its wealth.

- Vulnerability, which involves the strength of structures, including the issues of formal and informal developments, engineered vs. non-engineered structures, design and construction methods and technologies, material and construction standards, and the mechanism of failure (including flying debris and wind erosion)

PRINCIPLE OF THE PROPOSED RISK MODEL

The proposed risk model is under development. It will take into account and integrate all the above factors. It is important to note that from the wind climatic point of view it will be based on the spatial statistics of extreme wind events, which is important for adequate disaster planning. This is different from the threshold level of wind speeds derived on the basis of extreme value distribution of data from particular recording stations (Milford, 1987), which is typically used in design loading codes. Such an approach is more realistic and appropriate from a disaster-management point of view. Furthermore, for South African conditions it eliminates two basic shortcomings of the available wind data, namely
- it takes into account the climatic differences in the characteristics of extreme wind events, which are significant, and
- it eliminates the issue of poor coverage and reliability - the representativity of long-term reliable recording stations in South Africa (16 stations to cover an area of 1,2 million km^2).

A wind damage model for South Africa will not only contribute to disaster planning and management, but could also improve the rational basis for building codes and structural design.

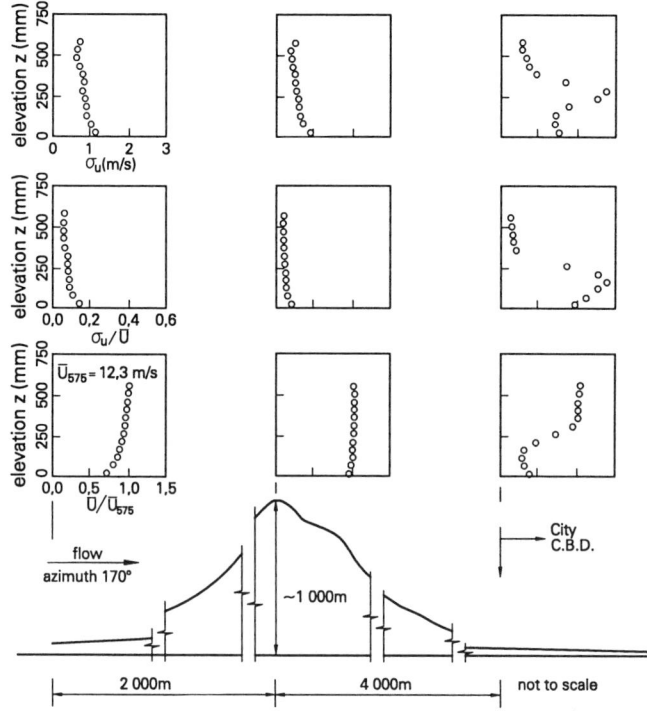

Figure 5: Wind characteristics influenced by dominant topography

REFERENCES

Berz G. (1991). The increasing significance of windstorms and the IDNDR. *Journal of Wind Engineering and Industrial Aerodynamics*, Vol. 41-44 (1992), 23-25.

Chiu G.L.F., Kiremidjian A S and Chiu A N L (1995). Damage hazard states for hurricane risk analysis. *Proc. 9th International Conference on Wind Engineering*, New Delhi, January 1995, 1973-1978.

Cohen J.S. (1999). Partners in mitigation: towards a national wind engineering policy. *Proc.International Conference; Wind Engineering into the 21st Century*, Lyngby, Denmark, June 1999, 1383-1388.

Cook N.J. (1984). The performance of buildings in the United Kingdom. *Proc. of the Conference on Design against Wind Induced Failure*, Bristol, January 1984.

CSIR/ESKOM (1992). *Transmission line loading due to narrow winds, Part 1: Tornado activity in South Africa 1905-1991; Appendix A: Significant tornadoes*, Report 550 23121, Division of Building Technology, CSIR, Pretoria, South Africa.

Davenport A.G. (1995). The role of wind engineering in the reduction of natural disasters. *Proc. 9th International Conference on Wind Engineering*, New Delhi, January 1995, Special Paper.

FEMA (1999). *Taking storm from the shelter: building a safe room inside your home*, Federal Emergency Agency, Mitigation Directorate, Washington DC, USA.

Gandemer J. and Moreau S.H. (1999). Wind loads and architectural design – application to tropical cyclonic dwelling. *Proc. International Conference; Wind Engineering into the 21st Century,* Lyngby, Denmark, June 1999, 1401-1408.

Goliger A.M., Waldeck J.L. and Milford R.V. (1992). A topographical study to investigate wind nuisance. *Journal of Wind Engineering and Industrial Aerodymanics,* Vol 44, Nos 1-3, pp 1693-1704.

Goliger A.M. (1996). *Hypothetical scenario of damage to houses due to a worst-case tornado event in South Africa,* Internal report BOU/I31, Division of Building Technology, CSIR, Pretoria, South Africa.

Goliger A.M. and De Coning E. (1999). *Tornado cuts through Cape Flats,* Publication of the Division of Building and Construction Technology, CSIR, with the co-operation of the SA Weather Bureau.

Goliger A.M. (2000). *Electronic database of wind damage in South Africa,* Division of Building Technology, CSIR, Pretoria, South Africa.

Holmes J.D. (1999), Modelling of extreme thunderstorm winds for loading of structures and risk assessment. *Proc. International Conference; Wind Engineering into the 21st Century,* Lyngby, Denmark, June 1999, 1409-1415.

INEL (1997). *The partnership for natural disaster reduction; Home saver project,* Brochure of the Idaho National Engineering Laboratory, July 9, 1997.

Leicester R.H., Bubb C.T.J., Dorman C. and Beresford F.D. (1979). An assessment of potential cyclone damage to dwellings in Australia. *Proc. 5th International Conference on Wind Engineering,* Fort Collins, Colorado, USA, July 1979, 23-36.

McDonald J.R. (1983). *A methodology for tornado hazard probability assessment,* Institute for Disaster Research, Texas Tech University, prepared for US Nuclear Regulatory Commission, NUREG/CR-3058.

Mdlekeza M., van Wyk A.T. and Goliger A.M. (1999). *Report on the severe storm damage in Umtata, 15 Dec. 1998,* Internal report BOU/I141, Division of Building Technology, CSIR, Pretoria, South Africa.

Melaragno M.(1982). *Wind in architectural and environmental design,* Van Nostrand Reinhold Company.

Milford R.V. (1987). Annual maximum wind speeds for South Africa. *The Civil Engineer in South Africa, January 1987,* 5-15.

Milford R.V. and Goliger A.M. (1997). Tornado risk model for transmission line design. *Journal of Wind Engineering and Industrial Aerodynamics,* Vol 72 (1997), 469-478.

Minor J.E. and Mehta K.C. (1979). Wind damage observations and implications. *Journal of the Structural Division ASCE,* November 1979, 2279-2291.

Munich Re (1994-1996). *Topics; Annual review of natural catastrophes,* Issues 1994, 1995 and 1996. Munich Reinsurance Company, D-800 München 40, Germany.

NIST (1998). BFRL investigates tornado damage. *NIST Building and fire research laboratory, US Department of Commerce, Research Update,* Vol 4, No 3, Fall 1998.

Reardon G.F. (1980). *Recommendations for the testing of roofs and walls to resist high winds,* Technical Report No 5, Cyclone Testing Station, James Cook University, Queensland, Australia.

Thompson L.E.(1994). Hurricane Andrew distruction of housing. *Housing Science,* Vol 18, No 4, pp.239-250.

Twisdale L.A.(1988). Probability of facility damage from extreme wind effects. *Journal of Structural Engineering,* Vol 114, No 10, October 1988.

Walker G.R., Reardon G.F. and Jancaukas E.D. (1987). Observed effects of topography on the wind field of cyclone Winifred. *Proc. 7th International Conference on Wind Engineering,* Aachen, West Germany, July 1987, Part 1, 139-148.

CONCRETE AND CONCRETE MATERIALS

STRUCTURAL ENGINEERING AND CONCRETE TECHNOLOGY IN DEVELOPING COUNTRIES: AN OVERVIEW

Ali S. Ngab

Department of Civil Engineering AlFateh University Tripoli, Libya

ABSTRACT

The twentieth century has witnessed an unprecedented growth in human population. During the last fifty years the world population has doubled from 3 to 6 billions. The majority of this increase is in the developing countries. More than 85 percent of the world population reside in these countries. By year 2050, more than 75 percent of the world's population will live in urban areas. To serve these needs and changes, large amounts of materials are needed for construction of houses, office buildings, roads and infrastructures required for decent living. The last century has also witnessed an unprecedented development in structural engineering where concrete played the predominant role in construction. The success of this material in the developed world is significant and there is a clear trend that such development will continue and shift to the developing world. Concrete consumption worldwide is in the order of 10 billion tons per year, next only to the total consumption of water. Thus, for obvious reasons concrete lends itself as possibly the only construction material to help solve these imminent problems of developing countries. However, many of these countries lack the funds and technology to face these great challenges. It is the objective of this paper to assess and evaluate the ability of concrete technology in solving many of structural engineering challenges in developing countries. An attempt will be made to review the conditions and demands of the construction sector related to concrete use and production. The basic requirements for sound concrete industry in developing countries will be addressed. The need for an appropriate technology will be discussed on light of the vast amount of knowledge available on structural engineering and concrete technology worldwide.

KEYWORDS

Concrete, technology, developing countries, structural engineering, appropriate technology

INTRODUCTION

Concrete consumption worldwide is in the order of 10 billion tons per year, next only to the total consumption of water. Developing countries have a significant share of this amount and the future of the concrete industry will shift towards the developing nations for many social and economical reasons. It is certain that the demand for housing and infrastructure is difficult to meet now and in the future in many of these countries. The housing shortage in South Africa for example is a bout one million and, in a small country such as Libya the shortage is estimated as fifty thousands units per year. In India the housing shortage is now estimated to be more than forty million unites. The situation in other African, Asian and South American countries is similar. Most of these countries became independent in the early fifties and sixties. After independence these nations were faced with a great demand for development. This demand to build includes residential, public and industrial buildings, roads, pavements, bridges, water structures, sewage treatment plants, power stations, dams etc. In order to carry out such great and difficult tasks, these counties relied on concrete as the only possible construction material to do the job. However, many lacked the basic ingredients to sustain such challenges. Many did not have the cementing material that is the basic element in concrete production. Lack of funds made importing this material from abroad very difficult. Thus, with the exception of the so-called oil rich counties, many developing countries had and still difficulty in providing the basic raw materials such as cement and reinforcing steel. Almost all developing countries have severe constraints as regards material and financial resources. India for example with its well-established construction industry, and highly developed technical manpower, lacks financial resources and funds (13). There is high correlation between construction and the gross domestic product (GDP). There are clear indicators that the construction industry is dependent on economic performance. Most developing countries especially in Africa have very low GDPs and hence correspondingly low construction expenditures. Even the so-called oil-rich countries lack qualified technical manpower. Other countries in Africa lack both technical manpower and also the needed funds for construction. Although in many countries the basic constituents of concrete as a man-made construction material are locally available; it took many decades to manufacture cement. In order for concrete to be able to service these needs as a primary material of construction, an assessment of its potential and limitations associated with its use in developing countries must be addressed. This paper will focus on the use of concrete in these counties with the objective of formulating an out line for an appropriate concrete technology for developing countries. This technology must be based on the enormous amount of knowledge currently available on concrete behavior in the industrialized world. This knowledge must be assessed and if possible modified to suit the local conditions under severe constraints that are encountered in the developing world. This is to be achieved without much sacrifice on the fundamental properties of sound concrete construction such as strength, durability and economy.

CURRENT TRENDS IN CONCRETE TECHNOLOGY

Developments in Concrete Knowledge

Cement being the most important ingredient for making concrete in now produced in many countries. Table (1) shows the cement production world wide for industrialized and developing countries alike. Cement consumption in the U.S. has averaged between 75 to 90 million ton China by far is the leading cement producer, the United States of America ranks third in cement production after Japan. Presently, India produces more than 58 million tons annually ranking as the fourth world cement producing country. Libya for example produces more than 5 million tons annually. World wide, cement production exceeded 1.25 billion tons in 1991, according to the U.S. Bureau of Mines. This production is estimated to reach 2 billion tons annually. Production and consumption of cement in developing countries is an indicator of the concrete industry activities. In Egypt for example, the local industry increased its production output and improved its quality to meet the increasing demand to build during the nineties. In 1997, 2.3 million tons of cement was imported to put the cement consumption in Egypt over 22 million ton per year. Under developed counties in the sub Sahara region such as Uganda suffer from shortage of cement that is imported from Kenya. Uganda's annual consumption, however, is only about half a million-ton. In Saudi Arabia the cement consumption grew rapidly after the 1991 Gulf war due to the construction boom that followed.

Table (1) World Production of Portland Cement (1995)
(in million tons)

United States	76
Germany	49
France	27
Italy	45
Japan	93
Russia	140
Spain	30
South Korea	45
Other countries	400

China	362
Brazil	31
India	64
Turkey	27
Egypt	21
South Africa	7
Algeria	12
Morocco	9
Libya	5

Table (2) Projection of Regional Cement Consumption, million tons,
(Year 2005, Adapted from World Cement V.27, no 5 May 1996)

Europe (including EU)	432
Asia	1082
Africa	77
North America	92
South America	142

The numbers shown in Table (1) and (2) indicate that cement production of the developing countries is equal to that of the developed world. When consumption of cement per capita is considered however, the developing world consumption is about 150 kg /capita while in the industrialized world the consumption is about 400 kg/capita. This is an indication of the volume of the concrete industry in both cases. According to Witmann (16), concrete's success is due to ingredients of worldwide availability and low cost, accessible technology, low energy consumption, feasibility of engineering its properties and performance to suit a wide variety of requirements. Compressive strengths as low as 0.1 MPa and as high as 800 MPa are feasible; densities between 100 kg/m3 and 5000 kg/m3 are also achievable. Concrete is the most versatile building material compared to steel, aluminum, timber, plastics and glass, and it is limited only in its tensile strength when unreinforced (16). Significant advances in concrete technology have been made worldwide. There are now more than sixteen types of concrete that are actually used in many construction sites. Table (3) shows the types and uses of the known types of concrete as of today. Researchers have investigated almost any thing one can think of related to concrete technology. The amount of available knowledge on concrete is numerous. Considering the strength of concrete as a measuring stick, it could be argued that an upper ceiling on concrete strength has been practically reached. Strength over 100 MPa in many situations could be practical. Moreover, significant advances in concrete durability has been also made, new terms such as High Performance Concrete (HPC) has been introduced in the late nineties as an extension of High Strength Concrete (HSC) of the seventies. Normal Strength Concrete (NSC), however, is still the dominant structural material in many parts of the world. These advancements in material behavior have been coupled with significant research in the design and maintenance aspects. The design procedures and specifications are well established and many countries are modifying their design and specification procedures. The amount of available literature and research results accumulated over the years is enormous. These results and work are somewhat oriented towards the needs of the industrialized countries. Unfortunately, the situation in the developing countries is not bright. There is still too much to be done. Temptation to follow the industrialized world is great, but the limitations are even greater. Concrete in many parts of the world is still a mere mixture of the basic materials of gravel, cement and water and the road a head is very long.

Table (3) different types of concrete in the industrialized countries (16)

Designation	Characteristics	Main Application
Flowable fill	Self-compaction, low strength, low shrinkage	Back fill of excavation
Permeable concrete	Moderate strength	Drainage pipes, slabs
Impermeable concrete	Low water cement ratio	Reservoirs
Foamed concrete	High shrinkage	Thermal insulation
Roller compacted	Dry consistency	Dams, Roads
Frost resistant	Entrained air, low permeability	Cold climates
Under water concrete	Flowing, highly cohesive	Place concrete under water
Mass concrete	Low heat, to avoid cracking	Dams, massive foundations
Shotcrete	Fast stiffening	Tunneling
Mortars	Low strength, low shrinkage	Masonry works
Colored concrete	White cement and pigments	Architectural purposes
Pumped concrete	Use retarders to delay setting	All kinds of construction
Precast concrete	High early strength	Concrete elements
Aggressive environment	Low permeability	Structure exposed to aggressive environment
Fiber concrete	Fibers added, durability	Industrial floors
High performance	Low w/c special materials	High performance elements

Future Trends and Developments

High Performance Concrete (HPC) is now under investigation for durability and strength considerations. Concrete as high as 800 MPa is being investigated, this new type of concrete known, as Reactive Powder Concrete (RPC) is produced using special concrete mixes with superplasticisers and steel fibers. Use of admixtures is increasing rapidly. Use of fly ash, silica fume, slag and other cement replacement is now under way in many countries. Chemical admixtures are now included to about 88% of concrete production in Canada, 85% in Australia, and 71% in the U.S. (8). Similar trends are reported for most developed countries. Use of admixtures in concrete in developing countries is very limited.

New research is shifting towards rehabilitation of deteriorated structures and efficient structural use of the newly developed high strength materials. The emerging construction technologies in these countries include also, Smart Concrete (SC), Self-Placing Concrete (SPC), and Autoclaved Aerated Concrete (AAC). New types of admixtures such as shrinkage reducing admixture, hardening accelerator admixture, and corrosion inhibitors for reinforced concrete are also under experimentation. These advances in concrete technology have been coupled with advancements in design refinements and computerization in addition to unification of the many concrete codes and specifications such as the Euro Code. In addition, calls to improvements in concrete education at the university level in the civil engineering departments are frequently being heard (9). These trends are continuing for better improvements and better understanding of concrete structures.

Concrete in Developing Countries

In most developing countries, the concrete sector just like any other sector of industry suffers from many constraints and shortcomings. The most dominant deficiencies of this industry include; lack of good quality materials; lack of well trained labors, shortage in qualified local engineers and technicians, deficiency in know- how technology and total or partial dependence on imported technology. These factors combined with low social development, brain drain and high demand for construction with limited allocated funds and resources put the whole industry in a state of instability.

With these many constraints, it is difficult to achieve the desired goals of good concrete practice. The principles of such practice, however, need to be clearly understood and appropriately applied if real progress is to be made.

In most of the developing countries, there is a huge shortage of houses and other basic infrastructures; traditional construction techniques that are currently in use in many of these countries can not cope with this great demand. Concrete in many developing countries is prepared at site by the laborers or by simple concrete equipment. Ready mixed concrete plants are not wide spread. Lack of funds also contributes to this imminent problem especially in countries in the sub African region and in many regions in Asia and South America. The problem of this shortage in housing and other buildings can only be solved through rapid methods of concrete construction such as prefabricated construction. However, large reinforced concrete panels, walls, floors and roofs used in Europe may not be suitable, due to the high cost of construction, huge capital investment, heavy machinery and equipment needed for production, transportation and erection of components. Therefore, special techniques have to be developed to suit the conditions of these countries. Such systems will result in savings in materials such as cement and steel and cost of construction. There is an urgent need for the ready mixed and precasting systems of concrete production to grow rapidly in the years to come to meet the increasing needs of development. However, the industrialization of the concrete sector may produce social problems of great magnitude in many heavily populated countries.

Appropriate Design

An appropriate design must fulfill certain standards. Designing for strength and durability as well as practicality and economy would be an excellent performance criterion for developing countries. Traditionally concrete is classified by its compressive strength, typically at 28 days; this is not convenient at certain circumstances. Due to long delays in construction in these countries design strength of 90 days may be appropriate. The consequence of such choice would increase the choice of cementing material and allows a more effective use of blended cement leading to more economic concrete mixes. The use of this cement enhances concrete durability also; Moreover, it helps in utilization of the wasted material such as fuel ash and blast furnace slag for better environment (6). Durability is an important design consideration in developing countries in particular. These countries can not afford to build in the first place. The cost of repair of deteriorated structures take up a significant portion of expenditure. Based on conditions of deteriorated structures in the developed countries and the cost of their repair, the author estimates that 50 percent of total construction output may be spent on repair and maintenance related construction in developing countries if the problems of durability are not considered at the out set. Many concrete durability problems are due to inadequate consideration at the design stage. Structural detailing and appropriate material selection are at core of the deterioration problem. Over congested reinforcement leading to difficulty in concerning will definitely lead to premature deterioration. Corrosion of reinforcing steel is the single most important cause of deterioration of concrete structures. Designers should avoid heavy steel reinforcement, especially in compression areas where steel in many cases in not necessary. The author has observed that in many concrete structures he reviewed or redesigned that the amount of access steel is more than one third of the normal acceptable reinforcement. This is a reflection of traditional conservative approach followed by many designers in developing countries. The consequences of such designs on safety of concrete structures are known. Concrete cover is another important design consideration that needs special attention for durable structures. In the Middle East in a study of more than 42 reinforced concrete structures, concrete deterioration was related to inadequate cover (7). A similar investigation carried out by the author (12) on the deterioration of concrete structures near the Mediterranean coast showed similar findings. The strength and durability objectives have to be coupled with practicality and economy of the structure in terms of material availability, labor skills and level of quality control, and initial as well as long term maintenance cost. This is of great importance in the developing counties that lack the financial resources and sophisticated technical know how.

Design Codes

Structural loads and environmental conditions are resisted by the structure that is actually built and not by the structure shown in the drawings. Therefore, no matter how good the design code and the design methods used, it is not possible to produce safe and durable structures unless buildings built properly under excellent construction techniques and good supervision. In many of the developing countries foreign consulting offices from different backgrounds carry out the design of structures. The consulting engineers come from very different backgrounds and different engineering schools. The construction companies that execute the jobs are from different nationalities belonging to different concrete practices. The situation becomes more complicated when the supervision is left to young, newly graduate, inexperienced engineers. The products of such conditions are predictable.

Current local concrete codes for developing countries in many cases are mere adaptation of those in the developed countries. The use of these codes in many situations become difficult and untrained engineers tend to follow them by the letter or avoid them all together and base their decisions on common sense and traditional construction practices. These practices in many cases ignore details and regulations. It is a matter of significance to point out that codes in these countries should be written in a very simple yet comprehensive manner with a few rules so that they could be used in practice. In addition it should be emphasized that risk factors in the developing counties should be somewhat different than in the industrialized countries for many good reasons. The task to write a unified concrete code for developing countries is extremely difficult. This code has to reflect many particularities. These would certainly include educational, technological, social and economical conditions that influence greatly the safety factors, which must be adapted for the local conditions. The engineers and workmanship capabilities have to be considered and taken into account. Design and construction methods used in the developing countries must be incorporated in the code. Design codes in these countries must emphasize, adequate detailing, adequate inspection and supervision, and adequate material control to insure safety, durability and economy.

TECHNOLOGY TRNSFER IN CONCRETE

Technology Transfer

For any country to impart on technology transfer in a certain field of science and technology, it must investigate the existing potentials and capabilities as well as the aims behind such transfer. This has to be done within a time and an economical frame. It may be thought at the outset that concrete technology is the easiest technology to be tackled by the developing countries. This is so because this technology depends mainly on utilization of local materials, local building technology, systems and labors. However, technology transfer is a concept that goes beyond the simple fact of importing technology or know-how. Technology transfer means grasping the essence and know how of the imported technology and gaining the knowledge of adapting it to local conditions, as well as having the ability to absorb any new innovations in the field.

Today many engineers in the developing countries treat concrete as an extension of the indigenous building material of the late twenties. The dominance of the industrialized concrete technology is vivid in the so-called oil rich countries more than others. The high rise concrete structures in Dubai, the great man-made prestressed concert pipe river over 4000 km in Libya and the King Fahd bridge in Saudi Arabia are only few examples which one can sight. These structures are built by foreign technology and know how. The local know-how technologies in such projects are almost non existent. Moreover, the transfer of technology via many projects in these counties did not succeed over the many years. On the contrary many countries are now totally dependent on this imported technology leading to a" fictitious" development condition.

Concrete Education and Research

On the educational level, most universities in the developing countries have good programs in concrete design. Graduates from these schools have the basic background identical to any western university graduate. This is possibly due in part to the staff of the universities who themselves are a product of the western advancements alluded to previously. However, once they graduate, engineers are faced with the actual concrete industry in the country. Many of these engineers can not find a well-established engineering firm and in many cases when they are hired, they are assigned jobs and positions of managers in the public companies. In many cases these managerial jobs are demanding and the newly graduate has no room for progress in what he or she has learnt in the many school years. The responsibility of supervising companies of advanced technology makes the engineer in an embarrassing situation. He has to approve or disapprove and make concrete decisions where he was never trained. Many of these engineers either become full time managers or go back to school. It has been suggested that poor quality of education in concrete technology in developed countries is probably responsible for the deterioration of concrete structures around the world (1,15). It was concluded that while a well-developed culture exits in the concrete industry for provisions of adequate strength, no similar culture exists for provisions of adequate durability. Moreover, Mheta and Gerwick (9) claim that even in the developed world there are serious problems with recent research in the area of concrete materials, construction and design at the prestigious civil engineering institutions all over the world. They call for the recognition of concrete engineering as a profession vital to development of new and better ways meeting structural and civil engineering works needs. They conclude that a lot of ongoing concrete research has only a little or no relevance to the primary needs of the concrete industry. And that most of the current research in concrete has a narrow focus and, therefore, is of limited value and as the industrialized world is now moving towards more and more standardized concrete, innovations and research are strangled. Mehta and Gerwick concluded that for concrete industry to continue meeting the needs of the society in the future, drastic changes in research and education would have to be made.

The situation in the developing countries is similar if not worse. On the educational level, concrete technology is not recognized as a profession. Courses are given on routine basis relying on curricula of western universities without any modifications or renewal. . On the research and development level the situation is not good at all. There is very little amount of money allocated to the concrete advancements in the developing countries compared to those in the industrialized nations. Even at research centers in the universities if they exit at all, the funds are insignificant and when allocated they are likely to be spent on any thing but research. Most research done at universities and research centers in these countries is insignificant and in many cases irrelevant and is not directly oriented towards appropriate technology for these countries. Changes are called for in education and research meet the development needs in these countries.

Appropriate Concrete Technology

The word appropriate technology may have different meaning to different audiences. In developing countries that are tangled in this dilemma of technology transfer for many decades, there are at least two main diverse ideas about appropriate technology. First, the concept that total transfer of advanced industrialized technology and mere adaptation of this technology would bring about the needed desired development. The second point of view is to concentrate on sustainable development based on local and traditional means of technologies using indigenous materials including concrete. The first approach ignores the local conditions, while the second approach embraces these conditions entirely. It is the authors' point of view that neither of these approaches would succeed to bring about the desired development. The first approach was applied in many oil-rich countries and after several decades, there was no real technological base for concrete in these countries. The second approach was applied in several poor African and Asian countries with limited success.

It is suggested that with regard to concrete technology, appropriateness means middle of the road technology. Neither do we advocate the use of advanced technology nor do we advocate use of primitive substitute construction technology. Our understanding of appropriate concrete technology is to take the benefits of both. For example; High Performance Concrete technology may be used to upgrade the low technical concrete production to levels where durability becomes central to concrete making. Upgrading the local technologies, quality of materials, local labors and training of engineers to supervise these developments will do this. Designs must not be a mere adaptation of other designs regardless of their sophistication, but should be in harmony with local and environmental conditions, designs which are simple yet comprehensive, easy to construct, utilizing local materials and using limited funds. The final objective of course, is to guarantee that concrete structures built in developing counties are durable, safe and affordable by the great majority of inhabitants.

Thus, there is a definite need for an appropriate concrete technology for developing countries. These countries need to establish the means for concrete technology transfer without committing the traditional mistakes of merely copying the developed technology or using very primitive indigenous materials that are not suitable for solving these imminent construction problems. Low cost substitute material such as vegetable fibers, wood, sugar cane baggasse and bamboo for reinforced concrete were actually tried as long ago as 1914 in many countries such as China, Japan, Philippians, India, Egypt, Columbia, Thailand and even in the US. (3). The use of these materials as a substitute for concrete is debatable. It is not the purpose of this paper to dismiss this approach without consideration; however, it is now believed that only through use of concrete technology, the construction industry in developing counties may progress.

CONCLUSIONS

Concrete started as a mere mix of three basic materials; cement, aggregate and water at the beginning of this century. Unfortunately, in many developing countries it remains to be so. By the end of the twentieth century, concrete has become highly complex mixture and significant improvement on strength, durability are made. Nevertheless, the basic principles of concrete making will continue in the years to come. In developing countries in particular, good concrete practice needs to be clearly understood and practiced and the new improvements need to be comprehended and when possible adapted, if concrete is to continue to service the human race effectively. Appropriate concrete technology does not mean the use of primitive indigenous materials as might one traditionally would think. On the contrary, it is believed that only through adaptation of advanced, yet appropriate methods would the developing world be able to face the great demands for concrete construction currently and in the years a head. Developing countries need to take advantage of the long experience of the developed world. Changes in the education system related to concrete and research is called for real progress.

In short, developing countries need huge amounts of concrete for development. They need a concrete industry capable of producing ready mixed, precast, durable, strong, self-flowing, self-cured, low shrinkage, and low cost concrete. Reinforced concrete with small amounts of steel reinforcement or other alternatives. Sustainable concrete that is suitable for hot weather and severe environment conditions. Education and research efforts in the future should be focused towards fulfilling these needs. Several specific steps could be considered necessary to facilitate and accelerate concrete progress in these counties, these include:

1) Updating educational programs where concrete technology becomes a specialization and a profession.
2) Establishing new research centers for appropriate concrete technology and allocating the needed funds for appropriate research.
3) Introduction of ready mixed concrete industry in many developing countries.
4) Unification of concrete codes in developing countries on regional basis.
5) Establishing new regional and local training centers for training of local labors as specialists in concrete technology
6) Adapting new emerging technologies without delay on research basis and provide programs and seminars for continuing education and holding conferences dedicated solely to such topics of appropriate technologies.

These efforts and hopes difficult as they might be- have to be considered seriously, if significant advancement and development are to be made.

REFERENCES

1. Aitcin, P; Neville,(1993) A High performance Concrete Demystified, National Sciences and Engineering Research Council of Canada.

2. EL-Aboud, M and Swamy,R.N(1993) ' Light weight Sandwich Panels for Low Cost Housing' Proceedings, of the Third International Conference on Concrete for Developing Countries', Tripoli, Libya, Editor, Ngab, A, S.

3. Fang, H.Y '(1981) " Low Cost Construction Materials and Foundation Structures' Second Australian Conference on Engineering Materials, Sydney.

4. Idron, G,(1994) "Modeling in Applied and Strategic Concrete Research, Keynote address at NATO/RILEM Workshop on the Modeling of Microstructure and Potential for Studying Transport Properties and Durability, Paris.

5. Koca,C. Abit, O (1996)" Development of Ready Mixed Concrete Industry in Turkey, Fourth International Conference on Concrete for Developing Counties, Gazimangusta, North Cyprus.

6. Malier,Yvea (1992) " High Performance Concrete from Material to Structure" .

7. Maslehudin, M. (1994) Concrete Durability in very Aggressive Environment Proc. ACISP.144 American Concrete Institute, Detroit, pp191-211.

8. Mehta,P.K. and Monteiro (1993).J.M. Concrete: Structure, Properties and materials Second Edition, Prentice Hall, p. 548.

9. Mehta, P.K and Gerwick, B.C (1996) '" Concrete in The Service Of The Modern World', Proceedings of the International Conference" On Concrete in the Service Of Mankind", University of Dundee, Editors, R. Dhir, and, M. McCarthy.

10. Morris, J. and Sephton, S.(1996) ' " the Role of Cement and Concrete Industry in Developing Countries." Proceedings of the International Conference" On Concrete In the Service of Mankind", University of Dundee. Editors, R. Dhir.

11. Newman, K. (1983) The Role of Ready Mixed Concrete in Developing Countries.

Proceedings of the First International Conference on Concrete Technology for Developing Countries. Amman.

12. Ngab, A. Barony, S. and Hatoush, Z, (1990)'" On Deterioration of Concrete Structures in Libya'", Proceedings of the International Conference on 15 Th. on Our World On Concrete and Structures, Singapore.

13. Trikha, D.N (1996) Ultra Structures –Indian Experience. Proceedings of the International Conference" On Concrete In The Service of Mankind", University of Dundee. Editors, R. Dhir, and, M. McCarthy.

14. Pomeroy, C.D (1991) Whither Concrete? . Second Australian conference on Engineering Materials, Sydney. Editor, D. J. Cook.

15. Wang, J, Knight, K. Swann, L (1996) " Appropriate Concrete Design" Proceedings of the International Conference" On Concrete in the Service of Mankind", University of Dundee. Editors, R. Dhir, and, M. McCarthy.

16. Wittmann, F, Gebauer, J. and Torrent, R.(1996) " The Versatility of Concrete" Proceedings Of the International Conference On Concrete In The Of Service Mankind, University Of Dundee., Editors, R. Dhir, and, M. McCarthy.

Strength Development of High Strength High-Performance Concrete at Early Ages

T. Yen [1], K.S.Pann [2], Y.L.Huang [3]

[1,3] Department of Civil Engineering, Chung-Hsiung University,
Taichung, Taiwan
[2] Department of Civil Engineering, Cheng-Hsiu Institute of Technology
Kaohsiung, Taiwan

ABSTRACT

This study aims to investigate the characteristics of the early strength development of HPC. Experimental tests were conducted on concrete with a 28 days' strength of 56 Mpa starting from 6 hours after mixing. The test methods include ultrasonic pulse velocity measurement and maturity method in addition to the standard compression test. Seven curing ages (6hrs, 12hrs, 1day, 2days, 3days, 7days and 28days) were selected for the tests.

Test results show that the increase rates of the strength growth of HPC are significantly different before and after 1 day of curing age. By selecting the curing age of 1 day as the separating point, a prediction model is proposed to describe the strength development in the two separated stages and is further verified by the tests. Moreover, compared to the test results, it is found that the prediction model releases better agreement than those obtained from the method of maturity.

KEYWORDS

high performance concrete, strength, curing temperature, early age, wave speed, maturity

INTRODUCTION

High-performance concrete, HPC, is a new generation of concrete. Its high strength and high flow ability are the two basic characteristics that are usually mentioned at the same time [1]. HPC has multiple advantages, such as improving environment and construction technique. It will become a major construction material in the modern trend for higher construction quality.

Construction work is usually required to speed up and remove shoring and form work sooner under the pressure of reducing cost. Yet construction disasters and engineering flaws may occur due to insufficient strength of concrete from these acts. HPC has high flow ability and high strength to last a long period, but the initial strength after casting is relatively small like conventional concrete. It is therefore quite important to understand and denominate the strength at early age.

This research aims to investigate the strength prediction model for concrete at early age using ultrasonic wave and maturity method. The speed of ultrasonic wave is quite related to the constitutions of concrete [2-3]. Since the concrete at early age has rapider hydration and the wave speed of the hydration product C-S-H is faster than that of water and void, the wave speed of the early-age concrete increases more quickly.

Temperature and time are two major control factors for the development of concrete strength, especially, the concrete strength at early age is greatly affected. The concrete under higher curing temperature will reach the maximum temperature earlier. The Maturity Method [4,5] was developed basing on this concept. Till now, almost all predictions of concrete strength use water-cement ratio as main factor [6]. The others such as gradation of aggregates, property of cement and content [7-9] are often regarded as minor affecting factors. Either water-cement theory or maturity prediction has focused on the long term strength basing on the variations of stress and elastic modulus to form prediction models [10]. Yet the research reports about the strength at early age have rarely been seen.

Right after concrete casting to about 1 day age, the hydration rate of cement is the most rapid. After one-day age, the formation of hydration heat slows down and the strength development of paste becomes slow, too. Recent researcher [11] has proposed a two-period model to predict the strength of conventional concrete at early age. According to the strength developing process, this paper establishes a strength prediction model of concrete at early age based on the maturity concept, covering concrete from mixing to 56 days age, different curing temperatures and ages of paste.

EXPERIMENTS

Materials

Type I Portland Cement is used, the maximum size of coarse aggregate is 1/2", the fineness modulus of fine aggregate is 2.9, and the superplasticizer of HICON HPC100 which meets the requirements of ASTM G-Type is used.

Variables

The variables considered in serial tests are shown in the following:
Curing temperature: $T_{cured}=10^0C$, 23^0C(ambient temperature), 40^0C
Water/binder ratio: $w/(c+p)$=0.33, 0.36, 0.40
Age: t=6, 12 hrs, 1 day, 2, 3, 7, 14, 28 and 56 days
Overfill ratio of paste: n=1.5, 1.6, 1.8
ratio of cement replaced by fly ash: 0%, 10%, 20%, 30%

Where w stands for the weight of water, c is the weight of cement, p is the weight of pozzolanic material, and n is the ratio of the volume of paste to the voids between aggregates. The mix proportions of HPC are shown in Table 1.

Specimens and Test Methods

The size of cylinder specimen for compression test is $\phi 10 \times 20$cm. They are tested according to ASTM C109 specifications. The variation of internal temperature was measured with a thermometer within 48 hours after concrete casting. Readings were taken every two hours to calculate the maturity and related compressive strength. The heights of the specimens were measured, the both ends were smeared with Vaseline and the wave speeds were detected before every compressive experiment was conducted.

TABLE 1
THE MIX PROPORTIONS OF HPC (kg/m^3)

w/(c+p)	0.33	0.36	0.36	0.36	0.40
Overfill ratio of paste (n)	1.6	1.8	1.6	1.5	1.6
Cement	395	435	369	333	334
Fly ash	158	147	157	163	158
Slag	17.2	22.8	19.4	17.5	17.6
Sand	633	589	629	652	633
Aggregate	967	901	962	996	967
Water	167	200	180	167	189
S.P.	20.8	18.3	16.4	17.9	15.3

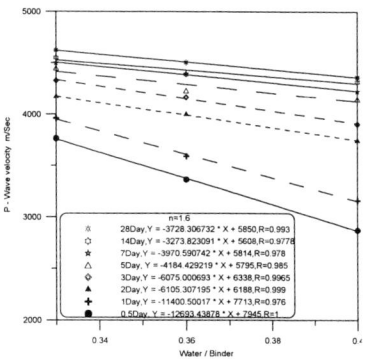

Fig.1 Relationship between w/(c+p) and wave speed

RESULTS AND DISCUSSIONS

Wave Speed of HPC

During early age, the hydration product, C-S-H, forms quickly to fill the voids in paste. Therefore, the increase rate of the wave speed at the early age is larger than in any later ages. As can be seen in Table 2, the largest increase rate of wave speed occurs at the age of one day. The relationship between curing age and wave speed of HPC with n equal to 1.6 and cured at 23°C is shown in Fig. 1. According to the slope of the regressive line in Fig. 1, an earlier-age concrete possesses a larger increase rate of wave speed whatever the water-paste ratio is and the maximum increase rate occurs at the age of one day or 0.5 day. Basically, the age of one day is the separating point at which the increase rate of wave speed changes.

TABLE 2
WAVE SPEED AND INCREASED RATE OF CONCRETE AT VARIOUS AGES

ages w/(c+p)	0.5 day	1 day	2 day	3 day	5 day	7 day	14 day	28 day
	Wave speed (m/sec)							
0.33	3763	3962	4170	4326	4448	4499	4550	4622
0.36	3365	3591	3996	4165	4280	4373	4450	4505
0.40	2873	3161	3683	3903	4147	4222	4315	4361
	Increased wave speed (m/day)							
0.33	---	398	207	156	61.1	25.3	7.30	5.15
0.36	---	452	405	169	57.5	46.5	11.0	3.92
0.40	---	577	521	219	122	37.6	13.3	3.24

Development of the Compressive Strength of HPC

The developments of compressive strength of HPC are summarized as shown in Fig. 2. By observing, it is found from the figures that before one-day age, except for the one with curing temperature of 10^{0}C, the strength of HPC is far greater than the increase rate after one-day age. In addition, the lower the water-cement is, the greater the strength gain becomes. At 0.5 day age and curing temperature of 23^0C and 10^0C, most strength increase rates do not exceed 1% of that at 56 days age. While for the curing temperature of 40^0C, the strength increase rate at 0.5 day age has reached about 2%. It means enhance the curing temperature can stimulate the development of strength at early age.

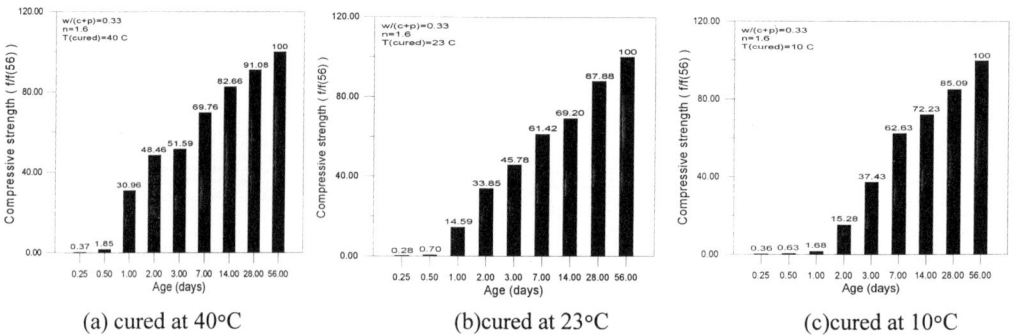

(a) cured at 40°C   (b) cured at 23°C   (c) cured at 10°C

Fig.2 Strength growth of HPC versus curing age

Effects of Curing Temperature on Strength of HPC

Fig. 3 indicates the strength development of HPC. It can be observed that all the strengths of HPC increase with age in a parabola. High and low curing temperatures have certain effects on the strength of HPC. Under low curing temperature of 10^0C, the strength of HPC obviously decreases as shown in Fig. 4. Under high curing temperature of 40^0C, the strength of HPC increased within early age but the strength after 7 days age does not increase. It is concluded that high curing temperature is in favor of strength of HPC for early age but not for the long age.

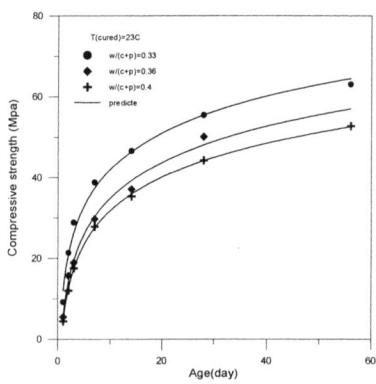

Fig.3 Development of compressive strength of HPC

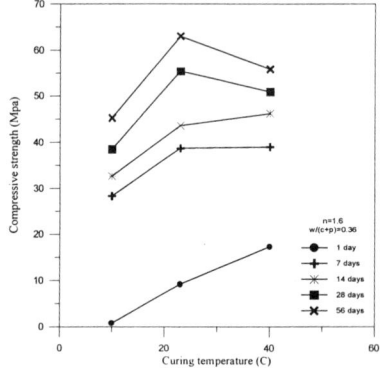

Fig. 4 Effects of curing temperatures on the compressive strength of HPC

Temperature Change inside Concrete

The temperature changes inside concrete from casting to 48 hours age under three curing temperatures for HPC with water/binder ratio=0.33 are shown in Fig. 5. In Fig. 5(a), the curing temperature is 23^0C, similar to the trend of hydration heat releasing curve of cement, the concrete has two peaks on the temperature curve. The second peak occurs within 20 to 24 hours after casting. Fig. 5(b) shows the temperature change under curing temperature of 40°C. It indicates the temperature inside concrete increases rapidly right after mixing due to high curing temperature. The second peak occurs within 12 to 20 hours after casting, which is earlier than that under curing temperature of 23°C. Fig. 5(c) shows the result of 10°C curing temperature. It differs from the former in the occurring of the second peak, which delays to within 35 to 45 hours after casting. It can be thus drawn that the higher the curing temperature is, the sooner the second peak reaches.

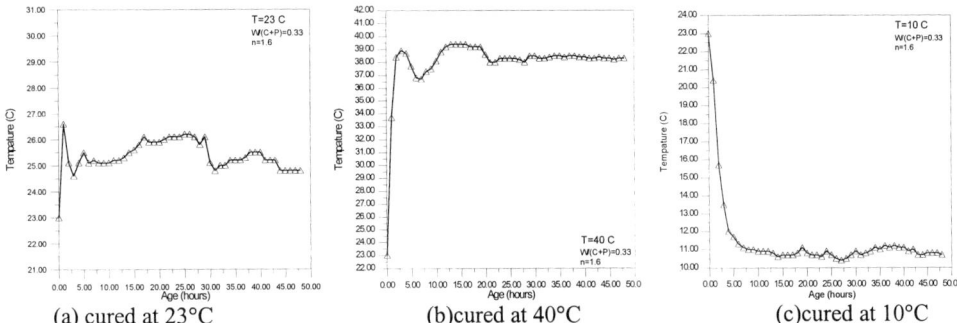

(a) cured at 23°C   (b) cured at 40°C   (c) cured at 10°C

Fig.5 Temperature change in the interior of concrete

Prediction by Maturity Method

According to Nurse-Saul Function method suggested by ASTM C 1074, the maturity of concrete can be determined by temperature and time factors as following equation:

$$M = \sum_{0}^{t}\left[(T - T_0) \times \Delta t\right] \quad (1)$$

where M = maturity when age is t
T = average temperature of concrete in the period of Δt
T_0 = basic temperature (when curing temperature is within 5 ~ 43°C, for Type I cement without any admixture, T_0 = -10°C; if curing temperature is not in this range, T_0 = 4.4°C)

According to Plowman[12], the relation between maturity and strength can be represented as follows:

$$\frac{S_e}{S_{28}} = A + B \times \log\frac{M}{1000} \quad (2)$$

where S_e = compressive strength when maturity is M
S_{28} = compressive strength of 28 days
A, B = constants
M = maturity (F-hr)

Prediction Model for Strength of HPC at Early Age

From the tests, it is found that except for the curing temperature of 10°C all other cases have relatively large slope change of curve at all ages before 1 day. The slope change between 1 day and 7 days reduces gradually. After 7 days, the slope change is even smaller. Most strength increase rates of HPC at before and after 1 day age have significant difference which is similar to the heat evolution of

Portland Cement [13]. Similar result is also observed in the wave speed detection. Accordingly, taking 1 day as the separating point is acceptable.

Summing up the above analysis and discussions with reference to the reports of the hydration of cement on strength development by Young[13] and Oluokun[14], it can be estimated roughly that the strength of HPC develops rapidly before 1day age. After 1 day, it reduces gradually so no any single curve can be used to reasonably predict the strength development of concrete. This paper divides the curing age from casting to 56 days into two periods. First period starts from 0.25 day to 1 day (2 days when curing temperature is 10^0C). The second period starts from 1 day (or 2 days) to 56 days. The first period is depicted by the equation (3) and the second period by the equation (4):

$$y = a \cdot x^b \qquad (3)$$

$$y = a \cdot (x-c)^b \qquad (4)$$

Substitute in all test data and run linear regression to get coefficients in the above equations as shown in Table 3.

TABLE 3
COEFFICIENTS OF A, B AND C IN Eqs. (3) AND (4)

Curing Temp. T(°C)		40	23	23	23	23	23		10
w/(c+p)		0.33	0.33	0.36	0.36	0.36	0.40		0.33
Overfill ratio of paste (n)		1.6	1.6	1.8	1.6	1.5	1.6		1.6
0.25 ¤ 1 day	A	14.173	6.430	4.953	5.268	3.873	3.390	0.25 day	1.295
	B	3.1999	2.8398	2.7493	2.4996	2.3336	2.5578	¤	1.7651
	C	0	0	0	0	0	0	2 day	0
	R^2	0.9762	0.9095	0.9519	0.9983	0.8949	0.9369		0.9132
1 ¤ 56 days	a	26.997	23.433	17.919	14.971	17.29	14.239	2 ¤ 56 day	16.375
	b	0.1887	0.2512	0.3371	0.364	0.2933	0.3336		0.2714
	c	0.9154	0.978	0.9546	0.9356	0.9834	0.9754		1.95
	R^2	0.9878	0.9931	0.9991	0.9981	0.9965	0.9911		0.9916

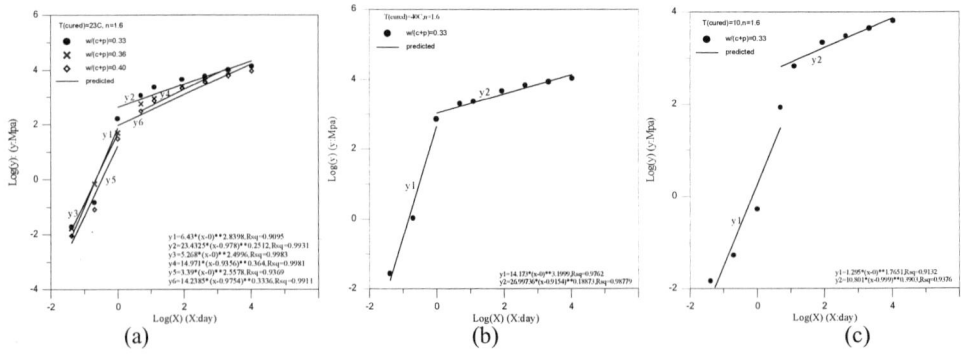

Fig.6 Strength Development under (a)23°C (b)40°C (c)10°C curing temperature

Fig.6 shows the prediction curve. In average, for the case under curing temperature of 10^0C, the predicted strength of concrete within 1 day age is much off. The other two curing temperature cases of 23^0C and 40^0C have more acceptable predicted results. For 1 day and over to 56 days age cases, all three have fairly reliable results.

Comparison of Measured Results with the Predicted Strength from Maturity Method and Prediction Model

The predicted strength from maturity method and the prediction model and the measured strengths are listed in Table 4. The strength percentage by maturity method shows a negative value before 6 hours age, which is obviously not an applicable result. After 12 hours age, under curing temperature of 10^0C, the predicted strength at 1 day age by maturity method still has large deviation. For curing temperature of 23^0C and 40^0C, the predicted strength at 1 day over to 2 days has smaller deviation. Comparing the predicted strengths of the prediction model and the maturity method with the measured strengths, we can found that the prediction model is obviously superior to maturity method, because the former has smaller deviation.

TABLE 4
COMPARISON OF THE STRENGTH OF MEASURED, MATURITY METHOD
AND PREDICTION MODEL

w/(c+p)	n	T(°C)	Age (hr)	Strength ratio (%)			
				Measured	Prediction	Maturity Method	
						M(°F-hr)	Se/S28
0.33	1.6	10	6	0.42	0.29	270.81	-2.78
			12	0.74	0.99	497.61	2.41
			24	1.98	3.36	945.99	7.88
			48	17.96	11.43	1846.71	13.58
		23	6	0.32	0.23	379.71	-5.49
			12	0.79	1.62	759.24	7.49
			24	16.61	11.60	1532.97	20.65
			48	38.52	40.04	3059.28	33.59
		40	6	0.40	0.33	499.14	-6.20
			12	2.03	3.03	1014.66	12.89
			24	33.99	27.83	2070.99	32.09
			48	53.20	53.37	4159.71	50.85

CONCLUSIONS

According to the tests, analysis and discussions, the following conclusions are drawn:

1. Earlier-age HPC has larger increase rate of wave speed, especially before the age of 1 day. After 3 day, the increase rate becomes smooth apparently.
2. For HPC with different water/binder ratios, under ambient curing temperature (23^0C), the strength develops the fastest within 1 day from casting. The development of strength then reduces obviously after 1 day.
3. The strength development of HPC is related to the water/binder ratio and is affected by the curing temperature. It is found that only curing temperature above the ambient temperature can adequately stimulate the development of strength at early age but not for the low curing temperature. However, this effect becomes less significant after 7 days.
4. For HPC under ambient curing temperature (23^0C) and 40^0C, the two-period prediction model

using one-day age as the separating point is proven to have fairly good reliability. As for curing temperature of 10^0C case, 2 days are taken as the separating point for the prediction model.
5. For the strength of HPC within one-day age, although the suggested two-period prediction model has relatively large deviation, it has rather better accuracy than the presently used maturity method.

REFERENCES

1. Stphan, D.E. (1994). High Performance Concrete, *Proceedings of International Workshop on HPC*, Bangkok. Thailand.

2. Lai C. P. (1998). *Effects of the Composition of Concrete on it's Flow Properties and the Ultrasonic Pulse Velocity in Concrete*, Ph.D. dissertation, National Chung-Hsiung University, Taiwan.

3. Stephen P.Pessiki and Matthew R. Johnson. (1996). Nondestructive Evaluation of Early-Age Concrete Strength in Plate Structures by the Impact-Echo Method, *ACI Materials Journal*, **Vol.93, No.3**, 260-271.

4. ASTM C1074-87. (1992). Standard Practice for Estimating Concrete Strength by the Maturity Method, *Annual Book of ASTM Standard*, **Vol.04.02**.

5. Nicholas J. Carino, *The Maturity Method*, CRC Handbook on Nondestructive Testing of Concrete, 101-143.

6. Gilkey, H.J. (1961). Water-Cement ratio versus strength another look, *J. of the ACI*, **V.57, No.10**, .1287-1312.

7. Walker.Snton and Bloom. (1960). Effect of Aggregate size on properties of concrete, *ACI Journal*, **V.57, No.9**, 283-298.

8. Aitcin,P.C. and Metha,P.K. (1990). Effect of coarse Aggregate characteristics on mechanical properties of high strength concrete, *ACI Materials Journal*, Mar.-Apr., 103-107.

9. Williams,R.I.T. (1962). Effect of cement content on the strength and elastic properties of dry lean concrete, Technical Report No.TRA/323, *Cement and Concrete Association*, London, Nov., 28.

10. Hirsch,T.J. (1962). Modulus of Elasticity of Concrete Affected by Elastic Modulus of Cement Paste Matrix and Aggregate, *Journal of American Concrete Institute*, **Vol.59, No.3**, 427-452.

11. Yen, T. (1993). The Development of Concrete in Early Age, *Proceedings of 17th National Conference on Theoretical and Applied Mechanics*, 817-821.

12. Plowman, J.M. (1956). Maturity and Strength of Concrete, *Magazine of Concrete Research* (London), **V. 8, No. 22**, Mar., 13-22.

13. Young, J.F. (1981). Hydration of Portland Cement, *EMMSE*, Material Research Lab. PA.

14. F.A. Oluokun, E.G. Burdette, J.H. Deatherage. (1990). Early Age Concrete Strength Prediction by Maturity-Another Look, *ACI Materials Journal*, Nov.-Dec., 565-572.

COMPRESSIVE BEHAVIOUR OF CONCRETE CONFINED BY ARAMID FIBRE SPIRALS

H. Y. Leung [1] and C. J. Burgoyne [2]

[1] Department of Building and Construction, City University of Hong Kong,
Tat Chee Avenue, Kowloon, Hong Kong
[2] Department of Engineering, University of Cambridge, Trumpington Street,
Cambridge, United Kingdom

ABSTRACT

The paper reports the experimental results of compressive tests on concrete cylinders wrapped by aramid fibre spirals. Both circular concrete cylinders with a single spiral and rectangular concrete specimens with two interlocking spirals are investigated. Concentric compression tests were carried out. It was found that the behaviour of confined concrete is influenced not only by the concrete strength, but also the spiral leg spacing and the degree of interlocking. Concrete cylinders with close spacing and a high degree of interlocking usually gave higher strength and ductility. Experimental results are compared with analytical data.

KEYWORDS

Concrete, compression, aramid fibre, spiral, interlocking spirals, passive confinement.

INTRODUCTION

It is generally accepted that plain concrete exhibits a brittle failure when it is compressed, which leads to a rapid loss of load-carrying capacity. The concept of using spiral reinforcement is to restrain the concrete from expansion and thus delay the failure. In the past, a lot of research has focused on using steel spirals and rings to confine concrete (Ahmad and Shah 1982, El-Dash and Ahmad 1995, and Mander *et al* 1988). The increase in ductility and strength were prominent. With the advent of composite materials, replacement of steel by fibre-reinforced-plastic (FRP) seems to be a rational method to solve the corrosion problems. In addition, due to the difference in the stress-strain behaviour of FRP and steel, the induced confining pressure is also different when subjected to compression. The stress of steel remains virtually constant after its yield point so the induced pressure cannot increase after yielding. On the other hand, FRP possesses linear elastic properties. The stress of the FRP confinement keeps on increasing with strain, and thus a monotonically increasing confining pressure is produced. The maximum confining pressure is obtained when the ultimate strength of the confinement is reached. When further load is applied, failure of the confined concrete often occurs as a result of fracture of the confinement reinforcement. In order to verify this, an experimental programme was carried out and the results are described in this paper.

EXPERIMENTAL DETAILS

Single Spiral

Kevlar 49 aramid fibre was used to make the spirals. Ten yarns were used to form a bundle which was then formed into a spiral by impregnating with resin and wrapping round a mandrel (Leung 2000). The mechanical properties of the aramid bundle are listed in Table 1.

TABLE 1
MECHANICAL PROPERTIES OF ARAMID SPIRALS

Material	Cross-sectional area	Ultimate load	Ultimate stress	Ultimate strain	Elastic modulus
Kevlar 49 aramid (ten yarns)	1.75 mm^2 (excluding resin)	2.42 kN	1.38 GPa	1.53 %	90.1 GPa

In order to study the effect of spiral pitch on confined concrete, single spirals were fabricated with different spacings, (10 mm, 20 mm 35 mm and 50 mm), but the total height was fixed at 194 mm. Typical spirals are shown in Figure 1. Three polythene rods, which have a negligible contribution due to their low elastic modulus (4 GPa), were used to hold the spiral in space.

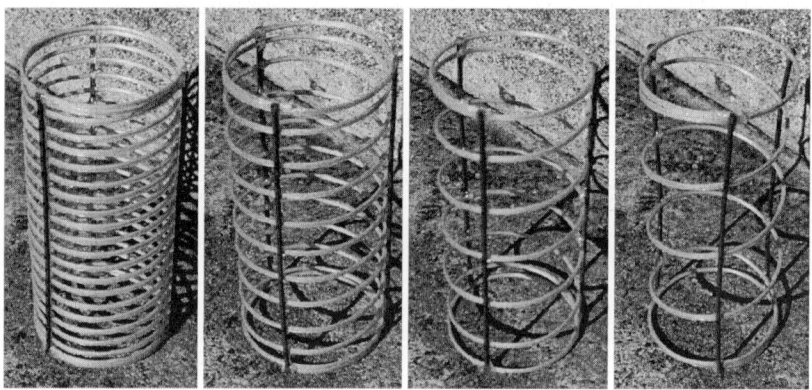

Figure 1: Aramid fibre single spirals

Double Interlocking Spirals

If the spirals are to be used to reinforce non-circular sections, some form of interlocking is required. To investigate this problem, two spirals with a fixed pitch of 20 mm, but with different centre-to-centre spacing were also tested. Typical interlocking spirals can be seen in Figure 2 which also shows the polythene rods used to maintain the necessary configuration.

With the diameter of the spiral (d_{sp}) fixed at 90 mm, five different centre-to-centre distances of adjacent spirals c_{sp} were utilized (36, 45, 57, 69 and 81 mm). These correspond to c_{sp}/d_{sp} ratios of 0.40, 0.50, 0.63, 0.77 and 0.90 respectively.

Concrete Specimens

The height of the concrete mould was 200 mm and the height of the spiral was 194 mm, so a thin 3 mm concrete cover could be provided at both ends. A 5 mm cover was provided around the outside giving an overall diameter of 100 mm. More details can be seen in Figure 3. It should be noted that the slenderness ratio adopted here is 2.0. This value was found to be the limiting value for the end effects to come into play (van Mier et al 1997).

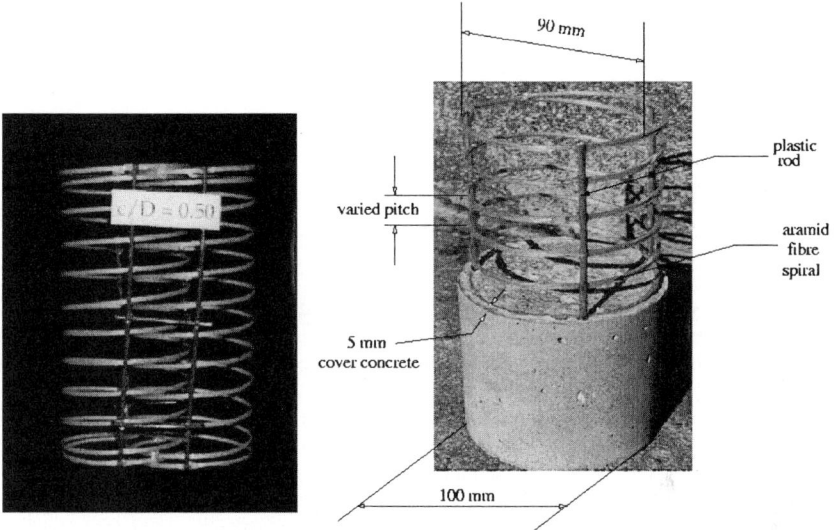

Figure 2 Double interlocking spirals   Figure 3: Details of specimen

Concrete cylinders with a single spiral and rectangular specimens with two interlocking spirals were cast. The concrete had a design strength of 40 MPa at 28-days. The concrete mix details are given in Table 2. The maximum coarse aggregate size was restricted to 5 mm, so that the aggregate particles could easily pass between the legs of the spiral. Ordinary Portland Cement was used with water/cement and aggregate/cement ratios of 0.45 and 4.5, respectively.

TABLE 2
CONCRETE MIX DESIGN

Material	Quantity, kg/m^3	Percentage by weight, %
aggregate (5 mm)	973	43.7
sand	703	31.6
OP cement	373	16.8
water	179	7.43

All the concrete specimens were cast in three lifts and each lift was followed by vibration on a shaking table. The specimens were then stripped from their moulds after one day and covered with plastic sheeting until the designated testing dates.

Forty two single spiral samples were tested and 12 with double interlocking spirals.

Loading Arrangements and Testing Procedures

All concrete specimens were tested in a 2000 kN concrete cube testing machine in a displacement controlled mode so that loading and displacement past the peak could be recorded. The loading rate adopted for all test was 0.3 mm/min. All specimens were preloaded with a load of 10 kN when the displacement reading in the data logger was adjusted to match that on the test machine. The test was continued until snapping of the spiral occurred.

For circular concrete specimens with a single spiral, tests were made at 8 days, 15 days and 28 days to obtain different compressive strengths. For the interlocking-spiral specimens, testing was carried out about 40 days after casting.

RESULTS AND DISCUSSIONS

Observations

For the concrete cylinders with a single spiral, it was observed that the longitudinal cracking of the cover concrete commenced near the mid-height of most samples. Such cracks gradually propagated to both ends with increasing load. When the load reached the concrete compressive strength, severe spalling of the cover occurred. However, it was noted that complete disintegration of the cover concrete did not occur. The cover failure was more noticeable in cylinders with closely-spaced spirals which physically separate the cover from the core. At the ultimate stage, fracture of the spiral, which occurred at mid-level of specimens, was sudden and explosive. Severe internal cracking was also noticed by examining the failed samples. It was observed that the core concrete has been reduced to fine rubble. Specimens after failure are shown in Figure 4.

Figure 4: Concrete specimens at failure (single spiral)

For concrete specimens with double interlocking spirals, the breaking of the interlocked spirals was found to be less explosive due to the shielding effect of the cover concrete which varied in thickness. As with the case of a single spiral, no change in the spiral was visible prior to fracture, but extensive cracking could be observed. Typical concrete specimens at failure are shown in Figure 5.

Figure 5: Concrete specimens at failure (double spirals)

Loading response

The experimental axial load-displacement profiles for single-spiral specimens are presented in Figure 6. All graphs are plotted to the same scale. The numerical results are also shown in Table 3. All data are averages of two or three specimens and the standard deviation (std.) of the data is also included. The theoretical prediction, which modifies the active confinement model (Kotsovos and Pavlovic 1995), is also given for comparison in the same table and figure (Leung 2000).

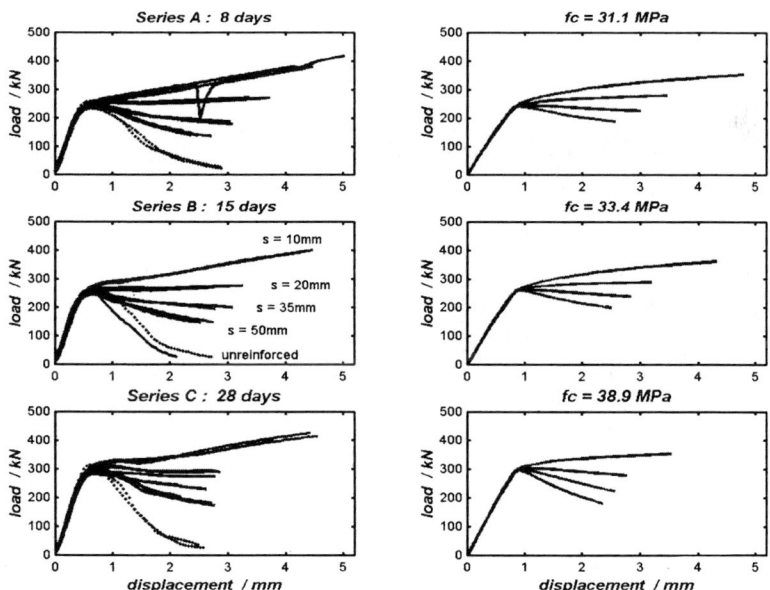

Figure 6: Experimental (on the left) and theoretical (on the right) results of concrete with a single spiral

TABLE 3
EXPERIMENTAL AND THEORETICAL DATA FOR CONCRETE CYLINDERS WITH SINGLE SPIRAL SUBJECTED TO UNIAXIAL COMPRESSION

Date of testing	Pitch mm	Maximum load kN	Displ. at max. load mm (std.)	Breaking load kN (std.)	Displ. when spiral breaks mm (std.)	Theoretical breaking displ. mm
8 days [A series]	plain	243.9	0.622 (0.02)			
	50	245.3		137.9 (1.39)	2.510 (0.18)	2.562
	35	242.4		184.4 (5.34)	2.990 (0.13)	3.007
	20	269.8		270.5 (2.07)	3.595 (0.20)	3.452
	10	392.9		392.9 (21.5)	4.555 (0.44)	4.790
15 days [B series]	plain	262.3	0.629 (0.05)			
	50	252.1		149.9 (4.65)	2.560 (0.17)	2.497
	35	249.9		197.8 (3.03)	2.877 (0.18)	2.851
	20	271.3		272.2 (3.70)	2.914 (0.30)	3.195
	10	389.2		391.0 (11.8)	4.170 (0.43)	4.303
28 days [C series]	plain	305.8	0.643 (0.05)			
	50	294.4		186.1 (16.4)	2.570 (0.34)	2.348
	35	287.9		237.7 (7.02)	2.637 (0.03)	2.555
	20	301.3		283.9 (9.00)	2.816 (0.04)	2.768
	10	412.8		412.2 (15.5)	4.240 (0.42)	3.543

Figure 6 shows that all specimens with a single spiral gave higher strength and strain values compared with unreinforced concrete cylinders, but the pre-peak behaviour is almost unaffected by the lateral reinforcement. Decreasing the spiral pitch increases the strain capacity of the specimens but only for the closest spacing is there a significant increase in failure stress. The lateral pressure created by the restraining effect of the spiral becomes larger at reduced pitch, so the confined concrete cylinder can sustain more loading before failure of the spiral occurs (see the last two columns of Table 4). However, this restraining effect does not become significant until the concrete cylinder approaches its ultimate axial compressive strength. The effect is delayed when higher concrete strengths are used; the specimen is stiffer, so the lateral expansion does not occur until later in the loading process. Thus, in series C with 10 mm pitch, there is a slight reduction in stress after the initial peak before the subsequent gain in strength caused by the spirals.

Supposing that the axial strain of plain concrete at peak load and the breaking strain of spirally confined concrete represent their ultimate values, the ultimate strain for confined concrete improves by about 4 times (for s = 50mm) and up to 7.5 times (for s = 10mm) compared with plain concrete (series A in Table 4). Figure 6 also shows that concrete strength strongly affects the magnitude of both the maximum strain and strength. When concrete with higher uniaxial strength is used, the strain magnification factor reduces slightly (see series C). Higher strength concrete cylinders display less ductility than the lower strength concrete cylinders. Lower strength concrete has a lower axial modulus and expands more transversely, which leads to longitudinal cracks and lower strength. But this behaviour allows more benefit to be achieved from the presence of spirals. In contrast, higher strength concrete usually gives lower ductility because high strength concrete itself is rather stiff and brittle. When high strength concrete cylinders are tested under uniaxial compression, comparatively little lateral expansion should be expected. A lower tensile force will then be generated in the spiral. When the high strength concrete cylinder reaches its ultimate state, disruption of the core concrete is found to be sudden and explosive due to the release of the large amount of energy inside the concrete matrix.

For the concrete specimens with interlocking spirals, the experimental data can be found in Table 4. The experimental results and the theoretical model are shown in Figure 7.

TABLE 4
EXPERIMENTAL AND THEORETICAL DATA FOR CONCRETE CYLINDERS WITH DOUBLE INTERLOCKING SPIRALS SUBJECTED TO UNIAXIAL COMPRESSION

c_{sp}/d_{sp}	Experimental breaking displacement (mm)	Theoretical breaking displacement (mm)
0.40	2.57	2.67
0.50	2.48	2.51
0.63	2.32	2.36
0.77	2.19	2.23
0.90	2.14	2.19

The specimen size increases with the centre-to-centre distance, so the increase in peak strength is purely due to change in specimen area, but the post-peak behaviour depends upon the spirals. It is found in Table 4 that, although the effect of c_{sp}/d_{sp} on the ultimate strain is rather small, it is worth noting that when the distance between the two spirals gets smaller, higher breaking strain values are recorded. So a high degree of overlapping leads to greater ductility. It should be understood that when two spirals are completely overlapped, the strain values would be enhanced as if the pitch is halved (see the curves for s = 10 mm and s = 20 mm in Figure 6). When c_{sp}/d_{sp} equals 1, there is no overlapping, so the ultimate strain would be identical to that of a single spiral.

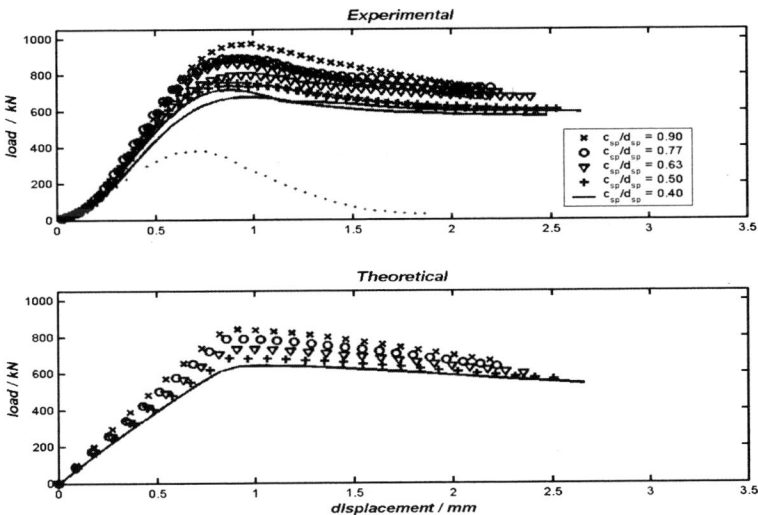

Figure 7: Experimental and theoretical results of concrete with double interlocking spirals

Good agreement between the experimental results and the theoretical prediction is found. Some general trends of the experimental data are also reflected in the theoretical results.

CONCLUSIONS

Concentric compression tests on short AFRP spirally-confined concrete cylinders have been described. The critical variables, including the concrete compressive strength, spiral spacing, arrangement of spirals and degree of interlocking, have been considered. A promising increase in strain is obtained experimentally and theoretically for a high degree of overlap. But a reverse situation is reported for high concrete strength and large spiral spacing. Bulging failure was found to be the governing mode for most uniaxially compressed concrete cylinders, which is usually followed by fracture of the spiral. On the analytical side, a fairly consistent prediction is made by the formulation reported elsewhere.

It is believed that by making use of such enhancements in the compression flange of beams, the strain and strength capacity of the beam should be increased.

REFERENCES

Ahmad, S.H. and Shah, S.P. (1982) "Stress-Strain Curves of Concrete Confined by Spiral Reinforcement", *ACI Journal*, **vol.79**, pp.484-490.

El-Dash, K.M. and Ahmad, S.H. (1995) "A model for stress-strain relationship of spirally confined normal and high-strength concrete columns", *Magazine of Concrete Research*, **vol.47**, pp.177-184.

Mander, J.B., Priestley, M.J.N and Park, R. (1988) "Observed stress-strain behaviour of confined concrete", *Journal of Structural Engineering, ASCE*, **vol 114, No.8**, pp.1827-1849.

Leung, H.Y. (2000) *Aramid fibre spirals to confine concrete in compression*, PhD thesis, University of Cambridge, 200pp.

van Mier, J.G.M., Shah, S.P., Arnaud, M., Balayssac, J.P., Bascoul, A., Choi, S., Dasenbrok, D., Ferrara, G., French, C., Gobbi, M.E., Karihaloo, B.L., König, G., Kotsovos, M.D., Labuz, J., Lange-Kornbak, D., Markeset, G., Pavlovic, M.N., Simsch, G., Thienel, K-C., Turatsinze, A., Ulmer, M., van Geel, H.J.G.M., van Vliet, M.R.A. and Zissopoulos, D., (1997) "Strain softening of concrete in uniaxial compression", *Materials and Structures*, **vol. 30**, pp.195-209.

Kotsovos, M.D. and Pavlovic, M.N. (1995) *Structure Concrete*, Thomas Telford, London.

BEHAVIOR AND ESTIMATION OF ULTRASONIC PULSE VELOCITY IN CONCRETE

Chao-Peng Lai[1], Yiching Lin[2] and Tsong Yen[2]

Department of Civil Engineering, Nan-Jeon Junior College of Technology and Commerce, Taiwan, R.O.C.
Department of Civil Engineering, National Chung-Hsing University, Taiwan, R.O.C.

ABSTRACT

Till now, the relationship between the pulse velocity and the strength of concrete is quite uncertain, it is widely recognized that the pulse velocity technique, in the present form, can not be used accurately for determining of concrete strength. The research is to investigate the fundamental behavior of ultrasonic pulse in concrete and to establish the mathematical models for estimation of concrete pulse velocity. Experimental studies are carried out to evaluate the models. The specimen compositions vary widely in aggregate content and water-cement ratio. The influences of mixture proportions as well as the pulse velocities of mortar and aggregate, on the concrete pulse velocity are analyzed. The results of the studies show that the pulse velocity of harden concrete can be estimated with errors less than 2.5%. In addition, changes in sand to aggregate ratio have little influences on the pulse velocity for various water-cement ratios. The pulse velocity of concrete is decreased by increasing the water-cement ratio.

KEYWORDS

Concrete, Mortar, Aggregate, Ultrasonic pulse velocity, Strength, Mixture proportions

INTRODUCTION

Application of ultrasonic pulse velocity (UPV) to nondestructive evaluation of concrete quality has been widely investigated for decades. Numerous data and the correlation relationships between strength and pulse velocity of concrete have been presented and proposed [1-5]. However, the measurements of the pulse velocity are influenced by many variables in concrete including mixture proportions, aggregate type and size, age of concrete, moisture content, and others [1]. Unfortunately, the factors affecting significantly the pulse velocity measurements might have little influences on strength of concrete. For instance, the pulse velocity of concrete having a water/cement ratio of more than 0.5 is drastically increased by increasing the aggregate content; at the same time, little change is observed in strength. As a result, the strength estimate by the pulse velocity method is not reliable if a pre-established calibration curve is not available [2].

Because the relationship between the strength and the pulse velocity is still quite uncertain, it is widely recognized that the pulse velocity technique, in the present form, cannot be used accurately enough for determination of concrete strength. Therefore, the objective of this paper is to investigate the fundamental behavior of ultrasonic pulse in concrete. Theoretical models for estimation of the pulse velocity are proposed. Experimental studies are carried out to evaluate the models. The specimen compositions vary widely in aggregate content and water-cement ratio. The influences of mixture proportions as well as the pulse velocities of mortar and aggregate, on the concrete pulse velocity are analyzed. The studies may result in reconsideration of the relationship between the concrete strength and the pulse velocity in a new light and then help plan future directions of studies to improve the application of pulse velocity to evaluation of concrete strength.

TEST PROGRAM

Specimens

Fifteen concrete mixture proportions are identified as Mixtures C1, C2 ...., and C15 in Table 1. The water-cement ratio (W/C) ranges from 0.3 to 0.7. Three volume ratios of sand to total aggregate (S/A) are considered to be 30%, 45%, and 60% for each water-cement ratio. These mixture proportions can be used to investigate the influence of changes in W/C and S/A on the pulse velocity of concrete.

TABLE1
MIXTURE PROPORTIONS AND PULSE VELOCITY OF CONCRETE

No.	W/C	S/A (%)	Average Pulse Velocity of Concrete Specimens (m/sec)				
			1 days	3 days	7 days	14 days	28 days
C1	0.7	30	3315	3705	4010	4084	4134
C2		45	3396	3772	4001	4104	4164
C3		60	3382	3771	4037	4084	4141
C4	0.6	30	3441	3888	4155	4240	4300
C5		45	3538	3847	4158	4219	4245
C6		60	3577	3855	4142	4213	4279
C7	0.5	30	3794	4093	4336	4376	4450
C8		45	3803	4133	4295	4358	4426
C9		60	3719	4033	4221	4291	4369
C10	0.4	30	4050	4307	4467	4517	4593
C11		45	4021	4305	4404	4481	4563
C12		60	3984	4240	4322	4453	4498
C13	0.3	30	4303	4477	4546	4609	4706
C14		45	4222	4452	4513	4593	4612
C15		60	4260	4414	4494	4583	4629

In addition, fifteen mortar mixture proportions, M1, M2, ..., and M15, corresponding to C1, C2, ..., and C15, respectively, were included in the studies. Each mortar mixture proportion was determined by proportionally increasing the mortar amount of the corresponding concrete to maintain a unit volume (1 m^3). These mortar and concrete specimens can be used to establish the model for estimation of concrete pulse velocity.

Three specimens were produced for each mixture proportion. All the specimens were cast in 100×200 mm cylindrical steel moulds and kept in their moulds for about 24 hours in the laboratory and de-moulded. Concrete cylinders were cured in water at 20°C and tested at 1, 3, 7, 14, and 28 days.

Experimental Equipment

The ultrasonic pulse velocities were measured by a commercially available pulse meter with an associated transducer pair. The principle of ultrasonic pulse velocity measurement involves sending an electro-acoustic pulse into concrete and measuring the travel time for the pulse to propagate through the concrete. The pulse is generated by a transmitter and received by a receiver. Knowing the path length, the measured travel time (Δt) can be used to calculate the pulse velocity (υ) as follows:

$$\upsilon = D/\Delta t \tag{1}$$

Where D is the depth of the cylinder. The concrete surface must be prepared in advance for a proper acoustic coupling. A small pressure is needed to ensure firm contact of the transducers against the concrete surface.

RESULTS AND DISCUSSIONS

Pulse Velocity of Specimens

The measurement results of concrete pulse velocity at ages of 1, 3, 7, 14, and 28 days are listed in Table 1. These values represent the average pulse velocity of three different specimens for each mixture proportion at a specific age.

Figure 1 shows the velocity development with age for C2, C5, C8, C11, and C14 (S/A=45%). In Fig. 1, one can find that the pulse velocity increases with a decreasing water-cement ratio. It is also observed that concrete gains the velocity rapidly at early ages for all the water-cement ratios. The pulse velocity of concrete at an age of 1 day can reach more than 80% of that for 28 days and more than 90% at 3 days.

Figure 2 shows the relationship between the pulse velocity of harden concrete at 28 days and the volume ratio of fine aggregate to total aggregate (S/A) for various water-cement ratios. It is clearly observed that changes in S/A have little influences on the pulse velocity for various water-cement ratios.

Table 2 tabulates the average pulse velocity measured from all mortar specimens. The pulse velocities of the mortar specimens are less than those of their corresponding concrete specimens.

Crushed river stones less than 13 mm in size were used as coarse aggregates in production of concrete specimens. To measure the pulse velocity, five cores with 75 mm in diameter and 150 mm in length were taken from the river stones. The measured pulse velocity of the stone cores ranges from 4908 to 5290 m/s. The average pulse velocity was calculated as 5175 m/s.

TABLE 2
AVERAGE PULSE VELOCITY OF MORTAR SPECIMENS

No.	Average Pulse Velocity of Mortar Specimens (m/sec)				
	1 days	3 days	7 days	14 days	28 days
M1	2797	3216	3412	3563	3598
M2	2751	3291	3544	3705	3724
M3	2755	3345	3640	3767	3809
M4	2944	3268	3598	3665	3731
M5	3050	3405	3772	3832	3890
M6	3101	3506	3799	3891	3957
M7	3170	3570	3806	3848	3910
M8	3205	3612	3864	3946	3995
M9	3239	3654	3904	3997	4073
M10	3513	3861	3953	4036	4077
M11	3652	3928	4045	4130	4191
M12	3747	4031	4160	4240	4294
M13	3919	4093	4160	4217	4295
M14	4034	4214	4269	4342	4406
M15	4088	4270	4334	4407	4478

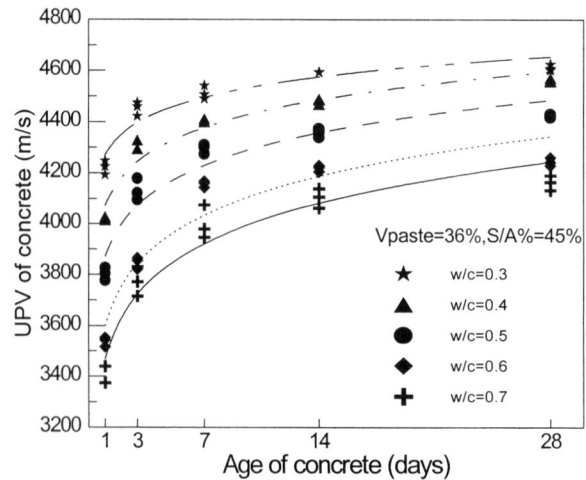

Figure 1: UPV development of concrete for different water-cement ratios

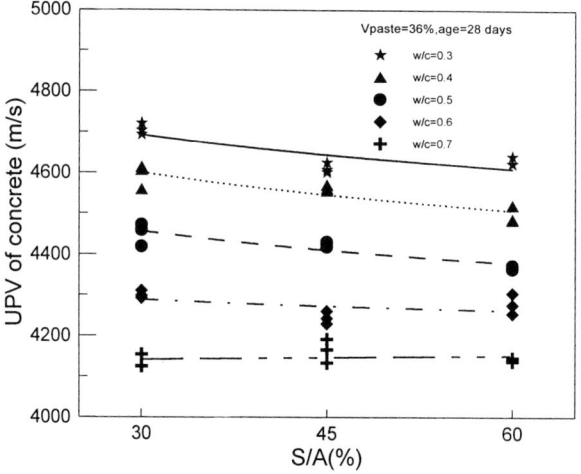

Figure 2: Relation between UPV of concrete and S/A for different water-cement ratios

Estimation Models For Concrete Pulse Velocity

The travel time (T_C) for a ultrasonic pulse that propagates through a composite material composed of N layers as shown in Fig. 3 can be expressed as:

$$T_C = T_1 + T_2 + \ldots + T_N \tag{2}$$

Where T_1, T_2, ... and T_N represent the elapse times needed for the pulse to propagate through layer 1, 2, ... and N, respectively.

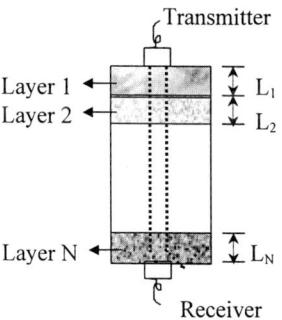

Figure 3: Schematic diagram of N-layered composite material

If the pulse velocities of each layer are known as υ_1, υ_2, ... and υ_N, Eq. (2) can be rewritten as follows:

$$\frac{L_C}{\upsilon_C} = \frac{L_1}{\upsilon_1} + \frac{L_2}{\upsilon_2} + \ldots + \frac{L_N}{\upsilon_N} \tag{3}$$

in which L_C and υ_C denote the total length and the pulse velocity of the N-layered composite material and L_1, L_2, ... and L_N denote the length of each individual layer from layer 1 to N, respectively. The variable of length (L) in Eq. (3) can be replaced with the variable of volume (V) based on the assumption that each layer has the same cross-sectional area. As a result, Eq. (3) can be further rewritten as:

$$\frac{1}{\upsilon_C} = \frac{(V_1/V_C)}{\upsilon_1} + \frac{(V_2/V_C)}{\upsilon_2} + \ldots + \frac{(V_N/V_C)}{\upsilon_N} \tag{4}$$

Eq. (4) is used to develop models for estimating the pulse velocity of concrete from its mixture proportion. From Eq. (4), a two-phase model for estimation of concrete pulse velocity is also proposed. This model assumes that concrete is composed of mortar (M) and coarse aggregate (CA). Accordingly, Eqs.(4) can be modified as Eqs. (5).

$$\frac{1}{\upsilon_C} = \frac{(V_M/V_C)}{\upsilon_M} + \frac{(V_{CA}/V_C)}{\upsilon_{CA}} \tag{5}$$

With the known mixture proportions of concrete and the previously measured pulse velocity of mortar and coarse aggregate, the concrete pulse velocity can be estimated using Eq. (5). The pulse velocities estimated by this two-phase model for all the fifteen concrete mixture proportions of various ages are compared to the measured results as shown in Fig. 4. It is found from Fig. 4 that the two-phase model can be used to estimate accurately the pulse velocity of concrete at ages after 7 days with estimation errors less than 2.5%.

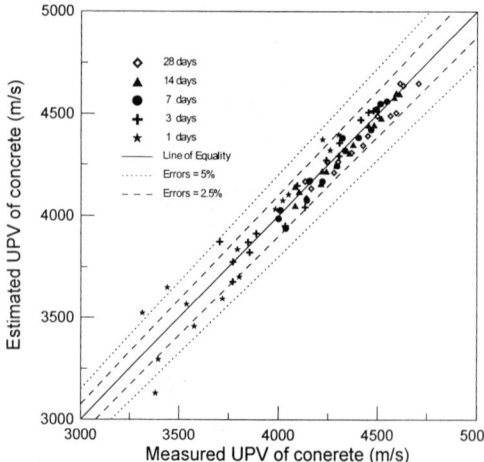

Figure 4: Comparison between estimated and measured UPV of concrete

The results show that two-phase models are suitable for pulse velocity estimation of concrete. The basic assumption of these models is to view concrete as a layered composite material as illustrated in Fig. 3. The two-phase model treats concrete as composed of one mortar layer and one aggregate layer as shown in Fig. 5(a), but for real concrete aggregate particles should be distributed in the mortar. This obvious difference between the model and the real concrete may result in errors as the model is used to estimate the concrete pulse velocity.

If the aggregate particles are uniformly located in the mortar as shown in Fig. 5(b), the transit time for a pulse propagating through the specimen is identical to the case of the two-layered model. In such an ideal case, the pulse velocity of concrete can be perfectly estimated by this model. For real concrete, the aggregate particles should distribute randomly. Consequently, the concrete pulse velocity estimated by the two-phase model may be over- or under-estimated depending on the distribution of aggregate particles.

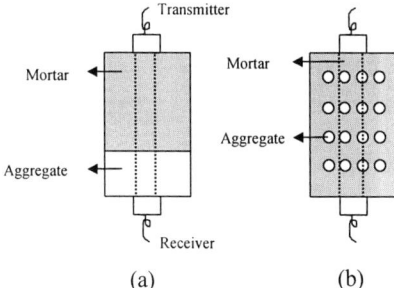

Figure 5: Schematic diagram of aggregate distribution in mortar

CONCLUSIONS

Several conclusions can be drawn as follows:

1. Changes in sand to all aggregate ratio have little influences on the pulse velocity of concrete for various water-cement ratios.

2. The pulse velocity of concrete is decreased by increasing the water-cement ratios.

3. The pulse velocity of harden concrete can be estimated with errors less than 2.5% by using the proposed models.

REFERENCES

1. Tanigawa, Y., Baba, K., and Mori, H. (1984). Estimation of Concrete Strength by Combined Nondestructive Testing Method. In Situ/Nondestructive Testing of Concrete. **ACI SP-82,** 57-76.

2. Sturrup, V.R., Vecchio, F.J. and Caratin, H. (1984). Pulse Velocity as a Measure of Concrete Compressive Strength. In Situ/Nondestructive Testing of Concrete. **ACI SP-82,** 201-227.

3. Popovics, S., Rose, J.L., and John S. Popovics, J.S. (1990). The Behavior of Ultrasonic Pulses in Concrete. *Cement and Concrete Research* **20:2,** 259-270.

4. Andrej Galan. (1967). Estimate of Concrete Strength by Ultrasonic Pulses Velocity and Damping Constant. *ACI Journal* **64:10,** 678-684.

5. Lin, Y., Changfan, H., and Hsiao, C. (1998). Estimation of High Performance Concrete Strength By Pulse Velocity. *Journal of the Chinese Institute of Engineer* **20: 6,** 661-668.

Study on the Rheological Behavior of Medium Strength High Performance Concrete

Chao-Wei Tang[1], Kuang-Hong Chen[2] and Tsong Yen[2]
[1] Department of Civil Engineering, Cheng-Shiu Institute of Technology
Kaohsiung, Taiwan
[2] Department of Civil Engineering, Chung-Hsiung University,
Taichung, Taiwan

ABSTRACT

A kind of high performance concrete (HPC) with medium strength, named TAICON in this article, has fine workability and segregation resistance. It is gradually widely applied in construction instead of ordinary concrete in Taiwan. In view of the currently utilized slump test as well as the other test methods that are not sufficient to clarify the flow behavior of TAICON, we used the two-point test procedure and the rheology test method proposed by Tattersall to investigate its flow properties. Test Results show that fresh TAICON confirms to Bingham model. Besides, the application of rheological method can lead to more stable results than the single value of slump test in describing the flowability of TAICON.

KEYWORDS

medium strength high performance concrete (TAICON), rheology, Bingham model, slump test

INTRODUCTION

As a construction material, HPC can flows easily when it is fresh and turns into a homogeneous mixture of all components as it hardens. Hence it has been seen that HPC gain worldwide popularity in the construction industry since 1980's. Likewise, in Taiwan, the use of HPC has been increasing gradually. However the normal concrete with low to medium strength is still the most common concrete for engineering purpose. In order to avoid the destruction of ecosystem due to the by-production of industry and reduce the consumption of limited energy in cement production, there is an urgent need to sufficiently utilize the mineral materials such as fly ash and slag. Consequently, to improve the properties of normal concrete and solve the environmental problems, a kind of medium strength HPC (25 MPa $\leq f_c' \leq$ 35 MPa, named as TAICON in this article) was actively promoted in Taiwan by industry, academy and government since 1997.

Comparing the constituents and properties of TAICON with that of normal concrete and HPC, as shown in Table 1, it reveals that the production of TAICON demands the mineral materials and superplasticizer. These materials will essentially affect the fresh properties of TAICON. So fresh TAICON can possesses fluid consistency and uniformity, and its slump value is usually more than 200 mm. However, the validity of slump test is generally recommended for concrete with a slump value ranging from 25 to 175 mm. In other words, the slump test method does not seem appropriate to characterize the workability of TAICON. Fortunately, over recent years, it has been well confirmed that the theory of rheology is much more sensitive to characterize the flow behavior of high strength HPC [1-3, 5]. Accordingly it is practicable to explore the rheological behavior of TAICON by using a rheological test method. In this research, the flowability of TAICON was investigated experimentally by using a self-established rheometer (TRM). The test results were additionally compared with the corresponding results of normal concrete and high strength HPC.

TABLE 1
COMPARISONS BETWEEN THE CONSTITUENT AND PROPERTIES OF NC, HPC AND TAICON

Type of concrete	Constituents of Material		Required properties	
	Material used	Restrict condition	Fresh stage	Hardened stage
Normal concrete (NC)	• Water • Cement • Aggregate	—	• Slump: 25-175 mm	• Compressive strength of 28 days: 14-35 MPa
High performance concrete (HPC✥)	• Water • Cement • Aggregate • Mineral materials • Superplasticizer	—	• Slump: 250 ± 15 mm (slump after 45min.: ≥ 235 mm)	• Compressive strength of 28 days: ≥ 55 MPa
Medium strength HPC (TAICON)	• Water • Cement • Aggregate • Mineral materials • Superplasticizer	• (W/B)○ ≥ 0.42 • Cement content < 300 kg/m^3 • Mineral materials content < 30%	• Slump: 230 ± 20 mm (slump after 45min.: ≥ 200 mm) • Slump-flow: 400-600 mm	• Compressive strength of 28 days: 25–35 MPa

Note ✥: According to the current definition of HPC in Taiwan; ○: W/B = Water/(Cement + Slag + Fly ash)

EXPERIMENT

Materials and Mixture Proportions

The test program started with materials selections. The selected ingredients included type I Portland cement, coarse aggregate with a maximum size of 12.7 mm, fine aggregate with a fineness of 2.90, G-type superplasticizer, and mineral materials of class F fly ash (2070 cm^2/g, LOI=4.0%) and slag (3990 cm^2/g, LOI≅0%). The superplasticizing used is HICON HPC 1000 confirming to ASTM C-494-Type-G. The main test variables include water/binder ratio (W/B) and binder combination. We chose four water/binder ratios of 0.45, 0.50, 0.55 and 0.60, and four cement/slag/fly-ash combinations; Series Aa - 7 : 2 : 1; Series Ba - 6 : 3 : 1; Series Bb - 6 : 2 : 2; and Series Bc - 6 : 1 : 3 by weight for the test program. The mixture proportions are summarized in Table 2.

Apparatus and Measurement

To measure the flowability of mixtures, the rheological measurements were conducted by using the TRM rheometer, in which the major components include synchronous drive motor, hydraulic gear, cross-shape spindle, sample container, and processing system. All the parts are mounted on a steel frame as shown in Figure 1. The authors have described its design and calibration in reference [5]. For each test specimen, the rheological measurement was executed, respectively, at 0, 30, and 60 min. after mixing. The test began with the maximum rotational speed of 50 rpm and the torque, T, was measured. In the same procedure the torque at rotational speed of 40, 30, 20, 10 and 5 rpm were measured. To ensure the stability of the results, a 10-second record was taken twenty seconds after a certain rotating speed is set. The relationship between the mean value of the lowest 10 torque measurements and the corresponding rotational speed is then obtained. In addition, a parallel slump test was also carried out for comparison.

TABLE 2
MIXTURE PROPORTIONS (kg/m^3)

Series	Mix no.	W/B	SP/B	W	C	SP	SL	F	S	CA
Aa series	45Aa	0.45	0.8%	212.6	330.8	4.73	94.51	47.25	591.8	1048
	50Aa	0.50	0.8%	222.8	311.9	3.56	89.10	44.55	591.8	1048
	55Aa	0.55	0.5%	232.0	295.3	2.11	84.36	42.18	591.8	1048
	60Aa	0.60	0.2%	240.3	280.4	0.80	80.10	40.05	591.8	1048
Ba series	45Ba	0.45	1.0%	211.9	282.6	4.71	141.28	47.09	591.8	1048
	45Bb	0.45	1.0%	209.6	279.5	4.66	93.15	93.15	591.8	1048
	45Bc	0.45	1.0%	207.3	276.4	4.61	46.07	138.22	591.8	1048
Bb series	50Ba	0.50	0.8%	222.0	266.4	3.55	133.22	44.41	591.8	1048
	50Bb	0.50	0.8%	219.6	263.6	3.51	87.86	87.86	591.8	1048
	50Bc	0.50	0.8%	217.5	261.0	3.48	43.50	130.50	591.8	1048
Bc series	55Ba	0.55	0.5%	231.3	252.3	2.10	126.16	42.05	591.8	1048
	55Bb	0.55	0.5%	229.0	249.8	2.08	83.28	83.28	591.8	1048
	55Bc	0.55	0.5%	226.8	247.4	2.06	41.24	123.71	591.8	1048

Note: C = Cement, B = Cement + Slag + Fly ash, SP = Superplasticizer, W = Water, SL = SLag, F = Fly ash, S = Sand, CA = Coarse Aggregate

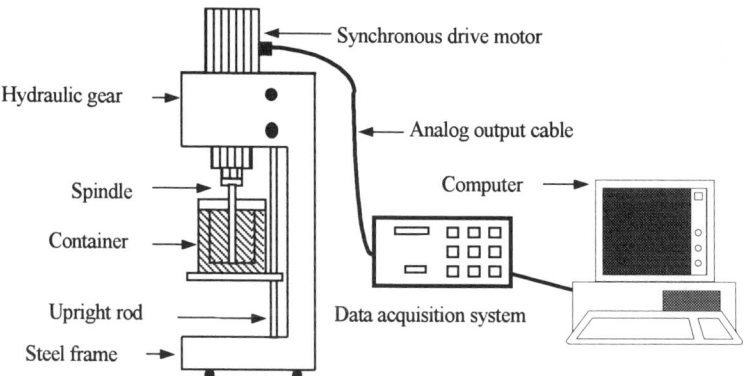

Figure 1: TRM rheometer apparatus

TEST RESULTS AND DISCUSSIONS

Test Results of Slump Test and Slump-Flow Test

According to the current definition of TAICON in Taiwan, the specified design strength at 28-day is at least 25 MPa, the slump is 230 mm ± 20 mm, and the retention value of slump is at least 200 mm after 45 minutes. On the whole, the results of Table 3 shows that the mixtures roughly agree with the definition of TAICON. However, it can be seen from Table 3 that the mixtures with approximately equal slump values can lead to a variety of slump-flow values. On the other hand, those with approximately equal slump-flow values do not result in agreeable slump values. These imply that the workability or consistency of fresh concrete cannot be sufficiently described by the results only from the slump test or slump-flow test.

TABLE 3
TEST RESULTS OF SLUMP TEST, SLUMP-FLOW TEST AND COMPRESSIVE STRENGTH

Mix no.	Slump (mm)				Slump-flow (mm)				f_c' (MPa)
	0 min.	30 min.	45 min.	60 min.	0 min.	30 min.	45 min.	60 min.	
45Aa	200	180	150	145	400	420	320	280	40.2
50Aa	195	190	180	170	405	380	380	350	29.8
55Aa	195	205	195	180	465	460	405	355	24.6
60Aa	205	210	210	195	420	405	370	370	22.9
45Ba	195	220	200	165	385	380	385	385	38.4
45Bb	195	210	195	195	400	390	390	370	37.0
45Bc	195	190	195	180	405	345	335	350	34.7
50Ba	205	180	180	165	440	425	385	375	25.6
50Bb	200	210	205	200	440	410	400	360	25.5
50Bc	210	200	185	165	420	410	345	360	25.9
55Ba	210	195	195	185	430	410	380	395	25.0
55Bb	200	185	185	175	420	380	360	380	20.9
55Bc	220	215	200	210	430	405	385	350	18.9

Test Results of Rheological Test

A typical result of measurements (mix no. 45Ba) using TRM rheometer is illustrated in Figure 2. From Figure 2 we can find in a good relationship of torque versus rotational speed (T-N) and the resulting values of g and h obtained by regression analysis are available. It means the Bingham model can well describe the flow properties of TAICON. So the linear relationship of T-N can be presented in the form of Equation (1). Where g is the intercept on the torque axis and h is the slope of the T-N line.

$$T = g + hN \tag{1}$$

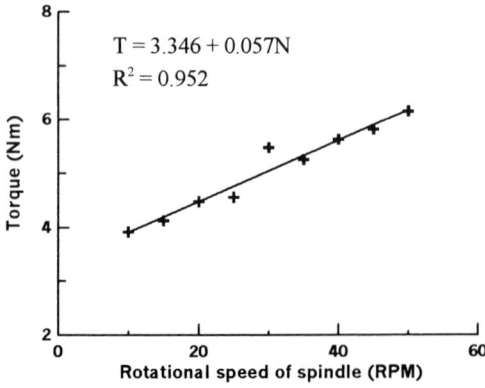

Figure 2: Relationship of torque versus rotational speed (T-N) (Sample 45Ba)

According to Equations (2) and (3), the units of g and h can be respectively transferred into the fundamental units of yield shear τ_0 (Pa) and plastic viscosity μ (Pa·s), as follows [5]:

$$\tau_0 = \frac{g}{\left[\dfrac{2\pi R_b^3}{3} + \dfrac{4\pi H \ln(R_c/R_b)}{(1/R_b^2 - 1/R_c^2)}\right]} \tag{2}$$

$$\mu = \frac{h}{\left[\dfrac{8\pi^2 H}{(1/R_b^2 - 1/R_c^2)} + \dfrac{\pi^2 R_b^4}{d}\right]} \tag{3}$$

Where H is the height of spindle, R_b is the radius of the spindle, R_c is the radius of the sample container, and d is the gap under the spindle. Table 4 gives the two Bingham constants (τ_0 and μ) of all the 13 concrete mixtures measured with the TRM rheometer, together with their slump values.

TABLE 4
COMPARISON OF SLUMP VALUE AND TWO BINGHAM CONSTANTS (τ_0 AND μ)

Mix no.	0 min.			30 min.			60 min.		
	Slump (mm)	τ_0 (Pa)	μ (Pa·s)	Slump (mm)	τ_0 (Pa)	μ (Pa·s)	Slump (mm)	τ_0 (Pa)	μ (Pa·s)
45Aa	200	563.9	19.3	180	684.2	22.5	145	906.3	20.7
50Aa	195	548.0	15.4	190	488.9	18.9	170	674.3	21.8
55Aa	195	588.0	6.8	205	335.3	11.1	180	486.6	13.9
60Aa	205	549.6	12.9	210	503.3	11.4	195	358.8	12.1
45Ba	195	416.2	16.8	220	407.1	21.8	165	591.0	25.4
45Bb	195	393.4	21.4	210	419.0	30.0	195	446.8	26.4
45Bc	195	382.9	26.8	190	380.9	26.4	180	469.5	26.8
50Ba	205	464.8	17.0	180	526.8	17.3	165	882.3	21.4
50Bb	200	427.7	21.9	210	415.8	23.4	200	487.4	22.6
50Bc	210	282.2	12.5	200	356.2	19.2	165	440.5	26.3
55Ba	210	459.7	13.7	195	394.2	16.5	185	374.6	13.6
55Bb	200	460.4	13.3	185	358.9	17.1	175	471.5	10.1
55Bc	220	279.3	11.1	215	294.2	10.2	210	338.3	14.3

Superiority of Rheological Test in Describing the Flow Behavior of TAICON

From Table 4 it is seen that the ranges of two Bingham constants, yield shear and plastic viscosity, vary from 279.3 to 906.3 Pa and 6.8 to 30.0 Pa·s, respectively, and the correlation of two Bingham constants, as shown in Figure 3, are independent of each other. This implies the inadequacy of single-point test (such as slump test), which can determine only one characteristic. For instance, in the case of 0 min. test results, it indicates in Table 4 that for mixtures having the same slump value, the corresponding τ_0 may differ from 1 to 1.6 times and the corresponding μ even differ from 1 to 3.9 times. This reveals that the rheological test is more sensitive than the slump test in demonstrating the flow behavior of fresh TAICON. In fact, the yield shear is the minimum shear needed to counteract the friction forces between the solid particles. If the applied shear is sufficient to overcome the threshold of shear, the concrete will begin flowing. While the plastic viscosity is the slope of the stress versus the shear rate after the stress exceeds the yield shear. This is the main reason for the discrepancy often encountered between the single-point test result and the perceived workability of given TAICON.

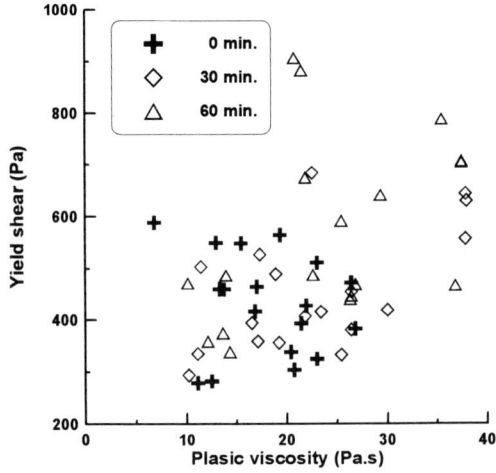

Figure 3: Relationship between the two Bingham constants (τ_0 and μ)

Evolution of Bingham Constants versus Time

To recognize the rheological properties of fresh concrete after a sufficiently long time, a proper characterization of the evolution of workability is required. Generally, the loss of workability is reflected by an increase of yield shear. However, the test results of Figure 4 shows that for mixtures having different water/binder ratios, the evolution curves of τ_0 versus time are different. It is due to the reason that in our test program the dosage of superplasticizer (SP) decrease with the increase of water/binder ratio (W/B). Thus the combined effect of SP and W/B will dilute the distinction of all mixtures. For instance, the test results of 60 min. shown in Figure 4 indicate that the efficiency of SP is lapsed, so the yield shear will be mainly controlled by the water/binder ratio. A similar tendency for the evolution curves of μ versus time can be also found in Figure 5.

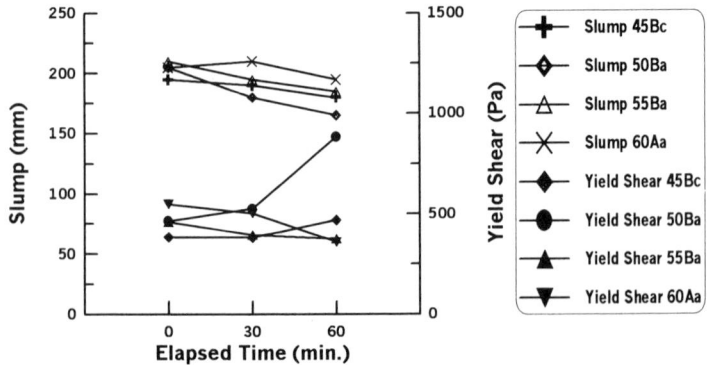

Figure 4: Evolutions of τ_0 versus time

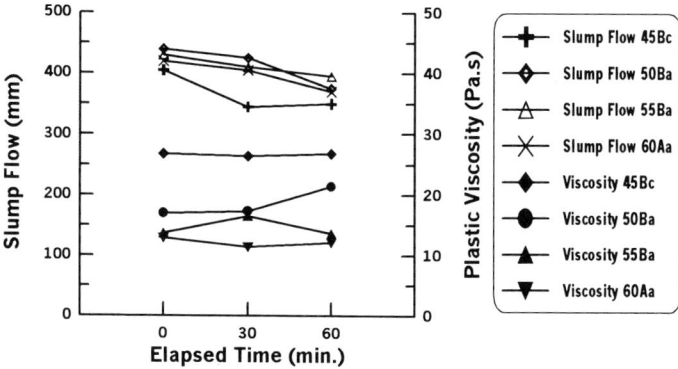

Figure 5: Evolutions of μ versus time

Effect of Binder Combination on the Flow Behavior of TAICON

Four series of binder combinations of cement, slag, and fly ash were selected to cover the mixture proportions of mineral admixtures used in construction practices. Figure 6 shows that the yield shear of 45Aa is larger than other mixtures at three different duration. It implies that cement content is the main factor to affect the yield shear of fresh concrete. On the other hand, Figure 7 demonstrates that the plastic viscosity at 0 min. is evidently increased from 45Ba to 45Bc. This may attribute to the greater amount of fly ash in 45Bb and 45Bc, since the lower density of fly ash will result in an increase of the volume of cementitious material. It leads to the situation that less free water is available for the lubrication of the particles, and thus the plastic viscosity increases with the increasing amount of fly ash. However, the plastic viscosity at 60 min. is almost invariable. The reason is that at this time period more free water is released from the slag particles and the distinction between each mixture will be diluted.

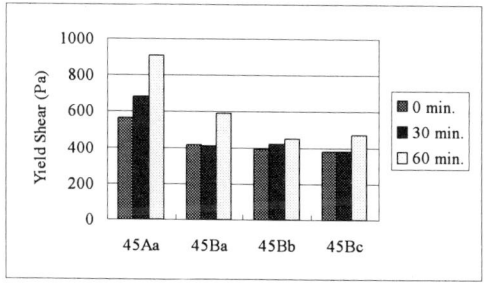
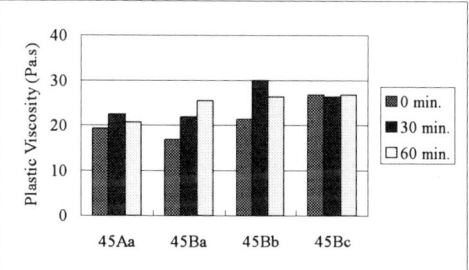

Figure 6: Effect of binder combination on τ_0 Figure 7: Effect of binder combination on μ

Comparison of the Bingham Constants for Various Concretes

Since the Bingham parameters are sensitive to mixture proportions and apparatus used, it is hard, if not impossible, to establish certain constants that are suitable to all test cases. However, the ranges of the 2 parameters obtained from the TRM rheometer tests are compared with that of normal concrete (NC) and HPC, as summarized in Table 5. It indicates that TAICON generally has the properties of HPC and thus possesses lower τ_0 and higher μ than normal concrete for which a low τ_0 value signifies a high flowability whereas a high μ value means a greater resistance to aggregate segregation.

TABLE 5
COMPARISON OF BINGHAM CONSTANTS BETWEEN NC, HPC AND TAICON

Concrete	Authors	τ_0 (Pa)	μ (Pa·s)
Normal concrete[4]	Murata & Kikukawa	150-300	5-50
	Sakata et al.	60-200	14-20
	Uzomaka	1750-7570	90-800
	Morinaga	600-6000	-
High strength HPC	Hu C. & Larrard[3]	50-2000	70-500
	Hu C., Larrard and Gjorv Odd E.[2]	0-1836	11-170
TAICON	Tang C.W., Yen T. and Chen K.H.	0 minute: 279.3-588.0	6.8-26.8
		30 minute: 294.2-684.2	10.2-30.0
		60 minute: 338.3-906.3	10.1-26.8
		Summarized: 279.3-906.3	6.8-30.0

CONCLUSIONS

Based on the test results in this research, the following conclusions are summarized:

1. Fresh TAICON confirms to Bingham model and thus the two Bingham constants (τ_0 and μ) together characterize its rheological behavior. Moreover, the rheological test is more sensitive than the slump test in demonstrating the flow behavior of fresh TAICON.

2. As seen in the rheological results, an increase of fly ash tends to result in larger plastic viscosity, which is corresponding to a worse workability and the yield shear is affected sensitively by cement content.

3. TAICON generally has the properties of HPC and thus possesses lower τ_0 and higher μ than normal concrete for which a low τ_0 value signifies a high flowability whereas a high μ value means a greater resistance to aggregate segregation.

Acknowledgements

This research is supported by the National Science Council, R.O.C., under Grant NSC87-2211-E-005-026.

References

1. De Larrard F., Hu C., Sedran T., Szitkar C., Joly M., Claux F. and Derkx F. (1997). A new rheometer for soft-to-fluid fresh concrete. ACI Materials Journal, **V. 94. No. 3,** 234-243.

2. Hu. C., De Larrard F., and Gjorv O.E. (1995). Rheological Testing and Modeling of Fresh High Performance Concrete. Materials and Structures, **V. 28, No. 175,** 1-7.

3. Hu C. and F. De Larrard. (1996). The Rheology of Fresh High-Performance Concrete. Cement and Concrete Research, **Vol. 26, No. 2,** 283-294.

4. Tattersall G.H. and Banfill P.F.G. (1983). The Rheology of Fresh Concrete, Boston: Pitman Books, 34-191.

5. Yen T., Tang C.W., Chang C.S. and Chen K.H. (1999). Flow Behavior of High Strength High Performance Concrete. Cement & Concrete Composites, **V. 21,** 413-424.

INFLUENCE OF ADMIXTURES AND QUARRY DUST ON THE PHYSICAL PROPERTIES OF FRESHLY MIXED HIGH PERFORMANCE CONCRETE

M. Fauzi M. Zain[1], Md. Safiuddin[2], T. K. Song[1], H.B.Mahmud[3], and Y. Matsufuji[4]

[1]Department of Civil and Structural Engineering, University Kebangsaan Malaysia
[2]Department of Civil and Environmental Engineering, University of Windsor, Ontario, Canada
[3]Department of Civil Engineering, University of Malaya, Malaysia
[4]Department of Architecture, Faculty of Engineering, Kyushu University, Japan

ABSTRACT

The effects of mineral and chemical admixtures, and quarry dust on the physical properties of freshly mixed high performance concrete were investigated. Silica fume and fly ash were used to replace 10% of cement by weight. They were used either separately or in combination with quarry dust in the concrete mixes. Quarry dust was incorporated to replace 20% and 40% of sand in the respective mixes. Superplasticizer and air entraining admixture were used as chemical admixtures to control the flowability of fresh concrete. The physical properties of fresh concrete mixes were monitored batch by batch. The properties of fresh mixes were determined in respect of slump, slump flow, V-funnel flow, air content, mix temperature and unit weight. It was observed that the properties, particularly the flow and the entrained air content, were influenced by the presence of mineral and chemical admixtures. In presence of superplasticizer, silica fume provided better flow properties than fly ash. Besides, the incorporation of quarry dust improved the flowability of fresh mixes. The outcome of the present study suggests the use of quarry dust as partial replacement of sand to augment the flowability of high performance concrete.

KEYWORDS

High Performance Concrete, Silica Fume, Fly Ash, Quarry Dust, Superplasticizer, Air Entraining Admixture, Slump, Slump Flow, V-Funnel Flow

INTRODUCTION

An admixture is defined as a material other than Portland cement, aggregate or water, that is added to concrete before or during mixing, in order to modify one or more of its properties in the fresh and hardened states (Dodson, 1990). Chemical and mineral admixtures are the two most widely used admixtures in the production of concrete. Chemical admixtures function by the dispersion of cement in the aqueous phase of concrete and increase the normal rate of cement hydration. On the other hand, mineral admixtures function through the reactions with by-products of hydrating cement or without any reaction with either the cement or its by-products.

Amongst the chemical admixtures, water reducing admixtures come first because they are used in concrete in large volume. The latest water reducing admixtures are superplasticizers, which are often called as super water reducers or supers. Chemically, it is the salt of sulfonated naphthalene formaldehyde condensate or sulfonated melamine formaldehyde condensate (Aitcin, 1992). This type of admixture causes deflocculation of cement grains and this is the process in which the cement grains in suspension can recover their initial grain size. The use of superplasticizer leads to an appreciable reduction in the quantity of mixing water because a lot of this water is no longer entrapped in the cement grain flakes. Superplasticizers can reduce the quantity of mixing water required to produce concrete of a given consistency by 12% to 20% (ASTM 1017-92, 1992). These admixtures also result in making additional surfaces of the cement particles available for early hydration. It should be noted that superplasticizers exert their action by decreasing the surface tension of water and by equi-directional charging of the cement particles. This property, coupled with the addition of mineral admixture, results in excellent flowability in the fresh concrete mixes (Nawy, 1996). The other chemical admixture, which is often used in cold weather concreting, is air entraining admixture. The entrainment of air has a number of beneficial effects. The most important of these is an increase in resistance to frost attack and to deterioration due to exposure to repeated freezing and thawing. The first mention of air entrainment as a possible cause for improved frost resistance of concrete was made in 1939 (Powers, 1939). The results of that work also indicated that entrained air reduced bleeding and improved the uniformity of concrete. Theoretically, there is no need of entrained air in concretes that are to be used in a hot country like Malaysia. However, in order to improve the handling, placeability and finishability of concrete, it is strongly suggested that a small amount of entrained air in fresh concrete be used (Zain et al., 1999). The present study reports the influence of superplasticizer and air entraining admixture on the physical properties of freshly mixed high performance concrete.

Mineral admixtures are siliceous or siliceous and aluminous materials which in themselves possess little or no cementitious value but will, in finely divided form and in the presence of moisture, chemically react with calcium hydroxide at ordinary temperatures to form compounds possessing cementitious properties (ACI SP-19, 1988). The literature of mineral admixture is quite rich. There are many publications available on the beneficial effects of mineral admixtures. Some studies have been made on the blending of mineral admixture with concrete as an alternative to Portland cement for the improvement of concrete properties (Malhotra, 1987). The production of high strength concrete using Class-C fly ash is an example of the development (Cook, 1982). Various types of reactive mineral admixture have also been used in Japan for obtaining super-workable concrete (Hayakawa et al., 1994). Very recently some mineral admixtures such as fly ash, slag, limestone and siliceous stone powders have been used as the partial replacement of sand (Uchikawa et al., 1995). The incorporation of mineral admixtures results in widening the range of grain size in the vicinity of cementitious materials. These admixtures, being extremely fine materials, fill the microvoids in grain packing and thereby improve the compactness of the concrete matrix and at the same time the rheological properties of the fresh mix. This study also reveals the influence of mineral admixtures, especially fly ash and silica fume on the physical properties of fresh high performance concrete.

Beside the literature of mineral admixtures, there are very few reports on the usage of vast quantities of wastes generated by mixing and quarrying industries. Small amount of the wastes generated by industrial and urban activities has been used in the construction of road and in the manufacture of building materials such as lightweight aggregates, bricks, tiles and autoclaved blocks. An attempt was also made to study the effect of rock dust as an alternative to sand on the strength and workability of concrete mixes (Nagaraj & Banu, 1996). A recent study has focussed the potential use of quarry dust in developing high performance concrete (Safiuddin et al., 2000). The effect of quarry dust on the flow properties of high performance concrete has been highlighted in this publication. The present study also reports the influence of quarry dust on the physical properties, especially the flowability of fresh high performance concrete mixes.

EXPERIMENTAL PROGRAM

Materials

Crushed granite stone, mining sand, quarry dust, Type I normal Portland cement, silica fume, fly ash, a naphthalene formaldehyde based superplasticizer and an air entraining admixture were used in this study. Tap water was used as mixing water. All the materials were locally available except silica fume and superplasticizer. Silica fume and fly ash were used as mineral admixtures whereas superplasticizer and air entraining admixture were added as chemical admixtures. Quarry dust was obtained from a local granite stone quarry. Mining sand and quarry dust took the role of fine aggregate. The physical properties of the materials are shown in Table 1.

Table 1: Physical Properties of Materials

Materials	Properties
Coarse Aggregate (CA): Crushed Granite Stone	Specific Gravity: 2.62 Absorption: 0.9% Size: ≤19 mm Fineness Modulus: 6.82
Fine Aggregate (FA): Mining Sand (MS)	Specific Gravity: 2.60 Absorption: 1.2% Size: ≤4.75 mm Fineness Modulus: 3.01
Fine Aggregate (FA): Quarry Dust (QD)	Specific Gravity: 2.63 Absorption: 0.6% Size: ≤4.75 mm Fineness Modulus: 4.20
Binder (B): Cement (C)	Specific Gravity: 3.15 Average Size: 10 µm
Binder (B): Silica Fume (SF)	Specific Gravity: 2.20 Average Size: 0.1 µm
Binder (B): Fly Ash (FA)	Specific Gravity: 2.26 Maximum Size: 53 µm
Superplasticizer (SP)	Specific Gravity: 1.21 Solid Content: 41%
Air Entraining Admixture (AEA)	Specific Gravity: 1.02 Solid Content: 8%
Water (W)	pH: 6.9 Dissolved Solids: < 2000 ppm

Mix Proportions

Normal Portland cement (NPC), silica fume (SF) and fly ash (FA) concretes were prepared with the water-binder ratios of 35% and 50%. Quarry dust (QD) was also used with normal Portland cement, silica fume and fly ash to prepare six types of concrete mixes which are designated as shown in Table 2. NPC concrete was considered as the control concrete. Quarry dust was used to replace 20% and 40% of sand respectively in preparing the concrete mixes containing quarry dust. Silica fume and fly ash were used as 10% replacement of cement in the concrete mixes containing silica fume and fly ash respectively. Sherbrooke mix design method was followed to get the mix proportions (Aitcin, 1997). The obtained proportions of crushed granite stone, mining sand and quarry dust were on the basis of saturated surface dry condition, but practically they were mixed in air dry condition. Hence, the amount of mixing water was corrected for the absorption of aggregates and quarry dust in conjunction with the consideration of water contributed by liquid superplasticizers. The details of mix proportions are given in Table 2.

Scope of Tests

Materials were weighed on air dry basis. The quantity of concrete prepared for each batch was at least 10% in excess of the requirement. The constituent materials were mixed in a rotating pan type mixer (capacity 0.05 m^3) at ambient temperature. Superplasticizer and air entraining admixture were added by two equal parts with the mix water before and during mixing.

Table 2: Mix Proportions

Type of Mix	CA	FA	QD	B			W	SP (%)	AEA (%)
				C	SF	FA			
Water-Binder Ratio = 35%									
NPC	1013	651	-	530	-	-	185.5	1.75	0.043
NPC20QD	1013	521	133	530	-	-	185.5	1.75	0.043
NPC40QD	1013	391	266	530	-	-	185.5	1.75	0.043
FA	1003	644	-	477	-	53	185.5	1.75	0.093
FA20QD	1003	515	132	477	-	53	185.5	1.75	0.093
FA40QD	1003	386	264	477	-	53	185.5	1.75	0.093
SF	1002	638	-	477	53	-	185.5	2.25	0.068
SF20QD	1002	510	131	477	53	-	185.5	2.25	0.068
SF40QD	1002	383	262	477	53	-	185.5	2.25	0.068
Water-Binder Ratio = 50%									
NPC	1055	690	-	400	-	-	200.0	1.00	0.055
NPC20QD	1055	552	141	400	-	-	200.0	1.00	0.055
NPC40QD	1055	414	282	400	-	-	200.0	1.00	0.055
FA	1047	685	-	360	-	40	200.0	1.00	0.089
FA20QD	1047	548	140	360	-	40	200.0	1.00	0.089
FA40QD	1047	411	280	360	-	40	200.0	1.00	0.089
SF	1046	678	-	360	40	-	200.0	2.00	0.067
SF20QD	1046	542	139	360	40	-	200.0	2.00	0.067
SF40QD	1046	407	278	360	40	-	200.0	2.00	0.067

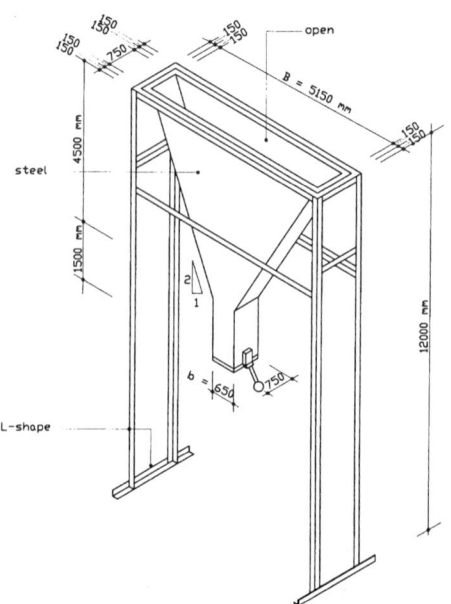

Figure 1: V-Funnel Apparatus

The mixing was carried out for about six minutes. Immediately after the completion of mixing, fresh concrete was sampled for the determination of slump, slump flow, V-funnel flow, mix temperature, entrained air content and unit weight. ASTM Standards, ASTM C143-90a (1990) and ASTM 1064-86 (1986), were adhered respectively to determine slump and slump flow, and the mix temperature. Entrained air content and unit weight were measured according to British Standards, BS 1881: Part 6 (1983) and BS 1881: Part 7 (1983). V-funnel flow was determined by an apparatus shown in Figure 1. This apparatus is similar in concept to the apparatus designed and developed by the Japanese researchers (Uomoto & Ozawa, 1994).

TEST RESULTS AND DISCUSSION

The freshly mixed high performance concrete was sampled and tested to determine some of its physical properties. These are given in Table 3. It was observed that the properties of fresh concrete are interrelated, especially the two characteristic flows and the air content. The relations between the two characteristic flows and the air content are shown in Figure 2 and Figure 3.

Table 3: Physical Properties of Freshly Mixed High Performance Concrete

Type of Mix	Slump (cm)	Slump Flow (cm)	V-Funnel Flow (l/s)	Air Content (%)	Unit Weight (kg/m^3)	Mix Temperature (^{0}C)
Water-Binder Ratio = 35%						
NPC	17.0	48.5	1.03	2.0	2409	25
NPC20QD	18.5	49.0	0.95	2.1	2398	25
NPC40QD	19.5	49.5	0.66	2.2	2399	26
FA	17.0	45.0	0.80	1.9	2366	28
FA20QD	18.5	47.5	0.70	1.9	2369	27
FA40QD	20.0	48.5	0.49	2.0	2384	28
SF	19.0	49.5	0.39	2.2	2360	26
SF20QD	19.5	50.5	0.35	2.3	2349	28
SF40QD	21.0	52.5	0.25	2.4	2336	28
Water-Binder Ratio = 50%						
NPC	17.5	50.5	1.82	1.6	2313	24
NPC20QD	20.0	51.0	1.74	1.7	2337	24
NPC40QD	21.0	51.0	1.41	1.8	2382	24
FA	19.0	46.0	1.56	1.8	2361	24
FA20QD	20.0	48.5	1.54	1.9	2352	25
FA40QD	21.5	50.0	1.51	2.0	2351	25
SF	20.0	51.5	1.49	1.6	2355	25
SF20QD	21.5	53.0	1.40	1.7	2326	25
SF40QD	23.0	53.5	1.29	1.8	2339	26

The obtained slump of the concrete mixes was between 17 and 23 cm. The slump flow of the mixes was in the range of 45 to 54 cm. The mixes with 35% water-binder ratio produced lower slump, slump flow and V-funnel flow than the mixes with 50% water-binder ratio. The mixes with 35% water-binder ratio possess higher volume concentration of cementitious materials but lower volume concentration of aggregates. Besides, the binder-aggregate ratio of the mix with 35% water-binder ratio is higher than that of the mixes with 50% water-binder ratio. Hence, according to earlier studies (Hobbs, 1976; Zain et al., 1999), the former mixes should provide higher flow properties than the latter. Exceptionally, the findings of the present study are not similar. This is because the flow properties do not depend upon the volume concentration of cementitious materials and aggregates alone. It is directed by the availability of water at a large extent and controlled by the dosages of superplasticizer and air entraining admixture. The available water in the mixes with 50% water-binder ratio was greater than that in the mixes with 35% water-binder ratio. The availability of sufficient water is the key factor to maintain the flowability in the fresh mixes. Moreover, high performance concrete usually shows a gluey behavior due to its cohesiveness, which plays a part in reducing the flow rate. This tendency was evidently higher in the mixes with 35% water-binder ratio.

It was also distinguishable that the mixes containing mineral admixtures and/or quarry dust produced higher slump and slump flow but lower V-funnel flow than NPC concrete mixes. The reduction in V-funnel flow is possibly due to the plastic behavior of mineral admixture and the arching of aggregates and/or quarry dust at the opening of the V-funnel. Besides, compared to FA concrete, V-funnel flow was lower in SF concrete. Silica fume is more plastic than fly ash in binding paste and becomes sticky to the sides of the V-funnel. Moreover, the friction between the vertical sides and the mix acts against the flow. As a result, the V-funnel flow was reduced considerably.

The incorporation of quarry dust in concrete mixes efficiently improved the flow properties of concrete mixes in terms of slump and slump flow. Conversely, it caused the reduction in V-funnel flow. The decrease in the viscosity and the increase in slump and slump flow in the concrete mixes with quarry dust are caused by the decrease in finer particles existent in NPC, FA and SF concrete mixes containing sand alone as fine aggregates. Besides, it was expected that the slump and slump flow of the concrete mixes containing silica fume and quarry dust in together would be reduced due to the sticky behavior and higher specific surface of silica fume. Instead, these mixes exhibited excellent slump and slump flow. Although SF20QD and SF40QD concrete mixes include 10% silica fume, it seems the total specific surface area of finer materials is reduced due to the 20% and 40 % replacement of sand respectively with the coarser quarry dust. As a result, it induces higher slump and slump flow. In opposition, the greater amount of coarser materials persuades the arching of aggregates at the neck of V-funnel. Quarry dust was coarser than sand with a higher fineness modulus of 4.20. Therefore, the amount of coarser materials in the concrete mixes incorporating quarry dust was higher which tempted the arching of aggregates and lowered the V-funnel flow.

Superplasticizer was used in all concrete mixes. It lessened the water demand and provided higher flowability. Furthermore, it improved the packing of the concrete mixes by improved grading in the range of the fines. As superplasticizer improved the packing of a concrete mix, a self-compacting effect evolved at higher consistencies. During self-compaction, the flowable binding paste tends to release the entrapped air and superplasticizer takes this entrapped air out of the concrete mix. The entrapped air is inclined to separate the component materials of concrete mix, which in turn tempts to decrease the flowability. Therefore, the removal of entrapped air is beneficial to the flowability.

The concrete mixes containing silica fume required higher dosage of superplasticizer than the mixes with normal Portland cement and fly ash. It follows that the concrete mixes with silica fume offer larger specific surfaces, which demand higher amount of superplasticizer to reach a chosen consistency. Silica fume particles, which cause the increase in the dosage of superplasticizer, are in the size range in which the flocculation forces act. The flocculation of the particles decreases the flowability of the mixes. Consequently, it causes the increase in the demand of superplasticizer. This explanation is supported by one earlier work in the related field (Durekovic & Popovic, 1990).

It is well known that the use of air entraining admixture results in the reduction of mixing water and gives more homogeneous, less permeable concrete by the distribution of microscopic fine spheroids of entrained air in the binding paste. The spheroids of entrained air would decrease the porosity by reducing the capillary water channel structure of the binding paste. The dosages of air entraining admixture largely depend on the mix composition. The presence of mineral admixture and superplasticizer controls the dosages of air entraining admixture. In the present study, the mixes containing silica fume consumed lower amount of air entraining admixture than the mixes with fly ash. Practically, the mixes with silica fume should need a higher dosage of air entraining admixture, but it also depends on the presence of superplasticizer. In presence of superplasticizer, the demand of air entraining admixture decreases (Durekovic & Popovic, 1990). As the mixes with silica fume consumed higher dosages of superplasticizer, they needed comparatively lower dosages of air entraining admixture.

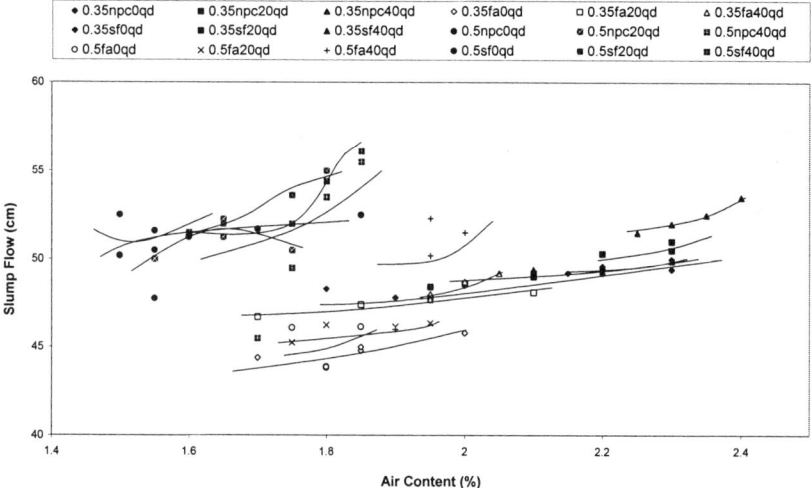

Figure 2: Relation between Slump Flow and Air Content

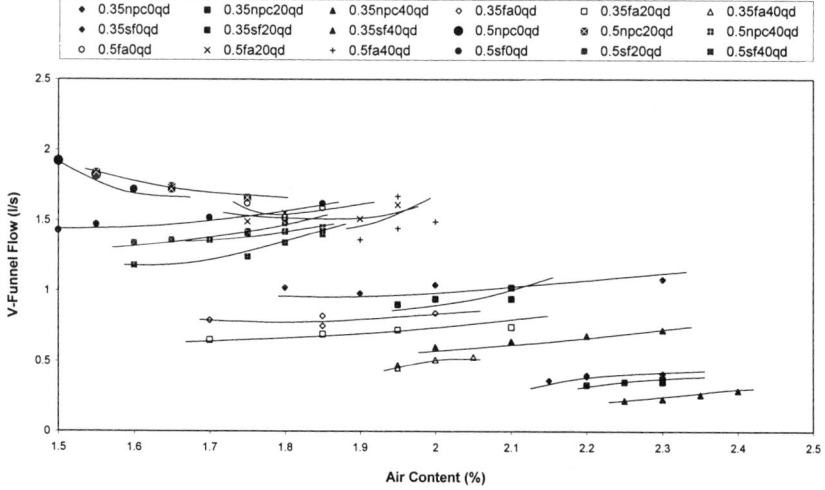

Figure 3: Relation between V-Funnel Flow and Air Content

The recommended amount of entrained air in concrete depends on the purpose for which the concrete is going to be used, the location and climatic condition, the maximum size of aggregate and the richness of the mix. Usually, the acceptable amount of entrained air ranges from 3% to 6% (Shetty, 1993). In the present study, the mixing was conducted in a non-freezing environment. Hence, a lower amount of entrained air was sufficient to facilitate the flowability. The entrained air content in the mixes was between 1.5% and 2.5%. It is true that entrained air improves the flowabilty of fresh concrete, but the excessive amount of entrained air does not favor the flowability. From the outcome of present study, it is evident that the higher amount of entrained air tends to decrease the flowability. The effect of entrained air content on slump flow and V-funnel flow has been shown in Figure 2 and Figure 3. It is suggested that an entrained air content between 1.5% and 2% is adequate to maintain excellent flowability in hot weather concreting.

The unit weight of concrete mixes was in the range of 2313 kg/m^3 to 2409 kg/m^3. The variations were mainly due to the differences in mix proportions and the presence of mineral and chemical admixtures. The mixes with 35% water-binder ratio produced higher unit weight than the mixes with 50% water-binder ratio. This is because the former mixes possess the higher content of cement and mineral admixtures. The higher volume concentration of cementitious materials makes the mixes with 35% water-binder ratio less porous and highly dense resulting in greater unit weight.

The flowability of concrete mixes is also a function of mix temperature. The evaporation from the mixes increases with the higher mix temperature and results in the greater loss of water and entrained air. This in turn tends to reduce the flowability. It is noticeable that the lower mix temperature was recorded in the mixes with 50% water-binder ratio. The mix temperature depends on the ambient temperature and the reactions that take place within the concrete mixes. As the ambient temperature was almost same for all mixes, the heat evolved during the reactions influenced the mix temperature. The greater amounts of cementitious materials and chemical admixtures extend the scope of chemical reactions that evolve higher heat energy. This might be a reason for the higher mix temperatures that have been recorded in the mixes with 35% water-binder ratio.

The mix temperature should be maintained within the range of 15^{0}C to 25^{0}C (Aitcin, 1993). It is well established that the control of the parameters of the air bubble system becomes rather difficult when concrete temperature is higher than 25^{0}C. In this study, the mixing was carried out at ambient temperature and the temperature of concrete mixes was in the range of 24^{0}C to 28^{0}C. Therefore, difficulties were experienced to maintain the required amount of entrained air in controlling the flowability. However, the amount of entrained air was maintained above the minimum air content of 1.5% by balancing the dosages of superplasticizer and air entraining admixture. It is recommended from the outcome of present study that an efficient control of the entrained air in concrete mixes is necessary to maintain the flowability during the production and delivery of high performance concrete in hot climate. Superplasticizer in combination with air entraining admixture can ensure this criterion.

CONCLUSIONS

The following conclusions can be drawn based on the results of the present study concerning the effects of admixtures and quarry dust on the physical properties of fresh high performance concrete:

1. Incorporation of mineral and chemical admixtures provides higher slump and slump flow in fresh mixes of high performance concrete.

2. Silica fume in company with superplasticizer produces higher slump and slump flow than fly ash in fresh concrete mixes.

3. Superplasticizer along with air entraining admixture provides the better control of entrained air and flowability in freshly mixed concrete.

4. Lower dosages of air entraining admixture between 1.5% and 2% are sufficient in hot weather concreting to enhance the flowability.

5. The amounts of cementitious materials and chemical admixture influences the mix temperature that takes an important role in the control of entrained air in fresh concrete mixes.

6. The partial replacement of sand by quarry dust increases the slump and slump flow of fresh concrete mixes.

7. V-funnel flow of the concrete mixes was not in harmony with the slump and slump flow. Further study is necessary to understand the mechanism of V-funnel flow and to identify the factors that affect the V-funnel flow.

ACKNOWLEDGEMENT

The authors would like to express sincere gratitude to University Kebangsaan Malaysia for providing the fund for the research. The authors also thank Ready Mixed Concrete (M) Sdn. Bhd., Sika Kimia (M) Sdn. Bhd. and Grace (M) Sdn. Bhd. for the supply of test materials and technical support throughout the research program.

REFERENCES

ACI SP-19 (1988). Cement and Concrete Terminology. *ACI Special Publication*, Vol. 19, American Concrete Institute, Detroit, Michigan, USA, p.107.

Aitcin, P.C. (1997). Sherbrooke Mix Design Method. *Proceedings of the One-day Short Course on Concrete Technology and High Performance Concrete: Properties and Durability*, University of Malaya, Kuala Lumpur, Malaysia, pp.49-83.

Aitcin, P.C. (1993). Durable Concrete: Current Practice and Future Trends. *Proceedings of V.M. Malhotra Symposium, ACI Special Publication*, Vol. 144, American Concrete Institute, Detroit, Michigan, USA, pp.85-104.

Aitcin, P.C. (1992). The Use of Superplasticizer in High Performance Concrete. *In Malier, Y. (Ed.), High Performance Concrete: From Material to Structure*, E & FN Spon, London, UK, pp.14-33.

ASTM C1017-92 (1992). Standard Specification for Chemical Admixtures for Use in Producing Flowing Concrete. *Annual Book of ASTM Standards*, Vol. 04.02, American Society for Testing and Materials, Philadelphia, USA.

ASTM C143-90a (1990). Standard Test Method for Slump of Hydraulic Cement Concrete. *Annual Book of ASTM Standards*, Vol. 04.02, American Society for Testing and Materials, Philadelphia, USA.

ASTM C1064-86 (1986). Standard Test Method for Temperature of Freshly Mixed Portland Cement Concrete. *Annual Book of ASTM Standards*, Vol. 04.02, American Society for Testing and Materials, Philadelphia, USA.

BS 1881: Part 106 (1983). Methods for Determination of Air Content of Fresh Concrete. *Testing Concrete*, British Standards Institution, London, UK.

BS 1881: Part 107 (1983). Methods for Determination of Density of Compacted Fresh Concrete. *Testing Concrete*, British Standards Institution, London, UK.

Cook, J.E. (1982). Research and Application of High Strength Concrete Using Class C Fly Ash. *Concrete International*, Vol. 4, No. 7, pp.72-80.

Dodson, V.H. (1990). *Concrete Admixtures*. Van Nostrand Reinhold, New York, USA.

Durekovic, A., and Popovic, K. (1990). Superplasticizer and Air-Entraining Agent in OPC Mortars Containing Silica Fume. *In Vazqueze. E. (Ed.), Admixtures for Concrete: Improvement of Properties, RILEM Proceedings 5*, Chapman and Hall, London, UK, pp.1-9.

Hayakawa, M., Matsouka, Y., and Shindoh, T. (1994). Development and Application of Super-workable Concrete. *In Bartos, P.J.M. (Ed.), Special Concretes: Workability and Mixing, RILEM Proceedings 24*, E & FN Spon, London, UK, pp.183-190.

Hobbs, D.W. (1997). Influence of Aggregate Volume Concentration upon the Workability of Concrete and Some Predictions Resulting from the Viscosity-Elasticity Analogy. *Magazine of Concrete Research*, Vol. 28, pp.191-202.

Malhotra, V.M. (1987). *Supplementary Cementing Materials for Concrete*. CANMET SP 86-8E, Ottawa, Canada.

Nagaraj, T.S., and Banu, Z. (1996). Efficient Utilization of Rock Dust and Pebbles as Aggregates in Portland Cement Concrete. *The Indian Concrete Journal*, Vol. 70, No. 1, pp.53-56.

Nawy, E.G. (1996). *Fundamentals of High Strength High Performance Concrete*. Longman Group Limited, London, UK.

Powers, T.C. (1993). The Bleeding of Portland Cement Paste, Mortar and Concrete. *Research Bulletin*, No. 2, Portland Cement Association, p.97.

Safiuddin, M., Zain, M.F.M., and Yusof, K.M. (2000). Utilization of Quarry Dust in Developing High Performance Concrete. *Proceedings of the 3rd Structural Specialty Conference of the Canadian Society for Civil Engineering*, London, Ontario, Canada, pp.378-384.

Shetty, M.S. (1993). *Concrete Technology*. S. Chand & Company Limited, India.

Uchikawa, H., Hanehara, S., and Hirao, H. (1995). Influence of Microstructure on the Physical Properties of Concrete Prepared by Substituting Mineral Powder for Part of Fine Aggregate. *Cement and Concrete Research*, Vol. 26, No. 1, pp.101-111.

Uomoto, T. and Ozawa, K. (1994). Committee Report on Super-workable Concrete. *JCI Proceedings*, Vol. 16, No. 1, Japanese Concrete Institute, pp.11-14.

Zain, M.F.M., Safiuddin, M., and Yusof, K.M. (1999). A Study on the Properties of Freshly Mixed High Performance Concrete. *Cement and Concrete Research*, Vol. 29, No. 9, pp.1427-1432.

TOWARDS SUSTAINABLE CONCRETE TECHNOLOGY IN AFRICA

Ali S. Ngab[1] and S.P.Bindra[2]

[1] Department of Civil Engineering, Al Fateh University, Tripoli, Libya
[2] Department of Civil Engineering, Engineering Academy Tajoura, Tripoli, Libya

ABSTRACT

Concrete buildings save significant energy and emit lesser emissions compared to buildings comprised of lighter materials. There are little or no defined parameters to fulfill the requirements of sustainable development. The paper focuses on problems and challenges for the concrete industry today and in the future with special reference to environmental concerns. It demonstrates that how these concerns can be effectively addressed in evaluation of concrete as a versatile building materials in the light of the ability to fulfill the requirements of sustainable development for both developed and developing countries in general and for African region in particular. It analyses the current state of concrete industry in its march towards the new millennium based upon the results of a major exercise in the form of regional survey initiated in Libya. The survey in the form of a comprehensive questionnaire involves considered views, best practice case studies, design methods, concrete as a building product etc. It probes issues like conservation of concrete making materials, enhancement of durability of new and existing structures and need for holistic education, training and research in concrete technology.

KEYWORDS

Sustainable, Concrete Technology, Indicators, Survey, Questionnaire, Africa

INTRODUCTION

Concrete industry today is the largest consumer of natural resources such as water, sand, gravel, and crushed rock. Recent estimates (3) show that it consumes around 8 billion tonnes of natural aggregates and over one billion tonnes of Portland cement. Each tonne of Portland cement is accompanied by releases of around one tonne of carbon dioxide which is one of the Green House Gases (GHG) responsible for global warming leading to climate change. Thus concrete industry is a large consumer of natural resources and energy consumption that has a big impact on natural environment. It has come under close scrutiny because of its being one of the most energy intensive and environmental polluting industry by Kyoto Protocol to the UN Framework Convention on Climate Change (Dec 1997).

Concrete production involves carefully controlled mixture of cement, sand, coarse aggregates and water which is mixed and placed in a manner to obtain concrete with good workability, durability, and strength. Expansive alkali reaction and chemical attack, freezing and thawing in cold climate regions, premature corrosion of embedded steel and inadequate structural design for seismic loading are some of the glaring examples of threat to long term performance and reduced service life of concrete structures. After a useful life of a few decades nearly all of it waste. With the fast depletion of raw natural material resources for concrete production, and public concern due to environmental pollution the construction industry is facing a crisis and will soon become unsustainable. Therefore, the concrete industry has an immense responsibility both for a more efficient utilisation of these resources, and taking measures to prevent environmental degradation and to create a better harmony with the natural environment.

What are alternatives to satisfy both developing countries and today's consumers? How can we avoid environmental problems and increase quality of life? Our assets are material, intellectual and human wealth to take on this challenge. Fair Share of Environmental Space (FSES) originally introduced by Green Peace as a challenging concept to share in resource use in each country based on population can be extended for concrete production. FSES sets an inspiring course for innovative businesses, local authorities and communities. FSES will be able to benefit the poor countries by joining the global economy without further straining the already overloaded ecological and life support systems of the planet. The state of concrete material production demonstrates that an increasing in concrete use is causing raw material resource depletion, air pollution, and creates risks of global warming. This trend is becoming progressively unacceptable in many countries of both developed and developing world. A number of national and local authorities are endeavouring to design highly efficient plants using better concrete production processes and management philosophies for concrete production in cities.

Case studies in Libya and its surrounds in Africa and Mediterranean region identifies key elements upon which the foundation of an environmental friendly sustainable technology for concrete technology can be built. After examining issues related to design, new technologies, controls, corporate culture, new processes, and conservation of concrete making materials vital to put concrete industry on sustainable path. Recommendation of specific strategies for application in Libya, Mediterranean countries, African region and possibly other parts of the world are given.

AN APPRAISAL OF CONCRETE CONSTRUCTION

Concrete construction utilising plain, reinforced and prestressed concrete depending upon the particular end use, plays a dominant role in the construction industries of both developed and developing countries. This is primarily because of the reason that concrete buildings save significant energy and emit lesser emissions compared to buildings comprised of lighter materials. Appraisal shows that construction industry cannot support several billion tonnes of radically poor energy intensive and wasteful construction materials most of which is concrete each year. Yet that is where our present development course will take us in 50 years. In recent years novel concepts like material effectiveness and longevity, minimum materials design and production, scrap recovery, reuse, recycling and materials savings through better quality, greater product effectiveness and smarter and intelligent design are gaining wide spread attention. Some of the new materials using closed loop principles are gaining novel properties from reprocessing to become a major material for modern construction. The case in point is "Environ" biocomposite- a decorative surface finish material, made from recycled paper and bioresin, that looks like stone, cuts like wood, is twice as hard as red oak, and has half the weight of granite but better abrasion resistance.

Experts believe that if entire spectrum of materials savings were systematically applied using smarter design, this would reduce the total flow of materials needed to sustain building industry by a factor much nearer to one hundred, or even more. Research and development studies show that it is possible to produce concrete, which reduces use of energy and raw material resources. As per an estimate, in recent years the energy used to produce concrete has been falling by typically a percent or two a year– faster when energy prices rise, slower when they fall. It is found that a host of opportunities still exists for doing more with much, much less. This would not only reduce waste and improve concrete quality as we develop new technologies and find better ways to apply them.

An overview of current data shows that more industrialized countries as main culprits.
- Global and regional ecological limits are already being exceeded by developed countries which Are the main culprit (5Tonnes of CO2 emission compared to 0.5 Tones /per person per year in developing countries).
- One billion i.e. 20% world population uses 80% of concrete, fossil fuel, metals, woods, minerals and other resources every year.
- Total consumption is still growing despite increasing the efficiency of using these resources.
- Population and landscaped destruction is increasing.

FSES is the total amount of energy, non-renewal resources, agricultural land and forests that we can use without causing irreversible environmental damage or depriving future generations of the resources they will need. This concept can be extended for concrete production to ensure that share of concrete production has to be increased for developing countries and reduced in developed countries in proportion to the population. The amount of environmental space is limited. This would require setting practical targets. To set clear and realistic direction, speed and order of magnitude of change is desperately needed to implement radical change in design of product and processes to realize close loop economy.

To be more responsible and set targets, Germany based Wuppertal Institute & International Panel on Climate Change (IPCC) has set limits for CO2 as 1.1 Tones per person per year for an optimistic scenario based upon projection of 10billion population & 1.7 Tones for 7 billion population. Currently Germany emits 12 Tones of CO2 and Europe 7 Tones of CO2 emission. This would require changes in the spatial organization of economy. This means we need Economic instruments and command and control measures (such as the ban on CFCs) to reduce and reallocate use of environmental space. It is critical that prices reflect the ecological truth. Subsidies for unsustainable agriculture and nuclear energy, for e.g. have to be removed. National governments should work, individually and together, to introduce a revenue-neutral 'ecological' tax reform, preferably at a pan-African, Pan European or EU level. By increasing taxes on resources, waste and pollution, with compensating measures for the least well off, consumption patterns can be made greener. Use the revenue raised to cut taxes on labor and to create new, environmentally sound job opportunities.

INITIATIVES AND METHODS FOR SUSTAINABLE CONCRETE TECHNOLOGY

Sustainable development requires that waste must be greatly reduced without compromising our well being. Any improvement that provides the same or a better performance from a lesser or no waste can produce sustainable concrete with less effort, transportation, waste, and cost. Therefore, sustainable technology for concrete production requires application of life cycle methodology to design, manufacture, construction, maintenance and the management of building projects. Through the principles of sustainability, sustainable concrete technology introduces resistance design –a design that needs to be expanded into durability design to include time as a new dimension in the design calculation. Thus it is a process and practice of meeting the needs of present without sacrificing the needs of future generation. It is an environmentally sound dynamic process - a commitment to rational exploitation of resources and environmental management with principle of equity and social justice. It not only helps to meet the challenges to protect our building industry against physical, biological and social evils but also provides tools, techniques and technology to preserve, protect, enhance, beautify and conserve it. Based upon best practice initiatives (2, 4) methods for development of sustainable concrete technology can be classified into the following six main categories which reinforce one another.

Design; New Technologies; Controls; Corporate Culture; New Processes, Saving Materials

Design

This involves applying holistic approach in all phases of integrated life cycle design of concrete structures for meeting multiple requirements of functionality, durability, aesthetics, health, economy and ecology. State of the art shows that virtually all the energy saving equipment used in concrete raw material and finished product production now in use were designed using rules of thumb that are not right. There is scope and need to achieve large energy savings in such common place equipment as valves, ducts, fans, dumpers, motors, wires, heat exchangers, insulation, and most other elements of technical design, in most of the technical systems that use energy, in most applications like crushing of boulder and grinding etc. This new efficiency revolution relies not so much on new technology as on the more intelligent application of existing technology. It is felt that best changes in the design are the simplest. For example, a costly task of removing contaminated air could be achieved by a new mechanical flow controller using a single moving part operated simply by gravity and airflow which can reduce energy by around 50-90 percent, reduce construction cost, and improve safety and performance. New geometries can quintuple the efficiency of aerators. Big savings in energy of the order of 40 percent escaping through the walls of cement manufacturing plant can be saved through super-insulation. Completely redesigning the process can drastically reduce the waste and effluents.

New Technologies

Trends show that new materials, design and fabrication techniques, electronics, and software can give integrated design of concrete structures and technology development patterns so that technologies become more powerful than the sum of their parts. This would require enhanced use of performance based design concept in the design of concrete structures in a unified manner. We can also use relutance motors in cement production (which can continuously adjust their software for peak efficiency under all operating conditions), smart materials to sophisticated sensors, rapid prototyping to ultra-precision fabrication, improved power-switching semiconductors to atomic scale manipulation, micro-fluidics, and micro-machines. As per an estimate, technologies available today can save about twice as much electricity in cement manufacture as was feasible 5 years ago, at only a third the real cost. Increased use of fuel cells and recovering waste heat can help to raise the useful work to more than 90 percent of the original final energy.

Controls

Enhanced use of information technology like using or writing a simple program to help optimizing the plant operation in concrete production can provide large savings. Better controls that measures the concrete quality actually coming out of batching plant and keep fine-tuning the process for desired results could cut the waste about in half. Measurement and control intelligence can be distributed into each piece of aggregate crushing, cement manufacturing and concrete production equipment so that each part of the process governs itself. Reactions can be kept at right temperatures and stone crushing machine tools fed to crush at the optimal rate. The more localized the control and feedback, the more precise the level of control. Use of ubiquitous microchips now permit not just such simple controls but also the construction of neural networks that learn, and use of fuzzy logic that helps make smart decisions. There is scope to employ distributed intelligence concept in self-organizing concrete production systems. It would help in many decentralized decision making, interpreting events, interacting with and learning from each other, and controlling their collective behavior through interaction of their diverse local decisions, much like an ecosystem works. Thus world of concrete production is to be controlled more by biologic- because biological systems already have evolved successful design solutions. Billions of years of design experience reflected in biological principles can be carried out in concrete process design and cement manufacturing using powerful computers equipped with artificial intelligence and genetic algorithms which evolve the fittest solutions by a mathematical vision of Darwinian natural selection.

Corporate Culture

There is need to take advantage of power tools for measurements, simulation, emulation and graphic display to turn the operating process from linear to cyclic which requires design, build, measure, analyze and improve. Improved maintenance and hardware typically yields savings approaching 50 percent with six months paybacks. Thus a substantial savings can be made to produce sustainable concrete by making the concrete production business as a learning organization rewarding measurement, monitoring, critical thought, and continuous improvement.

New Processes

Process innovations in concrete industry can help cut out steps, materials, and costs. They achieve better results using simpler and cheaper inputs. This requires improving processes by developing highly resource efficient materials, techniques, and equipment resulting into better output quality, less manufacturing time, less space, often less investment, and probably less total cost. High temperature process of cement manufacturing can be replaced by gentler, cheaper ones based on biological models that often involve microorganisms or enzymes. This can come by observing and imitating nature

Saving Materials

Good concrete should need less material to create a functional structure. This can be achieved by computer aided design which calculates stresses and strains and determines exactly how little material we make the structure just as strong as we want- but not stronger. Often this requires far less material by putting strength where it is needed. Thus changes in design can produce vastly better function. Another area of savings is the efficiency with which the raw material is converted into required concrete. Net shape and near net shape manufacturing can help concrete into the process where every molecule of material when fed in to the process emerge as a useful product. This unlocks a way to save materials consolidating many parts each individually fabricated into a single large part molded to net shape. Another ways to waste fewer concrete materials is to improve production quality and to improve design not merely of specific component but the entire process that uses it. Research needs to be carried out on repairs, reuse, upgrading, remanufacturing and recycling to keep the gift of good concrete and good work moving on to other users and other uses.

DISCUSSION OF RESPONSES FROM A TYPICAL CASE STUDY

A questionnaire was designed to know considered views, perceived problems and suggested solutions with regard to diverse aspects of concrete and sustainable development. The responses have been analyzed from the limited data obtained from both builders and users of concrete in Libya and surrounds. For the quality sustainable concrete check-off questions, weighted mean scores have been computed for each item and bar charts have been drawn to show the relative scores in decreasing order. Charts were presented in each case reflecting the views of the respondents. The results show the followings:

1. Most of respondents are married males with age varying between 25 to 60. Majority of them is full time government employees related to concrete industry. They produce, operate and use concrete from last over five years. They participate in concrete quality activities and have a few opportunities to present their ideas at sustainable concrete group meetings.
2. Majority of them feels that design, new technologies, controls, corporate culture, new processes and saving materials are important for development of sustainable concrete technology.
3. Nearly all of them feel that saving materials, enhancement of durability of concrete structures and holistic approach in concrete technology education, research and practice are essential elements for development of sustainable technology.
4. They also feel that at present consumer protection laws do not protect their interests.
5. Almost all the respondents feel the need for raising public awareness, visionary leadership from within concrete community, regulatory requirements, better energy management and quality consciousness culture to support the structure of an environment friendly concrete technology for sustainable development in Libya and its surrounds in Africa.

The analysis of opinions in the form of check-off questions which give factors that affect the sustainable concrete production management, construction operations and processes associated with environmentally critical areas, environmental pollution, national, regional & international cooperation, education and training, R&D and development policy, indicates that life in the future can be better for us all. The promises of business as usual economic growth will not deliver either a higher quality of life for the already well off or a higher standard of living for the poor in the West, East or south. We must start from environmental space to allow everyone to enjoy a good life.

Recommendations

1. The stored information in the form of database can be managed if identification of a sustainable concrete technology indicators is given based upon in depth R&D results on all types of concrete as a low cost building material, environmental impact assessments, remnant life assessment, and extended surveys from a broad spectrum of clientele from construction industry, governmental agencies and other organizations. This would require developing a package to perform the job of searching and sorting the number of measures in combination to resolve a problem. The computer package should be capable when asked to give integrated measures, which are specially needed in sustainable concrete production for quality buildings. Similarly one can ask computer to give integrated measures which have worked well in last 10 years or more based upon experience in a given area. In this way one can get the desired solution from the database quickly and easily.

2. A much larger sample is needed if we are to come to area specific conclusions. Many of the problems discussed are common to many other countries in Africa and may be similar other countries in the Mediterranean region.

3. Development of sustainable concrete technology requires adoption of holistic approach that has its roots in the idea that considers the needs and expectation of society as a whole and treats the concrete industry as a part of the whole. It needs extensive studies on other society needs like conservation of the earth's natural resources and safe disposal of polluting wastes in other industries.

4. With a view to ensure a more efficient interaction with society and their ultimate clients for promoting a sustainable development, there is an urgent need to re-examine the role, tasks, functions and their Modus Operandi of large number of existing local, national, regional and international societies, associations, bodies, networks and voluntary organisations related to concrete industry.

CONCLUDING REMARKS

Development of sustainable concrete technology requires an adoption of holistic approach that has its roots in the idea that considers the needs and expectation of society as a whole and treats the concrete industry as a part of the whole. An appraisal presented in this paper demonstrates that policies in most parts of world including Africa are for excessive concrete production leading to raw material depletion and pollution in their immediate surroundings. Bringing a shift in concrete production and construction trends in both developing and developed countries for enhancing the durability of concrete structures requires radical changes in design, new technologies, controls, corporate culture, new processes, and saving materials.

Based upon analysis of a survey using a comprehensive questionnaire has given an innovative approach to indicate that sustainable concrete technology in Africa is not a dream. It can be reality using a technology and practice which is capable of meeting the multiple requirements of the people, societies and nature in an optimal way during the life cycle of the concrete as a building material. This of course would require sound understanding of various sustainability factors like flexibility towards functional changes of the facility and high durability in the case of long target service life concrete structures and changeability, recycleability in the case of concrete structures with moderate or short target service life. Dealing with sustainability in concrete industry needs not only a concerted plan of action but also daring, caring and sharing approach. Daring without knowledge is risky. Knowledge without daring is fruitless.

REFERENCES

1. Bindra S.P. (1996). Eco-Friendly Construction Techniques in Malta: A Feasibility Report, Ministry of Public Works & Construction, Malta.
2. Bindra S.P. 2000. Transport Energy Conservation & Environment: Some Case Studies. Proc Energy Conservation in Industry and its Effect on the Environment Symposium, 19-21 September 2000; Tripoli, Libya.
3. Gjorv, Odd E and Sakai, K. (2000). Concrete Technology for a Sustainable Development in the 21st Century, E & FN Spon, London UK.
4. Hawken Paul, Lovins A & Lovins L (1999). Natural Capitalism - the Next Industrial Revolution, Earthscan Publications Ltd., London.

DEVELOPMENT OF ARTIFICIAL LIGHTWEIGHT AGGREGATES

S. Mohd[1], C. K. Wah[2] and P. Y. Lim[3]

[1] Assoc. Prof. School of Civil Engineering, Universiti Sains Malaysia, Perak Branch Campus, 31750, Perak, Malaysia
[2,3] Post Graduate Student School of Civil Engineering, Universiti Sains Malaysia, Perak Branch Campus, 31750, Perak, Malaysia
E-mail: cesabar@kcp.usm.my

ABSTRACT

Lightweight concrete has been used as building material for many decades throughout the world, especially in the countries like United States, Japan and Europe. The application ranges from lightweight partitions, walls and secondary structural components to the primary structural components. These concrete are either foamed type concrete or no fines, lightweight aggregates concrete. The compressive strength ranges from 7 N/mm2 for light partition to about 40 N/mm2 for the primary structural components. Recent development using artificially manufactured lightweight aggregates from various natural materials such as expanded clay, expanded shale, processed volcanic rocks, enable lightweight concrete to achieve compressive strength up to 70 N/mm2. With this kind of development, the use of lightweight concrete can be enhanced. This paper highlights a research program carried out by the authors to investigate the possibility of using industrial wastes; namely Pulverized Fuel Ash (PFA) obtained from Power Stations, mixed with quarry dusts from granite and limestone quarries, processed and manufactured into lightweight aggregates of various sizes. Investigation involves the determination of mix proportions and properties of the raw materials, production processes, and concrete mix design and the structural performance of the materials. Some preliminary results will be highlighted and discussed.

KEYWORDS

Artificial, aggregates, lightweight, high-strength, concrete, industrial-wastes, pulverized fuel ash, quarry-dust.

INTRODUCTION.

Concrete as a construction material is used for a wide range of application; from simple structures such as concrete drains to highly complicated advanced applications such as high rises, long span bridges, domes etc. It is highly durable and strong material. It can withstand almost any kind of environment and considerably less maintenance. With such a wide application, the level of technology required for the construction is also varied; from as simple as manually mixed on site by almost any level of knowledge and experience to produce concrete drains to advanced technology to produce long span slender structural components which requires a level of expertise to analyze, design and construct.

The technological development in the construction industry has obligatory the designers, engineers and people involved in the production of concrete to be more dynamic and adaptable to the changes. Range of application has increases. Construction are no longer only on natural grounds but on more difficult ground. Modern machineries, the use of Information Technology, the application of modularized prefab system are some of the technological changes in the construction industries. These changes will continue to takes place, forced the designers and engineers to find a more suitable alternative construction materials. One such material is Lightweight High Strength Concrete (LHSC).

Lightweight concrete has been used as building material for many decades throughout the world, especially in the countries like United States, Japan and Europe. The application ranges from lightweight partitions, walls and secondary structural components to the primary structural components. These concrete are either foamed type concrete or no fines, lightweight natural aggregates concrete. The compressive strength ranges from 7 N/mm2 for light partition to about 40 N/mm2 for the primary structural components. Recent development using artificially manufactured lightweight aggregates from various natural materials such as expanded clay, expanded shale, processed volcanic rocks, enable lightweight concrete to achieve compressive strength up to 70 N/mm2. With this kind of development, the use of lightweight concrete can be enhanced.

Lightweight Aggregates Concrete

Lightweight aggregates can be broadly classified into two categories; natural aggregates and artificially manufactured aggregates. Natural lightweight aggregates are normally obtained from the volcanic rocks such as pumice with density ranges from $500kg/m^3$ to $900kg/m^3$. It was formed from the sudden cooling of the volcanic lava. Concrete produced using natural lightweight aggregates in countries like UK, USA and Europe can achieve compressive strength up to about $40N/mm^2$ with density of about $1700N/mm^2$.

There are many types of artificially manufactured lightweight aggregate, amongst others are expanded clay, expanded shale, foamed slag, blast furnace slag, pulverized fuel ash and perlite. In UK, various types of lightweight aggregates are readily available in the market. Leca, Lytag, Pellite, Granulex and Liapor are some of the commercial names available in UK. The bulk dry density of these aggregates range from $425kg/m^3$ (expanded clay) to about $900kg/m^3$ (blast furnace slag) and can be used to produce concrete with compressive strength of about $40N/mm^2$ [R.L. Munn, 1993].

Lightweight Aggregates from PFA

Works on the use of PFA as the raw material for the production of lightweight aggregate can be traced back to the early 1930s. However, not until 1960s the first production of lightweight aggregate manufactured from PFA came into the market by a power plant in UK. This aggregate has been used quite extensively in UK especially for the pre-fabrication components [Torrey, 1978].

Aim and Objectives

The use of lightweight concrete as construction material in Malaysia is not common and can be considered at early stage. Range of application is limited to light partitions and secondary structural components. Type of lightweight concrete used is confined to the foamed type concrete with compressive strength ranges from as low as $7N/mm^2$ to about $12N/mm^2$ to produce pre-cast components. Natural aggregates are mostly from granite and limestone quarries, unlike in countries where lightweight natural aggregates can be found from, amongst other, the volcanic rocks.

In Japan, artificial manufactured aggregates from perlite, a type of volcanic rock, enable high strength lightweight concrete to be produced. These aggregates are having a specific gravity of around 0.8.

With the development of the steel industries and the construction of coal-fired power stations in Malaysia, the possibility of using wastes from these industries as artificially manufactured aggregates is high. Works on the processing fly ashes from the coal-fired power stations into artificial aggregates have been carried out for decades in countries like Europe and India. These aggregates can be used to produce concrete of compressive strength up to $50N/mm^2$ with density of around $1700N/mm^2$. The characteristics of the aggregates produced from the fly ashes varied, and very much dependent on the origin of the coal, type of firing system, and processing procedures. In principle, the type of aggregates intended can be designed and artificially produced.

This is the primary aim of this work, to produce artificial aggregates with relative gravity as low as possible using industrial wastes as the main raw material, namely fly ashes and steel slag. These aggregates are intended to be used in the production of high strength lightweight concrete for reinforced structural components as well as pre-stressed reinforced concrete structures. Ideally, the aim is to produce concrete with compressive strength of around $70N/mm^2$ at a density of around $1200N/mm^2$.

The objective of the work is in three folds. One is to reduce the self-weight of the components of a structure. With a density of about half of the normal weight concrete, the reduction in self-weight is significant. Generally speaking, about 70% of the service load on a foundation of a building at normal service condition is from dead load, which in turn about 70% is from the weight of concrete to form the structure. In some circumstances, especially the high rises; the cost of foundation can be as high as 40% of the total cost of the structure. Therefore, the use of lightweight concrete will reduce the foundation cost quite significantly.

Secondly is to reduce the transportation cost. The introduction of open industrialized modular system in building construction will change the traditional construction practice to the modern, industrialized manufacturing based construction system. When this happen, most of the structural components will be factory made. This will add to the existing factory made components such as concrete bridge girders, concrete poles, concrete manholes, etc. These components need to be transported to the site. Because of the limit on the weight by the highway authority for the trucks to pass on certain roads, the number of component can be carried per trip is limited, although the truck can take more. Therefore, the cost of transportation is high, in general is about 25% of the cost of the component. Therefore, the use of lightweight concrete will reduce the transportation cost by the increase in number of component carried per trip.

And, thirdly is to utilize waste materials and at the same time preserve natural resources. Slag from steel industries and pulverized fuel ash from coal-fired power stations are some of the industrial wastes. These waste materials are disposed of by land filling at open dumping sites within the premises of the industry. The accumulation of the material requires large dumping area and at the same time with the more stringent requirement by the authority on the characteristics of the wastes, forced the industries to find alternatives for the waste disposal.

METHODOLOGY

Raw materials from power station, steel industries and quarries were tested for its chemical composition, physical characteristics and loss-on-ignition. BS Standards and ASTM Standards were used where appropriate for all the tests. Sintering process was carried out for each raw material to evaluate the changes taken place during and after the sintering process. This initial sintering process was used to evaluate the suitability of the material for further investigation. These materials were then mixed at various combination and composition. Sintering process for each combination and composition were carried out to find the most suitable combination and composition of the raw materials, with PFA remains as the main raw material. Charcoal and clay were added as foaming agent and binder respectively. Sintered materials were tested for specific gravity, hardness and other physical characteristics. 75mm concrete cubes were tested for compressive strength at 28 days.

RESULTS AND DISCUSSION

Chemical composition and loss-On-Ignition of the raw materials used in this investigation are presented in Table 1. The chemical analysis of the raw materials shows that each of the raw materials do not contain any chemical that might adversely affect the properties of concrete. The chemical composition of Perlite given in Table 1 is used as comparison. Perlite has been used as main raw material to produce lightweight aggregates in Japan having a specific gravity of around 0.8. Carbon content in each of the raw materials is low as shown by the Loss on ignition results and suitable for use in concrete. Most importantly, the PFA used has shown having enough glassy minerals (SiO_2, Al_2O_3 and Fe_2O_3) to remain as main raw material. These minerals bloated at high temperature and remain glassy when cooled to form hard and dense material.

TABLE 1
CHEMICAL COMPOSITION OF RAW MATERIALS.

Mineral (%)	Raw Materials			
	PFA	Quarry Dust	Steel Slag	Perlite
SiO_2	50.04	75.0	38.8	74.2
Al_2O_3	24.0	14.0	6.55	13.2
Fe_2O_3	15.1	1.7	1.3	1.0
CaO	3.4	0.93	35.1	1.0
Na_2O	0.97	2.1	0.37	3.51
MgO	1.03	0.3	12.1	0.1
K_2O	2.12	6.0	0.47	4.35
SO_3	1.0	0.04	3.3	-
LOI	8.6	0.53	0.84	2.2

Table 2 shows the various combination and composition of the raw materials to form the test samples for the intended aggregates. PFA was used as the main raw material with amount ranges from 70% to about 50%. To improve its properties, quarry dust was added. Clay and charcoal was used as binder and foaming agents respectively. These samples were sintered at 1200°C.

As shown in Table 2, the average specific gravity of the aggregates produced ranges from as low as 1.167 to 1.414 with voids ranges from 51.6% to 41.7%. Average hardness using Rockwell Hardness test ranges from 64 to 86 using 60kg load.

(a) (b) (c)

Figure 1: Raw Materials – a) PFA; b) Quarry Dust; c) Clay

TABLE 2
COMPOSITION OF RAW MATERIALS

Sample	Raw material				Test Result		
	PFA	Quarry Dust	Clay	Charcoal	Specific Gravity	% Voids	Average Hardness
1	70	10	15	5	1.33	42.3	70
2	65	20	10	5	1.412	42.1	64
3	65	15	15	5	1.38	45.1	69
4	65	20	15	0	1.414	41.7	86
5	60	25	10	5	1.26	45.0	64
6	60	20	15	5	1.308	44.4	73
7	55	25	15	5	1.221	49.9	76
8	55	20	20	5	1.197	51.6	71
9	55	20	10	15	1.167	47.4	65
10	50	35	10	5	1.235	46.4	76

From these results, it was decided Sample 7 is the composition of raw materials to be investigated further. Three levels of maximum temperature were chosen; they are 1200°C, 1210°C and 1220°C.

Results and physical observation of the aggregates produced at different sintering maximum temperature are presented in Table 3. Results shows that pellets sintered at 1220°C could produce aggregates with low specific gravity and high crushing resistance. Rockwell Hardness test also show that the aggregates produced at 1220°C has the highest value. The aggregate crushing value (ACV) obtained from ACV test carried out in accordance to BS 812: part 110: 1990 was 14%. This indicates that the aggregates have high crushing resistance as compared to limestone aggregates with ACV ranges from 20% to 25%.

TABLE 3
PHYSICAL OBSERVATION OF SAMPLE 7

	Maximum temperature, ^{0}C		
	1200	1210	1220
Specific Gravity	1.22	1.21	1.20
Crushing resistance	Moderate	Moderate	High
% void	49.9	47	50
Rockwell Hardness	76	91	94
Colour	Brownish	Brownish	Brownish
Brittleness	Moderate	Moderate	Moderate

(a)   (b)   (c)

Figure 2 : a) Sintering Process; b) Artificial Aggregate samples; c) Aggregates from PFA alone

A series of 75mm concrete cubes were tested for compressive strength. The optimum mix design was established to produce the highest strength to density ratio. Various factors, which affect concrete strength, were investigated. In order to produce high strength cement paste, admixture was used as cement replacement material. The optimum amount of fine aggregates was also investigated, in conjunction with aggregate/cement ratio. The highest strength to density ratio obtained from this work was average cube strength of 61N/mm^2 at an average density of 1600kg/m^3. The corresponding mix proportion as given in Table 4.

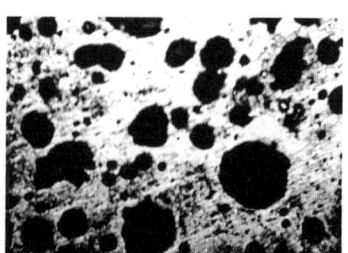 

Figure 3: Micro structure of an aggregate   Figure 4: Rockwell Hardness Test

TABLE 4
CONCRETE MIX PROPORTION

	Mix Proportion					Result		
	W/C	A:C	%Fines	% PFA	% Admixture	Compressive Strength (N/mm2)	Density (kg/m3)	Slump (mm)
Sample	0.37	3:1	40	20	1.5	61.0	1600	70

Figure 5: Concrete cubes tested for compressive strength

CONCLUDING REMARKS

The PFA obtained from coal-fired power station has the chemical composition suitable to be processed to produce artificially manufactured lightweight aggregates. Investigation has shown that its properties can be further improved by mixing it with granite quarry dust. The use of clay as binder agent further improved its palleting and sintering processes. Adding charcoal as foaming agent helps to produce voids to reduce its density.

Results obtained and presented here are preliminary. Further investigation is still being carried out to find the optimum composition of the raw material as well as the right sintering process.

Acknowledgement

The Author wish to thank the Ministry of Science Technology and Environment Malaysia for sponsoring this research program through IRPA Grant 1999, Zen Concrete Industries (M) Sdn bhd for jointly sponsoring the research and USM for providing the support needed.

References

ACI 211.2-91 (1994), *'Standard Practice For Selecting Proportions for Structural Lightweight Concrete'*

ACI 213R-87 (1994), *'Guide to Structural lightweight Aggregate Concrete'*

ACI 226.3R-87 (1994), *'Use of Fly Ash in Concrete'*

Ahlbeck (1982), *' lightweight Aggregates US patent'*, No 4,341,562

Andrew Short (1974), *'Lightweight Aggregate Concrete'*, Part1 Concrete July pp 47-48

Lydon F.D. (1975), *' Artificial Aggregates for Concrete'*, Concrete, September, pp49-52

Toray S (1978), *'Coal Ash Utilisation, Fly Ash, Bottom Ash and Slag'*, 1st Edition Noyes Data Corp., USA

Munn R.L. (1993), *'Lightweight aggregates and Lightweight Concrete'*, – Past, Present and Future, Australia

A COMPARATIVE STUDY OF NORMAL CONCRETE WITH CONCRETES CONTAINING GRANITE AND LATERITE FINE AGGREGATES

Funso Falade
Department of Civil Engineering
University of Lagos,
Nigeria
E-mail: ffalade@hotmail.com

ABSTRACT

Three types of fine aggregates are investigated for making concrete, namely sand for normal concrete, granite and laterite fines. The properties considered are compressive and tensile strengths. Three mix proportions of cement, fine and coarse aggregates are used: 1:1.5:3, 1:2:4 and 1:3:6 (cement: fine: coarse) with water/cement ratios of 0.62, 0.75 and 1.0 respectively. 100mm cubes and 150 x 300mm cylinders were prepared and cured in water at 21 ± 1^0 C. The specimens were tested for each mix at curing ages of 7, 14, 21, and 28 days. The results showed that concrete containing granite fines has the highest strength values both in compression and tension while concrete made with laterite fines has the lowest strength. For all the samples the strength characteristics decreased with increase in aggregate/cement ratio but increased with age irrespective of type of fine aggregate and mix proportion. Generally, the tensile strength was found to vary from 10 - 12% of the compressive strength values for 1:1.5:3 and 1:2:4 mixes and it was about 15% for 1:3:6 mix.

KEYWORDS

Concrete, Granite Chips, Granite Fines, Sand, Laterite Fines, Compressive Strength, Tensile Strength

INTRODUCTION

The strength of concrete is known to vary with some factors such as cement content, type of aggregates (fine and coarse), water/cement ratio, mix proportion, curing age, etc. Normal concrete has as its components cement, sand, and granite chips. Efforts have been made and continue to be made both within and outside Nigeria to replace these components especially cement and sand with a view to reducing construction cost without adverse effect on the strength characteristics of concrete. The use of cement offers the best method of stabilizing the constituents of concrete and its allied products. Falade (1) reported that the replacement of cement with sawdust ash in concrete reduced strength. He noted that the rate of gain in strength was more rapid at curing ages of 21 and 28 days in mixtures with high proportion of sawdust ash content than concrete with 0% ash. The results indicated that sawdust ash was not a good substitute for cement but that it possessed pozzolanic property. However it was recommended that some quantity of sawdust ash could be used in concrete as retarding agent without compromising its engineering properties. This is to delay the setting time of concrete thus compensating for the effect of hot weather both by helping to offset the lost of workability and minimize the increase in the amount of water to produce the required workability. Adepegba et al (2) studied the pozzolanic activities of six waste materials from mines and farms in Nigeria. They found that magnetite and limonite were heavy pozzolanas, while blast furnace slag, riget stone, riget coke and coal ash of palm bunch could be classified as normal weight which when treated could be used to blend the ordinary Portland cement for production of lime pozzolanic cement. Stroeven and Bui (3) studied the effect of rice husk ash on compressive strength of gap-graded concrete. They established that up to 40% of Portland cement could be replaced with rice husk ash in concrete. Stroeven and Vu (4) investigated the effect of partial replacement of Portland cement with Calcined Kaolin on cement pastes and mortars. They concluded that calcined kaolin is highly reactive pozzolan, which can be economically used to replace part of the Portland cement in concrete. They further observed that for moderate quantities of concrete in which the cement content would be around 300 or 325kg/m^3, the replacement percentage could be up to 30.

Sand is very expensive to procure and apart from this, depending on its source, it may require additional treatments such as washing to remove some undesirable substances and organic matters that may impair the strength of concrete. Investigations made on local raw materials have led to identification of laterite and granite fines as good substitutes for sand in concrete. Laterite fines are available in large quantity in tropical regions. The pioneer work on laterite in Nigeria was by Adepega (5) who compared the strength properties of normal concrete with those of 'laterized concrete'. He reported that laterite fines are good substitute for sand as component in concrete. It was reported by Balogun and Adepega (6) that when sand is mixed with laterite fines the most suitable mix for structural application is 1:1.5:3 with water/cement ratio of 0.65 provided that the laterite content is kept below 50%. Madu (7) observed that Eastern Nigeria laterite aggregates are good for road chippings and concrete but with slightly inferior results to those observed from igneous aggregates. Lasisi and Osunade (8) reported that laterite soils are highly heterogeneous and vary widely from one location to another, even with distance of few kilometers, hence empirical relationship derived from tests carried out in an area may not be necessarily applicable to soil in another location. In a related study, Falade (9) established that the location from where laterite is procured does not significantly affect its strength properties. Granite fines are a new material in the construction industry. They are produced as a by-product of granite chips in the quarries. The chips are widely used as coarse aggregate in concrete. Presently, only a small percentage of the granite fines is used for construction of blocks, electric poles, well and culvert rings.

The aim of this research is to investigate the strength characteristics of concrete made with three different fine aggregates using three mix proportions.

MATERIALS AND EXPERIMENTAL PROCEDURE

(i) *Materials*

The variables in this study are type of fine aggregate (sand, granite fines and laterite fines), mix proportion (1:1.5:3, 1:2:4 and 1:3:6) and curing ages (7,14,21 and 28 days). The cement used for this experiment is Ordinary Portland Cement from Ewekoro Cement Factory in Nigeria whose properties conform to BS 12 (10). The sand was obtained from Osun River Basin along the Ife-Ede Road. The granite fines were collected from Strabag quarry, Ibadan and the chips from a construction site at Obafemi Awolowo University. The laterite fines were collected from a borrow pit along the Ife-Ibadan road. The particle size of the fine aggregates used are those passing sieve No. 4 aperture 4.75mm but retained on sieve No. 220 aperture 0.063mm. The moisture contents of the fine aggregate are 3.40% for sand, 2.69% for laterite and 3.20% for granite fines while the specific gravities are 2.73, 2.62 and 2.70 for sand, laterite and granite fines respectively. Three mix proportions of cement, fine and coarse aggregates were used. These are 1:1.5:3, 1:2:4, 1:3:6 with water/cement ratios of 0.62, 0.75 and 1.0 respectively. The W/C ratios were kept constant for all the samples of the fine aggregate for each mix.

(ii) *Moulding of Specimens*

100mm cubes and 150 x 300mm cylinder specimens were prepared. Three specimens were made for each mix and age. Batching was by weight. Mixing was done mechanically in a mobile rotating drum mixer. The fine aggregate and cement were mixed together, and then the coarse was added. When the mixture had been thoroughly mixed, the required quantity of water was added and the mixing continued until an homogenous concrete mass was obtained. The specimens were made in accordance with BS 1881 (11). During the casting the moulds were intermittently shaken to ensure the dislodgement of trapped air so that a densified and non-entrained air concrete could be produced. The moulds with the cast specimens were left for $24\pm\frac{1}{2}$ hours with polythene cover before they were demoulded and transferred into a curing tank. The specimens were cured in water at 21 ± 1^0C

(iii) *Testing of Specimens*

Three specimens were tested at each curing age of 7, 14, 21, and 28 days. The strength characteristics of the specimens were tested using a loading rate of 120KN/min. on a 600KN Avery Denison Universal testing machine. The average of the load at which three specimens for each age failed was found and used to calculate the compressive and tensile strength values for each group.

RESULTS AND DISCUSSION

(i) *Results*

The particle size distribution of the fine and coarse aggregates is presented in Table 1. Particles passing sieve size 4.75mm but retained on 0.063mm were used. This ensured that the dust particles were removed from the mixes which otherwise might reduce the bond between the cement paste and coarse aggregate and this could result in lower strength. Figure 1-6 show the compressive and tensile strength values for the test specimens.

TABLE 1

RESULTS OF GRADATION OF AGGREGATES

SIEVE SIZE (mm)	PERCENTAGE PASSING (%)			
	Sand	Granite Fines	Laterite Fines	Gravel
25				100
20				94.29
14				10.56
10				4.2
4.75	99.52	96.6	99.8	
2.36	95	95.00	92.96	
1.10	92.88	41.54	53.60	
0.60	14.0	17.23	25.83	
0.425	11.84	13.47	16.72	
0.30	9.0	11.44	11.0	
0.15	1.52	9.29	6.80	
0.063	0.28	6.40	4.67	

Table 2 shows the composition of the fine aggregates as obtained from literature (12 – 14). Lack of chemicals hindered the determination of the actual oxides content of the fine aggregates used in this study but it is believed that the percentages of the various oxides are within the specified percentages contained in the referenced literature.

TABLE 2

OXIDE COMPOSITION OF THE FINE AGGREGATES

OXIDE	GRANITE* %	LATERITE** %	SAND*** %
SiO_2	71.05	54.53	62.45
Al_2O_3	15.30	20.89	17.42
Fe_2O_3	0.75	6.67	2.62
MgO	1.10	1.65	0.82
Na_2O	3.40	0.22	1.47
K_2O	4.90	3.71	2.94
TiO_2	0.25	0.92	0.43
P_2O_5	0.35	0.23	0.45
MnO	Trace	0.80	0.61
H_2O	0.06	-	-
CaO	0.90	0.54	0.82

Sources: *   Didier (12)
                **   Zdenek (13)
                ***   Geffrey (14).

(ii) *Discussion*

Generally, the results show that concrete containing low Aggregate/Cement ratio has higher strengths (compressive and tensile) than those with high Aggregate/Cement ratio. For example, in Fig 1 at 7- day curing age, the compressive strength of normal concrete having 1:1.5:3 (Agg/Cem = 4.5) is 15.2N/mm^2 and the corresponding tensile strength is 2.1N/mm^2 (Fig 2).

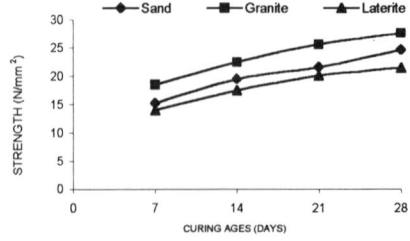

FIG.1 COMPRESSIVE STRENGTH FOR THE FINE AGGREGATES AT DIFFERENT CURING AGES (1:1 1/2:3 MIX)

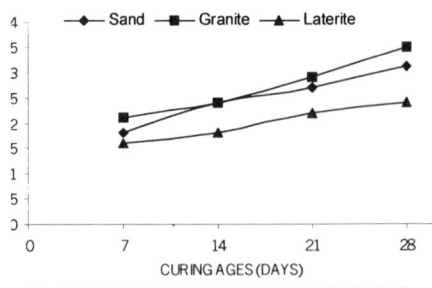

FIG. 2 TENSILE STRENGTH FOR THE FINE AGGREGATE AT DIFFERENT CURING AGES (1:1.5:3 MIX)

In Figures 3 and 4, for the same type of concrete and age the compressive and tensile strength values are 13.6N/mm² and 1.6N/mm² for 1:2:4 (Agg/Cem = 6)

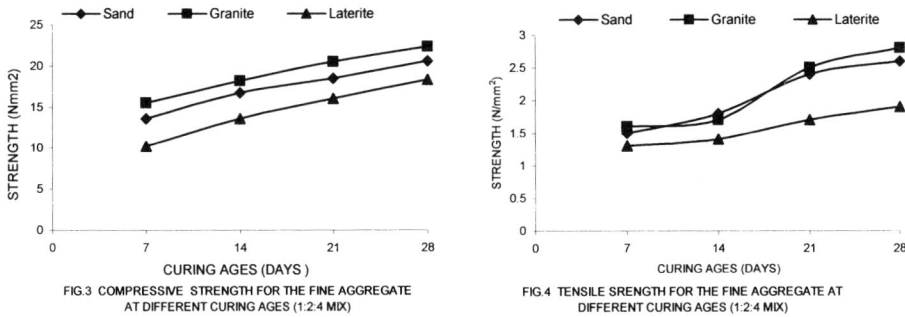

FIG.3 COMPRESSIVE STRENGTH FOR THE FINE AGGREGATE AT DIFFERENT CURING AGES (1:2:4 MIX)

FIG.4 TENSILE STRENGTH FOR THE FINE AGGREGATE AT DIFFERENT CURING AGES (1:2:4 MIX)

while in Figures 5 and 6, the compressive and tensile strength values are 9.1N/mm2 and 1.3N/mm2 for 1:3:6 (Agg/Cem = 9).

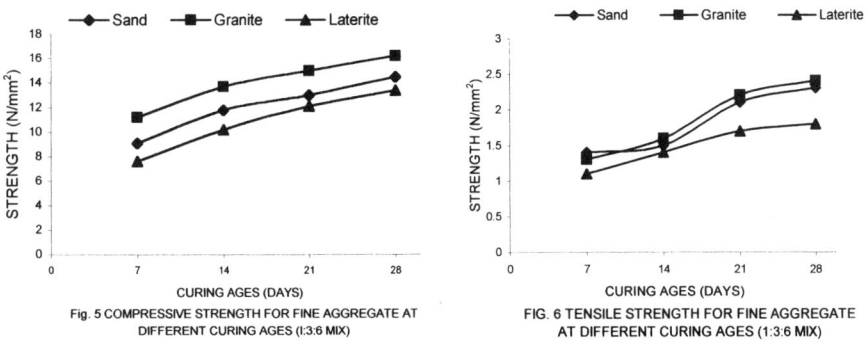

Fig. 5 COMPRESSIVE STRENGTH FOR FINE AGGREGATE AT DIFFERENT CURING AGES (1:3:6 MIX)

FIG. 6 TENSILE STRENGTH FOR FINE AGGREGATE AT DIFFERENT CURING AGES (1:3:6 MIX)

This observation is also true for concretes made with granite and laterite fines. The results further indicate that the strength characteristics vary with the type of fine aggregate. Concrete containing granite fines has the highest strength values followed by normal concrete while laterized concrete has the least. In Fig 1 for 1:1.5:3 mix proportion, the compressive strength at 7th day for concrete containing granite fine is 18.5N/mm² while for the same mix proportion and age the strength values are 15.2N/mm² and 14.0N/mm² for concrete containing sand and laterite fines (Figures 3 and 4) respectively. In Figures 2,4 and 6 for the same mix proportion and age the tensile strengths are 2.1N/mm², 1.8N/mm² and 1.6N/mm² for concrete containing Granite fines, Sand and Laterite fine. Strength increased with age irrespective of mix proportion and type of fine aggregate. For example, the strength values for concrete containing sand are 15.2N/mm², 19.5N/mm², 21.6N/mm² and 24.6N/mm² at 7, 14, 21 and 28 day for 1:1.5:3 mix proportion. For the same curing ages and mix proportion the strength values are 18.5N/mm², 22.5N/mm², 25.6N/mm² and 27.4N/mm² for concrete containing granite fines. The same trend was observed for concrete containing laterite fine. Figures 3,4,5,and 6 show that concrete having 1:2:4 and 1:3:6 mix proportions followed the same trend of increase in strength values with age. From Table 2, granite fines have silica (S_1O_2), content of 71.05%, sand has 62,45% and laterite fines has 54.53%. In the presence of moisture, lime (CaO) and silica (SiO_2) react to produce tricalcium silicate (C_3S) and dicalcium silicate (C_28). The hydration products of these two compounds are tobermorite gel and calcium hydroxide. The tobermorite particles are responsible for the cementing properties as well as other important engineering properties such as strength and shrinkage of concrete (Chong, 15). In the case under consideration, the reaction of lime (CaO, 64 - 67%) in cement (Neville, 16) with silica in the respective fine aggregates during the formation of cement paste may have been responsible for the difference in the development of strength since the silica content varies with type of fine aggregates. The higher the cementing property of a cement paste the higher the bond and strength.

It was found that the tensile strength varies from 10 - 12% of compressive strength for 1:1.5:3 and 1:2:4 mix proportions and about 15% for 1:1.5:3 mix. Other factors that may contribute to difference in strengths are water/cement ratio, moisture content and specific gravity of the fine aggregates. The moisture content for granite fines is 3.2, 3.4 for sand and 2.6 for laterite fines. While the specific gravities are 2.70, 2.73 and 2.62 for granite fines, sand and laterite fines respectively. The water/cement ratios used were obtained for concrete containing sand and they were used for concretes made with other fine aggregate without making adjustment for difference in moisture content and volume of fine aggregate as a result of difference in specific gravities since batching was by weight. The lower specific gravity of laterite fines shows that greater quantity of it will be needed on equal weight basis with sand and granite. This results in concrete containing laterite fine having particles with higher specific surface area than sand and granite fines thus requiring higher optimum W/C ratio at which maximum strength occurs, (Falade, 17).

CONCLUSIONS

The results of this study show that:

1. The three types of fine aggregates considered are suitable for making concrete.
2. The tensile strength was found to be between 10 - 12% of the compressive strength of 1: 1.5:3 and 1:2:4 mixes and about 15 & for 1:3:6 mix proportion.
3. The strengths of the specimen increase with age irrespective of type of fine aggregate and mix proportion.
4. The strength characteristics decrease with increase in aggregate/cement ration.

REFERENCES

1. F. Falade (1990), 'Effect of Sawdust Ash on the Strength of Laterized Concrete' West Indian Journal of Engineering. Vol. 15 No. 1 pp 71-84
2. D. Adepegba, J.A. Akinnawonu and A. Adejumo (1990), 'Exploitation of Pozzolanic Materials in Nigeria' Proc; Third Ansti Intl. Seminar of Civil Engineering Subnetwork, University of Mauritus, pp 97-107
3. P. Stroeven and D.O. Bui (1997), 'Effect of Rice Husk Ash on compressive Strength of Gap-graded Concrete Containing Very Fine Sand' procs; Structural Engineering Analysis and Modeling (SEAM4) Accra, Ghana, Vol. 1 pp 64-72.
4. P. Stroeven and D.D. Vu (1997), 'Effect of Kaolin on Characteristics of Portland Cement Pastes and Mortars' procs; Structural Engineering Analysis and Modeling (SEAM 4) Accra, Ghana, Vol. 1 pp 73-79.

5. D. Adepegba (1975), 'A Comparative Study of Normal Concrete with Concrete which contains Laterite Fine Instead of Sand'. Building Science, 10, 135 - 141

6. L.A. Balogun and D. Adepegba (1982) 'Effect of Varying Sand Content in Laterized Cone', Int. J. of Cement Compose Lightweight, 4, 235 - 241

7. R. M. Madu (1980): 'The Performance of Laterite Stone as Concrete Aggregates and Road Chippings; Bull. RILEM Mat. Struct. 13, 403-411.

8. F. Lasisi and J. A. Osunade (1984), 'Effect of Grain Size on the Strength of Cubes made from Lateritic Soils; Building and Environment. 19, 55-58.

9. F. Falade (1991), 'The Significance of Source of Laterite on the Strength of Cement-stabilized Lateritic Blocks, Journal of Housing Science, 15, 121-131.

10. British Standard Institution BS 12 (1971), 'Portland Cement (Ordinary and rapid hardening) Part 2.

11. British Standard Institution BS (1970), 'Method of Testing Concrete, Part 2.

12 J. Didier (1973), 'Granite and their enclave, McGraw Hill Book Company, New York

13. K. Zdenek (1971), 'Geology of Recent Sediments. McGraw Hill Book Company, New York .

14. G. Geffrey (1969), 'Composition and Analysis of Silicate rocks, Longman Publishing Company, London.

15. C. V. Y. Chong (1977), 'Properties of Materials, Macdonald & Evans Ltd., Plymouth.

16. A. N. Neville (1981), 'Properties of Concrete, English Language Book Society.

17. F. Falade (1994), 'Influence of Water/Cement Ratios and Mix proportions on workability and characteristic strength of concrete containing laterite fine aggregate' Building & Environment, Vol. 29, pp. 237-240.

THE INFLUENCE OF RECYCLED AGGREGATE ON THE EARLY COMPRESSIVE STRENGTH AND DRYING SHRINKAGE OF CONCRETE

A.R.M. Ridzuan[1], A.B.M. Diah[2], R. Hamir[2] and K.B. Kamarulzaman[2]

[1]Faculty of Civil Engineering, Mara University of Technology
40000 Shah Alam, Selangor, Malaysia
[2]School of Civil Engineering, Science University of Malaysia
Perak Branch Campus, 31750 Tronoh, Perak, Malaysia
cebakar@kcp.usm.my

ABSTRACT

Recycling waste concrete as source for the production of new concrete can help control environmental pollution and the problem of depleted natural aggregates. The effects of using crushed waste concrete as coarse aggregates upon strength and drying shrinkage of concrete were investigated. Waste concrete cubes which has been tested for compressive strength as a compliance of construction specification were crushed and utilize as coarse recycled aggregates in new concrete. It is important to mention that in order to simulate the real life condition waste concrete with very minimal information about its originality is used in its natural moisture condition. The recycled aggregates were tested for grading, specific gravity, bulk density, impact crushing and water absorption and the results compared with those for natural aggregate concrete. Concrete mixes of design strength of $20N/mm^2$, $25N/mm^2$, and $30N/mm^2$, were prepared using this recycled aggregates as coarse aggregates and tested. From the strength point of view the recycled aggregate concrete compared well with natural aggregate concrete. Therefore could consider for various potential applications. With respect to drying shrinkage the recycled aggregate concrete were found to have comparable shrinkage with the corresponding natural aggregate concrete.

KEYWORDS

Recycled, concrete, aggregate, compressive strength, drying shrinkage, waste concrete.

INTRODUCTION

Concrete is one of the widely use material in construction. Hence waste concrete is an unavoidable by product of concrete industry and construction. Waste concrete is derived from a number of sources. The most common are from demolished concrete structures, fresh batch leftovers, and waste concrete cubes that have been tested as compliance to building specification. This debris is usually thrown away causing environmental pollution. In some area in developed countries such as the United States, Europe and Japan natural aggregates is becoming scarce and bringing aggregates from far away places increases cost of concrete (Frondistou Yannas,1977). Since control of waste materials pollution is increasingly important, hence recycling these waste concrete materials seem inevitable solution. Although in Europe and other developed countries the recycling of buildings material started after the end of World War II (Buck 1977) it utilization was as means of solving an economic problem. Recycling as means of sustainable utilisation of materials did not actually start until fairly recently.

Researchers have tried to relate the quality of recycled aggregate concrete to the properties of the original concrete and paste, deterioration condition of the old concrete, crushing procedure, and the new mix composition; their findings have been extensively reviewed and discussed by Hansen (1992). It is generally accepted that the cement paste from original concrete that is adhered to the recycled aggregate plays an important role in determines the performance of recycled aggregate concrete. The qualities of the paste and the interface zones, as well as the paste content of the original concrete, thus influence the properties of the recycled aggregate concrete. Regarding the strength development of recycled aggregate concrete several contrasting results were obtained between investigator in the United States, Europe, and Japan. Among this Frondistou-Yannas (1977), Hansen and Narud (1983), Ravindrajah et al (1985, 1987) all found that the compressive strength of recycled concrete made from recycled coarse aggregate and natural sand have 0 – 35% lower compressive strength than natural aggregate concrete. Contrary to this Yoda et al (1985) and Diah et al (1998) found that the compressive strength of recycled aggregate concrete made from recycle coarse aggregate were either comparable or higher than natural aggregate concrete. From these findings it seems that more research is needed to understand the behaviour of the recycled aggregate concrete as structural material.

Most researchers have made the original concrete in the laboratory to have control of the properties and mix proportions of the original concrete and to be able to study some specific properties of recycled concrete under controlled conditions. However, this will lead to results that may differ from those obtained when field-demolished concrete or waste concrete, that has been exposed to climatic conditions, is used to produce recycled aggregate.

In Malaysia, very little is known about the use of recycled aggregate in manufacture of new concrete. This is probably due to the lack of knowledge about the behaviour of the material under local climatic conditions. Therefore, the main thrust of this investigation is to evaluate recycled aggregate concrete from waste concrete cubes that has been exposed to local climatic conditions for use as coarse aggregate in new concrete, as such material is likely to provide both environmental and economic advantages. In this study, an investigation has been carried out to quantify properties of concrete made by fully and partially replacing natural coarse aggregate with recycled aggregate. The properties investigated are aggregate characteristic, workability, compressive strength development and drying shrinkage at early age.

EXPERIMENTAL DETAILS

Materials

The materials used in this study are ordinary Portland cement in compliance with MS 522, river sand and coarse aggregate, which comply with BS 812. Waste concrete used in this study were concrete cubes brought from neighboring construction, which send concrete cubes for compressive strength testing at 7 and 28 days as a compliance to construction specification. The range of the design strength of these cubes is between 25 and 30 N/mm^2 and of age more than 2 years. These cubes were first broken into smaller pieces by the help of crushing machine and than into smaller pieces in a jaw crusher to produce recycled concrete aggregate of nominal maximum size of 20mm. The crushed product was than sieve in to 20mm and 10mm faction. The 20mm and 10mm coarse aggregate for the concrete mix were than combined in the ratio of 2:1 respectively for both natural and recycled aggregate concrete mix. The grading is as shown in Table (1 – 2) in accordance to BS 812 Part 3: 1985.The physical properties of all the aggregates in term of specific gravity, bulk density, aggregates impact and crushing values, and water absorption are presented in Table 3 in accordance to BS 1881.

Mix Design

The characteristic design strength chosen for the concrete mixes at 28 days were 20, 25, and 30 N/mm^2. This covered the range of design strength that is commonly used in practice. Two sets of mixes were prepared one for natural aggregates concrete and recycled aggregate concrete. The water cement ratio of the mixed range from 0.55 to 0.64. Since there is no existing standard method of designing concrete mixes incorporating recycled aggregate the recycled aggregates concrete mixes were derived simply by replacing the natural coarse aggregate proportion in the natural aggregate concrete mix, design using DOE method, with recycled coarse aggregate. All the concretes were mixed in a 0.04 m^3 horizontal pan mixer. The concrete mixing procedure were carried out in accordance with BS 1881: Part 108. The mix design proportion is shown in Table 4. Eight concrete mixes were prepared namely NAC 20, NAC 25, NAC 30, RAC 20, RAC 25, RAC 30, RAC 25 – 50 and RAC 25 – 75. The denotation depicted that for NAC 20 the concrete mix is of natural aggregate concrete of designed strength 20N/mm^2, for RAC 20 this means recycled aggregate concrete of designed strength 20N/mm^2, and for RAC 25 – 50 and RAC 25 – 75 these are recycled aggregate concrete mix of designed strength 25 N/mm^2 with 50% and 75% replacement of recycled coarse aggregate respectively by mass of natural aggregate. (i.e. 50% Recycled Aggregate with 50% Natural Aggregate and 75% Recycled Aggregate with 25% Natural Aggregate respectively).

Casting And Curing of Specimens

For each concrete mix, 18 nos. of 100mm cubes and 2 nos. of 25 mm x 75mm x 258mm prism were made in standard steel moulds. The cube was used to determine the compressive strength and the prisms were used for the drying shrinkage test. After 24 hours of molding, the cubes were removed form the mould and placed in water at 20°C until testing. The prism were kept in water for 7 days prior to drying in the laboratory environment of 23°C and 65% RH.

Compressive Strength Development Test and Shrinkage Measurement

Cube specimen (100mm side) were tested for compressive strength in accordance with BS 1881: Part 116 (1983). The shrinkage strain measurement was made on longitudinal side of a prism over 200mm-gauge length with a demountable mechanical strain gauge. At any time, the average of four reading taken in two identical prism. The shrinkage measurement was continuing for a period of 28 days.

RESULTS AND DISCUSSION

Workability

The freshly mixed concrete mixes were tested immediately after mixing for workability using slump tests. The test result showed that the natural aggregates concrete mixes show better workability (slump of 60mm) then recycled aggregate concrete mix (RAC) with slump measurement ranging between 35 – 40mm.

Compressive Strength Development

Table 5 show the compressive strength development of all the mixes. All the mixes attained the 28 days design strength of 20 N/mm^2, 25 N/mm^2, 30N/mm^2 and target strength at 28 days of 33.1 N/mm^2, 38 N/mm^2 and 43.1 N/mm^2 respectively. For the same design strength all the RAC mixes show higher compressive strength compared to the corresponding NAC mixes. The rate of compressive strength development is also higher for RAC compared to NAC mixes. The strength development of RAC and NAC show similar trend for compressive strength development.

It is observed that the compressive strength at the 3, 7 and 28 days shows increase in compressive strength as the design strength increases. The higher compressive strength may be attributed to the shape of the recycled aggregate which are more angular and rough in texture compared to natural aggregates..

Table 6 shows the relative compressive strength of recycled and natural aggregate concrete at 3, 7 and 28 days. It can be seen that at 3 days the compressive strength of RAC mix is 7 – 35% higher than the corresponding NAC mixes. At 7 days RAC mixes is 20 – 32% higher than NAC mixes and 28 days the RAC mixes is 2 – 20% higher than NAC mixes.

Effect of water cement ratio

Table 7 shows the effect of water cement ratio on compressive strength of the mixes. Generally the trend is similar for NAC and RAC mixes, such that the higher the water cement ratio the lower the compressive strength.

Effect of proportion of recycled aggregate

Table 5 illustrated the effect of percentage replacement of natural aggregate with recycled aggregate (mix NAC 25, RAC 25, RAC 25-75, RAC 25-50). It is observed that there was a tendency for slight increase in compressive strength the higher the proportion of Recycled Aggregate (RA) in the mixes at all ages of testing. This is most probably due to the angular shape of the RA that provides stronger bonding between the concrete matrix.

Age Factor of Recycled and Natural Aggregate Concrete

The age factor of concrete is the relative strength of concrete at any age to its 28 days. Table 8 shows the age factor for the concrete mixes at ages of 3, 7 and 14 days. As can be seen the age factor at 3 days varied from 0.43 – 0.47 for most mix except for mix RAC 25 and RAC 25 –50. At the age of 7 days the age factor for NAC mixes varied quite consistent from 0.58 – 0.59 and for RAC mix varied from 0.63 – 0.83. At age 14 days the age factor of the NAC mixes varied from 0.64 – 0.72 in contrast to RAC mix which varied quite consistently between 0.82 – 0.89. The results indicate that the utilisation of recycled coarse aggregate in concrete is advantageous to its compressive strength development and the RAC mixes seems to develop its strength faster than the NAC mixes.

Drying Shrinkage

Table 9 shows the results of shrinkage measurement and the development of shrinkage with time up to an early age of 28 days. The results indicated that all the mix started to reduce shrinking as early as 14 days after expose to drying. All the prism of various mix show similar trend of shrinkage development with the NAC mixes shows greater shrinkage development compared to RAC mixes. However in the case of NAC 25 and RAC 25 which is in contrast to the above deduction. From the results the shrinkage of the RAC mixes is 33 – 55% lower than the NAC mixes. This is in contrast to finding in Hansen (1992).

Effect of water cement ratio

From the results obtained the effect of water cement ratio indicate that the higher the water cement ratio the higher the shrinkage.

Effect of proportion of recycled aggregate

Table 10 shows the effect of different percentage replacement of coarse aggregate with recycled coarse aggregate. The results shows that for all RAC mixes of similar water cement ratio the shrinkage is less than the corresponding NAC mixes of similar water cement ratio concrete.

CONCLUSION

The replacement of natural coarse aggregates with recycled concrete aggregate was found to be advantageous to the compressive strength development of recycled aggregate concrete. For all mixes, the recycled aggregate concrete mixes manage to attain the required designed strength and have compressive strength 2-20% higher than the corresponding natural aggregate concrete. Regarding the drying shrinkage some of the recycled aggregate mix seems to show comparable shrinkage compared to the corresponding natural aggregate concrete.

From the results obtained in this study, the recycled aggregate concrete could be considered for various potential applications, such as slabs, foot pavements, drainage and medium strength structural concrete element. However further research is still needed as the durability aspect still needs to be fully quantify and understand.

References:

British Standard Institution BS 1881: Part 108: 1983, Method for Making Test Cubes from Fresh Concrete.

British Standard Institution BS 1881: Part 116: 1983, Method of Determination of Compressive Strength of Concrete Cubes.

British Standard Institution BS 812: Part 3: 1985, Method for Determination of Particle Size Distribution.

Buck, A.D.(1977), Recycled Concrete as Source of Aggregate. *American Concrete Institute Journal*, pp. 212 – 219.

Diah.A.B.M (1998) Utilisation of Waste Concrete as aggregat for environmental protection, *Malaysian Science & Technology Congress '98, Symposium D: biodiversity & Environmental Sciences*.

Frondistou – Y (1977), Waste Concrete as aggregate for new concrete. *Journal of the American Concrete Institute*, Proceedings, **74 : 8**, pp. 373 – 376.

Hansen T.C.(1992), *Recycled Aggregates and Recycled Aggregate Concrete. Third State – of – The Art Report 1949 – 1989, RILEM Report 6, Recycling of Demolished Concrete and Masonary; Spoon*.

Hansen T.C., and Narud. H. (1983), Strength of Recycled Concrete Made from Crushed Concrete Coarse Aggregate, *Concrete International – Design and Construction*, **5;1** pp. 79-83.

MS 522: Part 1: 1989, Specification for Portland Cement.

Ravindrajah R.S., Loo, Y.H and Tam, C.T.(1987), *Recycled Concrete as Fine and Coarse Aggregates* Magazine of Concrete Research **39; 141** pp. 214 – 220.

Yoda K., Yoshikane T., Nakashima Y., and Soshiroda T.(1985), Recycled Cement and recycled Concrete in Japan, *Proceedings of the Second International RILEM Symposium on Demolition and Reuse of Concrete and Masonry*, Chapman and Hall, London, pp 527 – 536.

TABLE 1
SIEVE ANALYSIS OF FINE AGGREGATE

Sieve Size	Percentage Passing %
5.00mm	96.2
2.36mm	82.5
1.18mm	50.9
600μm	20.9
300μm	3.6
150μm	0.7
pan	0.0

TABLE 2
SIEVE ANALYSIS OF RECYCLED AGGREGATE (RA) AND NATURAL AGGREGATE (NA) (GRANITE)

Sieve Size (mm)	RA (20mm) % Passing	NA (20mm) % Passing	RA (10mm) % Passing	NA (10mm) % Passing
37.5	100.0	100.0	100.0	100.0
20.0	100.0	99.92	100.0	100.0
14.0	65.0	97.83	69.35	99.41
10.0	32.5	80.27	30.96	98.47
5.0	2.5	30.36	2.36	28.74
2.8	0.0	18.94	0.09	8.42
pan	0.0	0.0	0.0	0.0

TABLE 3
PHYSICAL PROPERTIES OF AGGREGATE

Type of Aggregate	Specific Gravity			Loose Bulk density kg/m³	Aggregate Impact Value %	Aggregate Coasting Value %	Water absorption %
	SSD	Oven dried	App				
Natural Agg	2.55	2.52	2.60	1390	16.5	16.0	1.35
Recycled Agg.	2.31	2.23	2.41	1255	31.4	31.0	3.3

TABLE 4
MIX PROPORTIONS FOR RECYCLED AGGREGATE CONCRETE (RAC) AND NATURAL AGGREGATE CONCRETE (NAC)

Mix	Design Strength N/mm²	Cement Kg/m³	Water Kg/m³	Aggregate (kg/m³) 20mm		10mm		Sand	Water Cement
				NA	RA	NA	RA		
NAC20	20	320	205	537		268		1065	0.64
NAC25	25	355	205	550		275		1010	0.58
NAC30	30	375	205	520		285		960	0.55
RAC20	20	320	205		537		268	1065	0.64
RAC25	25	355	205		550		275	1010	0.58
RAC30	30	375	205		520		285	960	0.55
RAC 25 – 50	25	355	205	275	275	137.5	137.5	1010	0.58
RAC 25 – 75	25	355	205	137.5	412.5	68.9	206.3	1010	0.58

TABLE 5
COMPRESSIVE STRENGTH OF RECYCLED AGGREGATE CONCRETE (RAC) AND NATURAL AGGREGATE CONCRETE (NAC) (N/mm^2)

	Mix	1day	3days	7days	14days	21days	28days
W/C = 0.64	NAC 20	7.00	14.49	19.54	23.90	30.00	33.04
	RAC 20	7.98	17.00	27.10	34.10	36.60	38.00
	NAC 25	8.00	17.92	23.19	25.51	33.91	39.94
W/C = 0.58	RAC 25	10.00	27.20	34.00	39.10	42.00	45.00
	RAC 25–50	7.80	24.30	33.75	37.50	40.00	40.50
	RAC 25-75	10.00	21.50	33.25	38.50	40.75	43.50
W/C = 0.55	NAC 30	11.00	19.44	24.47	27.31	36.21	41.36
	RAC 30	14.00	20.80	30.40	39.00	43.00	48.00

TABLE 6
RELATIVE COMPRESSIVE STRENGTH OF RAC AND NAC (% OF RAC mix)

MIX	AGE (DAYS)		
	3	7	28
NAC 20	85	72	86
RAC 20	100	100	100
NAC 25	65	68	80
RAC 25	100	100	100
NAC 25	83	70	92
RAC 25 – 75	100	100	100
NAC 25	73	69	98
RAC 25 – 50	100	100	100
NAC 30	93	80	86
RAC 30	100	100	100

TABLE 7
EFFECT OF WATER CEMENT RATIO ON COMPRESSIVE STRENGTH

W/C	NAC			RAC		
	3d	7d	28d	3d	7d	28d
0.64	14.49	19.54	33.04	17.00	22.10	38.00
0.58	19.92	23.19	39.94	27.20	34.00	45.00
0.55	19.44	24.47	41.36	20.80	36.40	48.00

TABLE 8
AGE FACTOR OF RECYCLED AND NATURAL AGGREGATE CONCRETE

Mix	3 days	7 days	14 days	28 days
NAC 20	0.44	0.59	0.72	1.00
RAC 20	0.45	0.71	0.89	1.00
NAC 25	0.45	0.58	0.64	1.00
RAC 25	0.65	0.82	0.88	1.00
RAC 25-75	0.49	0.83	0.88	1.00
RAC 25-50	0.60	0.76	0.86	1.00
NAC 30	0.47	0.59	0.67	1.00
RAC 30	0.43	0.63	0.82	1.00

TABLE 9
DRYING SHRINKAGE mm x 10^{-3} OF RAC AND NAC

MIX	TIME (DAYS)				
	0	7	14	21	28
NAC 20	0	70	90	100	100
NAC 25	0	10	10	10	20
NAC 30	0	80	90	90	90
RAC 20	0	30	40	45	45
RAC 25	0	40	70	75	75
RAC 30	0	30	50	60	60
RAC 25-50	0	70	80	80	80
RAC 25-75	0	20	40	50	50

TABLE 10
EFFECT OF PROPORTION OF RECYCLED AGGREGATE ON DRYING SHRINKAGE AFTER 4 WEEKS

W/C	MIX	SHRINKAGE mm x 10^{-3}
0.58	NAC 25	90
	RAC 25	75
	RAC 25 – 75	50
	RA 25 - 50	80

NUMERICAL PREDICTION ON THE ELASTIC MODULUS OF AGGREGATE

How-Ji Chen[1] and Hui Chi Chan[2]

[1] Department of Civil Engineering, National Chung-Hsing University,
Taichung, Taiwan, R.O.C.
[2] Department of Civil Engineering, National Chung-Hsing University,
Taichung, Taiwan, R.O.C.

ABSTRACT

In this study, a cylindrical concrete model is established and analyzed by ANSYS program. The model is assumed that concrete is a composite material which combined mortar and aggregate to simulate the situation of aggregate and the distribution of stress. From the numerical results, the relationship among elastic modulus of concrete (E_c), elastic modulus of aggregate (E_a), elastic modulus of mortar (E_m) and volume content of aggregate (V_a) can be established. On the other hand, according to the numerical results, a regression formula of E_a was evaluated as

$$E_a = 6.3 E_c - 5.1 E_m - 9.1 V_a E_c + 8.7 V_a E_m$$

Furthermore, the experimental results show that the reliability of prediction formula Ea is acceptable in lightweight aggregates. By using simple test steps, the E value of lightweight aggregate can be effectively reached within 2 days.

KEYWORDS

numerical method, concrete, mortar, aggregate, elastic modulus, regression formula

INTRODUCTION

The theoretical model of E_c formulas were mostly established base on the composite materials' theories. They usually regard aggregate and mortar as two materials and use the E values of those two as parameters. Vogit, Reuss, Counto and many other researchers have proposed parallel connection, series connection and other ideal models to predict E_c [1-4]. However, they are confined under some ideal conditions such as constant strain mode, constant stress mode and shape of aggregate to establish ideal model and reach the theoretical solution. This ideal model, in fact, may have enormous difference with the stress distribution of aggregate and mortar inside concrete.

Lightweight aggregate is a porous material. There is a big difference between it and normal weight aggregate in composition. It has porous interior and more condensed outer surface. Unlike normal weight aggregate, it is not considered a homogeneous material[5,6]. Therefore, there is no feasible way to estimate E value of lightweight aggregate. It can be only estimated by theoretical method. Especially the lightweight concrete made up of lightweight aggregate is making it more difficult to predict E value. To predict E_c of lightweight concrete effectively, first, accurate E value of lightweight aggregate must be predicted. This is especially important for lightweight concrete.

NUMERICAL ANALYSIS

This study adopts numerical method to simulate the stress distribution of mortar and aggregate in concrete. First, we assume aggregate and mortar are both homogeneous materials and we set the size and position of aggregate inside concrete. By using the ANSYS program we find the Ec based on stress-strain curve of concrete and analyze the relationship among E_a, E_m, E_c and V_a.

Comparison of Theoretical and Numerical Solutions of E_c

To make sure the reliability of E_c from the numerical analysis, three theoretical models and formulas are used for comparison. They are Vogit model, Reuss model and Counto model as shown in Fig. 1.

Vogit model assumes two materials (aggregate and mortar) are in parallel combination. Under loading condition, the deformation behavior of two materials is equal to constant strain mode. If the E values (E_a and E_m) are known, the E_c of composite material is calculated from the following equation:

$$E_c = V_a E_a + V_m E_m$$

To emphasize the difference between elastic modulus of these two materials, this research takes $E_a = 1.0 \times 10^4$ MPa, $E_m = 4.0 \times 10^4$ MPa, $v=0.18$, various V_a from 10% to 90% and every 10% increment is a set so there are totally nine sets for analysis. The E_c values from all sets are compared with the results from theoretical formula as shown in Fig. 2(a). It is found that the E_c values from these two methods are almost the same. It indicates the numerical model by ANSYS program to calculate E_c value can reach a perfect result under constant strain mode.

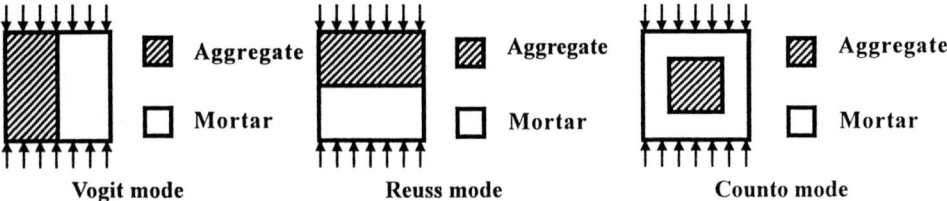

Figure 1: Theoretical model of concrete

Reuss model assumes two materials are in serial combination. Under loading condition, the stress behavior of two materials is equal to constant stress mode. If the E values (E_a and E_m) are known, the E_c of composite material is calculated from the following equation:

$$Ec = \frac{1}{\dfrac{V_a}{E_a} + \dfrac{V_m}{E_m}}$$

The comparison between E_c values from numerical analysis and from theoretical formula is shown in Fig. 2(b). It indicates the numerical model of this research has an acceptable result under constant stress mode.

The composite material model by Counto assumes a square shape aggregate inside the center of a hexahedron mortar. If the E values of the two materials are known, the E_c of composite material can be calculated by the following theoretical equation:

$$\frac{1}{E_c} = \frac{1-\sqrt{V_a}}{E_m} + \frac{\sqrt{V_a}}{\sqrt{V_a}E_a + (1-\sqrt{V_a})E_m}$$

The comparison between numerical analysis results and the theoretical calculation results is shown in Fig. 2(c). In the figure, it is found that, for $E_m/E_a=4$ (a big difference between E values of two materials), the error between numerical solution and theoretical solution is big, especially when V_a increases from 20% to 30%. There is an almost 7.0% error between numerical and theoretical solutions. The reason is that in the deriving process of the theoretical formula, the composite materials in parallel combination part are assumed as constant stress mode and those in serial combination part are assumed as constant strain mode. In real Counto model case, the stress and strain behavior are not in constant stress or constant strain modes. It can be proved from the stress distribution result of numerical analysis.

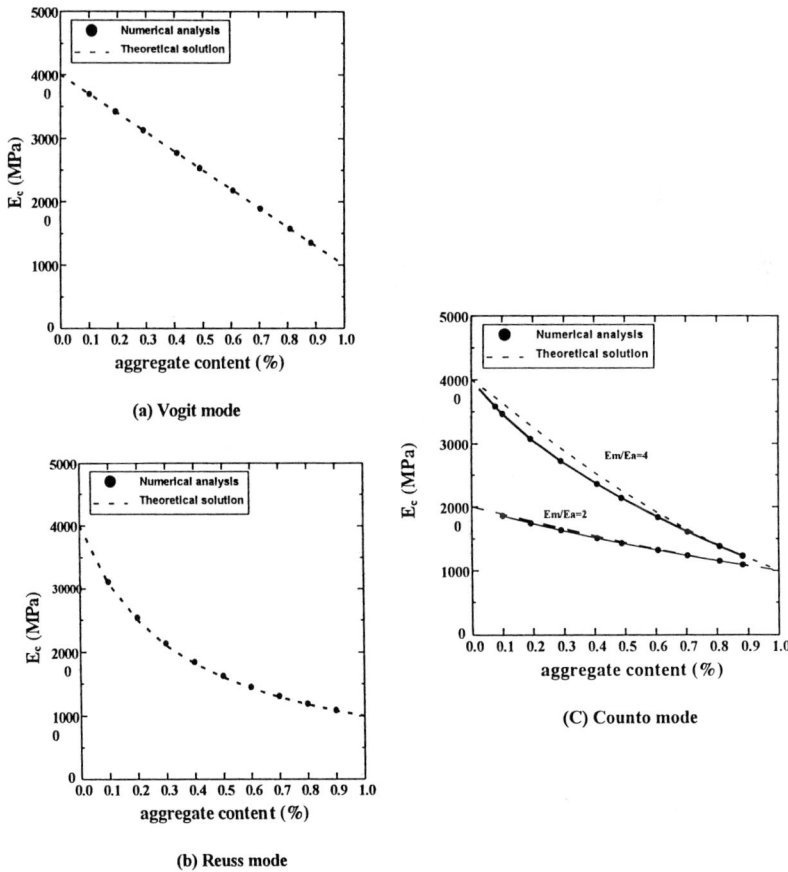

Figure 2 Comparison of theoretical and numerical solutions

As $E_m/E_a=2$, the maximum error is about 2.4%. It is concluded that when the E values of two materials are close, the assumption of composite material in constant stress and constant strain mode becomes more acceptable.

To sum up the above numerical analysis data and compare with theoretical solutions, the result shows concrete model established by numerical program ANSYS are acceptable.

Relation among E_a, E_m and E_c

To establish the relation among E_a, E_m and E_c, this study uses cylindrical specimen with V_a from 10% to 50% and ratio of E_m / E_a from 0.3 to 6.0. The results are shown in Fig. 3. Changing the size of specimen with a fix aggregate size (10mm) controls the volume content of aggregate. Aggregate is placed in the center of specimen. In $E_m / E_a < 1$ (upper half in figure) case, when E_m / E_a is fixed, E_c/E_m will increase with increasing V_a. When E_m / E_a decreases, the difference between E_c and E_m (or the ratio) becomes bigger. This is because that aggregate is a reinforcing particle; the E_c value will increase due to increase of high E material. It is more obvious as V_a increases. In $E_m / E_a > 1$ (lower half in figure) case, when E_m / E_a is fixed, E_c/E_m will decrease with increasing V_a. When E_m / E_a increases, the difference between E_c and E_m (or the ratio) becomes bigger. This is a lightweight aggregate concrete case; aggregate particle is a weak material, and the E_c value will decrease due to increase of low E material. It also is more obvious as V_a increases.

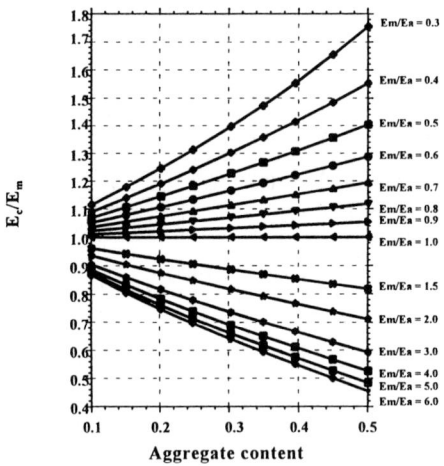

Figure 3: Relation among E_a, E_m, E_c and V_a

DETERMINATION OF E_a FORMULA

Based on the several theoretical formulas recommended in references, Va, Ec, and Em should be used as the main parameters, and the formula presented not more than quadratic polynomial is acceptable. Therefore, the completely quadratic polynomial of 3 variables mentioned above will be used to proceed the regression formula of Ea. The function is as follows:

$$E_a = f(V_a,\ E_c,\ E_m,\ V_a E_c,\ V_a E_m,\ E_c E_m,\ V_a^2,\ E_c^2,\ E_m^2)$$

The result of numerical analysis shown in Fig.3 is regarded as reference in selecting points used in the function. Refer to the mechanical properties of concrete, this study assumes 5 Ec values are 10000, 15000, 2000, 25000, and 30000MPa. The 7 Em/Ea values are 3.0, 2.0, 1.5, 1.0, 0.7, 0.5, and 0.4. The 6 Va values are 0.25, 0.3, 0.35, 0.4, 0.45, and 0.5. A total of 210 sets of sample data can be acquired. The Shazame program was used to analyze the selected sample points, the prediction formula can be determined as

$$E_a = 1251.2V_a + 6.67E_c - 5.4E_m - 8.8V_aE_c + 8.3V_aE_m - 0.59 \times 10^{-4}E_cE_m - 4479V_a^2 + 0.25 \times 10^{-4}E_c^2 + 0.33 \times 10^{-4}E_m^2 + 938.6$$

Statistical quantities related to the regressed relationship are illustrated in Table 1, in which the values of the standard error, T-ratio and P represent respectively the error, significance and the extent of concentration of data. It can be seen that the coefficients associated with the terms of Va and Va² correspond to extremely large standard errors, less significance and worse data concentration of the regression. Accordingly, it can then conclude that these two terms can be deleted with reasonable accuracy.

TABLE 1
RESULTS OF REGRESSION $E_a = f(V_a, E_c, E_m, V_aE_c, V_aE_m, E_cE_m, V_a^2, E_c^2, E_m^2)$

Parameter	Coefficient	Standard error	T-Ratio	P-value
V_a	1251.2	0.2458e5	0.5090e-1	0.959
E_c	6.6773	0.2736	24.40	0.000
E_m	-5.4031	0.2568	-21.04	0.000
V_aE_c	-8.7942	0.3279	-26.82	0.000
V_aE_m	8.3350	0.3009	27.70	0.000
E_cE_m	-0.58958e-4	0.7076e-5	-8.332	0.000
V_a^2	-4479.0	0.2971e5	-0.1507	0.880
E_c^2	0.24863e-4	0.5205e-5	4.777	0.000
E_m^2	0.32602e-4	0.2966e-5	10.99	0.000
constant	938.55	4898	0.1916	0.848

R-square=0.9956

In further analysis, it is assumed that the Ea formula has only other 7 terms and constants. Putting sample data into calculation, the formula is determined as follows:

$$E_a = 6.8E_c - 5.5E_m - 8.9V_aE_c + 8.4V_aE_m - 0.58 \times 10^{-4}E_cE_m + 0.24 \times 10^{-4}E_c^2 + 0.33 \times 10^{-4}E_m^2 + 732.58$$

Although the R-square is almost equal to 1, prediction formula of Ea is still complicated and has too many terms, which is inconvenient for practical use. In order to simplify the terms in prediction formula, the parameters are re-selected based on the theoretical equation. After several times of analysis, the formula can be determined as follows:

$$E_a = 6.3E_c - 5.1E_m - 9.1V_aE_c + 8.7V_aE_m$$

TABLE 2
RESULTS OF REGRESSION $E_a = f(E_c, E_m, V_a E_c, V_a E_m)$

Parameter	coefficient	Standard error	T-Ratio	P-value
E_c	6.3317	0.1391	45.66	0.000
E_m	-5.1219	0.1351	-37.91	0.000
$V_a E_c$	-9.1029	0.3123	-29.15	0.000
$V_a E_m$	8.7352	0.2990	29.28	0.000

R-square=0.9924

Table 2 shows each term has more accuracy, besides, comparing with the previous result of analysis, standard errors greatly decrease, T-ratio of each term adds up in the formula, and R-square of formula also reaches to 0.9924. This result shows the prediction formula has high reliability and is a simple formula good for practical use.

EXPERIMENTAL RESULTS

In order to find various E values of mortar and concrete in different compressive strength, this research measures different E_c and E_m at different curing ages of specimens. When it reaches desired curing age, we take the specimens to perform compression test under universal testing machine. Transducer is used to directly measure deformation and find E values.

Two mix proportions were used in mortar; the water-cement ratios (w/c) are 0.43 (Type I) and 0.48 (Type II). Specimens of 1 and 2 days curing age were tested for E values. The total amount of mortar and concrete specimens is 24. They were cured at constant temperature (23°C) and constant moisture (>95%) curing room. Mortar was blended firstly and then mixed with lightweight aggregate to make small cylindrical concrete specimens. In addition to making concrete specimen, mortar specimen is also made for comparison. The sizes of the concrete and mortar specimens were selected as 22ϕ × 22 mm for Type I and 30 ϕ × 30 mm for Type II. The lightweight aggregates used were expanded shale produced in Taiwan with a particle density of 800 kg/m3- 1000 kg/m^3, the sizes of aggregate are near 9 mm.

The compression test results of concrete and mortar are used to find E_c and E_m respectively. Substituting the two E values into prediction formula to find E_a. The results are shown in Table 3.

It is found from the results that E_a of four sets are from 5385 MP$_a$ to 7150 MP$_a$. The average is 6100 MP$_a$ and the variation stays within 18%. This indicates that determining E_a from the prediction formula and experimental results have consistency to some level. It means use the numerical method to simulate the situation of single aggregate in concrete and deriving prediction formula of E_a is reasonable and acceptable.

TABLE 3
DETERMINATION OF E_a BASED ON PERDITION FORMULA

Specimens	Curing days	V_a	$E_c(MP_a)$	$E_m(MP_a)$	$E_a(MP_a)$
Type I	1	36.5%	3510 * (4095) 4680	2710 3580(3087) 2970	6250
	2	36.5%	6010 * (6260) 6510	6840 7210(6890) 6620	5385
Type II	1	14.4%	2785 3435(2950) 2630	2360 2370(2365) *	5620
	2	14.4%	5280 5810(5523) 5480	4800 * (5305) 5810	7150

CONCLUSION

This study uses numerical method to simulate the situation of aggregate and stress distribution inside concrete, to establish E relation among aggregate, mortar and concrete. With verification of simple tests, the results are reached as followed:

(1) By comparing the numerical analysis result with theoretical solution, it is found in serial or parallel theory models, the numerical E_c solution is almost as the theoretical solution. In Counto model, when the E ratio of two materials is 4, the maximum error is near 7%. It is because the theory assumed constant stress and constant strain modes, which did not match the real case. This becomes more obvious when the E ratio of two materials increases.

(2) If E_a is bigger E_m, E_c will be bigger than E_m. If E_a is smaller E_m, E_c will be smaller than E_m. This trend becomes more obvious as V_a increases.

(3) This study indicates the practicability of using numerical method to establish the relationship among E_a, E_m, E_c and V_a. Substituting the experimental results into the prediction formula of E_a to find the E value of lightweight aggregate is proved to be a good method and with reliability.

(4) According to the numerical results, a prediction formula E_a of a cylindrical concrete model was evaluated as

$$E_a = 6.3 E_c - 5.1 E_m - 9.1 V_a E_c + 8.7 V_a E_m$$

REFERNCES

[1] Sidney Mindess, J. Francis Young. (1981). Concrete, *Prentice-Hall*, 339-344.
[2] Hansen, T. C. (1960). Strength, Elasticity, and Creep as related to the Internal Structure of Concrete, Chemistry of Cement, *Proceedings of the Fourth International Symposium*, Monograph 43, **Vol. 2**, Washington, 709-723.
[3] Hirsch, T. J. (1962). Modulus of Elasticity of Concrete Affected by Elastic Moduli of Cement Paste Matrix and aggregate, *ACI Journal*, Proceedings **Vol. 59**, Feb. 427-451.
[4] Popovics, S., and Erdey, M. (1970). Estimation of the Modulus of Elasticity of Concrete-Like Composite Materials, *Materials and Structures, Research and Testing* (RILEM, Paris), **Vol. 3, No. 16**, 253-260.
[5] J. Muller-Rochholz. (1979). Determination of the Elastic Properties of Lightweight Aggregate by

Ultrasonic Pulse Velocity Measurement, *The International Journal of Lightweight Concrete.* **Vol. 1, No.2**, 87-90.

[6] A. Ulrik Nilsen, Paulo J.M. Monteiro and Odd E. Gjorv. (1995). Estimation of the Elastic Moduli of Lightweight Aggregate, *Cement and Concrete Research*, **Vol.25, No. 2**, 276-280.

ESTIMATION OF CORROSION INITIATION TIME WITH OBSERVED DATA

Kanagasabai Ramachandran and Ali Karimi

Department of Civil and Environmental Engineering,
Imperial College of Science, Technology and Medicine, London, UK

ABSTRACT

Chlorination affects reinforced and pre-stressed concrete bridges, and methods are being developed to predict the remaining service life of bridges affected by chlorination. In the probabilistic estimation of service life of bridges, authors have used random field theory to model the variation of surface chloride concentration and effective diffusion coefficient. Random field theory requires the value of a parameter called the *'scale of fluctuation'* from experimental data. In this paper Kriging method has been applied to estimate the missing data on surface chloride concentration. These values are then used in the estimation of the scale of fluctuation of surface chloride concentration.

KEYWORDS

Random field, Reliability, Bridge assessment, Corrosion, Kriging, Probabilistic estimation.

INTRODUCTION

It is now well established that chlorination has affected the strength of a number of reinforced concrete bridges in Great Britain and USA. However the mechanism of chlorination process still lacks clear understanding and a number of possible chloride ion transportation processes have been suggested and are being investigated by a number of researchers from different fields. The most popular one appears to be the concept of diffusion of chloride ions through porous media of concrete. In spite of its deficiencies Fick's second law of diffusion is widely used to estimate and predict the chlorination levels at different depths of concrete. In order to use Fick's second law of diffusion, it is necessary to know two important parameters viz. surface chloride concentration and the effective or apparent diffusion coefficient. It is very difficult, if not impossible, to know these values exactly. For example, the surface chloride concentration on a motorway bridge not only varies from time to time at a particular location but also varies very significantly from point to point across the bridge. The diffusion coefficient also varies from point to point along the depth of concrete due to microstructure of concrete, water content to mention a few.

In the probabilistic estimation of corrosion initiation time by Fick's law the surface chloride concentration and diffusion coefficient are usually assumed as random variables. Since these parameters varies very significantly from point to point, it has been suggested by Karimi and Ramachandran (1999) that random field theory may have to be applied to give a more realistic representation of the variability of these two statistical quantities. In order to apply random field theory, it is necessary to estimate the 'scale of fluctuation' of these two parameters from experimental data on the chlorination levels at different levels and locations. Unfortunately measured/observed data available are scarce and some statistical methods have to be used to infer missing data on chlorination levels. In the study undertaken here, non-linear regression analyses were first performed to get the best estimate for the surface chloride concentration at different locations from the data available. It has been noticed that a fair number of values of surface chloride concentration are missing and they have to be guessed/assumed or estimated in a scientific way before proceeding with the estimation of 'scale of fluctuation' of the surface chloride concentration. In this study a method known as Kriging has been used to estimate the missing values of surface chloride concentration.

For clarity of presentation, the paper will include a very concise description of random field theory. Methods of estimating 'scale of fluctuation' will then be discussed and the use of Kriging method to estimate the missing values of surface chloride concentration will be described in detail with a case study.

PROBLEM DEFINITION

Probabilistic estimation of service life requires the evaluation of corrosion initiation time as well as propagation time. This paper will concentrate only on the initiation time. In our earlier work random field modelling of surface chloride concentration and effective diffusion coefficient has been discussed, Karimi and Ramachandran (1999). Random field theory requires a *variance function $\gamma(T)$* to determine the correlation between the local averages. The *variance function $\gamma(T)$*, the correlation function $\rho(\tau)$ and the *scale of fluctuation θ* are related by the equations:

$$\theta = \lim_{T \to \infty} T \, \gamma(T) \tag{1}$$

$$\gamma(T) = \frac{1}{T^2} \int_0^T \int_0^T \rho(t_1 - t_2) dt_1 dt_2 = \frac{2}{T} \int_0^T \left(1 - \frac{\tau}{T}\right) \rho(\tau) d\tau \tag{2}$$

$$\theta = 2 \int_0^\infty \rho(\tau) d\tau = \int_{-\infty}^\infty \rho(\tau) d\tau \tag{3}$$

Hence θ is equal to the area under the correlation function $\rho(\tau)$ ($\rho(\tau)$ may be exponential, triangular or Gaussian model). A correlation function is chosen and method of least squares is then used to find the scale of fluctuation θ, if there are sufficient suitable data. But it is difficult to collect sufficient data on the surface chloride concentration on bridges. Therefore some scientific way of estimating missing data is necessary to proceed with random field modeling of the surface chloride concentration. In our earlier paper, we assumed a value for the scale of fluctuation by inspection/intuition and performed a series of analyses with different scale of fluctuations to study the sensitivity of the scale of fluctuation on the reliability analysis. In this study a method of assessing the scale of fluctuation θ_c of surface chloride concentration C_s with insufficient data will be described. The use of this scale of fluctuation in the probabilistic analysis of corrosion initiation time has been already published, Karimi and Ramachandran (1999).

EVALUATION OF THE SCALE OF FLUCTUATION θ

Vanmarcke (1983) has proposed simple procedures for the estimation of the scale of fluctuation θ of a random field variable. The easiest method is to use the computed sample correlation coefficients from actual observed data sets. Considering a stationary random process $X(t)$ with mean μ_X and standard deviation σ_X, the *auto-covariance function* $C_X(\tau)$ for any interval length τ and the corresponding *correlation coefficient function* $\rho_X(\tau)$ are given by:

$$C_X(\tau) = \lim_{T \to \infty} \frac{1}{T} \int_0^T [x(t) - \mu_X][x(t+\tau) - \mu_X] dt$$

$$= E[(x(t) - \mu_X)(x(t+\tau) - \mu_X)]$$
(4)

$$\rho_X(\tau) = \frac{C_X(\tau)}{\sigma_X^2}$$
(5)

Hence for a given data set of finite length, the corresponding coefficients of correlation for various interval lengths τ can be evaluated using Eqn. 4. By using the measurements taken from surveys undertaken on reinforced concrete bridge cross-beams exposed to de-icing salts, Wallbank (1989), the scale of fluctuation θ_C for the surface chloride concentration C_s can be estimated by non-linear regression analysis.

Estimation of θ_c

In order to evaluate sample correlation coefficients from measured data, the data set needs to be of a suitable length and consist of sample measurements taken at reasonably frequent intervals. Many of the data sets compiled from the data collected from surveys on a number of bridge cross-beams are too short in length or do not contain sufficient data points. Also three cross-beams have only been found to be reasonably surveyed, and even in these cases sufficient data points are not available to evaluate sample correlation coefficients. A typical data set collected from one of these crossbeams is shown in Figure 1. However a set of comprehensive data is required to estimate θ_c. In this paper, measured values of the random variable C_s at given sample locations have been used to predict values of C_s at other required locations in order to produce more complete data sets. The technique by which this has been achieved is referred to as *kriging* and is briefly described below.

Kriging

Consider a spatially random process $Z(x)$ which varies continuously over a region R. Let z_i, where $i = 1, 2,...n$, be a series of observations of the random variable $Z(x)$ measured at the corresponding spatial coordinates x_i in R. One simple approach in modelling global or large scale variations in the mean value of a spatially random process is the use of *trend surface analysis*. A polynomial function is fitted through the points (z_i, x_i). The multiple regression model used may be expressed in vector notation as:

$$Z(x) = y^T(x)K + \varepsilon(x)$$
(6)

where $Z(x)$ is the value of the random variable at point x, $y^T(x)K$ is the trend and $\varepsilon(x)$ is a zero mean random variable representing fluctuations from this trend. The vector $y(x)$ represents the parameters to be estimated. Assuming that the errors $\varepsilon_i(x)$ at different locations have a constant variance and that they are independent, *ordinary least squares regression* analysis can be used to fit the proposed model to the observed data z_i yielding estimates of the model parameters K. Then

$$\hat{K} = (Y^T Y)^{-1} Y^T z \qquad (7)$$

where z is a column vector of observed values at sample sites x_i and Y is a (n×m) matrix with row vectors $y^T(x_i)$, where $i = 1, 2, ..., n$. The regression model proposed above ignores the second order effects that may be present in the random process $Z(x)$. Hence first order variations in the mean value of the process are only considered and the effect of any residual spatial dependence is neglected. This limitation in the model can be over come by relaxing the assumption of independence of residuals through the use of *generalised least squares regression*. Then

$$Z(x) = y^T(x)K + U(x) \qquad (8)$$

$$\hat{K} = (Y^T C^{-1} Y)^{-1} Y^T C^{-1} z \qquad (9)$$

where $U(x)$ are zero mean correlated random variable and C is a (n×n) matrix of covariances between the residuals $U(x_i)$ and $U(x_j)$ for each possible pair (i,j) of the n sample sites. The covariance matrix C has to be estimated from the data.

The covariance matrix C between residuals $U(x)$ can be estimated using the *covariance function* $C(\tau)$ which can be derived from the sample *variogram* $\gamma(\tau)$. The variogram $\gamma(\tau)$ is defined under the assumption of intrinsic stationarity. Then

$$E[Z(x+\tau) - Z(x)] = 0 \qquad (10)$$

$$VAR[Z(x+\tau) - Z(x)] = 2\gamma(\tau) \qquad (11)$$

$$\gamma(\tau) = \sigma^2 - C(\tau) \qquad (12)$$

where $\gamma(\tau)$ is the variogram, $C(\tau)$ is the covariance function and σ^2 is the variance of the random variable $Z(x)$. The natural sample estimator of the variogram is

$$\gamma(\tau) = \frac{1}{2n} \sum (z_i - z_j)^2 \qquad (13)$$

where n is the number of pairs of data separated by the interval τ. Therefore by estimating the sample variogram from the data set, an analytical model for the variogram $\gamma(\tau)$ can be fitted to the observed sample variogram and hence the form of the covariance function $C(\tau)$ can be defined using Equation (12). Once the covariance structure of a random process is known, it can be used to predict values of the random variable at other required locations by Eqn. 8. The local component $U(x)$ to be estimated may be expressed in terms of weighted linear combinations of the observed residuals u_i:

$$\hat{U}(x) = \sum_{i=1}^{n} \lambda_i U(x_i) \qquad (14)$$

where λ_i are the weights applied to each of the variables $U(x_i)$ and $\hat{U}(x)$ is the value of the local component $U(x)$ to be estimated at the point x. To obtain a suitable estimate of $U(x)$ consider the expected mean square error between the values of $U(x)$ and $\hat{U}(x)$:

$$E\left(\left[\hat{U}(x)-U(x)\right]^2\right) = E\left[\hat{U}^2(x)\right] + E\left[U^2(x)\right] - 2E\left[U(x)\hat{U}(x)\right]$$
$$= \lambda^T C \lambda + \sigma^2 - 2 \lambda^T c(x)$$
(15)

where C is the (n×n) matrix of co-variances between all possible pairs of n sample sites and $c(x)$ is the column vector of co-variances between the prediction point x and each of the n sample sites. The weights λ_i can be selected by minimising this mean square error. Hence differentiating with respect to the vector λ:

$$\lambda = C^{-1}c(x) \qquad (16)$$

Hence
$$\hat{U}(x) = \lambda^T U = c^T C^{-1} U \qquad (17)$$

where U is a column vector of the observed residuals $U(x_i)$ at each sample site x_i. The value of $\hat{u}(x)$ is then added to the known trend $y^T(x)K$ to obtain the predicted value $\hat{z}(x)$.

APPLICATION

The method described above will now be applied to the chlorination problem under investigation. A typical data set collected from one of the cross-beams is shown in Figure 2. The regression model $C_s(x) = \mu_C + U(x)$ is used in this study. By removing the mean from the observed values, the residuals $U(x_i)$ at the locations x_i have been obtained. Based on the residuals evaluated, the sample variogram has been estimated using Eqn.13 and is shown in Figure 1.

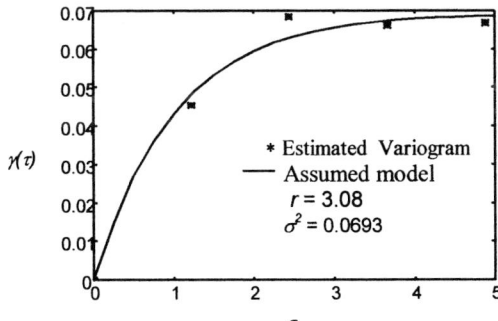

Figure 1: Estimated sample variogram for C_s

The sample variogram has been estimated from the measurements taken in the interval 5m to 20m along the beam as these data points are relatively closely spaced and hence can be used to describe the covariance structure of the random process reasonably well. The resulting covariance function $C(\tau)$ obtained has subsequently been used on a global scale. The magnitude of the parameters σ^2 and r in the exponential variogram have been estimated using least squares regression analysis to obtain the best fit of the variogram model on to the evaluated sample variogram. The resulting values obtained are indicated in Figure 1. Values of C_s at various locations have been predicted using the techniques described previously and the completed data set is shown in Figure 3.

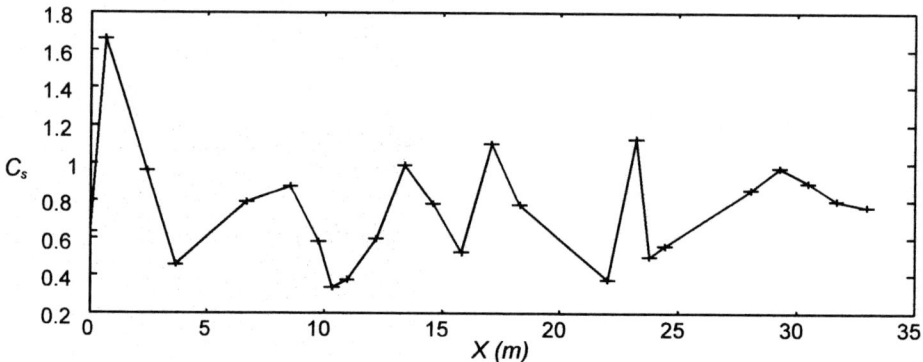

Figure 2: Initial set of measured values of C_s

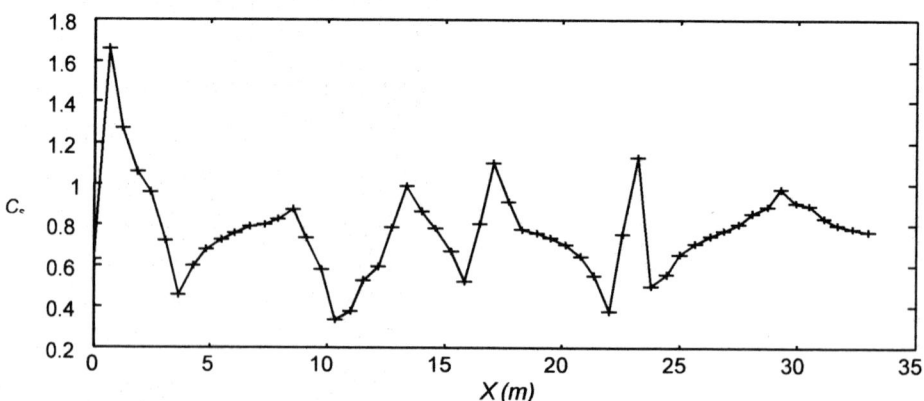

Figure 3: Complete set of data for C_s after Kriging

Estimation of θ_c:

As described in previous section the scale of fluctuation θ_C is estimated from the complete set of data produced by Kriging. The scale of fluctuation θ_C for the three beams are found to be 1.27, 2.68 and 2.12.

CONCLUSIONS

This paper discusses the problem of dealing with insufficient data on chlorination levels/surface chloride concentration in the estimation of chlorination initiation time in reinforced and pre-stressed bridges. Kriging method is used in geological sciences to predict missing data from geological surveys. Therefore it is suggested that Kriging method may be used to estimate missing data on surface chloride concentration in bridges. The complete data set can then be used to get a reliable value for the 'scale of fluctuation' of surface chloride concentration, a parameter required for random field analysis. A simple analysis has been undertaken on chlorination data collected on a bridge to explore the feasibility of this approach. The research is progressing and it appears that Kriging can be advantageously used in chlorination problems.

REFERENCES

Karimi, A and Ramachandran, K. (1999). Probabilistic estimation of corrosion in bridges due to chlorination.*ICASP8, Application of Statistics and Probability,* A.A. Balkema, Rotterdam.
Vanmarcke, E. H. (1983*). Random field: analysis and synthesis*. MIT Press.
Wallbank, E. J. (1989). The performance of concrete in bridges: A survey of 200 highway bridges. *G. Maunsell & Partners.*

CONSTRUCTION TECHNOLOGY AND METHODS

CONSTRUCTION AND INSTRUMENTATION OF A CONCRETE MODULAR BLOCK WALL

A. Kasa [1], F. H. Ali [2], N. Nasir [1]

[1] Department of Civil Engineering, Universiti Kebangsaan Malaysia,
Bangi, Malaysia
[2] Department of Civil Engineering, University of Malaya,
Kuala Lumpur, Malaysia

ABSTRACT

A full-scale model of 5 meter high and 10 meter wide test section of concrete modular block wall backfilled with cohesive residual soil and reinforced with geogrid was constructed at Nilai Industrial Park, Malaysia. Pisa II stonewall system was used for the facing unit while Miragrid geogrid was used for reinforcement. Performance during construction was monitored by incorporating extensive instrumentation including inclinometer, strain gauges, pressure cells, piezometer tubes and surface and wall settlement markers. This paper explains the procedure and methodology used for installation, shows the results obtained from this instrumentation and compares them with the results obtained from calculation.

KEYWORDS

Modular block wall, residual soil, instrumentation, construction, geogrid, reinforced earth structures.

INTRODUCTION

In Malaysia, the first reinforced structure was built in 1982 and since then it has been widely used. Most of retaining structures are retaining walls with average height of 8.0 meters. Other structures are bridge abutments, wing walls and side buttresses (Chiu et. al, 1987). The principle of reinforced earth is analogous to reinforced concrete. In the design of reinforced concrete, the bond stress between the bar and concrete and anchorage stress provide sufficient tensile stress to the concrete. However, in reinforced earth, the tensile stress is contributed by friction between the frictional backfill material and reinforcement strip (Ingold, 1982).

Backfill materials are usually specified as granular materials such as river sand and mining sand. However, sand is quite expensive nowadays. Another alternative is to use quarry dust or residual soil. Economically, residual soil has more advantage since it can be found abundantly in Malaysia at low cost. By using residual soil, contractors can cut the overall construction cost and time since no transportation of backfill material to sites is required. The most widely used reinforcing elements are steel strips. Although steel strips are well accepted, there are still uncertainties regarding its durability especially under wet soil condition as encountered in cohesive residual soil. It seems that the most practical reinforcing element to work in residual soil is geogrid, which is durable and friendly to the environments. In addition, in this case, geogrid is easier to handle, transport and install than steel strips.

A full-scale model of 5 meters high geogrid reinforced wall with residual soil as backfill was constructed at Lot PT1568, Nilai Industrial Park, Malaysia. The reasons to construct the retaining wall were to increase the usable space, to strengthen the slope and to study field behaviour of this reinforced soil wall system (Anuar & Faisal, 1997).

PROPERTIES

Fill Materials

Fill material used in this system is sedimentary slate residual soil, which is brownish in colour. It contains particles of various sizes ranging from pebble to clay. It consists of 34.3 % silt and clay, 31.1 % sand and 34.6 % gravel. According to United Soil Classification Systems (USCS), this soil can be classified as inorganic clayey sand and silty sand (SC - SM).

Reinforcement

Miragrid geogrid is a high-strength, flexible polyester geogrid specially designed to provide long lasting reinforcement and stability to reinforced earth structures. It consists of high tenacity, high molecular weight polyester (PET) yarns knitted and woven into a stable geometric configuration. The coated polyester yarns are bundled into machine (length) direction strands and space to provide uniform grid shaped opening. These grid apertures allow for easy soil penetration through the plane of the geogrid, forming a highly efficient tensile reinforcement element. Miragrid's flexibility and high long-term design strength make it an ideal reinforcement for reinforced structures (Nicolon Mirafi, 1997).

Modular Block

Pisa II blocks used in modular walls are manufactured from high strength concrete with minimum strength of 35 Mpa. They are available in many configurations. All of the blocks have keys, which provide a mechanical interlock with courses above and below any particular layer of blocks. They are also self-sloping and self-aligning. A standard configuration is illustrated in Figure 1.

Drainage

Groundwater infiltration of surface runoff can cause saturation of the reinforced soil that will substantially reduce soil strength and reduce the wall's factor of safety. To prevent the fill from becoming saturated by providing a good drainage system to the reinforced structure, free draining material such as granular soil is placed just behind the wall and at the end of geogrid layer. Geosynthetic pipes, which have tiny holes, are used to drain water accumulated at the base of the structure. The pipes are covered with special cloth to avoid soil particles from entering into the drainage system.

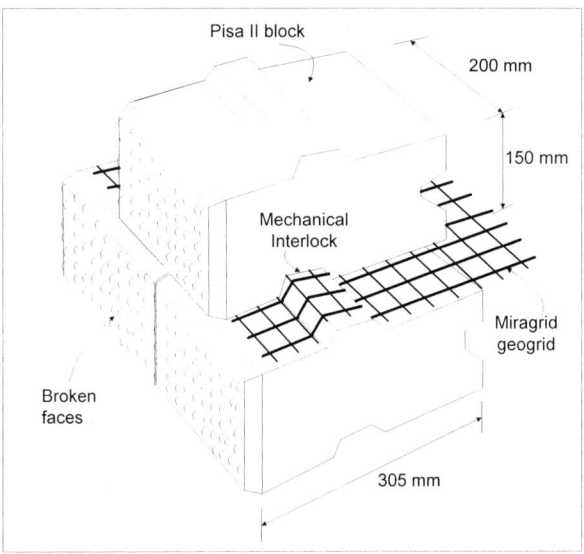

Figure 1: Pisa II block configuration and mechanical interlock between the blocks

INSTRUMENTATION AND CONSTRUCTION

An elevation plan to show the location of various types of instrumentation is shown in Figure 2. The purpose of instrumentation is to analyse the behaviour and the performance of this retaining wall. The instruments used in this study include strain gauges, inclinometer, standpipe and pneumatic piezometers, total pressure cells for horizontal and vertical pressure, surface and wall settlement markers.

Ten strain gauges are used to measure tensile forces in geogrid reinforcements. They are located at three different layers in the structure. Instrumented geogrid layers are N4, N6 and N8 as shown in Figure 2. The strain gauges are calibrated using tensile tests in the laboratory. Standpipe and pneumatic piezometers are applied to measure pore water pressures in order to check the workability of drainage system. The coefficients of lateral and vertical pressures at the base of the structure are measured by total pressure cells. An inclinometer is used to measure the horizontal movements while surface settlement markers are used to measure the overall settlements of the reinforced structure.

Construction was initiated by erecting a concrete levelling footing for accurate placing of the first row of modular blocks. Then, installation of geosynthetic drainpipe at the base of retaining wall was done. After that, foundation or base of retaining wall was prepared. However, because of rainy season, the installation of instruments on top of foundation was not done after the preparation of foundation. Backfilling of soil up to 300 mm was done prior to the installation of instruments to avoid instruments being damaged due to rainfall. The process of backfilling was performed by the use of a small bulldozer and shovels. Before compaction, the instruments were installed at their specified locations on foundation by excavation.

After all instruments were installed except instrumented geogrid layers, all connecting cables were accumulated at one location for ease of readings and compaction. Compaction equivalent to 95 % of Standard Proctor Density was specified. However, due to high frequency of rainfall or wet soil condition, it was difficult to achieve that level of compaction. A manually operated vibrating plate compactor was used to compact the backfill. Geogrid layers were then placed in position according to the design. During each stage, modular blocks were supplied by throwing them from the top of slope. Hammer was used to align and level the blocks. Normally after the placement of modular blocks, sand was placed and compacted behind the blocks.

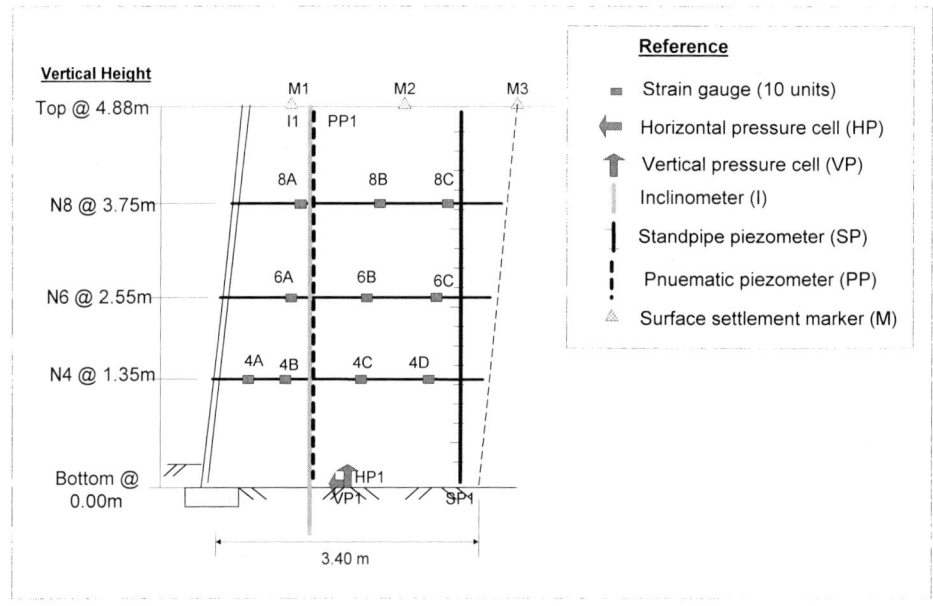

Figure 2: Location of various types of instrumentation

In this project, it was found that the rate of construction was largely controlled by the rate of supply and placement of fill material. It seemed that the drainage system was working properly because water was detected flowing through drainpipe during compaction. High frequency of rainfall during construction also played important role on the progress of work. It also had a major influence on properties of fill material and compaction level. Erosion, which took place during construction, had removed fine particles and exposed large particles. Optimum compaction level could not be achieved due to wet soil condition. Beside high frequency of rainfall, worst hazy condition had also slowed down the progress of construction since workers could not remain at site. The construction took two and half months to complete while in normal condition it could be finished in less than a month.

RESULTS AND DISCUSSION

Tensile Forces

It seemed that backfilling did not have much effect on tensile force. The highest change in tensile force recorded was 62 N/m, which was insignificant. During backfilling on geogrid N4, negative tensile forces were recorded by strain gauges 4A and 4D (See Figure 3). The reduction of tensile forces from initial readings could be due to large soil particles that hit the geogrid and caused local displacements near strain gauges. However, the values were small and unimportant.

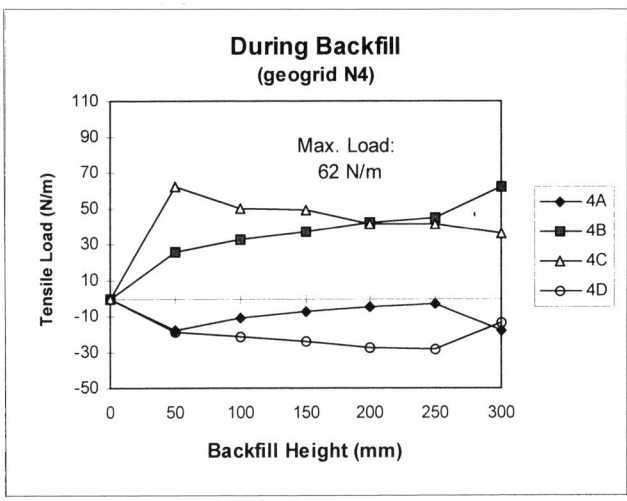

Figure 3: Changes in tensile force due to backfilling (N4)

The effects of compaction on tensile force are represented in Figure 4 and 5. The compaction was done using a manually operated vibrating plate compactor at 300 mm (12") lifts. For each location, there were 4 numbers of trips, which took approximately 5 minutes to finish. The relationship between tensile force and number of trip was plotted for geogrid N6 as shown in Figure 4. The highest change in tensile force recorded by strain gauges on geogrid N6 due to compaction was 129 N/m. This value was almost twice as much as the maximum value due to backfilling.

It was predicted that the higher the height of fill, the lesser the effect of compaction on tensile force. Figure 5 shows the changes in tensile force due to compaction at 1.5 meters above geogrid N4. Strain gauges on the geogrid recorded smaller changes in tensile force compared to the compaction at 0.3 meter above the same geogrid.

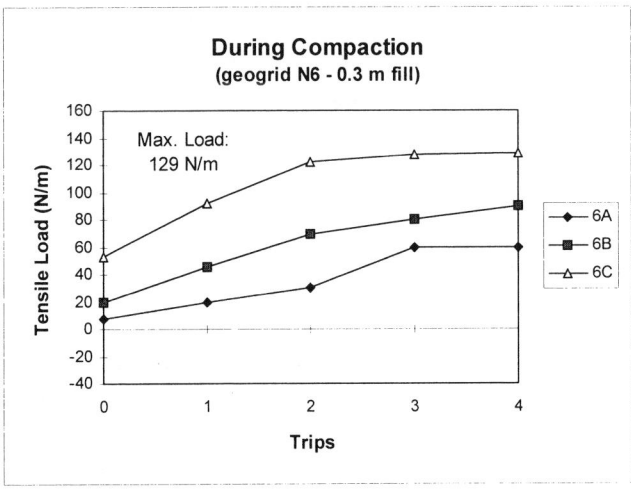

Figure 4: Changes in tensile force due to compaction (N6)

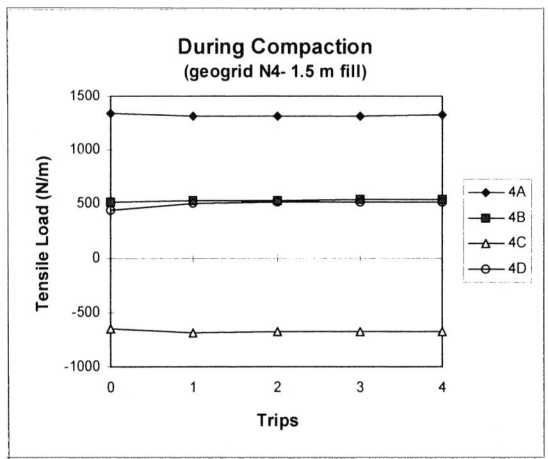

Figure 5: Changes in tensile force due to compaction after addition of 1.5 m of fill above N4

Geogrid N4 was installed at 1.35 meters above the bottom of wall. After installation and before backfilling, initial readings were taken for each strain gauge. After that, readings were taken at 2.55, 2.85, 3.15, 3.75, 4.05 and 4.88 meters as shown in Figure 6. It can be seen that tensile forces increased with the increase of construction height or increase of fill above geogrid except for strain gauge 4C. At the beginning, strain gauge 4C recorded a reduction in tensile force with the increase in construction height but after 3.15 meters, the force began to increase with the increase of construction height. For geogrid N4, maximum change in tensile force was recorded by strain gauge 4A, which was located at 0.3 meter (12") from concrete block wall/facing. The highest value recorded at 4.88 meters (experiencing 3.53 meters of compacted fill above the geogrid) was 3.624 kN/m. The second highest value was recorded by strain gauge 4D with tensile force of 1.639 kN/m at 4.88 meters. While strain gauge 4C recorded the lowest value with tensile force of - 0.543 kN/m at 3.15 meters.

The highest calculated tensile force using a computer program, Risiwall 4.0, was 9.387 kN/m at geogrid N4. So, the highest value measured during construction was 38.6 % of the calculated value.

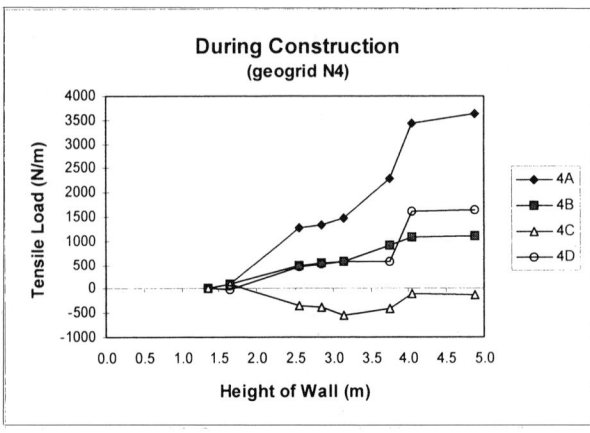

Figure 6: Changes in tensile force during construction for N4

Pore Water Pressure

No water was detected in the standpipe and pneumatic piezometers during construction, which indicated that the drainage system was working satisfactorily.

Horizontal And Vertical Earth Pressures

Changes in horizontal earth pressure during construction are shown in Figure 7. The results show that the stresses increased with the increase of fill above pressure cell or increase in construction height. At full height of wall, the pressure cell recorded 3.65 kN/m² of stress. While changes in vertical earth pressure during construction are shown in Figure 8. The results show that the stresses increased with the increase of fill material above pressure cell. At full height of wall, the pressure cell recorded 22.10 kN/m² of stress.

During construction, the measured coefficient of earth pressure, k for this retaining structure was 0.17 while the theoretical Rankine's active earth pressure coefficient, k_a was 0.32. It was 53.1 % of the theoretical k_a value. The measured k value can be obtained by using Eqn. 1.

$$k \text{ (measured)} = \frac{\text{Horizontal stress (measured)}}{\text{Vertical Stress (measured)}} \qquad (1)$$

While the theoretical k_a can be calculated by using Eqn. 2.

$$k_a \text{ (theoretical)} = \frac{1 - \sin \phi}{1 + \sin \phi} \qquad (2)$$

Where ϕ = drained friction angle.

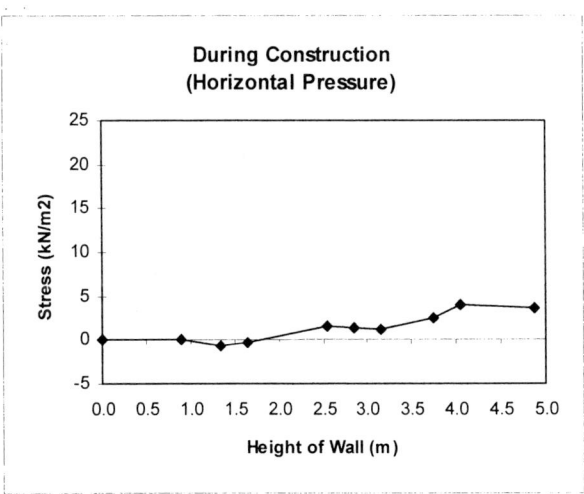

Figure 7: Changes in horizontal pressure during construction

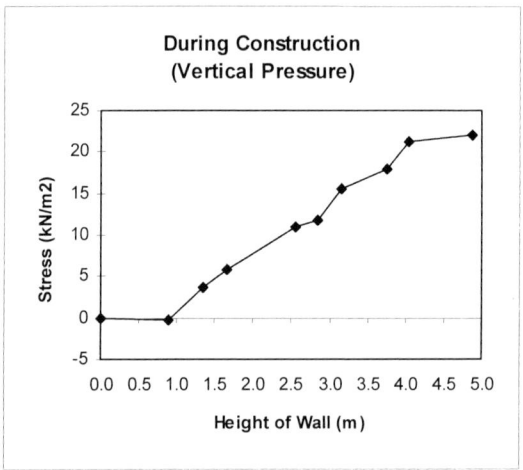

Figure 8: Changes in vertical pressure during construction

CONCLUSION

It is found that the tensile forces in the geogrid reinforcements as well as horizontal and vertical earth pressures were within the permissible limits. Thus it can be said that the structure has performed satisfactorily during construction. However, monitoring should be continued to see the performance of the structure after end of construction.

ACKNOWLEDGEMENT

The authors would like to express their thanks to the following for financial support and permission to use their materials and to publish details of the project included in this paper:
Risi Stone Systems (M) Sdn. Bhd. and Royal Ten Cate Regional Office.

REFERENCES

Anuar K. and Faisal H.A. (1997). Reinforced Modular Block Wall with Residual Soil as Backfill Material. *Geotropika '97* Johor Bahru, Malaysia, 425-431.

British Standards Institution (1995). *BS8006 Code of Practice for Strengthened/Reinforced Soils and Other Fills*, British Standards Institution, London.

Chiu, et al. (1987). Reinforced Earth Wall in Malaysia. *9th Southeast Asian Geotechnical Conference*, Bangkok, 8: 121-138.

Ingold, T. S. (1982). *Reinforced Earth*, Thomas Telford Ltd, London

Nicolon Mirafi (1997). *Miragrid Strong and Flexible Geogrids for Segmental Retaining Wall Reinforcement*, In-house Document, Nicolon Mirafi, Norcross, Georgia.

ANALYSIS OF MULTILAYER SYSTEM WITH GEOSYNTHETIC INSERTION – SENSITIVITY ANALYSIS

B.B. Budkowska and J.Yu

Department of Civil and Environmental Engineering
University of Windsor, Windsor, Ontario, Canada N9B 3P4

ABSTRACT

The inclusion of a geogrid in the angular aggregate results in the development of a new composite layer, the modulus of which is higher than the unreinforced material. That results from the gradual stiffening process of multi-layer structure by the function of interlocking when subject to repetitive loading. The analysis of this process is tackled in the framework of sensitivity theory by allowing for the development of a composite layer around the geogrid. The modified elastic continuum theory was utilized to develop the constitutive model. The concept of permanent modulus of deformation was incorporated. The analysis process required the deformed geometry updating. The theoretical formulation of sensitivity analysis employed the adjoint method. It allowed for determination of the permanent modulus of composite layer that was much higher as compared with unreinforced system.

KEYWORDS

Geogrid, interlocking, sensitivity analysis, permanent modulus, continuum theory, conjugated system method.

INTRODUCTION

In the last two decades an increasing interest has been observed in the behavior of flexible multi-layer systems that have found broad application in transportation area. The modes of pavement's failure comprise many forms of distress like fatigue cracking, rutting, thermal cracking and others. The detrimental effect of the distress caused by rutting can be reduced by application of the appropriate remedy which are offered by the newly developed materials. They include the broad family of geosynthetics, among which geogrid is mainly designed to function as reinforcement. The reinforcement feature offered by geogrid when inserted in a proper layer of a pavement system is of particular interest when rutting issue is of concern. Significant decrease of permanent deformation (rutting) has been observed after the application of geogrid into flexible pavement. The interlocking between geogrid and angular aggregates produces a new composite layer. It is obtained by incorporation of the adjacent rough granular material which can interpenetrate through large geogrid's openings.

The analysis of stiffening process of the layer containing geogrid subject to repetitive loading can be performed by means of homogenization method or allowing for development of a composite layer around the geogrid. In this paper the latter approach was applied in the framework of sensitivity theory. It allows for determination of the increase of permanent modulus of stiffer layer of known thickness. The numerical analysis is supported by the laboratory record of monitored permanent displacements of the selected points placed in the unreinforced and reinforced cross-sections (Nejad et. al. (1996)).

The constitutive model used in numerical identification process is based on the elastic continuum theory modified for the purpose of analysis in order to take into account the unrecoverable deformations. It is done by incorporation the concept of permanent modulus of deformation. Updating geometry after one sequence of load application is performed in order to take into account the unrecoverable deformation. The applied load of constant value is assumed to be distributed over the circular area. The parameters of laboratory model allow for considering the problem as axisymmetric. The theoretical formulation of sensitivity analysis employs the conjugated system method. It is performed in the scope of variational calculus. The first variation of the functional which represents the work done by the unit load applied at the control point on the decrement of permanent displacement caused by the change of modulus of permanent deformation of a composite layer is equal to the work of the stresses of the conjugated system on the corresponding variations of strains imposed on the primary (unreinforced) system. The changes of the permanent displacement of control point due to the placement of geogrid are provided by laboratory data. Combining this fact with known thickness of composite layer, the modulus of permanent deformation of composite is determined.

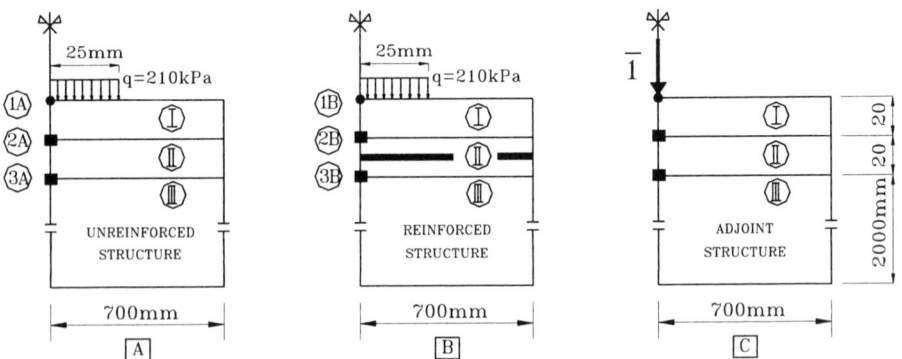

Figure 1: The structural cross-section (unreinforced and reinforced) and conjugated structure (not to scale)

DEVELOPMENT OF COMPOSITE LAYER

The cross-sections of laboratory multilayer model given by Nejad et al. (1996) are shown in Figure 1. It contains an unreinforced section A and reinforced section B. Section A is composed of three layers: a propriety cold-mix bituminous wearing course on the top, a crushed aggregate of basaltic origin in the middle, and sand subgrade made up of Sydney sand at the bottom. On the other hand, section B has a layer of geogrid included in the middle of the second layer. The accumulative permanent displacements are monitored at points 1, 2, 3 at each cross-section and are designated as 1A, 2A, 3A,1B, 2B, 3B, respectively. The system is subjected to repetitive loading provided by pneumatic tire moving with constant velocity which produces uniformly distributed pressure q = 210 kPa. It is reasonable to assume that the loading is distributed on the circular area having a radius r = 25 mm. The results of accumulated permanent displacements w_{1A} and w_{1B} at the cross-sections A and B for the point 1A and 1B (provided by Nejad et. al. (1996)) are shown in Figure 2.

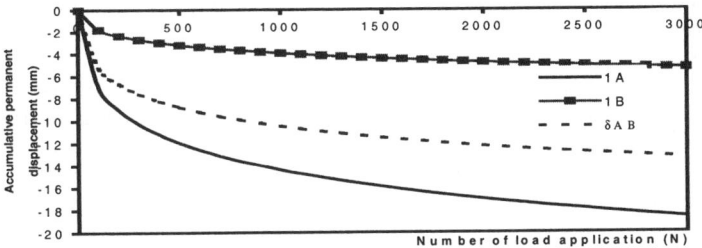

Figure 2: Monitored accumulated permanent deformations at point 1A and 1B

The permanent deformation of both points is found to change in cubic fashion with respect to number of load repetition (logarithmic scale) after regression analysis upon the lab data has been accomplished, and is listed below:

$$w_{1A}(N)[mm] = 0.0109 - 0.5376*\log_{10}(N) - 1.6435*\log_{10}^2(N) + 0.0741*\log_{10}^3(N) \quad (1)$$

$$w_{1B}(N)[mm] = 0.003 - 0.135*\log_{10}(N) - 0.3518*\log_{10}^2(N) - 0.0319*\log_{10}^3(N) \quad (2)$$

The difference of Eqn. 1 and Eqn. 2 denoted as δw_{AB} shown as dotted line in Figure 2 represents the decrease of the accumulated permanent displacement due to placing of geogrid in the mid layer.

$$\delta w_{AB}(N)[mm] = 0.0079 - 0.4026*\log_{10}(N) - 1.2917*\log_{10}^2(N) + 0.106*\log_{10}^3(N) \quad (3)$$

To identify the permanent modulus of each layer as the function of each sequence of loading application, the deformations of point 1A and 1B (denoted as δw_{1A}^S and δw_{1B}^S) corresponding to every single sequence of loading application are calculated and illustrated in Figure 3. The difference, defined as:

$$\delta w_{1S} = \delta w_{1A}^S - \delta w_{1B}^S \quad (4)$$

and denoted as δw is in Figure 3 included and will be considered in the sensitivity analysis.

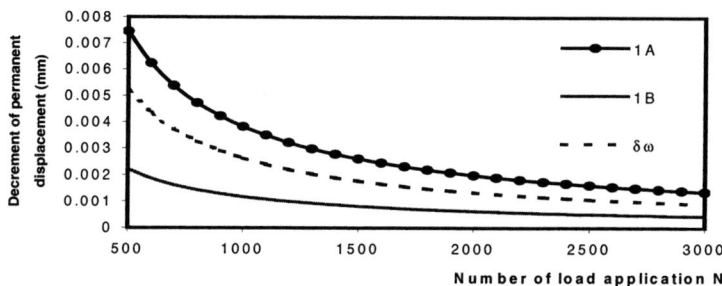

Figure 3: Permanent deformation corresponding to single load application

By using the displacement corresponding to single load application at point 1A, 2A, 3A as control parameters, the entire process of simulation of permanent deformations is performed with the aid of the finite element method (FEM). The suitable FEM package is provided by general analysis program ABAQUS (1998). The numerical simulation involves the geometry updating after every sequence of deformation. The permanent modulus of each layer (under unreinforced condition) was determined and verified by another multi-layer analysis program KENLAYER (Huang, (1993)). The results are summarized by the following equations (illustrated in Figure 4):

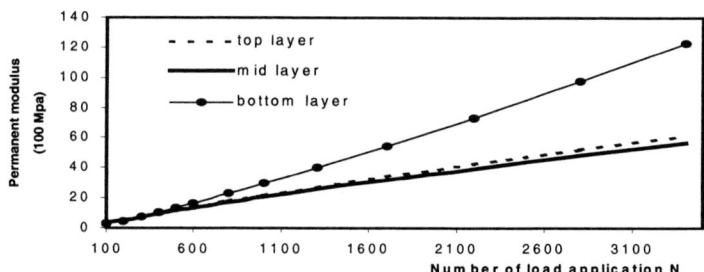

Figure 4: Distribution of permanent moduli for the unreinforced structure

$$E_{PT}(N) = 0.6887 + 0.0217 * N - (1.1885e - 6) * N^2 \tag{5}$$

$$E_{PM}(N) = 1.5296 + 0.02 * N - (1.2989e - 6) * N^2 \tag{6}$$

$$E_{PB}(N) = 0.9127 + 0.0283 * N + (2.3959e - 6) * N^2 \tag{7}$$

where: E_{PT}: the permanent modulus of top layer,
  E_{PM}: the permanent modulus of mid layer,
  E_{PB}: the permanent modulus of bottom layer,
  N: number of load application.

SENSITIVITY ANALYSIS

To identify the composite layer in the mid layer generated from the inclusion of geogrids, sensitivity analysis is incorporated. The adjoint structure is formed with the same geometry as the unreinforced structure but having the material property as the reinforced structure. Thus it deforms in the same fashion as the structure with geogrid in the mid layer. Unit load is applied to the monitored point 1 of the adjoint structure. The permanent modulus of the composite layer is taken as the design variable, while the permanent deformation produced by one single sequence of load application is considered as the state variable. The functional of permanent deformation is formulated in the scope of variational calculus. Its first variation is determined with involvement of adjoint system whereas some statically admissible variations of kinematic field are imposed on the primary system. Then, the following relationship is valid:

$$1\delta w_{1S} = \int_V (\overline{\sigma_z}\delta\varepsilon_z + \overline{\sigma_r}\delta\varepsilon_r + \overline{\sigma_\theta}\delta\varepsilon_\theta + \overline{\tau_{rz}}\delta\gamma_{rz})dV \qquad (8)$$

where: δw_{1S}: the decrement of permanent displacement at point 1 due to inclusion of geogrid for each single sequence of load application,

$\overline{\sigma_z}, \overline{\sigma_r}, \overline{\sigma_\theta}$: the normal vertical, radial and circumferential stresses respectively produced in conjugated system,

$\overline{\tau_{rz}}$: the shear stress in the conjugated system,

$\delta\varepsilon_z, \delta\varepsilon_r, \delta\varepsilon_\theta, \delta\gamma_{rz}$: the first variation of strains generated by the variations of kinematic field imposed on the primary system.

The constitutive model employed in analysis of primary and conjugated system satisfies the following relationships:

$$\sigma_z = \frac{E_p(1-v)}{(1+v)(1-2v)}\left[\varepsilon_z + \frac{v}{1-v}(\varepsilon_r + \varepsilon_\theta)\right] \qquad (9)$$

$$\sigma_r = \frac{E_p(1-v)}{(1+v)(1-2v)}\left[\varepsilon_r + \frac{v}{1-v}(\varepsilon_z + \varepsilon_\theta)\right] \qquad (10)$$

$$\sigma_\theta = \frac{E_p(1-v)}{(1+v)(1-2v)}\left[\varepsilon_\theta + \frac{v}{1-v}(\varepsilon_z + \varepsilon_r)\right] \qquad (11)$$

$$\tau_{rz} = \frac{E_p}{2(1+v)}\gamma_{rz} = G_p \gamma_{rz} \qquad (12)$$

where: E_p: modulus of permanent deformation,
G_p: shear modulus of permanent deformation,
v: Poisson's ratio.

The sensitivity theory allows for extension of the constitutive relationships on account of the design variables. Hence, the first variations of stresses defined by Eqn. 9 ~ Eqn.12 have the form:

$$\delta\sigma_z = \frac{1-v}{(1+v)(1-2v)}\left[\varepsilon_z + \frac{v}{1-v}(\varepsilon_r + \varepsilon_\theta)\right]\delta E_p + \frac{E_p(1-v)}{(1+v)(1-2v)}\left[\delta\varepsilon_z + \frac{v}{1-v}(\delta\varepsilon_r + \delta\varepsilon_\theta)\right] \quad (13)$$

$$\delta\sigma_r = \frac{1-v}{(1+v)(1-2v)}\left[\varepsilon_r + \frac{v}{1-v}(\varepsilon_z + \varepsilon_\theta)\right]\delta E_p + \frac{E_p(1-v)}{(1+v)(1-2v)}\left[\delta\varepsilon_r + \frac{v}{1-v}(\delta\varepsilon_z + \delta\varepsilon_\theta)\right] \quad (14)$$

$$\delta\sigma_\theta = \frac{1-v}{(1+v)(1-2v)}\left[\varepsilon_\theta + \frac{v}{1-v}(\varepsilon_r + \varepsilon_z)\right]\delta E_p + \frac{E_p(1-v)}{(1+v)(1-2v)}\left[\delta\varepsilon_\theta + \frac{v}{1-v}(\delta\varepsilon_z + \delta\varepsilon_r)\right] \quad (15)$$

$$\delta\tau_{rz} = \gamma_{rz}\delta G_p + G_p \delta\gamma_{rz} \quad (16)$$

Since the primary system is subjected to constant load, the variation of the statically admissible kinematic field imposed on the structure does not generate the increments of stresses. This means that left hand sides of Eqn. 13 ~ Eqn.16 are equal to zeros. Thus the sought variations of strains of primary system required by Eqn. 8 can be determined from the following equations:

$$\frac{\sigma_z}{E_p}\delta E_p = -\frac{E_p(1-v)}{(1+v)(1-2v)}\left[\delta\varepsilon_z + \frac{v}{1-v}(\delta\varepsilon_r + \delta\varepsilon_\theta)\right] \quad (17)$$

$$\frac{\sigma_r}{E_p}\delta E_p = -\frac{E_p(1-v)}{(1+v)(1-2v)}\left[\delta\varepsilon_r + \frac{v}{1-v}(\delta\varepsilon_z + \delta\varepsilon_\theta)\right] \quad (18)$$

$$\frac{\sigma_\theta}{E_p}\delta E_p = -\frac{E_p(1-v)}{(1+v)(1-2v)}\left[\delta\varepsilon_\theta + \frac{v}{1-v}(\delta\varepsilon_z + \delta\varepsilon_r)\right] \quad (19)$$

$$\delta\gamma_{rz} = -\frac{\gamma_{rz}}{G_p}\delta G_p \quad (20)$$

The solution of Eqn. 17 ~ Eqn. 20 gives:

$$\delta\varepsilon_z = -\frac{\varepsilon_z}{E_p}\delta E_p \quad (21)$$

$$\delta\varepsilon_r = -\frac{\varepsilon_r}{E_p}\delta E_p \quad (22)$$

$$\delta\varepsilon_\theta = -\frac{\varepsilon_\theta}{E_p}\delta E_p \quad (23)$$

$$\delta\gamma_{rz} = -\frac{\gamma_{rz}}{E_p}\delta E_p \quad (24)$$

Substitutions of the relationships (21) - (24) into Eqn. 8 results in:

$$\overline{1\delta w_{1S}} = \int_V \left(\overline{\sigma_z \varepsilon_z} + \overline{\sigma_r \varepsilon_r} + \overline{\sigma_\theta \varepsilon_\theta} + \frac{\overline{\tau_{rz} \gamma_{rz}} E_p}{G_p} \right) \frac{\delta E_p}{E_p} dV \quad (25)$$

For the axisymmetric problem, the infinitesimal volume dV is given as:

$$dV = 2\pi r (dr)(dh) \quad (26)$$

where: r, dr: the radial spatial variable and differential of r, respectively,
dh: infinitesimal increment of thickness of composite layer in vertical (z) direction.

To estimate δE_p for one single load application, Eqn. 26 can be discretized as follows:

$$\delta E_p = -\frac{\overline{1\delta w_{1S}} E_p}{2\pi \sum_i (\Delta h)_i \int_r \left(\overline{\sigma_z \varepsilon_z} + \overline{\sigma_r \varepsilon_r} + \overline{\sigma_\theta \varepsilon_\theta} + \frac{\overline{\tau_{rz} \gamma_{rz}} E_p}{G_p} \right) r dr} \quad (27)$$

where: Σ: the spatial discretization in vertical direction of the composite layer due to application of FEM,
Δh: the thickness of sublayer formed by FEM mesh within composite layer,
i: the number of sublayers formed within composite layer having thickness Δh.

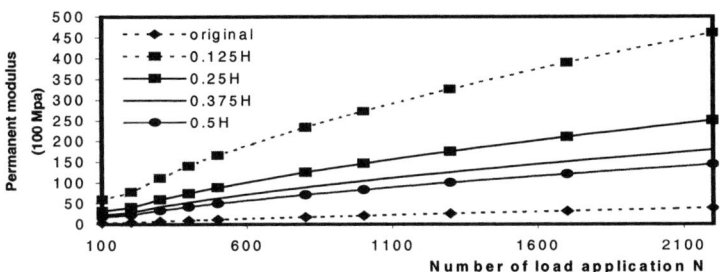

Figure 5: Distributions of modulus of permanent deformation E_p of composite layer of variable thickness

The results obtained by means of Eqn. 27 superposed to the initial value of E_p of the mid layer are presented in Figure 5 assuming that $\Delta h = 0.125 H; 0.25 H; 0.375 H,$ and $0.5H$, where H defines total thickness of the layer containing geogrid.

CONCLUSION

The presented formulation of sensitivity analysis allows for effective assessment of material characteristics of composite layer developed in the vicinity of stiffening element inserted in aggregate layer. The problem is of considerable importance in engineering practice with special attention to transportation area. The important aspect of analysis is embodied in the fact that the investigations are performed with respect to initial unreinforced system which allows for direct evaluation of advantages offered by new products when incorporated in traditional systems. Moreover, the variational approach of sensitivity analysis enables one to confine the exploration only to the regions where the composite layer can be developed. The constitutive model adopted in sensitivity investigation shows that the geogrid inclusion results in significant increase of stiffness of the original layer. The increase of stiffness is materialized through the improvement of modulus of permanent deformation of composite layer. The same effect, on the other hand, can be achieved through increase of thickness of an equivalent homogeneous layer having smaller value of modulus of permanent deformations.

ACKNOWLEDGEMENT

The authors acknowledge with thanks the support of the Natural Sciences and Engineering Research Council of Canada under Grant No. OGP 0110262, awarded to the first author, and the support of Ontario Graduate Scholarship of Science and Technology, awarded to the second author.

REFERENCES

ABAQUS Standard Users Manual, (1998). Version 5.8.5, Hibbitt, Karlsson & Sorensen, Inc., U.S.A

Huang, Y.H. (1993). *Pavement Analysis and Design*, Prentice Hall, Inc., New Jersey 07632

Nejad F.M. and Small J.C. (1996). Effect of Geogrid Reinforcement in Model Track Tests on Pavements, *Journal of Transportation Engineering*,. **122:6**, 468-474.

THE SCALE PITS OF SALDANHA STEEL

AN INNOVATIVE SOLUTION TO A COMPLEX PROBLEM

L. Fliss

Engineering Management Services (Pty) Ltd, Bedfordview 2008, Gauteng, South Africa

ABSTRACT

The design and construction of the scale pits of Saldanha Steel (SS) is a perfect example of how under extremely adverse conditions a critical problem could be solved by reverting from conventional to innovation. This is a case study where a two stage conventional construction of shoring and building was replaced by a one stage operation using the open caisson system, with the aim to save essential time and cost.

KEYWORDS

caisson, calcrete, concrete, hydrocyclone, pit, scale, shell, treatment, water.

INTRODUCTION

The steel production process requires large quantities of water for cooling and removal of scales, the iron oxide flakes produced during the continuous casting and hot rolling. As this process water has to be re-circulated, a water treatment plant is necessary to be provided to restore the quality of water to comply with the process requirements. The water treatment plant in steelworks is a major part of the whole project and includes primary retention of scales, filtration, removal of chemicals and cooling.

Scales retention takes place in hydrocyclone settling tanks, or scale pits, one for each cooling and de-scaling water stream. At SS there are two water streams, one for the continuous casting and another for the rolling mill. These water streams are running in flumes with steep falls of 6% to ensure the flushing of scales towards the scale pits. The flumes end with very steep 30% fall discharge pipes, tangentially connected into the circular hydrocyclones to create the cyclonic effect required for the fast settling process.

Due to the long steep slopes of the flumes and pipes, the discharge of water into the scale pits takes place at large depths, 10,7 m for the continuous caster and 13,7 m for the rolling mill, leading to a total pit depth of 14,3 m and 17,7 m respectively.

The coarse scales are settling on the bottom of the pit while the ascending clarified water passing through a filter is collected into a launder from where it flows into a sump, to be pumped further to the next phase of treatment. The scales will be removed from the bottom of the hydrocyclone by grub into trucks and recycled in the steel melting plant.

The basic dimensions of the two hydrocyclones required for the de-scaling are as per Table 1:

TABLE 1

Stream	Internal Dia	Depth	Feeding Pipeline Dia
Continuous caster	6,5 m	14,3 m	500 mm
Rolling Mill	16.0	17,7	1000

SITE CONDITIONS

Geology

The stratigraphy of SS site at the scale pits location consists of a sequence of hard pan (hard rock) and powder calcrete layers from ground level to 14m depth, underlain by stiff clayey silt of Varswater formation to a depth of 24 m, which becomes gradually a silty sand changing into very soft rock at -27m.

The water table was found at $-5,2$ m and an artesian second aquifer at -25 m.

Environment

The scale pits are located nearby the rolling mill buildings which were already erected and founded on shallow spread footings. Various installations of the water treatment plant were planed to be erected adjacent to the scale pits.

The environmental strict requirements did not allow any spillage from the pits, to prevent contamination of the aquifer.

Programme

The water treatment plant was on the critical path of the SS Project construction programme and the scale pits in particular due to their depth and difficulty were considered the most sensitive part of the whole programme.

The planned construction duration for the civil works of the scale pits was 10 months and meeting this target was of essence.

ORIGINAL DESIGN

The original design of the scale pits followed entirely the hydraulic scheme produced by the technology supplier and consisted of a circular hydrocylone with a rectangular underground pump station for each pit (Figure 1).

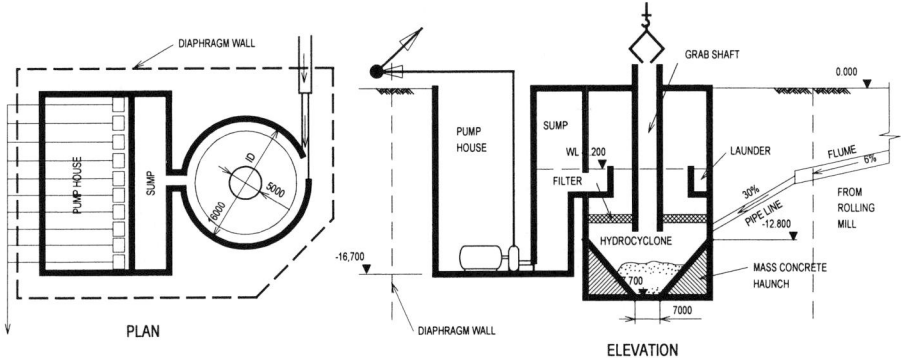

Figure 1: Scale Pit – Original Design

The construction method adopted for this design was to firstly construct two peripheral anchored diaphragm walls to shore the deep excavations and building afterwards conventionally the pits inside the de-watered enclosures. The shoring was required due to the proximity of existing shallow foundations and also to allow simultaneous construction of other plants around the pits.

The main problem with the diaphragm walls was that due to the difficult ground conditions caused by the hard rock layers existing within the first 14 m of depth, the rate of progress would have been extremely slow and unpredictable, therefore the required date of completion which was so critical for the whole project , could not be guaranteed.

Under these circumstances , the civil contractor proposed for the same lump sum price an alternative design, giving a firm guarantee for the completion date. The proposal was accepted and successfully carried out.

ALTERNATIVE DESIGN

As the construction of the diaphragm walls was the main obstacle in meeting the programme it was natural to look for a solution which eliminates this obstacle, using a totally different approach, namely the open caissons system.

Although the open caisson method is not an engineering novelty, being quite often used for bridge foundations or quay walls, the usage of caissons to create large underground chambers to accommodate reservoirs and pump stations is not at all common.

The advantages of using open caissons for the scale pits at SS were multiple :

- Speed of construction – as the sinking through hard rock was carried out using mechanical equipment such as woodpackers, jackhammers and excavators.

- Reduction in volume of excavation and keeping de-watering to a minimum.
- Cost saving through substantial reduction of reinforced concrete volume, as the caissons eliminated the costly diaphragm walls.

However, the caisson sinking operation being unusual for the normal contractors specialized in conventional industrial civil works, it would be required to go through a training programme and acquire some special equipment to achieve the expected rate of progress and quality of work. In addition more intense geotechnical field activity would be necessary to assist in controlling the de-watering of excavation and maintaining the uniform progress of caisson sinking.

The alternative design of the two scale pits included four circular caissons, varying in diameter from 6,6 to 16 m internally, two for hydrocyclones and the others for pump stations. The circular shape in plan for pump stations adopted in the alternative design, opposed to the original rectangular shape, is by far more economical for structures subjected to earth pressure around. The hydrocyclone and pump station for the rolling mill being much larger than those for the continuous caster, posed more difficult design and construction problems, therefore we will concentrate on describing these items (Figure 2).

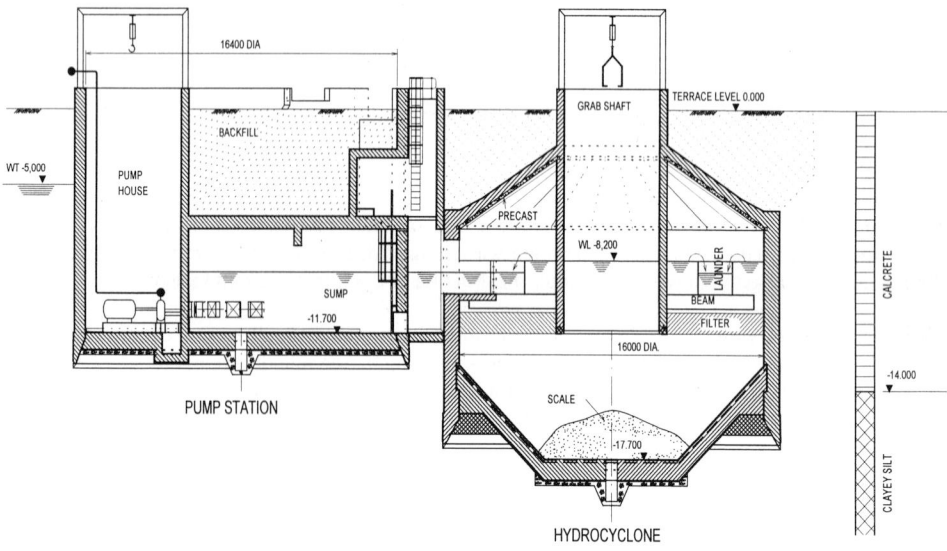

Figure 2: Section through Scale Pit- Alternative Design

The hydrocyclone pit was designed as a 16 m internal diameter caisson with an 800 mm thick wall to reach a depth of 16 m below terrace level. After passing through calcrete layers of various hardness it eventually penetrates some 2 m into a layer of stiff clay, to be stopped into a cast in-situ mass concrete ring.

The bottom part of the pit was designed as an inverted conical shell, a shape more economical than a flat raft with conical mass concrete haunching as it was provided in the original design. In addition, by adopting this shape for the pit bottom, the caisson could be shortened by 4 m saving thus cost and time. The water tightness of the pit bottom shell was achieved with water retaining concrete on top of which was applied a layer of 150 mm gunite.

Following the completion of the bottom shell, a temporary concrete tower was built on top to support on 8 jacks the 5 m diameter, 11,7 m high concrete grab shaft which eventually will be suspended from the pit roof. Meanwhile the grab shaft was providing temporary support during construction to the concrete roof structure, transferring its load to the temporary concrete tower. The grab shaft also provides support to the steel launder rested on 8 concrete precast beams which in turn will add stability to the shaft by propping it onto the caisson wall.

Supporting the launder on the radial beams has also the advantage of allowing it to be placed away from the caisson wall , creating thus overflow on both sides of the launder with the result of improving the hydrocyclone settling performance.

The pit roof was designed as a conical shell, opposed to the original flat slab and beams construction. The conical shell in addition to being the most economical structural shape to support the large circular load of the suspended grab shaft and the weight of the backfill on top, allows for a fast erection method, namely composite construction with precast structural permanent formwork and reinforced concrete topping, eliminating thus high scaffolds and complicated formwork.

After the roof was constructed and the backfill placed on top of it, the temporary supporting system was removed by firstly gradually disengaging the jacks and afterwards demolishing the temporary concrete tower.

Characteristic for the conceptual design of the hydrocyclone structure is the fact that all its main elements are multifunctional, performing simultaneously structural and hydro-mechanical roles. For example, the caisson wall is at the same time a water and earth retaining structure and also contributes to create the cyclonic movement of the water flow. Similarly, the pit conical bottom shell is not just a plug for the caisson but also contributes to the settling of the scales. The conical roof is supporting the grab shaft, the backfill and the radial beams which are supports for hydraulic equipment (launder and filter). The multifunctionality of the structural elements, combined with the optimised general shape of the hydrocyclone were the main factors contributing to the efficiency of the alternative design.

The pump station which was accommodated in a separate 16 m diameter caisson, included a pump house for 10 radially orientated pumps and a sump. While the sump is completely enclosed by a concrete roof with backfill on top, the pump house is open at the top to allow for the removal of the pumps for maintenance. A connecting chamber was constructed between the two caissons to allow the passage of water from the launder into the sump.

CONSTRUCTION

The construction of the two scale pits included a number of distinct activities but out of all, the sinking of the caissons was the most complex one due to the difficult natural conditions combined with a very tight programme.

It required a 24 hour/day operation to ensure the continuity of sinking, at an average rate of 300 mm/day, which was achieved through detailed planning of manpower and equipment.

Most important for maintaining the continuous progress and above all to prevent the caissons to get stuck, was to ensure the verticality of the caissons which was achieved by supporting the caissons at all times on 3 soil bearings at 120 degrees apart. The soil bearings were created in the form of ground pillars of natural soil initially about 3 m wide on which the cutting edge of the caisson was resting. The soil between the bearings was continually excavated and kept some 4 – 500 mm below the cutting edge while a new set of bearings was created at 200 mm below the cutting edge in between the active bearings supporting the caisson.

By reducing the width of the active bearings, the ground pressure on them increased to the level to which the soil under the cutting edge was collapsing and the caisson was sinking, landing eventually on the new set of 3 bearings. The operation was repeated until the caisson reached the designed depth where it was blocked. The controlled collapse of the active bearings was the actual mechanism of ensuring the verticality of the caisson.

Another difficult operation was the control of the water level inside the caisson, which was achieved by pumping the water from an internal central sump which was progressing in depth in parallel with the caisson, 2 m in advance of the cutting edge. The central de-watering sump was maintained active until the mass of the structure exceeded double the magnitude of the uplift force and then it was plugged.

The sinking of the smallest caisson of 6,6 m internal diameter posed a major problem, that of a very limited space for the operation of mechanical equipment together with manual excavation and de-watering, which was solved by using a mini excavator and training the team to work in confined conditions.

The four caissons were sunk simultaneously saving thus precious time and utilizing the equipment more efficiently. The civil works for the two double scale pits were completed in 8 months, two months ahead of schedule with substantial cost savings and to a high quality standard.

CONCLUSION

The success of the construction of the scale pits at Saldanha Steel is a perfect example of how a critical situation could be overcome, adopting an unconventional approach by a multidisciplinary team of designers and contractors, motivated to meet a difficult target.

The open caisson alternative proved to be the right solution, not only to solve the time schedule problem but also a more economical one than the original design. The saving in time and cost is the result of an innovative concept where, through an optimization process each element, particularly of the main structure, the large hydrocyclone pit, was designed to perform multiple functions.

This concept of multiple functionality of structural elements may find numerous applications particularly in the industrial civil engineering field as part of the general trend of multidisciplinary integration in industrial development.

UTILIZATION OF BAMBOO AS REINFORCEMENT IN CONCRETE FOR LOW-COST HOUSING

T.A.I Akeju and F. Falade
Department of Civil Engineering
University of Lagos
Akoka, Lagos
Nigeria

ABSTRACT

This paper presents the findings of an investigation into the performance of bamboo reinforced concrete beams in flexure. The variables are: reinforcement volume fractions, type of reinforcement, reinforcement surface conditions and curing ages. A designed mix proportion of 0.55:1:1.86:4.36 was used. The specimens were cured in water at $21\pm1^{\circ}C$. The results showed that the ultimate tensile strengths of bamboo splints and mild steel are $133.50 N/mm^2$ and $255.00 N/mm^2$ respectively with corresponding yield strengths of $68.75 N/mm^2$ and $207.50 N/mm^2$. The ultimate strengths of the beams increased with age and reinforcement volume fractions. The beams deflected less with increase in curing age for a given percentage of reinforcement content but deflection increased with increase in reinforcement content for a given curing age. All the test specimens exhibited flexural failure. The use of bitumen and sharp sand as coating on bamboo splints reduced the rate of water absorption from the concrete matrix by the splints and enabled the bamboo to maintain its initial strength without considerable reduction in strength with age. An increase in strength of up to 134.65% was recorded for beams containing 2.68% reinforcement volume fraction above the unreinforced concrete beams at 28-day curing age. This indicates that bamboo is a viable substitute for steel reinforcement in low-cost housing projects where limited load-carrying capacity is required.

KEYWORDS

Concrete matrix, Low-cost, Housing, Bamboo splints, Mild steel, Flexural strength, Tensile strength.

INTRODUCTION

In Nigeria, like other developing countries and also to a certain extent in industrialized ones, the housing shortage problem is assuming increasing dimension with consistent increase in the cost of building materials. Particularly affected is the low-cost housing sector. Concrete, widely used for construction work is weak in tension but strong in compression. Its tensile strength is about 10% of its compressive strength. When concrete structures are subjected to tensile stresses, reinforcement is provided in form of steel, synthetic and glass fibres to withstand the stresses. Generally, in developing countries particularly in Nigeria, steel reinforcement is commonly embedded in concrete to take the tensile stresses which plain concrete is incapable of resisting. Over the years, the price of steel has been increasing to the extent that the material is no longer affordable to persons in low and medium income groups.

The continuous increase in prices of steel has led to research investigation into viability of some local fibres as substitute, which would be comparable in reinforcing efficiency but lower in cost to steel in low-cost housing projects. Adetifa (1) reported that the ultimate load carrying capacity of fan-palm reinforced concrete beam increased with increase in percentage of reinforcement in concrete though not proportionately. Kankam (2) investigated the use of palm stalk as resources-saving fibre reinforcement in concrete. According to him, the use of palm stalk as fibre reinforcement offers greater restraint to shrinkage as compared to plain concrete. Mawenya and Mwamila (3) reported that sisal fibre is capable of imparting post cracking strength, ductility and toughness to concrete matrix and that experimental data revealed cracking strength improvement of up to 30% for 2% fibre volume content. Castro and Naaman (4) reported that natural fibres of the agave family have significant mechanical properties that make them eligible as potential reinforcement of cementitious matrixes. Balaguru and Foden (5) reported that addition of fibres improved the splitting strength and modulus of rupture of the composite. Niyogi and Dwarakanathan (6) showed that fibre reinforced beams compared to ones in plain concrete have higher load carrying capacities.

Bamboo is a natural fibre and widely available in Nigeria. It grows in natural vegetation among thick forest and in riverine areas. Its stem is jointed and divided into nodes where relatively soft material forms solid partition walls with the indication that the internodal fibres are discontinuous at the joints. Bamboo is used in its natural form in rural areas as columns (nakedly or as composite with mud). The culms are also used as flooring material, roof trusses and wall in temporary structures. In urban areas, they are predominantly used on construction sites as temporary support to formwork during concreting work and scaffolding during plastering and painting works.

The science and technique of using bamboo for various purposes particularly as reinforcement in structural concrete is well known in Asia, America and Australia. In Africa, efforts are being made by researchers to determine the engineering properties of bamboo species available in their localities. In Ghana, Kankam (7) has reported that mature bamboo culms can be used safely as substitute for steel in some structural members if adequately treated. Different loading arrangements on slab reinforced with bamboo have been investigated (8 – 9). The results showed that bamboo increased the load-carrying capacity of the slab. In Egypt, Youssef (10) conducted tests on Arundinaria Gigantea, a species of bamboo, he observed that the green bamboo with about 39% moisture content, possessed about 60% of the tensile strength 30% and 35% of the compressive and flexural strengths respectively of seasoned bamboo. In Nigeria, the extensive research work done to date on bamboo has been on its establishment, identification of different species and their locations. Omotoso (11) reported that there are seven species of bamboo in Nigeria. Among the species, Bambusa Vulgaris constitutes about 80%. Its ability to sustain appreciable load on construction sites led to the preliminary investigation on its engineering properties. Falade and Akeju (12) reported that Bambusa Vulgaris has ultimate strength of $133.50 N/mm^2$ and moduli of elasticity ranging from $15833.00 – 23843.00 N/mm^2$. They noted that the strength obtained was lower than that of steel reinforcement but higher than that of unreinforced concrete. These results necessitated the need to incorporate bamboo splints in concrete and assess the structural behavior of the composite.

The objectives of this study are (i) to compare the flexural strengths of concrete, which contain bamboo splints with unreinforced and steel reinforced concrete (ii) to apply surface treatment to the bamboo splints and evaluate its effect on the strength of the beams and (iii) to examine the behaviour of splints in concrete with age.

MATERIALS AND EXPERIMENTAL PROCEDURES

The bamboo culms varied from 17.65 - 21.10m in lengths, 102 - 110mm in external diameters and 300 - 400mm in internodal lengths. Visual observation and measurement with callipers showed that the diameter and the thickness of the bamboo culms were almost constant at the intermediate portion and subsequently decreased to the top. The age of the bamboo could not be determined but the texture of the fibres and the surface indicated that they were 'mature'. The culms were air-dried in the laboratory under ambient temperature to average moisture content of 12%. They were cut to the required splint size

(8x10x700mm) with circular sawing machine "Rapid". 10mm diameter mild steel was used. The tensile strengths of the bamboo and steel reinforcement were determined.

The particles of the fine aggregate (sharp sand) used were those passing sieve with aperture 4.75mm and retained on sieve with aperture 0.063mm. The coarse aggregate was from crushed granite having a size range of 10-20mm. The cement used is Ordinary Portland cement whose properties conform to British Standard BS12 (13). A designed mix of 0.55:1:86:4.34 (water: Cement: Sand: granite chips) for target strength of 40N/mm^2 was used. Six reinforcement volume fractions were considered. In the absence of any standard specifications for incorporating bamboo in concrete, reference was made to clauses 3.12.5 and 3.12.6 in British Standard BS8 110 (14), which stipulate a minimum of 0.24% and a maximum of 4% gross cross-sectional area for mild steel in beam. The volume fraction considered are 0, 0.67,1.34, 2.01 and 2.68 expressed as percentages of gross cross - sectional area (150 x 150mm) of the beam. The other variables are: type of reinforcement (bamboo splints and mild steel), reinforcement surface conditions (coated and uncoated splints) and curing ages (7,14,28 and 90 days). The number of splints corresponding to the selected percentages are 0, 2No, 4No, and 8No. Zero percent reinforcement was the control experiment. Fig. 1 shows the arrangement of the reinforcement in the beams.

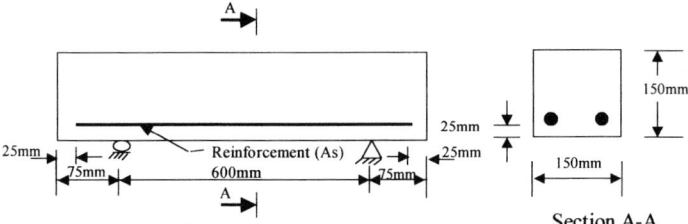

Fig. 1 Longitudinal Section of the Beam

At 0.67% reinforcement content, beams were cast with bamboo splints and steel (separately) and similarly with coated and uncoated splints. The beams were singly reinforced, the bamboo splints and mild steel have straight anchorage. Spacers were provided to hold the main reinforcement in position. The main reinforcement was tied to spacers with binding wire. A cover of 25mm was provided to prevent reinforcement from having direct access to water during curing.

MOULDING OF SPECIMENS

Two types of metal moulds were used: 150 x 150 x 150mm and 150 x 150 x 750mm for the cube and the beam specimens respectively. Batching was by weight. The mixture of cement, sand and granite chips were mixed mechanically in a mobile rotating drum mixer. When the constituents had been thoroughly mixed, the required quantity of mixing water was added gradually. The workability (using slump test) of the fresh concrete was determined immediately after the final mixing. The specimens were cast in accordance with British Standard BS 1881 (15). As the moulds were tapped, they were intermittently shaken to ensure the dislodgment of trapped air and more compaction. The moulds with the cast specimens were stored for 24±½ hours with polythene cover before they were demoulded. After demoulding, the specimens were transferred into a curing tank containing clean water where they were kept until the age of test. The average temperature of the curing water was 21 ± 1°C.

TESTING OF SPECIMENS

(i) *Compressive Strength*
The strength characteristic of each cube was determined using a loading rate of 120KN per minute on a 100kN Avery Dension Universal Testing Machine. A total of 54 cubes (150 x150 x 150mm) were tested. The specimens were tested immediately after removal from the curing tank.

(ii) *Flexural Strength*

A total of 90 beams (150 x150 x 750) were tested at curing ages of 7,14,28 and 90 days. Three specimens were tested for each age immediately after removal from the curing tank. Each test beam was simply supported over an effective span of 600mm on a pair of steel rods 32mm diameter placed at a distance 75mm from the edges of the beam. Two other similar rods were placed at a third point between the supports on top of the beam at a distance 200mm from each support. A spreader beam was positioned on the two rods while another rod was put across the spreader to transmit the load from the load cell. The beam was position such that the load was applied to the uppermost surface as cast in the mould. The loading arrangement of the specimen is shown in Fig. 2.

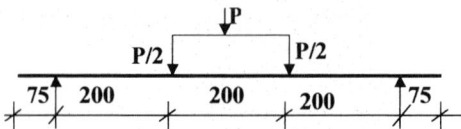

Fig. 2 Schematic Representation Of Loading Arrangement For Flexural Strength Test

With this arrangement, a constant maximum bending moment without shear was produced between the load points while the other portion of the beam was subjected to maximum shear with varying moment values. The moment between the load points induced compressive and tensile strengths at the top and bottom respectively. Deflections of the beams were measured using a dial gauge centrally positioned at the underside of each beam.

(iii) *Tensile Strength of Bamboo Splints Retrieved from Beams*

Tensile strength and moisture content tests were conducted on bamboo splints (coated and uncoated) that were retrieved from cured beams. At each age, a beam that had not been subjected to any loading was demolished to retrieve the embedded bamboo splints to determine their deterioration in concrete.

RESULTS AND DISCUSSION

(a) *Preliminary Tests on Reinforcing Materials*

The results of tensile strength test conducted on bamboo splints and mild steel showed that, the ultimate tensile strengths of bamboo splints and mild steel are $133.50 N/mm^2$ and $255.00 N/mm^2$ respectively with corresponding yield stresses of 68.75 and $207.54 N/mm^2$. The 24-hr absorption capacity test showed that bamboo absorbed 34.28% moisture. The fast water absorption is widely acknowledged as one of the drawbacks of using the bamboo with cementitious matrixes because they absorb part of its mixing water.

(b) *Plain and Reinforced Concrete Specimens*

(i) *Workability and Compressive Strength*

The workability of concrete batches showed an average slump value of 26.25mm. The cube specimens have average strength development of $24.26 N/mm^2$, $29.10 N/mm^2$, $35.55 N/mm^2$ and $37.93 N/mm^2$ at 7, 14, 28 and 90 days respectively.

(ii) *Flexural Strength*

The flexural strengths of the test beams varied with reinforcement volume fraction, type and surface conditions of reinforcement and curing age. When the percentage of reinforcement was increased, the flexural strength increased. In fig. 3, at 0% reinforcement content, the ultimate flexural strength at 7th day is 3.14 N/mm². For the same age, the strength values are 4.55, 5.05, 6.54 and 7.38N/mm² for 0.67% (uncoated bamboo splints), 1.34, 2.01 and 2.68 percentages of bamboo reinforcement respectively. These represent increases of 33.43%, 48.09%, 108.95% and 116.42%, above the strength of unreinforced concrete beams.

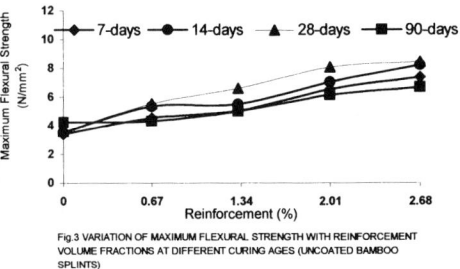

Fig.3 VARIATION OF MAXIMUM FLEXURAL STRENGTH WITH REINFORCEMENT VOLUME FRACTIONS AT DIFFERENT CURING AGES (UNCOATED BAMBOO SPLINTS)

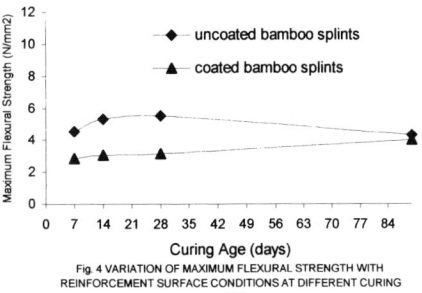

Fig. 4 VARIATION OF MAXIMUM FLEXURAL STRENGTH WITH REINFORCEMENT SURFACE CONDITIONS AT DIFFERENT CURING AGES

At 0.67% reinforcement volume fraction, beam containing steel reinforcement resisted more loads than those that contained bamboo splints. For example, in Fig. 4, at 7-day curing age, the average strength of beams containing bamboo splints is 4.55. N/mm² while that of steel reinforced beam is 6.79 N/mm² showing a strength ratio of approximately 1.50 above the beam that contained bamboo splints. Fig. 5 shows that the strengths of beams that contained uncoated bamboo splints were higher than those containing coated bamboo splints (bitumen + sharp sand).

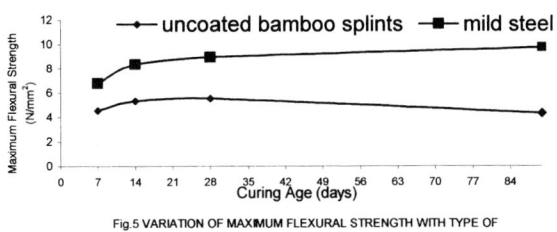

Fig.5 VARIATION OF MAXIMUM FLEXURAL STRENGTH WITH TYPE OF REINFORCEMENT AT DIFFERENT CURING AGES

The low ultimate strengths associated with beams that contained coated splints were due to bond failure as a result of loss of tension grip by concrete matrix and bamboo interface. The strength generally increased with age up to 90-day. At 90-day curing, all specimens that contained uncoated bamboo splints showed reduction in strength values. But unreinforced beams, beams containing coated splints and steel reinforced beams showed increases in strength values (figs. 3,4 and 5).

(iii) *Deflection*

Figure 6 shows typical load-deflection curves of the beams under gradually applied static loading at ages of 7, 14, 20 and 90 days for 0, 0.67, 1.34, 2.01 and 2.68% reinforcement volume fractions.
Deflections increased with increase in reinforcement content up to 28 days. At 90-day, the specimens showed reduction in deflection. The large deflections associated with beams containing bamboo splints as compared with those that contained mild steel might be attributed to its low modulus of elasticity (15.833-23.843 kN/mm²) compared to steel (210kN/mm²).

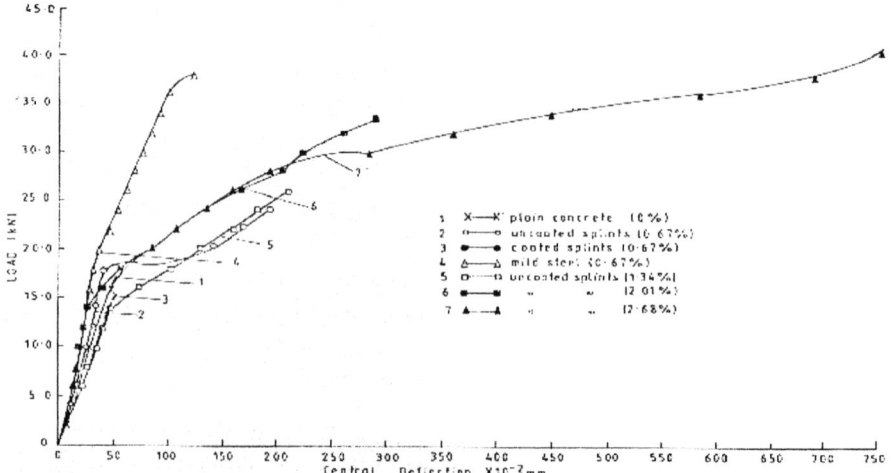

Fig. 6 Typical Load-Deflection Curves for Different Percentages of Reinforcement

(Iv) *Cracking and Yielding of the Composite*

The first crack load corresponds to the load at which micro cracks commenced on the beam while the yield is the load at which the beam momentarily refused to take load due to the yielding of the bamboo splints. Associated with yielding of bamboo is increase in deflection under constant load. After the yield load the relationship between stress and strain ceased to be linear. In most cases during the test the yield load coincided with the first crack load of the composite. When the bamboo has yielded, during the temporary refusal to take load by the test specimen, crack commenced at the soffit of the beams under the sustained yield load.

(v) *Mode of Failure*

Plain concrete specimens showed no ductility with brittle failure occurring shortly after the appearance of the first crack. After the final crack in the bamboo and steel reinforced concrete beams, the specimens exhibited drop in load –carrying capacity but showed some post-cracking resistance due to the presence of reinforcement that were bridging the cracks. Generally, the beams exhibited identical mode of failure (flexural) but the crack patterns varied. Crack usually started at the tension zone of the beams in the mid-third of the span where the greatest flexural strains occurred. The propagation of the cracks took different forms, depending upon the strength of the reinforcement –matrix interface as well as the brittleness of the reinforcement. Fig. 7 shows the crack patterns as the loading on the beams progressed.

Fig.7 Crack Patterns in Test Beams (Side View)

A comparison of first crack with failure loads showed that bamboo imparted post-cracking strength to beams. For example, the first crack loads are 17.10, 17.20, 20.00 and 22.00 kN at 7-day curing for 0.67, 1.34, 2.01 and 2.68% reinforcement volume fractions. For the same ages and percentages reinforcement, the ultimate loads are 24.20, 28.10, 34.50 and 37.60kN respectively. These show increase of 42.53%, 63.37%, 72.50% and 70.90% above the first crack loads. Other ages followed the same trend of higher failure loads than first crack loads.

(vi) *Behaviour of Bamboo Splints in Concrete*

At each curing age, an unloaded bamboo reinforced beam was carefully demolished to retrieve back the bamboo splints for moisture content and tensile strength test. Fig. 8 shows that the moisture in unloaded bamboo splints increased with age while those of coated splints, though increased but at a lower rate. At 90-day the average moisture content for the uncoated bamboo splints was 114.46% and 44.60% for the coated splints. The colour of the splints changed from light brownish (before embedding it in concrete) to deep yellowish (after retrieval from concrete). Fig. 9 shows that the tensile strength of the bamboo decreased progressively with age from the ultimate value of 133.50N/mm^2 before embedding the splints in concrete to average of 35.20N/mm^2 at 90-day for the uncoated splints and 82.06N/mm^2 at the same age for the coated splints. This trend can be attributed to chemical causes. Moisture caused progressive diffusion of hydration product into the interstices into the bamboo splints, the contaminated water in the fibres of the splints resulted in direct weakening of the fibres. The more the water in the fibre the lower the strength.

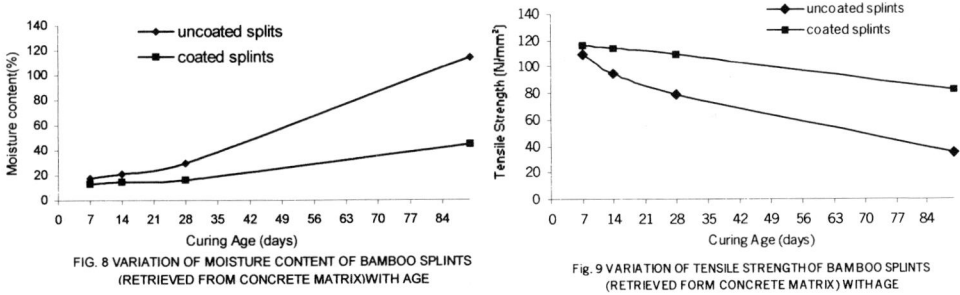

FIG. 8 VARIATION OF MOISTURE CONTENT OF BAMBOO SPLINTS (RETRIEVED FROM CONCRETE MATRIX)WITH AGE

Fig. 9 VARIATION OF TENSILE STRENGTH OF BAMBOO SPLINTS (RETRIEVED FORM CONCRETE MATRIX) WITH AGE

The coated bamboo splints retrieved from concrete matrix at each age showed that good bond existed between molten bitumen + sharp sand and concrete matrix while bamboo/molten bitumen interface showed ineffective bond initially but the bond improved with curing age. The improvement was made evident by the consistent increase in the strength of beams that contained coated bamboo splints at all curing ages considered in this study.

CONCLUSIONS

Based on the results presented in this paper, the following conclusion can be made:
i   The inclusion of bamboo splints in concrete beams increased the load carrying capacity of the beams but not proportionately.
ii   Strength was observed to improved by up to 134.65% above the strength of unreinforced beams at 28-day curing for 2.68% reinforcement volume fraction.
iii   For the same section and percentage reinforcement the failure load of mild steel reinforced beam was approximately 1.5 times that of its equivalent bamboo reinforced beams.
iv   Bamboo splints imparted post-cracking strength to concrete beams.

v   Bamboo splint deteriorated in concrete with age. Application of impervious surface coating like bitumen makes it more resistant to deterioration.

vi   The strength of bamboo reinforced beams is adequate for low-cost housing projects but further work needs to be done before final recommendation on the use of bamboo as reinforcement in concrete can be made.

RECOMMENDATIONS

The following on-going follow-up tests will better our understanding and confidence in the use of bamboo as reinforcement in concrete for low-cost housing projects:

i   Determination of durability, creep and shrinkage characteristics of bamboo reinforced concrete.

ii   Preservation and treatability of bamboo.

iii   Investigation of the behaviour of a prototype structure in which bamboo splints are used as reinforcement to assess the performance of bamboo reinforced concrete in service. The consistent curing of the specimens in water does not depict the practical situation.

REFERENCES

1. Adetifa, A.O. (1990). 'The behaviour of Fan-Palm Reinforced Concrete Beams under Flexural Loading, Proc. Second International Conference in Structural Engineering Analysis and Modeling (SEAM 2) pp 322-330.
2. Kankam, CK. (1990), 'Palm Stalk Fibre Reinforcement in Concrete to Control Shrinkage Stresses', Proc. SEAM 2 pp.357-370.
3. Mawenya, A.S. and Mwamila, B.L. (1990), 'Analysis and Behaviour of Sisal Fibre Reinforced Concrete' Proc. SEAM 2, pp. 333-349.
4. Castro, J and Naaman, (1981), Cement Mortar Reinforced with Natural Fibres; ACI Journal, Vol. 78, No. 1, pp. 69-78.
5. Balaguru, P and Foden, A (1996), 'Properties of Fibre Reinforced Structural Lightweight Concrete' ACI Structural Journal. Vol. 93, No.1 pp 62-78
6. Niyogi, S.K and Dwarakanathan, G.I (1985),'Fibre Reinforced Beam Under Moment and Shear'. Journal of Structural Engineering, Vol. 111, No.3 pp516-527
7. Kankam, J.A (1987), 'Bamboo as Reinforcement in Concrete' Proc. SEAM 1 pp 373-387.
8. Kankam, J.A., Perry, S.H. and Ben-George, M. (1986), Bamboo Reinforced Concrete One-way Slabs subjected to Line Loading' Int. Journal for Development Technology, vol. 4, pp. 1-9.
9. Kankam, J. A., Ben-George, M. and Perry, S.H., (1986), Bamboo Reinforced Concrete Two way Slabs subjected to Concentrated Central Loading' The structural Engineer, London, vol, 64B, No. 41, pp85-92.
10. Youssef, M.A.R. (1976), Bamboo as Substitute for Steel Reinforcement in Structural Concrete' New Horizon in Cost. Malts. Vol.1.
11. Omotosho, T.O, (1983), 'Bamboo in Nigeria' Independent Student Project, Forestry Research Inst., Ibadan.
12. Falade, F and Akeju, T.A.I., (1997), The Potentials of Bamboo as Constructions Materials (Accepted for Publication in SEAM Journal Ghana).
13. British Standard BS 12 (1971), Portland Cement (Ordinary and Rapid Gardening) Part 2, British Standard Institution, London.
14. British Standard BS 8110 (1985), The structural Use of Concrete. Part 1' Design Materials and Workmanship, British Standard Institution, London.
15. British Standard BS 1881 (1970), Methods of Testing Concrete, Part 2', British Standard Institution.

ENHANCING HOUSING DELIVERY USING A SIMPLE PRECAST CONSTRUCTION METHOD

J. Kanyemba

Building Research Unit, Rural Industries Innovation Centre (RIIC),
P. Bag 11, Kanye, Botswana

ABSTRACT

The provision of affordable housing has been a huge social issue facing many developing countries and increasingly it is becoming a political problem in some countries. Among the reasons for slow delivery of housing units has been lack of novelty in design and construction. This is evidenced by construction methods such as the brick and mortar method that has been used for a long time in most countries. On the other hand, methods like precast concrete units, which come in several designs that include wall panels, discrete slab panels, columns and roofing elements have not been widely adopted.

This paper describes a simple precast construction method that can be used to increase the number of housing units delivered per year. Precast panels, which can be manufactured at small yards, were used with steel lipped channels to construct a housing structure. The reinforced concrete panels are relatively light and thus they can be handled easily on site. The steel lipped channels are flexible and they can be fixed in several configurations to suit the architectural layout of the house.

The project proved that construction time was greatly reduced and unskilled labour could be utilised in some construction stages hence generating employment. It was also deduced that, repetitive units, which is a characteristic of low and medium cost houses, could be built within a reasonable period.

KEYWORDS

Housing delivery, precast concrete panels, lipped channels, employment creation.

INTRODUCTION

The provision of affordable housing to low-income groups is a major challenge facing many developing countries. The demand for houses continues to increase and the housing backlog swells everyday in most countries. In Botswana, the housing need has been estimated to be 12 000 units per year of which 60% are in the low-income category according to Min. of Lands & Housing (1999). There are several reasons for poor housing delivery in developing countries and among them is that there are fewer players in housing provision in most countries, poor policies, lack of finance and lack of incentives for innovation in providing solutions.

There are three main facets to housing delivery namely sites, services and structures. This paper has focused on the third element but the other two are also important in contributing towards effective delivery. There have been concerted efforts in most countries in upgrading semi-urban areas by constructing roads and sewer lines. In Botswana, this programme has been given high priority and it has been viewed as a solution to high urban land prices according to Min. of Finance & Development Planning (2000). In designing the housing structure control and innovation can be exercised and a suitable structure planned to suit the needs. The design could be made as cheap as possible and the construction duration can also be reduced by utilising fast-trek methods such as precast units.

It has been realised in several countries that the provision of a conventional house to each person who needs it cannot be achieved easily and this has called for non-conventional methods to be used according to Schlotfeldt (2000). Non-conventional methods are not poor quality systems but methods that are not specifically mentioned in building codes. Discrete precast concrete panel can be put in this category. The trend in a lot of countries is that these methods are eventually drafted into codes of practice after proving themselves over time as outlined by Knoetze et. al (2000). Tackling the housing shortage by encouraging non-conventional construction thus promotes novelty and would certainly attract participation from the private sector.

To alleviate the current problems, more housing units are needed annually and this calls for building systems that are quick, simple but of good quality. Studies on the social problems of low cost housing according to Schlotfeldt (2000) have indicated that solutions that involve the participation of beneficiaries in various forms are more successful. The precast construction method described in this paper utilises unskilled labour and there is an opportunity for recipients to participate and thus generate employment.

PRECAST STRUCTURE

To evaluate the cost advantages of the precast concrete panel and lipped channel building system a single-sized dwelling unit of area 13.5 m^2 was designed and constructed at RIIC's premises. The concept of precast houses is not new and several tests have been done on thermal performance, acoustic performance and condensation and the results showed good characteristics in several climatic zones in Southern Africa as set out in Agrément South Africa Mantag Certificate 1997/M46.

Design of Structure

A rectangular configuration was chosen for the layout of the unit and the inside dimensions were 3m x 4.5m. Steel lipped channels that were connected back-to-back or back-to-side were provided at 1.5m centres and the initial design was for a 75x50x10x2 mm section but a 100x50x20x2 mm section was used because the smaller channels were not available. Nevertheless, the bigger channels provided adequate lateral confinement to the 50 mm-thick precast concrete panels of size 300x1400x50 mm. The panels were made under factory conditions by a local concrete products company. The concrete panels were provided with 2x2 mm high-tension steel reinforcement primarily to resist lateral forces and to prevent cracking. The compound channel sections were designed to be spot welded at 500mm centres through their length of 3m but continuous welds were later used because minute gaps between the welded elements were found to be undesirable.

The columns were designed to resist lateral forces and vertical forces from the roof structure. An analysis of the stability of the system is shown in the Appendix. The panels that were used had vertical interlock and the lipped channels provided lateral restraint and for the purpose of analysis it was assumed that the discrete panels behaved as a continuous masonry wall.

Isolated footings of size 400x400x400 mm were used for each compound column that was cast in to a depth of 300 mm. The ground conditions consisted of a thin layer of alluvium sand overlying a clayey sand layer. A 5 mm thick mild steel base plate of size 300x300 mm was welded at the end of each column assembly for uniform load distribution.

Construction of the Precast Structure

The construction stages of the dwelling unit are summarised in Table 1, which also shows the resources used for the different stages. Also shown under duration is the computed time for a conventional structure.

TABLE 1
SUMMARY OF CONSTRUCTION ACTIVITIES

Activity	Duration (days)		Resources	Comments
	Precast	Conventional		
Site Preparation	5	6	2 casual labourers, technician	The work, which was planned for one day, took longer because of difficult ground conditions.
Foundation construction	4	8	2 casual labourers, technician	
Floor slab	2	2	1 casual labourer, technician officer	One day was spent on hardcore preparation and the other on concreting.
Wall Panels	2	4	2 casual labourers	This could have been done in half a day but time was lost in fitting the window frame.
Roof structure	3	3	1 casual labourer, technician	
Sheeting	2	2	1 casual labourer	
Finishes	4	4	1 casual labourer	This activity involved plastering and painting.

Two casual labourers and a technician with minimal building skill were involved in the construction of the unit. The project was completed within 22 days at a cost of about R5 500.00 that included material and labour. About 15% of the construction amount was spent on direct labour on the project.

The main construction stages of the unit are shown in the figures 1, 2 and 3 below.

Figure 1. Erection of compound channels.

Figure 2. Fixing of 'wall plate', door and window frames

Figure 3. The finished dwelling unit.

DISCUSSION

Savings in time

The main activities for the substructure that included erection of channels, hardcore preparation and slab casting were accomplished in five days. Although the casting of the channels was problematic because of waterlogged conditions, the problems would have been compounded if a conventional strip footing foundation had been used.

Seven days were spent on the superstructure whose main activities included the erection of the panels and the roof structure. The panels were erected within a day and the roof structure was also erected within two days and this included half a day that was spent making the trusses. The galvanised roof sheets and the ridge cap were installed within three days although this activity could have been done within two days but vital time was lost because of reworks. The finishes involved plastering and painting and the work was done within four days. This stage proved to be time consuming because the operatives were not adept at plastering.

The overall time used for construction was 22 days and a saving of at least seven days was realised when compared to the computed time for a conventional structure. The estimated time for the conventional structure was based on average rates for labour for different categories of construction from the local industry. The savings of construction time translates to savings in labour costs and the overall effect would be to reduce the cost of the project.

Economy

The cost of materials of the precast structure was 10% cheaper when compared to the conventional type and the main difference was at the substructure level. The reason for this was that strip footings are used with conventional construction compared to isolated footings employed in precast construction and they use larger volumes of excavation and concrete for the same foundation area compared to isolated footings and foundation brick walls are necessary for conventional structures.

The economy of precast structures was realised in labour and this was reflected in the small pilot project. At the substructure level in precast construction, skilled labour was only employed in setting out and the other activities such as erection of the channels utilised unskilled labour. In conventional brick construction, bricklayers would be required for the foundation brick walls and unskilled operatives are normally employed on activities such as foundation excavations and ferrying materials to required areas.

Unskilled labour was used for building the walls by stacking the panels, which in contrast with conventional brick construction would require, skilled labour. Fittings such as window frames and doorframes were done using unskilled labour in precast construction, whereas skilled labour would be required in brick construction.

Other options of substituting conventional materials or systems achieved economy in time and costs; for example, a steel channel was used as a wall plate in place of a timber system; lintels were not required for the structure.

Advantages

The precast concrete panel-lipped column construction system was found to have these advantages over conventional construction:

- Easy to construct,
- Short construction period,
- Special skills are not really needed,
- Labour costs are low,
- The finishes are simple and less costly,
- Quality control can be implemented at an early stage,
- Little maintenance is required with time.

There are some disadvantages with the system though such as the difficulty of constructing odd-shaped units and inadequate local small-scale capacity to manufacture the panels.

CONCLUSIONS AND RECOMMENDATIONS

It was proved that significant savings could be made in building materials, labour costs and construction time with the precast concrete panel building system compared to conventional brick and mortar construction. The following conclusions were drawn from the work:

- Precast construction is cost-effective in comparison to conventional methods of construction of houses.
- Since construction duration was significantly reduced for precast construction compared to other methods like brickwork construction, more units could be put up without compromising quality.

- The savings in labour costs utilising the precast system were enormous because unskilled operatives were used for several tasks when compared to conventional brickwork construction.
- The cost of materials for the precast unit were about 10% cheaper compared to the conventional unit and this could transform into huge savings if several units are being constructed.
- Housing delivery can be enhanced if precast building systems are utilised.

The following recommendations are intended to promote the building system as a means of providing affordable low-cost housing options:

- Governments should provide incentives to the private sector or individuals for developing innovative building systems.
- National building committees should be flexible in accepting new systems such as the described and procedures of incorporating new methods in codes should be improved.
- Small and medium enterprises in the building sector should be trained on non-conventional construction methods.

REFERENCES

Ministry of Lands and Housing, (1999), National Policy on Housing in Botswana, The Government Printer, Gaborone.

Ministry of Finance & Development Planning, (1997), National Development Plan 8, 1997/98 – 2002/03, The Government Printer, Gaborone.

Schlotfeldt, C., (ed), (2000), Housing is not about Houses, The Boutek Experience, CSIR Building and Construction Technology, Pretoria.

Knoetze, T. P., van Wamelen, J. and Kraayenbrink, E. A., (2000), Technical assessment of construction products in South Africa, Agrément News.

Agrément South Africa, Mantag Certificate M46, (1997), Matla housing system, Agrément SA.

BS 5628 : Part 1 : (1978), Code of practice for Use of masonry, Part 1. Structural use of unreinforced masonry, British Standards Institution.

BS 5950 : Part 1 : (1990), Code of practice for Use of structural steel, British Standards Institution.

APPENDIX

Stability Analysis of Precast Wall System

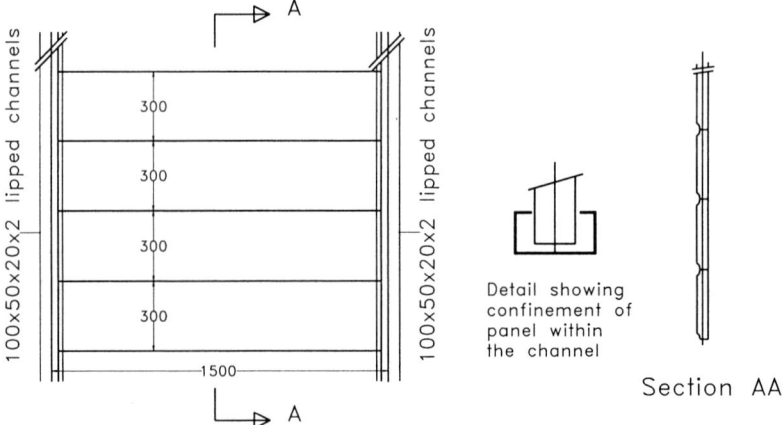

Figure 4. Elevation and details of precast wall system.

The panel system was treated as subjected to lateral loads since the load from the roof was carried by the lipped channels through a wall plate, which was welded at the top of the channels.

Check for lateral stability: BS 5628 (1980)

Check for limiting dimensions;- For the middle panel supported on three edges-:

height × length $\leq 1500 t_{ef}^2$

Where t_{ef} is the effective thickness of the wall
Data: height = 2.1 m, length = 1.4 m, thickness = 50 mm.
$RHS = 2.1 \times 1.4 = 2.94$

$LHS = 1500 \times 0.05^2 = 3.75$ OK!

Check for moment of resistance;

$$M_R = \frac{f_{kx}}{\gamma_m} Z$$

Where f_{kx} is the characteristic flexural strength
  γ_m is a partial factor of safety
  Z is the section modulus
$M_R = 5.35 \times 10^5$ Nmm = 0.535 KNm

The design moment in panels due to wind loading or accidental loads, M_w;

$$M_w = \alpha W_k \gamma_f L^2$$

Where α is a bending moment coefficient
  γ_f is a partial safety factor
  L is the length of the panel
  W_k is the characteristic load

$M_W = 0.0187$ KNm $< M_R$ OK!

Check for axial capacity of compound channels: (BS 5950 : Part 1 : 1990)

$$P_c = A_g p_c$$

Where P_c is the compression resistance
  A_g is the gross sectional area
  p_c is the compressive strength

For the back-to-back 100x50x20x2 section;
$I_{YY} = 5.75 \times 105$ mm^4
$r_{YY} = 24.9$ mm
Slenderness, λ = length/r_{YY} = 3100/24.9 = 124 < 180   OK

From Table 27(c), $p_c = 91$ N/mm^2

∴ P_c = 2x464x91 = 84.45 kN > 3.56 kN (load from roof).

PROCESS CHAINS –
A BASE FOR EFFECTIVE PROJECT MANAGEMENT

I. Weiser

Department of Civil Engineering, Bauhaus-University Weimar,
Marienstraße 5, D-99423 Weimar, GERMANY

ABSTRACT

The use of information and communication technologies in the construction industry did not achieve the desired advantages in terms of costs and quality of buildings, yet. Compared with other branches of industries information technology in civil engineering does not reach the desired extensiveness in its application. Planning and construction processes are characterised by work partitioning, document orientated methods and a lack of automation caused by the uniqueness of each individual item.

Within this paper the idea of a continuous flow of information and an integral view on separate building processes for a specific type of building, which is limited in its complexity, will be discussed. Therefore the method of process modelling will be introduced.

The modelled processes and applied methods need to be in relation to the building respectively to its digital product model. These relations need not only be maintained in an information and communication system but should also serve in facilitating project communication. The link between the communication and building model is implicitly realised through the content of the work described in the planning processes. These contents are nowadays described in documents. The transition from document oriented to model oriented work makes it possible for references between process and building model to be realised through a document linkage.

The combined view of process and product models enables the control of real processes as well as the simulation of planned sequences. Because of these opportunities substantial planning mistakes can be avoided and logistic conflicts can be detected and eliminated in advance.

KEYWORDS:

Building process, process modelling, industrial product house, integration, active process model, document linkage

BUILDING PROCESS

The building process contains all activities needed for planning, construction and use of a building. Concerning to the best result all these activities are linked in a logical sequence. This term implies the hole life cycle of a building. The beginning of the building process is the expression of necessity of premises. The end of the process is reached by the end of utilisation caused by demolition.

The use of information and communication technologies in the construction industry did not achieve the desired advantages in terms of costs and quality of buildings, yet. Compared with other branches of industries information technology in civil engineering does not reach the desired extensiveness in its application.

The building process is featured by a high degree of complexity and document orientated methods. Planning and construction processes are characterised by a great deal of work partitioning according to different crafts with different sizes and forms of organisation. The project work lacks of communication within a company and between project partners and a non-existing or inefficient project co-ordination. In addition there are many uncertainties and risks like weather and changing conditions on non-stationary sites and the usual changes of planning after work on site has already started. The lack of automation and standardisation caused by the uniqueness of each individual item and the mostly manual work causes a high planning effort for every building.

The organisation of companies is mostly function orientated. Each organisational unit has a own database which is developed and maintained by proprietary software systems. Each department is exclusively responsible for his own restricted part of the project. That's why discontinuities between departments inevitably occur. Such discontinuities induce mistakes, losses of information, interpretation failures and a slowdown of the overall process. Therefore the objective should be to reduce or at least to optimise these discontinuities in using a basic approach of integration.

ADVANTAGES OF PROCESS MODELLING

One approach to solve the mentioned problems of the building industry is seen in the understanding of the building process as an process for producing a customer focused industrial product. A well defined product according to objectives and guidelines needs to be in the centre of all activities. The co-operation of all project members must be organised in terms of an integration.

The scope of this work aims to the idea of a continuous flow of information and an integral view of separate building processes for one family houses, which are limited in their complexity, with a high degree of prefabrication and an increased use of standardised modules. By means of an effective planning and producing of this product "house", the quality of the product itself can be improved. To reach these objectives the method of process modelling will be introduced.

A model is the abstraction of a part of reality. It is based on the abstraction of characteristic properties of real objects by maintaining the basic structure and behaviour of the real object. [Scheer]
A focus on processes, or collection of tasks and activities that together transform inputs into outputs, allows organisations to view and manage materials, information and people in a more integrated way. [Garvin]

The process model of the building process for the one family house serves as the basis for process analysis and process optimisation. Process analysis means the analysis of information and material flow, the locating of changes of organisation units and used systems, the finding of discontinuities within the overall process and the discovering of deficits and critical processes during the building process. An optimisation of processes can be reached by changing process sequences, an improved use of human resources and the employment of modern communication technology.

Another objective of process modelling is the development of an process orientated thinking within the project team. Caused by the conjoint view of the same model the individual member of this team recognises his position within the overall process and the responsibility of all members for the collective product. The transparent representation in an easy to understand graphical model (fig. 1) enables the project members to realise the complexity of the overall process. The understanding of the process is the precondition for an optimisation of co-operated work and an improved communication between project members. An efficient management of a building company depends on a good communication and a continuous flow of information. In this context a process model has to be understood as a communication model as well. If it is possible to improve the handling with communication and information there will be an obvious economy of time and cost. Several investigations come to the conclusion that by improving the communication between project members up to 50% of building costs could be saved.

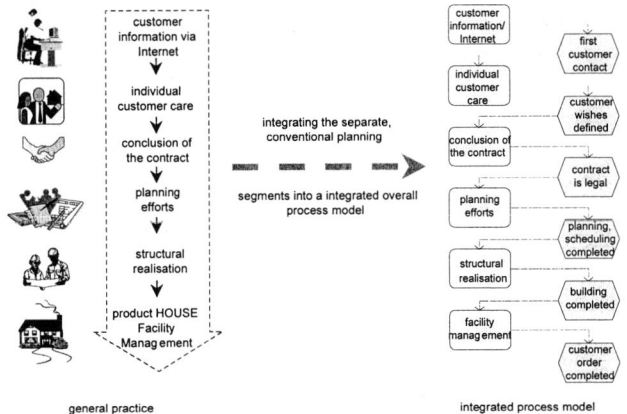

Figure 1: Mapping of construction activities into a formal model

PROCESS MODELLING

The process model shows different aspects of the building process like organisational structure, information or data structure and flows, events, processes, functions, relations and used applications or software systems. Within the scope of the presented work the modelling tool ARIS (Architecture of Integrated Information Systems) from IDS Scheer AG was used. Semiformal methods can be used for modelling of different aspects. The ARIS concept enables the modelling of all aspects of the building process in an integrated way. To model the overall process there are different modelling methods for each of the mentioned aspects and a set of different diagrams is required for the entire description of a process. One example for such a diagram is a organisation chart. The hole organisational structure for the building process of the one family house can be modelled in one organisation chart. Here one can find every member of the process from groups of building workers up to the executive engineer in a hierarchical order. Every person involved is defined by his name {Mr. Smith}, his responsibility [authorised to sign the contract] and his role {sales manager}. The integration of the different types of diagrams, which represent only different views on the same process, is reached by the use of only one internal representation of each object which can be referred to from many different diagrams. If the organisation object {Mr. Smith} is working in different sub-processes within the building process, e.g.

the conclusion of the contract or the accounting, for modelling these sub-processes only references of the object {Mr. Smith} from the organisation chart will be used.

It is also possible to integrate the different views e.g. the function, the organisational and the data view via a control view (fig. 2). Here the logical control view can be modelled, showing for the hole process which functions need to be carried out, as well as their logical sequence. The functions are identical to those of the referred view as well as the organisational units are. For this integrated view, the method of "Event-Driven Process Chains" (EPC) is used. The EPC consists basically of functions and events, as well as logical connectors {activates; is activated}, for defining the control flow. The models of the other views (e.g. organisational or data view) are integrated by connecting the elements of these models to the functions in the EPC

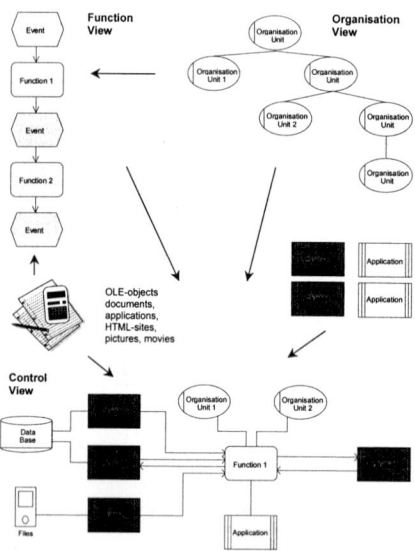

Figure 2: Modelling Method (control view)

The first objective of process modelling is to show a most possible realistic representation of the real process. That means, that every single process which is needed to plan and to build the house need to be defined and to be represented sequentially. Therefore in a first step the graphical representation of the real process will be modelled. The created model is now independent of the real process. The real process may change, but the model is a representation of this process as perceived at some point in time and will not change automatically. If things change then the model needs to be revised; a new model will be created. In [Greenwood] such a model is called a "passive model". This kind of model only helps to understand and to analyses a process. Within such a model it is already possible to integrate the different aspects of the building process, but it is not possible to model the dynamic behaviour of the real process.

With the chosen modelling tool it is possible to pass over the passive model to a "dynamic passive model" [Greenwood] by adding certain parameters to each object used in the model. Examples for such parameters could be the number of participating resources on a function, or the daily working hours of a resource. By changing these parameters it is possible to determine the behaviour of the process in varying circumstances. So if a function has to be repeated for any reason – the number of needed resources will rise. To simulate the process behaviour the model will be analysed via computer simulations based on mathematical methods. With the assistance of such simulations it is possible to make correct predictions for example for the time-dependent behaviour of the real process. The main value of dynamic passive models is, that with simulating real behaviour it is possible to compress time. The planning process for the one family house may take several weeks, but it can be simulated within only a few minutes. This enables a large number of different scenarios to be evaluated at a reasonable cost. But these models still have no direct relation to the real process.

To enable a selective simulation and to facilitate changes and additions to the very complex process model, it is important to provide an appropriate structure for the overall process model. The process model for the discussed one family house is structured according to figure 3. The model is subdivided in several levels of abstraction. Simulations are possible within one level or more than one level. Sub processes have to be modelled on the lowest level. The number of sub processes and the grade of detailing exert a big influence of the process model quality in terms of completeness and accurateness. For selective simulations the sub processes can be combined and linked on the different levels of abstraction.

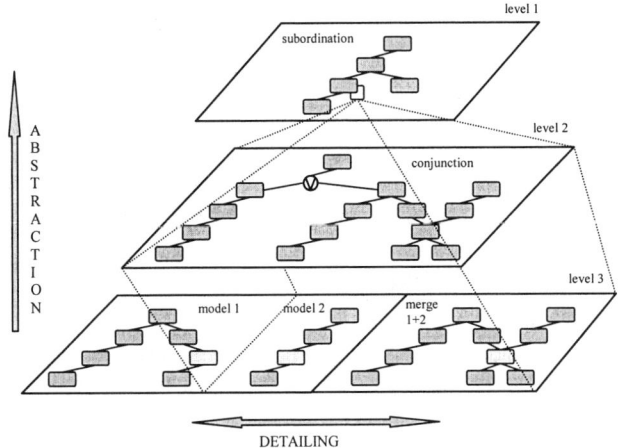

Figure 3: Levels of abstractions

To realise an integral view of the building process it is necessary to realise a direct relation between the real process and the process model. The created model and the real process need to be synchronised so that the model reflects the current state of the referred process at any time. In [Greenwood] such a model is described as an "active model". An active model should enable to be used for monitoring, that is to provide a high-level view of the current state of the real process.

RELATIONSHIP BETWEEN PROCESS AND PRODUCT MODEL

With the presented work a first approach for a realisation of an active model will be introduced. This approach in a first step only deals with the planning process. It appears to become much more difficult to realise an active model after work at building site already started. The feedback which is transferred from the site is mostly behind schedule and incorrect.

For the planning process approach the relationship between the process model and the real building process will be realised by a indirect relation to the building or correspondingly to its digital product model. The link between the communication and building model is implicitly realised through the content of the work described in the planning processes. These contents are nowadays described in documents.

Document is the term for a file, which includes information. It is known which application has to be applied to use these information [GLOSSAR]. The transition from document to model orientated work makes it possible for references between process and building model to be realised through a document linkage (fig. 4).

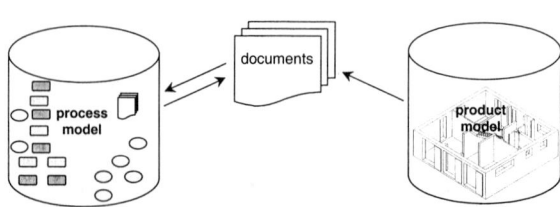

Figure 4: Linkage of process and product model

All project members work on the same building model and add their work to this model. The architect changes the design of the house using the same model as the engineer for the planning of all necessary house installations. The status quo of planning is represented within the digital 3D building model. For a better understanding and the possibility to give the planning results to a third party which has no access to the 3D model, the up-to-date information will be documented in e.g. drawings, bills of quantities or time schedules. During the building process these documents need to be updated frequently. If anything at the building model is changed, the changes need to be reflected in the actual documents. Via suitable mechanism like DMS and CSCW [Heinrich] project members need to be informed about document changes.

The base for such a integration via documents provides a central fileserver. All documents like drawings, lists, contracts, letters but also graphics and pictures are saved there digitally. Within the described work an existing document-management-system (DMS) is used. All project members are able to access the central document pool via a permanent linkage within a LAN / WAN or mobile via modem, ISDN or the internet. With the possibility of a world-wide access to the information, separately-working project partners are able to communicate and co-operate within a network. A scenario of a mobile usage of documents could be, that up-to-date information will be requested by a foreign partner or the architect is working at home and needs to update his work on the project. With every usage of the network time and costs can be saved.

All documents which are developed during the planning process need to be placed within the process model. The drawing of the groundfloor needs to be placed next to the event where all design issues are completed and the drawing can be given to other team members for their following work. In this way the flow of documents and information can be represented and analysed. Via an OLE-linkage it is possible for every project partner to get access to the documents. The access and the authorisation for documents is configured via roles.

All documents need to be saved in clear file structures which must be subdivided according to different criteria like project type or number, floors of a building or different versions of design. The file structures need to be adjustable corresponding to project requirements.

A process optimisation necessitates simplification and acceleration of document and data flow. This can be reached via a reasonable reduction of documents and a modified structure of them. All involved project members should be demanded to work within the 3D model and to generate special documents only to send them to building authorities or third persons. For a standardised, industrial product only a standardised amount and structure of documents is appropriate. Within this amount of data redundancies must be avoided and the consistence of information must be guaranteed.

SCENARIO

Using a 3D CAD-system the digital model of an one family house (fig. 5) is modelled. The preliminary design, for instance the ground floor (fig. 6) is presented as the plan of the ground floor in a drawing (fig. 7). This drawing is prepared according to predefined content and rules, so that every team member can find the appropriate information in an understandable form. The prepared drawing is saved in the structured file system of the document-management-system under a specific name [groundfloor_h1] and typical extension [dwg]. The same drawing can be found under the same name within the process model via an OLE-linkage at the corresponding position (fig. 8). Project members now are able, according to their authorisation, to work on this drawing by starting the appropriate application via selecting the symbol for the document in the process model. In this way several members are able to integrate their work and by reaching the final design the drawing can be transmitted to the building site for realisation.

Figure 5: One family house

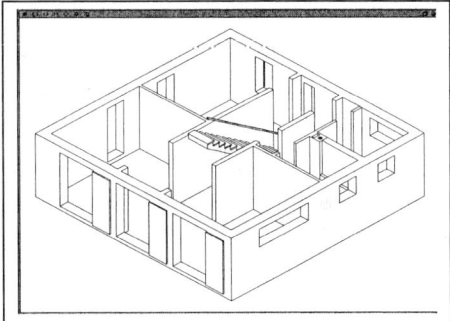

Figure 6: Digital model groundfloor

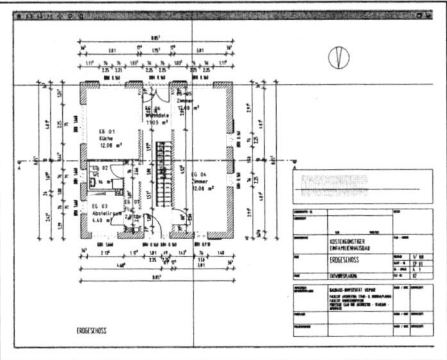

Figure 7: Drawing groundfloor

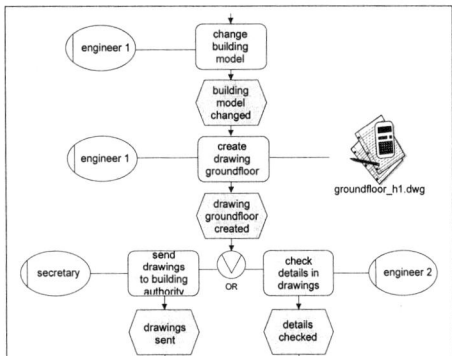

Figure 8: Process model

PERSPECTIVES

The combined view of process and product models enables the control of real processes as well as the simulation of planned sequences. Because of these opportunities a considerable amount of planning mistakes can be avoided and logistic conflicts can be detected and eliminated in advance.

A very important step for the use of electronic document-management-systems throughout the building process will be the world-wide launch and the legal approval of an electronic signature. Using this technology it will be possible to convert drawings, bills and contracts into legal documents via the internet or other networks.

REFERENCES

Allweyer T., Babin-Ebell T., Leinenbach S., Scheer A.-W. (1996). *Model-based Re-engineering in the European Construction Industry,* Construction on the Information Highway, Proceedings of CIB Workshop - W78 "Working Commission on Information Technology in Construction" and TG 10 "Task Group on Computer Representation of Design Standards and Building Codes", Bled (Slovenien), University of Ljubljana, Ljubljana, S. 21 - 31

Garvin D.A. (1998). *The process of organization and management*, Sloan Management Review 39:4. http://mitsloan.mit.edu/smr/past/1998/smr3943.html

Greenwood R.M., Robertson I., Snowdon R.A., Warboys B.C. (1995). *Proceedings 5th. Conference on Business Information Technology CBIT '95*, Department of Business Information Technology, Manchester Metropolitan University, Manchester, UK

GLOSSAR [Online] (2000)
http://www.GLOSSAR.de

Heinrich T., Hauschild Th., Grosche A. (2000). *Unterstützung von Kommunikation und Kooperation bei Revitalisierungsvorhaben,* Forum Bauinformatik, VDI Fortschritt-Berichte 4:163, Berlin, S. 59-67

Lindfors C.T. (2000). *Value chain management in construction: Modelling the process of house building,* Proceedings of International Conference for Construction Information Technology CIT 2000, Icelandic Building Research Institute, Reykjavik, Iceland, S. 575-583

Scheer A.-W. (1998). *ARIS – Vom Geschäftsprozeß zum Anwendungssystem*, 3. Auflage, Springer-Verlag, Berlin

ANALYSIS OF PLACEMENT ERRORS OF BARS IN REINFORCED CONCRETE CONSTRUCTION

J.O. Afolayan

Civil Engineering Department, Ahmadu Bello University
Zaria, NIGERIA

ABSTRACT

This paper reports an investigation into errors associated with the positioning of reinforcing bars in reinforced concrete elements. Data were collected from different construction sites on spacing of main and secondary bars in slabs, and spacing of links in beams and columns. Deviations from the values in the working structural drawings are noted as errors and are analysed to probe the quality control on the sites visited. The errors are also classified as fatal and non-fatal with their frequencies calculated. The results show that, on the average, the upper control limit (UCL) on an acceptable range for a good quality control is exceeded by 32% for all spacing errors in all the sites. The probability of human error related to placement of reinforcements is also estimated to be about 0.35.

KEYWORDS

Human error, reinforcement placement, error frequency, control limit, working drawing, workmanship

INTRODUCTION

Concrete as one of the principal materials for construction does not possess sufficient ductility to give adequate warning in case of failure. However, it possesses some properties in form of compressive strength, durability and fire resistance which are classified as good. A remedy for rendering concrete as a strong and durable building material is to reinforce it with steel bars. Thus, failure against major cracking and excessive deflection resulting from tensile, compressive, flexural and shear stresses is curtailed.

Structural failure might be considered to be the occurrence of one or more types of undesirable structural responses including the violation of predefined limit states. Fortunately, real structures only rarely fail in a serious manner but when they do it is often due to causes not directly related to the predicted nominal loading or strength probability distribution. Other factors such as human errors, negligence, poor workmanship or neglected loadings are most often involved. For instance, previous reports (Lew et at, 1982; Carino et al, 1983; Igba, 1996) attest to structural failures due to human errors, especially as they relate to reinforcement detailing. Meister (1962) quoted that 20 to 50% of all equipment failures are due to human errors.

This work analysis data on observations made at four different construction sites with emphasis on errors in positions of reinforcements in reinforced concrete slabs, beams and columns. The errors are classified into fatal and non-fatal for the purpose of characterizing the level of skill exhibited at the sites from which data were collected. The probability of human error is also estimated and compared with the global value (Matousek & Schneider, 1976) associated with poor workmanship and carelessness.

DATA ACQUISITION AND ANALYSIS

Four different construction sites were surveyed for data collection. The spacings of main and secondary bars for two hundred and twenty (220) slab units were measured. Similarly the spacings of links were measured on forty-nine (49) beams and on ninety-three (93) columns at varying points along their lengths. Tables 1 to 3 show samples of data recorded for some sites on slabs, beams and columns respectively. In these Tables several measurements which tally with the values in the working structural drawings are excluded.

TABLE 1
SPACING ERRORS FOR MAIN AND SECONDARY BARS IN SLABS
(RECOMMENDED SPACING = 200mm c/c)

Slab no.	Absolute measured error for main reinforcement (mm)	Absolute measured error for secondary reinforcement (mm)
1	5	5
2	20	10
3	15	17
4	4	10
5	5	5
6	5	4
7	20	8
8	10	20
9	5	17
10	10	15
11	20	15
12	10	5

TABLE 2
SPACING ERRORS FOR LINKS IN BEAMS
(RECOMMENDED SPACING = 175mm c/c)

Beam no.	Average absolute measured error (mm)	Beam no.	Average absolute measured error (mm)
1	8.8	8	9.7
2	8.3	9	10.0
3	13.3	10	6.6
4	5.0	11	8.6
5	11.7	12	10.0
6	5.3	13	10.5
7	9.3	14	10.0

TABLE 3
SPACING ERRORS FOR LINKS IN COLUMNS
(RECOMMEDED SPACING = 200mm c/c)

Column no.	Average absolute measured error (mm)	Column no.	Average absolute measured error (mm)
1	5.3	14	1.3
2	6.0	15	10.0
3	8.3	16	10.0
4	3.0	17	6.7
5	10.3	18	5.0
6	8.8	19	8.3
7	3.7	20	6.7
8	11.5	21	8.3
9	16.7	22	13.3
10	4.3	23	7.5
11	3.0	24	8.0
12	10.5	25	6.7
13	6.3	26	15.7

Of importance in this report are the statistical characters of the measured errors. Table 4 summarises the statistical analysis of the errors observed. It is clear that the levels of workmanship vary from site to site and from structural element to element. On the average, the variability associated with the positioning of links on columns is higher than for either beams or slabs. This observation can be understood considering the general layout of columns on sites. However, the mean error is greatest for the spacing of main and secondary bars in slabs than for the spacing of links in beams and columns.

TABLE 4
SUMMARY OF STATISTICS OF THE MEASURED ERRORS

Element	Mean Error (mm)				Coefficient of Variation			
	SITE				SITE			
	1	2	3	4	1	2	3	4
SLABS	10.67	10.35	9.67	10.47	0.55	0.38	0.53	0.53
BEAMS	7.93	10.91	10.73	9.24	0.61	0.49	0.51	0.46
COLUMNS	7.91	8.59	10.34	10.52	0.75	0.63	0.48	0.45

SIGNIFICANCE OF RESULTS

Attempts are made to establish the probability distribution for the measured errors and it is shown that in general a positively skewed normal distribution is followed as in Figure 1.

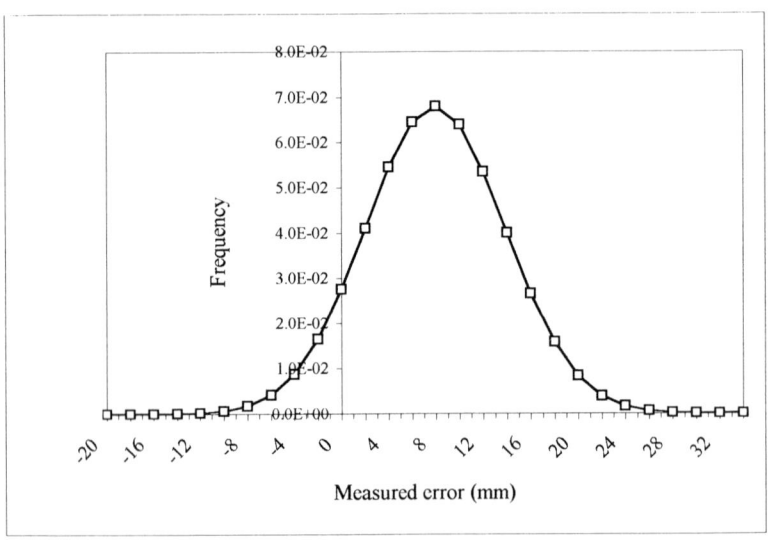

Figure 1: Error distribution for links in beams for a typical site

In quality control, one of the interests is the acceptable range for variations in measured variables. For this purpose, the acceptable range for the four sites is determined and it is found to be ±6.68mm. From all the data collected for all the sites an overall average for the spacing error measured is about 9.78mm. This implies that on the average for the sites the UCL is exceeded by about 32%. This is quite revealing with respect to the level of workmanship obtainable at those sites. It can be deduced that there is no satisfactory control on these sites in accordance with one of the practical working rules on the relationship between satisfactory control and the number of points falling outside limits (Grant, 1964). In short, the inspection effectiveness on these sites falls below average.

For all the elements surveyed, errors are classified as fatal and non-fatal. If the measured spacing coincides with the spacing in the structural working drawing, it is registered a non-fatal error. However, if the measured spacing is less or greater than the specified, it is recorded a fatal error. On this basis the summary for the fatal and non-fatal estimates are shown in Table 5.

TABLE 5
ERROR FREQUENCIES FOR REINFORCEMENT PLACEMENT

Site	Element	Fatal (%)	Non-fatal (%)
1	SLAB	30	70
	BEAM	37	63
	COLUMN	54	46
2	SLAB	33	67
	BEAM	35	65
	COLUMN	37	63
3	SLAB	38	62
	BEAM	30	70
	COLUMN	32	68
4	SLAB	31	69
	BEAM	34	66
	COLUMN	31	69

It is clear that proneness to error commission is highest for columns than for beams and slabs. On the average, a non-fatal frequency of 64.83% and 35.17% of fatal error frequency are obtained as far as placement of reinforcing bars is concerned. A step further is the determination of the probability of human error in relation to reinforcement detailing on site. Limiting one to the locality surveyed, this probability stays at 0.35. This value corresponds to the global value (Matousek & Schneider, 1976) associated with ignorance, carelessness and negligence and it is about 9% better than an earlier estimate (Afolayan & Igbozurike, 1997) for errors related to horizontal alignment of reinforced concrete columns at some other locality far from where the data for this study were obtained.

CONCLUDING REMARKS

The errors associated with the placement of reinforcing bars in reinforced concrete building components have been studied. Four different construction sites were visited for data collection. The results obtained show very clearly that workmanship at those sites which is a representative of the expectation for the locality is quite unsatisfactory. The estimated probability of human error in placing reinforcements is about 0.35. It is therefore indicative that the importance of adherence to structural detailing on sites has a very poor attendance in the study environment.

REFERENCES

Afolayan J.O. and Igbozurike S.I. (1997). Analysis of Human Error in Reinforced Concrete Column Construction, Submitted to *Building and Enviroment*.

Carino N. J., Woodword K.A., Leyendecker E.V. and Fattal S.G. (1983). A Review of the Skyline Plaza Collapse, *Concrete International*, 35-41.

Grant E.L. (1964). *Statistical Quality Control*, McGraw-Hill, Inc., New York

Igba P.O.E. (1996). Societal Mentality and Engineering Prcatice in Nigeria, *The Nigerian Engineer*, **34:2**, 24-32.

Lew H.S., Carino N.J. and Fattal, S.G. (1982). Cause of Condominium Collapse in Cocoa Beach, Florida, *Concrete International*, 64-67.

Matousek, M. and Schneider J. (1976). Untersuchungen zur Struktur des Sicherheitsproblems bei Bauwerken, *Bericht No 59*, Institut Fuer Baustatik und Construktion, Eidgenoessiche Technische Hochschule, Zurich.

Meister D. (1962). The Problem of Human initiated Failures, in *Proceedings of the 8th National Symposium on Reliability and Quality Control*, IEEE, New York, 234-239.

STRUCTURAL ENGINEERING EDUCATION

TEACHING STRUCTURAL ANALYSIS: A CURRICULUM FOR AN UNDER-GRADUATE CIVIL ENGINEERING DEGREE AND LEARNING ISSUES

B. W. J. van Rensburg

Department of Civil Engineering, University of Pretoria, Pretoria 0002, South Africa

ABSTRACT

Structural engineering is one of the disciplines contained in the under graduate civil engineering university degree. There is a growing demand to include more environmental, managerial and economics subjects and other humanitarian topics in the civil engineering course. There is thus pressure to reduce the structural engineering content of the course. This paper discusses a curriculum for the structural analysis part of the structural engineering component. Topics which are considered essential are discussed and proposals to make the learning process as time-effective as possible are outlined.

KEYWORDS

Structural analysis, teaching, under-graduate curriculum, learning issues.

INTRODUCTION

The environment within which the professional civil engineer practices the profession is ever changing. The ever-changing milieu demands that the curriculum of the civil engineering course at university level be continuously under scrutiny. There is a growing demand to include more environmental, managerial and economics subjects in the course. In addition, it is also considered necessary for the civil engineer to be able to communicate effectively and to be able to orchestrate public involvement in projects (especially in South Africa).

In South Africa there is no significant specialisation as far as the under-graduate civil engineering degree at university level is concerned. The civil engineer is expected to work in various diverse sub-disciplines of civil engineering during the initial part of his career before he decides to concentrate on a specific discipline or field in civil engineering. The civil engineering course does not have space to accommodate new subject matter unless other subjects are scaled down. This creates pressure to reduce the structural engineering content. The question now arises to what are the minimum requirements for the structural engineering component in the curriculum of a bachelor degree in civil engineering.

This paper more specifically discusses the structural analysis part of the structural engineering component of the curriculum. Topics which are considered essential and proposals to make the learning process as time-effective as possible are outlined.

STRUCTURAL ANALYSIS

The structural engineering content for a typical four-year bachelor degree in civil engineering in South Africa can be divided into:
- engineering mechanics: 2.5 %
- mechanics of /strength of materials: 2.5 %
- structural analysis: 3.75 %
- structural design: 8.75 %

Engineering mechanics (statics and dynamics) would include topics such as force systems; equilibrium; forces and moments in beams and simple structures; work and potential energy; kinematics; kinetics; and energy and momentum. Mechanics (or strength) of materials (Gere & Timoshenko, 1997) includes topics such as stress and strain; two-dimensional elasticity; bending and shear deformations; and buckling of columns. The structural design component includes topics such as structural loading, reinforced and pre-stressed concrete design, steel design, timber design and synthesis. The contents of the structural analysis subjects are the topic of this paper and will be covered in more detail.

The role of structural analysis is put into perspective by Norris et al (1991) (p.xvii): "(Structural) analysis is a means to an end - not the end itself - since the primary objective of the structural engineer is to design, not to analyse. Concentration on the study of analysis ... does allow the structural engineering student to develop the capacity for logical, rational, analytical thinking and to cultivate his or her appreciation, feeling, and intuition for structural behaviour... Of course, the student will never benefit fully from this educational experience in analytical thinking unless the analytical studies are correlated with a parallel study of structural design. By studying design, the student develops judgment, perception, imagination and creativity - in short the ability to synthesise".

TEACHING FOR UNDERSTANDING

The general principles of effective teaching (Felder & Brent, 1996) are outside the scope of this paper. However, factors that contribute to the mastering of a subject and need to be highlighted include:
- relating the subject material to the real world;
- relating the material covered in a course to the material covered in previous subjects; and
- integrating the material with the material in other courses done simultaneously (where appropriate).

Understanding, and the retention of knowledge are seriously hampered if structural analysis is taught strictly according to the many "clinical" structural analysis handbooks available.

SIGN AND DRAWING CONVENTIONS

It is common that conflicting sign and drawing conventions are used in the mechanics, strength of materials, structural design and structural analysis subjects in the same civil engineering curriculum. Even within the text of several handbooks the convention is not conforming. This lack of consistency creates unnecessary stumbling blocks for the student in the process of coming to grips with structural engineering.

The following consistent sign and drawing convention, in line with several prominent structural design handbooks (Reynolds & Steedman, 1988 and Steel Construction Institute, 1994) and some structural analysis textbooks, is proposed which should meet the requirements of both design and analysis subjects. The bending moment diagram should be drawn on the flexural tension side of frame members. For a horizontal beam this implies that for sagging bending the bending moment ordinates will be below the base line and for hogging moments the diagram will be above the base line. (If the structure is of reinforced concrete, the reinforcement to resist bending is required near the tension face. The bending moment diagram therefore indicates to the designer where the reinforcement is required.) With such a convention it is not necessary to indicate a sign for the ordinates of the bending moment diagram.

When the student initially studies bending moment and shear force diagrams a more rigorous mathematical approach is followed. The relationship between the shear force and bending moment functions is used and the sign convention is therefore relevant. Several important structural analysis handbooks adopt the sign convention that sagging moment is positive. Drawing the bending moment diagram on the tension side and choosing sagging bending as positive will be consistent with choosing the lateral axis as positive downwards for a beam.

BENDING MOMENT DIAGRAMS AND DEFLECTION CURVES

It is important in the process of solving problems always to draw the bending moment diagram and approximate deflection curve, so that students can develop an appreciation for the relationship between the loading, the bending moment and the deflection of the structure. The relationship between the bending moment and the curvature is utilised for this purpose. The student must then compare the deflected shape that correlates with the bending moment diagram, to the deflected shape he had visualised. This process should assist "towards an improved understanding of structural behaviour" (Brohn & Cowan, 1977).

DEFINITION OF SUPPORT CONDITIONS

It is important that symbols depicting various support conditions should be clearly defined. All possible situations should be defined, including those that could exist at an axis of symmetry. (The topic of symmetry considerations is discussed later in this paper.) That fact that these symbols indicate a theoretical situation should be emphasised. These theoretical conditions should be related to practical situations to assist the student in bridging the gap between analysis and synthesis. It is important for the study of symmetry to also define the support case when rotation and movement in one direction are prevented, but movement in the perpendicular direction is permitted.

FREE BODY DIAGRAMS AND VECTORIAL RESULTANTS AND COMPONENTS

The principles of free body diagrams and vectors are covered in the statics part of engineering mechanics courses. It is important to also cover and apply these topics in the structural analysis courses. The seemingly simple problem of calculating the axial and shear forces in an inclined member of a multi-member frame presents an unnecessary challenge if the mentioned principles are not applied. Similarly, the fact that the shear force and axial force in an inclined member supported by a vertical support roller, are the components of the resultant vertical reaction should be appreciated by the student. The application of vector mechanics is expedient in the analysis of some three-dimensional structures. (The topic of three-dimensional structures is discussed later.)

SHORTENING OF MEMBERS DUE TO FLEXURE

In order to correctly predict the behaviour of sway frames it is necessary to distinguish between the orders of magnitude of the following contributions to the joint displacements:
- joint displacements due to lateral bending or sway of the structure;
- joint displacements due to axial strains; and
- joint displacements due to shortening of a flexural member caused by the deflected shape.

Consider a rectangular frame ABCD, subjected to both horisontal and vertical loading, with supports at A and D, two columns AB and CD and a horisontal beam BC. Students generally appreciate the side sway deflection but frequently judge that the distance between nodes B and C should shorten significantly due to the flexure. It should be demonstrated that the magnitude of shortening due to flexure is an order smaller than the magnitude of the displacements of the nodes due to flexural sway displacements.

VALIDATION OF ANSWERS

Students must be strongly encouraged never to accept values obtained as answers to problems without scrutiny and only to use the given answers as a final check. The normal procedure to check the analysis of a problem is to trace the solution step by step. Another check is to do equilibrium calculations. Upper and lower bounds for answers could be obtained by adopting certain simplifying assumptions. For example by visualising the deflected shape, hinges could be introduced at assumed points of zero bending moment to make the structure statically determinate. Correlating the calculated bending moment diagram with the visualised deflected shape of the structure could also indicate inconsistencies, as discussed above. Developing techniques to obtain approximate answers is also important for computer analyses.

ANALYSING "REAL" STRUCTURES

When problems illustrating the principles of structural analysis are set, it motivates students appreciably if the problem is described in terms of "real" structures. Loads, dimensions and properties within a practical range will produce realistic answers for forces, moments and deflections. In this way students are assisted to bridge the gaps between analysis, design and synthesis. Hibbeler (1995), for example, sets problems to "depict realistic situations encountered in practice. It is hoped that this realism will both stimulate the student's interest in structural analysis and develop the skill to reduce any such problem from its physical description to a model or symbolic presentation ..."

PRACTICAL SESSIONS

Teaching structural behaviour is more effective if students are involved in different learning modes (according to their learning styles). Involving the students with studying the behaviour of models (both pre-manufactured and self constructed); visits to construction sites to study the structural systems and computer analyses graphically illustrating structural behaviour, are important learning aids. (Behr, 1996 and Hilson, 1972.) It is, however, important that such practical work is structured, managed and integrated with the analytical sessions.

VIRTUAL WORK

Specific important topics to include in the curriculum will now be discussed. Virtual work is a very important and general-purpose technique to study structural analysis. Topics that should be included, apart from the obvious items of calculating deflections and rotations due to axial, shear and bending deformations, are the topics of determining deflections (and forces) due to support movements, temperature, prestress and lack-of-fit. Numerical integration formulae for determining the integral of a product of two moment of force functions should be introduced at an early stage to simplify calculations and mathematics, so that the student can appreciating the principles involved.

METHOD OF SUPERPOSITION

The method of superposition, also known as the flexibility or force method, is an important building block in the pyramid of knowledge of the structural analyst. It is enlightening to determine the degrees of indeterminacy, apply the redundancies, determine the force and moment diagrams for each load case and to compile the set of superposition equations. It is useful specifically for continuous beams to computerise the process. Beams with varying cross-section and with prestressing cables could readily be handled with this method. Treating the complete computerisation of the flexibility method for a general frame is considered a low priority at under-graduate level; the process is very intricate, seldom used in available frame analysis programs and the insight gained by the student is limited.

METHOD OF SLOPE-DEFLECTION

The method of slope-deflection still has definite advantages: It provides an introduction to the method of moment-distribution and also a route for arriving at the member stiffness matrix. The method could also be developed (at graduate level) to do geometrical non-linear (second-order) and stability analyses of simple frame problems. The computer (and packages such as MATLAB®) should be utilised to solve the simultaneous equations that are compiled.

METHOD OF MOMENT-DISTRIBUTION

The method of moment-distribution also still has merit (in the age of computers). The method assists in teaching the concept of bending stiffness. In addition it affords the student the opportunity to do an approximate analysis, and as emphasised earlier, this is an important part of checking the reasonableness of the answers. Students should not be laboured with numerous cycles of distribution or with doing distributions for frames with sway.

SYMMETRY

Many real structures have some or other form of symmetry. The introductory study of symmetrical and anti-symmetrical behaviour is enlightening, especially when studying buckling behaviour. The student should practice introducing the appropriate boundary conditions on the axis of symmetry. At a later stage the principles learnt could be applied to the computer analysis of large frames or continuum structural problems (such as plates).

GRID STRUCTURES

Most textbooks illustrate structural analysis techniques with the aid of only plane frames. A grillage approximation can be used to model floor and bridge slabs. By including this topic in the curriculum several important principles are introduced. The plane grid member (with loads normal to the plane) also has three stress resultants. Similar to plane frames, two basic different approaches can also be applied to grid structures: the force method (superposition) or a displacement method (slope-deflection or moment-distribution). The principles of symmetrical behaviour could be utilised to reduce numerical work.

THREE-DIMENSIONAL FRAMED STRUCTURES

Many three-dimensional framed structures, such as transmission towers, cannot be approximated by plane frames. It is educational to study a few simple three dimensional structures. The student is challenged to consider the forces on, and behaviour of, a structural member in space. Again the force or displacement methods can be applied. The student has to transform forces and moments, and also displacements and rotations, in three dimensions when considering equilibrium and compatibility. Vector algebra could be put to good use in many problems.

THE MATRIX STIFFNESS METHOD

The matrix stiffness method is the basis for the finite element method and should be introduced by means of a simple method. Rectangular plane frames, neglecting axial deformations, should be dealt with first. Neglecting axial deformations reduces the number of degrees of freedom (the number of unknown nodal displacements). The solution of simple rectangular plane frames is rapid, transparent and easy to follow. In addition, the results obtained by utilising this method can easily be compared with those of obtained by slope-deflection and moment-distribution. The matrix stiffness method can now be generalised to include axial deformations and transformation of axes and expanded to include the third dimension.

INFLUENCE LINES

The principles of influence lines are applied in structural design. Covering this topic in the structural analysis curriculum is considered necessary. Betti's Law and the Mueller-Breslau principle can be derived utilising the principles of virtual work. The Mueller-Breslau principle capitalises on the student's intuitive feel for the deformation of a structure to determine the most critical imposed loading conditions.

GEOMETRIC NON-LINEAR BEHAVIOUR

It is often difficult for under-graduates to relate linear elastic analysis to structural instability and very often these topics are considered in isolation. Geometric non-linear (or second-order) analysis of plane frames assists in linking the initial near-linear behaviour with the eventual elastic instability. For this purpose it is necessary to introduce the stability functions. It is suggested that the slope-deflection method should be applied initially, before complete computerisation (at graduate level) is introduced. A package such as MATLAB® could again reduce numerical tedium.

The student gains considerable insight when doing a number of second-order analyses of a simple frame at different load levels. A structure with two degrees of freedom (one translational and one rotational, neglecting axial deformation), for example, could be considered. Making use of computers or tables for the stability functions the amount of calculation work is limited. Inspection of the load-deflection curve and the second-order bending moments is enlightening.

MATERIAL NON-LINEAR BEHAVIOUR

Another difficulty for students is to relate elastic behaviour to plastic behaviour. Again these topics of elastic and plastic behaviour (and analyses) are seen in isolation. Elastic-plastic analyses of continuous beams or plane frames assist in linking elastic and plastic behaviour. The student gains considerable insight when analysing a simple structure with increasing load and observing the elastic-plastic behaviour. To reduce the amount of calculation the student could make use of standard formulae or a computer program to quantify the various stages in the incremental loading process. An example could be a two-span continuous beam.

INTRODUCTORY TOPICS AT UNDERGRADUATE LEVEL

With structural engineering being a part of the curriculum of the bachelor degree in civil engineering, some structural analysis topics can only be introduced at under-graduate level. The following topics are cases in point:
- structural dynamics; and
- finite elements.

Students should be made aware of the basic concepts and application of these subject areas.

TOPICS THAT COULD BE OMITTED AT UNDERGRADUATE LEVEL

Bearing in mind the circumstances as stated previously, some topics (that could be educational and elegant to know) will have to be omitted. Some of the topics mentioned below appear in

structural textbooks and it is suggested that, given the topics that should be covered, the following topics have to be left out of the under-graduate curriculum:
- elastic weights analogy;
- three-moment equation;
- column analogy for fixed frames; and
- moment-area methods.

Problems, which are ideally suited to be solved be means of these methods, could be done just as effectively by virtual work combined with numerical integration of the products of two moment functions.

CONCLUSION

This paper outlines a minimum core curriculum for the structural analysis teaching within a first civil engineering degree. Aspects that should assist in effective learning and promote teaching towards understanding are also discussed. It is suggested that the methods of slope-deflection and moment-distribution should be retained for their educational value. It is further recommended that topics such as simplified treatment structural symmetry, grids, three-dimensional frames and non-linear analysis should form part of the curriculum. The change over the years (with the general availability of computers) should have been away from hand-numerical expedient methods to topics that assist the understanding of structural behaviour. Experience in approximate methods is necessary to check computer analyses.

REFERENCES

Behr R.A. (1996), Computer Simulations Versus Real Experiments in a Portable Structural Mechanics Laboratory, *Computer Applications in Engineering Education*, John Wiley, **4:1,** 9-18.

Brohn D.M. and Cowan, J. (1977). Teaching towards an improved understanding of structural behaviour, *The Structural Engineer*, **55:1**, January, 9-17.

Felder R.M. and Brent R. (1996). *Effective Teaching*, North Carolina State University, USA.

Gere J.M. and Timoshenko S.P. (1997). *Mechanics of Materials*, 4th Ed., PWS, Boston, USA.

Hibbeler R.C. (1995), *Structural Analysis*, 3rd Ed., Prentice Hall.

Hilson B. (1972), *Basic Structural Behaviour via Models*, Crosby Lockwood.

Norris C.H., Wilbur J.B. and Utku (1991). *Elementary Structural Analysis*, 4th Ed., McGraw-Hill.

Reynolds C.E. and Steedman J.C. (1988). *Reinforced Concrete Designer's Handbook*, 10th Ed., E.&F.N.Spon, London.

Steel Construction Institute (1994), *Steel Designers' Manual*, 5th Ed., Blackwell Scientific Publications, Oxford, Britian.

INTERACTIVE USE OF COMPUTERS TO PROMOTE A DEEPER LEARNING OF THE STRUCTURAL BEHAVIOUR

Rafiq M. Y., and Easterbrook D. J.
School of Civil and Structural Engineering University of Plymouth UK

ABSTRACT

An alternative approach to the teaching of Structural Engineering, which is based on the use of computer, is presented in this paper. In this approach emphasis has been shifted to a deeper understanding of the behaviour of structures for which the computer plays an important role. The authors believe that the introduction of suitable methodologies for the validation of computer results and the utilisation of the power of computers to promote deeper learning should be essential in teaching engineering students. Intelligent use of computers can help us achieve this goal.

The paper discusses success stories of the authors' achievements about the intelligent use of the computer in deeper learning of structural behaviour. The paper shows how a different approach, using computers to model real world problems, can allow the students to widely explore the design space and very quickly acquire a deeper knowledge, which has traditionally been the domain of the experienced engineer. In this paper, examples of the students' work, which clearly demonstrate deeper learning and better understanding of the behaviour of the structure is presented.

KEYWORDS

Deep learning, Intelligent use of the computer, Parametric studies, Graphic Use interface, Human Computer Interaction

INTRODUCTION

There can be little doubt that the electronic microcomputer is perhaps singular amongst recent tools in its capacity to revolutionise the design process by offering first-hand support to designers in their activities. The desktop microcomputer has become commonplace in the modern engineering practice.

Microcomputers are gaining a wider application as a result of the continuous improvements in performance-to-cost ratio. Concurrent advances in software and hardware have presented us with *enabling technologies* that helps to further the use of IT in supporting the design processes (as well as other fields) by making it easier to create more effective programs.

Norman [Norman 1986] emphasised the importance of Human Computer Interaction (HCI) for delivering effective support. Modern software has helped the user in this regard. Whilst ever faster computer processors are being developed which allow greater processing to be achieved in the same time span, graphic user interfaces (GUIs) have benefits which makes the process more transparent. Graphical interfaces help to simplify tasks and make the behaviour of programs more intuitive, and as such constitute ways in which software itself has become more empowering. The capability to allow the user to control the order in which actions are executed is critical to the effectiveness of certain software, to achieve the intended goals [Mathews J 2000].

GUI development tools and techniques have helped to make programs more *user-centred*, by enabling information to be clearly presented allowing the user interaction in a logical and uncomplicated manner. GUIs enable designers to carry out parametric studies to explore possible design solutions and to gain a better understanding of the process through interactive presentation of information.

MODE OF COMPUTER USE

Engineers have used computers for many years to assist them perform complicated analysis, design and drafting tasks. The introduction of AutoCad and other drafting and graphic packages has almost totally eradicated the use of drawing boards from design offices.

Recent developments in computer graphics, GUIs and the introduction of Windows based applications have greatly enhanced the use of computers, and computers have become much more user friendly. Software developers have utilised this opportunity to develop powerful computer packages, which assist designers in their activities.

Unfortunately in practice, computers are still used to perform routine tasks. The growing use of computers in design practices has also increased the danger of computer abuse by novice graduates. Too much reliance on computer results without proper understanding of the behaviour of the problem can lead to very expensive disasters.

The use of computers in education, similar to their use in the design practices, has been limited to routine tasks. The authors believe that it is time for a revolutionary shift in the way engineering students are educated. Emphasis must be shifted to a deeper understanding of the behaviour of the problem for which the computer is being used. The introduction of suitable methodologies for the validation of computer results helps in reducing the danger of computer abuse. The utilisation of the power of computers to promote deeper learning should be essential in teaching engineering students. Interactive use of computers can help us achieve this goal.

The authors contend that advances in computer hardware and software now allow a deferent approach to the role of the computer in design. In the University of Plymouth, we have adopted a different approach in the teaching of Structural Design. Our approach is more inline with the recent Standing Committee on Structural Safety (SCOSS) Report which identifies a need for more guidance on understanding structural behaviour and its proper modelling for computer analysis.

In the past students were taught Structural Analysis and Structural Design separately. Students often had difficulties in establishing links between analysis and design. Topics covered in structural analysis were generally limited to hand calculation of member forces of frames and trusses. Analysis of complicated structures such as multi-storey buildings, plates and shells were generally limited to approximate analysis and linear elastic theoretical analysis of plates using tables developed through solving differential equations (see for example Theory of Plates and Shells by Timoshenko). Students had very little opportunity to develop a full understanding of these theories.

With the introduction of the computer in teaching, students were taught to use computers as an alternative method to hand calculations. They often used computers to analyse trusses and rigid frames and compared the computer results with their hand calculations. Hence the use of computers were generally limited to the analysis of frames.

There is no doubt that the computer has provided the graduate with an excellent means of reducing computational time, but often, it is only used to perform full calculations faster for a single solution to the problem.

Some institutions have introduced finite element analysis into their undergraduate courses. At the start of the FEA use the learning curve is extremely steep. Once learned, it is often limited to the analysis of simple problems.

Another major problem with FEA is the amount of data generated by the computer. The process and extraction of useful information for design is a daunting task for the students.

The other, more interesting, module in an undergraduate course is Structural Design. This module is traditionally based on the applications of Codes of Practice. The majority of the rules within the Codes of Practice are empirical based. Students find little link between Structural Analysis, which is a more theoretical based module, and Structural Design, which is a mainly empirical based module. Students prefer to use Tables from the Codes of Practice for Design and Tables of bending moment, shear force and deflection formulae in design guides such as for example The Steel Designer's Manual [Constrado, 1981], and Reinforced Concrete Designers' Handbook [Reynold, 1981], rather than using results of finite element analysis.

It is considered that the way computers are used presently, in the teaching of structural analysis and design, students have very little opportunity for deeper learning of the basic concepts and the behaviour of the structure. In the current approach, students have little opportunity for innovative thinking, as the scope of teaching is rigidly constrained.

Once students have graduated from university and are employed in design practices, they lack confidence as they often have little to offer to the practice. In design practices, graduates undergo another stage of training and they slowly merge into the troop of traditional designers.

In the University of Plymouth, Structural Analysis is combined with Structural Design. Students are using computers extensively to analyse complicated structures. They are taught how to appropriately model various types of structures and to use approximate methods to validate the results of the computer analysis. Various approximate analysis techniques have been developed for the validation of computer results. They also learn how to use checking models (MecLeod 1992) to set up alternative, simplified models to check results produced by the FEA.

Computers are used as a vehicle for deep learning. Students are presented with open-ended briefs. In some cases students are presented with a choice of problems for which they work in groups of four or five. They study the problem thoroughly and within their group they formulate a design brief. The tutor acts as a supervisor to guide the students and discuss with them their problems. An important part of any project is extensive parametric study, for better and deeper understanding of structural behaviour. The importance of establishing links between analysis and design, by extracting useful information from large amount of computer results, is greatly stressed. Another important part of the project task is validation of the computer model and thorough the verification of computer results, using approximate analysis methods.

In the following case study an example of the students work is presented.

CASE STUDY

The problem given to the students was to study the effects of the beam dimensions on the behaviour of in-situ reinforced concrete two-way spanning slabs, as defined in the British Code of Practice BS8110 [BSI, 1997]. Students were asked to use LUSAS [FEA, 1999] finite element analysis to analyse the slab and to present what they learnt their peers.

The main points in the brief, submitted by students, are summarised as follows:

- They used an interior slab as defined in table 3.14 of BS8110.
- The slab dimension was selected to be 6m by 8m and 250mm deep.

- A single uniformly distributed load was considered (this is the most common type of load on slabs).
- It was stated that the beam depth is the most important parameter that affects the behaviour of the floor system as a whole.
- It was proposed that a parametric study would be conducted by changing the depth of the beam from 250 mm (a shallow beam) to 1300 mm in order to study the effect of changing the beam depth on the overall behaviour of the structure (floor system)
- Two different analytical models were proposed for the FEA:
    (i)   A Plate Bending Element to model the slab plate and an Engineering Grillage element to model the beam part of the floor system.
    (ii) A Thin Shell Element type to model the slab plate and a 3D Thick Beam element to model the beam part of the floor system
- A symmetrical quarter model of the slab was considered for the analysis.
- BS8110 Table 3.14, Yield Line Analysis and Timoshenko Tables were used for validation of the computer results.

Students proposed analytical models were discussed and finalised with the help of tutor. This was one of the areas in which tutor input was essential, as the students had no prior experience with analytical modelling of structures.

During the consultation process, the tutor guided the students with the various general aspects of modelling techniques, related to various types of structures. In this case students were introduced to Grill, 3D beam, Thin Plate, and Thin Shell elements. General techniques of mixing compatible elements were discussed. The concept of mesh refinement and convergence criteria was introduced.

Students used this knowledge to propose their own models for the problem related to their projects. The proposed models were discussed and refined by the tutor and students, until a satisfactory model was achieved.

The projects were mainly student centred and the role of the tutor was the active supervision of the process and the assessment and critique of the student work during the project.

At stages during the process, students presented their work to their peers for information dissemination and obtaining feed back from their peers.

Parametric Studies

A systematic approach is needed to give the designer the means to examine the effect of different constraints and parameter variations upon the problem to be investigated. In this way, the designer will be able to concentrate investigation in specific regions of interest.

BS 8110 has given guidelines for the design of two types of floor systems:
    (i)   In-situ reinforced concrete slabs supported by surrounding beams.
    (ii) Flat slabs, supported only by columns

In the first case, the assumption behind the BS8110 design is that the beams are sufficiently stiff so that the beam deflection is much smaller than the deflection at the centre of the slab. In this case the slab can be considered to be continuous over the supports, for the proposed slab in this case study. The slab is divided into regions of mid-strip and edge strips in each direction and it is assumed that the majority of the bending moment is resisted by the mid-strip regions in each direction.

In the second case, the only supports provided for the slab are the columns at the four corners of the slab. The punching shear capacity of the slabs at the column supports is sometimes enhanced by the provision of column heads.

In this case, the majority of the moments are concentrated around the column and strips joining each of the two adjacent columns, carries the majority of the bending moment.

Unfortunately BS8110 does not give any guidance about the relative stiffness of the beam in comparison with the slab. In practice, if the slab is supported on beams with depth greather than the slab, then it is considered as a two-way spanning slab and the reinforcement within the slab is calculated on the basis of the Table 3.14 of the BS8110. A similar procedure is adopted by Codes of Practices in other countries.

The results of the parametric study, carried out by students, is presented in Figures 1.

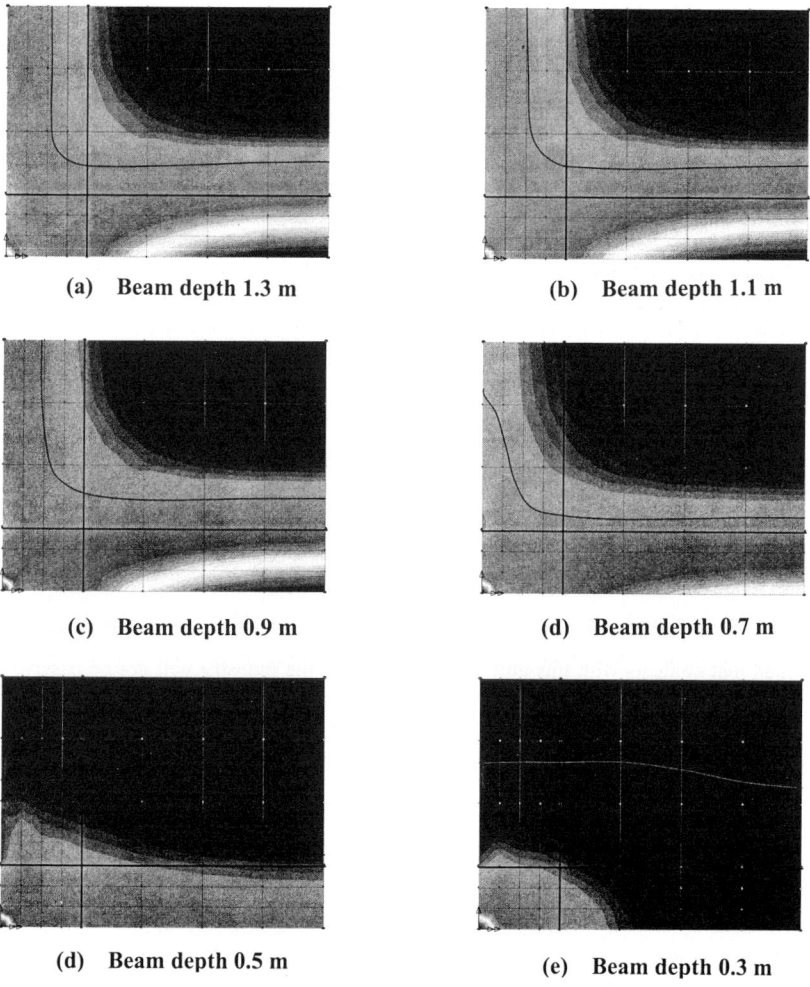

(a) Beam depth 1.3 m (b) Beam depth 1.1 m

(c) Beam depth 0.9 m (d) Beam depth 0.7 m

(d) Beam depth 0.5 m (e) Beam depth 0.3 m

Figure 1 Contour Plots of Bending Moment for Bars Parallel to Y-axis

DISCUSSION OF RESULTS

It is clear that the distribution of bending moments and hence the reinforcement distribution within the slab is very sensitive to the beam depth. Figure 1 shows that for a beam depth up to 500 mm the concentration of the bending moments is around the column i.e. in the region of the slab edge strip, as defined in BS8110. In this case the slab behaves more like a flat slab rather than a two-way slab spanning. If the slab reinforcement is detailed in accordance with the recommendations of BS8110 for two-way spanning slabs, only minimum reinforcement is provided in this region and the majority of the reinforcement bars are placed in the mid-strip. This will result in insufficient reinforcement bars in this region, which may result in excessive cracking of the slab in this area.

The result of this parametric study also demonstrates that when the depth of the beam is above 500 mm, the slab starts behaving like a two-way slab and the BS8110 design rules for two-way slabs would be valid.

Use of Graphic Interface

Figures 1 (a) to (e) show contour plots of bending moments for reinforcement bars parallel to the y-axis. It is clear that by using a Graphic User Interface (GUI) it is possible to exactly follow what is happening when one of the design parameters are changed. It clearly shows how changing the depth of the supporting beams affects the behaviour of the floor system.

Students used Microsoft Power Point for their presentation and they combined the animation power of Microsoft Power Point with the powerful graphic capabilities of LUSAS. Slides were animated to demonstrate the flow of bending moment contours from the mid-strip of the slab towards the column, as the depth of the beam was reduced.

Lessons learned by the students can be summarised as:

- Students were able to fully appreciate the effect of the changes in the depth of the beam to the performance of the structure.
- They were able to demonstrate the deficiencies inherent in the Codes of Practice.
- They were able to pinpoint the existence of practical pitfalls, caused by misinterpretation of the Codes of Practice.
- They were able to take home the fact that the responsibility of the design always lies with the engineer and the Codes of Practice are merely a good practice guide which can not cover every possible design scenario.
- It gives students a better understanding of the concepts and helps them gain confidence.
- It is hoped that students with this attitude, when joining the industry will not be passive learners, but they will be able to contribute efficiently to the practice.

Without the use of the computer in this reflective mode it would have not been possible for the students to appreciate the full behaviour of the structure and hence promote their deep learning.

Link Between Analysis and Design

Once the computer analysis was completed the students had the major task of verifying the computer result. They also had the task of extracting information for design. The main objective of this exercise was that the students should be able to:

- Use their knowledge, learned during the analysis process, to critically review the computer output and select limited useful design information in critical location on the structure.
- Appreciate link between analysis and design by extracting concise and useful information from the pile of computer output to be used in design.

- Appreciate that the analysis is not the end of the story. If during the design any major changes are proposed to satisfy essential design requirements, the students must go back and re-analyse the structure to reflect this change.
- View analysis and design as an integrated process.

CONCLUSIONS

The authors have found that industry is increasingly expecting structural engineering graduates to be instantly productive when starting their employment. This has also been noted for students commencing an industrial placement year. Industry does not have the resource to spend time in training graduates who have been only taught Code of Practices in their rigid format. They are expected to be aware of real world practice.

Interactive use of computers in exploring many aspects of the problem and conducting a parametric study by changing the important design parameters can greatly enhance the learning process.

Using GUIs, to visually view different aspects of structural behaviour, is a powerful tool that should be encouraged more widely in teaching of structural design.

Promoting student centred learning, in which students take control of their learning and the tutor can act as an active supervisor to guide students in their learning, gives students confidence in their work and promotes interest in the activity which they are engaged.

There is little doubt that computers are an essential part of life in any profession. The interactive use of the computer, in which it is used as a decision support tool, can enhance the decision making role of the designer.

Proper use of computers in teaching can promote deep learning and give confidence to graduates, which better meets industry requirements.

References

BS8110: *"Structural use of concrete: Part 1"*, London, British Standards Institutions, 1997.

Constrado, (1981), *Steel Designers' Manual*, Constructional Steel research and Development Organisation (CONSTRADO), Colins, UK.

Lusas Finite Element System. FEA Ltd. UK 1999.

Mathews J. D, (2000), *Optimisation and Decision Support During the Conceptual Stage of Building Design – New Techniques Based on the Genetic Algorithm,* PhD thesis, University of Plymouth, UK.

Norman D.A., (1986) ,"Cognitive Engineering" in "User Centred System Design", Norman D.A. & Draper S.W. (eds.), Lawrence Erlbaum Associates Publishers, New Jersey and London, pp.31-61.

Reynolds and Steedman, (1981), *Reinforce Concrete Designers' Handbook*, A Viewers Publication.

Timoshenko, S., and Woinowsky-Krieger, (1981), *Theory of Plates and Shells,* McGraw-Hill Publication.

LATE PAPERS

EXPERIMENTAL INVESTIGATION OF COLD-FORMED STEEL BEAM-COLUMN SUB-FRAMES : ENHANCED PERFORMANCE

K F Chung[1] and M F Wong[2]

[1]Associate Professor, [2]Research student
Department of Civil and Structural Engineering, the Hong Kong Polytechnic University,
Hong Kong, China.

ABSTRACT

This paper presents the findings of an experimental investigation on cold-formed steel beam-column sub-frames with bolted connections engineered for high structural efficiency and buildability. Based on two previous investigations on a total of twelve beam-column sub-frame tests, the basic connection configurations of bolted moment connections with specific ranges of sizes of cold-formed steel sections and hot rolled steel gusset plates was established. In order to increase the structural performance of the proposed connections, another eight beam-column sub-frames with hot rolled steel gusset plates of engineered shape were executed. It was found from the tests that for connections with thick gusset plates, flexural failure of cold-formed steel sections was critical and the moment resistance of the connections was found to be over 85% of the moment capacity of the cold-formed steel sections. It was thus demonstrated that through rational design and construction, effective moment connections between cold-formed steel sections may be readily achieved for practical building applications.

KEYWORDS

Cold-formed steel moment connections, beam-column sub-frames, experimental investigation, C-sections back-to-back

INTRODUCTION

In order to extend the effective use of cold-formed steel in building application, a research project on the structural performance of bolted moment connections in beam-column sub-frames was undertaken. Two previous experimental investigations on a total of twelve beam-column sub-frame tests were executed by Wong and Chung (2000a and 2000b), and the basic connection configurations of bolted moment connections with specific ranges of sizes of cold-formed steel sections and hot rolled steel gusset plates was established as follows:

- All structural members such as beams and columns are formed with two lipped C- sections back-to-back with interconnections.

- Moment connections between beams and columns are formed with hot-rolled steel gusset plates.
- Only the webs of lipped C-sections are bolted onto gusset plates.
- Four bolts per member are used.

The connection details are rationalised after considering ease of fabrication and installation. In general, the proposed moment connections are not able to develop full moment capacity of the connected members due to discontinuity of load paths along section flanges in the sections. It is aimed to develop bolted moment connections with effective use of materials to mobilise at least 75% of the moment capacity of the connected sections. Experimental investigations into connections between cold-formed steel sections are also reported by Bryan (1993), Zhao and Hancock (1991), Chung and Lau (1999), and also Wheeler, Clarke and Hancock (1999).

Based on the previous two experimental investigations on bolted moment connections on both internal and external beam-column sub-frames under both lateral and gravity loads, the maximum moment resistance of the proposed connections was found to range from 75% to 85% of the moment capacities of the connected sections. This paper presents the findings of another experimental investigation on the structural performance of engineered bolted moment connections. A total of eight internal and external beam-column sub-frames with hot rolled steel gusset plates of 50 mm chambers were executed. Comparisons among the test results together with recommendations on bolted connections between cold-formed steel sections were also presented.

TEST PROGRAM AND TEST PROCEDURES

In order to examine the structural performance of bolted moment connections against bolt pitches, two internal beam-column sub-frame tests with different pitches of 180 mm and 240 mm were carried out, i.e. *S180D1* and *S240D1*; the thickness of the gusset plates is 10 mm. Two external beam-column sub-frame tests with similar connection configurations were also carried out for comparison, i.e. *E180D1* and *E240D1*. Furthermore, in order to examine the structural performance of bolted moment connections against gusset plate thickness, two internal beam-column sub-frame tests with gusset plates of 6 mm thick were carried out, i.e. *S180D4* and *S240D4*. Two additional external beam-column sub-frames with a gusset plate of 6 mm thick were also carried out for comparison, i.e. *S180D4* and *E240D4*.

In the present investigation, all the test specimens are constructed according to the proposed basic configuration. All bolts are 16mm in diameter and of Grade 8.8. A double section of two lipped C-sections back-to-back with a section depth of 150 mm, a flange width of 60 mm and a thickness of 2.0 mm is used in all tests; the sections are designated as C15020 DS. The yield strength of the sections is 450 N/mm^2 which is designated as G450. The moment capacity of the double section obtained from four point load tests is 21.36 kNm. Details of the test series are summarised in Table 1. Figure 1 illustrates the general arrangements of the test set-up together with the proposed connection configurations for both the internal and the external beam-column sub-frames. Both the applied load and the displacements of each member of the test specimens were measured during the entire deformation history. The tests were terminated when either section failure or member buckling occurred, or deformation of the test specimens became excessive, i.e. over 250 mm.

TEST RESULTS

The results of all the eight tests are summarised in Table 1 together with the measured material properties and dimensions of both the cold-formed steel sections and the hot rolled steel gusset plates. Among the eight tests, two different modes of failure were identified:

- *FFcs*   Flexural failure of cold-formed steel section, as shown in Figure 2
- *LTBgp* Lateral torsional buckling of hot-rolled steel gusset plate, as shown in Figure 3

The maximum moment resistances of the test specimens may be evaluated at two locations, namely, at the centreline and at the failure position of the connections for different purposes. The centreline evaluation enables easy comparison of the moment resistances of the connections against applied moments obtained directly from conventional structural analysis while the failure position evaluation is required for connection design. In the data analysis, the moment resistances are first evaluated at the centreline of the connections, and level arm coefficients are then applied according to the associated failure modes to give the moment resistances of the connections at the failure positions. For test specimens with excessive deformation under testing, the moment resistance of the connection is restricted to be the applied moment at a connection rotation of 0.05 radian.

COMPARISONS AND DISCUSSIONS

After data analysis, the moment rotation curves of the internal and the external beam-column sub-frames are presented in Figure 4. In order to assess the effectiveness of the bolted moment connections, a moment resistance ratio, Ψ, is established which is defined as follows:

$$\Psi = \frac{\text{Measured moment resistance of a connection}}{\text{Measured moment capacity of connected section}}$$

The results of all the test specimens are compared among each other and the findings are presented as follows:

a) Tests with a bolt pitch of 180 mm

   Tests *S180D1* and *E180D1* - 10 mm thick gusset plates
   In both tests, flexural failure in the connected cold-formed steel sections was apparent. The moment resistance ratios, ψ, at the failure position of the connections in tests *S180D1* and *E180D1* were found to be 0.87 and 0.92 respectively.

   Tests *S180D4* and *E180D4* - 6 mm thick gusset plates
   In both tests, lateral torsional buckling of the hot rolled steel gusset plates was critical. The moment resistance ratios, ψ, at the failure position of the connections in tests *S180D4* and *E180D4* were found to be 0.65 and 0.64 respectively.

b) Tests with a bolt pitch of 240 mm

   Tests *S240D1* and *E240D1* - 10 mm thick gusset plates
   In both tests, flexural failure in the connected cold-formed steel sections was apparent. The moment resistance ratios, ψ, at the failure position of the connections in tests *S240D1* and *E240D1* were found to be 0.92 and 0.87 respectively.

   Tests *S240D4* and *E240D4* - 6 mm thick gusset plates
   In both tests, lateral torsional buckling of the hot rolled steel gusset plates was critical. The moment resistance ratios, ψ, at the failure position of the connections in tests *S240D4* and *E240D4* were found to be 0.67 and 0.62 respectively.

c) For connections with thick gusset plates, it is shown that flexural failure of the connected cold-formed steel sections is always critical, and the corresponding moment resistance ratio of the connections, ψ, is 0.85. This is regarded as a favourable mode of failure with high structural efficiency. For connections with thin gusset plates, lateral torsional buckling of the hot rolled steel gusset plates is always critical, and the corresponding moment resistance ratio, ψ, of the connections is about 0.60. It is regarded to be inefficient with low structural efficiency.

d) For connections with similar configurations but with different bolt pitches, it is shown that there is little difference in the moment resistance ratios. Thus, the bolt pitch of 180 mm may be considered as an optimal value for effective moment connections between sections C15020 DS.

CONCLUSIONS

Based on the findings of the experimental investigation, it is concluded that bolted moment connections between cold-formed steel sections are readily achieved using the proposed connection configurations. The bolted moment connections are demonstrated to be effective in transmitting moment between the connected sections, and thus enabling effective moment framings in cold-formed steel structures. Engineers are encouraged to build light-weight low to medium rise moment frames with cold-formed steel sections.

ACKNOWLEDGEMENTS

The research project leading to the publication of this paper is supported by the Research Grants Council of the Hong Kong Government of the Special Administrative Region (Project No. PolyU5031/98E), and also by the Research Committee of the Hong Kong Polytechnic University Research (Project No. G-V750). The tests were carried out at the Heavy Structure Laboratory of the Department of Civil and Structural Engineering, the Hong Kong Polytechnic University. The authors would like to express their gratitude to Mr Liu S. Y. and also to the technicians of the Heavy Structure Laboratory for the execution of the tests. The test specimens were supplied and fabricated by the P & Ls' Engineering Co. Ltd.

REFERENCES

1. Wong, M.F. and Chung, K.F. (2000a) Experimental investigation of cold-formed steel beam-column sub-frames: Pilot Study, *Proceedings of the Fifteenth International Specialty Conference on Cold-Formed Steel Structures*, St. Louis, Missouri U.S.A., October 19-20, 2000, 607-618.
2. Wong, M.F. and Chung, K.F. (2000b) Experimental investigation of cold-formed steel beam-column sub-frames: Comparative study, *Proceedings of the First International Conference on Structural Stability and Dynamics*, Taipei, December 2000, 587-592.
3. Bryan, E.R. (1993) The design of bolted joints in cold-formed steel sections, *Thin Walled Structures*, 1993, Vol. 16, 239-262.
4. Zhao, X.L. and Hancock, G.J (1991) T-joints in rectangular hollow sections subject to combined actions, *Journal of Structural Engineering*, ASCE, Vol. 117, No. 8 August 1991.
5. Chung, K. F. and Lau, L. (1999) Experimental investigation on bolted moment connections among cold-formed steel members, *Engineering Structures*, Vol. 21, 898-911.
6. Wheeler, A.T., Clarke, M.J. and Hancock, G.J. (1999) Finite element modelling of eight bolt rectangular hollow section bolted moment end plate connections, *Proceeding of the Second International Conference on Advances in Steel Structures*, Hong Kong, December 1999, 237-244.

Table 1 Summary of test program and test data

Test	Section	Maximum applied force (kN)	Failure mode	Maximum moment resistance		Moment resistance at 0.05 rad. (kNm)	Member		Gusset plate		Centreline of connection		Failure position of connection		
				Moment (kNm)	Rotation (rad.)		Thickness (mm)	Yield strength (N/mm^2)	Thickness (mm)	Yield strength (N/mm^2)	Normalised moment (kNm)	Ψ	Normalised moment (kNm)	Ψ (CFS)	Ψ (HRS)
S180D1	C15020DS	21.47	FFcs (CFS)	23.75	0.040	-	2.02	486	9.96	302	21.77	1.02	18.61	0.87	-
E180D1	C15020DS	22.70	FFcs (CFS)	25.11	0.043	-	2.06	476	9.96	302	23.05	1.08	19.71	0.92	-
S180D4	C15020DS	8.91	LTBgp	19.71	0.099	17.30	2.02	490	5.96	329	14.32	0.67	13.78	0.65	0.88
E180D4	C15020DS	9.35	LTBgp	20.68	0.130	17.14	2.02	477	5.96	329	14.18	0.66	13.65	0.64	0.87
S240D1	C15020DS	22.91	FFcs (CFS)	25.34	0.036	-	2.01	478	9.96	302	23.74	1.11	19.59	0.92	-
E240D1	C15020DS	20.86	FFcs (CFS)	23.07	0.038	-	2.04	450	9.96	302	22.62	1.06	18.66	0.87	-
S240D4	C15020DS	8.41	LTBgp	18.60	0.058	18.08	2.03	468	5.96	329	14.89	0.70	14.33	0.67	0.92
E240D4	C15020DS	7.75	LTBgp	16.70	0.020	-	2.03	482	5.96	329	13.75	0.64	13.23	0.62	0.85

Notes:

- S   denotes an internal beam-column sub-frame with a 'cross' shaped gusset plate under lateral load
- E   denotes an external beam-column sub-frame with a 'tee' shaped gusset plate under lateral load
- 180 denotes a bolt pitch of 180 mm
- 240 denotes a bolt pitch of 240 mm
- D   denotes 4 bolts per member with 50 mm chamfers in gusset plate
- 1   denotes a 10 mm thick gusset plate
- 4   denotes a 6 mm thick gusset plate

The measured moment capacity of C15020DS G450 is 21.36 kNm

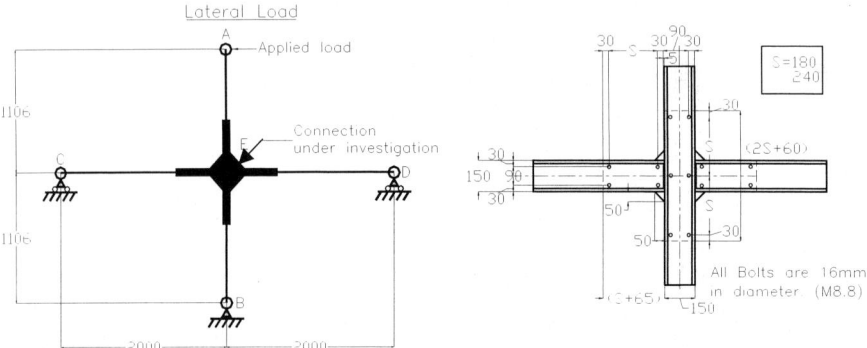

Internal beam-column sub-frame tests with 50 mm chamfers in hot-rolled steel 'cross' shaped gusset plates.

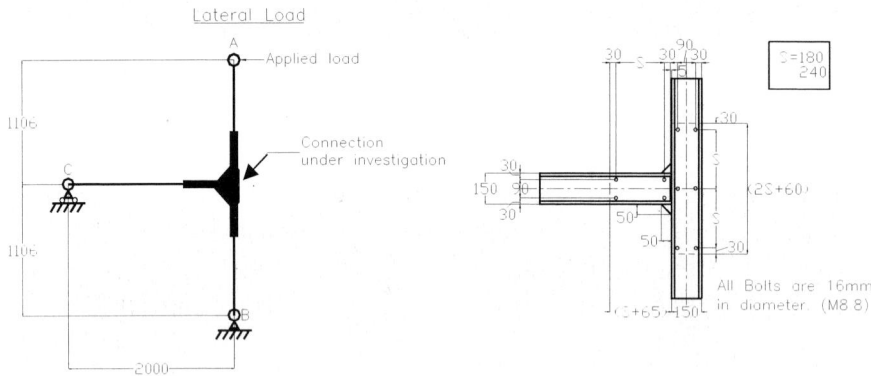

External beam-column sub-frame tests with 50 mm chamfers in hot-rolled steel 'tee' shaped gusset plates.

Figure 1   General set-up of beam-column sub-frame tests.

Figure 2 Flexural failiure of connected cold-formed steel sections.

Figure 3 Lateral torsional buckling of hot-rolled steel gusset plate.

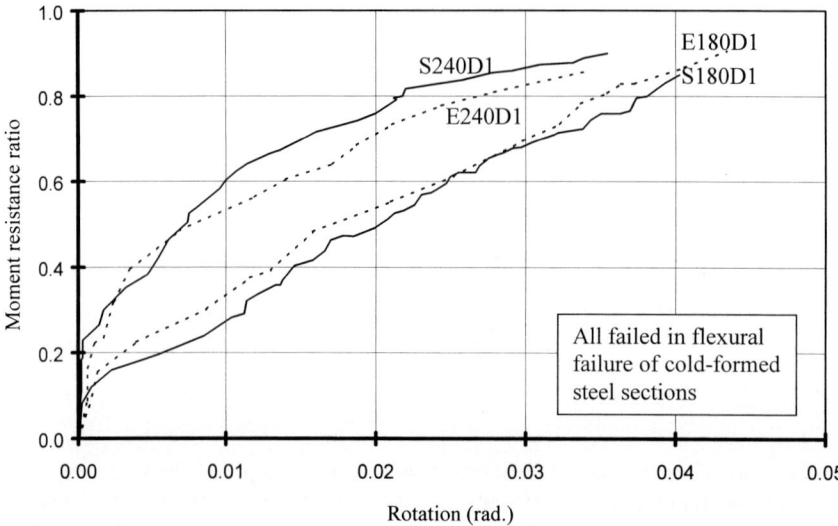

Figure 4a Moment rotation curves of beam-column sub-frames tests with 10 mm thick gusset plate.

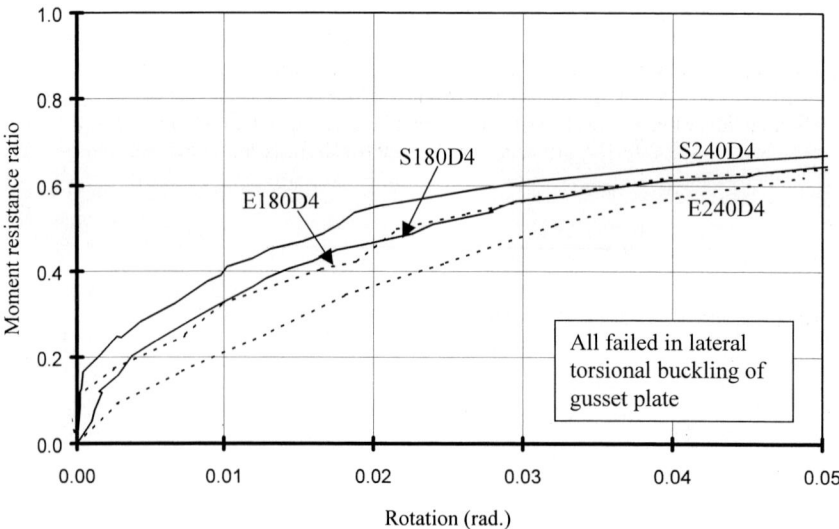

Figure 4b Moment rotation curves of beam-column sub-frames tests with 6 mm thick gusset plate.

MODIFIED PLASTICITY THEORY FOR REINFORCED CONCRETE SLAB STRUCTURES OF LIMITED DUCTILITY

C.T.Morley and S.R.Denton

University Engineering Department, Trumpington Street,
Cambridge CB2 1PZ, Britain

ABSTRACT

The paper presents recent work, mainly for a PhD dissertation, on approaches to the analysis of bending failure in concrete slabs with limited ductility, due to a high percentage of bending reinforcement or some other cause. The aim is to make comparatively simple extension of the well-known Yield Line Theory for slabs, rather than employ non-linear finite element analysis. The paper advocates use of the Galerkin method, with comparatively few assumed modes of deflection, which reduces to the upper bound kinematic approach for perfectly-plastic materials. When using the method it proves advantageous to include mechanisms involving 'wide yield lines' of constant curvature over a certain width (which also prove useful in predicting the effect of reduced strength in anticlastic curvature). Examples of predictions by this method are given, with approximate load-deflection curves for slabs of varying ductility, as well as outlines of the further development required before the method could be used for structural assessment in practice. The method seems promising, but needs further work, especially on including likely patterns and ranges of residual stress at zero applied load.

KEYWORDS

Plasticity, Galerkin, concrete, reinforcement, slabs, ductility, strain-softening.

INTRODUCTION

Computer programs based on the upper bound theorem of plasticity theory are increasingly being used for assessment of the strength of existing concrete structures (as distinct from designing new structures, where the lower bound theorem may well be more appropriate). In particular, programs (e.g. Middleton (1997)) based on yield line theory (YLT) are in use for assessment of reinforced concrete highway bridge slabs. Such slabs frequently have comparatively large reinforcement percentages. Then, quite apart from such problems as making a proper estimate of membrane action and large-deflection effects, finding a satisfactory collapse mechanism giving a low upper bound on the collapse load, and assessing the likelihood of failure in shear rather than bending (as assumed in YLT), the question may arise whether the slab has sufficient ductility to allow the use of plasticity theory and YLT at all. One obvious approach is simply to disallow the use of YLT when a given ductility criterion is violated – often the depth of concrete in compression, when evaluating the ultimate moment in a yield line, is limited to a certain proportion, say one quarter, of the effective depth to the tension reinforcement. This approach leaves an awkward gap in such assessment procedures – a range of bridge slabs which cannot be assessed easily.

The aim of this paper is to help fill that gap. An obvious possibility would be to carry out full non-linear finite element analyses in such cases of limited ductility. However, several NLFE analyses would presumably be needed, for each slab and load case, with different initial conditions so as to cover the range of residual stress, support settlement etc. likely to have occurred during the life of the slab before it is loaded towards collapse. This could be rather cumbersome. So this paper explores the scope for modifying YLT, in a hopefully simple way, to cope with cases of limited ductility. Of course, as soon as one departs from strict plasticity theory, some of the useful consequences of that theory will fall away too – we shall thus have to cope with residual stress, but perhaps this can be done in a simple manner also.

Another motive for some of the work, again of relevance to slabs with high steel proportions, is to allow, within plasticity theory, for the reduced strength of slab elements in anticlastic bending with principal curvature directions not coinciding with the (orthogonal) reinforcement directions.

PROPOSED BASIC APPROACH

The basic approach put forward is to use the Galerkin method to predict an approximate load-deflection curve, from which the maximum load is read. In this method, the (assumed small) deflections of the slab are taken to be representable by the sum of a small number of predetermined mode shapes w_i, each with a magnitude which varies during loading. Thus geometric compatibility, and any kinematic boundary conditions, are satisfied at all stages. For an assumed set of mode-shape magnitudes, the resulting curvatures and strains lead, via the given material properties, to stress-resultants (e.g. moments and shearing forces) – and thence to a set of external loads p and boundary forces in equilibrium with the stress-resultants. Unfortunately, these loads p will normally not be in the same pattern as the required external loading λq, where λ is a load factor. So in the Galerkin method, equilibrium under λq is satisfied in an overall way, by making p and λq do the same virtual work on each of the assumed mode shapes w_i.
Thus we require

$$\Sigma (p - \lambda q) w_i = O, \quad \text{for all i} \qquad (1)$$

where the sum is taken over all relevant terms and areas, using a suitable computer program to evaluate numerically all required integrals. In our program, we use virtual work to convert the external equilibrium set p, doing work on each w_i, into an integral over the slab area of products of moment and curvature.

For each λ, the magnitudes of the mode-shapes have to be chosen to satisfy Eqn.1, within a certain tolerance, and we have an incremental method (Denton (2001)) for homing in on the solution.

The mode-shapes selected for the method include the mode of elastic deformation (with uncracked concrete) under the given applied load pattern, and a few other modes (such as modes for elastic behaviour with cracked concrete, and possible plastic modes). In particular we include the plastic collapse mechanism thought to give a low upper bound for the equivalent slab with unlimited ductility. There are strong similarities between the Galerkin method and the work-equation approach in the upper-bound theorem of plasticity. According to Denton and Morley (2000), this Galerkin method, applied in this way to an elastic-perfectly-plastic structure, will always give an upper bound on the true collapse load, rather like the usual work equation for a mechanism. So, if we use the expected YLT mechanism among the modes for the Galerkin method, for a slab with unlimited ductility we can expect to obtain the YLT estimate of the collapse load. In effect, this Galerkin method fills in (approximately) the load-deflection curve between the initial uncracked behaviour and the final stage of plastic collapse. Then, for cases of limited ductility, especially ductility not far short of unlimited, we can hope that this Galerkin method will also give useful predictions of the collapse load.

Incidentally, a postulated theorem for bodies of limited ductility - that an overestimate of the peak strength may be obtained by following an assumed mechanism and equating the *maximum* energy dissipation rate to the rate at which external loads do work - may be disproved by counter-example (Denton and Morley (2000)).

WIDE YIELD LINES

One immediate problem with the Galerkin approach set out above is that the YLT mechanisms, which we wish to incorporate as Galerkin modes, involve discontinuities of slope, at the yield lines. So how can they properly be added to continuous elastic modes? And how can the material properties, usually expressed in terms of strain and curvature rather than concentrated displacement or rotation, then be used to obtain stress-resultants such as bending moments?

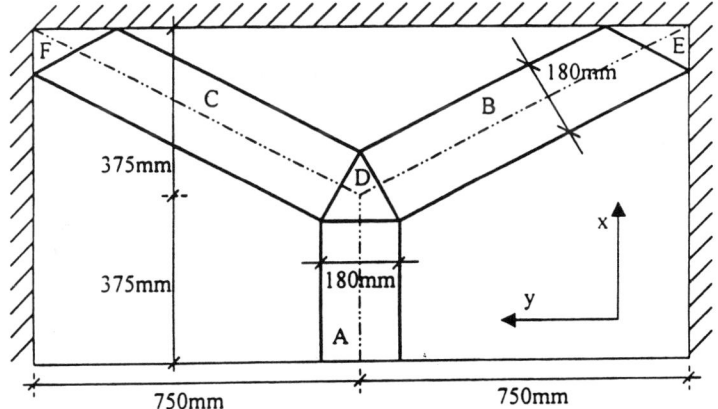

Figure 1: Typical 'wide yield line' mechanism

The solution is to develop modified YLT mechanisms, in which the rotation in each yield line is spread out as uniform cylindrical curvature over a narrow zone (centred on the original yield line position). The width of these zones, or 'wide yield lines' is to be chosen later, perhaps in the light of experiment, but will probably be a small multiple of the slab effective depth. This width need not be the same for every yield line. A typical example of such a mechanism, for a rectangular slab simply-supported on three sides and free on the fourth, is shown in Figure 1.

The geometry of mechanisms formed from such 'wide yield lines' has been studied (Denton (2001)). It turns out that many of the equations are rather similar to those for the 'strut-and-tie' method of lower-bound analysis for reinforced concrete walls loaded in plane, with bands of uniaxial compression meeting at nodal regions of biaxial compression. In particular, across nodal regions such as regions D,E and F in Figure 1 and D in Figure 2, the curvatures are in fact constant, and can be determined from the curvatures κ_a etc in the intersecting wide yield lines.

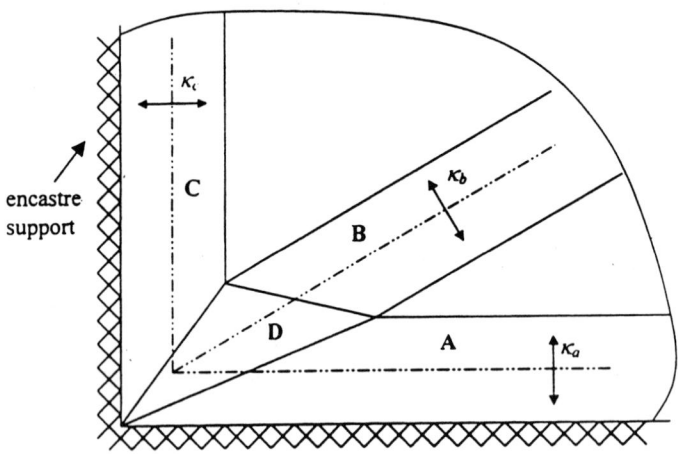

Figure 2 : Nodal region at intersecting fixed edges

Although wide yield lines are mainly intended for use in cases of limited ductility, they do have application with perfect plasticity. This is to tackle the problem of reduced bending strength in anticlastic curvature, when the principal curvature directions are at an angle to the orthogonal reinforcement. This reduction, from the bending strength predicted on yield lines using the well-known Johansen square yield criterion, is zero for very low steel proportions (when the concrete effectively has infinite compressive strength). However, the reduction can become quite marked (perhaps 30%) for high proportions of steel (in which case limited ductility would also be an issue). Unfortunately, the effect of this reduction for a given slab cannot be calculated using ordinary YLT with concentrated yield lines, because YLT mechanisms only involve uniaxial curvature. However, with wide yield lines a nodal region such as D in Figure 2 will involve anticlastic curvature. If such yield lines are made wide enough (Figure 3 with dimension a increased), appreciable areas of slab are in anticlastic bending. This opens the way to allowing for reduced strength in such areas, by treating the width a as a variable in an upper-bound plastic analysis, and using it to minimise the predicted upper bound on the collapse load.

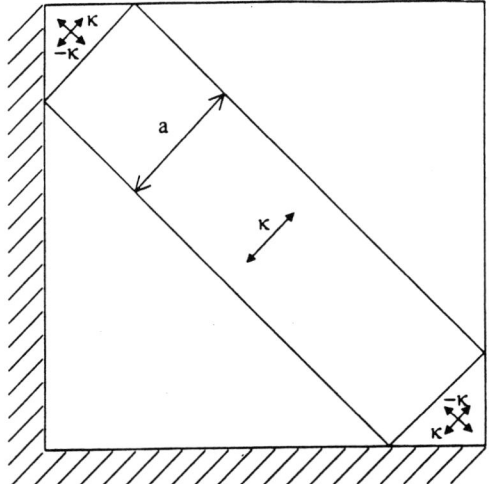

Figure 3 : Wide yield line across a balcony slab

EXAMPLE OF THE GALERKIN METHOD

A circular slab of radius R, under uniform load p per unit area, has been analysed (Denton and Morley (2000)), with various limits on ductility. The moment-curvature relationship, assumed the same in radial and circumferential directions, had elastic behaviour up to κ_y, then plastic behaviour at constant

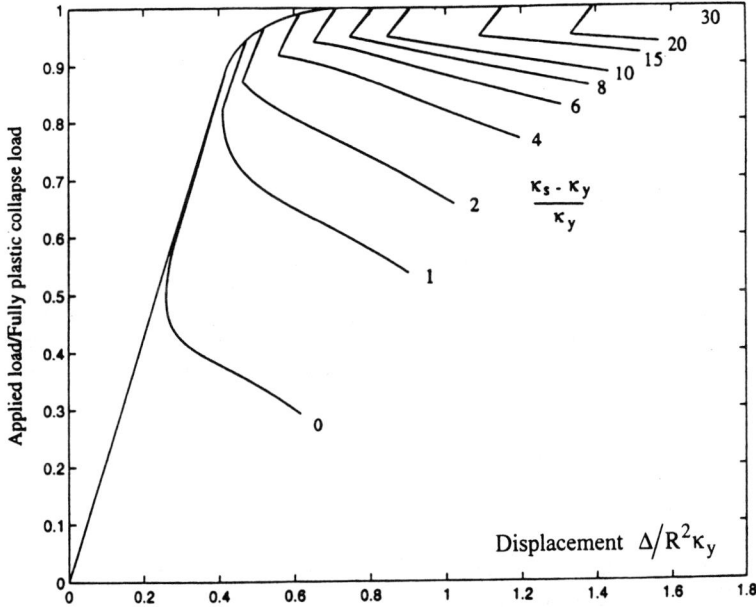

Figure 4 : Load-deflection plots for varying ductility

fully-plastic moment M_p up to κ_s, at which point the moment dropped immediately to zero (i.e. showing total strain-softening). The Galerkin method did not strictly speaking employ a wide yield line in this case, but as well as the linear-elastic deflection pattern it considered a conical mechanism, intended for plastic collapse, with a spherical central region out to radius R/20 (i.e. the equivalent of a wide yield line).

Predictions of the analysis are shown in Figure 4. This gives graphs of load against deflection, with the applied load as a proportion of the fully-plastic collapse load plotted against a dimensionless measure of central deflection. The graphs are plotted for different values of the length of the plastic plateau, as a multiple of the curvature at first yield. Because of the sudden strain-softening at the end of the plastic plateau, the graphs show very peaky behaviour at low ductility. However, if the plastic plateau length is at least twice the curvature at first yield, the predicted peak load is within 5% of the value for perfect plasticity.

EXPERIMENTS

A short series of experiments on yield lines with limited ductility are described by Denton (2001) and in Denton and Morley (1999). Wide rectangular slabs, simply supported along the two shorter sides and free on the others, were loaded with two patch loads near midspan, so that the expected failure mode was a single full-width yield line at midspan. Extra reinforcement was provided to ensure failure in this mode, and the two loads were far enough apart not to interfere with behaviour on this yield line. However, the loads were placed towards one free edge, to provoke the near part of the yield line possibly to strain-soften before the far end reached the expected peak moment.

The two test slabs were 105 mm thick and 750 mm wide, and had span 1500 mm between supports. The main longitudinal reinforcement was designed so that one specimen at its theoretical ultimate moment would have a neutral axis depth about one third of the steel effective depth – close to the usually-suggested limit for plastic analysis. The second slab had more reinforcement, with a neutral axis depth about half the effective depth – well beyond the usual limit, so that rather brittle behaviour was expected. Each slab had a companion beam with similar properties, but only 200 mm wide, loaded symmetrically.

The Galerkin method with two modes, elastic, and plastic with a central 'wide yield line', was applied to the companion beams. The modelling of the material properties was then adjusted slightly to give reasonable agreement between experiment and prediction, for both the comparatively ductile and the rather brittle cases. This modelling was then applied to the wide slabs, using in the Galerkin analysis five deflection modes – two based on linear-elastic analysis for isotropic and post-cracking stiffness, and three patterns with wide yield lines, including of course one mechanism with a single central wide yield line. For all wide yield lines, a constant width of 180 mm, twice the effective depth, was chosen. The bending moments (M_x, M_y, M_{xy}) at points across the slab surface were calculated from the corresponding curvatures using a special computer program based on a layering method, with the mid-depth strains calculated so as to give zero membrane forces (steps had been taken to eliminate lateral forces at the slab edges). Of course, these computed mid-depth strains might not themselves form a compatible system, and in a situation with significant membrane effects in-plane displacement modes could be included in the Galerkin method (see Olonisakin (1996)).

Predicted load-deflection curves for the two slabs, along with experimental results, are plotted in Figure 5. Both slabs were found to achieve (just) the theoretical YLT load, but there were some discrepancies with the companion beam tests. The analysis matches fairly well the key features of the test results, but the experimental post-peak behaviour is rather more ductile than the analysis predicts.

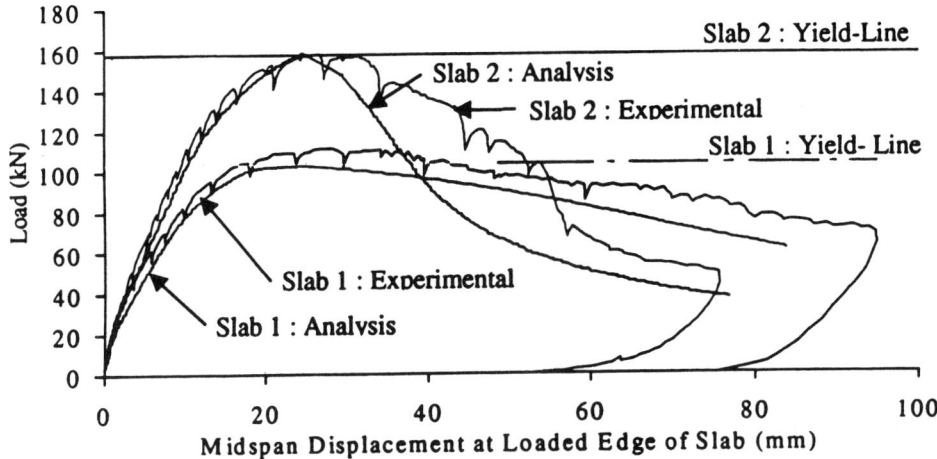

Figure 5 : Results for wide slabs loaded eccentrically

REQUIRED FURTHER DEVELOPMENTS

One of the most profound and useful corollaries of plasticity theory for structures made of perfectly-plastic materials is that residual stresses – caused for example by other loading cases taken beyond first yield, or by support settlement – have no effect on the predicted collapse load. So in ordinary YLT there is no consideration of such residual stresses, which have no effect on the work equation at full plasticity for a further increment of deflection in the assumed mode.

Unfortunately, as soon as one departs from strictly perfect plasticity, these corollaries of plastic theory fall away. If the proposed Galerkin method is to be used for structural assessment in practice, we must contemplate analysing a structure several times over, with different values and patterns of residual stress representing the extremes of the range likely to develop in practice, during the life of the structure before it is finally loaded to collapse. Thus far we have not done any such analyses – all of the examples presented had zero residual stress at the zero-load condition. However, it may well be possible to adapt the Galerkin approach to take account of such residual stresses – the problem will be to predict reasonable patterns and magnitudes for residual stress.

Another consideration is that the greater the percentage of bending reinforcement, giving rise to fears about limited ductility, the greater the likelihood that failure will be by shear rather than bending. So consideration of shear strength and shear failures should continue in parallel with any further developments of this method.

CONCLUSIONS

Approaches to the bending analysis of reinforced concrete slabs with limited ductility have been presented, intended to constitute a comparatively simple extension of the well-known method of Yield Line Theory. The Galerkin method appears to have some promise, but further development is needed, especially to consider likely ranges of residual stress. The 'wide yield lines' introduced to assist the Galerkin approach, also have some application in predicting the effect of reduced strength in anticlastic bending.

REFERENCES

Denton S.R. (2001). PhD Dissertation, University of Cambridge (in preparation).

Denton S.R. and Morley C.T. (1999) Accounting for limited ductility in concrete structures. FIB Symposium, Prague, 1999. Session: Modelling of concrete structures.

Denton S.R. and Morley C.T. (2000). Limit analysis and strain-softening structures. *International Journal of Mechanical Sciences* **42**, 503-522

Middleton C.R. (1997). Concrete bridge assessment: an alternative approach. *The Structural Engineer* **75:23-24**, 403-409.

Olonisakin A.A. (1996). Reinforced concrete slabs with partial lateral edge restraint. PhD Dissertation, University of Cambridge.

ACCIDENTAL DAMAGE ON UNREINFORCED MASONRY STRUCTURES

F.S. Crofts [1] and J.W Lane [2]

[1] Department of Civil Engineering, Technikon Pretoria, Private Bag X680, Pretoria, South Africa
[2] Masonry Consultant, Sandton, South Africa

ABSTACT

Design guidance is given for partial safety factors at the limit state of accidental damage and accidental actions to be considered for incorporation in the South African loading code, SABS 0160 - 1989 for the design of unreinforced masonry and other structures.

KEYWORDS

Accidental damage, accidental actions, partial safety load factors, residual strength, stability, progressive collapse, limit state of accidental damage, unreinforced masonry.

INTRODUCTION

In the 1980's, a comprehensive limit state formulation of South African structural codes was developed by a working group of the Structural Division of the South African Institution of Civil Engineering (SAICE). The proposals were accepted by the SABS and the loading code SABS 0160 was amended to reflect the limit state principles. Subsequently all issues and revisions of design codes for structural materials in South Africa, wherever feasible (Crofts & Lane, 2000) have been expressed in limit states format and conforms to a set of partial load factors and a uniform system for defining load combinations (Kemp et al, 1987).

The structural masonry code SABS 0164, Part 1 for Unreinforced Masonry is the one code that has not yet been revised accordingly. This Code is based on BS5628, Part 1 and reflects partial load factors based on British practice in the materials code. The structural masonry code SABS 0164, Part 2 for Reinforced and Prestressed Masonry has been revised and is used in conjunction with SABS 0160.

In September 1998 a South African National Conference on Loading was held in Midrand. The objective of this Conference was to re-activate the loading code working group and consider the impact of many international developments that have taken place since the inception of SABS 0160. Currently the SAICE working group on the South African Loading Code, representing the South African Bureau of Standards (SABS) Technical Committee 5120:SC2 – General Procedures for Loading in Buildings and Industrial Structures has been reconvened and meets on a regular basis. In the event of Eurocodes (which are consistent with SABS and essentially with ANSI/ASCE provisions) not being adopted and thus necessitating updating of South African Loading Codes, this paper will give guidelines for partial safety factors to be used for accidental damage in the design of unreinfoced masonry and accidental actions to be considered for incorporation in the South African loading code.

ACCIDENTAL LOADS VERSUS ACCIDENTAL DAMAGE

Introduction

Accidental damage is caused by an accidental action. Design to allow for accidental actions (loads or forces) is just a particular case of design for stability. Rules introduced following the partial collapse of the large panel precast concrete structure at Ronan Point in the United Kingdom in 1968 apply to buildings of five storeys and above; these rules were made to guard against gas explosions and do not in themselves ensure safety against all hazards. Accidental damage (as a result of an accidental action) can be limited by planned sacrifice or by greater structural strength. The choice is a design matter although the solution must satisfy the broad functional requirements as set out in SABS 0164 - 1.

Accidental Loads

SABS 0164-1 states, "In addition to designing the building to support loads arising from normal use, there should be a reasonable probability that it will not collapse catastrophically under the effect of misuse or accident. No building can be expected to be resistant to the excessive loads, that could arise because of an extreme cause, but it should not be damaged to an extent disproportionate to the original cause."

When, because of the layout of a structure there is a potential hazard, such as a brick pier/column or wall in a basement/garage, liable to vehicle impact, the design should ensure that, in the event of an accident, there is an acceptable probability that despite serious damage the structure will remain standing.

In Table 2, SABS 0160 and draft international ISO/DIS 9652-1 Masonry – Part 1 Unreinforced Masonry Design, the partial load factor for accidental loads is 1,0 x accidental load. Table 1 shows an extract of table 2 SABS 0160 indicating the partial load factors for accidental loading. This table indicates that the accidental load design is undertaken at ultimate limit state. There is no serviceability criteria and this accidental load acts in isolation with any other imposed load, thus having the load combination factor = 0. Accidental actions, loads or forces are largely unpredictable.

TABLE 1
PARTIAL LOAD FACTORS AND LOAD COMBINATION FACTORS

Type of load	Partial load factor γ_l		Load combination factor
	Ultimate limit state	Serviceability limit state	ψ_i
k) Accidental loads	1,0	-	0

Furthermore, owing to the nature of a particular occupancy or use of a structure, it may be necessary in the design concept or a design appraisal to consider the effect of a particular hazard and to ensure that, in the event of an accident, there is an acceptable probability of the structure remaining after the event, even in a damaged condition. Where there is a probability of vehicles running into and damaging or removing vital loadbearing members of the structure in the ground floor (or basement), the provision of bollards, protective walls, retaining earth banks, etc. should be considered.

Accidental Damage

Analysis of SABS 0164, written in the format of limit state of resistance was developed to ensure that margins of structural safety, gave similar results of codes for other materials.

After the Ronan Point disaster in the UK there were changes in the British Building Regulations to deal specifically with damage due to an accident such as a gas explosion for buildings of 5 storeys and more. The 1976 regulations required that the building should be so constructed that structural failure consequent on removal of any one member in a storey should be localised and limited in extent to a certain area of that storey. If this was not possible, that member should be capable of sustaining, without structural failure, an additional load or action corresponding to 34 kN/m^2. Subsequent revisions were aimed at reducing the sensitivity of a building to disproportionate collapse. One current design option is to provide effective horizontal and vertical ties in accordance with the structural codes of practice, requirements for design strengths related to 34 kN/m^2 loading. Where such provisions cannot be made, it is recommended that the structure should be able to bridge over a loss of an untied member and that the area of collapse be limited and localised.

Finally, if it not possible to bridge over the missing member, such a member should be designed as a protected element capable of sustaining additional loads related to a pressure of 34 kN/m^2. In practice the 34 kN/m^2 is used to determine a notional load that is applied sequentially to key elements and not as a specific overpressure that would result from a gas explosion. According to Ellis & Currie (1998), these requirements are considered to result in more robust structures which will generally be more resistant to disproportional failure due to various causes.

SABS 0164 - 1 (appendix E) gives the provisions for the aforementioned design aspects for accidental damage. As in the British code a key or protected member is a member that can withstand without collapse an accidental load of 34 kN/m^2. This value seems to be incongruous in the South African environment, since this value is based on British practice where town gas is used. In the 1996 version of the IStructE publication "Appraisal of existing structures" it is stated 'if it is certain that such a building will remain without a piped gas supply, the loading for the assessment of key elements may be taken as 17 kN/m^2"; this value would therefore be more appropriate for South African conditions.

The design for accidental damage is not only applicable to unreinforced masonry but is a measure to ensure the structural integrity of buildings. With the design of unreinforced masonry the serviceability limit states of deflection and cracking are seldom if ever relevant. Designers of masonry structures should be aware that checking for accidental damage as different to designing for an accidental action The wide variety of form of structural masonry and the brittle nature of unreinforced masonry make their use even more of a design matter, needing more individual judgement, than almost any other material.

CODE PROVISION FOR ACCIDENTAL DAMAGE

Design Code Provision

SABS 0164 - 1, currently gives partial safety load factors for loading that one would expect to find in SABS 0160. These values are based on British Practice.

Design values given for accidental damage (as well as for normal design) at ultimate limit state are similar to those found in BS 5628 - 1 (and differs from those in SABS 0160); these partial safety factors are as follows:.

- Design dead load = 0,95 Gk or 1,05 Gk (1)
- Design imposed load = 0,35 Qk except that in the case of buildings used predominantly for storage or where the imposed load is of permanent nature, use 1,05 Qk. (2)
- Design wind load = 0,35 Wk (3)

Gk, Qk and Wk refer to the characteristic dead, imposed and wind load. The characteristic values used in British codes of practice corresponds with the 'nominal load' contribution Dn, Ln and Wn in SABS 0160.

The origin of these partial safety factors values in BS 5628 - 1 for accidental damage is not well documented, but according to Morton (1987) when the effect of the accidental removal of a wall is being considered, the design dead load is changed to 0,95 Gk or 1,05 Gk. Qk and Wk are reduced to approximately a third of their value. This is done on the basis that the accidental event is unlikely to occur when all the worst condition prevail.

Loading Code Provision

The South African loading code SABS 0160

The design load effects pertaining to both ultimate and serviceability limit states is obtained from Eqn. 4 or 5 where the effects of the nominal loads multiplied by partial load factors given in column 2 or column 3 of Table 2, SABS 0160, as applicable, and the relevant load combination factors given in column 4 or interpreted from Table 3, for time dependent additional loads in correlation to the dominant load.

design load effect $Q = \gamma_i D_n$ (4)
and $= \gamma_i D_n + \gamma_i Q_{nj} + \Sigma(\psi_i \gamma_i Q_{ni})$ (5)
$\qquad\qquad i \neq j$
in which

- γ_i the partial load factors given in Table 2, SABS 0160
- D_n the nominal permanent load effect
- Q_{nj} the dominant imposed load effect for the load combinations and limit state under consideration
- Q_{ni} additional imposed load effects relevant and significant to the load combination and limit state under consideration
- ψ_i the load combination factors given in tables 2 & 3, SABS 0160

and the 'nominal load' contribution for a material with time-dependent characteristics shall be obtained from Eqn 6 for the design point in time value, Q_{dp}

$$Q_{dp} = \gamma_D D_n + \Sigma(\psi_i \gamma_i Q_{ni})$$ (6)

The design point-in-time load effect obtained from Eqn 6 may be required in the following design situations:
- determination of the sustained load contribution for analysis of the time-dependant behaviour of materials at serviceability limit state

- analysis of stability of structures which have localised accidental damage at the limit state of accidental damage
- analysis of residual strength of structures at a limit state of progressive collapse.

At the serviceability limit state the following combinations may be used:

$$1{,}1D_n + 1{,}0L_n \qquad (7)$$
$$1{,}1D_n + 0{,}3L_n + 0{,}6W_n \qquad (8)$$

with some exceptions namely that the imposed load factor 0,3 shall be replaced by 0,6 for garages, filing and storage areas and taken as 0 for roof loads. Note that the 0,6 serviceability wind load factor is applicable to the 50 year mean return period of wind speed and shall be modified for other return periods. The sustained portion of the loads at the serviceability limit state is obtained from

$$1{,}1D_n + 0{,}3L_n \qquad (9)$$

except as modified as noted above for the factor 0,3.

Discussion on SABS 0160

Mention is made of the use of the design point-in-time load effect obtained from Eqn. 6 for the analysis of the time-dependant behaviour of materials at serviceability limit state, the analysis of stability of structures which have localised accidental damage and the analysis of residual strength of structures at a limit state of progressive collapse.

Except for the serviceability criteria covered in the Code the latter two criteria are just listed but not addressed adequately in the loading code, i.e. the conditions of limit states of accidental damage and progressive collapse.

Current European proposals

ENV 1991–2-7 on accidental actions due to impact and explosions has been published as a European prestandard in 1998 and has to be available with a National Application Document (to be drafted) for trial use in the UK (Ellis & Currie, 1998). The ENV proposes three approaches to design for accidental actions, each with assigned to a different category of accidental design situations. The three categories varies from limited to large consequences.

The ENV includes a more detailed approach to design for accidental situations than current UK practice, in that it provides guidance for specific hazards, rather than providing a general level of robustness.

The accidental loading requirement has now been included in the 1996 BS loading code with the following text "When an accidental load is required for a key or protected element approach to design, that load shall be taken as 34 kN/m^2; the appropriate material Code shall be used".

The draft international standard ISO 9652-1 has the following partial load factors for accidental damage (Ellis & Currie, 1998):
- Design dead load = 1,0 Gk (10)
- Design imposed load = 0,35 Qk except that, in the case of buildings used predominantly for storage, or where the imposed load is of permanent nature, 1,0 Qk should be used. (11)
- Design wind load = 0,35 Wk (12)

In EC 6 (1995) (masonry code), there is a clause stating: "A structure shall be designed in such a way it will not be damaged by events like explosions, impact or consequences of human error, to an extent disproportionate to the original cause". There are no partial load factors for accidental damage.

Future development of UK and European guidance

With the responsibility for categorisation resting with national authorities, there will be an opportunity for the Eurocode to reflect any position that the UK wishes to adopt in relation to disproportionate collapse procedures (Ellis & Currie, 1998).

Structural Codes in Canada, the United States (Ellingwood, 1999)

Conditional or accidental limit states:

Load effects arising from rare accidental events, such as fires, explosions of natural gas, detonation of bombs, or vehicular impact, may initiate a catastrophic collapse of a structure. Here, the primary concern is life safety rather than damage prevention, and the design philosophy is to create a robust structural system in which damage is contained rather than prevented entirely. As a consequence, while the conditional limit states are ultimate or strength related, in the general sense described above, sources of load-carrying capacity not ordinarily permitted might be mobilised in the design analysis.

United States: ASCE 7/ANSI A 58 first introduced a requirement for general structural integrity against unforeseen events in 1972, shortly after the Ronan Point collapse. ASCE 7-95 contains a performance requirement that a building be designed to sustain local damage, with the structural system as a whole remaining stable. It recommends the following load combination for checking for the ability of a damaged structure to maintain overall stability for a short time following an accidental load event:

$$(0,9 \text{ or } 1,2) D + 0,5 L + 0,2 W \quad (13)$$

This check requires the notional removal of load-bearing elements. If certain key elements in the structure must be designed to withstand the effects of the accident (perhaps to allow the development of alternate load paths), they should be designed using the combination,

$$(0,9 \text{ or } 1,2) D + A_k + (0,5 L \text{ or } 0,2 S \text{ or } 0,2 W) \quad (14)$$

in which A_k is the postulated action due to the abnormal load and S is the load due to snow. Normally, only the main load-bearing structure would need to be checked.. Building code officials in the US are not enthusiastic about provisions related to general structural integrity because they are difficult to cast in prescriptive code language. Most building codes in the United States do not contain such provisions.

Canada: The National Building Code of Canada (NBCC) requires structures to be designed for sufficient structural integrity to withstand all effects that may reasonably be expected to occur during the service life. Structural integrity is defined as "the ability of the structure to absorb local failure without widespread collapse". Designers are advised to consider and take measures against severe accidents with probabilities of occurrence of approximately 10^{-4} /years or more. Several general approaches are suggested and date back to the 1970 edition of the NBCC, issued not long after the Ronan Point collapse.

Structural Code in Australia

In AS 1170.1 the following approach is followed: "Accidental and misuse are normally unforeseen events therefore it is impracticable to design for them. However, precaution should be taken to limit the effects of local collapses caused by such actions i.e. to prevent progressive collapse".

RECOMMENDATIONS

Introduction

The recommendations in SABS 0164 - 1 based on the British experience have proved to be satisfactory and therefore the order of magnitude of these values should still apply.

In the proposal document for a comprehensive limit state formulation for the South African Loading Code by Kemp et al (1987) gave the partial load factors for the serviceability as in Eqns 7, 8 and 9.

There was however the following sentence in the original publication:

"For analysis of residual strength and stability of structures with localised accidental damage (limit state of progressive collapse or accidental damage), load factors for the serviceability limit state should be used".

These serviceability service state values compare favourably with the values given for accidental damage found in BS 5628 - 1; i.e. Eqns. 1, 2 and 3.

Comparing these factors with the design point-in-time load effect obtained from $Q_{d\,p} = \gamma_D D_n + \Sigma(\psi_i \gamma_i Q_{ni})$ in SABS 0160, and equating $D_n = G_k$ and $L_n = Q_k$ and $W_n = W_k$, these equations become:

$$\cong 1{,}1D_n \text{, or} \qquad \text{(similar to (1))}$$
$$\cong 1{,}1D_n + 1{,}0L_n \text{ or } 1{,}1D_n + 0{,}35L_n \text{ or} \qquad \text{(similar to (2))}$$
$$\cong 1{,}1D_n + (0{,}35 \text{ or } 1{,}05) L_n + 0{,}35W \qquad \text{(similar to (3))}$$

Considering the design point-in-time load effect for the analysis of the time-dependant behaviour of materials at serviceability limit state in SABS 0160, Eqns 7, 8 & 9 for load combinations thus become:

$1{,}1D_n$ for dead load or self-weight only, (10)
$1{,}1D_n + 1{,}0L_n$ for dead load and imposed load (7)
$1{,}1D_n + 0{,}3L_n$ for dead load and sustained imposed load (9)
$1{,}1D_n + 0{,}3L_n + 0{,}6W_n$ for dead load, imposed load and windload (8)

When comparing the serviceability limit state partial safety factors in SABS 0160, Eqns. 7, 8, 9 & 10 compare favourably with the partial safety factors for accidental damage in BS 5628-1 (Eqns. 1,2, & 3) and ISO 9652-1 (Eqns 10,11, & 12), except for Eqn. 8 which has a higher factor for windloading, i.e. 0,6 thus making this equation more onerous.

When designing for accidental damage much is left to the individual designer, for example to determine the probable design point-in-time load for accidental damage with the unlikelyhood of all the worst conditions prevailing at a given time.

Design Guidance on Accidental Loads

It is proposed that when an accidental load is required for a key or protected element approach to design, that load shall be taken as 17 kN/m^2.

Design Guidance on Partial Safety Factors for determining Residual Strength of Structures at a Limit State of Accidental Damage

It is proposed that the following sentence is included in the South African loading code:

For analysis of residual strength and stability of structures with localised accidental damage (limit state of progressive collapse or accidental damage), load factors for the serviceability limit state should be used;
or
For analysis of residual strength and stability of structures with localised accidental damage (limit state of progressive collapse or accidental damage) the load factors indicated in eqns. 7, 8 9 and 10 should be used.

REFERENCES

AS1170.1 Australian Loading Code Part 1

British Standards Institute. Use of masonry. Part 1. *Structural use of unreinforced masonry*. BS 5628: 1978. Part 1

Crofts, F.S; Lane, J.W. *Structural concrete masonry. A design guide*. Concrete Manufacturers Association, 2000.

Draft International Standard ISO 9652-1: Masonry – Part 1: *Unreinforced masonry design by calculation*.

Ellingwood B.R. *A comparison of general design load requirements in structural codes in Canada, Mexico and the United States*. Engineering. Journal, AISC, 2nd quarter 1999.

Elllis B.R; Currie D.M. *Gas explosions in buildings in the UK: Regulations and risk*. The Structural Engineer, Volume 76/No 10, 6 October 1998.

Kemp A.R; Milford R.V; Laurie J.A.P; *Proposals for a comprehensive limit states formulation for South African structural codes*. The Civil Engineer in South Africa, September 1987.

London, Structural Engineers Trading Organisation Ltd. *Appraisal of existing structures*, 2nd ed., October 1996.

Morton J. *Limit state philosophy. Partial safety factors and the design of walls for compression and shear.* Brick Development Association. London, December 1987.

Proceedings Eurocode 6: Design of masonry structures – Part1-: *General rules for buildings – Rules for reinforced and unreinforced masonry*. ENV 196 – 1-1 : 1995.

South African Bureau of Standards. *The general procedures and loadings to be adopted in the design of buildings*. SABS 0160:1989 (amended 1990, 1991 & 1993).

South African Bureau of Standards. The structural use of masonry. Part 1. *Unreinforced masonry walling*. SABS 0164: Part 1:1980

South African Bureau of Standards. The structural use of masonry. Part 2. *Structural design and requirements for reinforced and prestressed masonry*. SABS 0164: Part 2:1992.

South African National Conference on Loading. *Towards the development of a unified approach to design loading on civil and industrial structures for South Africa*. Eskom Conference Centre, Midrand. September 1998.

A CONTACT BASED DYNAMIC DELAMINATION BUCKLING ANALYSIS OF COMPOSITES

S. Mohammadi[1] and A. Asadollahi[2]

Department of Civil Engineering, University of Tehran

ABSTRACT

In this study, a combined finite/discrete element algorithm is developed to simulate delamination and inplane fracture in laminated composites subjected to dynamic loadings. The application of the finite/discrete element strategy to modelling of dynamic loading of composites is innovative, and will provide a significant advance in comparison to presently available capabilities of numerical modelling of this complex physical problem. In this method of modelling of composites, the possible delaminated region is modeled using a discrete element mesh, and the rest of the structure is modelled by a standard finite element mesh. Each group of similar plies is modelled by one discrete element. Each discrete element will be discretized by a finite element mesh and might have material or geometric nonlinearities. The interlaminar behaviour of discrete elements is governed by bonding laws which include contact and friction interactions for the post delamination phase. Once two layers are delaminated, the corresponding interface will still be capable of further contact and friction interaction. The performance of the model to correctly simulate the physical behaviour of composites subjected to impact loadings will be assessed by solving several test cases available in the literature.

KEYWORDS

Discrete element method, Dynamic delamination buckling, Crack propagation, Composites

INTRODUCTION

It is evident that impact loading can cause severe damage in composite laminates. The phenomenon of failure by catastrophic crack propagation poses problems in all applications, particularly in the aerospace industry in which safety is of paramount importance, but where over-design carries heavy penalties in terms of excess weight. Therefore, the development of reliable models for determining the failure behaviour of growing advanced materials are vitally important.

---

[1] Assistant Professor, corresponding author.
[2] Postgraduate student.

In general, according to the orthotropic laminated nature of composites, the failure modes may be classified into four different types : matrix failure, delamination, shear cracking, and erosion damage. There is, however, agreement that the most dominant causes of damage during impact are matrix cracking coupled strongly with complex mode delamination mechanisms Ambur(1995).

Numerical simulation of arbitrary shaped components is traditionally performed by the finite element techniques, which is rooted in the concepts of continuum mechanics and is not suited to general fracture propagation and fragmentation problems. In contrast, the discrete element method (DEM) is specifically designed to solve problems that exhibit strong discontinuities in material and geometric behaviour Munjiza (1995). The discrete element method idealizes the whole medium into an assemblage of individual bodies, which in addition to their own deformable response, interact with each other (through a contact type interaction) to capture the characteristics of the discontinuum and to perform the same response as the medium itself.

In this paper, some of the main aspects of modelling of composites by DEM are discussed. The final summary and conclusions will follow some representative results of the numerical tests for assessing the performance of the method.

DISCRETE ELEMENT MODELLING OF COMPOSITES

Figure 1 shows a typical combined FE/DE mesh for a quarter of a composite plate subjected to concentrated central loading. In a combined FE/DE method, the fractured region is modelled using a discrete element mesh and the remainder of the specimen is modelled by a standard finite element mesh. A combined mesh enables us to prevent unnecessary contact detection and interaction calculations which comprise a major part of the analysis time, Mohammadi (1997_3).

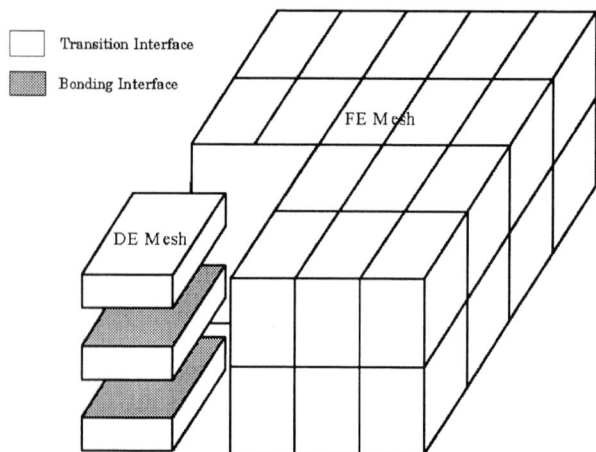

Figure 1: Combined FE/DE mesh

Each group of similar layers is modelled by one discrete element and each discrete element is discretized by a standard finite element mesh. The interlaminar behaviour of discrete elements is governed by bonding laws, including contact and friction interactions for the post delamination phase. Interactions between finite elements and discrete elements are modelled by transition interfaces (Figure 1).

Inplane fracture may result in the creation of new discrete bodies which are in contact and friction interaction with neighbouring bodies. A special remeshing algorithm is adopted to maintain compatibility conditions in newly fractured regions.

From the computational point of view, the discrete element procedure comprises three steps: object representation, contact detection, and contact interaction. The first two steps are closely associated to each other and are usually discussed within the framework of the contact detection algorithms.

Contact detection

An Alternating Digital Tree, Bonet (1991) contact detection algorithm is employed to detect the possibility of contact between discrete elements. In this method, each object is represented by a bounding box. Each bounding box is then represented by a space bisection algorithm resulting in a binary tree database structure. Figure 2 illustrates the space bisection procedure and the associated binary tree for a typical problem. The use of binary tree structure would dramatically increase the performance of contact search, because once one node of the tree is found to be sufficiently far from an object, all its descendant nodes will be eliminated from the contact search of that object.

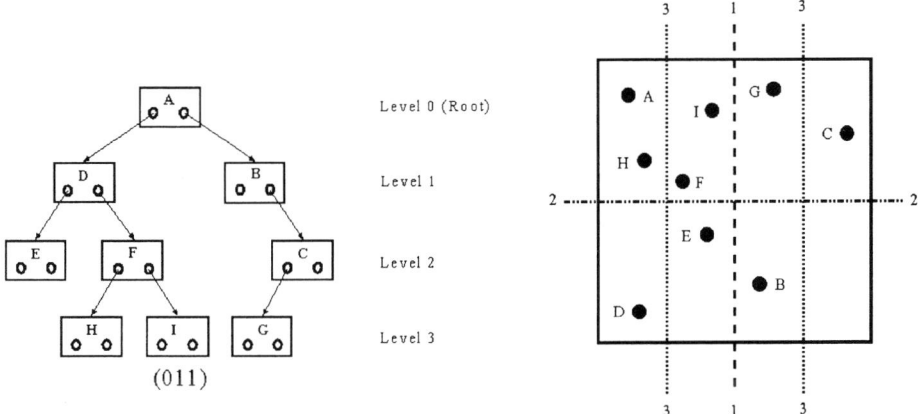

Figure 2: The space bisection approach, and its associated binary tree data structure.

Contact interaction

Once the possibility of contact between discrete bodies is detected, another method has to be used to satisfy the impenetrability condition of the bodies. Many different methods have been developed for enforcing a constraint condition on the governing equation of a well established physical behaviour. Among them the penalty method is likely to be the most appropriate scheme for adopting into an explicit contact analysis. In this method, penetration of the contactor object is used to establish the contact forces between contacting objects at any given time (See Figure 3).

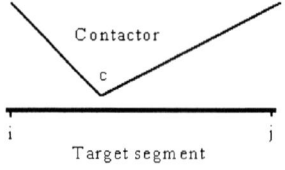
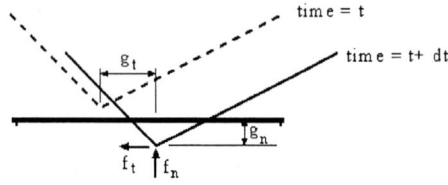

a) Before contact b) Possible normal and tangential gaps

Figure 3: Contact force based on impenetrability.

According to the weak form of the boundary value problem, the component form of the virtual work of the contact forces associated to the contact node is given by Mohammadi (1998_1), Schonauer (1993):

$$\delta W^c = f_k^{\ c}\ \delta g_k = f^c_k\ \frac{\partial g_k}{\partial u_i^s}\ \delta u_i^s$$

where k=n,t and i=x,y, and u_i^s is the i-component of the displacement vector at node s, $\mathbf{g} = (g_n, g_t)$ is the relative motion vector, and $\mathbf{f}^c$ is the contact force vector over the contact area $\mathbf{A}^c$,

$$\mathbf{f}^c = \mathbf{A}^c\ \mathbf{\sigma}^c$$

$$\mathbf{\sigma} = \mathbf{\alpha}\ \mathbf{g} = \begin{bmatrix} \alpha_n & 0 \\ 0 & \alpha_t \end{bmatrix} \begin{bmatrix} g_n \\ g_t \end{bmatrix}$$

where α is the penalty term matrix, which may vary between single contact nodes. The corresponding recovered residual force is then evaluated as:

$$r_i^s = f^c_k\ \frac{\partial g_k}{\partial u_i^s}$$

The calculated contact force has to be distributed to the target and the contactor nodes.

DELAMINATION INITIATION

The Chang-Springer criterion may be properly used for predicting the initiation of delamination. Chang and Springer (1986_2):

$$\left(\frac{\sigma_z^{\ 2}}{N^2}\right) + \left(\frac{\sigma_{xz}^{\ 2} + \sigma_{yz}^{\ 2}}{T^2}\right) = d^2$$

where N and T are the unidirectional normal and tangential strengths of the bonding material, respectively and failure occurs when d becomes greater than or equal to one.

MATERIAL MODEL

The imminence of material failure is monitored by the orthotropic Hoffman criterion, where a geometric yield surface is constructed from three tensile yield strengths σ_T, three compressive strengths σ_C, and three shear strengths σ_S. It may be defined as:

$$\Phi = \tfrac{1}{2}\sigma^T \mathbf{P} \sigma + \sigma^T \mathbf{p} - \bar{\sigma}^2(\kappa)$$

where the projection matrix **P**, and the projection vector **p** are defined based on the nine material yield strengths and a normalized yield strength $\bar{\sigma}$ (see Schellekens et al. (1990)), and κ is a softening/hardening parameter.

A bilinear local softening model is also adopted in this study to account for release of energy and redistribution of forces which caused the formation of a crack. It may properly avoid the mesh dependency of the results by introducing a length scale into the softening material model ,Mohammadi (1997_3).

The additivity postulate of computational plasticity is used to formulate the rate form of the stress return algorithm. The integration of the flow rule in a finite step is then performed by the backward Euler method coupled with the Newton-Raphson iterative scheme.

NUMERICAL SIMULATIONS

The author has previously published a number of papers on verifying the performance of the approach in modelling the complex behaviour of progressive cracking in composites (Mohammadi (1997_3,1998_1), Owen(1998)). Therefore, in this paper only further numerical simulations of some engineering applications are presesnted.

Fracture and delamination buckling analysis of an orthotropic composite specimen is considered. The material properties used in the calculations are listed in Table 1 (Liu (1993)). The composite [90_n, 0_n, 90_n] ply layout is assigned to the beam with (LHW=10.16,0.249,2.54cm) geometric descriptions. The specimen is subjected to quasi-static concentrated loading P=2300 KN applied at its centre line. An eight layer finite/discrete element mesh was used to model half of the beam.

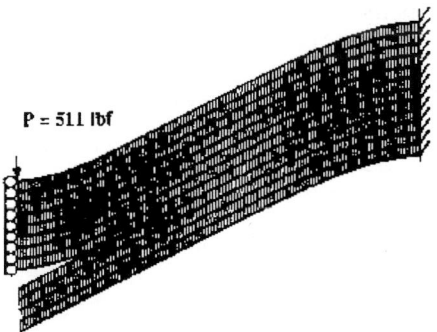

Figure 4:Delaminated deformed shape of the beam observed from the experiments repoted by Liu

TABLE 1
MATERIAL PROPERTIES OF THE ORTHOTROPIC COMPOSITE BEAM

$$E_{xx} = 139200 \text{ M Pa}, \quad G_{xx} = 5580 \text{ M Pa}$$
$$E_{yy} = 9700 \text{ M Pa}, \quad G_{yz} = 3760 \text{ M Pa}$$
$$v_{xy} = v_{yz} = .3, \quad \rho = (1.38\text{-}2)\frac{kg}{m^3}$$
$$X_t = 1150 \text{ M Pa}, \quad X_c = 1120 \text{ M Pa}$$
$$Y_t = 40 \text{ M Pa}, \quad Y_c = 170 \text{ M Pa}$$
$$S = 100 \text{ M Pa}$$

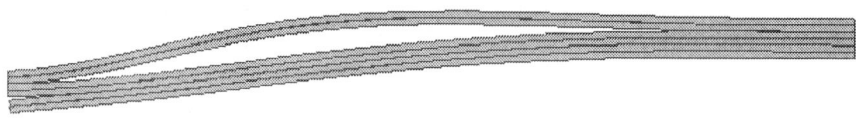

(a) Buckling of top delaminated layer.}

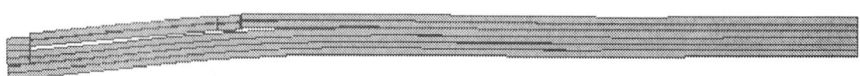

(b) Fracture patterns

Figure 5: Deformed shape and crack patterns of the [90 $_n$,0 $_n$,90 $_n$] composite specimen.

Figure 5a illustrates the buckling mode of delaminated layer. Although Liu (1993) reported the same delamination patterns, however, no buckling mode was reported at the top delaminated interface.

A fracture analysis was also performed to predict the real damage loading and damage mode of the beam (Figure 5b). Matrix cracking across the thickness of the top layer of the specimen prevents the formation of a buckling mode and the overall behaviour of the specimen reduces nearly to an unbonded multi-layer beam.

Impact loading of a composite plate - delamination analysis

A numerical simulation is undertaken to assess the performance of the method for dealing with progressive debonding phenomena (no material fracture) in a laminated composite plate which is subjected to a high velocity impact at its centre (based on the experiments undertaken by Worswick (1995).

Because of symmetry, only one quarter of the plate is modelled. Also, only the central region of this model is meshed by a DE mesh (See Figure 6). The composite ply pattern is set to $[90_n,0_n,90_n,0_n,90_n]$. The impact loading is simulated by a triangular load applied from 0 to 5 μsec with a peak force of 5 kN.

Material properties and other necessary information are given in Table 2.

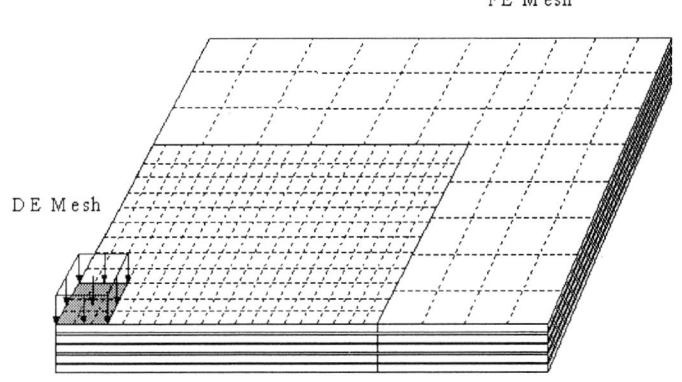

Figure 6: FE/DE mesh of the composite plate.

TABLE 2
MATERIAL PROPERTIES FOR T800/P2302-19 GRAPHITE RESIN

Model size = 0.0762 × 0.0508 × 0.0044 m
DE region = 0.05 × .035 m
Ply layout [90, 0, 90, 0, 90]
E_{xx} = 152.4e3 M Pa , E_{yy} = 10.7e3 M Pa
v = 0.35 , ρ = 1.55 e3 $\frac{kg}{m^3}$
X_t = 2772 M Pa , X_c = 3100 M Pa
Y_t = 79.3 M Pa , Y_c = 231 M Pa
S = 132.8 M Pa

Figures 8 and 9 illustrate the debonding patterns at different layer interfaces for two different stages of the loading. Delamination patterns are clearly developing from the central region of the plate, i.e. the impacted zone, towards the edges of the plate.

Figure 7 depicts the comparison of the displacement history of the centre of the plate for this mesh and a coarser mesh. The comparisons are made for both the top and bottom point (Series 1:Coarse–

Top,Series 2 :Fine-Top ,Series 3 : Coarse – Bot,Series 4 :Fine- Top) across the thickness of the plate, and clearly shows the mesh independency of the results.

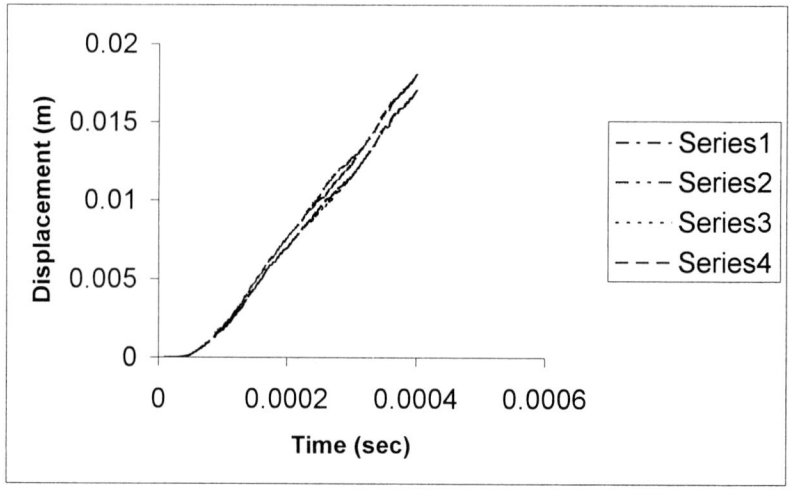

Figure 7: Displacement history of the central point

CONCLUSIONS

The combined finite/discrete element has proved to be an efficient algorithm for dealing with multi-fracture and fragmentation processes, which frequently arise from impact loadings on structures. An alternating digital tree method is adopted to reduce the extensive numerical costs of the contact detection phase. A local remeshing scheme is introduced for geometric modelling of the cracks, which plays an important role in avoiding the excess distortions of the finite elements in the vicinity of cracks. Several numerical tests have been used to assess the performance of the method.

ACKNOWLEDGEMENTS

The authors would like to appreciate Prof. D.R.J. Owen and Dr. D. Peric, from University of Wales Swansea, who were closely involved in the developing stage of the work.

REFERENCES

Ambur.D.R and Starnes.J.H (1995). Low- speed impact damage-initiation characteristics of selected laminated composite plates. *AIAA Journal*, **33:10**,1919-1925.

Bicanic.N, Mungiza .A , Owen.D.R.J, and Petrinic.N .(1995). From continua to Discontinua- a combined finite element / discrete element modelling in civil engineering . In Topping.B.H.V , editor , *Developments in Computational Techniques for Structural Engineering*, pages 45-58 ,Civil-CompPress.

Bonet.J and Peraire.J.(1991). An alternating digital tree (ADT) algoritm for 3D geometric searching and intersection problems . *International Journal for Numerical Methods in*

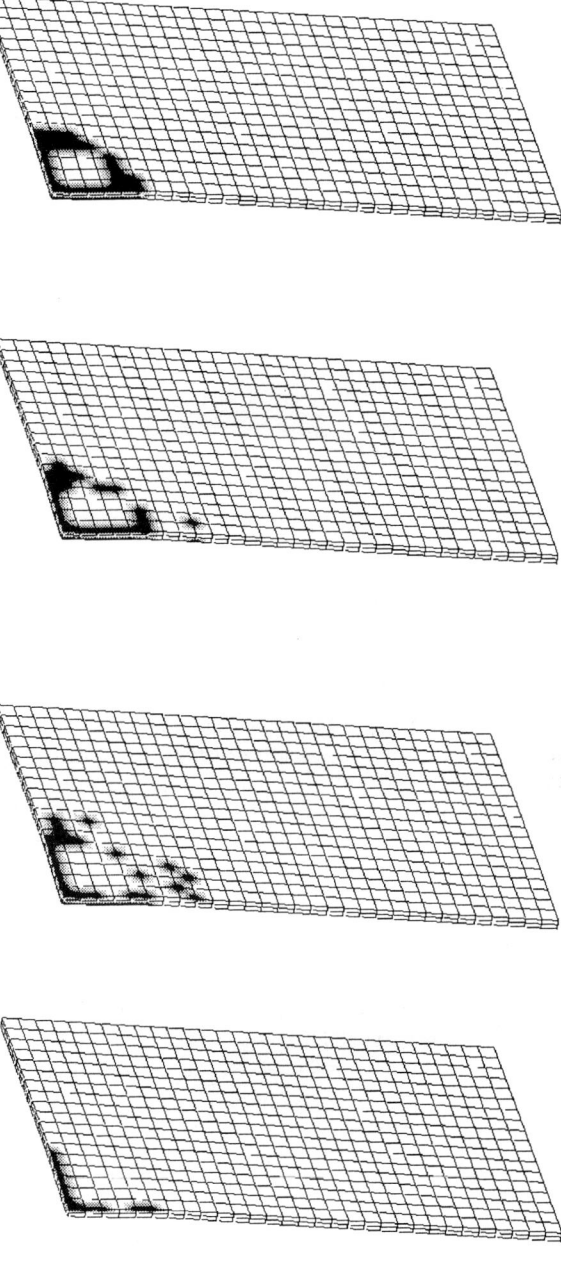

Figure 8: Delamination patterns at layer interfaces at T=0.00006 sec.

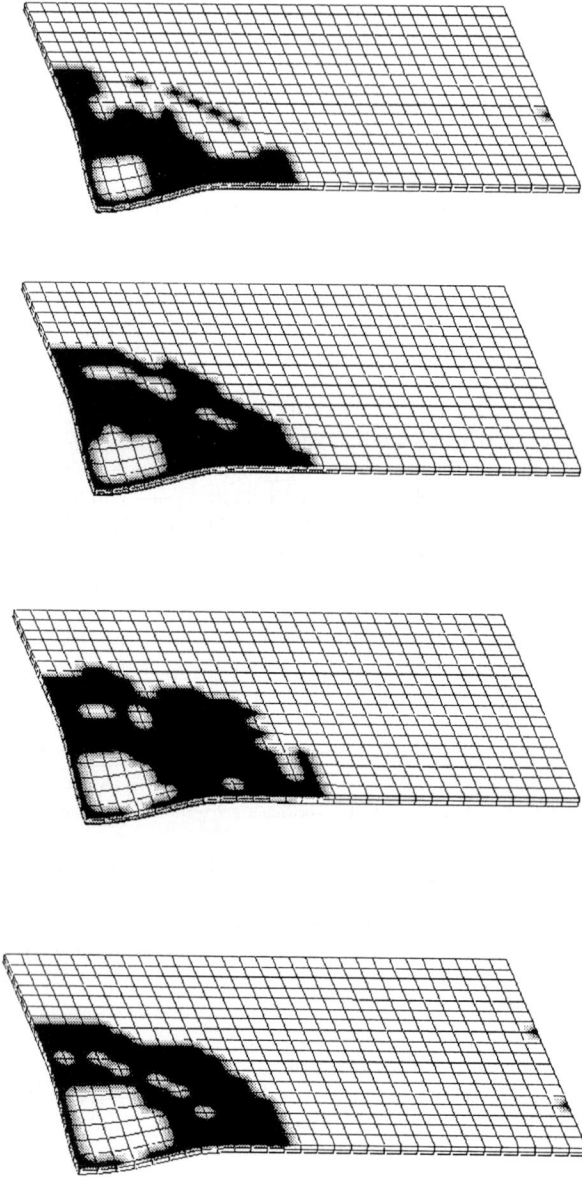

Figure 9: Delamination patterns at layer interfaces at T=0.00012 sec.

Engineering **31,**1-17.

Chang .F.K and Springer.G.S. (1986) .The strength of fiber reinforced composite bends. *Composite Materials* **20:1,** 30-45 .

Hallquist. J.O , Goudreau.G.L, and Benson.D.J .(1985). Sliding interfaces with contact-impact in large scale lagrangian computations .Computer Methods in Applied Mechanics and Engineering,**51.**

Lagace.P.A and Wolf.E. (1995) . Impact damage resistance of several laminated material systems. *AIAA Journal,* **33:6,** 1106-1113.

Liu.S, Kutlu.Z, and Chang.F.K.(1993). Matrix cracking-induced delamination propagation in graphite/epoxy laminated composites due to a transverse concentrated load . In Stichcomb.W.W and Ashbange.N.E, editors , Composite Materials : Fatique and Fracture, ASTM STP 1159, volume 4, pages 86-101 .

Mohammadi. S, Owen.D.R.J , and Peric.D .(1997). Discontinuum approach for damage analysis of composites . In Crisfield.M . A , editor, Computational Mechanics in UK- *5 th ACME Conference* , pages 40-44 ,London,UK.

Mohammadi. S, Owen.D.R.J , and Peric.D.(1998). A combined finite / discrete element algoritm for delamination analysis of composites. *Finite Elements in Analysis and Design*,**28,**321,336.

Mohammadi . S , Owen.D.R.J , and Peric.D.(1997) Delamination analysis of composites by discrete element method. In Owen.D.R.J, Onate.E , and Hinton .E, editors , Computational Plasticity , COMPLAS V, pages 1206-1213, March 1997. Barcelano ,Spain.

Mohammadi . S , Owen.D.R.J and Peric.D .(1997).3D fracture analysis of composites by DEM. In M.S. Shephard.M.S, editor , *Fourth US National Congeress on Compputational Mechanics,*page 539. USACM, SanFrancisco,USA.

Munjiza. A , Owen .D.R.J, and Bicanic.N. (1995). A combined finite -discrete element method in transient dynamics of fracturing solids . *Engineering Computations,***12,**145-174.

Owen. D.R.J , Peric.D , and Mohammadi . S . (1998) . Discrete element modeling of multi – fracturing solids and structures. In Foreign Object Impact and Energy Absorbing Structure, pages 50-67. Institution of Mechanical Engineers (IMechE) , Aerospace Industries Division.London.

Petrinic .N.(1996) Aspects of Discrete Element Modeling Involving Facet-to-Facet contact Detection and Interaction . *PhD thesis*, Department of Civil Engineering,Univercity of Weles Swansea,UK.

Rowlands.R.E.(1985). Strength (failure) theories and their experimental correlation .In Sih .G.C and Skudra.A.M , editors, Handbook of Composites,Vol. 3,*Failure Mechanics of Composites* , chapter 2,pages 71-125 .Elsevier Science Publishers B.V.

Schellekens .J.C.J and de Borest.R.(1990). The use of hoffman yield criterion in finite element analysis of anisotropic composites. *Computers and Structures* **37:6,**1087-1096.

Schonauer.M, Rodic.T, and Owen.D.R.J . (1993). Numerical modelling of thermomechanical prosseses related to bulk forming operations . *Journal De Physics IV* **3,**1199-1209.

Worswick .M.J, Strazincky.P.V, and Majeed .O.(1995).Dynamic fracture of fiber reinforced composite coupons. In Sun.C.T, Sankar.B.V, and Rajapakse .Y.D.S, editors, Dynamic Response and Behavior of Composites , *ASME,* AD-**46**, 29-41 .

Modelling of Cable Vibration Effects of Cable-stayed Bridges

S.H.Cheng   D.T.Lau

Ottawa-Carleton Bridge Research Institute, Department of Civil and Environmental Engineering,
Carleton University, Ottawa, Ontario, Cananda, K1S 5B6

ABSTRACT: The dynamic responses of a cable-stayed bridge may be significantly affected by the vibration of the stay cables. In previous studies, the interaction effects between the vibration motions of the bridge deck and the pylons and the transverse vibration of the stay cables are often ignored. In the present study, a three-node cable model has been developed to represent the transverse motion of the cable. The interactions between the vibrating stay cables and other structural components of the bridge superstructure are considered by the application of the dynamic cable stiffness. Numerical examples are presented to demonstrate the validity and efficiency of the proposed model. The impact of cable vibration behaviour on the dynamic characteristics of the cable-stayed bridges is discussed.

KEYWORDS: cable-stayed bridge, cable vibration, finite element models, dynamic cable stiffness, long-span structure, numerical, modal analysis

INTRODUCTION

The dynamic responses of a cable-stayed bridge under typical dynamic load conditions involve motions of the deck and the pylons, as well as that of the stay cables. Thus, the dynamic behaviour of this type of long span, three-dimensional flexible structures are complex. In most of the previous static and dynamic numerical studies of cable-stayed bridges (Fleming & Egeseli,1980; Parvez & Wieland, 1987; Wethyavivorn, 1987;Abdel-Ghaffar & Nazmy, 1991; Nazmy & Abdel-Ghaffar, 1992), each individual stay cable is usually modelled by a truss element. Although the single-truss element model can give accurate results in static analysis of cable-stayed bridges, it cannot account the lateral vibration motions of the cables on the dynamic responses of the bridge. Recent observations have shown that natural frequencies of the high strength steel cables in typical cable-stayed bridges may lie in the frequency range of the first several vibration modes of the bridge structure. Thus, under certain dynamic load conditions, the vibration of the stay cables and its interaction with the other parts of the bridge superstructure can be significant. Recently, some researches have been carried out to improve the modelling of the interaction effect of the stay cables. Causevic and Screckovic(1987) have suggested using linear springs and lumped masses to represent the stay cables in the analysis model. In other studies (Baron & Lien, 1973; Yiu, 1982; Tuladhar & Brotton, 1987; Abdel-Ghaffar & Khalifa, 1991; Tuladhar et al, 1995), a multi-link model, which discretizes each stay cable into several elements, has been adopted by several researchers to analyze the dynamic behaviour of the cable-stayed bridges due to wind and seismic loads. However, the multi-link modelling scheme significantly increases the number of degree-of-freedom of the analytical model, especially when analyzing modern multi-stay cable-stayed bridges. Although comparisons have been made between the results obtained from the single-truss model and the multi-link model in some of the recent research (Abdel-Ghaffar & Khalifa, 1991; Tuladhar et al, 1995), the impact of cable vibration effect on the dynamic responses of the cable-stayed bridges is still not clear.

In this paper, a new and more convenient cable model of using 3-node cable element is developed. It can accurately model the vibrational characterisitcs of the stay cables more efficiently with less number of degree-of-freedom than the multi-link model. The interaction behaviour between the cables, the pylons and the bridge deck are considered by adopting the dynamic stiffness of the cable. The accuracy and reliability of this new cable model are verified by numerical examples.

DISPLACEMENT FUNCTIONS

Transverse motion of the stay cable

Considering small amplitude transverse vibration of a horizontal string fixed at both ends, the transverse vibration frequency of the string can be determined as (Cheng, 1999) $f_n = \frac{n}{2l}\sqrt{S/\mu}(n = 1, 2, 3, \cdots)$, where S is the tension in the string, μ is mass per unit length of the string, and l is the length of the string. The corresponding mode shape is $Y_n(x) = \frac{\omega l}{n\pi}\sin\frac{n\pi x}{l}$, where ω is the circular frequency.

The transverse displacements at an arbitrary point on a stay cable as shown in Figure 1 can be decomposed into two parts. The first part is the linear displacements related to the nodal displacements at the two ends of the cable connected to the girder and the pylon. The second part can be treated as the displacements contributed from the transverse vibration of the cable fixed at both ends, which can be expressed as the summation of the Sine series following the mode superposition method. Thus, the total transverse displacements of the cable can be expressed as follows

$$\bar{v}(\bar{x}) = \frac{\bar{x}}{L}\bar{v}_i + \left(1 - \frac{\bar{x}}{L}\right)\bar{v}_j + \sum_n a_{\bar{y}_n}\sin\frac{n\pi\bar{x}}{L} \qquad (1)$$

$$\bar{w}(\bar{x}) = \frac{\bar{x}}{L}\bar{w}_i + \left(1 - \frac{\bar{x}}{L}\right)\bar{w}_j + \sum_n a_{\bar{z}_n}\sin\frac{n\pi\bar{x}}{L} \qquad (2)$$

where $\bar{v}$, $\bar{w}$ are the displacements along the $\bar{Y}$, $\bar{Z}$ axes in the local coordinate system, L is the length of the cable, $\bar{v}_i$, $\bar{w}_i$, $\bar{v}_j$, and $\bar{w}_j$ are the nodal displacements at the node i and node j along the $\bar{Y}$, $\bar{Z}$ axes in the local coordinate system, and $a_{\bar{y}_n}$, $a_{\bar{z}_n}$ are the generalized coordinates corresponding to the nth sine shape function term. The number of harmonic terms to be taken in the summation of Eqns.1 and 2 are determined by the accuracy requirement of the analysis.

Longitudinal motion of the stay cable

The extensional behaviour of the stay cable is another important component of its dynamic motion. In practice, the nonlinear static stretching stiffness of the stay cable can be expressed as $K_c = \frac{K_e K_g}{K_e + K_g}$, where $K_e = \frac{EA}{L}$ is the elastic stiffness, $K_g = \frac{12T^3}{W^2 L_h^3}$ is the geometric stiffness, E is the elastic Young's modulus, A is the cross-sectional area of the cable, L is the chord length of the cablel, L_h is the length of the horizontal projection of the cable, T is the mean tension in the cable, and W is the weight per unit length.

When the bridge is subjected to dynamic loads, the stay cables resist the movements of the deck and pylon at the two end supports. The longitudinal and transverse motions in the plane of the stay cable are dynamically coupled. In the model developed here, the geometric nonlinearity of the stay cable due to the coupling between the motions in the transverse and the longitudinal directions of the cable in the dynamic responses is considered by the dynamic cable stiffness. The dynamic extensional cable stiffness at the frequency ω of the structure has been derived by Davenport(1994), which has the form as follows

$$K(\omega) = \frac{K_e\left[1-\left(\frac{\omega}{\omega_0}\right)^2\right]}{1+\left(\frac{K_e}{K_g}\right)-\left(\frac{\omega}{\omega_0}\right)^2} \qquad (3)$$

where $\omega_0 = \frac{\pi}{L}\sqrt{\frac{Tg}{W}}$ is the fundamental frequency of the "taut wire". The dynamic cable stiffness concept has been applied in the analysis of guy cable structures.

Studies on the behaviour of massive guy cables(Davenport & Steels, 1965) have shown that the dynamic stiffness effect has a significant impact on the dynamic behaviour of the guy cable structures. On the other hand, the evaluation of the influence of the cable vibration on the dynamic behaviour of cable-stayed bridges by dynamic cable stiffness has not been reported. It is of practical importance to determine how significant are the differences in the bridge responses as evaluated from of the dynamic stiffness and the static stiffness, and the possibility when the effective stiffness of the cable becomes negative. In order to more accurately account for the cable behaviour in the modelling of the dynamic responses of the cable-stayed bridges, the formulation of the dynamic cable stiffness is included in the cable model developed in the current study. The impact of the dynamic stiffness is investigated in details in the case study of the Nanpu Bridge presented in a companion paper(Cheng & Lau, 2001).

A spring-mass-spring system as shown in Figure 2 is adopted to derive of the dynamic stiffness of the stay cable. The cable mass which participates in the longitudinal translational motion is assumed to be lumped at the center point m of the cable model. Appropriate constraints are applied to the motion of the lumped mass when considering the longitudinal motion of the stay cable. The stiffnesses of the springs at the two sides of the mass are the geometric stiffness K_g and the elastic stiffness K_e, respectively. The in-plane displacement $\bar{u}(\bar{x})$ of an arbitrary point $\bar{x}$ on the cable relative to the displacement at the cable support point of the pylon can be expressed as follows

$$\bar{u}(\bar{x}) = \begin{cases} \frac{2\bar{x}}{L}(\bar{u}_m - \bar{u}_j), & 0 \leq \bar{x} \leq \frac{L}{2} \\ \frac{2(L-\bar{x})}{L}(\bar{u}_m - \bar{u}_j) + \frac{2\bar{x}-L}{L}(\bar{u}_i - \bar{u}_j), & \frac{L}{2} < \bar{x} \leq L \end{cases} \qquad (4)$$

where $\bar{u}_i$, $\bar{u}_m$ and $\bar{u}_j$ are respectively the longitudinal displacements at points i, m and j for the 3-node cable model. The stiffness matrix and mass matrix of the 3-node cable model can be derived following the standard finite element procedure (Cheng, 1999).

NUMERICAL EXAMPLE

A numerical example of a three-span cable-stayed bridge is presented to demonstrate the validity and reliability of the proposed 3-node cable model. This example bridge has been used by Tuladhar et al(1995) to study the effect of cable vibration on the modal analysis results and seismic response of the cable-stayed bridge. The geometry and details of the finite element mesh of the example bridge are shown in Figure 3. In the present research, the emphasis is placed on the comparison of the different modelling schemes of the transverse motion of the cable, and the influence of the cable vibration effect on the modal analysis results.

Three different analytical models are employed in the comparison study. The bridge deck and pylons in all three models are modelled by 3D beam elements. The stay cables are respectively modelled by the proposed 3-node cable elements, the multi-link elements and the single-truss elements in separate analytical models. The tension forces in the cables are obtained from the dead load condition. The motion of the bridge is assumed to be restricted in the plane of the structure, i.e. only three degrees-of-freedoms (two translations and one rotation) are considered at each node of the numerical model of the bridge.

Comparison between 3 − node cable model and single − truss cable model

Results obtained from the proposed 3-node cable model and the single-truss cable model are found to be close with regard to the basic modes of the bridge itself. Numerous pure cable vibration modes are identified by using the 3-node model. These local cable vibration modes show up as additional vibration modes independent of the structural modes with frequencies between those of the natural modes of the bridge structure itself. The mode shapes obtained from the 3-node cable model include the motions not only of the deck and the pylons, but also of the stay cables as well. In the case of the single-truss cable model, the cables in the mode shapes remain straight as the vibrational behaviour of the cables are ignored. The modal frequencies obtained from the proposed 3-node cable model and the single-truss cable model are plotted against the modal order in Figure 4. The flat regions on the curve of the 3-node cable model result indicate the pure cable modes. They show up between the basic modes of the bridge itself.

Although the two ends of the stay cable in cable-stayed bridge may move with the deck and the pylon(s) under dynamic loads, results show that expression presented earlier in the paper to calculate the natural frequencies and the associated mode shapes of the transverse motion of a string fixed at both ends can still be used with reasonable accuracy to give the estimates of the transverse motion frequencies of the stay cables. The natural frequencies of the stay cables in the example bridge, obtained from both the analytical expressions and the numerical modal analysis results using the 3-node cable elements, are presented in Table 1. Good agreement is observed in the fundamental natural frequencies of all the cables, which demonstrates the validity and accuracy of the proposed model.

Comparison between 3 − node cable model and multi − link cable model

To further demonstrate the efficiency of the proposed model, the example bridge is also analyzed using multi-link elements to represent the cables. Each cable is divided respectively into two, four and eight links in three separate analyses. The fundamental cable frequencies obtained from the multi-link models are also presented in Table 1. Results show that for the example here, each cable should be divided into at least eight links to achieve accurate results. The first twenty-two natural frequencies obtained from the 3-node model and the eight-link model, including the bridge frequencies and all the cable fundamental frequencies, are presented in Table 2. The agreement between these two sets of results is excellent. The total number of degrees-of-freedom of the 3-node cable model is 171, whereas that of the eight-link model is 387. For modern long-span multi-stay cable-stayed bridges, there can be more than a hundred stay cables in a bridge. If the multi-link model is employed to model the stay cables, the number of degree-of-freedom of the analytical model is therefore significantly higher than that of the proposed cable model. It not only requires more effort in data preparation, but also increases significantly the costs in computer time and storage.

CONCLUSIONS

The effect of cable vibration behaviour on the dynamic behaviour of the cable-stayed bridges is studied in this paper. A 3-node cable model, which can represent accurately not only the transverse motion of the cable, but the interaction between the motion of the cables and the other parts of the bridge superstructure by introducing the dynamic stiffness in the longitudinal direction, has been developed. It is computationally more efficient than the multi-link approach, especially in the analysis of the modern multi-stay cable-stayed bridges. The validity and accuracy of the proposed model are verified by the numerical example. Results show that considering the cable vibration behaviour, the natural frequencies, the mode shapes and the order of the vibration modes of the cable-stayed bridges are affected. A large number of pure cable vibration modes as well as new cable-deck coupled modes are identified by using the proposed 3-node cable model. These modes cannot be predicted if the cable is modelled by the single-truss element. The natural frequencies of the conventional high strength steel cables in the modern cable-stayed bridges fall in the frequency range of the first several basic bridge modes. Thus, the transverse motions of the cables will influence the dynamic response of the bridge and should be considered in the dynamic response analysis.

References

1. Abdel-Ghaffar, A.M. and Khalifa, M.A.(1991) Importance of Cable Vibrations in Dynamics of Cable-stayed Bridges. *Journal of Engineering Mechanics, ASCE.* **Vol. 117, No.11:** 2571-2589.
2. Abdel-Ghaffar, A.M. and Nazmy, A.S.(1991). 3-D Non-linear Seismic Behaviour of cable-stayed bridges. *Journal of Engineering Mechanics, ASCE* **Vol. 117, No.11:** 3456-3476.
3. Baron, F., and Lien, S.(1973) Analytical Studies of a Cable-stayed Bridge. *Computers and Structures* **Vol. 3:** 443-465.
4. Causevic, M.S., and Sreckovic, G.(1987) Modelling of Cable-stayed Bridge Cables: Effects on Bridge Vibrations. *Proceedings of International Conference on Cable-stayed Bridges, Bangkok, Thailand*, 407-420.
5. Cheng, S.H. (1999) Structural and Aerodynamic Stability Analysis of Long-span Cable-stayed Bridges. *Ph.D. thesis, Department of Civil and Environment Engineering, Carleton University, Ottawa, Canada.*
6. Cheng, S.H., and Lau, D.T.(2001) Parametric Study of Cable Vibration Effects on the Dynamic Response of Cable-stayed Bridges. *Proceedings of International Conference on Structural Engineering, Mechanics and Computation, Cape Town, South Africa, 2-4 April*, Paper Ref: SEMC2001/211.
7. Davenport, A.G.(1994) A Simple Representation of the Dynamics of a Massive Stay Cable in Wind. *Proceeding of International Conference on Cable-Stayed and Suspension Bridges, Deauville. October* **Vol.2:** 427-438.
8. Davenport, A.G., and Steels, G.N.(1965) Dynamic Behaviour of Massive Guy Cables. *Journal of Structure Division, ASCE* **Vol.91.ST2:** 43.
9. Fleming, J.F., and Egeseli, E.A.(1980). Dynamic behaviour of a cable-stayed bridge. *Earthquake Engineering and Structural Dynamics* **8(1):** 1-16.
10. Nazmy, A.S., and Abdel-Ghaffar, A.M.(1992) Effects of ground motion spatial variability on the response of cable-stayed bridges. *Earthquake Engineering and Structural Dynamics* **Vol. 21:** 1-20.
11. Parvez, S.M., and Wieland, M.(1987). Earthquake behaviour of proposed multi-span cable-stayed bridge over river Jamuna in Bangladesh. *Proceedings of International Conference on Cable-stayed Bridges, Bangkok, Thailand*, 479-489.
12. Tuladhar R., and Brotton, D.M.(1987) A Computer Program for Nonlinear Dynamic Analysis of Cable-stayed Bridges under Seismic Loading. *Proceedings of International Conference on Cable-stayed Bridges, Bangkok, Thailand*, 315-326.
13. Tuladhar R., Dilger, W.H. and Elbadry, M.M.(1995) Influence of Cable Vibration on Seismic Response of Cable-stayed Bridges. *Canadian Journal of Civil Engineering* **Vol. 22:** 1001-1020.
14. Wethyavivorn, B.(1987) Dynamic behaviour of Cable-stayed Bridges. *Ph.D. thesis, Department of Civil Engineering, The University of Pittsburgh, Pittsburgh, Pa.*
15. Yiu, P.K.A.(1982) Static and Dynamic Behaviour of Cable Assisted Bridges. *PhD. thesis, Department of Civil Engineering, the University of Manchester, Manchester, United Kingdom.*

Table 1: COMPARISON OF THE FUNDAMENTAL FREQUENCIES OF THE STAY CABLES IN THE EXAMPLE BRIDGE

Cable No.	Analytical (Hz)	Three-Node Cable Element (Hz)	Multi-link Element (Hz)		
			2 links	4 links	8 links
1	0.4838	0.4844	0.4348-0.4364	0.4633-0.4810	0.4813-0.4876
2	0.7601	0.7527-0.7588	0.6827-0.6860	0.7364	0.7496
3	1.1332	1.1342	1.0208	1.1013-1.1083	1.1277
4	1.1913	1.1913	1.0653-1.0714	1.1623	1.1842
5	0.8218	0.8206	0.7391-0.7415	0.8013	0.8163
6	0.5117	0.5091-0.5147	0.4589	0.4975-0.5029	0.5067-0.5125
7	0.4853	0.4888	0.4348-0.4364	0.4633-0.4810	0.4813-0.4876
8	0.7622	0.7527-0.7588	0.6827-0.6860	0.7408	0.7546
9	1.1402	1.1410	1.0208-1.0267	1.1013-1.1176	1.1347
10	1.1849	1.1847	1.0653-1.0714	1.1554	1.1775
11	0.8212	0.8131-0.8289	0.7391-0.7415	0.7991	0.8139
12	0.5122	0.5091-0.5147	0.4623	0.4975-0.5029	0.5067-0.5125

Table 2: COMPARISON OF 3-NODE CABLE MODEL AND MULTI-LINK CABLE MODEL MODAL ANALYSIS RESULTS

Mode No.	Natural Frequency (Hz)		Mode Shape
	Proposed Three-Node Cable Element	Multi-link Element (8 links per cable)	
1	0.3481	0.3465	Heave(girder, Sym.)
2	0.4615	0.4584	Heave(girder, Anti-Sym.)
3-6	0.4844-0.5148	0.4813-0.5125	Pure cable motion
7	0.5892	0.5895	Heave(girder, Anti-Sym.)
8-10	0.7526-0.8131	0.7496-0.8139	Pure cable motion
11	0.8206	0.8259	Heave(girder, Sym.)
12	0.8289	0.8306	Pure cable motion
13	0.9240	0.9204	Heave(girder, Anti-Sym.)
14-17	0.9677-1.0250	0.9530-0.9997	Pure cable motion
18	1.1017	1.1110	Heave(girder, Sym.)
19-22	1.1342-1.1913	1.1277-1.1842	Pure cable motion

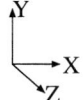
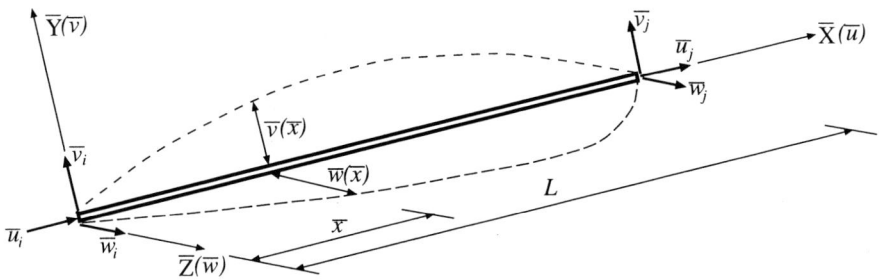

Figure 1: Transverse vibration of the stay cable

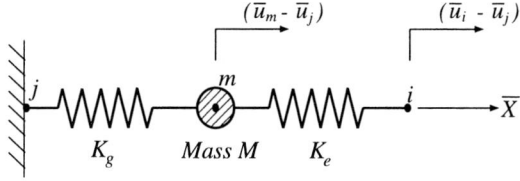
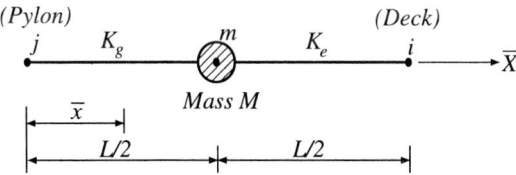

Figure 2: Spring-mass-spring system representing the dynamic behaviour of the stay cable in the longitudinal direction

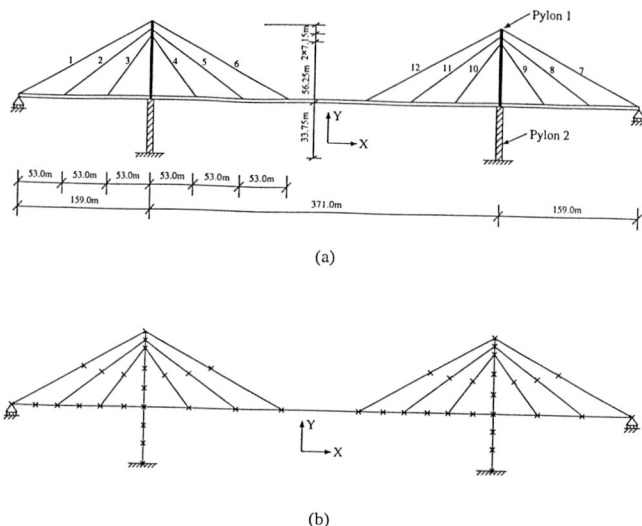

Figure 3: Geometry and mesh details of the example bridge (a) Geometry of the example bridge (b) Mesh details of the example bridge

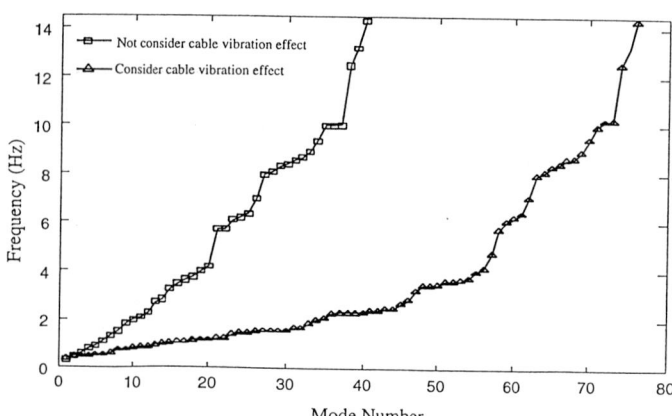

Figure 4: Impact of cable vibration effect on the dynamic characteristics of the example bridge

Parametric Study of Cable Vibration Effects on the Dynamic Response of Cable-stayed Bridges

S.H.Cheng   D.T.Lau

Ottawa-Carleton Bridge Research Institute, Department of Civil and Environmental Engineering, Carleton University, Ottawa, Ontario, Cananda, K1S 5B6

ABSTRACT: In this paper, a 3-node cable model is employed to investigate the cable vibration effects on the dynamic responses of cable-stayed bridges. The transverse vibration of the stay cables in cable-stayed bridges under dynamic loads, and the interaction between the motions of the cables and the other parts of the bridge superstructure are considered. A comprehensive parametric study of eight example bridges is carried out. The parameters considered in the study include the span length, the pylon shape and the cable arrangement. The impact of considering the dynamic cable stiffness in the dynamic analysis of cable-stayed bridges is also discussed.

KEYWORDS: cable-stayed bridge, cable vibration, finite element models, dynamic cable stiffness, long-span structure, numerical, modal analysis

INTRODUCTION

The common practice of modelling the stay cables in cable-stayed bridges by single-truss elements (Fleming & Egeseli,1980; Parvez & Wieland, 1987; Wethyavivorn, 1987;Abdel-Ghaffar & Nazmy, 1991; Nazmy & Abdel-Ghaffar, 1992) cannot represent the lateral motions of the cables and the interaction effects with other parts of the bridge superstructure. The multi-link approach proposed in some recent studies (Baron & Lien, 1973; Yiu, 1982; Tuladhar & Brotton, 1987; Abdel-Ghaffar & Khalifa, 1991; Tuladhar et al, 1995) improves the accuracy in the modelling of the cable behaviour. However, by dividing each cable into multiple elements, the size of the analytical model is increased significantly. A 3-node cable model has been proposed in a recent study by Cheng and Lau (2001). The new cable model can simulate the transverse vibration of the cable and cable-bridge interaction effects. It is more computationally efficient than the multi-link model. A comprehensive parametric study using the 3-node cable model is carried out in the present study. Eight example bridges are analyzed to investigate the impact of considering cable vibration and the use of the dynamic cable stiffness on the dynamic responses of the cable-stayed bridges. Other parameters considered in the investigation include the span length, the pylon shape and the cable arrangement.

CASE STUDY OF NANPU BRIDGE

Nanpu Bridge is a long-span cable-stayed bridge located in Shanghai, China. It has a center span of $423m$, as shown in Figure 1. The bridge girder has a composite deck cross-section comprised of a prestressed concrete slab supported by two I-shaped steel beams. Two different numerical models are employed to study the cable vibration effect on the dynamic characteristics of the Nanpu Bridge. In these two models, the main girder of the bridge and the two H-shaped pylon are modelled by the 3D beam finite elements (Cheng, 1999). The difference between these two models is in the way the stay cables are modelled. In the first model, the stay cables are modelled by single-truss elements, whereas in the second model, the cables are modelled by the proposed 3-node cable elements.

Modal analysis

The modal analysis results of the Nanpu Bridge are presented in Table 1. The results show that considering the cable vibration effect in the dynamic model of the Nanpu Bridge, additional pure cable vibration modes and new cable-deck coupled vibration modes are identified in the modal analysis. These additional modes are found in the frequency spectrum between the previously determined modes of the bridge without considering the cable vibration effect. The new cable-deck coupled modes involve the transverse motions of the stay cables in the cable plane and the torsional vibration of the main girder in the center span, with the cable motions as the dominant component. These coupled vibration behaviour as well as the pure cable vibration modes cannot be identified by the single-truss model. Comparing the results obtained from the two models, it is noted that the natural frequencies of the corresponding bridge modes are close to each other, with those obtained from the 3-node cable model slightly lower. This is because in the 3-node cable model, the oscillations of the cables increase the vibrating mass of the bridge system. With approximately the same stiffness but a greater mass, the natural frequencies of the vibration modes obtained from the 3-node cable model are therefore lower. In the same frequency range up to $0.8Hz$, there are 14 modes extracted from the single-truss cable model, whereas more than 200 modes are identified by the 3-node cable model. The additional cable vibration modes may have a significant impact on the dynamic performance and long-term safety of the bridge system. The mode shapes of some typical bridge vibration modes obtained by using the 3-node cable model are shown in Figure 2. The interaction of the cable vibration behaviour with the motion of the main girder and the pylons is observed.

The fundamental frequency of the longest cable in the Nanpu Bridge obtained from the modal analysis is $0.476Hz$, whereas that of the shortest is $1.579Hz$. The frequencies of the cable vibration modes fall in the frequency range between modes 5 to 45 of the bridge structure, i.e. between $0.426Hz$ and $1.607Hz$. For dynamic motions of the bridge in this frequency range, the vibration of the cables and their interaction with the motions of the deck and the pylons can be significant, which can severely affect the response characteristics of the bridge.

Dynamic cable stiffness

In the proposed 3-node cable model (Cheng, 1999; Cheng & Lau, 2001), the interaction effect between the cable vibration and that of the other parts of the bridge superstructure is represented by the dynamic cable stiffness $K(\omega) = \dfrac{K_e\left[1-\left(\frac{\omega}{\omega_0}\right)^2\right]}{1+\left(\frac{K_e}{K_g}\right)-\left(\frac{\omega}{\omega_0}\right)^2}$, where K_e is the elastic stiffness, K_g is the geometric stiffness, ω is the circular frequency of the structure, $\omega_0 = \frac{\pi}{L}\sqrt{\frac{Tg}{W}}$ is the fundamental frequency of the "taut wire". By modifying the longitudinal stiffness, the extensional behaviour of the cable is influenced.

In recent years, a number of cases of cable replacements have been reported because of the corrosion and fatigue problems in the cables. These two phenomena occur simultaneously and reinforce each other. The tension in the cable can drop significantly due to the damage to the wires inside the stay cable. To study the effect of the tension loss on the dynamic behaviour of the stay cables, the dynamic stiffness of the longest and the shortest cables of the Nanpu Bridge are investigated by modifying the tension in the cable taken as $0.75T_{DL}$, $0.50T_{DL}$ and $0.25T_{DL}$ respectively, where T_{DL} is the initial cable tension caused by the dead load. From the results presented in Table 2, it can be noticed that the value of the dynamic cable stiffness depends more on the tension in the cable than on the external exciting frequency of the bridge motion. A negative value of the extensional stiffness of the longest cable occurs under the condition when the tension force is reduced to $0.25T_{DL}$ and the oscillating frequency of the bridge is $1.0Hz$. Physically, this negative stiffness indicates that instead of resisting the enlongation of the cable, the deformation tend to increase. The cable then is under an unstable condition. If the cable tension is reduced to $0.50T_{DL}$, when the bridge vibrates with a frequency of $0.2Hz$, the extensional stiffness of the longest cable considering the dynamic interaction effect decreases by 15.18% to $0.837 \times 10^7 N/m$. The effect of the dynamic interaction between the cables and the other parts of the superstructure on the extensional stiffness of the shortest cable is much less than that on the longest cable.

Assuming there is no tension loss due to corrosion or fatigue problems of the cable, the dynamic cable stiffness of the longest and the shortest cables in the Nanpu Bridge are determined for the bridge natural frequencies at $0.1Hz$, $0.2Hz$, $0.5Hz$ and $1.0Hz$. No significant changes in the extensional stiffness of the cables are found. If the dynamic interaction between the cable and the other parts of the superstructure is neglected, the extensional cable stiffness obtained is determined to be $0.9862 \times 10^7 N/m$. Considering the dynamic stiffness for the bridge structural frequency of $0.5Hz$, the extensional cable stiffness is changed to $0.9861 \times 10^7 N/m$ with only a difference of 0.01%. For the case of the extensional stiffness $K(\omega)$ of the cable approaches the critical state of zero, the corresponding bridge frequency for the longest cable is found to be $7.76Hz$, and for the shortest cable $105.6Hz$. These frequencies are obviously beyond the bridge frequency range that are typical of cable-stayed bridges excited by earthequakes, wind and traffic. For the case of the shortest cable, the dynamic stiffness effect is obviously negligible. Thus, if there is no tension loss due to corrosion or fatigue problem, the high pre-tension and relatively low mass of the cables reduce the significance of impact of the dynamic stiffness on the extensional behaviour of the stay cables in cable-stayed bridges. The effect observed here is not as significant as that reported by Davenport and Steels(1965) on the behaviour of the guy cables, which typically have a lower initial tension and a greater massthan the stay cables in cable-stayed bridges. However, in reality the stay cables are usually subjected to deterioration due to the corrosion and fatigue problems, the tension in the cable may be reduced significantly. The dynamic stiffness effect may become more significant and thus should then be considered in the analysis, especially for the long stay cables in bridges.

PARAMETRIC STUDY

To better understand the cable vibration behaviour on the dynamic responses of modern cable-stayed bridges, parametric studies are carried out to investigate its impact on bridges of different structural layouts in terms of span length, pylon shape and cable arrangement type. Four different center span lengths are considered, which are $300m$, $423m$ (span length of the Nanpu Bridge), $500m$ and $600m$. The range covers the typical span length of modern medium- to long-span cable-stayed bridges. Two cases of typical cable arrangement in modern cable-stayed bridges are considered, i.e. the parallel cable-plane arrangement and the inclined cable-plane arrangement. Correspondingly, the associated pylon shapes are taken as the H shape as in the Nanpu Bridge, and the inverted-Y shape as shown in Figure 3. The combinations of these structural parameters yield eight different example bridges.

Pylon shape

For the example bridge considered here, the span length, the cross-section of the main girder, the cross-sectional area of the stay cables and the boundary conditions are taken the same as those of the Nanpu Bridge. In the parametric studies, the shape of the pylons is changed to an inverted-Y shape, as shown in Figure 4. Consequently, the parallel cable-plane arrangement of the Nanpu Bridge is changed to the inclined cable-plane arrangement in some of the parametric cases.

Two sets of modal analysis results, of which the stay cables are modeled respectively by the single-truss element and the 3-node element, are compared. As discussed befored, numerous pure cable vibration modes are extracted by using the 3-node cable model. Corresponding to the first ten bridge structural vibration modes obtained from the single-truss cable model, 144 new pure cable vibration modes are identified by using the 3-node cable model. No cable-deck coupled mode is identified in this case. The torsional frequencies are increased when considering the influence of the cable vibration behaviour, whereas for the other bridge modes, the modal frequencies are slightly decreased. This is different from the case of the Nanpu Bridge having H-shaped pylons, of which the bridge natural frequencies, including the torsional modes, all decrease due to the influence of the transverse motion of the cables.

Comparing the modal analysis results with those obtained from the Nanpu Bridge, it can be observed that by changing the pylon shape from H to the inverted-Y, the frequency of the floating mode is slightly decreased, whereas those for the heave modes are slightly increased. A 3.94% increase in the frequency of the sway mode is also observed. For the torsional modes, the impact is more significant. The frequency of the lowest symmetric torsional mode is increased by 32% from $0.495Hz$ to $0.654Hz$, whereas that of the lowest anti-symmetric torsional mode is increased by 20.2% from $0.607Hz$ to $0.730Hz$. This is because that the inverted-Y shape pylon has significantly higher torsional resistance than the H-shaped one, thus can suppress more effectively the torsional motion of the girder.

Variation of span length

The impact of varying the span length on the natural frequencies of cable-stayed bridges with parallel cable-plane and the inclined cable-plane are shown in Figures 4 and 5, respectively. The frequencies of six typical low order modes are plotted against span length. The change of the frequencies of the fundamental floating mode (floating), the lowest symmetric sway mode (1-S-L), the lowest symmetric heave (1-S-V) and anti-symmetric heave (1-A-V) modes, and the lowest symmetric torsional (1-S-T) and anti-symmetric torsional (1-A-T) modes are presented. Two curves are drawn for each mode. The solid lines represent the results of neglecting the cable vibration behaviour and the dotted lines represent the results of considering the cable vibration behaviour.

In all the cases analyzed, the frequencies drop more rapidly when the span length changes from a medium span of $300m$ to a long span of $423m$. Changing from the span length of $423m$ to $600m$, the changes in frequencies are more gentle. It is observed that the variation in the span length has a significant impact on the sway behaviour of the bridge girder. When the center span length is doubled from $300m$ to $600m$, the sway frequency is reduced by 75.5%. The reduction percentage of the sway mode in the case of the inclined cable-plane arrangement is found to be almost the same. As shown in these two figures, the natural frequencies of the floating, the sway and the heave modes are hardly affected by the cable vibration behaviour, whereas the frequencies of the torsional modes are slightly changed due to the transverse motion of the cables. No obvious relationship is identified between the cable vibration effect on changes of the modal frequencies and the variation of the span length.

Cable arrangement type

Comparing the results shown in Figures 4 and 5, the frequencies of the floating, the sway and the heave modes are not significantly affected by the change in the cable arrangement type. However, the frequencies of the torsional modes are significantly increased when the type of cable arrangement is changed from the parallel to the inclined. For the lowest symmetric torsional mode, a 32.8% increase is observed for the case of span length of $300m$, and a 37.58% increase is observed for the case of span length $600m$. These increases in torsional frequencyare due to the fact that the inclined cable-plane arrangement type is more efficient in resisting the torsional response motion in the bridge system.

The cable vibration behaviour hardly affects the frequencies of the floating, the sway and the heave modes of both types of cable arrangement. By taking into account the transverse cable motions, the obtained frequencies of these modes are slightly lower. For the torsional modes, the same trend of decrease in the frequency values due to the cable vibration effect is observed when the bridge has a parallel cable-plane arrangement, whereas the torsional frequencies are found to slightly increase in the cases when the inclined cable-plane arrangement is adopted for the bridge.

SUMMARY

The effects of cable vibration behaviour on the dynamic characteristics of the cable-stayed bridges are studied in this paper. A 3-node cable model, which can represent accurately not only the transverse motion of the cable, but the interaction between the motion of the cables and the other parts of the superstructure by the dynamic stiffness in the longitudinal direction, is employed in the analysis. Results obtained from the case study of Nanpu Bridge show that the natural frequencies, the mode shapes and the modal order of the cable-stayed bridges are affected by considering the cable vibration behaviour. A large number of pure cable vibration modes as well as new cable-deck coupled modes are identified. The dynamic cable stiffness depends predominantly on the cable tension. Parametric study of eight example bridges show that the inverted-Y shape pylon and the inclined cable-plane arrangement result in higher torsional rigidity for the bridge system than the H-shaped pylon and the parallel cable-plane arrangement. For medium to long-span cable-stayed bridges, the frequencies of the floating, the sway and the heave modes are not much affected by the cable motion, whereas the torsional frequencies are slightly changed. The cable motions tend to decrease the modal frequencies, except for the torsional modes of the bridges having inclined cable-plane arrangement.

References

1. Abdel-Ghaffar, A.M. and Khalifa, M.A.(1991) Importance of Cable Vibrations in Dynamics of Cable-stayed Bridges. *Journal of Engineering Mechanics, ASCE.* **Vol. 117, No.11:** 2571-2589.
2. Abdel-Ghaffar, A.M. and Nazmy, A.S.(1991). 3-D Non-linear Seismic Behaviour of cable-stayed bridges. *Journal of Engineering Mechanics, ASCE* **Vol. 117, No.11:** 3456-3476.
3. Baron, F., and Lien, S.(1973) Analytical Studies of a Cable-stayed Bridge. *Computers and Structures* **Vol. 3:** 443-465.
4. Cheng, S.H. (1999) Structural and Aerodynamic Stability Analysis of Long-span Cable-stayed Bridges. *Ph.D. thesis, Department of Civil and Environment Engineering, Carleton University, Ottawa, Canada.*
5. Cheng, S.H., and Lau, D.T.(2001) Parametric Study of Cable Vibration Effects on the Dynamic Response of Cable-stayed Bridges. *Proceedings of International Conference on Structural Engineering, Mechanics and Computation, Cape Town, South Africa, 2-4 April*, Paper Ref: SEMC2001/211.
6. Davenport, A.G.(1994) A Simple Representation of the Dynamics of a Massive Stay Cable in Wind. *Proceeding of International Conference on Cable-Stayed and Suspension Bridges, Deauville. October* **Vol.2:** 427-438.
7. Davenport, A.G., and Steels, G.N.(1965) Dynamic Behaviour of Massive Guy Cables. *Journal of Structure Division,ASCE* **Vol.91.ST2:** 43.
8. Fleming, J.F., and Egeseli, E.A.(1980). Dynamic behaviour of a cable-stayed bridge. *Earthquake Engineering and Structural Dynamics* **8(1):** 1-16.
9. Nazmy, A.S., and Abdel-Ghaffar, A.M.(1992) Effects of ground motion spatial variability on the response of cable-stayed bridges. *Earthquake Engineering and Structural Dynamics* **Vol. 21:** 1-20.
10. Parvez, S.M., and Wieland, M.(1987). Earthquake behaviour of proposed multi-span cable-stayed bridge over river Jamuna in Bangladesh. *Proceedings of International Conference on Cable-stayed Bridges, Bangkok, Thailand,* 479-489.
11. Tuladhar R., and Brotton, D.M.(1987) A Computer Program for Nonlinear Dynamic Analysis of Cable-stayed Bridges under Seismic Loading. *Proceedings of International Conference on Cable-stayed Bridges, Bangkok, Thailand,* 315-326.
12. Tuladhar R., Dilger, W.H. and Elbadry, M.M.(1995) Influence of Cable Vibration on Seismic Response of Cable-stayed Bridges. *Canadian Journal of Civil Engineering* **Vol. 22:** 1001-1020.
13. Wethyavivorn, B.(1987) Dynamic behaviour of Cable-stayed Bridges. *Ph.D. thesis, Department of Civil Engineering, The University of Pittsburgh, Pittsburgh, Pa.*
14. Yiu, P.K.A.(1982) Static and Dynamic Behaviour of Cable Assisted Bridges. *PhD. thesis, Department of Civil Engineering, the University of Manchester, Manchester, United Kingdom.*

Table 1: IMPACT OF CABLE VIBRATION BEHAVIOUR ON THE DYNAMIC CHARACTERISTICS OF NANPU BRIDGE

Single-Truss Cable Element		Proposed Three-Node Cable Element		Mode Shape
Mode No.	Frequency (Hz)	Mode No.	Frequency (Hz)	
1	0.1216	1	0.1216	Floating(girder, Anti-Sym.)
2	0.3433	2	0.3422	Heave(girder, Sym.)
3	0.3687	3	0.3679	Sway(girder, Sym.)
4	0.4255	4	0.4246	Heave(girder, Anti-Sym.)
		5	0.4696	New mode *
		6-12	0.4758-0.4767	Pure cable motion ($\#C22$)
		13	0.4829	New mode *
		14-20	0.4890-0.4901	Pure cable motion ($\#C21$)
5	0.4992	21	0.4951	Torsion(girder, Sym.)
6	0.5069	22	0.5042	Bending(pylon, Anti-Sym.)
		23-30	0.5256-0.5262	Pure cable motion ($\#S22$)
7	0.5392	31	0.5376	Bending(pylon, Sym.)
		32-38	0.5407-0.5419	Pure cable motion ($\#C20$)
		39	0.5448	New mode *
		40-47	0.5464-0.5474	Pure cable motion ($\#S21$)
		48	0.5537	New mode *
		49-54	0.5571-0.5577	Pure cable motion ($\#C19$)
		55	0.5599	New mode *
		56-63	0.6048-0.6058	Pure cable motion ($\#S20$)
8	0.6106	64	0.6073	Torsion(girder, Anti-Sym.)

* New mode: Dominant cable vibration coupled with torsion of main girder at the center span.

Table 2: IMPACT OF CABLE TENSION VARIATION ON THE DYNAMIC CABLE STIFFNESS

Tension (N)	ω (rad/s)	Dynamic Cable Stiffness $K(\omega)$ (N/m)	
		Longest Cable	Shortest Cable
$0.75T_{DL}$	$0.1 \times 2\pi$	0.9530×10^7	3.0847×10^7
	$0.2 \times 2\pi$	0.9529×10^7	3.0847×10^7
	$0.5 \times 2\pi$	0.9524×10^7	3.0847×10^7
	$1.0 \times 2\pi$	0.9507×10^7	3.0847×10^7
$0.50T_{DL}$	$0.1 \times 2\pi$	0.8370×10^7	3.0620×10^7
	$0.2 \times 2\pi$	0.8365×10^7	3.0620×10^7
	$0.5 \times 2\pi$	0.8323×10^7	3.0620×10^7
	$1.0 \times 2\pi$	0.8156×10^7	3.0620×10^7
$0.25T_{DL}$	$0.1 \times 2\pi$	0.3766×10^7	2.8538×10^7
	$0.2 \times 2\pi$	0.3689×10^7	2.8538×10^7
	$0.5 \times 2\pi$	0.3092×10^7	2.8535×10^7
	$1.0 \times 2\pi$	-0.3964×10^7	2.8525×10^7

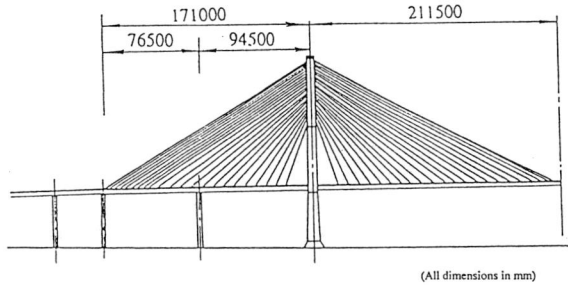

Figure 1: Structural layout of Nanpu Bridge(half)

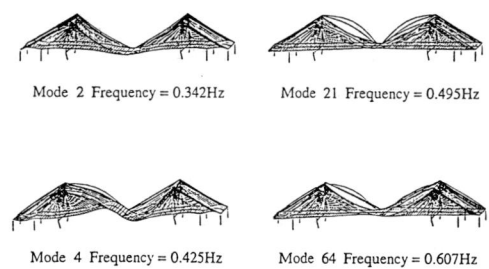

Figure 2: Mode shapes of the Nanpu Bridge (Three-node cable model)

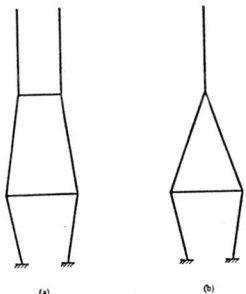

Figure 3: Modification of the pylon shape(a) Pylon of the original Nanpu Bridge (b) Modified pylon of the example bridge

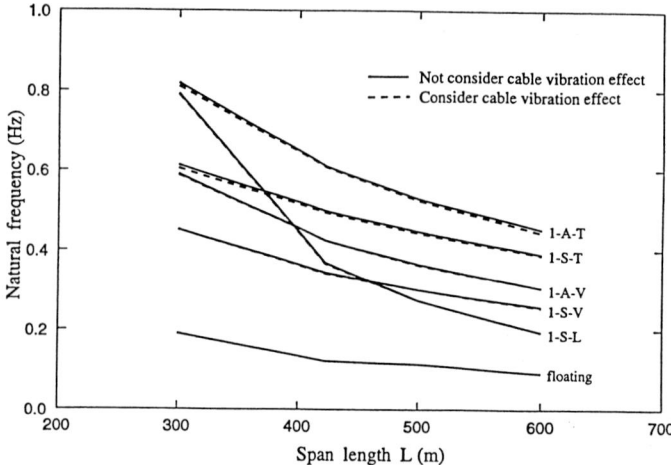

Figure 4: Impact of span length variation on the natural frequency of parallel cable-plane cable-stayed bridges

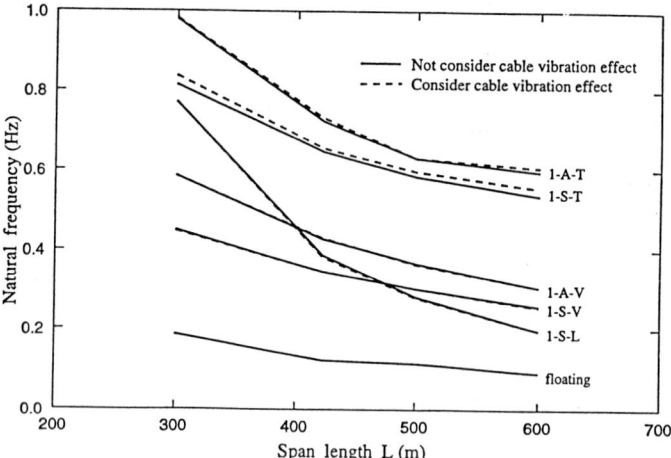

Figure 5: Impact of span length variation on the natural frequency of inclined cable-plane cable-stayed bridges

FINITE ELEMENT MODELLING ON VIERENDEEL MECHANISM IN STEEL BEAMS WITH LARGE CIRCULAR WEB OPENINGS

[1]Liu, T.C.H., [2]Chung, K.F., and [2]Ko, A.C.H.

[1]Manchester School of Engineering, Manchester, England.
[2]Department of Civil and Structural Engineering, the Hong Kong Polytechnic University, Hong Kong, China.

ABSTRACT

The paper presents a finite element investigation on the Vierendeel mechanism in steel beams with large circular web openings. A finite element model is established with both material and geometrical non-linearity so that the moment capacities of tee sections above and below the web openings may be accurately evaluated in the presence of co-existing axial and shear forces. Moreover, load re-distribution across the web openings may be directly incorporated during the assessment of the load carrying capacities of steel beams with large circular web openings. An empirical shear moment interaction curve at the perforated sections is then proposed for practical design of steel beams with medium to large circular web openings against the Vierendeel mechanism. It is found that shear yielding in steel beams with large circular web openings is very important as it reduces the effectiveness of load re-distribution across the web openings. Consequently, the plastic moment capacities of tee sections at the low moment side of the web openings may not be mobilized fully.

KEYWORDS

Circular web opening, finite element analysis, material and geometric non-linearity, design development, shear moment interaction curves

INTRODUCTION

In modern buildings, openings are frequently required to be provided in structural members so that building services may be incorporated into structural zones for simplified layout and installation. Moreover, the overall depth of the construction zone may be reduced accordingly, and this is beneficial for multi-storey buildings with large headroom requirement. At present, there is a tendency to use large water pipes and air ducts, and openings of dimensions up to 75% of the depth of floor beams are often required. As reported by Lawson (1987), Darwin (1990), Redwood (1993), and Oehlers and Bradford (1995), the presence of large web openings may have a severe penalty on the load carrying capacities of floor beams, depending on the shapes, the sizes, and the locations of the openings. Both rectangular and circular openings are commonly used, and reinforcements around the web openings may be provided as necessary through rational design.

The presence of web openings in steel beams introduces three different modes of failure at the perforated sections:

(a) Shear failure due to reduced shear capacity,
(b) Flexural failure due to reduced moment capacity, and
(c) 'Vierendeel' mechanism due to the formation of four plastic hinges in the tee-sections above and below the web openings.

In a design method recommended by Ward (1990), the load carrying capacities of the beams with circular web openings is assumed to be attained after the formation of plastic hinges at the low moment side of the web openings. Moreover, a linear interaction formula is used to assess the moment capacities of the tee sections above and below the web openings under co-existing axial and shear forces. The method is subsequently adopted into Eurocode 3: Annex N (1993). In general, the method is regarded to be conservative as the beams are capable to carry additional load until all four plastic hinges at critical locations of the perforated sections are developed to form the Vierendeel mechanism. Moreover, the reduction in the moment capacity of the tee sections under coexisting axial and shear forces is less severe than that anticipated in the linear interaction formula.

SCOPE OF WORK

In order to provide accurate assessment on the load carrying capacities of steel beams with large circular web openings, a finite element model is established with both material and geometrical non-linearity by Liu (1996), Liu and Chung (1999), and Chung, Liu and Ko (2000). The moment capacities of tee sections above and below the web openings may thus be accurately evaluated in the presence of co-existing axial and shear forces. Moreover, load re-distribution across the web openings after initial yielding may be incorporated during the assessment of the load carrying capacities of steel beams. The yield patterns both at first yield and at failure will provide understanding on the structural behaviour of the steel beams. After calibration against test data, an empirical shear moment interaction curve at the perforated sections is also suggested for practical design of steel beams with medium to large circular web openings against the Vierendeel mechanism.

It should be noted that in the present investigation, all hot rolled steel I sections are of Class 1 or 2 (plastic or compact). All web openings are concentric to the mid-height of the sections with diameters, d_o, between 0.5 h and 0.75 h where h is the depth of the sections. No reinforcement around the web openings is considered. The formulation is presented in accordance with Eurocode 3 for easy reference. It should be noted that both the bending moment, M_{Sd}, and the shear force, V_{Sd}, due to global actions are evaluated at the centre of the web openings, as shown in Figure 1.

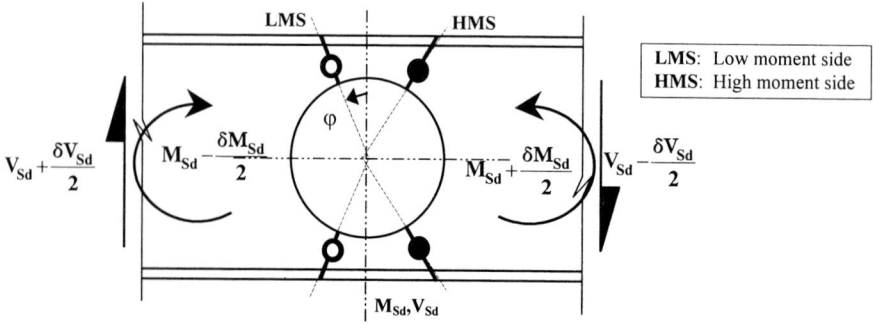

Figure 1   Vierendeel mechanism around circular web opening

FINITE ELEMENT MODELLING

Figure 2 presents a finite element model where the flanges and the web of a steel beam are discretized with iso-parametric eight-noded shell elements. A circular opening is formed in the web with refined mesh configuration. After sensitivity study on both the configuration and the density of the finite element mesh, it is found that over 750 shell elements are required to model the flanges and the web of the beam with almost half of the shell elements located around the web opening. Full Newton-Raphson solution procedure is used to analyse the beam during the entire deformation history. It should be noted that during the numerical investigation, it is necessary to ensure that the finite element model only fails at the perforated section, and failure in other parts of the beam including overall instability are prohibited. Moreover, the web opening is free from any boundary effect or point load.

The finite element model was calibrated against the test data of two steel I beams with single circular web opening reported by Redwood and McCutcheon (1968). Figure 2 also presents details of the geometrical dimensions and the material data of the steel beams. Moreover, the deformed finite element mesh under the Vierendeel mechanism of the perforated section at failure is also illustrated. For both tests, the load-deflection curves obtained from the finite element modelling are plotted in Figure 3 together with the measured test data for direct comparison. It is shown that both the maximum moment capacities of the perforated sections and the deformation characteristics of the beams are modelled satisfactorily.

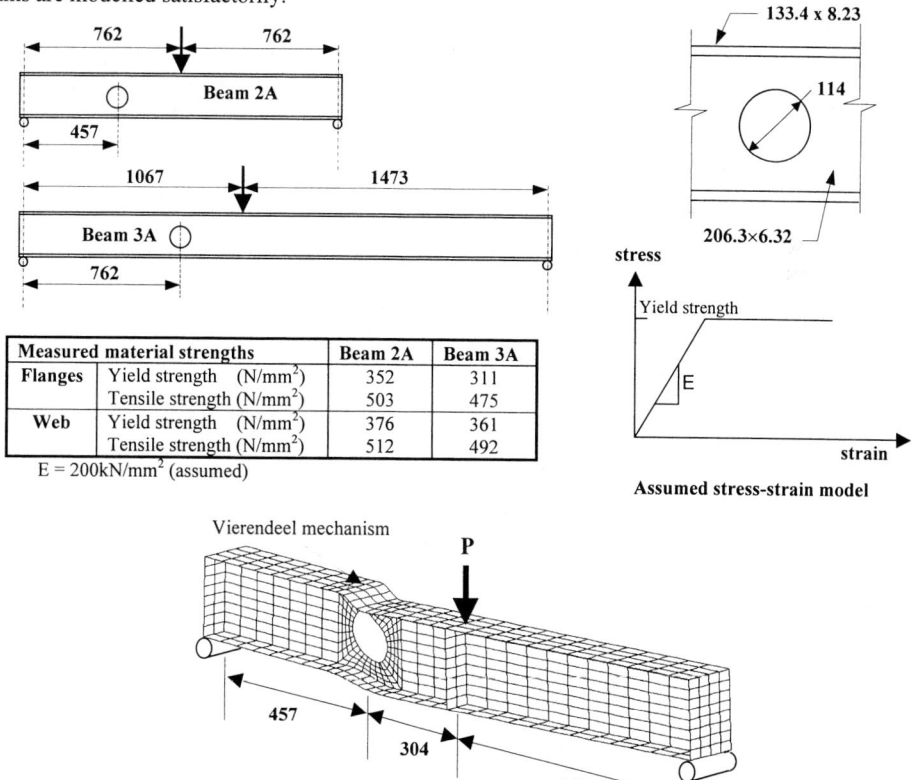

Figure 2   Details of Beams 2A and 3A with finite element model

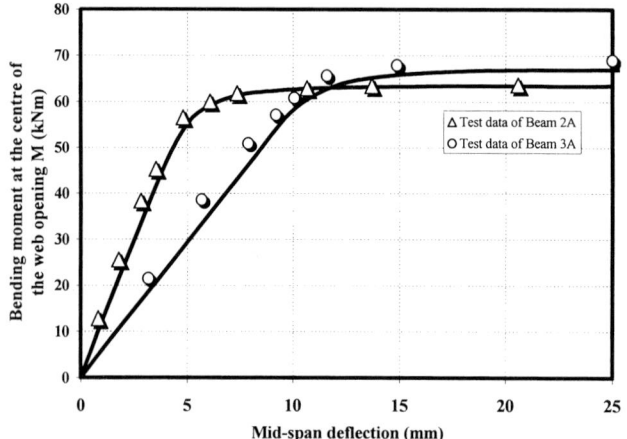

Figure 3   Load-deflection curves of Beams 2A and 3A

It is interesting to examine the stress distribution of the perforated section at both first yield and failure conditions of Beam 2A, as shown in Figure 4. First yield appears in the web of the tee sections at cross sections with $\varphi = 30°$ and $\varphi = -45°$. At the same time, shear yielding in the web of the tee sections at cross sections with $\varphi = \pm 45°$ and $0°$ are also apparent. However, this does not materialize a collapse mechanism, and the beam continues to carry additional loading until the Vierendeel moment is large enough to cause extensive yielding in the tee sections.

At failure, both the webs and the flanges of the tee sections at the HMS are extensively yielded. Furthermore, there is also extensive shear yielding in the webs of the tee sections with minimum web depth (i.e. $\varphi = 0°$). However, at the LMS, only the webs of the tee sections are yielded while the stress level of the flanges reaches only 60% of the yield strength. It should also be noted that the yield patterns at failure are very different from those assumed positions in the current design method as shown in Figure 1.

For Beam 3A, while the beam section and the opening are similar to those in Beam 2A, the location of the opening gives rise to a different shear to moment ratio, and thus a different behaviour from that in Beam 2A is observed. The load-deflection curve of Beam 3A is also presented in Figure 3 together with the measured test data for direct comparison. It is found that Beam 3A starts to yield much later than Beam 2A and the top flange at the HMS of the perforated section buckles locally under large global bending action.

In the finite element models of both Beams 2A and 3A, extensive shear yielding is found at the tee sections with minimum web depth at failure which reduces the effectiveness of load re-distribution across the web openings. Consequently, the tee sections at the HMS are found to be fully yielded while the tee sections at the LMS are only partially yielded at failure. This agrees well with the finding of a complementary analytical study by Ko and Chung (2000) on existing design methods that not all four plastic hinges are fully developed at failure.

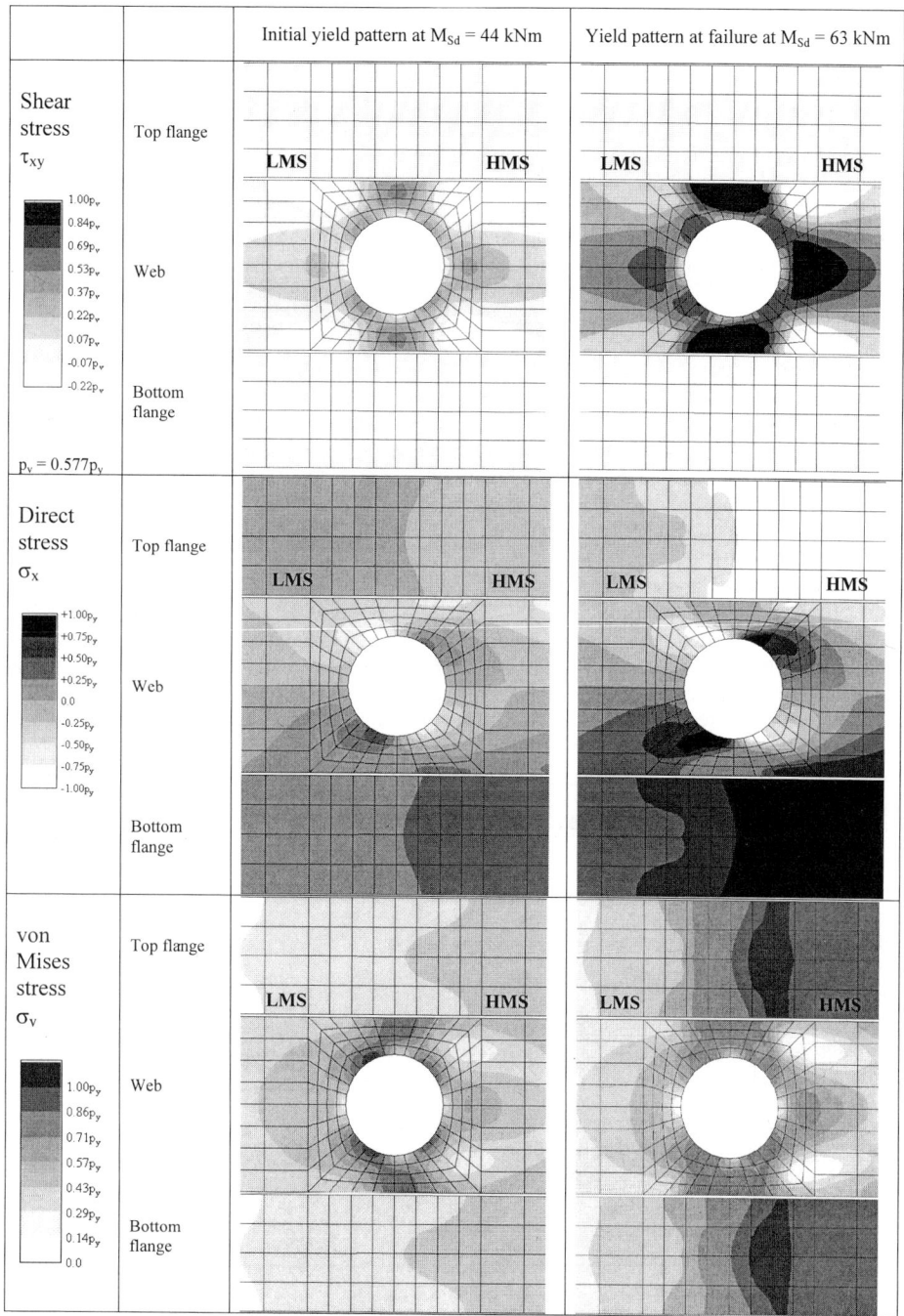

Figure 4   Stress distribution at the perforated section in Beam 2A
(Stresses are shown at the mid-thickness of the steel elements.)

SHEAR MOMENT INTERACTION CURVES

For practical design, it is highly desirable to develop simple design methods using shear moment interaction curves to assess the load carrying capacity of steel beams with circular web openings. A parametric study is carried out using similar finite element models as shown in Figure 2 to generate shear moment interaction curves of the perforated sections in four simply supported steel beams of different section sizes with circular web openings under uniformly distributed load. By varying the positions of the openings along the beam spans, four non-dimensional interaction curves relating the shear utilization ratio v, (or $V_{Sd} / V_{o,Rd}$) and the moment utilization ratio m, (or $M_{Sd} / M_{o,Rd}$) are obtained, and they are plotted altogether in Figure 5 for direct comparison. (Refer to Appendix for the design formulae of $M_{o,Rd}$ and $V_{o,Rd}$.) It is shown that at perforated sections, the shear moment interaction is not severe in general, and strength reduction is only significant at large value of either v or m.

It should also be noted that when the web openings are placed at the mid-span of the beams with $v = 0$, all beams fail in flexure with $m = 1$, i.e. at a bending moment very close to their respective plastic moment capacities. However, for beams with same serial sizes but with different weights per unit length, the curves are of similar shapes, but with different y-intercepts, i.e. different values of v. This is due to the under-estimation of the shear capacities of the perforated sections given by the design formula for sections with thick flanges, when compared with the finite element results.

It is interesting to compare the finite element results with other empirical but yet simple interaction curves which are also plotted in Figure 5. While the first order line may be too conservative and the third order curve is unconservative, the second order curve is considered to be simple and practical to the design of the Vierendeel mechanism in steel beams with circular web openings. With accurate prediction on the shear capacity, $V_{o,Rd}$, the non-linear curve with a power of 2.5 seems to be a good fit to the finite element results for all four steel beams with circular web openings of diameter d_o between 0.5 h and 0.75 h.

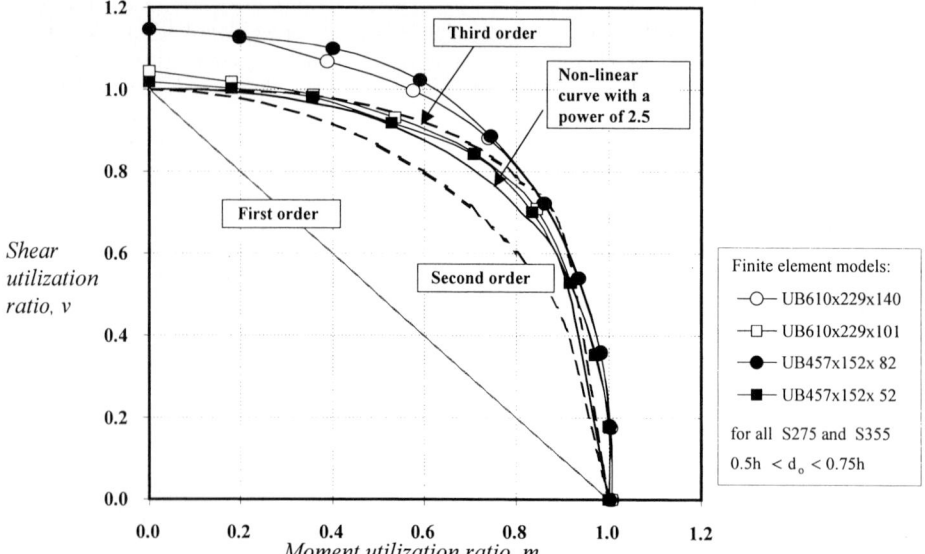

Figure 5   Shear moment interaction curves of four steel beams

SIMPLE DESIGN RULE

In order to provide simple design rule for practical design of steel beams with circular web openings, an empirical shear moment interaction curve (Eqn.1) is proposed as follows:

$$\left(\frac{V_{Sd}}{V_{o,Rd}}\right)^{2.5} + \left(\frac{M_{Sd}}{M_{o,Rd}}\right)^{2.5} \leq 1 \qquad \text{(Eqn. 1)}$$

The non-linear interaction formula may be re-presented to give a simple empirical design formula (Eqn.2) to assess the moment capacity of the perforated section, $M_{vo,Rd}$, under global shear force, V_{Sd}, against the Vierendeel mechanism as follows:

$$M_{vo,Rd} = M_{o,Rd} \left[1 - \left(\frac{V_{Sd}}{V_{o,Rd}}\right)^{2.5}\right]^{0.4} \geq M_{Sd} \qquad \text{(Eqn. 2)}$$

With the use of the proposed empirical design formula, all the design rules against flexural failure, shear failure and Vierendeel mechanism are based on perforated section with significant reduction in calculation effort. Furthermore, there is no need to evaluate section capacities of the tee sections.

CONCLUSIONS

A numerical investigation on the load carrying capacities of simply supported steel beams with large circular web openings is presented. A finite element model is established with both material and geometrical non-linearity so that the moment capacities of tee sections above and below the web openings may be accurately evaluated in the presence of co-existing axial and shear forces. Moreover, load re-distribution across the web openings may be directly incorporated during the assessment the load carrying capacity of steel beams with large circular web openings. Calibration of the finite element model against test data is also carried out. The finite element model is thus regarded to be applicable in predicting the general structural behaviour of steel beams with large circular web openings against various modes of failure. A parametric study on four steel beams with large circular web openings is also performed.

In general, there is extensive shear yielding at the tee sections with minimum web depth at failure which reduces the effectiveness of load re-distribution across the web opening after initial yielding. Furthermore, at failure, the tee sections at the HMS are found to be fully yielded while the tee sections at the LMS are only partially yielded. Moreover, accurate prediction on the load carrying capacities of steel beams with large circular web openings depends on the accuracy of the shear capacity assessment of the perforated sections. As shown in the shear moment interaction curves, the interaction due to co-existing shear and moment at the perforated sections is found not to be severe in general.

ACKNOWLEDGEMENTS

The research project leading to the publication of the paper is supported by the Research Grants Council of the Government of Hong Kong SAR (Project No. PolyU5085/97E). The financial support to the first author as a Honorary Research Fellowship to the Department of Civil and Structural Engineering of the Hong Kong Polytechnic University is also gratefully acknowledged.

REFERENCES

1) Lawson R.M. (1987) *Design for openings in the webs of composite beams*, CIRIA Special Publication and SCI Publication 068, CIRIA / Steel Construction Institute.

2) Darwin, D. (1990) *Steel and composite beams with web openings*, American Institute of Steel Construction, Steel Design Guide Series No. 2, Chicago, IL, USA.

3) Redwood, R.G. and Cho, S.H. (1993) *Design of steel and composite beams with web openings*, Journal of Constructional Steel Research, 25, 23-41.

4) Oehlers, D.J. and Bradford, M.A.(1995) *Composite steel and concrete structural members: Fundamental behaviour*, Pergamon.

5) Ward, J.K. (1990), *Design of composite and non-composite cellular beams*, The Steel Construction Institute.

6) *ENV 1993-1-3, Eurocode 3: Design of steel structures: Part 1.1 General rules and rules for buildings*, 1992, and *Amendment A2 of Eurocode 3: Annex N 'Openings in webs'*, 1998, British Standards Institution.

7) Liu T.C.H. and Chung, K.F. (1999) *Practical design of universal steel beams with single web openings of different shapes,* Proceedings of the Second European Conference on Steel Structures '*Eurosteel 99*', May 1999, Prague, 59-62.

8) Chung, K.F., Liu, T.C.H. and Ko, A.C.H. (2000) *Investigation on Vierendeel mechanism in steel beams with circular web openings*, Journal of Constructional Steel Research (in press).

9) Redwood, R.G. and McCutcheon, J.O. (1968) *Beam tests with un-reinforced web openings*, Journal of Structural Division, Proceeding of American Society of Civil Engineers, 94:ST1, 1-17.

10) Ko, C.H. and Chung, K.F. (2000) *A comparative study on existing design rules for steel beams with circular web openings,* Proceedings of the First International Conference on Structural Stability and Dynamics, ed. by Yang, Y.B., Leu, L.L. and Hsieh, S.H., December 2000, Taipei, 733-738.

Appendix   Design formulae

The global bending moment and the global shear force at the centre of the web opening are denoted as M_{Sd} and V_{Sd} respectively. The moment capacity and the shear capacity of the perforated section, $M_{o,Rd}$, and $V_{o,Rd}$ are given as follows:

$$M_{o,Rd} = f_y W_{o,pl} \qquad V_{o,Rd} = f_v A_{vo}$$

$$W_{o,pl} = W_{pl} - \frac{d_o^2 t_w}{4} \qquad A_{vo} = A_v - d_o t_w$$

where f_v is the shear strength of the steel taken as $0.577 f_y / \gamma_{Mo}$,
f_y is the design yield strength of the steel,
W_{pl} is the plastic modulus of the un-perforated section,
A_v is the shear area of the un-perforated section, or equal to $h \times t_w$ conservatively,
h is the overall depth of steel beam,
d_o is the diameter of the opening, and
t_w is the web thickness.

Loading Parameters at Cracks and Notches

Riad A. Fodhail
Mechanical Engineering Department
Faculty of Engineering, University of Aden, Yemen
P.O.Box: 5243, Maalla- Aden. Fax: 967-2-246611
E-mail: eng.aden.@y.net ye

Abstract

The stress and strain fields near cracks and sharp notches are highly inhomogeneous and multi-axial in nature. Therefore, the description of strength phenomena of the structures containing sharp notches or cracks requires loading parameters, which adequately describe the local loading situation for the material at hand. In this paper the loading parameters are derived from averaging the local stress and strain distribution at the notch root or crack tip over some microstructure related size scale. The behavior of micro, small and macrocracks as well as the behavior of notches is described by the obtained loading parameters.

Introduction

Fracture of a structural component may have different causes. They depend considerably on the kind of loading, on the state of local stresses and on the material properties. The fracture of a component mostly means also the breakdown of the whole structure. In linear elastic fracture mechanics, which is based upon the assumption of linear elasticity and infinitesimal strain, a macrocrack is assumed as a defect in the structural component and it was successfully shown that the stress intensity factor k [1] is a suitable global parameter. Through it the stress field at macrocracks is characterized. This also holds good for small scale yielding [2]. The assumption of crack-like flaw in engineering components allows describing and understanding the several damage phenomena. Thus by means of fracture mechanics considerable progress has been made in the safety and damage assessment of components. Not in every case, however, does the fracture proceed from the crack-like flaw. The assessment of state of local stress at notches in crack-free components is equally important. The local fracture process usually starts at areas of highly localized inhomogeneous stresses such as at cracks and sharp notches. Therefore, for deciding the load carrying capacity of structural components with such defects, the knowledge of the local stress field at these defects is required. The loading parameters (strength parameters), in this paper, are obtained by the concept of micro- support action. Neuber basically develops this concept [3]. It is based on the abstraction that the local fracture depends not only on the continuum mechanically calculated highest local stress but also on that in the neighbourhoods zones (Neuber's micro-support action). Hence a mean value (S_{em}) of a suitable equivalent stress or strain (S_e) over a characteristic volume V^* is built up starting from the notch root or the crack tip. V^* depends on the microstructure of the material and the used S_e. Mathematically the loading parameter S_{em} can be written in the following form:

$$S_{em} = \frac{1}{V^*} \int_{(V^*)} S_e(x,y,z) dV \tag{1}$$

S_{em} as an integral model is independent of the notch geometry and the lows of material behaviors.

If the state of stress in V* is largely homogenous in other coordinate directions (y, z), as the case of plane cracks and sharp notches, then it is sufficient to build S_{em} over a dimensional characteristic length d* of the volume element V*.

Therefore Eq. 1 becomes:

$$S_{em} = \frac{1}{d^*} \int_{(d^*)} S_e(x)dx \quad (2)$$

The value of d* for a particular material can be determined by comparing the experimental values of the endurance limits for two differently notched or cracked specimens as follows:

$$\frac{\Delta S_{em1}}{\Delta S_{em2}} = \frac{\int_{(d^*)} \Delta S_e(\rho_1, x)dx}{\int_{(d^*)} \Delta S_e(\rho_2, x)dx} = 1 \quad (3)$$

Where ρ_1 and ρ_2 are the radii of curvature of the two notched components.

Generally for steels, d* is in the range of few μm to few 100 μm, depending on the microstructure of the material and the used suitable equivalent stress or strain hypothesis.

S_{em} can be considered as a general global loading parameter for fracture in both homogeneous and inhomogeneous stress fields. It can be compared with the material characteristic from the tensile test for the same fracture mode.

Local stress field

Figs.1 and 2 show an elliptical notch and a linear crack in infinite plates under one dimensional tensile stress field.

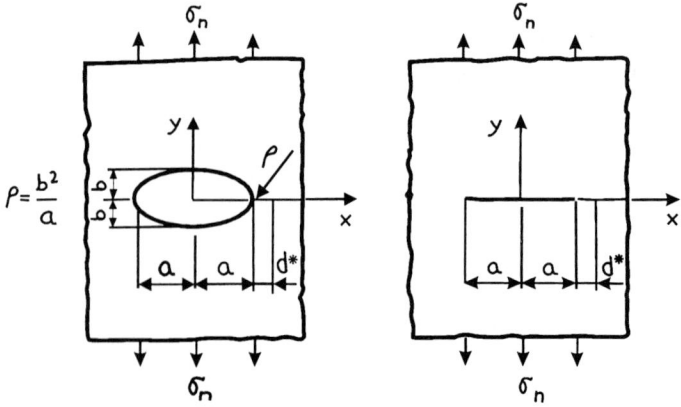

Fig. 1 Elliptical notch   Fig. 2 linear crack

The exact local stress field in the ligament of the elliptical notch for linear elastic stress-strain relationship is given in the following form [4]

$$\sigma_x = \sigma_n \left\{ \frac{1+2\sqrt{\rho/a}}{\sqrt{(x/a)^2 -1+\rho/a}[x/a+\sqrt{(x/a)^2 -1+\rho/a}]} - \right.$$

$$\left. \rho/a \frac{(1+\sqrt{\rho/a})[x/a+2\sqrt{(x/a)^2 -1+\rho/a}]}{[(x/a)^2 -1+\rho/a]^{3/2}[x/a+\sqrt{(x/a)^2 -1+\rho/a}]^2} \right\} \quad (4)$$

$$\sigma_y = \sigma_n \left\{ \frac{(x/a)^2 +(x/a)\sqrt{[(x/a)^2 -1]+\rho/a}+\rho/a}{\sqrt{(x/a)^2 -1+\rho/a}[x/a+\sqrt{(x/a)^2 -1+\rho/a}]} + \right.$$

$$\left. \rho/a \frac{(1+\sqrt{\rho/a})[x/a+2\sqrt{(x/a)^2 -1+\rho/a}]}{[(x/a)^2 -1+\rho/a]^{3/2}[x/a+\sqrt{(x/a)^2 -1+\rho/a}]^2} \right\} \quad (5)$$

By setting $\rho = 0$ in Eqs. (4) and (5), the following exact local stress field in the ligament of the crack is obtained:

$$\sigma_x = \sigma_n \left[\frac{x/a}{\sqrt{(x/a)^2 -1}} -1 \right] \quad (6)$$

$$\sigma_y = \sigma_x \frac{x/a}{\sqrt{(x/a)^2 -1}} \quad (7)$$

where

σ_n - nominal stress
$2a$ - notch or crack length
ρ - radius of curvature of the elliptical notch $= b^2/a$ at $x = a$
$\sigma_z = 0$ for plane stress case
$\sigma_z = \nu\,(\sigma_x +\sigma_y)$ for plane strain case
ν - Poisson's ratio

Loading parameters

In the vicinity of the notch root or the crack tip a multi-axial state of stress exists. For such a case of combined stresses, strength hypotheses are used to find the equivalent stress and the equivalent strain.

In the ligament of the notch or of the crack, the normal stresses and strains represent the principal stresses and strains.

By using maximum principal stress hypothesis ($S_e = \sigma_1 = \sigma_y$), the loading parameter given by Eq. (2) can be expressed for notches as shown below:

$$S_{em} = {}^N\sigma_{1m} = {}^N\sigma_{ym} = \frac{1}{d*}\int_a^{a+d*}\sigma_1(x)dx = \frac{1}{d*}\int_a^{a+d*}\sigma_y(x)dx \qquad (8)$$

By using maximum principal strain hypothesis ($S_e = \varepsilon_1 = \varepsilon_y$), the loading parameter given by Eq. (2) can be expressed for notches as shown below

$$S_{em} = {}^N\varepsilon_{1m} = {}^N\varepsilon_{ym} = \frac{1}{d*}\int_a^{a+d*}\varepsilon_1(x)dx = \frac{1}{d*}\int_a^{a+d*}\varepsilon_y(x)dx \qquad (9)$$

- for plane stress case

$$S_{em} = {}^N\varepsilon_{1m} = {}^N\varepsilon_{ym} = \frac{1}{d*}\int_a^{a+d*}\frac{1}{E}[\sigma_y(x) - \nu\sigma_x(x)]dx \qquad (9a)$$

- for plane strain case

$$S_{em} = {}^N\varepsilon_{1m} = {}^N\varepsilon_{ym} = \frac{1}{d*}\int_a^{a+d*}\frac{1}{E}\{\sigma_y(x) - \nu[\sigma_x(x) + \sigma_z(x)]\}dx \qquad (9b)$$

where E is the material's modulus of elasticity.

The results of integration of Eq. (8) and Eqs. (9a) and (9b) are obtained in a closed lengthy analytical form after the substitution of Eqs. (4) and (5) in them. The values of the loading parameters ${}^N\sigma_{1m}$ and ${}^N\varepsilon_{1m}$ for plane stress and plane strain cases are plotted against ρ/a for different values of $d*/a$ as shown in Figs. 3, 4, and 5 respectively.

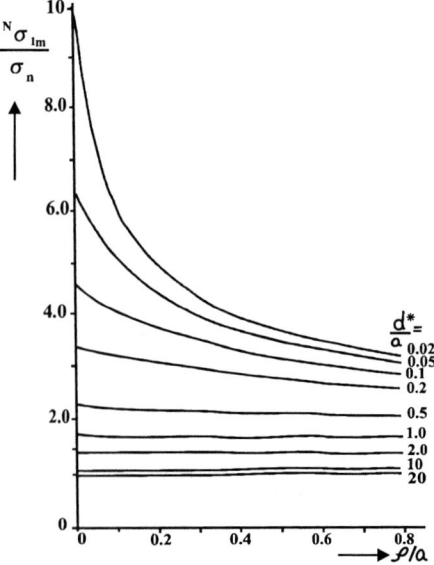

Fig. 3 ${}^N\sigma_{1m}$ in the notch ligament

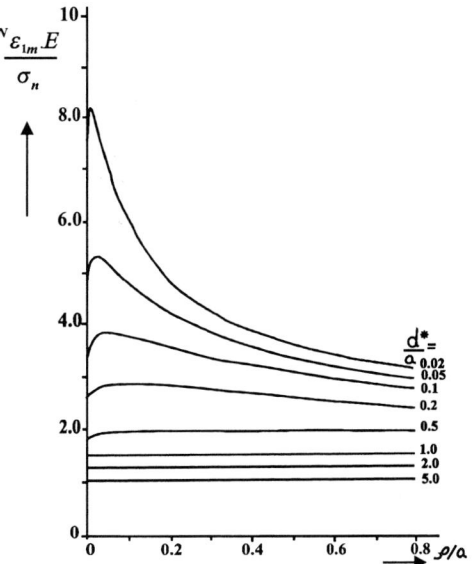

Fig. 4 $^N\varepsilon_{1m}$ in the notch ligament
for plane stress case, $\nu = 0.3$

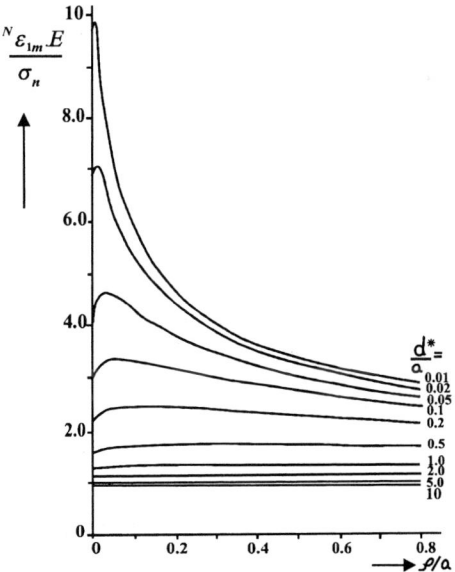

Fig. 5 $^N\varepsilon_{1m}$ in the notch ligament
for plane strain case, $\nu = 0.3$

For linear elastic behavior of materials, the stress and the strain fields at cracks have the same singularity. Any field can therefore be used. Applying the stress field approach for maximum principal stress hypothesis and using Eqs. (2) and (7), the loading parameters for cracks will be as follows:

$$S_{em} = {}^C\sigma_{1m} = \frac{1}{d*}\int_a^{a+d*}\sigma_y(x)dx = \frac{1}{d*}\int_a^{a+d*}[\sigma_n \frac{x/a}{\sqrt{(x/a)^2-1}}]dx$$

$$= \sigma_n\sqrt{1+\frac{2a}{d*}} \qquad (10)$$

By defining

- microcracks; $a \ll d*$
- small cracks; $a \approx d*$
- macrocracks; $a \gg d*$

This definition is also valid for micro, small, and macronotches.

For microcracks, $a \ll d*$

$${}^C\sigma_{1m} = \sigma_n \qquad (11)$$

It means that the nominal stress characterizes the strength of structural members with microcracks.

For macrocracks, $a \gg d*$

$${}^C\sigma_{1m} = \sigma_n\sqrt{\frac{2a}{d*}} = K_I\sqrt{\frac{2}{\pi d*}} \qquad (12)$$

K_I is the stress intensity factor for mode one fracture ($K_I = \sigma_n\sqrt{\pi a}$).
In Eq. (12), ${}^C\sigma_{1m}$ is proportional to K_I. This means that the stress intensity factor for linear elastic materials characterizes the stress field at macrocracks.

For small cracks, $a \approx d*$

$${}^C\sigma_{1m} = \sigma_n\sqrt{1+\frac{2a}{d*}} = K_I\sqrt{\frac{2}{\pi d*}}\sqrt{1+\frac{d*}{2a}} \qquad (13)$$

Eq. (13) shows that the loading parameter for small cracks is larger than that for macrocracks by the factor $\sqrt{1+d*/2a}$. The loading parameter at micro, small, and macrocracks have been found in a closed analytical form. The loading parameters at micro, small, and macronotches can also be represented in a closed analytical form, but are complicated.

Referring to Fig. 3, the maximum principal stress hypothesis, as already known from fracture mechanics, gives loading parameters for cracks greater than those for notches. As the radius of curvature is increased, the loading parameter is decreased. For micronotches (a << d*), the nominal stress σ_n is the loading parameter as in the case of microcracks. For d*/a > 0.5, the loading parameter for small notches and small cracks are approximately the same. Therefore Eq. (13) can be used for both the cases. The radius of curvature of the notch has no influence on the loading parameters in this range.

Compared to maximum principal stress hypothesis, the maximum principal strain hypothesis shows that the loading parameters for some sharp notches are greater than those at cracks (Figs. 4 and 5). The loading parameter found by using the maximum principal strain hypothesis depends essentially on d*/a. It increases as d*/a becomes smaller and is maximum for $\rho \approx d^*$.

Conclusion

In this paper the loading parameters are obtained by averaging the local stress and local strain distribution over a dimensional characteristic length d*, that depends on the microstructure of the material and the used equivalent strength hypothesis. In some papers the loading parameter is built over a definite distance from the crack tip such as in [5] over two grain sizes and in [6] over a distance of 1-3 grain sizes.

The loading parameters and the derived characteristic parameters are dependent on the equivalent strength hypothesis, notch geometry, the characteristic length d*, and the material behavior. The maximum principal stress hypothesis always gives larger loading parameters for cracks ($\rho = 0$) than for notches ($\rho \neq 0$). The maximum principal strain hypothesis gives larger loading parameters for some sharp notches than for cracks and they are maximum for $\rho \approx d^*$.

The nominal stress can be used as a loading parameter for structural members with microcracks and micronotches (a << d*). For structural members with macrocracks (a >> d*) the stress intensity factor can be used as a loading parameter. For small cracks the loading parameter is greater than that for macrocracks by the factor $\sqrt{1 + d^*/2a}$. The loading parameters for small cracks can also be used for small notches in case of d*/a > 0.5. The radius of curvature of the notch has no influence in this range.

References

[1] IRWIN, G. R., "Relation of a Stress near a Crack to Crack Extension Force", Int. Mech. Appl., Brussels, (1956).
[2] RICE, J. R. and ROSENGREN, G. F., "Plane Stain Deformation near a Crack Tip in a Power Hardening Material". T. Mech. Phys. Solids, Vol. 16, (1968).
[3] NEUBER, H., "Kerbspannungslehre" Akademie-Verlag, Berlin, 3. Auflage, (1985).
[4] SÄHN, S. and GÖLDNER, H., "Bruch and Beurteilungskriterien in der Festigkeitskehre" Fachbuchverlag. Leipzig (1989).
[5] RITCHIE, R. C., KNOTT, J. F. and RICE, J. R., Mech. Phys. Solids, Vol. 21, pp.395, (1973)
[6] RAWAL, S. P. and GURLAND, J., Proceeding of the second International Conference on the Mechanical Behavior of materials, Boston, pp. 1154 (1976), Federation of Material Societies, Dearborn, Michigan.

INVESTIGATIONS ON LOCAL STABILITY OF COMPRESSED WALL OF HOLLOW REINFORCED CONCRETE BRIDGE PIER WITH RECTANGULAR CROSS SECTION

E. P. Djelebov

Department of Public Works, Roads and Transport
Mpumalanga Province, Nelspruit, South Africa

ABSTRACT

The problem of stability of high slender reinforced concrete bridge piers faces the designer of modern bridges and it could be considered from two points of view:
- the problem for the overall stability of the slender RC bridge pier;
- the problem for the local stability of the different elements of the slender bridge RC pier and in particular the compressed wall of the pier.

Solutions of the first problem have been given in numerous theoretical and practical investigations. An attempt for solving of the second problem is given in this paper which is a result of diploma thesis for completion of post-graduate specialization (1979 – 80) in Chair "Solid Structures", Higher Institute of Architecture and Civil Engineering – HIACE), Sofia, Bulgaria. The paper discusses the possible methods of solving the problem for local stability of compressed wall of slender, RC, hollow bridge pier and gives practical theoretical solution of the problem, using simplifying, realistic assumptions and taking into account the actual behavior of material. The designer is offered an easy, straightforward method for checking of the adequacy of the thickness of the compressed wall of the slender RC bridge pier ensuring the local stability. The method allows simple computer programs or graphs to be created for the purpose of checking local stability of the compressed wall. By discussing the theoretical background for the proposed method, the paper also contributes to the better understanding of the stability problems by structural engineers and designers of structural concrete, especially in relation to non – linear behavior of the structural systems and real materials.

KEYWORDS

Stability, local stability, buckling, slender bridge piers, hollow slender bridge piers.

FORMULATION OF THE PROBLEM, ASSUMPTIONS AND METHODS FOR SOLVING

Formulation of the problem

A high slender reinforced concrete bridge pier with rectangular (box) cross section is given (Fig 1). Usually transverse diaphragms are provided for technological purposes (storage of materials for the climbing formwork) and to ensure the local stability of the compressed wall.

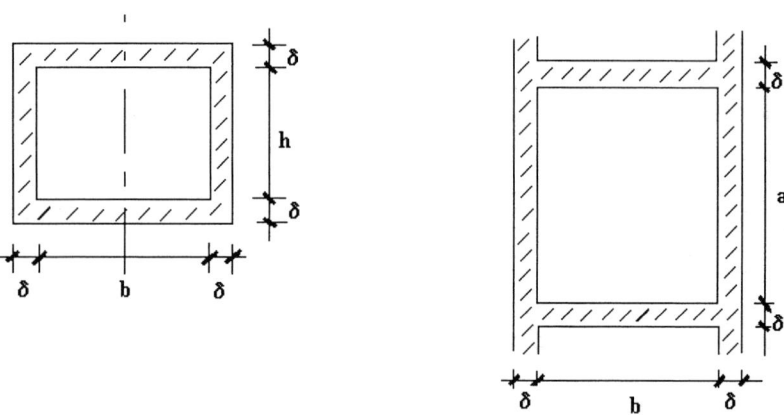

Figure 1 (a) and (b)

Assumptions

The following assumptions are adopted:
- the thickness of the pier wall is constant and equal to δ ;
- the compressive stresses in two adjacent cross sections at the proposed diaphragms are equal – in fact the moment diagram is not linear due to the geometrical non – linearity – moments of second degree are added (Fig. 3) and this allows as to use the available solutions for stability of plates;

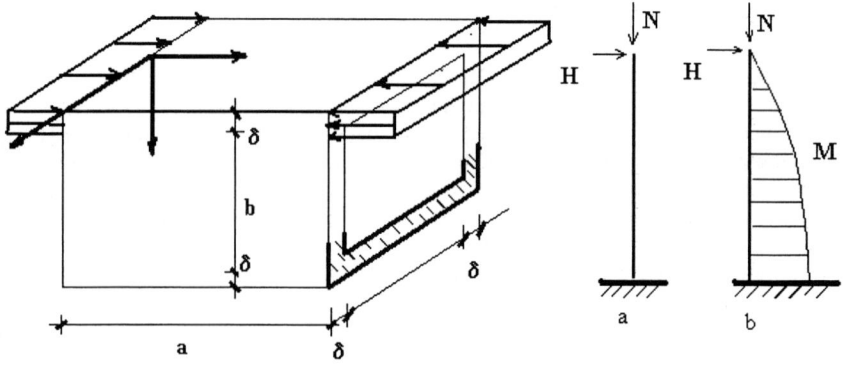

Figure 2   Figure 3

- if the theory of plates is used an error will occur, since due to the restraint, the plate has to be loaded with additional forces over the longitudinal edges (the cross section is not ideally rigid – i.e. it is geometrically non – linear system – see Fig. 4). As it will be shown the additional transverse forces are negligible and if not taken into account (i.e. if geometrical linearity is considered) then the results are on the safe side since these forces in fact increase the rigidity of the plate ;

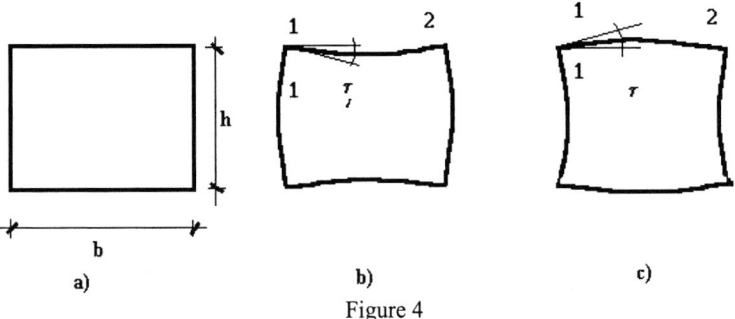

Figure 4

- the adopted for concrete non–linear stress–strain curve will be as shown on Fig. 5. Concrete is not a linear material and assuming physical (material) linearity will result in significant error.

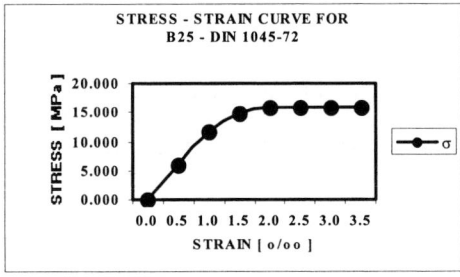

Figure 5

Methods for Solving of the Problem

The problem is approached by using the available investigations on the Theory of Plates and application of the Theory of Plasticity as well as different investigations on stability problems. The following approaches to the system were used in [1] for solution :
1. as geometrically and physically linear (Theory of Elasticity);
2. as geometrically non-linear and physically linear;
3. as geometrically linear and physically non-linear;
4. as geometrically and physically non-linear.

SOLUTION AS GEOMETRICALLY AND PHYSICALLY LINEAR SYSTEM

The theoretical solution is given in the Theory of Elasticity (see [5],[6]). The solution is based on the differential equation for indifferent equilibrium of plates:

$$\frac{\partial^4 \omega}{\partial x^4} + 2\frac{\partial^4 \omega}{\partial x^2 \partial y^2} + \frac{\partial^4 \omega}{\partial y^4} + P\frac{\partial^2 \omega}{\partial x^2} = 0 \qquad (1)$$

After taking into account of the properties of the material and the boundary conditions, as it is shown in [2], the following equation for determining of the critical stress/force is obtained from the homogenous system of two equations:

$$\sqrt{\sqrt{UV}+V}\, th\frac{\sqrt{\sqrt{UV}+V}}{2} + \sqrt{\sqrt{UV}-V}\, tg\frac{\sqrt{\sqrt{UV}-V}}{2} =$$
$$= \frac{b}{1.92}\left(2\frac{h}{b}+1\right)\frac{2}{b^2}\sqrt{UV} \qquad (2)$$

Where $U = P b^2 / D$; $V = (a/b)^2 m^2 \pi^2$; $D = E \delta^3 / 11.52$; $P_{cr} = U P / b^2$ and $\sigma_{cr} = U P / b^2 \delta$ or $\sigma_{cr} = U E \delta^3 / 12 (1 - \mu^2) b^2$ from where the critical stress/required thickness of the wall could be obtained. The results depend on the two relations – **a/b** and **h/b**. By substituting with fixed ratios **a/b** and **h/b** and with $m = 1, 2, ..., n$ it is possible to draw graphs for U. In fact interested is the first realistic form of indifferent equilibrium for $m = 1$ and the lowest P_{cr} (σ_{cr}). The respective calculations/graphs have been prepared for the practical values of **h/b = 1 / 3, 1 / 2, 1 / 1.5** and **1 / 1** and various ratios for **a / b**, but in these paper due to the restrictions, only one is shown on Fig. 7.

Figure 7

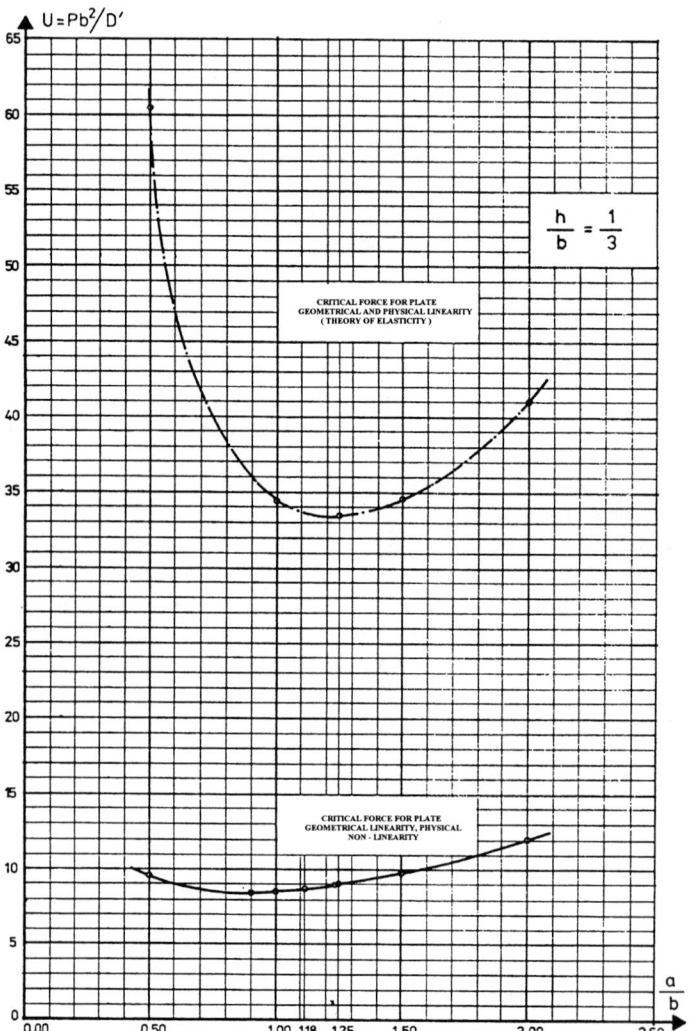

In order to simplify the work for checking of the local stability the following consideration could be done; in fact the designer will be satisfied if the thickness of the wall δ is such that the stability is ensured at the highest allowable stress/force. The **min** δ required for the local stability could be obtained from the above formulae and by solving them for δ and substituting of the critical stress with the maximum allowed stress in the concrete.

SOLUTION AS GEOMETRICALLY NON-LINEAR AND PHYSICALLY LINEAR SYSTEM

The solution is based on the Non – Linear Theory of Plates - see [3]. The deflections at each point is a function of the three co-ordinates (presented in McLoren's series):

$$u(x,y,z)=u_0(x,y,0)+\left(\frac{\partial u}{\partial z}\right)_{z=0}z+\frac{1}{2}\left(\frac{\partial^2 u}{\partial z^2}\right)_{z=0}z^2+...$$
$$v=v(x,y,z)=v_0(x,y,0)+\left(\frac{\partial v}{\partial z}\right)_{z=0}z+\frac{1}{2}\left(\frac{\partial^2 v}{\partial z^2}\right)_{z=0}z^2+ \quad (3)$$
$$w=w(x,y,z)=w_0(x,y,0)+\left(\frac{\partial w}{\partial z}\right)_{z=0}z+\frac{1}{2}\left(\frac{\partial^2 w}{\partial w^2}\right)_{z=0}z^2+...$$

But taking into account some simplifying assumptions for plates the differential equations representing the equilibrium of the geometrically non-linear plate are – see [3]:

$$\frac{1}{E\delta}\nabla^2\nabla^2\Phi=-(w,w)$$
$$D\nabla^2\nabla^2 w-\Delta(w,\Phi)-\Delta(\Phi,w)=q \quad (4)$$

where

W = w (x,y,z) is the function of the deflection and Φ = Φ (x, y) is function of the stresses. In our case the equations for the indifferent equilibrium of the geometrically non-linear plate in full are:

$$\frac{\partial^4\Phi}{\partial x^4}+\frac{\partial^4\Phi}{\partial x^2\partial y^2}+\frac{\partial^4\Phi}{\partial y^4}=\left(\frac{\partial^2 w}{\partial x\partial y}\right)^2-\frac{\partial^2 w}{\partial x^2}\frac{\partial^2 w}{\partial y^2}$$
$$D\left(\frac{\partial^4 w}{\partial x^4}+2\frac{\partial^4 w}{\partial x^2\partial y^2}+\frac{\partial^4 w}{\partial y^4}\right)-\frac{\partial^2\Phi}{\partial y^2}\frac{\partial^2 w}{\partial x^2}-\frac{\partial^2\Phi}{\partial x^2}\frac{\partial^2\Phi}{\partial y^2}+2\frac{\partial^2\Phi}{\partial x\partial y}\frac{\partial^2 w}{\partial x\partial y}=q$$

In [9] is shown that the additional axial forces in the plate (result of the geometrical non-linearity) in transverse direction P_y are very small comparing to the existing in the plate due to normal loading longitudinal axial forces P_x (P_y / P_x = **0.0048**) and hence it is not promising to follow that way for obtaining of solution for the problem.

SOLUTION AS GEOMETRICALLY LINEAR AND PHYSICALLY NON - LINEAR SYSTEM

Assumptions

The assumption that the concrete behaves as linear material will result in a serious error (see Fig. 5) which gives reasons to turn to the Non – Linear Theory of Elasticity (see [1] and [3]). The relation between the specific potential energy and tensor - deformation according to that theory is

W = W (ε_x, ε_y, ε_z, γ_{xy}, γ_{yz}, γ_{yz})

or   W = W (I_1, I_2, I_3)   (5)

where I_1, I_2 and I_3 are the three invariants of the tensor deformation.

In order to define the relationship between the stresses and strains, the theory of non-linear elasticity adopts the following assumptions:

1. the derivative of the specific potential energy for the third invariant of the tensor-deformation is equal to zero; the last means that one of the main deformations is equal to zero or close to zero and is relevant to the plates due to the plane deformation state. Also if in a system only the variations of the form are considered, then all stress states could be summarized with one Generalized Stress σ_i (to which corresponds a Generalized Strain ε_i) and then the relation stress - strain could be expressed with equations, similar to these from to the Linear Theory of Elasticity.

2. The second assumption is that the volume deformation is linear and elastic. The last allows to express the relation stress – strain formally in the same way as in the linear theory of elasticity. It should not be forgotten that in these expressions the sectional module E_c and the Poisson's constant μ for the non-linear theory of elasticity are variables but not constants and it is important to know the stress – strain curve for the Generalized Stress and Strain (which could be obtained from any loading - i.e. from compression of concrete).

After the expression of the relationship stress – strain as in the Linear Theory of elasticity two options could be followed:
- to follow the Engesser – Karman suggestion for loading and unloading zones and to work with Initial Modulus of deformation E_0 (unloading) and Sectional Modulus of deformation (loading) E_c or
- to follow Shanley suggestion without zones of loading and unloading and to work with one Sectional Modulus E_c and Poisson's constant μ different from **0.5** (which is correct for concrete – shrinking material with $\mu = $ **0.16 to 0.20**).

The second option is investigated in [2] since it is considered that the second option follows more closely the behavior of the real system – for more see [2]. Due to the similarity with the Linear Theory of Elasticity, similar to (1) differential equation for indifferent equilibrium of plates is obtained:

$$W=W(\varepsilon_x,\varepsilon_y,\varepsilon_z,\gamma_{xy},\gamma_{yz},\gamma_{xz}) \quad or \quad W=W(I_1,I_2,I_3) \qquad (5)$$

When the real stress – strain curve from Fig. 5 is used the following is obtained in [2]:

$$\bar{a}_{11}\frac{\partial^4 w}{\partial x^4}+2(\bar{a}_{12}+\bar{a}_{33})\frac{\partial^4 w}{\partial x^2 \partial y^2}+\bar{a}_{22}\frac{\partial^4 w}{\partial y^4}+\frac{12\sigma_x}{E\delta^2}\frac{\partial^2 w}{\partial x^2}=0 \qquad (6)$$

And

$$0.07\frac{\partial^4 w}{\partial x^4}+0.66\frac{\partial^4 w}{\partial x^2 \partial y^2}+0.3\frac{\partial^4 w}{\partial y^4}+\frac{P}{D}\frac{\partial^2 w}{\partial x^2}=0$$

Using the same considerations as for the geometrically and physically linear system above the following equation is valid:

$$\sqrt{\sqrt{0.98V^2+3.33UV}+1.1V}=A;$$
$$\sqrt{\sqrt{0.98V^2+3.33UV}-1.1V}=B;$$
$$(Ath\tfrac{A}{2}+Btg\tfrac{B}{2})=(2\tfrac{h}{b}+1)\sqrt{0.98V^2+3.33U} \qquad (7)$$

Where $U = P b^2 / D'$; $V = (a/b)^2 m^2 \pi^2$; $D' = E \delta^3 / 12$; $P_{cr} = U P / b^2$ and $\sigma_{cr} = U P / b^2 \delta$

The same graphs could be drawn and compared with the graphs from the solution as geometrically and physically linear elastic system. The conclusions above for the minimum required thickness δ are valid also and then **min** δ could be obtained by substituting of the critical stress with the maximum allowed stress in the concrete. The last is done for one case only on Fig. 6.

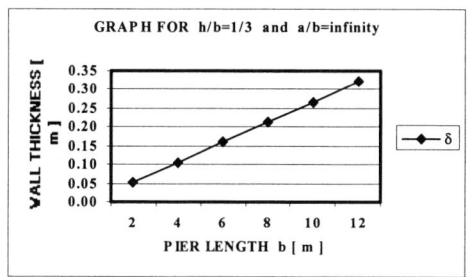

Figure 6

If it is supposed that **a/ b** approaches **infinity** then **min** δ to ensure the local stability of the wall withot diaphragms could be obtained for a range of realistic values of **b (4 to 12m)**:

$$\min \delta = b \sqrt{\frac{12 \sigma_{cr}}{E \min U}} = b \sqrt{\frac{0.006}{\min U}} \tag{8}$$

As it could be observed the relationship between δ and **b** is linear for **h/b = constant** and graphs as that shown on Fig. 6 could be drawn. By simple entering with the width of the cross section **b** and the proposed wall thickness δ in the graph for the relevant **h/b** the designer could check whether the requirements for local stability are fulfilled. The area above the line ensures stability and the area below the line represents instability.

SOLUTION AS GEOMETRICALLY NON - LINEAR AND PHYSICALLY NON - LINEAR SYSTEM

In this case is very difficult to give simple, rigorous solution and methods based on potential energy take precedence. In [2] some of the difficulties for obtaining of the solution are given and further investigations in that direction were not pursued due to the limited aims of the diploma thesis.

CONCLUSIONS

The following conclusions could be drawn as result of the investigation over the local stability of the compressed wall of hollow reinforced concrete bridge pier with rectangular cross section:

From the four known theoretical methods for solving of the problem
- the linear elastic approach gives the highest values for the critical stress – about 3.5 times:
- the geometric non-linearity of the system does not affect considerably the solution;
- the physical non-linearity has a significant influence on the solution of the problem since concrete is a material with well expressed non-linear behavior;

- the investigation of the system as a geometrically and physically non-linear involves sophisticated mathematical apparatus and will be time consuming;

1. the proposed solution of the problem as geometrically linear and physically non-linear leads to practical and way of checking the stability requirements by means of graphs or with small computer program for solving of transcendental equation. Shenley assumption seems satisfy the properties of concrete and to be closer to the modern understanding about stability problems.

REFERENCES:

1. Bezuhov N. I. (1968). *Basic Theory of Elasticity, Plasticity and Flow,*
2. Djelebov E. P. (1980). *Investigations on Local Stability of Compressed Wall of Hollow, Reinforced Concrete Bridge Pier with Rectangular Cross Section,* Diploma Thesis, Annual Proceedings of the Higher Institute of Architecture and Civil Engineering, Sofia, Bulgaria.
3. Lukash P. A. (1978). *Basic Non – Linear Mechanics,* Construction Publisher, Moscow, USSR
4. Volmir A. S. (1967). *Stability of Deformable Systems,*
5. Vurbanov C. P. (1975). *Stability and Dynamics of Elastic Systems,*
6. Vurbanov C. P. (1976). *Theory of Elasticity,*
7. Panovko Y. G. (1964). *Stability and Vibration of Elastic Systems – Modern Concepts, Paradoxes and Errors,*
8. Timoshenko S. P. (1971). *Stability of Bars, Plates and Shells,*
9. Umanskiy A.A. at al (1973), *Industrial, dwelling and Public Buildings and Structures Designer's Manual, Calculus and Theory, Part II,* Construction Literature Publisher, Moscow, USSR

STRENGTHENING OF REINFORCED CONCRETE CONTINUOUS BEAMS WITH CFRP COMPOSITES

S. A. El-Refaie, A. F. Ashour and S. W. Garrity

Department of Civil and Environmental Engineering, University of Bradford,
Bradford, West Yorkshire, BD7 1DP, UK

ABSTRACT

This paper reports the testing of five reinforced concrete continuous beams strengthened with externally bonded carbon fibre reinforced polymer (CFRP) composites. All beams had the same geometrical dimensions and internal steel reinforcement. The main parameters studied were the position and form of the CFRP composites. Three of the beams were strengthened using different arrangements of CFRP plate reinforcement and one was strengthened using CFRP sheets. The performance of the CFRP strengthened beams was compared with that of an unstrengthened control beam. Peeling failure was the dominant mode of failure for all the strengthened beams tested. The beam strengthened with both top and bottom CFRP plates produced the highest load capacity.

KEYWORDS

Reinforced Concrete, Peeling Failure, Strengthening, CFRP Composites, Continuous Beams, Deflections, Reactions, Strains, Capacity.

INTRODUCTION

Bonding plates to the external surfaces of existing reinforced concrete elements has proved to be an effective and a practical means of increasing strength and stiffness. Recently, fibre reinforced polymer composites (FRP) have been used as an alternative to steel plate external reinforcement. FRP composites have many advantages over steel plates including a high strength to weight ratio, corrosion resistance, availability in any length and ease of handling. On the other hand, FRP composites are costly and have no ductility which could lead to undesirable brittle failure of the strengthened elements.

Several research studies have been conducted on the strengthening and repair of simply supported beams (Norris et al. (1997), Swamy and Mukhopadhyaya (1999), Arduini et al. (1997) and Ritchie et al. (1991)) but very little seems to have been done on continuous beams. Sharma (1992) repaired eight two span reinforced concrete beams that had previously failed in flexure by replacing the concrete and steel reinforcement in the damaged regions with either normal concrete and new steel reinforcement or a new composite material called fibro-ferrocrete, comprised of ferrocement and fibre reinforced concrete. He concluded that the response of the repaired beams was similar to that of original beams and the performance of the beams repaired with fibro-ferrocrete was superior to and more consistent than that of beams repaired with normal concrete and reinforcement. The authors (El-Refaie et al. 2000) carried out an experimental investigation of two span reinforced concrete beams to study the effect of hogging strengthening using CFRP sheets on their strength and ductility. It was shown that increasing the length and number of CFRP layers produced a higher load capacity up to a certain limit beyond which no further improvement could be achieved. Overall, the beam ductility was reduced. This paper presents test results of five reinforced concrete continuous beams strengthened with CFRP plates and sheets. Unlike the previous tests by the authors, the steel reinforcement over the central support of the test specimens was the same as that provided at the mid-spans. Two locations of CFRP composites are examined.

TEST PROGRAMME

Five reinforced concrete two span beams were tested to failure. Test specimen geometry and reinforcement, as well as the loading and support arrangement are illustrated in Figure 1. Each beam was 8500mm long x 150mm wide x 250mm deep. The longitudinal reinforcement consisting of 4 bars of 16mm diameter was the same in each beam. Vertical links of 6mm bar diameter at 100mm centres were provided throughout each beam length in order to prevent shear failure.

The position and form of CFRP laminates were the main parameters investigated. Table 1 presents details of the external CFRP composites used. The CFRP laminates applied to the top face of the beams were 2500mm long and placed symmetrically about the line of the central support and those applied to the bottom face of the beam were 3500mm long and positioned symmetrically about the centres of both spans. Beam CB1 had no external CFRP laminates and was used as a control specimen. The top of the central region of beam CB5 was strengthened with CFRP sheets of same strength as that of the CFRP plate used for beam CB2.

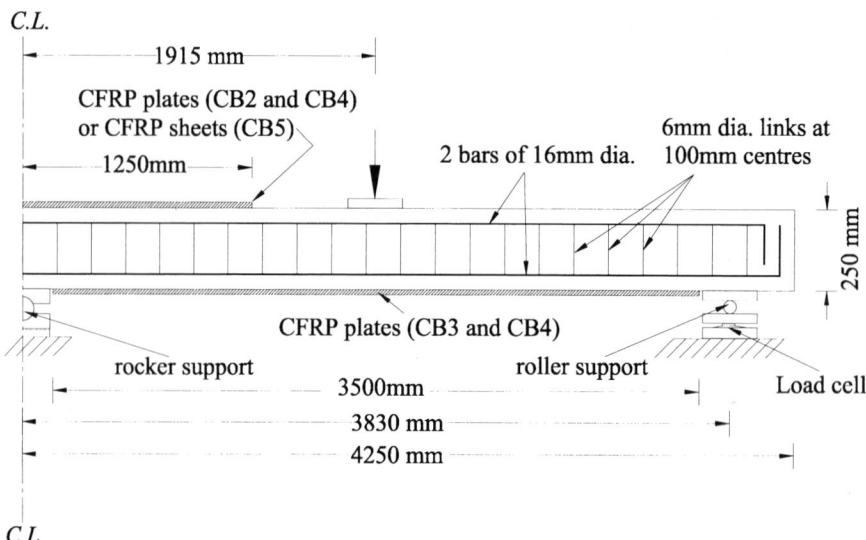

Figure 1: Geometrical dimensions, reinforcement and test rig of test specimen

MATERIAL PROPERTIES

The unidirectional CFRP sheets, CFRP plates, structural and epoxy bonding adhesives were provided by Weber and Broutin (UK) Ltd; details of the mechanical properties of these materials, taken from the manufacturers data sheets, are summarised in Table 1. The effective width of CFRP sheets was 110 mm whereas that of CFRP plates was 100 mm. The thickness of CFRP plates used was 1.2mm. For beam CB5, 6 layers of CFRP sheets of total thickness 0.702mm were used; those have an equivalent tensile strength to the CFRP plate used in CB2.

Ordinary Portland cement, sand and gravel coarse aggregate (10mm maximum size) were used to produce a concrete mix with a compressive strength of 30 N/mm² at 28 days. Three 100mm cubes and three 100x100x500mm prisms were made to provide values of the cube strength f_{cu} and the modulus of rupture f_r, respectively. The average values of those control specimens are given in Table 2. The yield strength f_y and modulus of elasticity E_s of the internal steel reinforcing bars are also shown in Table 2.

STRENGTHENING PROCESS

The concrete substrate was initially roughened by sand blasting and a 300mm straight edge was used to check that the surface deviation was within the acceptable 1mm limit recommended by the manufacturers. The quality of the substrate was then assessed by using a pull-off bond test to verify that the tensile strength was at least 1 N/mm² for applying sheets and 1.5 N/mm² for applying plates. The CFRP laminates were then applied using the appropriate bonding adhesive in accordance with the manufacturer's recommendations.

TABLE 1
DETAILS AND PROPERTIES OF CFRP COMPOSITES USED IN THE TEST SPECIMENS

Beam No.	Type	Position	f_f^* N/mm²	E_f^* kN/mm²	Bonding adhesive used	$f_a^\#$ N/mm²	$E_a^\#$ kN/mm²
CB1	none	none	none	none	none	none	none
CB2	plate	Top face	2500	150	Epoxy plus structural adhesive	19	9.8
CB3	plate	Bottom face					
CB4	plate	Both top and bottom faces					
CB5	sheet	Top face	3900	240	Epoxy plus bonding adhesive	17	5

* f_f and E_f = tensile strength and modulus of elasticity of CFRP composites, respectively.
\# f_a and E_a = tensile strength and modulus of elasticity of bonding adhesives used, respectively.

TABLE 2
PROPERTIES OF CONCRETE AND STEEL REINFORCEMENT USED IN THE TEST SPECIMENS

Beam No.	Concrete		Internal steel reinforcement					
			16mm dia. Longitudinal bars			Vertical stirrups		
	f_{cu} N/mm²	f_r N/mm²	No.	f_y N/mm²	E_s kN/mm²	No.	f_y N/mm²	E_s kN/mm²
CB1	24.0	3.0	2 bars (top) + 2 bars (bottom)	520	201	6mm closed stirrups at 100mm	308	200
CB2	43.6	4.6						
CB3	47.8	4.4						
CB4	46.1	4.4						
CB5	44.7	4.8						

TEST RIG AND RESULTS

Each test specimen comprised two equal spans; each test span of 3830mm was loaded at its mid point as shown in Figure 1. Load cells were used to measure the end support reactions and electrical resistance strain (ERS) gauges were attached to the longitudinal steel bars at the bottom mid-spans and the top over the central support to measure their surface strains. The mid-span deflections were measured using linear variable differential transformers (LVDTs). Load cell, ERS gauge and LVDT readings were recorded automatically, at each increment of the applied load, using data logging equipment.

Mid-Span Deflections

Figure 2 shows the total applied load versus mid-span deflection relationship for all test specimens. At the early stages of loading, all beams showed very similar stiffnesses to each other. After the occurrence of the first crack, beam CB3, strengthened at the mid-span soffit, exhibited higher stiffness than those strengthened over the central support (CB2 and CB5). Obviously, beam CB4 strengthened at mid-span soffit and top over the central support had the greatest stiffness of all the beams. Although the CFRP plate had a lesser Young's modulus than the CFRP sheet (see Table 1), the stiffness of the beam strengthened with CFRP sheets (beam CB5) was less than that of the beam strengthened with CFRP plate (beam CB2). This may be attributed to the lesser Young's modulus of the epoxy adhesive used to bond the sheets compared to that of the structural adhesive used to bond the plates with the concrete (see Table 1).

End support reactions

Figure 3 shows the amount of the load transferred to the end support plotted against the total applied load. As the results recorded for the two end support reactions were similar, only one end-support reaction is plotted in Figure 3. At the early stages of loading, the end support reactions of all the beams

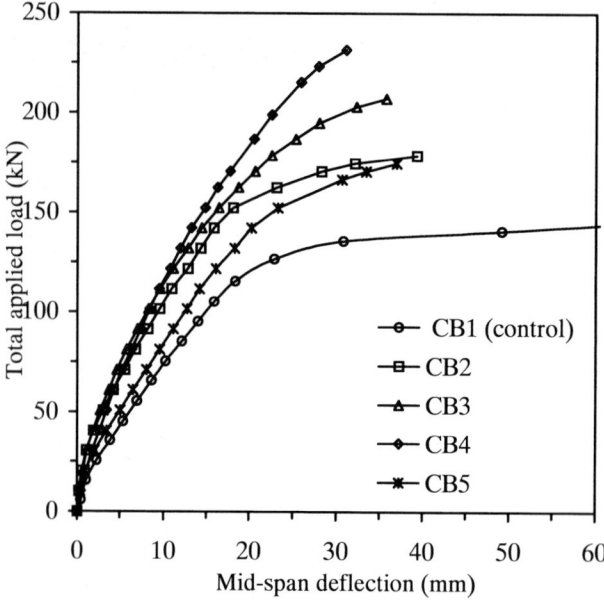

Figure 2: Total applied load against mid-span deflections

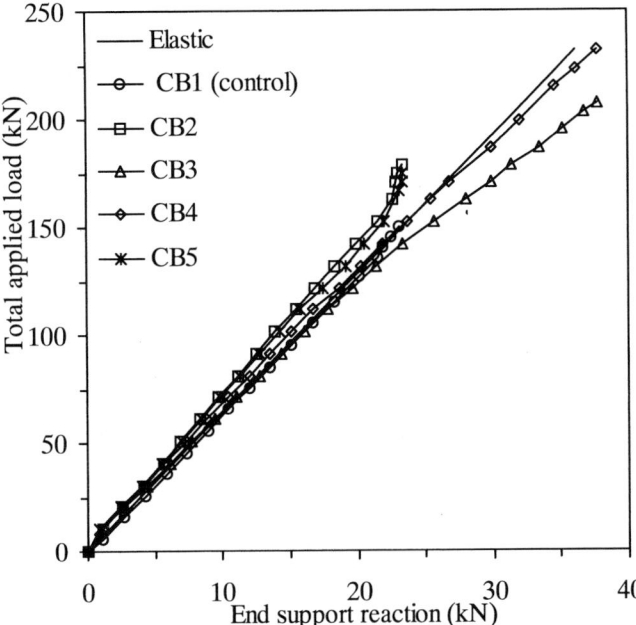

Figure 3: Total applied load against end support reaction

tested were very similar and were close to that obtained from an elastic analysis. Beams CB1 and CB4, which had uniform stiffness along the beam length, exhibited closer end support reactions to that obtained from linear elastic analysis up to failure. For the beams strengthened with CFRP composites over the central support (beams CB2 and CB5), their end support reactions were less than that of the control beam CB1. On the other hand, the end support reaction of beam CB3, which had CFRP plate at the beam soffit, was higher than that of the control beam CB1. The end support reactions of beams CB2 and CB5 strengthened with CFRP plate and sheets of equivalent strength, respectively, were very similar.

Load and moment capacities

The control beam CB1 failed in a conventional ductile flexural failure mode due to yielding of the internal steel bars in tension followed by crushing of concrete in compression over the central support and then the mid-span sections, respectively. The other four strengthened beams failed as a result of a peeling failure of the concrete cover adjacent to the external CFRP reinforcement as shown in Figure 4. The peeling failure mode was brittle, sudden and explosive and was accompanied by a loud noise.

Table 3 summarises the total failure load P_t (the two mid-span point loads) and the failure load enhancement ratio (ξ) which is the ratio of the ultimate load of an externally strengthened beam to that of the control beam. Strengthening the mid-span soffit (beam CB3) was found to give a higher ultimate load enhancement ratio than strengthening the top of the beam over the central support (beams CB2 and CB5). Beam CB4 strengthened with central support and mid-span CFRP plates failed at the highest load capacity and therefore showed the largest load enhancement ratio of all the strengthened beams tested. Using CFRP sheets of equivalent strength to CFRP plates produced nearly the same load capacity (beams CB2 and CB5).

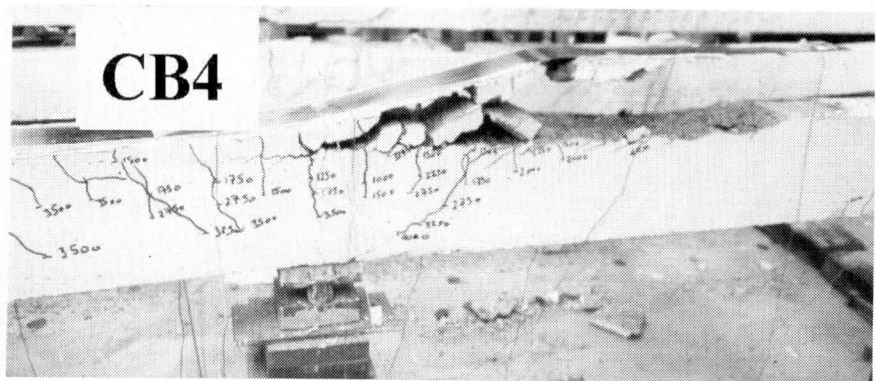

Figure 4: Typical peeling failure of CFRP composites over the central support (beam CB4)

Figure 5 shows the total applied load plotted against the hogging and sagging bending moments for the beams tested. The bending moment was calculated from the equilibrium considerations of beams using the measured end support reaction and mid-span applied load. The behaviour of all the beams at early load levels was nearly elastic. With increasing the applied load, many cracks occurred and the steel reinforcement yielded and consequently the bending moment was different from the elastic moment.

Table 3 presents the ultimate moment enhancement ratio which is the ratio of the ultimate moment of strengthened sections (hogging or sagging sections) to that of unstrengthened sections. In general, all strengthened sections resisted higher moment than the corresponding unstrengthened sections of the control beam. The ultimate moment enhancement ratio of beam CB2 is nearly the same as that of beam CB5 strengthened with CFRP sheets of equivalent strength to that of CB2 plate. By comparing the ultimate load and moment enhancement ratios of the test specimens, it appears that the ultimate moment enhancement ratio is nearly the same (beam CB4) or higher (beams CB2, CB3, CB5). The moment capacity of all the strengthened sections is increased by approximately 50% of the original (control) moment capacity. Unstrengthened sections exhibited no increase in the moment capacity.

TABLE 3

ULTIMATE LOAD (P_T), ULTIMATE LOAD ENHANCEMENT RATIO (ξ) AND ULTIMATE MOMENT ENHANCEMENT RATIO (η)

Beam No.	P_t (kN)	ξ	η	
			sagging	hogging
CB1	149.67	1.00	1.00	1.00
CB2	178.64	1.19	1.0	1.52
CB3	207.06	1.38	1.57	1.05
CB4	231.42	1.55	1.57	1.51
CB5	174.58	1.17	0.99	1.48

Internal reinforcement strains

Figure 6 presents the total applied load plotted against the tensile strains in the top steel bars over the central support and bottom steel bars at mid-spans. The yield load of the internal steel bars of the strengthened beams, that is the value of the applied load at which yielding of steel reinforcement occurred, is increased compared to that of the corresponding control beam. The yield load of the tensile steel bars was governed by the position of the external CFRP composites. Where external CFRP

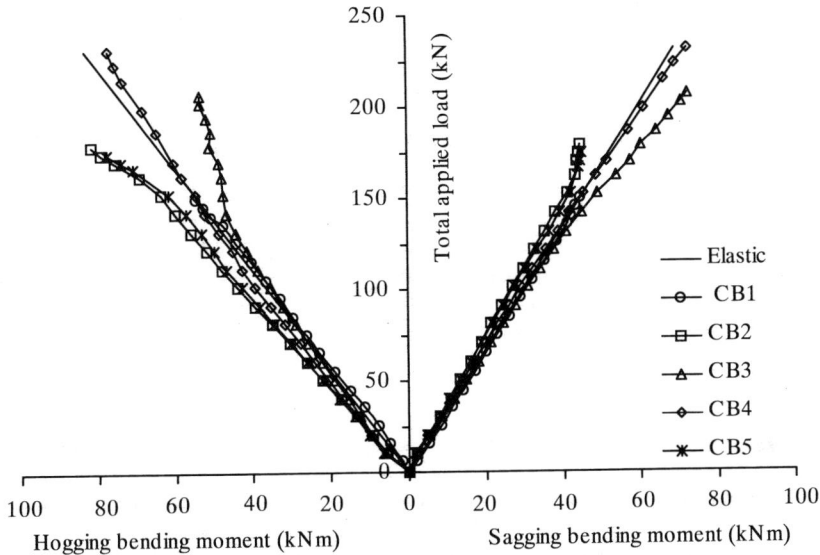

Figure 5 Load vs. bending moment of test specimens

laminates were provided, they participated in carrying tensile stresses and that reduced stresses in internal steel bars. For example, the yield load of the top steel bars over the central support is similar to that of the bottom steel bars at mid-spans for beams CB1 and CB4, but the yield load of the top steel bars over the central support is higher than that of the bottom steel bars at mid-spans for beams CB2 and CB5. Alternatively, the yield load of the bottom steel bars at mid-spans is higher than that of the top steel bars over the central support for beam CB3. CB2 steel strains in the top bars over the central

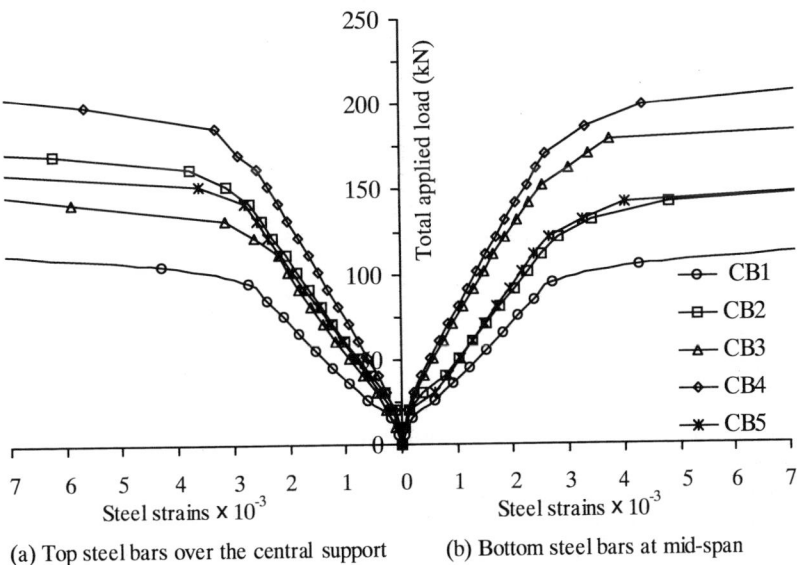

(a) Top steel bars over the central support   (b) Bottom steel bars at mid-span

Figure 6 Load vs. tensile steel bar strain of test specimens

support and bottom bars at mid-spans were very similar to those of CB5. Although most steel bars yielded, failure of the strengthened beams mainly occurred due to peeling of the CFRP composites.

CONCLUSIONS

Based on the experimental investigation presented in this paper, the following conclusions may be drawn:
- Using the CFRP composites to strengthen reinforced concrete continuous beams is an effective method; the load and moment capacities were increased by up to 55% and 57%, respectively.
- Peeling failure of the concrete cover adjacent to the CFRP composites was the dominant mode of failure for all the strengthened beams tested.
- Strengthening the mid-span soffit resulted in an increase in the end support reaction compared with that of the control beam. Conversely, the end support reaction of beams strengthened at the hogging zone over the central support was less than that of the control beam.
- Strengthening reinforced concrete continuous beams resulted in a reduction in the stresses in the steel bars compared with those in the control beam.
- The performance of the beams strengthened with plates or sheets of equivalent strength was very similar.
- Beams with mid-span soffit reinforcement had a higher load capacity than those with central support strengthening.
- Mid-span soffit and central support strengthening was found to be the most effective arrangement to give the highest stiffness and capacity.

ACKNOWLEDGEMENT

The experimental work described in this paper was conducted in the Heavy Structures Laboratory in the University of Bradford; the assistance of the laboratory staff is acknowledged. The authors are grateful to Weber and Broutin (UK) Ltd. for providing the CFRP reinforcement and the associated priming and bonding materials for the research.

REFERENCES

1. Arduini M., Di Tommaso A. and Nanni A. (1997). Brittle Failure in FRP Plate and Sheet Bonded Beams. ACI Structural Journal, 94:4, 363-370.
2. El-Refaie S. A., Ashour A. F., and Garrity S. W. (2000). Tests of Reinforced Concrete Continuous Beams Strengthened with Carbon Fibre Sheets. Proceedings of the 10th BCA Annual Conference on Higher Education and Concrete Industry, Birmingham, UK, 187-198.
3. Norris T., Saadatmanesh H. and Ehsani M. R. (1997). Shear and Flexural Strengthening of R/C Beams with Carbon Fiber Sheets. Journal of Structural Engineering, ASCE, 123:7, 903-911.
4. Ritchie P. A., Thomas D. A., Lu L. and Connelly G. M. (1991). External Reinforcement of Concrete Beams Using Fiber Reinforced Plastics. ACI Structural Journal, 88:4, 490-499.
5. Sharma A. K. (1992). Tests of Reinforced Concrete Continuous Beams Repaired with and without Fibre-Ferrocement. Concrete International, ACI 14:3, 36-40.
6. Swamy R. N. and Mukhopadhyaya P. (1999). Debonding of carbon-fibre-reinforced polymer plate from concrete beams. Proceedings of the Institution of Civil Engineers, Structures & Buildings, 134:11, 301-317.

ASSESSMENT OF A RANGE OF DESIGN MODELS FOR PREDICTING CREEP IN CONCRETE

G. C. Fanourakis[1] and Y. Ballim[2]

[1] School of Civil Engineering, Technikon Witwatersrand, P. O. Box 17011, Doornfontein, 2028, Johannesburg, South Africa.
[2] School of Civil and Environmental Engineering, University of the Witwatersrand, Private Bag 3, WITS, 2050, Johannesburg, South Africa.

ABSTRACT

Creep of concrete is a complex phenomenon which has proven difficult to model. Nevertheless, for many reinforced and prestressed concrete applications, a reasonably accurate prediction of the magnitude and rate of creep strain is an important requirement of the design process. Although laboratory tests may be undertaken to determine the deformation properties of materials, these are time consuming, often expensive and generally not a practical option. National design codes therefore rely on empirical prediction models to estimate the magnitude and development of the creep strain.

This paper considers the suitability of seven "code type" creep prediction models when compared with the actual strains measured on a range of concretes under laboratory control conditions. The concretes tested incorporate three aggregate types and two strength grades for each aggregate type. The results are compared with the predictions of creep using models contained in SABS 0100 (1992), ACI 209 (1992), AS 3600 (1988), CEB-FIP (1970, 1978 & 1990), the RILEM Model B3 (1995) methods.

The results indicate that the current SABS0100 and the CEB-FIP (1970) methods provide suitably accurate predictions over all the concretes tested. These methods gave and average coefficient of variation of 18% and a maximum value of 31%.

KEYWORDS

Concrete, creep, prediction, models, design codes, aggregates, w/c ratio

INTRODUCTION

The magnitude of creep in concrete is a design consideration which is of importance for the durability, long-term serviceability and, in the case of prestressed concrete, the load carrying capacity of structures. Its importance has been magnified by the recent tendency towards design highly stressed and slender members.

The magnitude of creep, which is required for design purposes, can be estimated at various levels. The choice of level depends on the type of structure and the quality of the data available for the design. In cases where only a rough estimate of the creep is required, which is suitable only for approximate calculations, an estimate can be made on the basis of a few parameters such as relative humidity, age of concrete and member dimensions. On the other extreme, in the case of deformation-sensitive structures, estimates are based on comprehensive laboratory testing and mathematical and computer analyses. Ideally, a compromise has to be sought between the simplicity of the prediction procedure and the accuracy of results obtained.

At the design stage, when often the only information available is the compressive strength of the concrete, the general environmental conditions of exposure and the member sizes, the designer has to rely on a design code model to estimate the extent and rate of creep strains. Given their nature, these models are not able to account for the full range of factors that are known to influence the creep deformation in concrete and simplicity of application is usually demanded by the users of the model. Nevertheless, the users of the model require some confidence as to the accuracy of the predictions as well as the range of error of the prediction.

This paper presents an assessment of seven of the more commonly used international code-type models (except for RILEM B3) which are used to predict creep strains, without the need for creep tests. The accuracy of these models was assessed by comparing experimental total creep values against those predicted at the corresponding ages by the SABS 0100 (1992), ACI 209 (1992), AS 3600 (1988), CEB-FIP (1970), CEB-FIP (1978), CEB-FIP (1990) and the RILEM Model B3 (1995) methods. The models were assessed against the strains measured on six different concretes, incorporating combinations of three aggregate types and two w/c ratios.

BASIS OF THE MODELS CONSIDERED

The models considered vary widely in their techniques, are empirically based and do not require any results from laboratory tests as input. However, certain intrinsic and/or extrinsic variables, such as mix proportions, material properties and age of loading are required as input to these models

With the exception of the RILEM Model B3 (1995), the above models derive from structural design codes of practice and express creep strain as the product of the elastic deformation of the concrete (at the time of loading) and the creep coefficient. The creep coefficient accounts for the effect of one or more intrinsic and/or extrinsic variables. The RILEM Model B3 (1995) is, by comparison, significantly more complex than the design code models and has a different structure as it enables the calculation of separate compliance functions for the basic creep and drying creep.

The SABS 0100 (1992) model is based on the British Standard method, BS 8110 (1985), with minor modifications arising from research conducted by Davis and Alexander (1992). These modifications entail the application of a Relative Creep Coefficient and the use of a different equation for the determination of the elastic modulus, depending on the aggregate type included in the concrete.

EXPERIMENTAL DETAILS

Materials

A single batch of CEM I 42,5 cement from the Dudfield factory of Alpha Cement was used for all the tests carried out in this investigation. A quartzite (Q) from the Ferro quarry in Pretoria, granite (G) from the Jukskei quarry in Midrand and andesite (A) from the Eikenhof quarry in Johannesburg were used as both the stone and sand aggregates for the concrete.

Laboratory Procedures

Mix proportions

A total of six mixes were prepared, using water/cement (w/c) ratios of 0.56 and 0.4, for each of the three aggregate types included in the investigation. For each mix, a constant water content of 195 l/m^3 was used. The w/c ratios of 0.56 and 0.4 were chosen to respectively represent typical medium and high strength concretes used in practice. Table 1 shows the mix proportions of the six concretes as well as the slump values obtained for the concretes.

TABLE 1
MIX PROPORTIONS AND SLUMP AND COMPRESSIVE STRENGTH TEST RESULTS OF THE CONCRETE USED IN THIS INVESTIGATION

Aggregate Type	Quartzite		Granite		Andesite	
Mix Number	Q1	Q2	G1	G2	A1	A2
Water (l/m^3)	195	195	195	195	195	195
Cement (kg/m^3)	348	488	348	488	348	488
19 mm Stone (kg/m^3)	1015	1015	965	965	1135	1135
Crusher Sand (kg/m^3)	810	695	880	765	860	732
w/c Ratio	0.56	0.4	0.56	0.4	0.56	0.4
a/c Ratio	5.24	3.50	5.30	3.55	5.73	3.83
Slump (mm)	90	50	115	70	95	55
Compressive Strength (MPa)	37	65	38	65	48	74

Preparation of specimens

Six 100 mm cubes were cast for each of the six mixes. In the case of each mix, three cubes were tested at seven days and three at 28 days after casting. The 28 day strength of each concrete, which is shown in Table 1, was taken as the average of the three compressive strength tests at that age. For each concrete type, six prisms, measuring 101.6 x 101.6 x 200 mm, were prepared for the creep and shrinkage testing.

Creep and associated shrinkage tests

The prisms were removed from the curing bath at age of 28 days after casting and, for each mix, three prisms were used for determining the total deformation under load. The other three prisms were used for monitoring the drying shrinkage strains in the same environment as the creep samples but in an unloaded condition. Initial elastic strains were determined by obtaining strain measurements on each of the loaded samples within 10 minutes of the application of the full load on the samples.

The loading frames are described in more detail by Ballim (1983) and are based on the ASTM C512-76 (1976) creep frame, except the load is applied by a hydraulic flat jack instead of a compressed spring. By means of an air conditioner and humidifier, the temperature and relative humidity in the room in which the frames were house was kept between 23 ± 3 °C and 65 ± 5 °C, respectively. The prisms in each of the six creep frames were subjected to a constant stress equal to 25 per cent of the 28 day compressive strength of the relevant mix. The stresses were maintained to an accuracy of ± 0,5 MPa for a period of six months.

In both the creep and shrinkage tests, strains were measured using a 100 mm Demec gauge across steel targets which had been glued onto opposite faces of the test prism. The Demec gauge is accurate to approximately 17 microstrain. Total strains were determined daily for one week, weekly until the end of one month, and approximately monthly thereafter for a total period of six months.

At each measuring period, the strain of each prism was taken as the average of the strains measured on the two opposite faces of the prism. The strain of each group of three prisms was taken as the average of the strains of the prisms in that group.

RESULTS AND DISCUSSION

The creep strain at any time was determined as:

$$\varepsilon_c(t) = \varepsilon(t) - \varepsilon_e - \varepsilon_{sh}(t) \quad (1)$$

where,
- $\varepsilon_c(t)$ = creep strain at any time t
- $\varepsilon(t)$ = measured strain on the loaded samples at any time t
- ε_e = average instantaneous elastic strain recorded immediately after loading
- $\varepsilon_{sh}(t)$ = drying shrinkage strain at any time t (from companion samples)

In order to provide a basis for comparing the creep strains of concretes with different strengths and different applied loads, the results are presented in the form of specific creep (C_c), which is defined as:

$$C_c = \frac{\varepsilon_c(t)}{\sigma} \quad (2)$$

The creep prediction methods mentioned earlier were used to predict the specific creep at the same ages at which measurements were taken for the concrete of each of the six mixes used in this investigation. It should be noted that the high-strength concretes tested (w/c =0.4) were outside of the allowable range for the use of the Australian model AS 3600 (1988). This model was therefore only used for the low strength group of concretes. Figures 1 and 2 show a comparison between the measured results for the concretes with w/c=0.56 and w/c=0.4 respectively, together with the corresponding strains predicted by the different models.

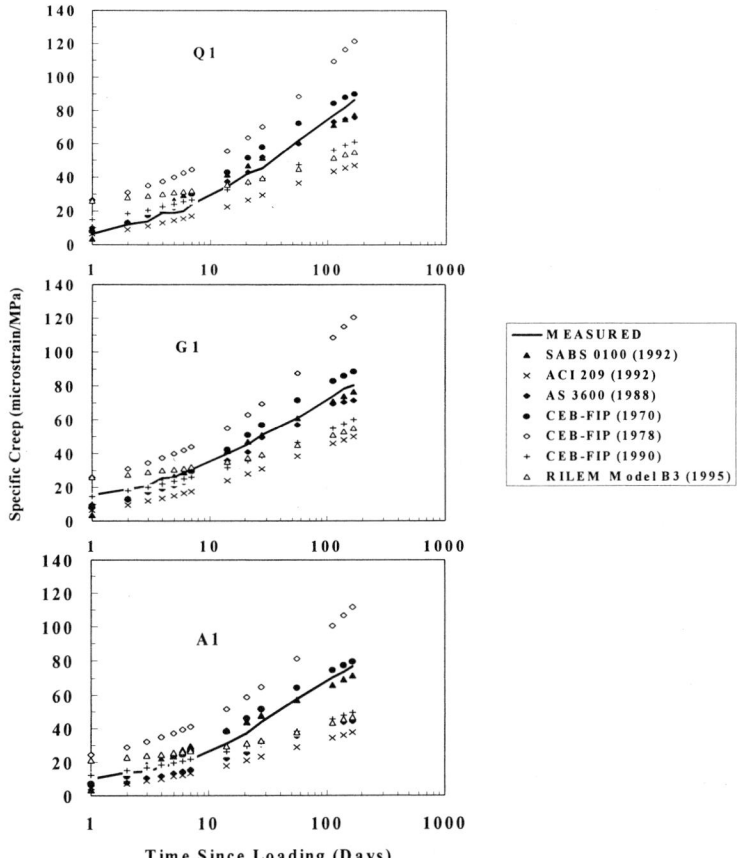

Figure 1: Comparison between the measured and predicted creep strains for the low-strength concretes

In order to provide a statistical basis for comparing the results of creep prediction methods, Bazant and Panula (1979) define a coefficient of variation of errors (ω_j) for single data sets as well for an number of data sets compared against the same prediction model (ω_{all}). The more accurate the prediction, the lower the value of ω_j. The calculated values of ω_j and ω_{all} for the different models assessed are shown in Table 2 below.

Assumptions Made in Analyses

In the case of both the SABS 0100 (1992) and BS 8110 (1985) methods, a final (30 year) creep coefficient (ϕ^*) is determined from a particular nomograph which accounts for relative humidity, age of loading and the effective thickness of the member. The lowest effective thickness shown in this nomograph is 150 mm whereas the effective thickness of the samples tested was 50 mm. A more accurate value of ϕ^* was therefore obtained by using an equation deriving from the CEB-FIP (1970) method.

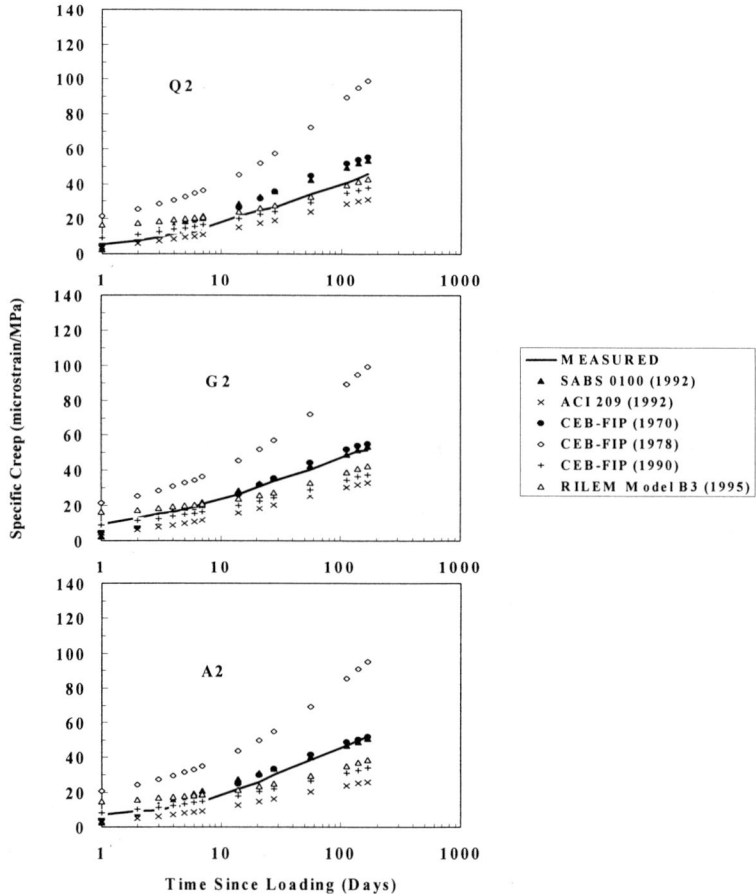

Figure 2: Comparison between the measured and predicted creep strains for the high-strength concretes

Furthermore, the SABS 0100 (1992) method allows for the estimation of a final (30 year) creep strain (ε_{cc}) of which 40%, 60% and 80% may be assumed to develop during the first month, 6 months and 30 months of loading, respectively. These values were fitted to a logarithmic curve in order to predict the creep strain at the ages at which measurements were taken on the samples. This curve takes the form:

$$\varepsilon_c(t) = \varepsilon_{cc}(0.258[\log_{10}(t)] + 0.0286) \quad (3)$$

The results indicate that the methods which yielded the most accurate predictions were the SABS 0100 (1992) and the CEB-FIP (1970) methods, both giving the lowest overall coefficients of variation (ω_{all}) of approximately 18%.

TABLE 2
COEFFICIENTS OF VARIATION FOR SPECIFIC CREEP

Prediction Method	Coefficients of Variation (ω_j)						ω_{all}
	Mix Q1	Mix Q2	Mix G1	Mix G2	Mix A1	Mix A2	
SABS 0100(1992)	16.5	31.7	9.6	9.0	15.8	17.0	18.2
ACI 209 (1992)	52.6	36.3	45.7	45.1	60.8	58.4	50.5
AS 3600 (1988)	12.5	n/a	13.4	n/a	47.2	n/a	29.2
CEB - FIP (1970)	18.1	31.3	15.0	12.3	13.9	9.9	18.1
CEB - FIP (1978)	66.0	148.6	53.9	95.1	65.6	112.8	96.1
CEB - FIP (1990)	32.7	19.8	27.7	31.2	39.6	38.3	32.2
RILEM Model B3 (1995)	45.5	29.2	33.0	21.9	45.3	32.6	35.6

The CEB-FIP (1978) showed very poor accuracy of prediction and this is probably the reason that it was superceded by the 1990 version. Nevertheless, the 1990 version does not seem to be an improvement on the 1970 version of the model. The ACI 209 model also showed poor prediction accuracy and, based on the present results, should not be recommended for use with South African concretes. In general, the methods that significantly over- or under-predicted the creep strains showed increasing deviation from the measured results with increasing time under load.

Although applied to the low-strength concretes only, the Australian model showed good prediction accuracy, except for Mix A1. Further investigation is necessary to establish if this is the result of an aggregate-type effect.

Finally, the RILEM B3 model seems to predict the creep performance of the high strength concretes better that it does that of the low strength concretes. This effect was also noted in a separate investigation by Ballim (2000). Also, the relatively poor overall coefficient of variation for this model (35.6%) does not seem to justify the significant complexity involved in its application.

CONCLUSIONS

- The seven models assessed show significant and wide variation in the magnitude of specific creep predicted over the time period considered. At the extremes, the CEB-FIP (1978) model showed the largest over-prediction, while the ACI 209 model showed the largest under-prediction.
- For the range of concretes tested, the SABS 0100 (1992) and the CEB-FIP (1970) models gave the most accurate predictions, with overall coefficients of variation of 18%.
- The Australian model, AS 3600, gave reasonable predictions for the low strength concretes while the RILEM B3 model showed better prediction accuracy for the high strength concretes tested.
- In its application to the present concretes, both the 1978 and 1990 versions of the CEB-FIP model appear not to be as accurate as the 1970 version of the same model.

REFERENCES

American Concrete Institute (ACI) (1992), ACI Committee 209, Subcommittee II. *Prediction of Creep, Shrinkage and Temperature Effects in Concrete Structures,* Report ACI 209R-92, Detroit, March, pp. 1- 12.

AS 3600 (1988) *Concrete Structures - AS 3600-1988,* Standards Association of Australia, North Sydney, pp. 8-14, 32-34.

ASTM C512-76 (1976) *Standard Method of Test for Creep of Concrete in Compression,* ASTM Book of Standards, Part 14, Philadelphia: American Society for Testing and Materials.

Ballim, Y. (1983) *The Concrete Making Properties of the Andesite Lavas from the Langeleven Formation of the Ventersdorp Supergroup,* MSc thesis, University of the Witwatersrand, Johannesburg, pp. 22-31, 67, 71-73, 80.

Ballim, Y (2000). The effect of shale in quartzite aggregate on the creep and shrinkage of concrete – A comparison with RILEM model B3. *Materials and Structures,* 2000, Vol. 33, May, pp. 235-242.

Bazant, Z.P. and Panula, L. (1979) Practical prediction of time dependent deformations of concrete, Parts I-VI, *Materials and Structures,* Vol. 12, pp. 169-183.

BS 8110 (1985) *Structural Use of Concrete, Part 2, Code of Practice for Design and Construction,* London, British Standards Institution.

CEB-FIP (1970) Comité Europeen du Béton - Federation Internationale De La Precontrainte, *International Recommendations for the Design and Construction of Concrete Structures, Principles and Recommendations,* FIP Sixth Congress, Prague, pp. 27-28.

CEB-FIP (1978) Comité Euro-International du Béton - Federation Internationale De La Precontrainte, *International System of Unified Standard Codes of Practice for Structures, Volume II - CEB-FIP Model Code for Concrete Structures,* 3rd ed. Lausanne, pp. 56, 331-344.

CEB-FIP (1990) Comité Euro-International du Béton, *CEB-FIP Model Code 1990, First Draft,* Lausanne, Mar., pp. 2-3, 2-28 to 2-40 (Information Bulletin No. 195).

Davis and Alexander (1992) *Properties of Aggregates in Concrete (Part 2),* Hippo Quarries Technical Publication, Sandton, South Africa, Mar., pp. 1-27, 42-43, 46-47.

RILEM Model B3 (1995) Creep and shrinkage model for analysis and design of concrete structures - model B3, draft RILEM Recommendation, prepared by Bazant, Z.P. and Baweja, S., *Materials and Structures,* Vol. 28, pp. 357-365, 415-430, 488-495, with Errata in Vol. 29 (1996) pp. 126.

SABS 0100 (1992) *Code of Practice for the Structural Use of Concrete, Parts 1 & 2, Part 1 : Design. Part 2 : Materials*; Pretoria: South African Bureau of Standards.

EFFECTS OF CEMENT TYPE AND WATER TO CEMENT RATIO ON CONCRETE EXPANSION CAUSED BY SULFATE ATTACK

Seb J. Ficcadenti[1]

[1] Ficcadenti and Waggoner Consulting Structural Engineers
Irvine, California 92614, USA

ABSTRACT

When groundwater or soils containing concentrations of sulfates come in contact with concrete, a reaction takes place that may be harmful to the concrete. Specifically, the sulfate reacts with the hydrated tricalcium aluminate in the cement to form crystals of ettringite that occupy a larger volume than the reactants. The forming of these crystals can cause expansion and distress in the concrete. Previous research has suggested that cements with limited amounts of tricalcium aluminate and low water to cement ratios are both necessary to control this reaction and extend the service life of concrete in sulfate environments. Much of this research relies on data collected over short periods of time, extrapolated over the service life of the concrete using linear models of the relationship between sulfate attack expansion and time.

Data collected by the Bureau of Reclamation, United States Department of the Interior, over 18 years of accelerated testing, designed to model a period of time exceeding 100 years, were analyzed as part of this investigation. The analysis of this data demonstrates that the water to cement ratio of the concrete has only a limited impact of the sulfate attack expansion of the concrete for each type of cement tested. The service life of concrete in sulfate environments made with sulfate resistant cements do not necessarily benefit from the use of low water to cement ratios. The analysis of the data demonstrates that the relationship between the sulfate attack expansion and time is not linear. This relationship varies from an exponential curve for non-sulfate resistant cements with low water to cement ratios, to an S-curve for sulfate resistant cements with high water to cement ratios.

KEYWORDS

Water to cement ratio, sulfate attack, expansion, cement type, ettringite, tricalcium aluminate.

INTRODUCTION

In 1954 the United States Bureau of Reclamation started a set of long term, accelerated tests to determine the effects of cement type and water to cement ratio on the resistance of concrete to chemical sulfate attack. These tests were conducted at their research facilities in Denver, Colorado, in the United States of America. The early results of these tests were reported in 1958 after completing the first three and one half years of measurements (Smith, 1958). The testing continued for a total of 18 years, simulating a period in excess of 100 years. However, the findings of the completed tests have never been published. Fortunately, the Bureau of Reclamation has maintained a complete set of records of these tests and they are available to the public.

The scope of this investigation has been to review and analyze all the test results obtained from the 600 test specimens included in the above study. The specimens were tested for changes in their properties of expansion, compression strength, modulus of elasticity, dynamic modulus and weight as they were exposed to very severe levels of sulfates in solution under an accelerated test protocol. The specimens reserved for compression and modulus of elasticity testing were used up in the first three years and were reported in the 1958 publication. However, the remaining specimens continued to be tested for expansion, dynamic modulus and weight until 1972. The measurements of expansion are the primary focus of this investigation.

The test results have considerable value considering the full range of concrete mix designs investgated in the specimens tested, the sophisticated equipment used to conduct the tests and the extended period of time over which the test results were obtained. The specimens tested consisted of a full test matrix of concretes composed of mix designs containing combinations of the five major cement types and four water to cement ratios. The equipment and facilities necessary to properly conduct these tests were developed by the Bureau of Reclamation and maintained over the extended testing period.

BACKGROUND

The American Concrete Institute publication ACI-201, entitled "Guide to Durable Concrete," recommends criteria for the design of concrete exposed to sulfate containing soils and solutions (ACI, 1992). These criteria consist of specifying mix designs containing sulfate resistant cements, such as types II and V, and low water to cement ratios. These criteria are in large part based on research conducted by the Bureau of Reclamation, the Portland Cement Association, and the Corps of Engineers (Kalousek, 1976). However, the findings of each of these research projects rely heavily on linear extrapolation in order to predict long-term behavior (Kalousek, 1972). Recent studies to develop empirical models to predict concrete expansion caused by sulfate attack also rely on a linear relationship between the expansion of concrete exposed to sulfate containing solutions and time (Kurtis, 2000). It is this assumption of linearity between concrete expansion and time that provides the foundation for the design criteria in ACI-201 recommended for concrete exposed to sulfate environments.

The process of chemical sulfate attack of concrete can be explained simply as three reaction steps (Mindess, 1999). First, the soluble sulfate ions penetrate into the concrete from the surrounding environment. Second, the sulfate ions react with the calcium hydroxide to form gypsum. Third, the gypsum reacts with the tricalcium aluminate to form the expansive compound, ettringite.

The process by which soluble sulfates enter hardened concrete and react with the available tricalcium aluminate to form ettringite is controlled in large part by the finite amount of tricalcium aluminate available in the cement for this reaction. The reaction can reasonably be modeled as the depletion of this fixed resource. The behavior of the depletion of a fixed resource is seldom linear. Most often the depletion of a fixed resource exhibits an accelerating or exponential behavior at first as the exposure to the reactants becomes wide spread, then the reaction slows as the fixed resource becomes scarce, eventually stopping altogether as the fixed resource is consumed. This type of reaction typically follows an S-curve relationship rather than a linear one.

The assessment of this behavior is complicated by the fact that expansion measurements can only be taken until the expansion causes the concrete to break apart. For example, when non-sulfate resistant cements are used the concrete can expand rapidly and break apart while the reaction is still accelerating. In this case only the early stages of the S-curve behavior can be measured.

The accelerated tests that are the subject of this investigation confirm that concretes exposed to sulfate containing solutions do not exhibit a linear relationship between expansion and time. Instead, their behavior follows an S-curve relationship, consistent with the depletion of a fixed resource model.

TEST MATRIX

The test matrix for this study consisted of 20 different mix designs, each represented by 30 test specimens. The total of 600 test specimens were divided into five groups of 120 specimens each. Each of the groups was made from concrete using one of the five basic cement types. The physical and chemical properties of each of the cement types are given in Table 1. Each specimen consisted of a cylinder of concrete three inches in diameter by six inches long. Each group of 120 specimens was composed of 4 sub-groups. The water to cement ratios used in the mix design of the four sub-groups were 0.45, 0.52, 0.58 and 0.65. The constituents of the twenty mix designs used in the study are given in Table 2.

Table 1: Physical and Chemical Properties of Cements

Cement Types	Type I	Type II	Type III	Type IV	Type V
Physical Properties					
Laboratory No.	M-1763	M-1700	M-2028	4495	6358
Specific Gravity	3.16	3.19	3.29	3.22	3.18
Normal Consistency, %	27.0	26.2	27.8	21.8	23.2
Air in Mortar, %	12.0	10.0	11.0	6.3	5.0
Passing No. 325 Screen, %	90.7	93.4	95.9	90.6	90.92
Blaine Surface, cm^2/gram	3524	3314	4531	3880	---
Chemical Analysis					
SiO_2, percent	21.43	22.27	20.54	22.92	24.44
Al_2O_3, percent	6.16	5.44	6.33	4.89	3.32
Fe_2O_3, percent	2.35	3.80	2.19	4.92	3.21
CaO, percent	65.03	63.46	63.41	59.40	63.32
MgO, percent	1.17	1.08	1.66	3.19	1.40
Na_2O, percent	0.07	0.07	0.08	0.13	0.12
K_2O, percent	0.47	0.48	0.67	1.17	0.13
SO_3, percent	2.37	1.72	3.09	1.93	1.42
Chemical Compounds					
C_3S, percent	50.3	42.1	47.5	20.3	40.9
C_2S, percent	23.6	32.1	23.1	50.5	39.3
C_4AF, percent	7.1	11.6	6.7	15.0	9.8
C_3A, percent	12.4	8.0	13.1	4.6	3.4
$CaSO_4$, percent	4.0	2.9	5.2	3.3	2.4

Each sub-group of 30 specimens was cured three different ways. Fourteen specimens were cured in a fog room for 14 days and then exposed to 50 percent relative humidity for 14 days, prior to exposure to the accelerated sulfate test protocol. Six specimens were cured for 28 days in the fog room, prior to exposure to the accelerated sulfate test protocol. The remaining ten specimens were cured continuously in the fog room and tested for compression strength at the same ages as the companion cylinders subjected to the accelerated sulfate test.

Six specimens of each design mix were used to measure expansion, dynamic modulus and weight properties throughout the test period, unless the specimens became so damaged by expansion that the measurements could no longer be made. Three of these specimens of each design mix were cured for 14 days and three were cured for 28 days in the fog room. The remaining specimens were used for compression testing to evaluate changes in strength and modulus of elasticity.

TEST PROTOCOL

To measure the relative resistance of concrete to sulfate attack, the Bureau of Reclamation employed two testing protocols at the time these tests were conducted, the continuous soaking test protocol and the accelerated soaking and drying test protocol. Standard three-inch by six-inch cylindrical concrete specimens, with inserts to permit length measurements, were employed in both types of tests.

Table 2: Constituent Makeup of Each Mix Design Tested

Mix Design Number	Cement Type	Water to Cement Ratio	Weight of Cement (N/m^3)	Weight of Water (N/m^3)	Weight of Sand (N/m^3)	Weight of Gravel (N/m^3)
31	I	0.45	335	151	740	1120
37	I	0.52	292	152	768	1105
38	I	0.58	262	152	796	1108
32	I	0.65	222	145	834	1115
39	II	0.45	351	158	807	1042
40	II	0.52	298	155	831	1024
41	II	0.58	265	153	864	1033
42	II	0.65	232	151	875	1019
47	III	0.45	348	156	797	1029
51	III	0.52	303	158	826	1018
49	III	0.58	266	155	854	1022
50	III	0.65	237	154	871	1015
52	IV	0.45	327	147	825	1061
53	IV	0.52	284	147	877	1050
45	IV	0.58	260	151	855	1023
46	IV	0.65	234	152	882	1027
33	V	0.45	356	160	746	1127
35	V	0.52	298	155	765	1099
36	V	0.58	266	154	792	1099
34	V	0.65	238	155	828	1105

In the continuous soaking test protocol, the concrete cylinders were continuously submerged in a 2.1 percent sodium sulfate solution at room temperature. Although this test readily identifies non-sulfate resistant concretes, it requires many years to prove the relative merits of comparatively resistant concretes. For this reason, the Bureau of Reclamation developed the accelerated soaking and drying test protocol. The accelerated test protocol was used for the tests in this investigation.

The accelerated soaking and drying test protocol requires approximately one-sixth the time of the continuous soaking tests. In this protocol, test specimens are alternately subjected to 16 hours of soaking in a 2.1 percent sodium sulfate solution maintained at 73 degrees Fahrenheit and 8 hours of drying in air under forced draft at 130 degrees Fahrenheit. The accelerated tests were run using automated equipment consisting of an elevated sulfate solution tank, an insulated specimen storage cabinet equipped with time controls for pumping the sulfate solution to and from the elevated tank, and an electric heater and blower for circulating warm air through the specimen cabinet.

For both test protocols, test specimens were measured periodically in a surface-dry condition to determine changes in length, weight and dynamic modulus of elasticity. For both test protocols, failure was defined as an expansion of 0.5 percent or a 40 percent reduction in dynamic modulus, whichever occurred first.

TEST RESULTS

The results of the expansion testing were consistent for both concretes fog room cured for 14 days and concretes fog room cured for 28 days. The longer curing period only delayed the expansion of the concrete slightly. The results of the expansion testing for the concrete cured for 28 days with water to cement ratios of

0.45 and 0.65 are provided in Figure 1. For clarity, the test results for concretes with type III and type IV cements are not shown. The concretes with type III cement behaved similarly to the concretes with type I cement and the concretes with type IV cement behaved similarly to the concretes with type II cement.

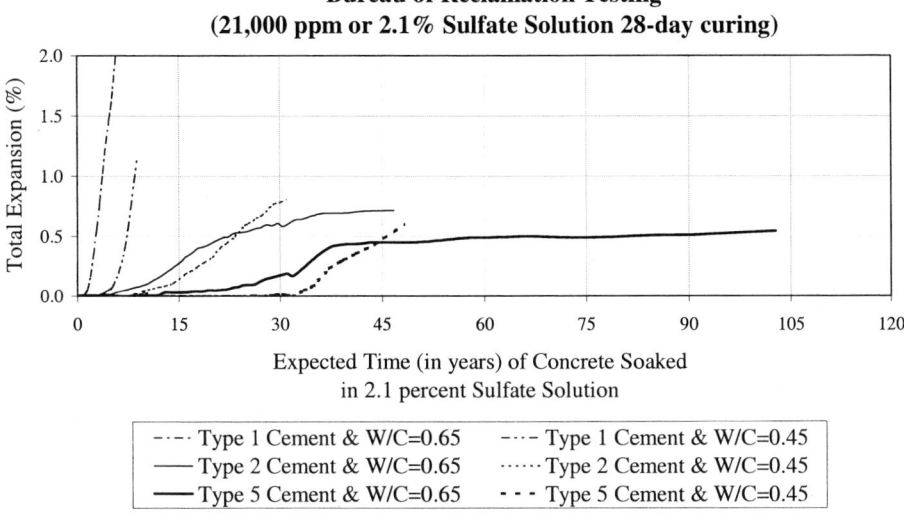

Figure 1: Expansion Test Results from Accelerated Soaking and Drying Protocol

The concretes with cement types I and III expanded very rapidly and broke apart before the S-curve behavior could be confirmed. However, the concretes with cement types II, IV and V all displayed the S-curve behavior. The concretes with type V cement, containing the least amount of tricalcium aluminate, exhibited the least amount of expansion and the expansion took place over the greatest time period. These results confirm that the expansion potential of the concrete is determined primarily by the quantity of tricalcium aluminate in the cement. In addition, these results confirm that an S-curve behavior is exhibited as the fixed amount of tricalcium aluminate is depleted in the reaction to form ettringite, not a linear behavior as previously assumed.

The test results indicate that water to cement ratio has only a limited impact on the expansion of the concrete exposed to sulfate solutions. Figure 2 provides the results of expansion testing on concretes made with type II cement at the four different ratios of water to cement. Water to cement ratio had little impact on controlling the expansion of these concretes exposed to sulfate solutions. The comparative characteristics of these curves are that the lower water to cement ratio concretes expanded slower at first and more rapidly later in the test as compared to the higher water to cement ratio concretes.

Similarly, Figure 3 provides the results of expansion testing on concretes made with type V cement at the four different ratios of water to cement. Again, water to cement ratio had little impact on controlling the expansion of these concretes exposed to sulfate solutions. As was the case with the concretes made with type II cement, the comparative characteristics of these curves are that the lower water to cement ratio concretes expanded slower at first and more rapidly later in the test as compared to the higher water to cement ratio concretes. However, it is interesting to note that the concretes with type V cement and the highest water to cement ratios exhibited the least amount of expansion and maintained their integrity for the entire testing period.

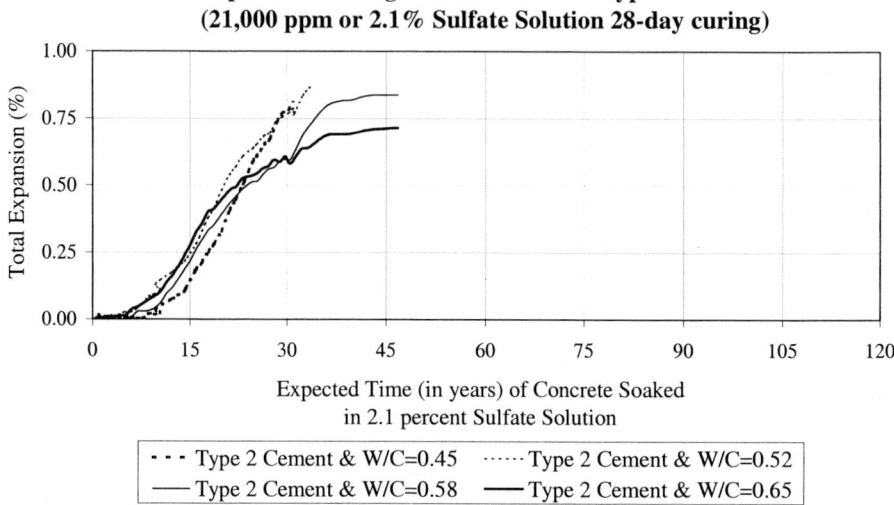

Figure 2: Expansion Test Results for Concretes with Type II Cements

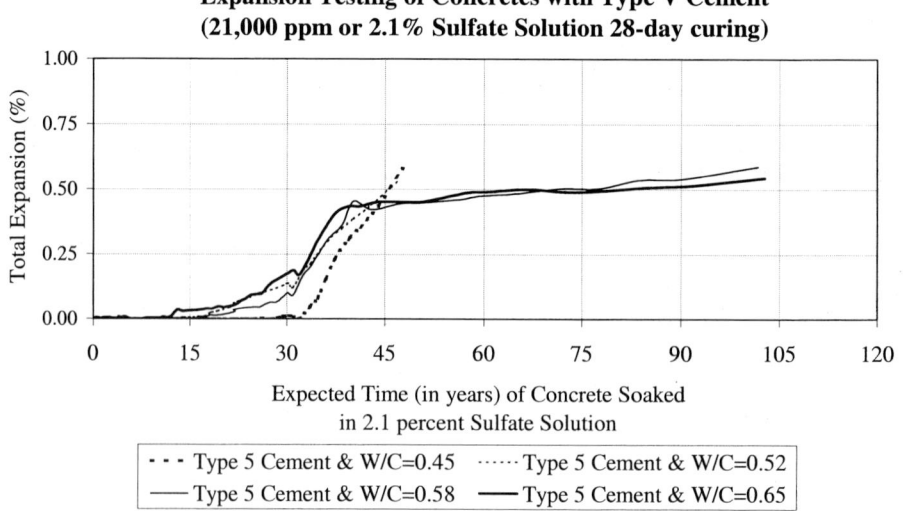

Figure 3: Expansion Test Results for Concretes with Type V Cements

CONCLUSIONS

The S-curve behavior of the test results shown in Figures 2 and 3 are consistent with the behavior of a depleting fixed resource system. The depletion of the tricalcium aluminate in the reaction that forms ettringite, and the resulting expansion, is slow at first as the sulfate solution penetrates the concrete. As the sulfate solution becomes more widespread within the concrete the rate of expansion increases. As the bulk of the tricalcium aluminate is consumed, the chemical becomes scarce and the expansion slows. Once the available chemical is used up, then the expansion stops.

It has been difficult to observe this behavior because test specimens with low water to cement ratios and specimens with cements containing higher amounts of tricalcium aluminate break apart before the expansion slows. Therefore, measurements of the slowing expansion could not be made on such test specimens. This difficulty has contributed to previous investigators falsely concluding that the behavior of sulfate attack expansion with time is linear.

The test results analyzed in this study investigated a wide variety of concrete mix designs, including those made with sulfate resistant cements and high water to cement ratios. Concretes with cements containing low quantities of tricalcium aluminate and possessing higher void spaces in the cement paste are best able to absorb their expansion potential and maintain their integrity until the sulfate attack reaction slows. This phenomenon is well illustrated in the test results analyzed in this investigation.

The water to cement ratio of the concrete has little impact on chemical sulfate attack expansion over time. These test results indicate that concretes with the same cement type exhibit similar behavior regardless of water to cement ratio. The only difference noted is that concretes made with lower water to cement ratios, when exposed to sulfate attack, expand slightly slower in the short term and slightly faster over the long term, as compared to comparable concretes with higher water to cement ratios.

REFERENCES

Smith, F. L. (1958), Effect of Cement Type on the Resistance of Concrete to Sulfate Attack. *United States Department Of the Interior Bureau of Reclamation Report No. C-828.*

Kalousek, G. L., Porter, L. C., & Benton, E. J., (1972). Concrete for Long Term Service in Sulfatre Environments. *Cement and Concrete Research* **2**, 79-89.

Kalousek, G. L., Porter, L. C., & Harboe, E. M., (1976). Past, Present, and Potential Developments of Sulfate-Resisting Concretes. *Journal of Testing and Evaluation* **4:5**, 347-354.

Amercian Concrete Institute Committee 201, (1992). *Guide to Durable Concrete*. American Concrete Institute, Michigan, USA.

Young, J.F., Mindess, S., Gray R. J., & Bentur, A., (1998). *The Science and Technology of Civil Engineering Materials*, Prentice Hall, New Jersey, USA.

Kurtis, K. E., Monteiro, P. J. M., & Madanat, S. M., (2000). Emperical Models to Predict Concrete Expansion Caused by Sulfate Attack. *American Concrete Institute Materials Journal* 97:2, 156-162.

IMPROVED MODELLING OF ELECTRICAL SUBSTATION EQUIPMENT FOR SEISMIC LOADS

G.C. Pardoen[1], R. Villaverde[1], R. Tavares[1], S. Carnalla[1]

[1]Department of Civil & Environmental Engineering, University of California, Irvine
Irvine, CA 92697-2175 USA

ABSTRACT

Electrical utilities use different types of substation equipment that have dynamic characteristics that are not well known for many classes of equipment. Manufacturer differences in design, method of structural support, physical dimensions, voltage class differences as well as differences at the component level are just some of the features that introduce variations of dynamic behavior. Although utilities have been specifying that equipment be seismically qualified by analysis for many years, there is uncertainty in the properties and methods of modeling used in these analyses.

This paper suggests improvements in the methods used for modeling electric substation equipment through a combination of experimental and analytical studies.

The experimental studies were a combination of in-situ field measurements of this equipment at four sites selected from Pacific Gas & Electric facilities and laboratory experimental modal analyses. The in-situ tests considered the transmissibility of the ambient and forced vibration ground motion to the support locations of the equipment as well as the low level response of the equipment itself due to force-calibrated hammer excitation. The experimental modal analyses considered the linear response of the equipment (frequency, mode shape, damping) under low level, force-calibrated hammer excitation. Improved experimental procedures are noted.

The analytical studies interpreted the vibration data using commercially available software in order to define the frequency, mode shape and damping characteristics of the equipment. Multiple degree-of-freedom, lumped parameter modal models were developed for the equipment. Simple analytical models were developed that closely matched the experimental data in order to predict the transformers' dynamic response under dynamic loading. These modal models can be integrated with standard finite elements to model structural response modifications due to mass or stiffness changes resulting from support structures. The analytical studies recommend improved methods for modeling equipment and major components.

KEYWORDS

Experimental Vibrations, Substation Equipment, Seismic Qualification, Finite Element Modeling

BACKGROUND

Electrical utilities use different types of substation equipment that have dynamic characteristics that are not well known for many classes of equipment. Manufacturer differences in design, method of structural support, physical dimensions, voltage class differences as well as differences at the component level are just some of the features that introduce variations of dynamic behavior. Although utilities have been specifying that equipment be seismically qualified by analysis for many years, there is uncertainty in the properties and methods of modeling used in these analyses.

An analytical and experimental study was conducted to improve the analysis capability when assessing the seismic qualification of specific electrical substation equipment. One goal of the experimental study was to quantify the transmissibility between the mounting base of electrical substation bushings and the foundation of the substation transformer that typically supports one or more of these bushings. See Figure 1. Another goal was to obtain the in-situ vibration characteristics of the bushings themselves when mounted to a substation's transformer. Based upon the experimental investigations, simple analytical models were developed, which closely matched the experimental data, in order to predict the transformers' dynamic response under dynamic loading.

Figure 1 – Electrical Substation Transformer and Bushing

The study included the field vibration tests of typical 230-kV and 500-kV transformers at four sites selected from representative Pacific Gas & Electric facilities as well as corroborative finite element analyses of the substation equipment. Additionally an experimental modal analysis was conducted at the Richmond Field Station of the University of California, Berkeley on a bushing mounted on a shake table mounting-stand.

FIELD VIBRATION TESTS

The experimental vibration tests obtained (a) the frequencies and damping characteristics for a variety of bushings as well as (b) the transmissibility between the transformer's base and each bushing's base.

The bushings' vibration characteristics were obtained from the experimentally derived frequency response functions using the ME'Scope modal identification software [1]. The frequency response functions were obtained at strategic positions along the length of a bushing by (a) measuring the acceleration response at these locations due to (b) the excitation provided by an APS 100-lb seismic shaker. The time domain data were recorded directly to a 16-channel, Hewlett Packard 3565 Dynamic Signal Analyzer which, in turn, were converted to frequency domain functions such as power spectra, coherence functions, frequency response functions, etc. An output channel of the HP 3565 DSA provided the band-limited, pure random input the horizontal shaker.

The transmissibility was calculated from the frequency domain quotient of the bushing's base response to the transformer's ground-level base response. Figure 2 depicts the Kinemetrics Ranger SS-1 seismometers used in the transmissibility study. The transmissibility study was undertaken to assess the adequacy of the amplification factor of 2.0 specified by the Institute of Electrical and Electronic Engineers in Standard 693-1997 for the seismic qualification of transformer bushings [2].

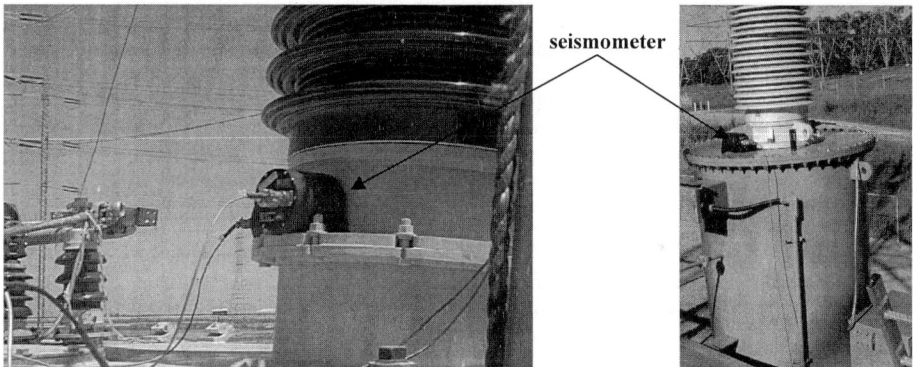

Figure 2 – Transmissibility Study Using Kinemetric Ranger SS-1 Seismometers

To supplement the field vibration tests, an experimental modal analysis was performed at the Richmond Field Station of the University of California, Berkeley on a bushing that had been previously subjected to shake table tests. See Figure 3. The thrust of these analyses was to use an alternative excitation source to the APS seismic shaker – a force-calibrated hammer – to obtain the frequency response functions of a representative bushing. In particular, it was anticipated – and the laboratory experiment confirmed – that the impact excitation would provide 'better' frequency response functions than those obtained using the APS seismic shaker as the excitation.

Using the coherence function as the measure of frequency response function quality, the impact technique not only provided better quality measurements at the Ignacio, and fourth, field substation test but subsequently at the Richmond Field Station. While Pacific Gas & Electric's management had been reluctant throughout the test program to use an impact device on a field-operational bushing for fear of the hammer cracking the ceramic 'skirts', the use of the force calibrated hammer at the last substation provided the evidence that the hammer was superior to the shaker for input and that better quality measurements, with not visible damage to the bushing itself, resulted.

 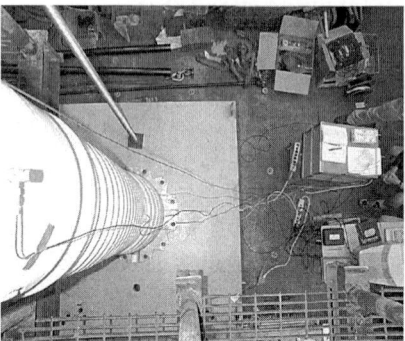

Figure 3 – Experimental Modal Analysis of Bushing at Richmond Field Station

RESULTS

Relatively simple analytical models were constructed of the transformer (which acted much like a large, rectangular prism rigid body) and the bushing in order to assess the bushing's dynamic response to different earthquake excitations. Amplification factors were then computed for each of the excitations considered, defining these amplification factors as the peak shear force at the base of the bushings over the corresponding shear force obtained when the bushings were assumed to be mounted directly to the ground. It was found that the amplification factors for the 500-kV bushings were within the amplification factor of 2.0 specified by Standard IEEE 693, but not those of the 230-kV bushings, which exceeded the IEEE standard by as much as a factor of two.

Representative analytical models for two 500-kV transformer/bushing substations are shown in Figures 4 and 6. The analytical/experimental corroborative data for fundamental frequencies for these two substations are shown in Figures 5 and 7.

IMPROVED MODELING

Transformers in electric power substations have experienced severe damage during past earthquakes. For example, transformers in substations owned by Pacific Gas & Electric were damaged during the 1989 Loma Prieta earthquake and, more recently, extensive substation damage was reported after the 1999 Taiwan earthquake. Most of the damage in the Loma Prieta earthquake was to the 230-kV and 500-kV transformers; the damage consisted of cracked porcelain bushings, anchorage failure, and development of leaks [3]. Damage to substations owned by Southern California Edison following the 1994 Northridge earthquake was characterized by transformer problems with their bushings, anchorage, radiators, lightning arrestors and conservator tanks [4].

Clearly any improvement in modeling of the complex electrical substation hardware must entail a combination of experimental and analytical techniques. While the experimental/analytical study conducted for the four Pacific Gas & Electric substations for provided reasonable data for modeling of bushings, it was clear that improvements could be in subsequent experiments. While the following experimental improvements have been submitted to Pacific Gas & Electric for subsequent testing of substation equipment, the follow-on project has yet to be funded.

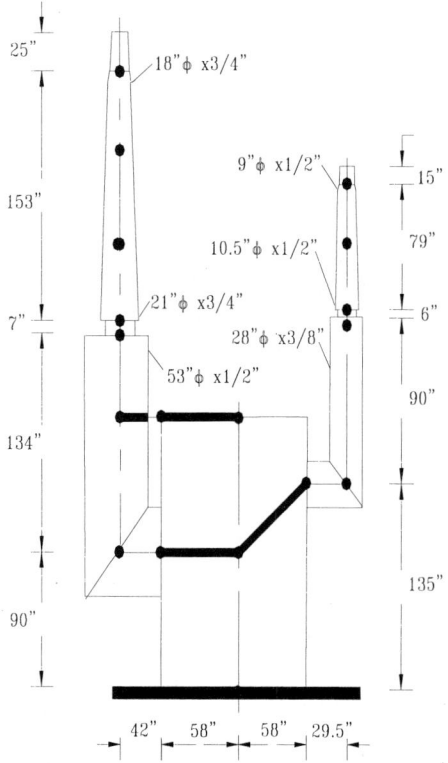

Figure 4 – Analytical Model of 500-kV Pauwels Transformer

Mode	Direction	Experimental Frequency (Hz)	Analytical Frequency (Hz)
1	Longitudinal	3.4	3.5
2	Transverse	3.4	3.5
3	Transverse	5.1	5.1
4	Longitudinal	5.6	5.5
5	Transverse	5.8	5.5
6	Longitudinal	6.4	6.4
7	Longitudinal	13.1	18.6
8	Transverse	-	18.8
9	Transverse	-	21.2
10	Longitudinal	-	23.1
11	Transverse	-	24.9
12	Longitudinal	-	30.5

Figure 5 – Analytical/Experimental Results of 500-kV Pauwels Transformer

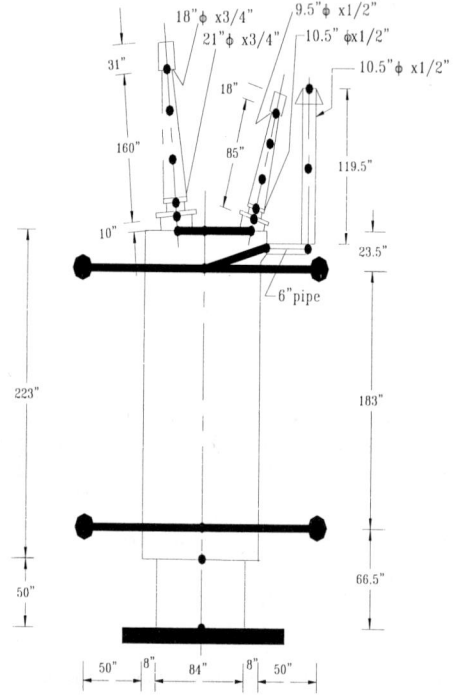

Figure 6 – Analytical Model of 500-kV Westinghouse Transformer

Mode	Direction	Experimental Frequency (Hz)	Analytical Frequency (Hz)
1	Both	2.4	2.4
2	Both	2.4	2.5
3	Longitudinal	3.1	3.6
4	Transverse	3.6	3.6
5	Longitudinal	4.6	6.0
6	Transverse	6.2	6.0
7	Longitudinal	8.2	7.5
8	Transverse	9.2	9.3
9	Both	10.1	11.1
10	Both	12.0	11.5
11	Both	19.1	15.3
12	Both	20.2	20.3

Figure 7 – Analytical/Experimental Results of 500-kV Westinghouse Transformer

IMPROVED EXPERIMENTAL/ANALYTICAL INVESTIGATION OF BUSHINGS

An improved experimental vibration approach requires a combination of laboratory and field-testing of substation equipment. The laboratory tests should consist of both force-calibrated hammer and shake table excitation whereas the field-testing should use ambient and eccentric mass shaker excitation. The experimental work is purposely divided into laboratory and field studies so that the vibration characteristics of the substation equipment, such as bushings, can be investigated from two different perspectives. Laboratory testing provides more in-depth, iterative investigations using approximate support conditions whereas field-testing usually provides the correct support conditions for the substation equipment but the nature of the experimentation does not lend itself to in-depth investigations.

Laboratory Tests: Force-calibrated Hammer – Based upon the availability of various types of equipment for laboratory testing, low-level excitation techniques can be used to determine their linear vibration characteristics from a force-calibrated hammer. Although commercially available 'hammers' range from ball-peen to sledge, it is anticipated that the force-calibrated mallet should suffice for most electrical substation equipment. Uni-axial and multi-axis accelerometers can measure the response and provide input to a multi-channel dynamic signal analyzer. Five to ten averaged responses should be sufficient to get 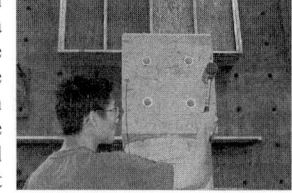 the required coherence for good modal identification from frequency response functions such as those obtained at the Ignacio Substation and the Richmond Field Station. The ME'scope software can provide the modal parameters from the frequency response functions.

Laboratory Tests: Shake Table – Uni-axial shake table tests should be conducted on the same substation equipment that was subjected to the force-calibrated hammer tests. Although appropriate boundary conditions can be a technical issue that must be decided a-priori, the intention of the shake table tests is to further investigate the substation equipment's vibration characteristics due to laboratory controlled base excitations using periodic random or burst random input. Additionally, the shake table's amplitude level can be gradually increased to examine the change in modal parameters due to excitation levels. The controlled increase in base excitation amplitude should provide some guidance as to the use of linear and nonlinear analytical models. The base-excitation correction factors should be applied to the shake table response data to in order to represent it in the standard frequency response function format. Uni-axial and multi-axis accelerometers should be used to measure the response and provide input to a multi-channel dynamic signal analyzer.

Field Vibration Tests: Force-calibrated Hammer – In-situ vibration characteristics of the substation equipment can be determined from low level force-calibrated hammer tests. The in-situ force calibrated hammer results can provide a link to those determined in the laboratory under different boundary conditions. As was shown in the tests conducted at the Ignacio substation, these low level forces do not cause any damage to the equipment including the brittle ceramic bushing skirts.

Field Vibration Tests: Eccentric Mass Shaker – Previous studies determined that a 100-lb shaker mounted to the top of transformer did not provide as good coherence (input/output correlation factor) for the frequency response function measurements as the force-calibrated hammer. The purpose of the eccentric mass shakers is to provide a decent, sinusoidal ground motion that will provide good data for the transmissibility calculation between the base excitation and the excitation at the attachment point of the substation equipment. Response measurements on the substation equipment itself can also be acquired during the eccentric mass shaker tests. Three sacrificial concrete mounting pads (4' x 4' x 1.5' with dual sonotube shear keys) can be constructed adjacent to the substation equipment. Pads 1 and 2 can be used to mount the eccentric mass shakers for N-S excitation whereas pads 2 and 3 can be used to mount the eccentric mass shakers for E-W excitation. The concrete pads can be constructed for easy removal after the forced vibration tests.

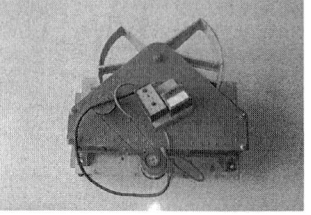

Improved Analytical Models - Improved analytical models should consist of both experimentally-derived, lumped parameter modal models as well as finite element models. ME'scope is an industry proven analytical platform that can determine the lumped parameter, modal model characteristics from the experimental frequency response functions. ME'scope provides estimates of frequency, mode shape and percent critical damping from the calibrated frequency response functions. The output consists of printed and animated vibration response. A 'movie' of the animation (as an *.avi file) can be created for each mode of vibration. The significance of the experimentally-derived, modal model serves at least three purposes:

- estimates of damping, derived from the in-situ hardware, can be incorporated into finite element models,
- sophisticated finite element models can be used to corroborate the results of these relatively simple, lumped parameter modal models, and
- 'structural dynamic modification' (what if queries) and 'forced response simulation' analyses can be easily performed with the confidence that the lumped parameter model accurately reflects the in-situ hardware.

CONCLUSIONS

A study was conducted to quantify the ground motion amplification that takes place at the base of bushings mounted on electrical substation transformers. The study included field vibration tests as well as analytical modeling. It was found that the ground amplification factors are as much as 1.89 for 500-kV bushings and 3.95 for 230-kV bushings. Thus the 500-kV bushings are in accordance with IEEE 693 recommendations whereas the 230-kV bushings have an amplification factor nearly double what is specified. While the field vibration tests help calibrate finite element models, obvious improvement in the experimental procedure were noted. An improved experimental/analytical program is proved for the improved modeling of electrical substation equipment.

REFERENCES

1. "ME'scope Visual Engineering Series", Vibrant Technology, Inc., Jamestown, CA, 2000, http://www.vibetech.com
2. Bellorini, S., Bettinali, F., Salvetti, M., Zafferani, G., "Seismic Qualification of Transformer High Voltage Bushings", IEEE Transactions on Power Delivery, 1998; 13(4): 1208-1213.
3. Schiff, A.J., "Northridge Earthquake: Lifeline Performance and Post-Earthquake Response", Technical Council on Lifeline Earthquake Engineering, Monograph 8, ASCE, August 1995.
4. Ibanez, P, Vasudevan, R., Vineberg, E.J., "A Comparison of Experimental Methods for Seismic Testing of Equipment", Nuclear Engineering and Design, 1973; 25: 150-162.

AUTHOR INDEX

Abdullah, M. 887
Afolayan, J.O. 1129, 1137, 1489
Ahmed, T.U. 861
Akeju, T.A.I. 1463
Alexander, M.G. 351
Alexander, N.A. 919
Ali, F.H. 1441
Al-Jabri, K.S. 1087
Alshebani, M.M. 437
Amin, M.S. 1185
Anderson, D. 251
Ansari, F. 1175
Apalak, M.K. 711
Arya, C. 1243
Asadollahi, A. 1539
Ashour, A.F. 1591
Atimtay, E. 1009
Au, F.T.K. 3
Austin, P.D. 499

Ballim, Y. 1599
Banerjee, P.K. 655
Barik, M. 639
Barnard, H., 1321
Basoenondo, E.A. 419
Bayoumi, S. 771
Beirow, B. 629
Benjeddou, A. 549
Bensafi, M. 1001
Beushausen, H.D. 351
Bhattacharyya, S.K. 565, 787, 793, 861
Bignell, P. 1037
Bindra, S.P. 1391
Bjelajac, N. 821
Boazu, R. 573
Bocchio, N. 463
Boswell, L.F. 359, 377
Budkowska, B.B. 755, 1449
Bullen, F. 1037
Burgess, I.W. 1087
Burgoyne, C.J. 335, 1357
Burkhardt, A. 267
Butler, D. 499
Byfield, M.P. 259, 393

Caddemi., S. 1113
Carnalla, S. 1615
Chan, H.C. 1423
Chanerley, A.A. 919
Chau, K.T. 977

Chen, H.J. 1423
Chen, K.H. 1373
Cheng, S.H. 1551, 1559
Cheng, Y.S. 3
Cheung, Y.K. 3
Choi, F.C. 1203
Chowdhury, S.H. 327
Chung, K.F. 1515, 1567
Clarke, J.L. 1243
Clifton, G.C. 1063
Colajanni, P. 1113
Comninos, K. 1043
Cousins, B.F. 763
Crawford, R.J. 259
Crofts, F.S. 1531

Davies, J.M. 221
de Andrade Filho, J.C. 913
de Andrade, C.F. 913
de Andrade, J.C. 913
de Felice, G. 411
De Matteis, G. 937, 955, 965
de Villiers, P.J. 1297
Della Corte, G. 955
Denton, S.R. 1523
Diah, A.B.M. 1415
Djafour, M. 835
Djelebov, E.P. 1583
du Plooy, N.F. 905
du Toit, F. 1043
Dunaiski, P.E. 1297, 1321
Dundu, M. 1155

Easterbrook, D.J. 1505
Ehl, O. 27
Elachachi, S.M. 1001
El-Lobody, E. 401
Elmarakbi, A. 755
El-Rafaie, S.A. 1591
Erbatur, F. 1147

Faggiano, B. 965
Falade, F. 1407, 1463
Fanourakis, G.C. 1599
Farjoodi, J. 853, 867
Feng, Y.T. 135
Ficcadenti, S.J. 473, 1607
Fliss, L. 1457
Fodhail, R.A. 1575

Galjaard, H.J.C. 385
Garrity, S.W. 1591
Gentile, C. 591
Gergess, A.N. 243
Geyer, S. 647
Ghobarah, A. 1235
Giżejowski, M. 275
Giżejowski, M.A. 695
Giles, R.S. 419
Gluck, J. 897
Goliger, A.M. 1305, 1329
Goorvadoo, N. 919
Gould, P.L. 39
Groenwold, A.A. 647
Grosskurth, K.P. 1251
Gunes, R. 711

Hamilton, C.H. 993
Hamir, R. 1415
Hancock, G.J. 213
Harte, R. 87
Hasebe, N. 483
Henry, A. 887
Henry, Jr., D.P. 655
Heyns, P.S. 905
Hobbelman, G.J. 447, 455
Hoon, K.H. 1019
Hose, Y.D. 993
Huang, Y.L. 1349
Humar, J.L. 1185, 1313

Imran, I. 947
Iordanescu, M. 621
Irawan, P. 985
Iskhakov, I. 367

Jarquio, R.V. 319, 343
Jing, J.J. 613
Jun, G.J. 613

Kahraman, M. 1147
Kamarulzaman, K.B. 1415
Kanyemba, J. 1471
Karakas, E.S. 711
Karczewski, J.A. 275, 695
Karimi, A. 1431
Kasa, A. 1441
Katsikadelis, J.T. 99
Kay, E.A. 1243
Kazanjy, R.P. 473, 993
Kemp, A.R. 51, 1289

Kennedy, J.B. 583, 599
Keršner, Z. 737
Kerdal, D. 835
Khan, M.M. 291, 1217
Klerck, P.A. 135
Knésl, Z. 737
Ko, A.C.H. 1567
Ko, J.M. 111
Komann, S. 173
Kratz, R.D. 351, 1051
Krätzig, W.B. 87
Krige, G. 1321
Krige, G.J. 291, 1217
Kruger, P.J. 1043
Kumar, V.A. 875

Lai, C.P. 1365
Lam, D. 401
Lam, S.S.E. 977
Landolfo, R. 937, 955, 965
Lane, J.W. 1531
Lau, D.T. 1551, 1559
Lau, N.C. 1071, 1079, 1103
LeBlanc, B. 1195
Letombe, S. 549
Leung, A.Y.T. 845
Leung, H.Y. 335, 1357
Li, Q.S. 845
Liang, X. 613
Lim, C.H. 985
Lim, J.K.T. 1019
Lim, K.L. 1019
Lim, P.Y. 1399
Lin, Y. 1365
Liu, T.C.H. 1567
Lo, S.H. 703
Low, K.H. 1019, 1163
Luo, Y. 829

Mackenzie, R. 1321
Mahachi, J. 1155, 1305
Mahgoub, M.A. 1313
Mahmud, H.B. 1381
Maity, D. 787
Mandara, A. 75
Mansur, M.A. 311
Marzouck, M.H. 599
Masarira, A. 283
Masih, R. 1225
Matsufuji, Y. 1381
Matsunaga, H. 541
Mazzolani, F.M. 75, 937, 955, 965
McBride, A.T. 719

McKinley, B. 377
Megnounif, A. 835
Melerski, E.S. 507, 763
Milford, R.V. 1305
Moestopo, M. 947
Mohammadi, S. 1539
Mohammed, G.A. 771, 779
Mohd, S. 1399
Morley, C.T. 1523
Moss, P.J. 1063
Mukhopadhyay, M. 639
Mullen, C.L. 1195
Munirudrappa, N. 875
Muscolino, G. 1113
Mwakali, J.A. 301

Náhlík, L. 737
Nasir, A.M. 499
Nasir, N. 1441
Nedjar, D. 1001
Nethercot, D.A. 15
Ngab, A.S. 1339, 1391
Ngu, T.L. 1079, 1095
Ni, Y.Q. 111
Noor, F.A. 919
Nurick, G.N. 1029

Obrębski, J.B. 161
Ocholi, A. 1129
Onyejekwe, O.O. 793
O'Regan, P.D. 1243
Osterrieder, P. 629, 803, 1279
Othman, I. 607
Owen, D.R.J. 135

Pahl, P.J. 63
Pan, T.C. 985
Pann, K.S. 1349
Pardoen, G.C. 473, 993, 1615
Parhi, P.K. 565
Pasternak, H. 173
Perbix, W. 1251
Plank, R.J. 1087
Postek, E. 275, 695
Purnomo, H. 419

Rafiq, M.Y. 1505
Rahman, M.K. 1259
Ramachandra, L.S. 861
Ramachandran, K. 1121, 1431
Razak, H.A. 1203

Reddy, B.D. 27
Redekop, D. 491, 687
Renansiva, R. 947
Resan, K.H. 607
Retief, J.V. 1289, 1297, 1329
Ribakov, Y. 897
Richardson, A. 887
Ridzuan, A.R.M. 1415
Ronca, P. 463
Ruess, M. 63

Safiuddin, Md. 1381
Samaan, M. 583
Scheele, F. 719
Schilling, S. 173
Secu, A. 573
Sen, R. 243
Sennah, K.M. 583, 599
Şensoy, S. 811
Severn, R.T. 195
Sha, W. 1071, 1079, 1095, 1103
Shalouf, F. 533
Sheikh, S.A. 123
Shin, H. 491
Shrestha, S. 1195
Shrivastava, S.C. 557
Sin, H.P. 1163
Sinha, P.K. 565
Song, D. 655
Song, T.K. 1381
Soroushian, A. 853, 867
Stefanescu, D.P. 573
Studnicka, J. 231
Sudarsono, A. 947
Sun, Z.G. 111

Taban-Wani, G. 675
Tahir, M.M. 251
Tan, K.H. 311
Tang, C.W. 1373
Tavares, R. 1615
Teh, L.H. 213
Ter Haar, T.R. 1289
Thambiratnam, D. 1037
Thambiratnam, D.P. 419, 499
Tickodri-Togboa, S.S. 675
Timashev, S.A. 147
Ting, C.N. 455
Trinchero, P.E. 525
Tuladhar, P. 1195
Tuna, M.E. 1009

Unay, A.I. 1269
Uzoegbo, H.C. 427

van der Kuilen, J.W.G. 463
van der Ploeg, J.A. 447
van Rensburg, B.W.J. 1497
van Zijl, G.P.A.G. 455, 729, 745
Vassileva, S.T. 927
Veer, F.A. 447, 455
Villaverde, R. 1615
Vrouwenvelder, T. 183

Wah, C.K. 1399
Walraven, J.C. 385
Wang, H. 655
Wang, Q. 829
Wang, X.F. 483
Wang, Z.Y. 977
Weiser, I. 1481
Weng, W. 311
Werner, F. 1279
Wierbicki, S. 275, 695

Wong, M.F. 1515
Wong, Y.L. 977
Wu, B. 977

Yang, A. 1019
Yang, L.F. 845
Yen, T. 1349, 1365, 1373
Yin, H. 985
Yu, J. 135, 1449
Yu, X.J. 687
Yuen, S.C.K. 1029

Zain, M.F.M. 1381
Zhang, J.Z. 845
Zhao, M. 1175
Zhao, Y. 1175
Zhu, J. 803
Zingoni, A. 515, 663

KEYWORD INDEX

Abnormal load, 607
Acceptability criterion, 913
Accidental actions, 1531
Accidental damage, 1531
Acid-resistance, 87
Active process model, 1481
Adaptive three phase search approach, 1147
Advanced analysis, 213
Advanced finite element model, 695
Africa, 1391
Aggregate, 737, 1365, 1415, 1423
Aggregates, 1399, 1599
Air entraining admixtures, 1381
Airship hangar, 173
Algorithm, 675
Allowable axial load, 343
Aluminium alloys, 75
Ambient vibration test, 591
Analog equation method, 99
Analysis, 311, 1051, 1279
Analytical continuation, 483
Analytical method in reinforced concrete, 343
Analytical modeling, 993, 1269
ANSYS, 711
Antenna, 621
Appropriate technology, 1339
Aramid fibre, 1357
Aramid, 335
Arbitrarily shaped hole, 483
Arching action, 1043
Architecture, 455
Artificial earthquake accelerogram, 927
Artificial, 1399
Assumed stress, 647
Asymmetric beam, 1071, 1103
Axial, 301
Axisymmetric shell structures, 499

Bamboo splints, 1463
Bar forces, 319
Base shear, 1313
Batten plate, 835
Beam columns, 213
Beam, 1087, 1163
Beam-column joint, 1235
Beam-column joints, 985
Beam-column sub-frames, 1515
Beams, 123, 311, 327, 393
Bearing capacity, 367
Bed joint, 437

Bending strength, 259, 463
Bending theory of shells, 515
Bi-axial bending, 343
Bi-axial, 919
Bifurcation, 557, 811
Bimaterial, 483
Bingham model, 1373
Bi-steel, 377
Blast load, 1029
Blast walls, 1043
Blast, 1051
Bolted connection, 275
Bond deterioration, 985
Bond strength, 1251
Boundary conditions, 591
Boundary element method, 99
Box girder, 583
Bracings, 599
Brick wall, 419
Bridge assessment, 1431
Bridge deck, 607
Bridge vibration, 3
Bridges, 231, 583, 599, 993, 1175
Brittle, 301
Bucking, 377
Buckling analysis, 213
Buckling capacity, 803
Buckling, 301, 557, 639, 1583
Buffer, 1019
Building process, 1481
Buildings, 231
Built-up column, 835
Buntons, 291, 1163, 1217

C++, 875
Cable vibration, 1551, 1559
Cable, 621
Cable-stayed bridge, 111, 1551, 1559
Caisson, 1457
Calcrete, 1457
Calculation model, 267
Capacity, 1591
Cement paste, 737
Cement type, 1607
Cementitious repair material, 1259
CFRP composites, 1591
CFRP, 1175
Chlorination, 1431
Circular column, 343
Circular tanks, 507

Circular web opening, 1567
Cladding panels, 937
Classification, 1321
Closed form, 483
Closed two-dimensional solution, 549
Code calibration, 1289
Code comparison, 1297
Code, 1279
Codification, 1305
Cold-formed steel moment connections, 1515
Collapse, 301, 955
Column axial load, 985
Column capacity axis, 343
Column capacity charts, 343
Column web, 251
Columns, 15, 123
Column-supported silo, 525
Combined finite/discrete element approach, 135
Complex stress function, 483
Composite bars, 161
Composite beam, 1071, 1095, 1103
Composite concrete construction, 351
Composite construction, 15
Composite structures, 1063
Composite, 393, 401, 541, 573
Composites, 565, 1539
Compressible fluid, 784
Compression, 301, 335, 1357
Compressive strength, 1407, 1415
Computer program, 1129
Concentrated, 243
Concentrically loaded, 301
Concrete bridge, 1225
Concrete forces, 319
Concrete matrix, 1463
Concrete stress and strain, 319
Concrete technology, 1391
Concrete, 327, 335, 401, 729, 737, 1175, 1243, 1251, 1339, 1357, 1365, 1399, 1407, 1415, 1423, 1457, 1523, 1599
Condition monitoring, 1175
Confinement, 335
Conjugated system method, 1449
Connections, 15, 955, 1087
Consistent-mass matrices, 663
Construction, 927, 1441
Contact effect, 695
Containment structures, 515
Containment walls, 1043
Continuous beams, 1591
Continuum theory, 1449
Control limit, 1489
Convergence, 63, 867
Cooling tower, 87

Coordinate transformation, 639
Core radius, 301
Corrosion, 1203, 1217
Cost analysis, 1137
Coupled problem, 784
Crack damage, 87
Crack injection, 1251
Crack propagation, 1539
Crack spacing, 327
Crack width, 327
Cracked frame, 779
Cracking, 311, 729
Cranes, 1321
Creep, 1599
Critical load, 283
Critical loading, 821
Critical processes, 1259
Critical stress, 737
Cross-ply, 541
C-sections back-to-back, 1515
Cubic elements, 213
Curing temperature, 1349
Curvature ductility, 947, 1009
Curvature, 499
Curved boundary, 639
Curved, 583, 599
Cyclic loads, 437, 473
Cylinders, 525
Czech Republic, 231

Damage assessment, 1225
Damage detection, 111
Damage, 243, 955
Decomposition, 63
Deep learning, 1505
Deflation, 63
Deflection, 1163
Deflections, 1591
Deformation, 275
Degenerate, 811
Delamination, 565
Demolition, 1225
Design codes, 1321, 1599
Design concepts, 803
Design development, 1567
Design lode, 1297
Design requirements, 1129
Design, 173, 447, 455, 1051, 1279
Detailing, 1051
Developing countries, 1339
Diaphragm, 473
Diaphragms, 599
Differential quadrature method, 491
Diffusion analysis, 1259

Diffusivity, 1259
Dilatancy, 745
Dimensioning, 161
Discharge tube, 533
Discontinuous deformation analysis, 719
Discontinuous media, 719
Discrete cracking, 745
Discrete element method, 1539
Discretisation, 675
Displacement ductility, 947, 1009
Displacement-field decomposition, 663
DMEM, 161
Document linkage, 1481
Door segments, 173
Double skin composite, 377
Dowels, 385
Drilling d.o.f., 647
Drop modeling/simulation, 1019
Drop/impact behaviour, 1019
Drying shrinkage, 1415
Dry-stack system, 427
Dual system, 1009
Ductility formulation, 947
Ductility, 123, 301, 919, 1523
Durability, 87, 183
Duration of load, 463
Dynamic analyses, 965
Dynamic analysis, 99, 499
Dynamic behavior, 913
Dynamic cable stiffness, 1551, 1559
Dynamic delamination buckling, 1539
Dynamic domain decomposition, 135
Dynamic loads, 1321
Dynamic pressure, 533
Dynamic properties, 687
Dynamic stability, 829
Dynamic, 583, 853, 867
Dynamics, 161

Early age, 1349
Earthquake design, 887
Earthquake engineering, 993
Earthquake resistant, 1009
Earthquake, 919
Eccentrically loaded, 301
Egg-shaped sludge digesters, 515
Eigenstates, 63
Eigenvalues, 63
Eigenvectors, 63
Elastic analysis, 507, 763
Elastic characteristics, 573
Elastic foundation, 755
Elastic half-space, 507
Elastic membranes, 99

Elastic mismatch, 737
Elastic modulus, 1423
Elastic stability, 835
Elasticity, 411
Elastic-plastic analysis, 221
Elasto-plastic phenomena, 275
Elasto-plastic, 861
Elastoplastic, 867
Elastoplasticity, 655
Elasto-visco-plastic, 675
Element matrices, 663
Elimination, 887
Employment creation, 1471
Energy absorption, 1037
Energy dissipation, 75
Enhanced strains, 27
Envelope process, 1113
Epoxy adhesive, 711
Epoxy resins, 1251
Equilibrium paths, 821
Equivalent parallelogram, 27
Equivalent rectangle, 319
Error frequency, 1489
Ettringite, 1607
Excited vibration, 829
Expansion, 1607
Experimental investigation, 1515
Experimental testing, 993
Experimental verification, 695
Experimental vibrations, 1615
Experimental, 1235
External column, 251
Externally bonded FRP, 1243

Fabric reinforced laminae, 573
Failure probability, 267
Failure, 771, 1029
Far-boundary, 793
Fatigue life, 629
FE-analysis, 173
Feed back control, 887
FEM analysis, 655
Fiber optic sensors, 1175
Fiber reinforced polymer composites, 1175
Fibre composites, 1243
Fibre reinforced polymers (FRP), 123
Field test, 1195
Finite and infinite element, 771
Finite difference, 821
Finite differences, 161
Finite element analysis, 1037, 1269, 1567
Finite Element Method (FEM), 793
Finite element method, 3, 491, 565, 737, 755, 784

Finite element modeling, 1615
Finite element models, 1551, 1559
Finite element technique, 861
Finite element, 39, 499, 639, 647, 779
Finite elements, 455, 663, 687, 745, 729
Finite strip method, 3
Finite-element, 401, 583, 599
Fire behaviour, 1063
Fire engineering, 1071, 1079, 1103
Fire protection, 1095
Fire resistance, 1071, 1079, 1095, 1103
Fire resistant steel, 1071, 1079, 1103
Fire, 15, 1071, 1087, 1095, 1103
Flange width-thickness ratio, 947
Flat shell, 647
Flexible end-plate, 1087
Flexible structure, 784
Flexible, 771
Flexural strength, 1463
Flexural strengthening, 1243
Flexural torsional buckling, 1279
Floor vibrations, 913
Flow, 675
Flowability, 1381
Fluid-structure interaction, 784, 793
Flush endplate, 1087
Flush end-plate, 251
Flutter, 811
Formula, 327
Fourier spectrum of the earthquake accelerogram, 927
Fracture mechanics, 737
Frame joints, 283
Frame, 1087
Frames, 51, 161, 221
Framing clip, 473
Free vibration test, 591
Free vibration, 499, 639, 875
Free-free beam, 861
Friction effect, 695
Friction welded shear connector, 377
Friction, 275
Frontal protection structures, 1037
FRP, 1175, 1235
Full interaction, 393
Full-scale silo tests, 525
Full-scale testing, 591
Fuzzy mathematics, 687

Gains, 887
Galerkin, 1523
Genetic algorithm, 1155
Genetic algorithms, 1147
Geogrid, 1441, 1449

Geometric nonlinearity, 621
Geometrical non-linearity, 711
G-force, 1019
Girder, 243
Glass structure, 455
Glass structures, 447
GPS, 613
Grading, 463
Grain, 533
Granite chips, 1407
Granite fines, 1407
Graphic use interface, 1505
Green Element Method (GEM), 793
Green's function, 483
Group theory, 663
Guard rails, 463
Guides, 291
Guyed towers, 621

Headed-studs, 385
Heat flux, 1095
Heat transfer, 1071, 1095, 1103
Heavy construction equipment, 1225
High performance concrete, 87, 1349, 1381
High strength concrete, 385
Higher mode effects, 1313
Higher-order theory, 541
High-strength, 1399
Historical structures, 1269
Hollow slender bridge piers, 1583
Homogeneous elastic half-space, 763
Homogenization, 411
Housing delivery, 1471
Housing, 1463
Human computer interaction, 1505
Human error, 1489
Human tolerance, 913
Hybrid method, 793
Hybrid solid hexahedral elements, 703
Hydrocyclone, 1457
Hygrothermal, 565
Hyperbolic shell, 499
Hysteretic, 919

Ideally elastic-plastic model, 367
Idempotents, 663
Identification, 591
I-girder, 599
Impact strength, 463
Impact, 861
Impact-induced effects, 1019
Imperfect shell, 39
Imperfection sensitive, 821

Imposed load, 1297
Inclusion, 483
Indeterminate structure, 1163
Indicators, 1391
Industrial product house, 1481
Industrial-wastes, 1399
Inelastic, 51, 243
Initial stiffness, 251
Inspection strategy, 183
Instability, 811
Instrumentation, 1441
Integration, 1481
Integrity, 147
Intelligent use of the computer, 1505
Interaction buckling, 803
Interaction detection, 135
Interaction laws, 135
Interface shear, 351
Interface, 1251
Interlaminar stress, 541
Interlocking spirals, 1357
Interlocking, 1449
Interpolation, 655
Intumescent coating, 1071, 1095, 1103
I-shaped steel member, 947
Isosafety, 1129
Iteration, 63

Joints, 447

Knowledge based systems, 927
Kriging, 1431

Lagrangian, 675
Laminate, 455
Laminated plate, 541
Large deflections, 1029
Large facilities, 195
Large rotation, 655
Large strain, 655
Larssen steel piles, 259
Lateral confinement, 977
Lateral load, 419, 427
Lateral torsional buckling, 283
Laterally loaded piles, 755, 763
Laterite fines, 1407
Learning issues, 1497
Lightly reinforced, 985
Lightweight aggregate concrete, 385
Lightweight, 1399
Limit state design, 183, 301
Limit state of accidental damage, 1531

Limit states, 1129
Limiting depth of the compression zone, 367
Lipped channels, 1471
Liquid inertia absorber, 905
Liquid storage tanks, 491
Live load, 607
Load carrying capacity, 1203
Load combinations, 267, 1289
Load distribution, 583, 599
Load factors, 1289
Load ratio, 1087
Loading code, 1297
Loading, 1051
Local and global buckling, 803
Local stability, 1583
Local-global, 39
Long-span structure, 1551, 1559
Loss of support, 779
Low-cost, 1463

Macro-elements, 1001
Maintenance, 147
Mapping, 639
Masonry strength, 1269
Masonry structures, 1269
Masonry, 411, 419, 427, 437, 729, 745
Material and geometric non-linearity, 1567
Matrix update method, 1185
Maturity, 1349
Maximal reinforcement cross-section area, 367
Mean upcrossing rate, 1113
Measurement problems, 1225
Medium strength high performance concrete (TAICON), 1373
Membrane action, 1043
Membrane hypothesis, 515
Mesh adaptivity, 135
Metal materials, 75
Metal systems, 75
Micro-alloyed steel, 1079
Micromechanics, 411
Mild steel, 301, 1463
Mindlin solution, 763
Mindlin-Reissner plates, 27
Mine shafts, 291
Minimum load eccentricity, 343
Minimum reinforcement, 1009
Minor axis, 251
Mixed-flexibility method, 51
Mixture proportions, 1365
Modal analysis, 111, 613, 1195, 1551, 1559
Modal residual vector, 1185
Modal shapes, 875

Modal test, 1203
Mode shapes, 1203
Model modification, 687
Modelling, 401, 447, 455, 955
Models, 1599
Modular block wall, 1441
Moisture profile, 1259
Moisture, 565
Moment adjustment factor, 1313
Moment resistance, 251, 1071, 1079
Moment resisting frame, 1235
Moment resisting frames, 955
Monitoring, 629
Monumental constructions, 75
Morphology, 1251
Mortar, 737, 1365, 1423
Moving trains, 3
Moving vehicles, 3
Multi-fracture, 135
Multi-stage identification, 111

Natural frequencies, 875
Natural frequency, 1203
Nau's method, 853, 867
Near-fault ground motion, 993
Neural network, 111, 927
Neural networks, 629
Newmark's method, 867
Non linear behavior, 1001
Non linear, 1043
Non-homogeneous bodies, 99
Nonlinear FE simulations, 1195
Nonlinear finite element, 1259
Nonlinear, 51, 853, 867
Non-linear, 919
Non-linearity, 87
Non-seismic detailing 977
Non-uniform stress, 525
Nuclear reactor containment, 231
Numerical method, 1423
Numerical modelling, 719
Numerical simulation method, 573
Numerical, 1551, 1559
Nylon, 447

Objectionable vibrations, 913
Ogival shells, 515
On-line monitoring, 613
Openings, 311
Operational changes, 629
Optical fibers, 1175
Optimal control, 147
Optimal passive control, 897

Optimal placement, 887
Optimisation, 1147, 1155
Orientation of the shear reinforcement, 367
Orthotropic, 647
Oscillations, 811
Overhang, 243
Overhead, 1321
Overstrength, 937, 965
Overturning moment, 1313

Parabolic stress method, 319, 343
Parallelisation, 135
Parametric studies, 1505
Parametric study, 875, 1163
Partial factors, 183
Partial safety load factors, 1531
Partially prestressed, 327
Passive confinement, 1357
Patch repair, 1259
Peeling failure, 1591
Perfobondstrip, 385
Performance based code, 183
Performance-based design, 965
Periodic, 811
Permanent deformation, 1019
Permanent modulus, 1449
Perturbation technique, 845
Perturbation, 811
Philosophy, 1051
Photoelastic, 779
Physical models, 719
Piezoelectric materials, 549
Piling, 259
Pin-jointed, 1155
Pipe, 771
Pipeline segments, 147
Pit, 1457
Plastic buckling, 557
Plastic design, 1279
Plastic, 437, 675
Plasticity, 301, 1523
Plate bending, 639
Plates, 1019
Point dislocation, 483
Polycarbonate, 455
Post-critical behaviour, 821
Potential energy, 763
Potential loss, 1137
Precast concrete panels, 1471
Precast elements, 351
Predicting ground motion, 927
Prediction, 1599
Pre-envelope covariances, 1113
Prestressed, 275, 327

Probabilistic analysis, 267
Probabilistic assessment, 1121
Probabilistic estimation, 1431
Probability of failure, 1113, 1129, 1137
Probability of survival, 1113
Probability, 1121
Process modelling, 1481
Profiled matrices, 63
Progressive collapse, 1531
Pulsation, 533
Pulse load, 861
Pulverized fuel ash, 1399
Push-off test, 401

q-factor, 965
Quadratic electric potential, 549
Quadratic nonlinear optimisation, 1185
Quarry dust, 1381, 1399
Quasi-isotropic model, 367
Quasi-periodic, 811
Questionnaire, 1391

Random field, 1431
Random variables, 1137
Rate dependence, 729
Rational mapping function, 483
RC bridge, 1195
RC element section, 367
Reaction walls, 195
Reactions, 1591
Rectangular and square column, 343
Rectangular column, 977
Recycled, 1415
Reduced modulus action, 259
Reduction, 533
Regression formula, 1423
Rehabilitation, 1235
Reinforced concrete columns, 993
Reinforced concrete, 311, 977, 1001, 1175, 1203, 1235, 1591
Reinforced earth structures, 1441
Reinforced gunite, 1043
Reinforced telecommunication tower, 629
Reinforced, 327
Reinforcement placement, 1489
Reinforcement, 1523
Reinforcing steel development length, 319
Reliability bounds, 1121
Reliability index, 1129
Reliability, 147, 183, 1431
Remeshing, 655
Renovation, 1225
Repair prioritization, 147

Repair, 123, 243, 1175, 1235, 1251
Representation theory, 663
Residual deflection, 243
Residual soil, 1441
Residual strength, 1531
Residual, 437
Resistant, 1051
Response spectra, 919
Retrofitting, 123
Reversed cyclic loading, 985
Rheology, 1373
Riemann-Hilbert problem, 483
Ringbeam, 525
Risk analysis, 183
Risk model, 1329
Rock mechanics, 291
Rotational ductility, 947
Rotational shell, 39

Safety concept, 267
Safety targets, 1289
SAISC silo guideline, 525
Sand washing plant, 905
Sand, 1407
Sandwich panels, 937
Sandwich plates, 549
Sandwich shells, 557
Scale, 1457
Screed, 351
Second-order analysis, 213
Second-order effects, 221
Second-order, 51
Sectional profiles, 1129
Seismic analysis, 1001
Seismic behaviour, 955
Seismic design, 937, 965, 1313
Seismic protection, 75
Seismic qualification, 1615
Seismic resistant design, 947
Seismic response, 1195
Seismic upgrade, 1235
Seismic, 123, 919
Selfweight, 243
Semi-rigid joint, 695
Semi-rigid joints, 51
Sensitivity analysis, 755, 1449
Serviceability limit state, 937, 965
Serviceability, 629, 1321
Shaft steelwork, 1217
Shakedown, 275
Shaking tables, 195
Shape memory alloys, 75
Shear adjustment factor, 1313
Shear deformation layerwise theory, 549

Shear effect, 755
Shear locking, 27
Shear moment interaction curves, 1567
Shear span, 977
Shear strength, 1009
Shear strengthening, 1243
Shear studs, 393, 401
Shear transfer, 473
Shear wall, 473
Shear walls, 937, 1009
Shear, 1235
Shear-connectors, 385
Shell analysis, 515
Shell of revolution, 39
Shell structrures, 515
Shell, 1457
Shells of revolution, 515
Silica fume, 1381
Silo, 525, 533
Simplified method, 1001
Simultaneous optimum design, 1147
Skew bridge, 875
Slabs, 123, 393, 1523
Slant legged, 875
Slender bridge piers, 1583
Slenderness ratio, 947
Slide, 275
Slim floor, 1071, 1095, 1103
Slump test, 1373
Smart structures, 549
Software, 621
Soil-cement blocks, 427
Soil-structure interaction, 771, 779
Space frames, 213
Space truss, 301
Special devices, 75
Spectral moments, 1113
Spherical shells, 557
Spiral, 1357
Spirals, 335
Spline finite strip, 835
Stability design, 803
Stability, 51, 221, 283, 301, 811, 821, 853, 1531, 1583
Stainless steels, 75
Static and dynamic response, 1037
Static deflection, 875
Static, 161
Steel adherends, 711
Steel arches, 173
Steel concrete sandwich construction, 377
Steel construction, 1063, 1279
Steel frames, 283, 937, 965
Steel pile walls, 259

Steel structures, 15, 231, 259, 955, 1079, 1095
Steel, 401
Steel-concrete composite structures, 385
Stiffened plate, 639
Stiffened plates, 1029
Stiffeners, 599
Stiffness-matrix, 675
Stochastic finite element method, 845
Stochastic load distributions, 267
Stochastic, 845
Stopper, 1163
Straightening, 243
Strain hardening, 221
Strain measurement, 1225
Strain, 437
Strains, 1591
Strain-softening, 1523
Strength, 311, 1349, 1365
Strengthening, 1243, 1591
Stress reduction, 1163
Stressed skin design, 937
Structural analysis, 703, 1497
Structural assessment, 1121
Structural categories, 1051
Structural codes, 955
Structural design, 15, 401, 1297, 1305
Structural dynamic response, 629
Structural dynamics, 845
Structural engineering, 1339
Structural health monitoring, 1185
Structural input loading, 905
Structural layout optimisation, 1147
Structural rehabilitation, 75
Structural reliability, 1121, 1289
Structural response, 419, 1063
Structural safety, 1121, 1269
Structural steel, 393
Structures, 147, 565, 675, 1175
Sub cells method, 573
Subgrade reaction, 755
Subspaces, 663
Substation equipment, 1615
Subterranean, 675
Sulfate attack, 1607
Superelement, 621
Superplasticizer, 1381
Surface roughness, 351
Surface texture, 351
Survey, 1391
Suspension bridge, 613
Sustainable, 1391
Symmetry groups, 663
System analysis, 183
System bounds, 1121

Tall buildings, 161, 887
Tanks, 231
Teaching, 1497
Tearing, 1029
Technology, 1339
Tee joint, 711
Tee-section, 301
Temperature effects, 507
Temperature, 565, 1087
Tensile strength, 1407, 1463
Test results, 385
Textile fabric, 173
Thermal conduction, 1095
Thermal conductivity, 1095
Thermal response, 1063
Thermal stress analysis, 711
Thin plate bending, 483
Thin plate, 639
Thin shell element, 695
Thin-walled member, 829
Three dimensions, 703
Timber, 463, 473
Time domain reliability, 1113
Time integration, 853, 867
Time scale, 729
To plasticize, 275
Toe-nail, 473
Tolerance, 853, 867
Topping, 351
Torsion, 283, 599, 607, 919
Towers, 231
Transient variational principle, 845
Transition junction, 525
Transparency, 455
Transverse shear effects, 557
Treatment, 1457
Tricalcium aluminate, 1607
Truncation boundary, 784
Truncation error, 853, 867
Truss model, 311
Truss system, 1137
Tuned absorber, 905
Two-phase material, 737

Ultimate drift ratio, 977
Ultimate strength of beams and T-beams, 319
Ultimate strength, 343

Ultimate, 583
Ultrasonic pulse velocity, 1365
Under-graduate curriculum, 1497
Underground, 675
Uniform hazard spectrum, 1313
Unreinforced interfaces, 351
Unreinforced masonry, 1531

Variation stiffness, 829
Vibration absorber, 905
Vibration, 541
Vibration-based damage detection, 1185
Vibrations, 63, 99, 549
Vibratory screen, 905
Virtual displacements, 845
Virtual, 675
Viscous damping system, 897

w/c ratio, 1599
Walking activity, 913
Wall, 533
Warping stiffness, 283
Waste concrete, 1415
Water to cement ratio, 1607
Water, 1457
Wave speed, 1349
Welded joints, 1137
Wet cracks, 1251
Wind damage, 1329
Wind disaster, 1329
Wind loading, 1305
Wind tunnel test, 629
Winkler model of soil, 507
Winkler soil, 763
Work, 675
Working drawing, 1489
Workmanship, 1489

Y beam, 607
Yield line, 1043
Yield strength of reinforcing steel, 319, 343